Geometric Approximation Algorithms

Mathematical
Surveys
and
Monographs

Volume 173

Geometric Approximation Algorithms

Sariel Har-Peled

American Mathematical Society
Providence, Rhode Island

2010 *Mathematics Subject Classification.* Primary 68U05, 68W25;
Secondary 68P05, 52Cxx.

For additional information and updates on this book, visit
www.ams.org/bookpages/surv-173

Library of Congress Cataloging-in-Publication Data

Har-Peled, Sariel, 1971–
Geometric approximation algorithms / Sariel Har-Peled.
p. cm. — (Mathematical surveys and monographs ; v. 173)
Includes bibliographical references and index.
ISBN 978-0-8218-4911-8 (alk. paper)
1. Approximation algorithms. 2. Geometry—Data processing. 3. Computer graphics. 4. Discrete geometry. I. Title.

QA448.D38H377 2011
516′.11—dc22

2011002940

∞ The paper used in this book is acid-free and falls within the guidelines
established to ensure permanence and durability.
Visit the AMS home page at http://www.ams.org/

10 9 8 7 6 5 4 3 2 1 16 15 14 13 12 11

Contents

Preface

> Finally: It was stated at the outset, that this system would not be here, and at once, perfected. You cannot but plainly see that I have kept my word. But I now leave my cetological system standing thus unfinished, even as the great Cathedral of Cologne was left, with the crane still standing upon the top of the uncompleted tower. For small erections may be finished by their first architects; grand ones, true ones, ever leave the copestone to posterity. God keep me from ever completing anything. This whole book is but a draft – nay, but the draft of a draft. Oh, Time, Strength, Cash, and Patience!
>
> – Moby Dick, Herman Melville.

This book started as a collection of class notes on geometric approximation algorithms that was expanded to cover some additional topics. As the book title suggests, the target audience of this book is people interested in geometric approximation algorithms.

What is covered. The book describes some key techniques in geometric approximation algorithms. In addition, more traditional computational geometry techniques are described in detail (sampling, linear programming, etc.) as they are widely used in developing geometric approximation algorithms. The chapters are relatively independent and try to provide a concise introduction to their respective topics. In particular, certain topics are covered only to the extent needed to present specific results that are of interest. I also tried to provide detailed bibliographical notes at the end of each chapter.

Generally speaking, I tried to cover all the topics that I believe a person working on geometric approximation algorithms should at least know about. Naturally, the selection reflects my own personal taste and the topics I care about. While I tried to cover many of the basic techniques, the field of geometric approximation algorithms is too large (and grows too quickly) to be covered by a single book. For an exact list of the topics covered, see the table of contents.

Naturally, the field of geometric approximation algorithms is a subfield of both computational geometry and approximation algorithms. A more general treatment of approximation algorithms is provided by Williamson and Shmoys [**WS11**] and Vazirani [**Vaz01**]. As for computational geometry, a good introduction is provided by de Berg et al. [**dBCvKO08**].

What to cover? The material in this book is too much to cover in one semester. I have tried to explicitly point out in the text the parts that are more advanced and that can be skipped. In particular, the first 12 chapters of this book (skipping Chapter 7) provide (I hope) a reasonable introduction to modern techniques in computational geometry.

Intellectual merit. I have tried to do several things that I consider to be different than other texts on computational geometry:

(A) Unified several data-structures to use compressed quadtrees as the basic building block and in the process provided a detailed description of compressed quadtrees.

(B) Provided a more elaborate introduction to VC dimension, since I find this topic to be somewhat mysterious.
(C) Covered some worthy topics that are not part of traditional computational geometry (for example, locality sensitive hashing and metric space partitions).
(D) Embedded numerous color figures into the text to illustrate proofs and ideas.

Prerequisites. The text assumes minimal familiarity with some concepts in computational geometry including arrangements, Delaunay triangulations, Voronoi diagrams, point-location, etc. A reader unfamiliar with these concepts would probably benefit from skimming or reading the fine material available online on these topics (i.e., Wikipedia) as necessary. Tail inequalities (i.e., Chernoff's inequality) are described in detail in Chapter 27. Some specific prerequisites are discussed in Chapter 28.

Cross-references. For the convenience of the reader, text cross-references to theorems, lemmas, etc., often have a subscript giving the page location of the theorem, lemma, etc., being referenced. One would look like the following: Theorem 19.3_{p257}.

Acknowledgments. I had the benefit of interacting with numerous people during the work on this book. In particular, I would like to thank the students that took the class (based on earlier versions of this book) for their input, which helped in discovering numerous typos and errors in the manuscript. Furthermore, the content was greatly affected by numerous insightful discussions with Jeff Erickson and Edgar Ramos. Other people who provided comments and insights, or who answered nagging emails from me, for which I am thankful, include Bernard Chazelle, Chandra Chekuri, John Fischer, Samuel Hornus, Piotr Indyk, Mira Lee, Jirka Matoušek, and Manor Mendel.

I would especially like to thank Benjamin Raichel for painstakingly reading the text and pointing out numerous errors and typos and for giving guidance on what needed improvement. His work has significantly improved the quality of the text.

I am sure that there are other people who have contributed to this work, whom I have forgotten to mention – they have my thanks and apologies.

A significant portion of the work on this book was done during my sabbatical (taken during 2006/2007). I thank the people that hosted me during the sabbatical for their hospitality and help. Specifically, I would like to thank Lars Arge (Aarhus, Denmark), Sandeep Sen (IIT, New Delhi, India), and Otfried Cheong (KAIST, Daejeon, South Korea).

Work on this book was also partially supported by NSF (CAREER) award CCR-0132901, and AF award CCF-0915984.

Errors. There are without doubt errors and mistakes in the text and I would like to know about them. Please email me about any of them that you find.

Sariel Har-Peled
sariel@uiuc.edu
Urbana, IL
April 2011

CHAPTER 1

The Power of Grids – Closest Pair and Smallest Enclosing Disk

In this chapter, we are going to discuss two basic geometric algorithms. The first one computes the closest pair among a set of n points in linear time. This is a beautiful and surprising result that exposes the computational power of using grids for geometric computation. Next, we discuss a simple algorithm for approximating the smallest enclosing ball that contains k points of the input. This at first looks like a bizarre problem but turns out to be a key ingredient to our later discussion.

1.1. Preliminaries

For a real positive number α and a point $\mathsf{p} = (x, y)$ in $\mathbb{R}^2$, define $\mathsf{G}_\alpha(\mathsf{p})$ to be the grid point $(\lfloor x/\alpha \rfloor \alpha, \lfloor y/\alpha \rfloor \alpha)$. We call α the ***width*** or ***sidelength*** of the ***grid*** G_α. Observe that G_α partitions the plane into square regions, which we call grid ***cells***. Formally, for any $i, j \in \mathbb{Z}$, the intersection of the halfplanes $x \geq \alpha i$, $x < \alpha(i+1)$, $y \geq \alpha j$, and $y < \alpha(j+1)$ is said to be a grid ***cell***. Further we define a ***grid cluster*** as a block of 3×3 contiguous grid cells.

Note that every grid cell $\square$ of G_α has a unique ID; indeed, let $\mathsf{p} = (x, y)$ be any point in $\square$, and consider the pair of integer numbers $\mathrm{id}_\square = \mathrm{id}(\mathsf{p}) = (\lfloor x/\alpha \rfloor, \lfloor y/\alpha \rfloor)$. Clearly, only points inside $\square$ are going to be mapped to $\mathrm{id}_\square$. We can use this to store a set P of points inside a grid efficiently. Indeed, given a point p, compute its $\mathrm{id}(\mathsf{p})$. We associate with each unique id a data-structure (e.g., a linked list) that stores all the points of P falling into this grid cell (of course, we do not maintain such data-structures for grid cells which are empty). So, once we have computed $\mathrm{id}(\mathsf{p})$, we fetch the data-structure associated with this cell by using hashing. Namely, we store pointers to all those data-structures in a hash table, where each such data-structure is indexed by its unique id. Since the ids are integer numbers, we can do the hashing in constant time.

Assumption 1.1. Throughout the discourse, we assume that every hashing operation takes (worst case) constant time. This is quite a reasonable assumption when true randomness is available (using for example perfect hashing [**CLRS01**]).

Assumption 1.2. Our computation model is the unit cost RAM model, where every operation on real numbers takes constant time, including log and $\lfloor \cdot \rfloor$ operations. We will (mostly) ignore numerical issues and assume exact computation.

Definition 1.3. For a point set P and a parameter α, the ***partition*** of P into subsets by the grid G_α is denoted by $\mathsf{G}_\alpha(\mathsf{P})$. More formally, two points $\mathsf{p}, \mathsf{q} \in \mathsf{P}$ belong to the same set in the partition $\mathsf{G}_\alpha(\mathsf{P})$ if both points are being mapped to the same grid point or equivalently belong to the same grid cell; that is, $\mathrm{id}(\mathsf{p}) = \mathrm{id}(\mathsf{q})$.

1.2. Closest pair

We are interested in solving the following problem:

Problem 1.4. Given a set P of n points in the plane, find the pair of points closest to each other. Formally, return the pair of points realizing $\mathcal{CP}(\mathsf{P}) = \min_{\mathsf{p} \neq \mathsf{q},\, \mathsf{p},\mathsf{q} \in \mathsf{P}} \|\mathsf{p} - \mathsf{q}\|$.

The following is an easy standard ***packing argument*** that underlines, under various disguises, many algorithms in computational geometry.

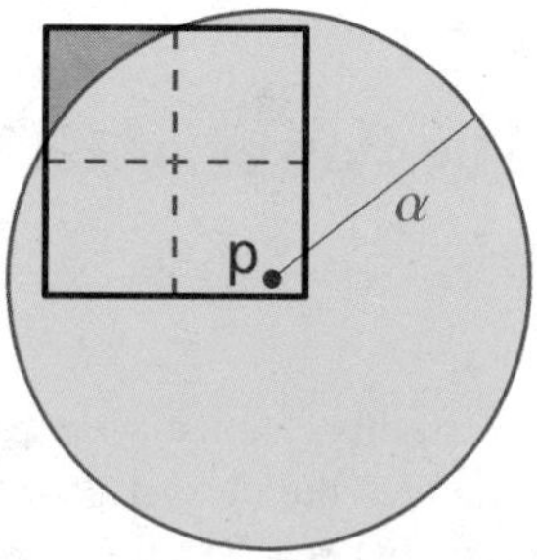

Lemma 1.5. *Let* P *be a set of points contained inside a square* $\square$*, such that the sidelength of* $\square$ *is* $\mathcal{CP}(\mathsf{P})$*. Then* $|\mathsf{P}| \leq 4$.

Proof. Partition P into four equal squares $\square_1, \ldots, \square_4$, and observe that each of these squares has diameter $\sqrt{2}\alpha/2 < \alpha$, and as such each can contain at most one point of P; that is, the disk of radius α centered at a point $\mathsf{p} \in \mathsf{P}$ completely covers the subsquare containing it; see the figure on the right.

Note that the set P can have four points if it is the four corners of $\square$. ∎

Lemma 1.6. *Given a set* P *of* n *points in the plane and a distance* α*, one can verify in linear time whether* $\mathcal{CP}(\mathsf{P}) < \alpha$*,* $\mathcal{CP}(\mathsf{P}) = \alpha$*, or* $\mathcal{CP}(\mathsf{P}) > \alpha$.

Proof. Indeed, store the points of P in the grid G_α. For every non-empty grid cell, we maintain a linked list of the points inside it. Thus, adding a new point p takes constant time. Specifically, compute id(p), check if id(p) already appears in the hash table, if not, create a new linked list for the cell with this ID number, and store p in it. If a linked list already exists for id(p), just add p to it. This takes $O(n)$ time overall.

Now, if any grid cell in $\mathsf{G}_\alpha(\mathsf{P})$ contains more than, say, 4 points of P, then it must be that the $\mathcal{CP}(\mathsf{P}) < \alpha$, by Lemma 1.5.

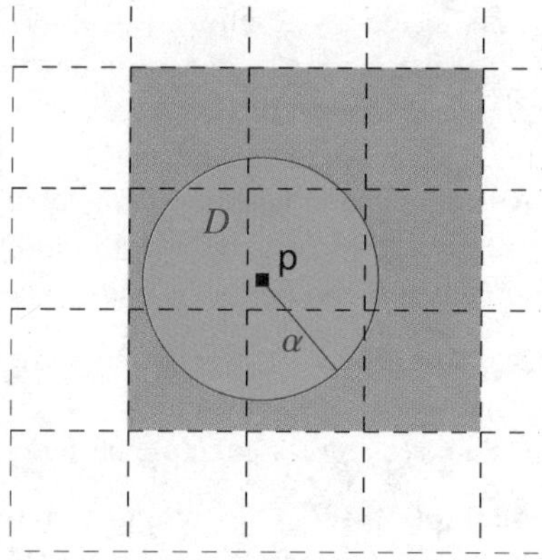

Thus, when we insert a point p, we can fetch all the points of P that were already inserted in the cell of p and the 8 adjacent cells (i.e., all the points stored in the cluster of p); that is, these are the cells of the grid G_α that intersects the disk $D = \mathrm{disk}(\mathsf{p}, \alpha)$ centered at p with radius α; see the figure on the right. If there is a point closer to p than α that was already inserted, then it must be stored in one of these 9 cells (since it must be inside D). Now, each one of those cells must contain at most 4 points of P by Lemma 1.5 (otherwise, we would already have stopped since the $\mathcal{CP}(\cdot)$ of the inserted points is smaller than α). Let S be the set of all those points, and observe that $|S| \leq 9 \cdot 4 = O(1)$. Thus, we can compute, by brute force, the closest point to p in S. This takes $O(1)$ time. If $\mathbf{d}(\mathsf{p}, S) < \alpha$, we stop; otherwise, we continue to the next point.

Overall, this takes at most linear time.

As for correctness, observe that the algorithm returns '$\mathcal{CP}(\mathsf{P}) < \alpha$' only after finding a pair of points of P with distance smaller than α. So, assume that p and q are the pair of points of P realizing the closest pair and that $\|\mathsf{p} - \mathsf{q}\| = \mathcal{CP}(\mathsf{P}) < \alpha$. Clearly, when the later point (say p) is being inserted, the set S would contain q, and as such the algorithm would stop and return '$\mathcal{CP}(\mathsf{P}) < \alpha$'. Similar argumentation works for the case that $\mathcal{CP}(\mathsf{P}) = \alpha$. Thus if the algorithm returns '$\mathcal{CP}(\mathsf{P}) > \alpha$', it must be that $\mathcal{CP}(\mathsf{P})$ is not smaller than α or equal to it. Namely, it must be larger. Thus, the algorithm output is correct. ∎

Remark 1.7. Assume that $\mathcal{CP}(\mathsf{P} \setminus \{\mathsf{p}\}) \geq \alpha$, but $\mathcal{CP}(\mathsf{P}) < \alpha$. Furthermore, assume that we use Lemma 1.6 on P, where $\mathsf{p} \in \mathsf{P}$ is the last point to be inserted. When p is being inserted, not only do we discover that $\mathcal{CP}(\mathsf{P}) < \alpha$, but in fact, by checking the distance of p to all the points stored in its cluster, we can compute the closest point to p in $\mathsf{P} \setminus \{\mathsf{p}\}$ and denote this point by q. Clearly, pq is the closest pair in P, and this last insertion still takes only constant time.

Slow algorithm. Lemma 1.6 provides a natural way of computing $\mathcal{CP}(\mathsf{P})$. Indeed, permute the points of P in an arbitrary fashion, and let $\mathsf{P} = \langle \mathsf{p}_1, \ldots, \mathsf{p}_n \rangle$. Next, let $\alpha_{i-1} = \mathcal{CP}(\{\mathsf{p}_1, \ldots, \mathsf{p}_{i-1}\})$. We can check if $\alpha_i < \alpha_{i-1}$ by using the algorithm of Lemma 1.6 on P_i and α_{i-1}. In fact, if $\alpha_i < \alpha_{i-1}$, the algorithm of Lemma 1.6 would return '$\mathcal{CP}(\mathsf{P}_i) < \alpha_{i-1}$' and the two points of P_i realizing α_i.

So, consider the "good" case, where $\alpha_i = \alpha_{i-1}$; that is, the length of the shortest pair does not change when p_i is being inserted. In this case, we do not need to rebuild the data-structure of Lemma 1.6 to store $\mathsf{P}_i = \langle \mathsf{p}_1, \ldots, \mathsf{p}_i \rangle$. We can just reuse the data-structure from the previous iteration that was used by P_{i-1} by inserting p_i into it. Thus, inserting a single point takes constant time, as long as the closest pair does not change.

Things become problematic when $\alpha_i < \alpha_{i-1}$, because then we need to rebuild the grid data-structure and reinsert all the points of $\mathsf{P}_i = \langle \mathsf{p}_1, \ldots, \mathsf{p}_i \rangle$ into the new grid $\mathsf{G}_{\alpha_i}(\mathsf{P}_i)$. This takes $O(i)$ time.

In the end of this process, we output the number α_n, together with the two points of P that realize the closest pair.

Observation 1.8. *If the closest pair distance, in the sequence $\alpha_1, \ldots, \alpha_n$, changes only t times, then the running time of our algorithm would be $O(nt + n)$. Naturally, t might be $\Omega(n)$, so this algorithm might take quadratic time in the worst case.*

Linear time algorithm. Surprisingly[①], we can speed up the above algorithm to have linear running time by spicing it up using randomization.

We pick a random permutation of the points of P and let $\langle \mathsf{p}_1, \ldots, \mathsf{p}_n \rangle$ be this permutation. Let $\alpha_2 = \|\mathsf{p}_1 - \mathsf{p}_2\|$, and start inserting the points into the data-structure of Lemma 1.6. We will keep the invariant that α_i would be the closest pair distance in the set P_i, for $i = 2, \ldots, n$.

In the ith iteration, if $\alpha_i = \alpha_{i-1}$, then this insertion takes constant time. If $\alpha_i < \alpha_{i-1}$, then we know what is the new closest pair distance α_i (see Remark 1.7), rebuild the grid, and reinsert the i points of P_i from scratch into the grid G_{α_i}. This rebuilding of $\mathsf{G}_{\alpha_i}(\mathsf{P}_i)$ takes $O(i)$ time.

Finally, the algorithm returns the number α_n and the two points of P_n realizing it, as the closest pair in P.

Lemma 1.9. *Let t be the number of different values in the sequence $\alpha_2, \alpha_3, \ldots, \alpha_n$. Then $\mathbf{E}[t] = O(\log n)$. As such, in expectation, the above algorithm rebuilds the grid $O(\log n)$ times.*

[①]Surprise in the eyes of the beholder. The reader might not be surprised at all and might be mildly annoyed by the whole affair. In this case, the reader should read any occurrence of "surprisingly" in the text as being "mildly annoying".

Proof. For $i \geq 3$, let X_i be an indicator variable that is one if and only if $\alpha_i < \alpha_{i-1}$. Observe that $\mathbf{E}[X_i] = \mathbf{Pr}[X_i = 1]$ (as X_i is an indicator variable) and $t = \sum_{i=3}^{n} X_i$.

To bound $\mathbf{Pr}[X_i = 1] = \mathbf{Pr}[\alpha_i < \alpha_{i-1}]$, we (conceptually) fix the points of P_i and randomly permute them. A point $\mathsf{q} \in \mathsf{P}_i$ is ***critical*** if $\mathcal{CP}(\mathsf{P}_i \setminus \{\mathsf{q}\}) > \mathcal{CP}(\mathsf{P}_i)$. If there are no critical points, then $\alpha_{i-1} = \alpha_i$ and then $\mathbf{Pr}[X_i = 1] = 0$ (this happens, for example, if there are two pairs of points realizing the closest distance in P_i). If there is one critical point, then $\mathbf{Pr}[X_i = 1] = 1/i$, as this is the probability that this critical point would be the last point in the random permutation of P_i.

Assume there are two critical points and let p, q be this unique pair of points of P_i realizing $\mathcal{CP}(\mathsf{P}_i)$. The quantity α_i is smaller than α_{i-1} only if either p or q is p_i. The probability for that is $2/i$ (i.e., the probability in a random permutation of i objects that one of two marked objects would be the last element in the permutation).

Observe that there cannot be more than two critical points. Indeed, if p and q are two points that realize the closest distance, then if there is a third critical point s, then $\mathcal{CP}(\mathsf{P}_i \setminus \{\mathsf{s}\}) = \|\mathsf{p} - \mathsf{q}\|$, and hence the point s is not critical.

Thus, $\mathbf{Pr}[X_i = 1] = \mathbf{Pr}[\alpha_i < \alpha_{i-1}] \leq 2/i$, and by linearity of expectations, we have that $\mathbf{E}[t] = \mathbf{E}\left[\sum_{i=3}^{n} X_i\right] = \sum_{i=3}^{n} \mathbf{E}[X_i] \leq \sum_{i=3}^{n} 2/i = O(\log n)$. ■

Lemma 1.9 implies that, in expectation, the algorithm rebuilds the grid $O(\log n)$ times. By Observation 1.8, the running time of this algorithm, in expectation, is $O(n \log n)$. However, we can do better than that. Intuitively, rebuilding the grid in early iterations of the algorithm is cheap, and only late rebuilds (when $i = \Omega(n)$) are expensive, but the number of such expensive rebuilds is small (in fact, in expectation it is a constant).

Theorem 1.10. *For set* P *of* n *points in the plane, one can compute the closest pair of* P *in expected linear time.*

Proof. The algorithm is described above. As above, let X_i be the indicator variable which is 1 if $\alpha_i \neq \alpha_{i-1}$, and 0 otherwise. Clearly, the running time is proportional to

$$R = 1 + \sum_{i=3}^{n} (1 + X_i \cdot i).$$

Thus, the expected running time is proportional to

$$\mathbf{E}\Big[R\Big] = \mathbf{E}\left[1 + \sum_{i=3}^{n} (1 + X_i \cdot i)\right] \leq n + \sum_{i=3}^{n} \mathbf{E}\Big[X_i\Big] \cdot i \leq n + \sum_{i=3}^{n} i \cdot \mathbf{Pr}\Big[X_i = 1\Big]$$
$$\leq n + \sum_{i=3}^{n} i \cdot \frac{2}{i} \leq 3n,$$

by linearity of expectation and since $\mathbf{E}[X_i] = \mathbf{Pr}[X_i = 1]$ and since $\mathbf{Pr}[X_i = 1] \leq 2/i$ (as shown in the proof of Lemma 1.9). Thus, the expected running time of the algorithm is $O(\mathbf{E}[R]) = O(n)$. ■

Theorem 1.10 is a surprising result, since it implies that uniqueness (i.e., deciding if n real numbers are all distinct) can be solved in linear time. Indeed, compute the distance of the closest pair of the given numbers (think about the numbers as points on the x-axis). If this distance is zero, then clearly they are not all unique.

However, there is a lower bound of $\Omega(n \log n)$ on the running time to solve uniqueness, using the comparison model. This "reality dysfunction" can be easily explained once one realizes that the computation model of Theorem 1.10 is considerably stronger, using hashing, randomization, and the floor function.

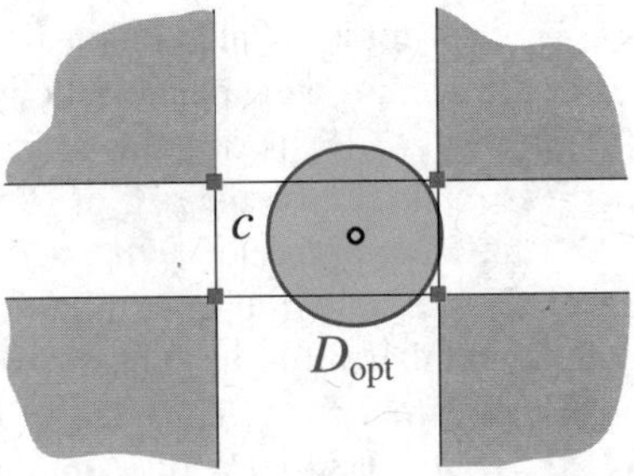

FIGURE 1.1. If the disk $D_{opt}(\mathsf{P}, k)$ does not contain any vertex of the cell c, then it does not cover any shaded area. As such, it can contain at most $k/2$ points, since the vertical and horizontal strips containing c each has at most $k/4$ points of P inside it.

1.3. A slow 2-approximation algorithm for the k-enclosing disk

For a disk D, we denote by radius(D) the ***radius*** of D. Let $D_{opt}(\mathsf{P}, k)$ be a disk of minimum radius which contains k points of P, and let $r_{opt}(\mathsf{P}, k)$ denote the radius of $D_{opt}(\mathsf{P}, k)$. For $k = 2$, this is equivalent to computing the closest pair of points of P. As such, the problem of computing $D_{opt}(\mathsf{P}, k)$ is a generalization of the problem studied in the previous section.

Here, we study the easier problem of approximating $r_{opt}(\mathsf{P}, k)$.

Observation 1.11. *Given a set P of n points in the plane, a point q, and a parameter k, one can compute the k closest points in P to q in $O(n)$ time. To do so, compute for each point of P its distance to q. Next, using a selection algorithm, compute the k smallest numbers among these n distances. These k numbers corresponds to the desired k points. The running time is $O(n)$ as selection can be done in linear time.*

The algorithm algDCoverSlow(P, k). Let P be a set of n points in the plane. Compute a set of $m = O(n/k)$ horizontal lines $h_1, \ldots, h_m$ such that between two consecutive horizontal lines, there are at most $k/4$ points of P in the strip they define. To this end, consider the points of P sorted in increasing y ordering, and create a horizontal line through the points of rank $i(k/4)$ in this order, for $i = 1, \ldots, \lfloor n/(k/4) \rfloor$. This can be done in $O(n \log(n/k))$ time using deterministic median selection together with recursion.[2] Similarly, compute a set of vertical lines $v_1, \ldots, v_m$, such that between two consecutive lines, there are at most $k/4$ points of P. (Note that all these lines pass through points of P.)

Consider the (non-uniform) grid G induced by the lines $h_1, \ldots, h_m$ and $v_1, \ldots, v_m$. Let X be the set of all the intersection points of these lines; that is, X is the set of vertices of G. For every point $\mathsf{p} \in X$, compute (in linear time) the smallest disk centered at p that contains k points of P, and return the smallest disk computed.

Lemma 1.12. *Given a set P of n points in the plane and parameter k, the algorithm* **algDCoverSlow**(P, k) *computes, in $O(n(n/k)^2)$ deterministic time, a circle D that contains k points of P and* radius$(D) \le 2r_{opt}(\mathsf{P}, k)$, *where $r_{opt}(\mathsf{P}, k)$ is the radius of the smallest disk in the plane containing k points of P.*

---

[2] Indeed, compute the median in the y-order of the points of P, split P into two sets, and recurse on each set, till the number of points in a subproblem is of size $\le k/4$. We have $T(n) = O(n) + 2T(n/2)$, and the recursion stops for subproblems of size $\le k/4$. Thus, the recursion tree has depth $O(\log(n/k))$, which implies running time $O(n \log(n/k))$.

Proof. Since $|X| = O((n/k)^2)$ and for each such point finding the smallest disk containing k points takes $O(n)$ time, the running time bound follows.

As for correctness, we claim that $D_{opt}(\mathsf{P}, k)$ contains at least one point of X. Indeed, consider the center u of $D_{opt}(\mathsf{P}, k)$, and let c be the cell of G that contains u. Clearly, if $D_{opt}(\mathsf{P}, k)$ does not cover any of the four vertices of c, then it can cover only points in the vertical and horizontal strips of G that contain c. See Figure 1.1. However, each such strip contains at most $k/4$ points, and there are two such strips. It follows that $D_{opt}(\mathsf{P}, k)$ contains at most $k/2$ points of P, a contradiction. Thus, $D_{opt}(\mathsf{P}, k)$ must contain a point of X. Clearly, for a point $\mathsf{q} \in X \cap D_{opt}(\mathsf{P}, k)$, this yields the required 2-approximation. Indeed, the disk of radius $2r_{opt}(\mathsf{P}, k)$ centered at q contains at least k points of P since it also covers $D_{opt}(\mathsf{P}, k)$. ∎

Corollary 1.13. *Given a set* P *of* n *points and a parameter* $k = \Omega(n)$*, one can compute in linear time a circle* D *that contains* k *points of* P *and* $\mathrm{radius}(D) \leq 2r_{opt}(\mathsf{P}, k)$.

Remark 1.14. If **algDCoverSlow**(P, k) is applied to a point set P of size smaller than k, then the algorithm picks an arbitrary point p of P and outputs the minimum radius disk centered at p containing P. This takes $O(|\mathsf{P}|)$ time.

Remark 1.15. One can sometime encode the output of a geometric algorithm in terms of the input objects that define it. This might be useful for handling numerical issues, where this might prevent numerical errors. In our case, we will use this to argue about the expected running time of an algorithm that uses **algDCoverSlow** in a black-box fashion.

So, consider the disk D output by the algorithm **algDCoverSlow**(P, k). It is centered at point p of radius r. The point p is the intersection of two grid lines. Each of these grid lines, by construction, passes through two input points $\mathsf{q}, \mathsf{s} \in \mathsf{P}$. Similarly, the radius of D is the distance of the intersection point p to a point $\mathsf{t} \in \mathsf{P}$. Namely, the disk D can be uniquely specified by a triple of points $(\mathsf{q}, \mathsf{s}, \mathsf{t})$ of P, where the first (resp. second) point q (resp. s) specifies the vertical (resp. horizontal) line of the grid that passes through this point (thus the two points specify the center of the disk), and the third point specifies the points on the boundary of the disk; see the figure on the right.

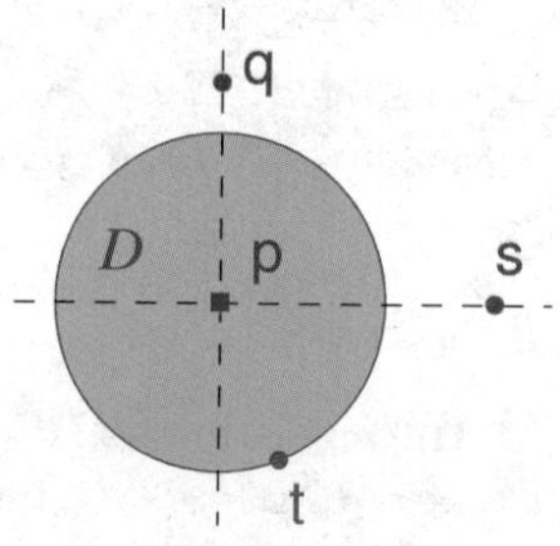

Now, think about running **algDCoverSlow** on all possible subsets $\mathsf{Q} \subseteq \mathsf{P}$. The above argument implies that although there the number of different inputs considered is exponential, the algorithm always outputs one of n^3 possible outputs, where $n = |\mathsf{P}|$.

Lemma 1.12 can be easily extended to higher dimensions. We get the following result.

Lemma 1.16. *Given a set* P *of* n *points in* $\mathbb{R}^d$ *and parameter* k*, one can compute, in* $O\left(n(n/k)^d\right)$ *deterministic time, a ball* $\mathbf{b}$ *that contains* k *points of* P *and its radius* $\mathrm{radius}(\mathbf{b}) \leq 2r_{opt}(\mathsf{P}, k)$*, where* $r_{opt}(\mathsf{P}, k)$ *is the radius of the smallest ball in* $\mathbb{R}^d$ *containing* k *points of* P.

1.4. A linear time 2-approximation for the k-enclosing disk

In the following, we present a linear time algorithm for approximating the minimum enclosing disk. While interesting in their own right, the results here are not used later and can be skipped on first, second, or third (or any other) reading.

As in the previous sections, we construct a grid which partitions the points into small ($O(k)$ sized) groups. The key idea behind speeding up the grid computation is to construct

algGrow(P_i,α_{i-1},k)
  Output: α_i
begin
  $\mathsf{G}_{i-1} \leftarrow \mathsf{G}_{\alpha_{i-1}}(\mathsf{P}_i)$
  for every grid cluster $c \in \mathsf{G}_{i-1}$
      with $|c \cap \mathsf{P}_i| \geq k$ **do**
    $\mathsf{P}_c \leftarrow c \cap \mathsf{P}_i$
    $\alpha_c \leftarrow$ **algDCoverSlow**(P_c, k)
    // $r_{\mathrm{opt}}(\mathsf{P}_c, k) \leq \alpha_c \leq 2r_{\mathrm{opt}}(\mathsf{P}_c, k)$

  return minimum α_c computed.
end

algDCover(P, k)
  Output: 2-approx. to $r_{\mathrm{opt}}(\mathsf{P}, k)$
begin
  Compute a gradation of P:
    $\{\mathsf{P}_1, \ldots, \mathsf{P}_m\}$
  $\alpha_1 \leftarrow$ **algDCoverSlow**(P_1, k)
  for $i \leftarrow 2$ to m **do**
    $\alpha_i \leftarrow$ **algGrow**(P_i, α_{i-1}, k)
  return α_m
end

FIGURE 1.2. The linear time 2-approximation algorithm **algDCover** for the smallest disk containing k points of P. Here, **algDCoverSlow** denotes the "slow" 2-approximation algorithm of Lemma 1.16.

the appropriate grid over several rounds. Specifically, we start with a small set of points as a seed and construct a suitable grid for this subset. Next, we incrementally insert the remaining points, while adjusting the grid width appropriately at each step.

1.4.1. The algorithm.

1.4.1.1. *Preliminaries.* We remind the reader that **algDCoverSlow**(P, k) is the "slow" algorithm of Lemma 1.16.

Definition 1.17. Given a set P of n points, a ***k-gradation*** $(\mathsf{P}_1, \ldots, \mathsf{P}_m)$ of P is a sequence of subsets of P, such that (i) $\mathsf{P}_m = \mathsf{P}$, (ii) P_i is formed by picking each point of P_{i+1} with probability $1/2$, (iii) $|\mathsf{P}_1| \leq k$, and (iv) $|\mathsf{P}_2| > k$.

We remind the reader that a ***grid cluster*** is a block of 3×3 contiguous grid cells.

Definition 1.18. Let $\mathsf{gd}_\alpha(\mathsf{P})$ denote the maximum number of points of P mapped to a single grid cell by the partition $\mathsf{G}_\alpha = \mathsf{G}_\alpha(\mathsf{P})$; see Definition 1.3.

Also, let $\mathtt{depth}(\mathsf{P}, \alpha)$ be the maximum number of points of P that a circle of radius α can contain.

Lemma 1.19. *For any point set* P *and* $\alpha > 0$, *we have that if* $\alpha \leq 2r_{\mathrm{opt}}(\mathsf{P}, k)$, *then any cell of the grid* G_α *contains at most* $5k$ *points; that is,* $\mathsf{gd}_\alpha(\mathsf{P}) \leq 5k$.

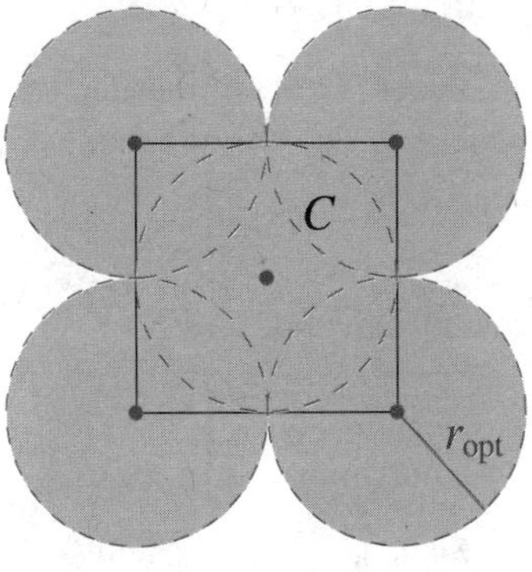

PROOF. Let C be the grid cell of G_α that realizes $\mathsf{gd}_\alpha(\mathsf{P})$. Place 4 points at the corners of C and one point in the center of C. Placing at each of those points a disk of radius $r_{\mathrm{opt}}(\mathsf{P}, k)$ completely covers C, as can be easily verified (since the sidelength of C is at most $2r_{\mathrm{opt}}(\mathsf{P}, k)$). Thus, $|\mathsf{P} \cap C| = \mathsf{gd}_\alpha(\mathsf{P}) \leq 5\,\mathtt{depth}(\mathsf{P}, r_{\mathrm{opt}}(\mathsf{P}, k)) = 5k$; see the figure on the right. ■

1.4.1.2. *The algorithm.* We first compute a gradation $\{P_1, \ldots, P_m\}$ of P. We use **algDCoverSlow** on P_1 to get the desired approximation for P_1. Next, we iteratively refine this approximation moving from one set in the gradation to the next one. Specifically, motivated by the above, we will maintain the following invariants at the end of the ith round: (i) there is a distance α_i such that $\mathsf{gd}_{\alpha_i}(P_i) \leq 5k$, (ii) there is a grid cluster in G_{α_i} containing k or more points of P_i, and (iii) $r_{\text{opt}}(P_i, k) \leq \alpha_i$.

At the ith round, the algorithm constructs a grid G_{i-1} for points in P_i using α_{i-1} as the grid width. We use the slow algorithm of Lemma 1.16 (i.e., **algDCoverSlow**), on each of the non-empty grid clusters, to compute α_i.

In the end of this process, the number α_m is the desired approximation.

The new algorithm **algDCover** is depicted in Figure 1.2.

Intuition. During its execution, the algorithm never maps too many points to a single grid cell being used. Indeed, we know that there is no grid cell of G_{i-1} containing more than $5k$ points of P_{i-1}. We expect every cell of G_{i-1} to contain at most $10k$ points of P_i, since $P_{i-1} \subseteq P_i$ was formed by choosing each point of P_i with probability $1/2$. (This of course is too good to be true, but something slightly weaker does hold.)

Note that the algorithm of Lemma 1.16 runs in linear time if the number of points in the set is $O(k)$. In the optimistic scenario, every cluster of $G_{i-1}(P_i)$ has $O(k)$ points, and computing the clustering radius for this point set takes $O(k)$ time. Now, every point participates in a constant number of clusters, and it follows that this fixing stage takes (overall) linear time in the size of P_i.

1.4.2. Analysis – correctness.

Lemma 1.20. *For $i = 1, \ldots, m$, we have $r_{\text{opt}}(P_i, k) \leq \alpha_i \leq 2r_{\text{opt}}(P_i, k)$, and the heaviest cell in $G_{\alpha_i}(P_i)$ contains at most $5k$ points of P_i.*

Proof. Consider the optimal disk D_i that realizes $r_{\text{opt}}(P_i, k)$. Observe that there is a cluster c of $G_{\alpha_{i-1}}$ that contains D_i, as $\alpha_{i-1} \geq \alpha_i$. Thus, when **algGrow** handles the cluster c, we have $D_i \cap P_i \subseteq c$. The first part of the lemma then follows from the correctness of the algorithm of Lemma 1.16.

The second part follows by Lemma 1.19. ■

Lemma 1.20 implies immediately the correctness of the algorithm by applying it for $i = m$.

1.4.3. Running time analysis.

Lemma 1.21. *Given a set P, a gradation of P can be computed in expected linear time.*

Proof. Observe that the sampling time is $O\left(\sum_{i=1}^m |P_i|\right)$, where m is the length of the sequence. Also observe that $\mathbf{E}[|P_m|] = |P_m| = n$ and

$$\mathbf{E}\Big[|P_i|\Big] = \mathbf{E}\Big[\mathbf{E}\Big[|P_i| \,\Big|\, |P_{i+1}|\Big]\Big] = \mathbf{E}\left[\frac{|P_{i+1}|}{2}\right] = \frac{1}{2}\mathbf{E}\Big[|P_{i+1}|\Big].$$

Now by induction, we get

$$\mathbf{E}\Big[|P_{m-i}|\Big] = \frac{n}{2^i}.$$

Thus, the running time is $O\left(\mathbf{E}\left[\sum_{i=1}^m |P_i|\right]\right) = O\left(\sum_{i=1}^m n/2^i\right) = O(n)$. ■

Since $|P_1| \leq k$, the call $\alpha_1 \leftarrow$ **algDCoverSlow**(P_1, k) in **algDCover** takes $O(k)$ time, by Remark 1.14.

1.4.3.1. *Excess and why it is low.* Now we proceed to upper-bound the number of cells of $\mathsf{G}_{\alpha_{i-1}}$ that contains "too many" points of P_i. Since each point of P_{i-1} was chosen from P_i with probability 1/2, we can express this bound as a sum of independent random variables, and we can bound this using tail-bounds.

Definition 1.22. For a point set P and parameters k and α, the ***excess*** of $\mathsf{G}_\alpha(\mathsf{P})$ is

$$\mathcal{E}(\mathsf{P}, \alpha) = \sum_{c \in \mathrm{Cells}(\mathsf{G}_\alpha)} \left\lfloor \frac{|c \cap \mathsf{P}|}{50k} \right\rfloor \leq \frac{|\mathsf{P}|}{50k},$$

where $\mathrm{Cells}(\mathsf{G}_\alpha)$ is the set of cells of the grid G_α.

The quantity $100k \cdot \mathcal{E}(\mathsf{P}, \alpha)$ is an upper bound on the number of points of P in heavy cells of $\mathsf{G}_\alpha(\mathsf{P})$, where a cell of $\mathsf{G}_\alpha(\mathsf{P})$ is ***heavy*** if it contains at least $50k$ points.

For a point set P of n points, the radius α returned by the algorithm of Lemma 1.16 is the distance between a vertex q of the non-uniform grid (i.e., a point of the set X) and a point of P. Remark 1.15 implies that such a number α is defined by a triple of points of P (i.e., they specify the three points that α is computed from). As such, α can be one of at most $O(n^3)$ different values. This implies that throughout the execution of **algDCover** the only grids that would be considered would be one of (at most) n^3 possible grids. In particular, let $\mathfrak{G} = \mathfrak{G}(\mathsf{P})$ be this set of possible grids.

Lemma 1.23. *For any $t > 0$, let $\gamma = \lceil 3 \ln n \rceil / 8k$ and $\beta = t + \gamma$. The probability that $\mathsf{G}_{\alpha_{i-1}}(\mathsf{P}_i)$ has excess $\mathcal{E}(\mathsf{P}_i, \alpha_{i-1}) \geq \beta$ is at most* $\exp(-8kt)$.

PROOF. Let $\mathfrak{G} = \mathfrak{G}(\mathsf{P})$ (see above), and fix a grid $\mathsf{G} \in \mathfrak{G}$ with excess $\mathcal{E}(\mathsf{P}_i, \varkappa(\mathsf{G}))$ at least β, where $\varkappa(\mathsf{G})$ is the sidelength of a cell of G.

Let $U = \left\{ \{\mathsf{P}_i \cap c\} \;\middle|\; c \in \mathsf{G}, |\mathsf{P}_i \cap c| \geq 50k \right\}$ be the sets of points stored in the heavy cells of G. Furthermore, let $\mathcal{V} = \bigcup_{X \in U} \Pi(X, 50k)$, where $\Pi(X, \nu)$ denotes an arbitrary partition of the set X into disjoint subsets such that each one of them contains ν points, except maybe the last subset that might contain between ν and $2\nu - 1$ points.

This partitions the points inside each heavy cell into groups of size at least $50k$. Now, since each such group lies inside a single cell of G, for G to be the grid computed for P_{i-1}, it must be that every such group "promoted" at most $5k$ points from P_i to P_{i-1}, by Lemma 1.20.

Now, it is clear that $|\mathcal{V}| = \mathcal{E}(\mathsf{P}_i, \varkappa(\mathsf{G}))$, and for any $S \in \mathcal{V}$, we have that $\mu = \mathbf{E}[|S \cap \mathsf{P}_{i-1}|] \geq 25k$. Indeed, we promote each point of $S \subseteq \mathsf{P}_i$, independently, with probability 1/2, to be in P_{i-1} and $|S| \geq 50k$. As such, by the Chernoff inequality (Theorem 27.18$_{\text{p341}}$), for $\delta = 4/5$, we have

$$\begin{aligned}\mathbf{Pr}\Big[|S \cap \mathsf{P}_{i-1}| \leq 5k\Big] &\leq \mathbf{Pr}\Big[|S \cap \mathsf{P}_{i-1}| \leq (1-\delta)\mu\Big] < \exp\left(-\mu\frac{\delta^2}{2}\right) \leq \exp\left(-\frac{25k(4/5)^2}{2}\right)\\ &= \exp(-8k).\end{aligned}$$

Furthermore, since $\mathsf{G} = \mathsf{G}_{\alpha_{i-1}}$, this imply that each cell of G contains at most $5k$ points of P_{i-1}, by Lemma 1.20. Thus we have

$$\begin{aligned}\mathbf{Pr}\Big[\mathsf{G}_{\alpha_{i-1}} = \mathsf{G}\Big] &\leq \prod_{S \in \mathcal{V}} \mathbf{Pr}\Big[|S \cap \mathsf{P}_{i-1}| \leq 5k\Big] \leq \exp(-8k\,|\mathcal{V}|) = \exp(-8k\mathcal{E}(\mathsf{P}_i, \varkappa(\mathsf{G})))\\ &\leq \exp(-8k\beta).\end{aligned}$$

Since there are at most n^3 different grids in $\mathfrak{G}$, we have

$$\mathbf{Pr}\left[\mathcal{E}(\mathsf{P}_i, \alpha_{i-1}) \geq \beta\right] = \mathbf{Pr}\left[\bigcup_{\substack{\mathsf{G}\in\mathfrak{G},\\ \mathcal{E}(\mathsf{P}_i,\varkappa(\mathsf{G}))\geq\beta}} (\mathsf{G} = \mathsf{G}_{\alpha_{i-1}})\right] \leq \sum_{\substack{\mathsf{G}\in\mathfrak{G},\\ \mathcal{E}(\mathsf{P}_i,\varkappa(\mathsf{G}))\geq\beta}} \mathbf{Pr}[\mathsf{G} = \mathsf{G}_{\alpha_{i-1}}]$$

$$\leq n^3 \exp(-8k\beta) \leq \exp\left(3\ln n - 8k\left(t + \frac{3\lceil \ln n \rceil}{8k}\right)\right) \leq \exp(-8kt). \qquad \blacksquare$$

1.4.3.2. *Bounding the running time.*

Lemma 1.24. *In expectation, the running time of the algorithm in the ith iteration is* $O\left(|\mathsf{P}_i| + \gamma^4 k\right)$, *where* $\gamma = \lceil 3 \ln n \rceil / 8k$.

Proof. Let Y be the random variable which is the excess of $\mathsf{G}_{\alpha_{i-1}}(\mathsf{P}_i)$. In this case, there are at most Y cells which are heavy in $\mathsf{G}_{\alpha_{i-1}}(\mathsf{P}_i)$, and each such cell contains at most $O(Yk)$ points. Thus, invoking the algorithm of **algDCoverSlow** on such a heavy cell takes $O(Yk \cdot ((Yk)/k)^2) = O(Y^3 k)$ time. Overall, the running time of **algGrow**, in the ith iteration, is $T(Y) = O\left(|\mathsf{P}_i| + Y \cdot Y^3 k\right) = O\left(|\mathsf{P}_i| + Y^4 k\right)$.

Set $\gamma = \lceil 3 \ln n \rceil / 8k$, and observe that the expected running time in the ith stage is

$$R_i = O\left(\sum_{t=0}^{n/k} \mathbf{Pr}[Y = t] \left(|\mathsf{P}_i| + Y^4 k\right)\right)$$

$$= O\left(|\mathsf{P}_i| + \mathbf{Pr}[Y \leq \gamma]\, \gamma^4 k + \sum_{t=1}^{n/k} \mathbf{Pr}[Y = t + \gamma](t + \gamma)^4 k\right)$$

$$= O\left(|\mathsf{P}_i| + \gamma^4 k + \sum_{t=1}^{n/k} \exp(-8kt) \cdot (t + \gamma)^4 k\right) = O\left(|\mathsf{P}_i| + \gamma^4 k\right),$$

by Lemma 1.23 and since the summation is bounded by

$$O\left(\sum_{t=1}^{n/k} 16 t^4 \gamma^4 k \exp(-8kt)\right) = O\left(\gamma^4 k \sum_{t=1}^{n/k} t^4 \exp(-8kt)\right) = O\left(\gamma^4 k\right). \qquad \blacksquare$$

Thus, by Lemma 1.24 and by Lemma 1.21, the total expected running time of **algDCover** inside the inner loop is

$$O\left(\sum_{i=1}^{m} \left(|\mathsf{P}_i| + \gamma^4 k\right)\right) = O\left(n + \left(\frac{3\ln n}{8k}\right)^4 km\right) = O(n),$$

since $m = O(\log n)$, with high probability, as can be easily verified.

Theorem 1.25. *Given a set* P *of n points in the plane and a parameter k, the algorithm* **algDCover** *computes, in expected linear time, a radius* α, *such that* $r_{\mathrm{opt}}(\mathsf{P}, k) \leq \alpha \leq 2r_{\mathrm{opt}}(\mathsf{P}, k)$, *where* $r_{\mathrm{opt}}(\mathsf{P}, k)$ *is the minimum radius of a disk covering k points of* P.

1.5. Bibliographical notes

Our closest-pair algorithm follows Golin et al. [**GRSS95**]. This is in turn a simplification of a result of Rabin [**Rab76**]. Smid provides a survey of such algorithms [**Smi00**].

The minimum disk approximation algorithm is a simplification of the work of Har-Peled and Mazumdar [**HM05**]. Note that this algorithm can be easily adapted to any point set in constant dimension (with the same running time). Exercise 1.3 is also taken from there.

Computing the exact minimum disk containing k points. By plugging the algorithm of Theorem 1.25 into the exact algorithm of Matoušek [**Mat95a**], one gets an $O(nk)$ time algorithm that computes the minimum disk containing k points. It is conjectured that any exact algorithm for this problem requires $\Omega(nk)$ time.

1.6. Exercises

Exercise 1.1 (Packing argument and the curse of dimensionality). One of the reasons why computational problems in geometry become harder as the dimension increases is that packing arguments (see Lemma 1.5) provide bounds that are exponential in the dimension, and even for moderately small dimension (say, $d = 16$) the bounds they provide are too large to be useful.

As a concrete example, consider a maximum cardinality point set P contained inside the unit length cube C in $\mathbb{R}^d$ (i.e., the ***unit hypercube***), such that $\mathcal{CP}(P) = 1$. Prove that

$$\left(\left\lfloor\sqrt{d}\right\rfloor + 1\right)^d \leq |P| \leq \left(\left\lceil\sqrt{d}\right\rceil + 1\right)^d.$$

Exercise 1.2 (Compute clustering radius). Let C and P be two given sets of points in the plane, such that $k = |C|$ and $n = |P|$. Let $r = \max_{p \in P} \min_{\overline{c} \in C} \left\|\overline{c} - p\right\|$ be the ***covering radius*** of P by C (i.e., if we place a disk of radius r around each point of C, all those disks cover the points of P).

(A) Give an $O(n + k \log n)$ expected time algorithm that outputs a number α, such that $r \leq \alpha \leq 10r$.

(B) For $\varepsilon > 0$ a prescribed parameter, give an $O(n + k\varepsilon^{-2} \log n)$ expected time algorithm that outputs a number α, such that $\alpha \leq r \leq (1 + \varepsilon)\alpha$.

Exercise 1.3 (Randomized k-enclosing disk). Given a set P of n points in the plane and parameter k, present a (simple) randomized algorithm that computes, in expected $O(n(n/k))$ time, a circle D that contains k points of P and $\text{radius}(D) \leq 2r_{\text{opt}}(P, k)$.

(This is a faster and simpler algorithm than the one presented in Lemma 1.16.)

CHAPTER 2

Quadtrees – Hierarchical Grids

In this chapter, we discuss quadtrees, arguably one of the simplest and most powerful geometric data-structures. We begin in Section 2.1 by defining quadtrees and giving a simple application. We also describe a clever way for performing point-location queries quickly in such quadtrees. In Section 2.2, we describe how such quadtrees can be compressed, constructed quickly, and be used for point-location queries. In Section 2.3, we show how to dynamically maintain a compressed quadtree under insertions and deletions of points.

2.1. Quadtrees – a simple point-location data-structure

Let P_{map} be a planar map. To be more concrete, let P_{map} be a partition of the unit square into polygons. The partition P_{map} can represent any planar map where a region in the map might be composed of several polygons (or triangles). For the sake of simplicity, assume that every vertex in P_{map} appears in a constant number of polygons.

We want to preprocess P_{map} for point-location queries, so that given a query point, we can figure out which polygon contains the query (see the figure on the right). Of course, there are numerous data-structures that can do this, but let us consider the following simple solution (which in the worst case, can be quite bad).

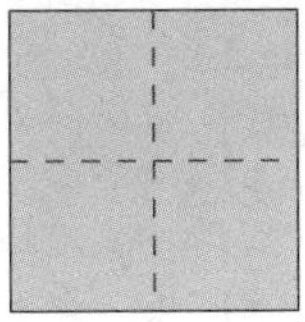

Build a tree $\mathcal{T}$, where every node $v \in T$ corresponds to a cell $\square_v$ (i.e., a square) and the root corresponds to the unit square. Each node has four children that correspond to the four equal sized squares formed by splitting $\square_v$ by horizontal and vertical cuts; see the figure on the right.

The construction is recursive, and we start from $v = \mathrm{root}_{\mathcal{T}}$. The ***conflict list*** of the square $\square_v$ (i.e., the square associated with v) is a list of all the polygons of P_{map} that intersect $\square_v$. If the current node's conflict list has more than, say, nine[1] polygons, we create its children nodes, and we call recursively on each child. We compute each child's conflict list from its parent list. As such, we stop at a leaf if its conflict list is of size at most nine. For each constructed leaf, we store in it its conflict list (but we do not store the conflict list for internal nodes).

Given a query point q, in the unit square, we can compute the polygon of P_{map} containing q by traversing down $\mathcal{T}$ from the root, repeatedly going into the child of the current node, whose square contains q. We stop as soon as we reach a leaf, and then we scan the leaf's conflict list and check which of the polygons in this list contains q.

[1]The constant here is arbitrary. It just has to be at least the number of polygons of P_{map} meeting in one common corner.

An example is depicted on the right, where the nodes on the search path of the point-location query (i.e., the nodes accessed during the query execution) are shown. The marked polygons are all the polygons in the conflict list of the leaf containing the query point.

Of course, in the worst case, if the polygons are long and skinny, this quadtree might have unbounded complexity. However, for reasonable inputs (say, the polygons are all fat triangles), then the quadtree would have linear complexity in the input size (see Exercise 2.2). The major advantage of quadtrees, of course, is their simplicity. In a lot of cases, quadtrees would be a sufficient solution, and seeing how to solve a problem using quadtrees might be a first insight into a problem.

2.1.1. Fast point-location in a quadtree. One possible interpretation of quadtrees is that they are a multi-grid representation of a point set.

Definition 2.1 (Canonical squares and canonical grids)**.** A square is a ***canonical square*** if it is contained inside the unit square, it is a cell in a grid G_r, and r is a power of two (i.e., it might correspond to a node in a quadtree). We will refer to such a grid G_r as a ***canonical grid***.

Consider a node v of a quadtree of depth i (the root has depth 0) and its associated square $\square_v$. The sidelength of $\square_v$ is 2^{-i}, and it is a canonical square in the canonical grid $\mathsf{G}_{2^{-i}}$. As such, we will refer to $\ell(v) = -i$ as the ***level*** of v. However, a cell in a grid has a unique ID made out of two integer numbers. Thus, a node v of a quadtree is uniquely defined by the triple $\mathrm{id}(v) = (\ell(v), \lfloor x/r \rfloor, \lfloor y/r \rfloor)$, where (x, y) is any point in $\square_v$ and $r = 2^{\ell(v)}$.

```
QTFastPLI(T, q, l, h).
    m ← ⌊(l + h)/2⌋
    v ← QTGetNode(T, q, m)
    if v = null then
        return QTFastPLI(T, q, l, m − 1).
    w ← Child(v, q)
            //w is the child of v containing q.
    If w = null then
        return v
    return QTFastPLI(T, q, m + 1, h)
```

FIGURE 2.1. One can perform point-location in a quadtree $\mathcal{T}$ by calling **QTFastPLI**$(T, q, 0, \mathrm{height}(T))$.

Furthermore, given a query point q and a desired level ℓ, we can compute the ID of the quadtree cell of this level that contains q in constant time. Thus, this suggests a natural algorithm for doing a point-location in a quadtree: Store all the IDs of nodes in the quadtree in a hash-table, and also compute the maximal depth h of the quadtree. Given a query point q, we now have access to any node along the point-location path of q in $\mathcal{T}$, in constant time. In particular, we want to find the point in $\mathcal{T}$ where this path "falls off" the quadtree (i.e., reaches the leaf). This we can find by performing a binary search for the leaf.

Let **QTGetNode**(T, q, d) denote the procedure that, in constant time, returns the node v of depth d in the quadtree $\mathcal{T}$ such that $\square_v$ contains the point q. Given a query point q, we can perform point-location in $\mathcal{T}$ (i.e., find the leaf containing the query) by calling **QTFastPLI**$(T, \mathsf{q}, 0, \mathrm{height}(T))$. See Figure 2.1 for the pseudo-code for **QTFastPLI**.

Lemma 2.2. *Given a quadtree $\mathcal{T}$ of size n and of height h, one can preprocess it (using hashing), in linear time, such that one can perform a point-location query in $\mathcal{T}$ in $O(\log h)$*

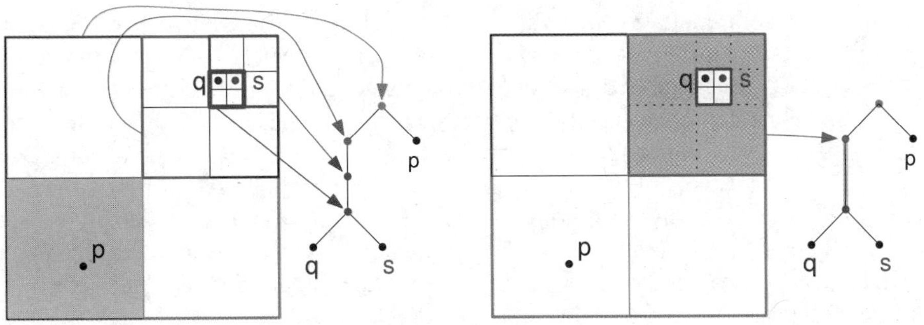

FIGURE 2.2. A point set, its quadtree and its compressed quadtree. Note that each node is associated with a canonical square. For example, the node in the above quadtree marked by p would store the gray square shown on the left and also would store the point p in this node.

time. In particular, if the quadtree has height $O(\log n)$ (i.e., it is "balanced"), then one can perform a point-location query in $\mathcal{T}$ in $O(\log \log n)$ time.

2.2. Compressed quadtrees

2.2.1. Definition.

Definition 2.3 (Spread). For a set P of n points in a metric space, let

$$\Phi(\mathsf{P}) = \frac{\max_{\mathsf{p},\mathsf{q}\in\mathsf{P}} \|\mathsf{p}-\mathsf{q}\|}{\min_{\mathsf{p},\mathsf{q}\in\mathsf{P},\mathsf{p}\neq\mathsf{q}} \|\mathsf{p}-\mathsf{q}\|} \qquad (2.1)$$

be the ***spread*** of P. In words, the spread of P is the ratio between the diameter of P and the distance between the two closest points in P. Intuitively, the spread tells us the range of distances that P possesses.

One can build a quadtree $\mathcal{T}$ for P, storing the points of P in the leaves of $\mathcal{T}$, where one keeps splitting a node as long as it contains more than one point of P. During this recursive construction, if a leaf contains no points of P, we save space by not creating this leaf and instead creating a null pointer in the parent node for this child.

Lemma 2.4. *Let* P *be a set of n points contained in the unit square, such that* $\mathrm{diam}(\mathsf{P}) = \max_{\mathsf{p},\mathsf{q}\in\mathsf{P}} \|\mathsf{p}-\mathsf{q}\| \geq 1/2$. *Let $\mathcal{T}$ be a quadtree of* P *constructed over the unit square, where no leaf contains more than one point of* P. *Then, the depth of $\mathcal{T}$ is bounded by $O(\log \Phi)$, it can be constructed in $O(n \log \Phi)$ time, and the total size of $\mathcal{T}$ is $O(n \log \Phi)$, where $\Phi = \Phi(\mathsf{P})$.*

PROOF. The construction is done by a straightforward recursive algorithm as described above.

Let us bound the depth of $\mathcal{T}$. Consider any two points $\mathsf{p}, \mathsf{q} \in \mathsf{P}$, and observe that a node v of $\mathcal{T}$ of level $u = \lfloor \lg \|\mathsf{p}-\mathsf{q}\| \rfloor - 1$ containing p must not contain q (we remind the reader that $\lg n = \log_2 n$). Indeed, the diameter of $\square_v$ is smaller than $\sqrt{1^2+1^2}2^u = \sqrt{2}2^u \leq \sqrt{2}\|\mathsf{p}-\mathsf{q}\|/2 < \|\mathsf{p}-\mathsf{q}\|$. Thus, $\square_v$ cannot contain both p and q. In particular, any node u of $\mathcal{T}$ of level $r = -\lceil \lg \Phi \rceil - 2$ can contain at most one point of P. Since a node of the quadtree containing a single point of P is a leaf of the quadtree, it follows that all the nodes of $\mathcal{T}$ are of depth $O(\log \Phi)$.

Since the construction algorithm spends $O(n)$ time at each level of $\mathcal{T}$, it follows that the construction time is $O(n \log \Phi)$, and this also bounds the size of the quadtree $\mathcal{T}$. ■

The bounds of Lemma 2.4 are tight, as one can easily verify; see Exercise 2.3. But in fact, if you inspect a quadtree generated by Lemma 2.4, you would realize that there are a lot of nodes of $\mathcal{T}$ which are of degree one (the *degree* of a node is the number of children it has). See Figure 2.2 for an example. Indeed, a node v of $\mathcal{T}$ has more than one child only if it has at least two children x and y, such that both $\square_x$ and $\square_y$ contain points of P. Let P_v be the subset of points of P stored in the subtree of v, and observe that $\mathsf{P}_x \cup \mathsf{P}_y \subseteq \mathsf{P}_v$ and $\mathsf{P}_x \cap \mathsf{P}_y = \emptyset$. Namely, such a node v splits P_v into at least two non-empty subsets and globally there can be only $n - 1$ such splitting nodes. Thus, a regular quadtree might have a large number of "useless" nodes that one should be able to get rid of and get a more compact data-structure.

If you compress them, they will fit in memory. We can replace such a sequence of edges by a single edge. To this end, we will store inside each quadtree node v its square $\square_v$ and its level $\ell(v)$. Given a path of vertices in the quadtree that are all of degree one, we will replace them with a single vertex v that corresponds to the first vertex in this path, and its only child would be the last vertex in this path (this is the first node of degree larger than one). This ***compressed node*** v has a single child, and the region rg_v that v "controls" is an annulus, seen as the gray area in the figure on the right.

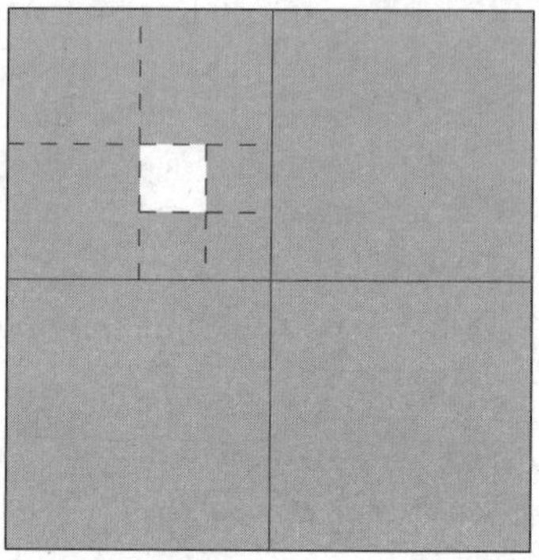

Otherwise, if v is not compressed, the ***region*** that v is in charge of is a square $\mathsf{rg}_v = \square_v$. Specifically, an uncompressed node v is either a leaf or a node that splits the point set into two (or more) non-empty subsets (i.e., two or more of the children of v have points stored in their subtrees). For a compressed node v, its sole child corresponds to the inner square, which contains all the points of P_v. We call the resulting tree a ***compressed quadtree***.

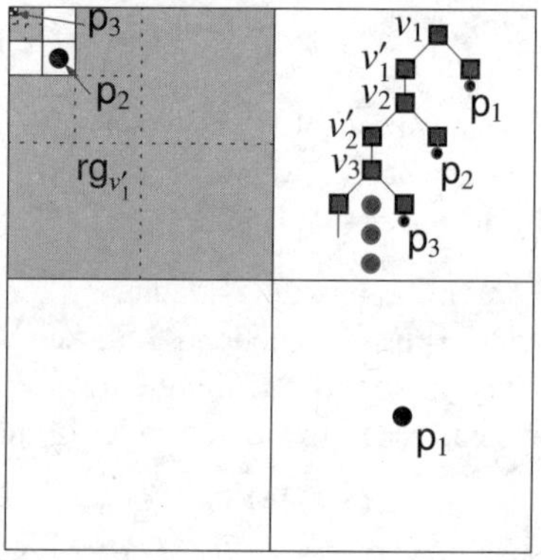

FIGURE 2.3.

Observe that the only nodes in a compressed quadtree with a single child are compressed (they might be trivially compressed, the child being one of the four subsquares of the parent). In particular, since any node that has only a single child is compressed, we can charge it to its parent, which has two (or more) children. There are at most $n-1$ internal nodes in the new compressed quadtree that have degree larger than one, since one can split a set of size n at most $n - 1$ times till ending up with singletons. As such, a compressed quadtree has linear size (however, it still can have linear depth in the worst case).

See Figure 2.2 for an example of a compressed quadtree.

Example 2.5. As an application for compressed quadtrees, consider the problem of reporting the points that are inside a query rectangle r. We start from the root of the quadtree and recursively traverse it, going down a node only if its region intersects the query rectangle. Clearly, we will report all the points contained inside r. Of course, we have no guarantee about the query time, but in practice, this might be fast enough.

Note that one can perform the same task using a regular quadtree. However, in this case, such a quadtree might require unbounded space. Indeed, the spread of the point set might be arbitrarily large, and as such the depth (and thus size) of the quadtree can be arbitrarily large. On the other hand, a compressed quadtree would use only $O(n)$ space, where n is the number of points stored in it.

Example 2.6. Consider the point set $\mathsf{P} = \{\mathsf{p}_1, \ldots, \mathsf{p}_n\}$, where $\mathsf{p}_i = (3/4, 3/4)/8^{i-1}$, for $i = 1, \ldots, n$. The compressed quadtree for this point set is depicted in Figure 2.3.

Here, we have $\square_{v_i} = [0,1]^2/8^{i-1}$ and $\square_{v'_i} = [0,1]^2/(2 \cdot 8^{i-1})$. As such, v'_i is a compressed node, where

$$\mathsf{rg}_{v'_i} = \square_{v'_i} \setminus \square_{v_{i+1}}.$$

Note that this compressed quadtree has depth and size $\Theta(n)$. In particular, in this case, the compressed quadtree looks like a linked list storing the points.

2.2.2. Efficient construction of compressed quadtrees.

2.2.2.1. *Bit twiddling and compressed quadtrees.* Unfortunately, to be able to efficiently build compressed quadtrees, one requires a somewhat bizarre computational model. We are assuming implicitly the unit RAM model, where one can store and manipulate arbitrarily large real numbers in constant time. To work with grids efficiently, we need to be able to compute quickly (i.e., constant time) $\lg(x)$ (i.e., $\log_2$), 2^x, and $\lfloor x \rfloor$. Strangely, computing a compressed quadtree efficiently is equivalent to the following operation.

Definition 2.7 (Bit index)**.** Let $\alpha, \beta \in [0,1)$ be two real numbers. Assume these numbers in base two are written as $\alpha = 0.\alpha_1\alpha_2\ldots$ and $\beta = 0.\beta_1\beta_2\ldots$. Let $\mathrm{bit}_\Delta(\alpha,\beta)$ be the index of the first bit after the period in which they differ.

For example, $\mathrm{bit}_\Delta(1/4 = 0.01_2, 3/4 = 0.11_2) = 1$ and $\mathrm{bit}_\Delta(7/8 = 0.111_2, 3/4 = 0.110_2) = 3$.

Lemma 2.8. *If one can compute a compressed quadtree of two points in constant time, then one can compute* $\mathrm{bit}_\Delta(\alpha,\beta)$ *in constant time.*

Proof. Given α and β, we need to compute $\mathrm{bit}_\Delta(\alpha,\beta)$. Now, if $|\alpha - \beta| > 1/128$, then we can compute $\mathrm{bit}_\Delta(\alpha,\beta)$ in constant time. Otherwise, build a one-dimensional compressed quadtree (i.e., a compressed trie) for the set $\{\alpha, \beta\}$. The root is a compressed node in this tree, and its only child[②] v has sidelength 2^{-i}; that is, $\ell(v) = -i$. A quadtree node stores the canonical grid cell it corresponds to, and as such $\ell(v)$ is available from the compressed quadtree. As such, α and β are identical in the first i bits of their binary representation, but clearly they differ at the $(i+1)$st bit, as the next level of the quadtree splits them into two different subtrees. As such, if one can compute a compressed trie of two numbers in constant time, then one can compute $\mathrm{bit}_\Delta(\alpha,\beta)$ in constant time.

If the reader is uncomfortable with building a one-dimensional compressed quadtree, then use the point set $\mathsf{P} = \{(\alpha, 1/3), (\beta, 1/3)\}$, and compute a compressed quadtree $\mathcal{T}$ for P having the unit square as the root. Similar argumentation would apply in this case. ■

Interestingly, once one has such an operation at hand, it is quite easy to compute a compressed quadtree efficiently via "linearization". The idea is to define an order on the nodes of a compressed quadtree and maintain the points sorted in this order; see Section 2.3 below for details. Given the points sorted in this order, one can build the compressed quadtree in linear time using, essentially, scanning.

[②]The root is Chinese and is not allowed to have more children at this point in time.

However, the resulting algorithm is somewhat counterintuitive. As a first step, we suggest a direct construction algorithm.

2.2.2.2. *A construction algorithm.* Let P be a set of n points in the unit square, with unbounded spread. We are interested in computing the compressed quadtree of P. The regular algorithm for computing a quadtree when applied to P might require unbounded time (but in practice it might be fast enough). Modifying it so that it requires only quadratic time is an easy exercise. Getting down to $O(n \log n)$ time requires some cleverness.

Theorem 2.9. *Given a set* P *of n points in the plane, one can compute a compressed quadtree of* P *in $O(n \log n)$ deterministic time.*

Proof. Compute, in linear time, a disk D of radius r, which contains at least $n/10$ of the points of P, such that $r \leq 2r_{\mathrm{opt}}(\mathsf{P}, n/10)$, where $r_{\mathrm{opt}}(\mathsf{P}, n/10)$ denotes the radius of the smallest disk containing $n/10$ points of P. Computing D can be done in linear time, by a rather simple algorithm (see Lemma 1.16_{p6}).

Let $\alpha = 2^{\lfloor \lg r \rfloor}$. Consider the grid G_α. It has a cell that contains at least $(n/10)/25$ points of P (since D is covered by $5 \times 5 = 25$ grid cells of G_α and $\alpha \geq r/2$), and no grid cell contains more than $5(n/10)$ points, by Lemma 1.19_{p7}. Thus, compute $\mathsf{G}_\alpha(\mathsf{P})$, and find the cell $\square$ containing the largest number of points. Let P_{in} be the set of points inside this cell $\square$, and let $\mathsf{P}_{\mathrm{out}}$ be the set of points outside this cell. Specifically, we have

$$\mathsf{P}_{\mathrm{in}} = \mathsf{P} \cap \square \quad \text{and} \quad \mathsf{P}_{\mathrm{out}} = \mathsf{P} \setminus \square = \mathsf{P} \setminus \mathsf{P}_{\mathrm{in}}.$$

We know that $\left|\mathsf{P}_{\mathrm{in}}\right| \geq n/250$ and $\left|\mathsf{P}_{\mathrm{out}}\right| \geq n/2$.

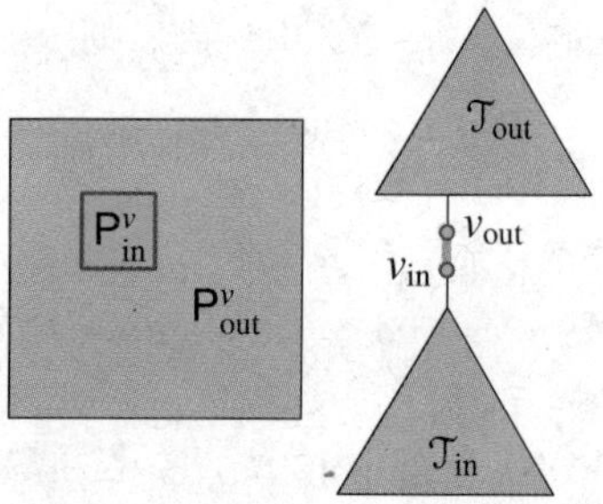

Next, compute (recursively) the compressed quadtrees for P_{in} and $\mathsf{P}_{\mathrm{out}}$, respectively, and let $\mathcal{T}_{\mathrm{in}}$ and $\mathcal{T}_{\mathrm{out}}$ denote the respective quadtrees. Create a node in both quadtrees that corresponds to $\square$. For $\mathcal{T}_{\mathrm{in}}$ this would just be the root node v_{in}, since $\mathsf{P}_{\mathrm{in}} \subseteq \square$. For $\mathcal{T}_{\mathrm{out}}$ this would be a new leaf node v_{out}, since $\mathsf{P}_{\mathrm{out}} \cap \square = \emptyset$. Note that inserting this new node might require linear time, but it requires only changing a constant number of nodes and pointers in both quadtrees. Now, hang v_{out} on v_{in} creating a compressed quadtree for P; see the figure on the right. Observe that the new glued node (i.e., v_{out} and v_{in}) might be redundant because of compression, and it can be removed if necessary in constant time.

The overall construction time is $T(n) = O(n) + T\left(\left|\mathsf{P}_{\mathrm{in}}\right|\right) + T\left(\left|\mathsf{P}_{\mathrm{out}}\right|\right) = O(n \log n)$. ■

Remark 2.10. The reader might wonder where the Tweedledee and Tweedledum operation (i.e., $\mathrm{bit}_\Delta(\cdot,\cdot)$) gets used in the algorithm of Theorem 2.9. Observe that the hanging stage might require such an operation if we hang $\mathcal{T}_{\mathrm{in}}$ from a compressed node of $\mathcal{T}_{\mathrm{out}}$, as computing the new compressed node requires exactly this kind of operation.

Compressed quadtree from squares. It is sometime useful to be able to construct a compressed quadtree from a list of squares that must appear in it as nodes.

Lemma 2.11. *Given a list C of n canonical squares, all lying inside the unit square, one can construct a (minimal) compressed quadtree $\mathcal{T}$ such that for any square $c \in C$, there exists a node $v \in T$, such that $\square_v = c$. The construction time is $O(n \log n)$.*

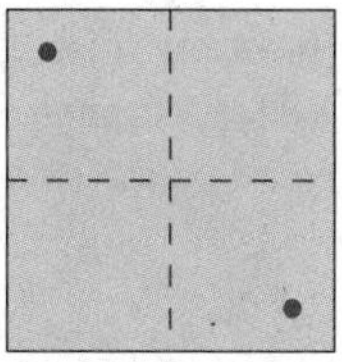

PROOF. For every *canonical* square $\square \in C$, we place two points into a set P, such that any quadtree for P must have $\square$ in it. This is done by putting two points in $\square$ such that they belong to two different subsquares of $\square$. See the figure on the right.

The resulting point set P has $2n$ points, and we can compute its compressed quadtree $\mathcal{T}$ in $O(n \log n)$ time using Theorem 2.9. Observe that any cell of C is an internal node of $\mathcal{T}$. Thus trimming away all the leaves of the quadtree results in a minimal quadtree that contains all the cells of C as nodes. ■

2.2.3. Fingering a compressed quadtree – fast point-location. Let $\mathcal{T}$ be a compressed quadtree of size n. We would like to preprocess it so that given a query point, we can find the lowest node of $\mathcal{T}$ whose cell contains a query point q. As before, we can perform this by traversing down the quadtree, but this might require $\Omega(n)$ time. Since the range of levels of the quadtree nodes is unbounded, we can no longer use a binary search on the levels of $\mathcal{T}$ to answer the query.

Instead, we are going to use a rebalancing technique on $\mathcal{T}$. Namely, we are going to build a balanced tree $\mathcal{T}'$, which would have cross pointers (i.e., fingers) into $\mathcal{T}$. The search would be performed on $\mathcal{T}'$ instead of on $\mathcal{T}$. In the literature, the tree $\mathcal{T}'$ is known as a ***finger tree***.

Definition 2.12. Let $\mathcal{T}$ be a tree with n nodes. A ***separator*** in $\mathcal{T}$ is a node v, such that if we remove v from $\mathcal{T}$, we remain with a forest, such that every tree in the forest has at most $\lceil n/2 \rceil$ vertices.

Lemma 2.13. *Every tree has a separator, and it can be computed in linear time.*

PROOF. Consider $\mathcal{T}$ to be a rooted tree. We precompute for every node in the tree the number of nodes in its subtree by a bottom-up computation that takes linear time. Set v to be the lowest node in $\mathcal{T}$ such that its subtree has $\geq \lceil n/2 \rceil$ nodes in it, where n is the number of nodes of $\mathcal{T}$. This node can be found by performing a walk from the root of $\mathcal{T}$ down to the child with a sufficiently large subtree till this walk gets "stuck". Indeed, let v_1 be the root of $\mathcal{T}$, and let v_i be the child of v_{i-1} with the largest number of nodes in its subtree. Let $s(v_i)$ be the number of nodes in the subtree rooted at v_i. Clearly, there exists a k such that $s(v_k) \geq n/2$ and $s(v_{k+1}) < n/2$.

Clearly, all the subtrees of the children of v_k have size at most $n/2$. Similarly, if we remove v_k and its subtree from $\mathcal{T}$, we remain with a tree with at most $n/2$ nodes. As such, v_k is the required separator. ■

This suggests a natural way for processing a compressed quadtree for point-location queries. Find a separator $v \in T$, and create a root node f_v for $\mathcal{T}'$ which has a pointer to v; now recursively build a finger tree for each tree of $T \setminus \{v\}$. Hang the resulting finger trees on f_v. The resulting tree is the required finger tree $\mathcal{T}'$.

Given a query point q, we traverse $\mathcal{T}'$, where at node $f_v \in \mathcal{T}'$, we check whether the query point $q \in \square_v$, where v is the corresponding node of $\mathcal{T}$. If $q \notin \square_v$, we continue the search into the child of f_v, which corresponds to the connected component outside $\square_v$ that was hung on f_v. Otherwise, we continue into the child that contains q; naturally, we have to check out the $O(1)$ children of v to decide which one we should continue the search into. This takes constant time per node. As for the depth for the finger tree $\mathcal{T}'$, observe $D(n) \leq 1 + D(\lceil n/2 \rceil) = O(\log n)$. Thus, a point-location query in $\mathcal{T}'$ takes logarithmic time.

Theorem 2.14. *Given a compressed quadtree $\mathcal{T}$ of size n, one can preprocess it, in time $O(n \log n)$, such that given a query point q, one can return the lowest node in $\mathcal{T}$ whose region contains q in $O(\log n)$ time.*

PROOF. We just need to bound the preprocessing time. Observe that it is $T(n) = O(n) + \sum_{i=1}^{t} T(n_i)$, where $n_1, \ldots, n_t$ are the sizes of the subtrees formed by removing the separator node from $\mathcal{T}$. We know that $t = O(1)$ and $n_i \leq \lceil n/2 \rceil$, for all i. As such, $T(n) = O(n \log n)$. ■

2.3. Dynamic quadtrees

Here we show how to store compressed quadtrees using any data-structure for ordered sets. The idea is to define an order on the compressed quadtree nodes and then store the compressed nodes in a data-structure for ordered sets. We show how to perform basic operations using this representation. In particular, we show how to perform insertions and deletions. This provides us with a simple implementation of dynamic compressed quadtrees.

We start our discussion with regular quadtrees and later on discuss how to handle compressed quadtrees.

2.3.1. Ordering of nodes and points. Consider a regular quadtree $\mathcal{T}$ and a **DFS** traversal of $\mathcal{T}$, where the **DFS** always traverse the children of a node in the same relative order (i.e., say, first the bottom-left child, then the bottom-right child, top-left child, and top-right child).

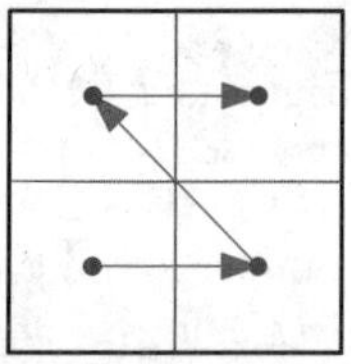

Consider any two canonical squares $\square$ and $\widehat{\square}$, and imagine a quadtree $\mathcal{T}$ that contains both squares (i.e., there are nodes in $\mathcal{T}$ with these squares as their cells). Notice that the above **DFS** would always visit these two nodes in a specific order, independent of the structure of the rest of the quadtree. Thus, if $\square$ gets visited before $\widehat{\square}$, we denote this fact by $\square \prec \widehat{\square}$. This defines a total ordering over all canonical squares. It is useful to extend this ordering to also include points. Thus, consider a point p and a canonical square $\square$. If $\mathsf{p} \in \square$, then we will say that $\square \prec \mathsf{p}$. Otherwise, if $\square \in \mathsf{G}_i$, let $\widehat{\square}$ be the cell in G_i that contains p. We have that $\square \prec p$ if and only if $\square \prec \widehat{\square}$. Next, consider two points p and q, and let G_i be a grid fine enough such that p and q lie in two different cells, say, $\square_\mathsf{p}$ and $\square_\mathsf{q}$, respectively. Then $\mathsf{p} \prec \mathsf{q}$ if and only if $\square_\mathsf{p} \prec \square_\mathsf{q}$.

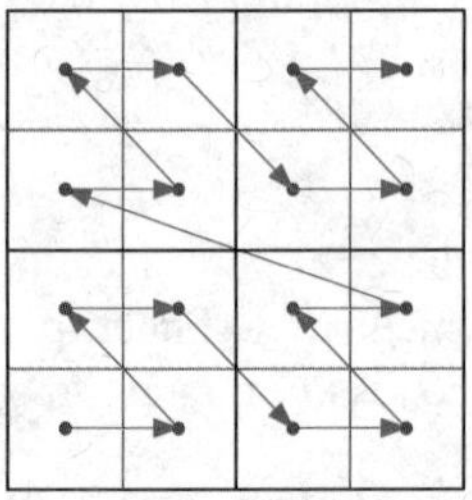

We will refer to the ordering induced by $\prec$ as the $\mathcal{Q}$-*order*. The ordering $\prec$ when restricted only to points is the ordering along a space filling mapping that is induced by the quadtree **DFS**. This ordering is know as the $\mathcal{Z}$-*order*. Specifically, there is a bijection z from the unit interval $[0, 1)$ to the unit square $[0, 1)^2$ such that the ordering along the resulting "curve" is the $\mathcal{Z}$-order. Note, however, that since we allow comparing cells to cells and cells to points, the $\mathcal{Q}$-order no longer has this exact interpretation. Unfortunately, unlike the Peano or Hilbert space filling curves, the $\mathcal{Z}$-order mapping z is not continuous. Nevertheless, the $\mathcal{Z}$-order mapping has the advantage of being easy to define. Indeed, given a real number $\alpha \in [0, 1)$, with the binary expansion $\alpha = 0.x_1x_2x_3\ldots$ (i.e., $\alpha = \sum_{i=1}^{\infty} x_i 2^{-i}$), the $\mathcal{Z}$-order mapping of α is the point $z(\alpha) = (0.x_2x_4x_6\ldots, 0.x_1x_3x_5\ldots)$.

To see that this property indeed holds, think about performing a point-location query in an infinite quadtree for a point $\mathsf{p} \in [0,1)^2$. In the top level of the quadtree we have four possibilities to continue the search. These four possibilities can be encoded as a binary string of length 2; see the figure on the right. Now, if $y(\mathsf{p}) \in [0, 1/2)$, then the first bit in the encoding is 0, and if $y(\mathsf{p}) \in [1/2, 1)$, the first bit output is 1.

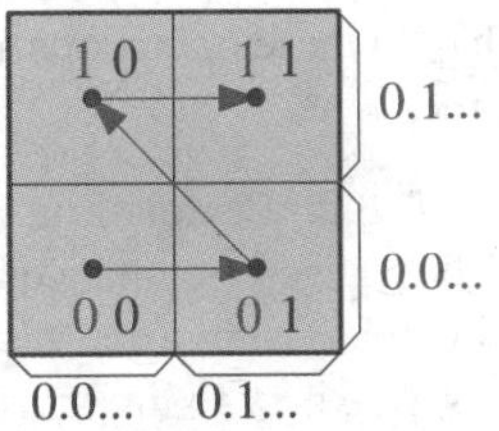

Similarly, the second bit in the encoding is the first bit in the binary representation of $x(\mathsf{p})$. Now, after resolving the point-location query in the top level, we continue this search in the tree. Every level in this traversal generates two bits, and these two bits corresponds to the relevant two bits in the binary representation of $y(\mathsf{p})$ and $x(\mathsf{p})$. In particular, the $2i+1$ and $2i+2$ bits in the encoding of p as a single real number (in binary), are the ith bits of (the binary representation of) $y(\mathsf{p})$ and $x(\mathsf{p})$, respectively. In particular, for a point $\mathsf{p} \in [0,1)^d$, let $\mathrm{enc}_{\prec}(\mathsf{p})$ denote the number in the range $[0, 1)$ encoded by this process.

Claim 2.15. *For any two points* $\mathsf{p}, \mathsf{q} \in [0,1)^2$, *we have that* $\mathsf{p} \prec \mathsf{q}$ *if and only if* $\mathrm{enc}_{\prec}(\mathsf{p}) < \mathrm{enc}_{\prec}(\mathsf{q})$.

2.3.1.1. *Computing the $\mathcal{Q}$-order quickly.* For our algorithmic applications, we need to be able to find the ordering according to $\prec$ between any two given cells/points quickly.

Definition 2.16. For any two points $\mathsf{p}, \mathsf{q} \in [0,1]^2$, let $\mathrm{lca}(\mathsf{p}, \mathsf{q})$ denote the smallest canonical square that contains both p and q. It intuitively corresponds to the node that must be in any quadtree storing p and q.

To compute $\mathrm{lca}(\mathsf{p}, \mathsf{q})$, we revisit the Tweedledee and Tweedledum operation $\mathrm{bit}_\Delta(\alpha, \beta)$ (see Definition 2.7) that for two real numbers $\alpha, \beta \in [0, 1)$ returns the index of the first bit in which α and β differ (in base two). The level ℓ of $\square = \mathrm{lca}(\mathsf{p}, \mathsf{q})$ is equal to

$$\ell = 1 - \min(\mathrm{bit}_\Delta(x_{\mathsf{p}}, x_{\mathsf{q}}), \mathrm{bit}_\Delta(y_{\mathsf{p}}, y_{\mathsf{q}})),$$

where x_{p} and y_{p} denote the x and y coordinates of p, respectively. Thus, the sidelength of $\square = \mathrm{lca}(\mathsf{p}, \mathsf{q})$ is $\Delta = 2^\ell$. Let $x' = \Delta \lfloor x/\Delta \rfloor$ and $y' = \Delta \lfloor y/\Delta \rfloor$. Thus,

$$\mathrm{lca}(\mathsf{p}, \mathsf{q}) = [x', x' + \Delta) \times [y', y' + \Delta).$$

We also define the lca of two cells to be the lca of their centers.

Now, given two cells $\square$ and $\widehat{\square}$, we would like to determine their $\mathcal{Q}$-order. If $\square \subseteq \widehat{\square}$, then $\widehat{\square} \prec \square$. If $\widehat{\square} \subseteq \square$, then $\square \prec \widehat{\square}$. Otherwise, let $\overline{\square} = \mathrm{lca}(\square, \widehat{\square})$. We can now determine which children of $\overline{\square}$ contain these two cells, and since we know the traversal ordering among children of a node in a quadtree, we can now resolve this query in constant time.

Corollary 2.17. *Assuming that the* bit_Δ *operation and the* $\lfloor \cdot \rfloor$ *operation can be performed in constant time, then one can compute the* lca *of two points (or cells) in constant time. Similarly, their $\mathcal{Q}$-order can be resolved in constant time.*

Computing bit_Δ **efficiently.** It seems somewhat suspicious that one assumes that the bit_Δ operations can be done in constant time on a classical RAM machine. However, it is a reasonable assumption on a real world computer. Indeed, in floating point representation, once you are given a number, it is easy to access its mantissa and exponent in constant time. If the exponents are different, then bit_Δ can be computed in constant time. Otherwise, we can easily $\oplus_{\mathrm{xor}}$ the mantissas of both numbers and compute the most significant bit that is one. This can be done in constant time by converting the resulting mantissa into a floating point number and computing its $\log_2$ (some CPUs have this command built in). Observe that all these operations are implemented in hardware in the CPU and require only constant time.

2.3.2. Performing operations on a (regular) quadtree stored using $\mathcal{Q}$-order. Let $\mathcal{T}$ be a given (regular) quadtree, with its nodes stored in an ordered-set data-structure (for example, using a red-black tree or a skip-list), using the $\mathcal{Q}$-order over the cells. We next describe how to implement some basic operations on this quadtree.

2.3.2.1. *Performing a point-location in a quadtree.* Given a query point $q \in [0,1]^2$, we would like to find the leaf v of $\mathcal{T}$ such that its cell contains q.

To answer the query, we first find the two consecutive cells in this ordered list such that q lies between them. Formally, let $\square$ be the last cell in this list such that $\square \prec q$. It is now easy to verify that $\square$ must be the quadtree leaf containing q. Indeed, let $\square_q$ be the leaf of $\mathcal{T}$ whose cell contains q. By definition, we have that $\square_q \prec q$. Thus, the only bad scenario is that $\square_q \prec \square \prec q$. But this implies, by the definition of the $\mathcal{Q}$-order, that $\square$ must be contained inside $\square_q$, contradicting our assumption that $\square_q$ is a leaf of the quadtree.

Lemma 2.18. *Given a quadtree $\mathcal{T}$ of size n, with its leaves stored in an ordered-set data-structure $\mathcal{D}$ according to the $\mathcal{Q}$-order, then one can perform a point-location query in $O(Q(n))$ time, where $Q(n)$ is the time to perform a search query in $\mathcal{D}$.*

2.3.2.2. *Overlaying two quadtrees.* Given two quadtrees $\mathcal{T}'$ and $\mathcal{T}''$, we would like to overlay them to compute their combined quadtree. This is the minimal quadtree such that every cell of $\mathcal{T}'$ and $\mathcal{T}''$ appears in it. Observe that if the two quadtrees are given as sorted lists of their cells (ordered by the $\mathcal{Q}$-order), then their overlay is just the merged list, with replication removed.

Lemma 2.19. *Given two quadtrees $\mathcal{T}'$ and $\mathcal{T}''$, given as sorted lists of their nodes, one can compute the merged quadtree in linear time (in the total size of the lists representing them) by merging the two sorted lists and removing duplicates.*

2.3.3. Performing operations on a compressed quadtree stored using the $\mathcal{Q}$-order. Let $\mathcal{T}$ be a compressed quadtree whose nodes are stored in an ordered-set data-structure using the $\mathcal{Q}$-order. Note that the nodes are sorted according to the canonical square that they correspond to. As such, a node $v \in \mathcal{T}$ is sorted according to $\square_v$. The subtlety here is that a compressed node v is stored according to the square $\square_v$ in this representation, but in fact it is in charge of the compressed region rg_v. This will make things slightly more complicated.

2.3.3.1. *Point-location in a compressed quadtree.* Performing a point-location query is a bit subtle in this case. Indeed, if the query point q is contained inside a leaf of $\mathcal{T}$, then a simple binary search for the predecessor of q in the $\mathcal{Q}$-order sorted list of the cells of $\mathcal{T}$ would return this leaf.

However, if the query is contained inside the region of a compressed node, the predecessor to q in this list (sorted by the $\mathcal{Q}$-order) might be some arbitrary leaf that is contained inside the compressed node of interest. As a concrete example, consider the figure on the right of a compressed node w. Because of the $\mathcal{Q}$-order, the predecessor query on q would return a leaf v that is stored in the subtree of w but does not contain the query point.

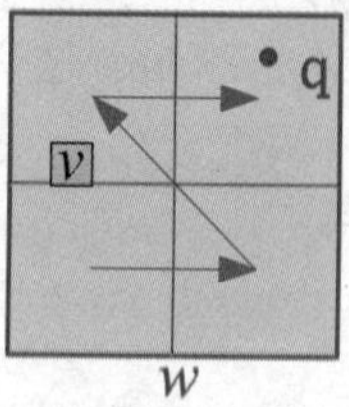

In such a case, we need to find the LCA of the query point and the leaf returned. This returned cell $\square$ could be in the quadtree itself. In this case, we are done as $\square$ corresponds to the compressed node that contains q. The other possibility is that $\square$ is contained inside the cell $\square_w$ of a compressed node w that is the parent of v. Again, we can now find this parent node using a single predecessor query.

The code for the point-location procedure is depicted on the right. The query point is q. The query point is not necessarily stored in the quadtree, and as such the cell that contains it might be a compressed node (as described above). As such, the required node v has $\mathsf{q} \in \mathsf{rg}_v$ and is either a leaf of the quadtree or a compressed node.

```
algPntLoc_Qorder(𝒯, q).
    □_v = predecessor_{𝒬-order}(𝒯, q).
    // □_v last node in 𝒯
    // s.t.  □_v ≤ q.
    if q ∈ □_v then
        return v
    □ = lca(□_v, q)
    □_w = predecessor_{𝒬-order}(𝒯, □).
    return w
```

Lemma 2.20. *We have that: (i) if* $\mathsf{q} \in \square_v$*, then* $\mathsf{q} \in \mathsf{rg}_v$ *and (ii) if the region of* $\mathcal{T}$ *containing the query point* q *is a leaf, then* v *is the required leaf.*

PROOF. (i) Otherwise, there exists a node $x \in \mathcal{T}$ such that $\mathsf{q} \in \mathsf{rg}_x \subseteq \square_x$ and $\square_x \subsetneq \square_v$, but that would imply that $\square_v \prec \square_x \prec \mathsf{q}$, contradicting how $\square_v$ was computed.

(ii) Since q is contained in a leaf, then the last square in the $\mathcal{Q}$-order that is before q is the cell of this leaf, that is, the node v. ■

Now, if $\mathsf{q} \notin \square_v$, then our search "failed" to find a node that contains the query point. So, consider the node w that should be returned. It must be a compressed node, and by the above, the search returned a node v that is a descendant of w. As a first step, we compute the square $\square = \mathrm{lca}(\square_v, \mathsf{q})$. Next, we compute the last cell $\square_u$ such that $\square_u \preceq \square$.

Now, if $\square_u = \square$, then we are done as $\mathsf{q} \in \mathsf{rg}_u$. Otherwise, if $\square$ is not in the tree, then u is a compressed node, and consider its child u'. We have that $\square$ is contained inside $\square_u$ and contains $\square_{u'}$. Namely, $\mathsf{q} \in \square_u \setminus \square_{u'} = \mathsf{rg}_u$. Now, u' is the only child of u (since it is a compressed node), which implies that $\square_u$ and $\square_{u'}$ are consecutive in $\mathcal{T}$ according to the $\mathcal{Q}$-order, and $\square$ lies in between these two cells in this order. As such, $u \prec \square \prec u'$ and as such u is the required node that contains the query point.

We get the following result.

Lemma 2.21. *Given a compressed quadtree* $\mathcal{T}$ *of size* n*, with its nodes stored in an ordered-set data-structure* $\mathcal{D}$ *according to the* $\mathcal{Q}$*-order, then point-location queries can be performed using* $\mathcal{T}$ *in* $O(Q(n))$ *time, where* $Q(n)$ *is the time to perform a search query in* $\mathcal{D}$*.*

2.3.3.2. *Insertions and deletions.* Let q be a point to be inserted into the quadtree, and let w be the node of the compressed quadtree such that $\mathsf{q} \in \mathsf{rg}_w$. There are several possibilities:

(1) The node w is a leaf, and there is no point associated with it. Then store p at w, and return.
(2) The node w is a leaf, and there is a point p already stored in w. In this case, let $\square = \mathrm{lca}(\mathsf{p}, \mathsf{q})$, and insert $\square$ into the compressed quadtree. This is done by creating a new node z in the quadtree, with the cell of z being $\square$. We hang z below w if $\square \neq \square_w$ (this turns w into a compressed node). (If $\square = \square_w$, we do not need to introduce a new cell and set $z = w$.) Furthermore, split $\square$ into its children, and also insert the children into the compressed quadtree. Finally, associate p with the new leaf that contains it, and associate q with the leaf that contains it. Note that because of the insertion w becomes a compressed node if $\square_w \neq \square$, and it becomes a regular internal node otherwise.
(3) The node w is a compressed node. Let z be the child of w, and consider $\square = \mathrm{lca}(\square_z, \mathsf{q})$. Insert $\square$ into the compressed quadtree if $\square \neq \square_w$ (note that in this

case w would still be a compressed node, but with a larger "hole"). Also insert all the children of $\square$ into the quadtree, and store p in the appropriate child. Hang $\square_z$ from the appropriate child, and turn this child into a compressed node.

In all three cases, the insertion requires a constant number of search/insert operations on the ordered-set data-structure.

Deletion is done in a similar fashion. We delete the point from the node that contains it, and then we trim away nodes that are no longer necessary.

Theorem 2.22. *Assuming one can compute the $\mathcal{Q}$-order in constant time, then one can maintain a compressed quadtree of a set of points in $O(\log n)$ time per operation, where insertions, deletions and point-location queries are supported. Furthermore, this can implemented using any data-structure for ordered-set that supports an operation in logarithmic time.*

2.3.4. Compressed quadtrees in high dimension. Naively, the constants used in the compressed quadtree are exponential in the dimension d. However, one can be more careful in the implementation. The first problem, for a node v in the compressed quadtree, is to store all the children of v in an efficient way so that we can access them efficiently. To this end, each child of v can be encoded as a binary string of length d, and we build a trie inside v for storing all the strings defining the children of v. It is easy, given a point, to figure out what the binary string encoding the child containing this point is. As such, in $O(\mathsf{d})$ time, one can retrieve the relevant child. (One can also use hashing to this end, but it is not necessary here.)

Now, we build the compressed quadtree using the algorithm described above using the $\mathcal{Q}$-order. It is not too hard to verify that all the basic operations can be computed in $O(\mathsf{d})$ time. Specifically, comparing two points in the $\mathcal{Q}$-order takes $O(\mathsf{d})$ time. One technicality that matters when d is large (but this issue can be ignored in low dimensions) is that if a node has only a few children that are not empty, we create only these nodes and not the other children. Putting everything together, we get the following result.

Theorem 2.23. *Assuming one can compute the $\mathcal{Q}$-order in $O(\mathsf{d})$ time for two points $\mathbb{R}^{\mathsf{d}}$, then one can maintain a compressed quadtree of a set of points in $\mathbb{R}^{\mathsf{d}}$, in $O(\mathsf{d}\log n)$ time per operation. The operations of insertion, deletion, and point-location query are supported. Furthermore, this can be implemented using any data-structure for ordered-set that supports insertion/deletion/predecessor operations in logarithmic time.*

In particular, one can construct a compressed quadtree of a set of n points in $\mathbb{R}^{\mathsf{d}}$ in $O(\mathsf{d}n\log n)$ time.

2.4. Bibliographical notes

The authoritative text on quadtrees is the book by Samet [**Sam89**]. He also has a more recent book that provides a comprehensive survey of various tree-like data-structures [**Sam05**] (our treatment is naturally more theoretically oriented than his). The idea of using hashing in quadtrees is a variant of an idea due to Van Emde Boas and is also used in performing fast lookup in IP routing (using PATRICIA tries which are one-dimensional quadtrees [**WVTP97**]), among a lot of other applications.

The algorithm described in Section 2.2.2 for the efficient construction of compressed quadtrees is new, as far as the author knows. The classical algorithms for computing compressed quadtrees efficiently achieve the same running time but require considerably more careful implementation and paying careful attention to details [**CK95, AMN⁺98**]. The idea of fingering is used in [**AMN$^+$98**] (although their presentation is different than ours) but the idea is probably much older.

The idea of storing a quadtree in an ordered set by using the $\mathcal{Q}$-order on the nodes (or even only on the leaves) is due to Gargantini [**Gar82**], and it is referred to as ***linear quadtrees*** in the literature. The idea was used repeatedly for getting good performance in practice from quadtrees.

Our presentation of the dynamic quadtrees (i.e., Section 2.3) follows (very roughly) the work de Berg et al. [**dHTT07**].

It is maybe beneficial to emphasize that if one does not require the internal nodes of the compressed quadtree for the application, then one can avoid storing them in the data-structure. If only the points are needed, then one can even skip storing the leaves themselves, and then the compressed quadtree just becomes a data-structure that stores the points according to their $\mathcal{Z}$-order. This approach can be used for example to construct a data-structure for approximate nearest neighbor [**Cha02**] (however, this data-structure is still inferior, in practice, to the more optimized but more complicated data-structure of Arya et al. [**AMN$^+$98**]). The author finds that thinking about such data-structures as compressed quadtrees (with the whole additional unnecessary information) is more intuitive, but the reader might disagree[3].

$\mathcal{Z}$-order and space filling curves. The idea of using $\mathcal{Z}$-order for speeding up spatial data-structures can be traced back to the above work of Gargantini [**Gar82**], and it is widely used in databases and seems to improve performance in practice [**KF93**]. The $\mathcal{Z}$-order can be viewed as a mapping from the unit interval to the unit square, by splitting the sequence of bits representing a real number $\alpha \in [0, 1)$, where the odd bits are the x-coordinate and the even bits are the y-coordinate of the mapped point. While this mapping is simple to define, it is not continuous. Somewhat surprisingly, one can find a continuous mapping that maps the unit interval to the unit square; see Exercise 2.5. A large family of such mappings is known by now; see Sagan [**Sag94**] for an accessible book on the topic.

But is it really practical? Quadtrees seem to be widely used in practice and perform quite well. Compressed quadtrees seem to be less widely used, but they have the benefit of being much simpler than their relatives which seems to be more practical but theoretically equivalent.

Compressed quadtrees require strange operations. Lemma 2.8 might be new, although it seems natural to assume that it was known before. It implies that computing compressed quadtrees requires at least one "strange" operation in the computation model. Once one comes to term with this imperfect situation, the use of the $\mathcal{Q}$-order seems natural and yields a reasonably simple algorithm for the dynamic maintenance of quadtrees. For example, if we maintain such a compressed quadtree by using skip-list on the $\mathcal{Q}$-order, we will essentially get the skip-quadtree of Eppstein et al. [**EGS05**].

Generalized compressed quadtrees. Har-Peled and Mendel [**HM06**] have shown how to extend compressed quadtrees to the more general settings of doubling metrics. Note that this variant of compressed quadtrees no longer requires strange bit operations. However, in the process, one loses the canonical grid structures that a compressed quadtree has, which is such a useful property.

---

[3] The author reserves the right to disagree with himself on this topic in the future if the need arises.

Why compressed quadtrees? The reader might wonder why we are presenting compressed quadtrees when in practice people use different data-structures (that can also be analyzed). Our main motivation is that compressed quadtrees seem to be a universal data-structure, in the sense that they can be used for many different tasks, they are conceptually and algorithmically simple, and they provide a clear route to solving a problem: Solve your problem initially on a quadtree for a point set with bounded spread. If this works, try to solve it on a compressed quadtree. If you want something practical, try some more practical variants like *kd*-trees [**dBCvKO08**].

2.5. Exercises

Exercise 2.1 (Geometric obesity is good)**.** For a constant $\alpha \geq 1$, a planar convex region T is ***α-fat*** if $R(T)/r(T) \leq \alpha$, where $R(T)$ and $r(T)$ are the radii of the smallest disk containing T and the largest disk contained in T, respectively.

(A) Prove that if a triangle $\triangle$ has all angles larger than β, then $\triangle$ is $\dfrac{1}{\sin(\beta/2)}$ fat. We will refer to such a triangle as being ***fat***.

(B) Let S be a set of interior disjoint α-fat shapes all intersecting a common square $\square$. Furthermore, for every $\triangle \in \mathsf{S}$ we have that $\operatorname{diam}(\triangle) \geq c \cdot \operatorname{diam}(\square)$, where c is some constant. Prove that $|\mathsf{S}| = O(1)$. Here, the constant depends on c and α, and the reader is encouraged to be less lazy than the author and to figure out the exact value of this constant as a function of α and c.

Exercise 2.2 (Quadtree for fat triangles)**.** Let P be a triangular planar map of the unit square (i.e., each face is a triangle but it is not necessarily a triangulation), where all the triangles are fat and the total number of triangles is n.

(A) Show how to build a compressed quadtree storing the triangles, such that every node of the quadtree stores only a constant number of triangles (the same triangle might be stored in several nodes) and given a query point p, the triangle containing p is stored somewhere along the point-location query path of p in the quadtree.

(Hint: Think about what is the right resolution to store each triangle.)

(B) Show how to build a compressed quadtree for P that stores triangles only in the leaves, and such that every leaf contains only a constant number of triangles and the total size of the quadtree is $O(n)$.

Hint: Using the construction from the previous part, push down the triangles to the leaves, storing in every leaf all the triangles that intersect it. Use Exercise 2.1 to argue that no leaf stores more than a constant number of triangles.

(C) (Tricky but not hard) Show that one must use a compressed quadtree in the worst case if one wants linear space.

(D) (Hard) We remind the reader that a ***triangulation*** is a triangular planar map that is compatible. That is, the intersection of triangles in the triangulation is either empty, a vertex of both triangles, or an edge of both triangles (this is also known as a ***simplicial complex***). Prove that a fat triangulation with n triangles can be stored in a regular (i.e., non-compressed) quadtree of size $O(n)$.

Exercise 2.3 (Quadtree construction is tight)**.** Prove that the bounds of Lemma 2.4 are tight. Namely, show that for any $r > 2$ and any positive integer $n > 2$, there exists a set of n points with diameter $\Omega(1)$ and spread $\Phi(\mathsf{P}) = \Theta(r)$ and such that its quadtree has size $\Omega(n \log \Phi(\mathsf{P}))$.

Exercise 2.4 (Cell queries). Let $\widehat{\square}$ be a canonical grid cell. Given a compressed quadtree $\widehat{T}$, we would like to find the *single* node $v \in \widehat{T}$, such that $\mathsf{P} \cap \widehat{\square} = \mathsf{P}_v$. We will refer to such a query as a ***cell query***. Show how to support cell queries in a compressed quadtree in logarithmic time per query.

Exercise 2.5 (Space filling curve). The ***Peano curve*** $\sigma : [0, 1) \to [0, 1)^2$ maps a number $\alpha = 0.t_1 t_2 t_3 \ldots$ (the expansion is in base 3) to the point $\sigma(\alpha) = (0.x_1 x_2 x_3 \ldots, \; 0.y_1 y_2 \ldots)$, where $x_1 = t_1$, $x_i = \phi(t_{2i-1}, t_2 + t_4 + \cdots + t_{2i-2})$, for $i \geq 1$. Here, $\phi(a, b) = a$ if b is even and $\phi(a, b) = 2 - a$ if b is odd. Similarly, $y_i = \phi(t_{2i}, t_1 + t_3 + \cdots + t_{2i-1})$, for $i \geq 1$.

(A) Prove that the mapping σ covers all the points in the open square $[0, 1)^2$, and it is one-to-one.

(B) Prove that σ is continuous.

CHAPTER 3

Well-Separated Pair Decomposition

In this chapter, we will investigate how to represent distances between points efficiently. Naturally, an explicit description of the distances between n points requires listing all the $\binom{n}{2}$ distances. Here we will show that there is a considerably more compact representation which is sufficient if all we care about are approximate distances. This representation would have many nice applications.

3.1. Well-separated pair decomposition (WSPD)

Let P be a set of n points in $\mathbb{R}^{\mathsf{d}}$, and let $1/4 > \varepsilon > 0$ be a parameter. One can represent all distances between points of P by explicitly listing the $\binom{n}{2}$ pairwise distances. Of course, the listing of the coordinates of each point gives us an alternative, more compact representation (of size $\mathsf{d}n$), but its not a very informative representation. We are interested in a representation that will capture the structure of the distances between the points.

As a concrete example, consider the three points on the right. We would like to have a representation that captures the fact that p has similar distance to q and s, and furthermore, that q and s are close together as far as p is concerned. As such, if we are interested in the closest pair among the three points, we will only check the distance between q and s, since they are the only pair (among the three) that might realize the closest pair.

FIGURE 3.1.

Denote by $A \otimes B = \left\{ \{x, y\} \,\middle|\, x \in A,\ y \in B, x \neq y \right\}$ the set of all the (unordered) pairs of points formed by the sets A and B. We will be informal and refer to $A \otimes B$ as a ***pair*** of the sets A and B. Here, we are interested in schemes that cover all possible pairs of points of P by a small collection of such pairs.

Definition 3.1 (Pair decomposition). For a point set P, a ***pair decomposition*** of P is a set of pairs

$$\mathcal{W} = \left\{ \{A_1, B_1\}, \ldots, \{A_s, B_s\} \right\},$$

such that (I) $A_i, B_i \subset \mathsf{P}$ for every i, (II) $A_i \cap B_i = \emptyset$ for every i, and (III) $\bigcup_{i=1}^{s} A_i \otimes B_i = \mathsf{P} \otimes \mathsf{P}$.

Translation: For any pair of distinct points $\mathsf{p}, \mathsf{q} \in \mathsf{P}$, there is at least one (and usually exactly one) pair $\{A_i, B_i\} \in \mathcal{W}$ such that $\mathsf{p} \in A_i$ and $\mathsf{q} \in B_i$.

Definition 3.2. The pair Q and R is $(1/\varepsilon)$-***separated*** if

$$\max(\mathrm{diam}(\mathsf{Q}), \mathrm{diam}(\mathsf{R})) \leq \varepsilon \cdot \mathbf{d}(\mathsf{Q}, \mathsf{R}),$$

where $\mathbf{d}(\mathsf{Q}, \mathsf{R}) = \min_{\mathsf{q} \in \mathsf{Q}, \mathsf{s} \in \mathsf{R}} \|\mathsf{q} - \mathsf{s}\|$.

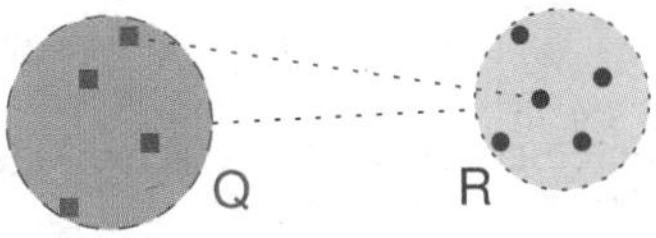

Intuitively, the pair $\mathsf{Q} \otimes \mathsf{R}$ is $(1/\varepsilon)$-***separated*** if all the points of Q have roughly the same distance to the points of R. Alternatively, imagine covering the two point sets with

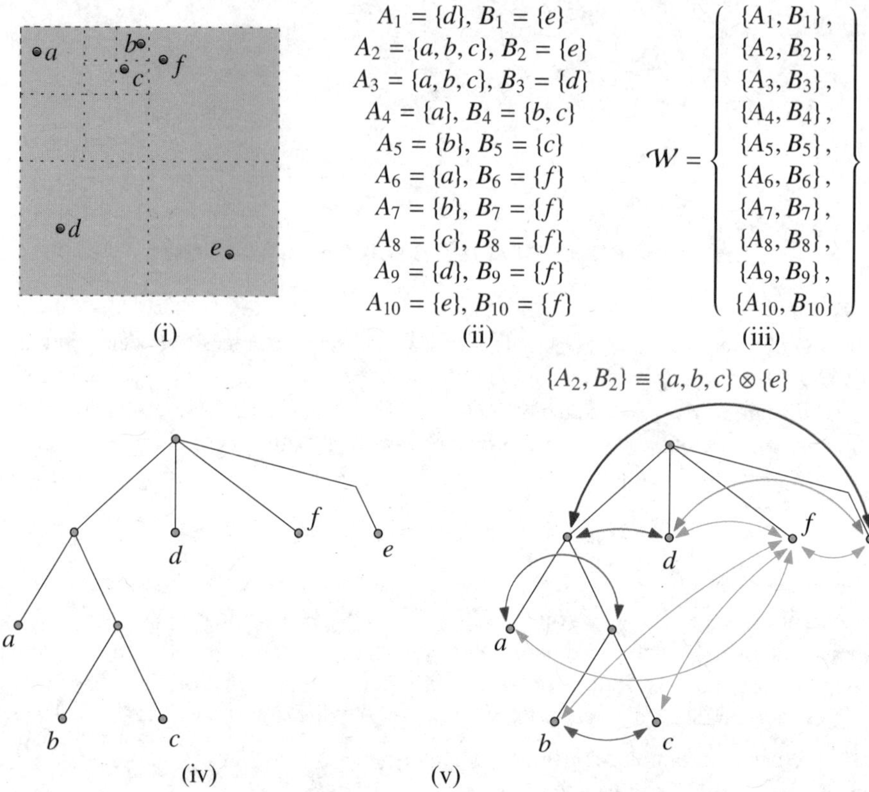

FIGURE 3.2. (i) A point set $\mathsf{P} = \{a, b, c, d, e, f\}$. (ii) The decomposition into pairs. (iii) The respective (1/2)-WSPD. For example, the pair of points b and e (and their distance) is represented by $\{A_2, B_2\}$ as $b \in A_2$ and $e \in B_2$. (iv) The quadtree $\mathcal{T}$ representing the point set P. (v) The WSPD as defined by pairs of vertices of $\mathcal{T}$.

two balls of minimum size, and require that the distance between the two balls is at least $2/\varepsilon$ times the radius of the larger of the two.

Thus, for the three points of Figure 3.1, the pairs $\{\mathsf{p}\} \otimes \{\mathsf{q}, \mathsf{s}\}$ and $\{\mathsf{q}\} \otimes \{\mathsf{s}\}$ are (say) 2-separated and describe all the distances among these three points. (The gain here is quite marginal, as we replaced the distance description, made out of three pairs of points, by the distance between two pairs of sets. But stay tuned – exciting things are about to unfold.)

Motivated by the above example, a well-separated pair decomposition is a way to describe a metric by such "well-separated" pairs of sets.

Definition 3.3 (WSPD). For a point set P, a ***well-separated pair decomposition*** (***WSPD***) of P with parameter $1/\varepsilon$ is a pair decomposition of P with a set of pairs

$$\mathcal{W} = \left\{ \{A_1, B_1\}, \ldots, \{A_s, B_s\} \right\},$$

such that, for any i, the sets A_i and B_i are ε^{-1}-separated.

For a concrete example of a WSPD, see Figure 3.2.

Instead of maintaining such a decomposition explicitly, it is convenient to construct a tree $\mathcal{T}$ having the points of P as leaves. Now every pair (A_i, B_i) is just a pair of nodes (v_i, u_i)

```
algWSPD(u, v)
  if u = v and Δ(u) = 0 then
      return            // Do not pair a leaf with itself
  if Δ(u) < Δ(v) then
      Exchange u and v
  If Δ(u) ≤ ε · d(u, v) then
      return { {u, v} }

  // u_1, ..., u_r – the children of u
  return ⋃_{i=1}^{r} algWSPD(u_i, v).
```

FIGURE 3.3. The algorithm **algWSPD** for computing well-separated pair decomposition. The nodes u and v belong to a compressed quadtree $\mathcal{T}$ of P.

of $\mathcal{T}$, such that $A_i = \mathsf{P}_{v_i}$ and $B_i = \mathsf{P}_{u_i}$, where P_v denotes the points of P stored in the subtree of v (here v is a node of $\mathcal{T}$). Naturally, in our case, the tree we would use is a compressed quadtree of P, but any tree that decomposes the points such that the diameter of a point set stored in a node drops quickly as we go down the tree might work. Naturally, even when the underlying tree is specified, there are many possible WSPDs that can be represented using this tree. Naturally, we will try to find a WSPD that is "minimal".

This WSPD representation using a tree gives us a compact representation of the distances of the point set.

Corollary 3.4. *For an ε^{-1}-WSPD $\mathcal{W}$, it holds, for any pair $\{u, v\} \in \mathcal{W}$, that*

$$\forall \mathsf{q} \in \mathsf{P}_u, \mathsf{s} \in \mathsf{P}_v \quad \max\Bigl(\mathrm{diam}(\mathsf{P}_u), \mathrm{diam}(\mathsf{P}_v)\Bigr) \leq \varepsilon \left\| \mathsf{q} - \mathsf{s} \right\|.$$

It would usually be convenient to associate with each set P_u in the WSPD an arbitrary representative point $\mathrm{rep}_u \in \mathsf{P}$. Selecting and assigning these representative points can always be done by a simple **DFS** traversal of the tree used to represent the WSPD.

3.1.1. The construction algorithm. Given the point set P in $\mathbb{R}^{\mathsf{d}}$, the algorithm first computes the compressed quadtree $\mathcal{T}$ of P. Next, the algorithm works by being greedy. It tries to put into the WSPD pairs of nodes in the tree that are as high as possible. In particular, if a pair $\{u, v\}$ would be generated, then the pair formed by the parents of this pair of nodes will not be well separated. As such, the algorithm starts from the root and tries to separate it from itself. If the current pair is not well separated, then we replace the bigger node of the pair by its children (i.e., thus replacing a single pair by several pairs). Clearly, sooner or later this refinement process would reach well-separated pairs, which it would output. Since it considers all possible distances up front (i.e., trying to separate the root from itself), it would generate a WSPD covering all pairs of points.

Let $\Delta(v)$ denote the diameter of the cell associated with a node v of the quadtree $\mathcal{T}$. We tweak this definition a bit so that if a node contains a single point or is empty, then it is zero. This would make our algorithm easier to describe.

Formally, $\Delta(v) = 0$ if P_v is either empty or a single point. Otherwise, it is the diameter of the region associated with v; that is, $\Delta(v) = \mathrm{diam}(\square_v)$, where $\square_v$ (we remind the reader) is the quadtree cell associated with the node v. Note that since $\mathcal{T}$ is a compressed quadtree, we can always decide if $|\mathsf{P}_v| > 1$ by just checking if the subtree rooted at v has more than one node (since then this subtree must store more than one point).

We define the ***geometric distance*** between two nodes u and v of $\mathcal{T}$ to be

$$\mathbf{d}(u, v) = \mathbf{d}(\Box_u, \Box_v) = \min_{\mathsf{p} \in \Box_u, \mathsf{q} \in \Box_v} \|\mathsf{p} - \mathsf{q}\| .$$

We compute the compressed quadtree $\mathcal{T}$ of P in $O(n \log n)$ time. Next, we compute the WSPD by calling **algWSPD**$(u_0, u_0, \mathcal{T})$, where u_0 is the root of $\mathcal{T}$ and **algWSPD** is depicted in Figure 3.3.

3.1.1.1. *Analysis.* The following lemma is implied by an easy packing argument.

Lemma 3.5. *Let $\Box$ be a cell of a grid G of $\mathbb{R}^{\mathsf{d}}$ with cell diameter x. For $y \geq x$, the number of cells in G at distance at most y from $\Box$ is $O\big((y/x)^{\mathsf{d}}\big)$. (The $O(\cdot)$ notation here, and in the rest of the chapter, hides a constant that depends exponentially on d.)*

Lemma 3.6. **algWSPD** *terminates and computes a valid pair decomposition.*

Proof. By induction, it follows that every pair of points of P is covered by a pair of subsets $\{\mathsf{P}_u, \mathsf{P}_v\}$ output by the **algWSPD** algorithm. Note that **algWSPD** always stops if both u and v are leafs, which implies that **algWSPD** always terminates.

Now, observe that if $\{u, v\}$ is in the output pair and either P_u or P_v is not a single point, then $\alpha = \max(\mathrm{diam}(\mathsf{P}_u), \mathrm{diam}(\mathsf{P}_v)) > 0$. This implies that $\mathbf{d}(\mathsf{P}_u, \mathsf{P}_v) \geq \mathbf{d}(u, v) \geq \Delta(u)/\varepsilon \geq \alpha/\varepsilon > 0$. Namely, $\mathsf{P}_u \cap \mathsf{P}_v = \emptyset$. ■

Lemma 3.7. *For the WSPD generated by* **algWSPD**, *we have that for any pair $\{u, v\}$ in the WSPD,*

$$\max\big(\mathrm{diam}(\mathsf{P}_u), \mathrm{diam}(\mathsf{P}_v)\big) \leq \varepsilon \cdot \mathbf{d}(u, v) \quad \textit{and} \quad \mathbf{d}(u, v) \leq \|\mathsf{q} - \mathsf{s}\|$$

hold for any $\mathsf{q} \in \mathsf{P}_u$ and $\mathsf{s} \in \mathsf{P}_v$.

Proof. For every output pair $\{u, v\}$, we have by the design of the algorithm that

$$\max\big(\mathrm{diam}(\mathsf{P}_u), \mathrm{diam}(\mathsf{P}_v)\big) \leq \max\big(\Delta(u), \Delta(v)\big) \leq \varepsilon \cdot \mathbf{d}(u, v).$$

Also, for any $\mathsf{q} \in \mathsf{P}_u$ and $\mathsf{s} \in \mathsf{P}_v$, we have $\mathbf{d}(u, v) = \mathbf{d}(\Box_u, \Box_v) \leq \mathbf{d}(\mathsf{P}_u, \mathsf{P}_v) \leq \mathbf{d}(\mathsf{q}, \mathsf{s})$, since $\mathsf{P}_u \subseteq \Box_u$ and $\mathsf{P}_v \subseteq \Box_v$. ■

Lemma 3.8. *For a pair $\{u, v\} \in \mathcal{W}$, computed by* **algWSPD**, *we have that*

$$\max\big(\Delta(u),\ \Delta(v)\big) \leq \min\big(\Delta(\overline{\mathsf{p}}(u)),\ \Delta(\overline{\mathsf{p}}(v))\big),$$

where $\overline{\mathsf{p}}(x)$ denotes the parent node of the node x in the tree $\mathcal{T}$.

Proof. We trivially have that $\Delta(u) < \Delta(\overline{\mathsf{p}}(u))$ and $\Delta(v) < \Delta(\overline{\mathsf{p}}(v))$.

A pair $\{u, v\}$ is generated because of a sequence of recursive calls **algWSPD**(u_0, u_0), **algWSPD**(u_1, v_1), …, **algWSPD**(u_s, v_s), where $u_s = u$, $v_s = v$, and u_0 is the root of $\mathcal{T}$. Assume that $u_{s-1} = u$ and $v_{s-1} = \overline{\mathsf{p}}(v)$. Then $\Delta(u) \leq \Delta(\overline{\mathsf{p}}(v))$, since the algorithm always refines the larger cell.

Similarly, let t be the last index such that $u_{t-1} = \overline{\mathsf{p}}(u)$ (namely, $u_{t-1} \neq u_t = u$ and $v_{t-1} = v_t$). Then, since v is a descendant of v_{t-1}, it holds that

$$\Delta(v) \leq \Delta(v_t) = \Delta(v_{t-1}) \leq \Delta(u_{t-1}) = \Delta(\overline{\mathsf{p}}(u)),$$

since (again) the algorithm always refines the larger cell in the pair $\{u_{t-1}, v_{t-1}\}$. ■

Lemma 3.9. *The number of pairs in the computed WSPD is $O(n/\varepsilon^{\mathsf{d}})$.*

PROOF. Let $\{u, v\}$ be a pair appearing in the output. Consider the sequence (i.e., stack) of recursive calls that led to this output. In particular, assume that the last recursive call to **algWSPD**(u, v) was issued by **algWSPD**(u, v'), where $v' = \overline{\mathrm{p}}(v)$ is the parent of v in $\mathcal{T}$. Then

$$\Delta(\overline{\mathrm{p}}(u)) \geq \Delta(v') \geq \Delta(u),$$

by Lemma 3.8.

We charge the pair $\{u, v\}$ to the node v' and claim that each node of $\mathcal{T}$ is charged at most $O(\varepsilon^{-\mathrm{d}})$ times. To this end, fix a node $v' \in V(\mathcal{T})$, where $V(\mathcal{T})$ is the set of vertices of $\mathcal{T}$. Since the pair $\{u, v'\}$ was not output by **algWSPD** (despite being considered), we conclude that $\Delta(v') > \varepsilon \cdot \mathbf{d}(u, v')$ and as such $\mathbf{d}(u, v') < r = \Delta(v') / \varepsilon$. Now, there are several possibilities:

(i) $\Delta(v') = \Delta(u)$. But there are at most $O\left((r/\Delta(v'))^{\mathrm{d}}\right) = O(1/\varepsilon^{\mathrm{d}})$ nodes that have the same level (i.e., diameter) as v' such that their cells are within a distance at most r from it, by Lemma 3.5. Thus, this type of charge can happen at most $O(2^{\mathrm{d}} \cdot (1/\varepsilon^{\mathrm{d}}))$ times, since v' has at most 2^{d} children.

(ii) $\Delta(\overline{\mathrm{p}}(u)) = \Delta(v')$. By the same argumentation as above $\mathbf{d}(\overline{\mathrm{p}}(u), v') \leq \mathbf{d}(u, v') < r$. There are at most $O(1/\varepsilon^{\mathrm{d}})$ such nodes $\overline{\mathrm{p}}(u)$. Since the node $\overline{\mathrm{p}}(u)$ has at most 2^{d} children, it follows that the number of such charges is at most $O\left(2^{\mathrm{d}} \cdot 2^{\mathrm{d}} \cdot (1/\varepsilon^{\mathrm{d}})\right)$.

(iii) $\Delta(\overline{\mathrm{p}}(u)) > \Delta(v') > \Delta(u)$. Consider the canonical grid G having $\square_{v'}$ as one of its cells (see Definition 2.1_{p14}). Let $\widehat{\square}$ be the cell in G containing $\square_u$. Observe that $\square_u \subsetneq \widehat{\square} \subsetneq \square_{\overline{\mathrm{p}}(u)}$. In addition, $\mathbf{d}\left(\widehat{\square}, \square_{v'}\right) \leq \mathbf{d}(\square_u, \square_{v'}) = \mathbf{d}(u, v') < r$. It follows that there are at most $O(1/\varepsilon^{\mathrm{d}})$ cells like $\widehat{\square}$ that might participate in charging v', and as such, the total number of charges is $O(2^{\mathrm{d}}/\varepsilon^{\mathrm{d}})$, as claimed.

As such, v' can be charged at most $O\left(2^{2d}/\varepsilon^{\mathrm{d}}\right) = O\left(1/\varepsilon^{\mathrm{d}}\right)$ times[①]. This implies that the total number of pairs generated by the algorithm is $O\left(n\varepsilon^{-\mathrm{d}}\right)$, since the number of nodes in $\mathcal{T}$ is $O(n)$. ■

Since the running time of **algWSPD** is clearly linear in the output size, we have the following result.

Theorem 3.10. *For* $1 \geq \varepsilon > 0$, *one can construct an* ε^{-1}-*WSPD of size* $n\varepsilon^{-\mathrm{d}}$, *and the construction time is* $O\left(n \log n + n\varepsilon^{-\mathrm{d}}\right)$.

3.2. Applications of WSPD

3.2.1. Spanners. It is sometime beneficial to describe distances between n points by using a sparse graph to encode the distances.

Definition 3.11. For a weighted graph G and any two vertices p and q of G, we will denote by $\mathsf{d}_{\mathsf{G}}(\mathsf{q}, \mathsf{s})$ the ***graph distance*** between q and s. Formally, $\mathsf{d}_{\mathsf{G}}(\mathsf{q}, \mathsf{s})$ is the length of the shortest path in G between q and s. It is easy to verify that d_{G} is a metric; that is, it complies with the triangle inequality. Naturally, if q and s belongs to two different connected components of G, then $\mathsf{d}_{\mathsf{G}}(\mathsf{q}, \mathsf{s}) = \infty$.

[①]We remind the reader that we will usually consider the dimension d to be a constant, and the O notation would happily consume any constants that depend only on d. Conceptually, you can think about the O as being a black hole for such constants, as its gravitational force tears such constants away. The shrieks of horror of these constants as they are being swallowed alive by the black hOle can be heard every time you look at the O.

Such a graph might be useful algorithmically if it captures the distances we are interested in while being sparse (i.e., having few edges). In particular, if such a graph G over n vertices has only $O(n)$ edges, then it can be manipulated efficiently, and it is a compact implicit representation of the $\binom{n}{2}$ distances between all the pairs of vertices of G.

A ***t-spanner*** of a set of points $\mathsf{P} \subset \mathbb{R}^{\mathsf{d}}$ is a weighted graph G whose vertices are the points of P, and for any $\mathsf{q}, \mathsf{s} \in \mathsf{P}$, we have

$$\|\mathsf{q} - \mathsf{s}\| \leq \mathsf{d}_{\mathsf{G}}(\mathsf{q}, \mathsf{s}) \leq t \|\mathsf{q} - \mathsf{s}\|.$$

The ratio $\mathsf{d}_{\mathsf{G}}(\mathsf{q}, \mathsf{s}) / \|\mathsf{q} - \mathsf{s}\|$ is the ***stretch*** of q and s in G. The ***stretch*** of G is the maximum stretch of any pair of points of P.

3.2.1.1. *Construction.* We are given a set of n points in $\mathbb{R}^{\mathsf{d}}$ and a parameter $1 \geq \varepsilon > 0$. We will construct a spanner as follows.

Let $c \geq 16$ be an arbitrary constant, and set $\delta = \varepsilon / c$. Compute a δ^{-1}-WSPD decomposition of P using the algorithm of Theorem 3.10. For any vertex u in the quadtree $\mathcal{T}$ (used in computing the WSPD), let rep_u be an arbitrary point of P_u. For every pair $\{u, v\} \in \mathcal{W}$, add an edge between $\{\mathrm{rep}_u, \mathrm{rep}_v\}$ with weight $\|\mathrm{rep}_u - \mathrm{rep}_v\|$, and let G be the resulting graph.

One can prove that the resulting graph is connected and that it contains all the points of P. We will not prove this explicitly as this is implied by the analysis below.

3.2.1.2. *Analysis.* Observe that by the triangle inequality, we have that $\mathsf{d}_{\mathsf{G}}(\mathsf{q}, \mathsf{s}) \geq \|\mathsf{q} - \mathsf{s}\|$, for any $\mathsf{q}, \mathsf{s} \in \mathsf{P}$.

Theorem 3.12. *Given a set* P *of* n *points in* $\mathbb{R}^{\mathsf{d}}$ *and a parameter* $1 \geq \varepsilon > 0$, *one can compute a* $(1+\varepsilon)$*-spanner of* P *with* $O\left(n\varepsilon^{-\mathsf{d}}\right)$ *edges, in* $O\left(n \log n + n\varepsilon^{-\mathsf{d}}\right)$ *time.*

PROOF. The construction is described above. The upper bound on the stretch is proved by induction on the length of pairs in the WSPD. So, fix a pair $x, y \in \mathsf{P}$, and assume that by the induction hypothesis, for any pair $z, w \in \mathsf{P}$ such that $\|z - w\| < \|x - y\|$, it follows that $\mathbf{d}_{\mathsf{G}}(z, w) \leq (1+\varepsilon)\|z - w\|$.

The pair x, y must appear in some pair $\{u, v\} \in \mathcal{W}$, where $x \in \mathsf{P}_u$ and $y \in \mathsf{P}_v$. Thus, by construction

$$\|\mathrm{rep}_u - \mathrm{rep}_v\| \leq \mathbf{d}(u, v) + \Delta(u) + \Delta(v) \leq (1 + 2\delta)\mathbf{d}(u, v) \leq (1 + 2\delta)\|x - y\| \tag{3.1}$$

and this implies

$$\begin{aligned}\max\left(\|\mathrm{rep}_u - x\|, \|\mathrm{rep}_v - y\|\right) &\leq \max(\Delta(u), \Delta(v)) \leq \delta \cdot \mathbf{d}(u, v) \leq \delta \|\mathrm{rep}_u - \mathrm{rep}_v\| \\ &\leq \delta(1 + 2\delta)\|x - y\| < \frac{1}{4}\|x - y\|,\end{aligned}$$

by Theorem 3.10 and since $\delta \leq 1/16$. As such, we can apply the induction hypothesis (twice) to the pairs of points rep_u, x and rep_v, y, implying that

$$\mathsf{d}_{\mathsf{G}}(x, \mathrm{rep}_u) \leq (1+\varepsilon)\|\mathrm{rep}_u - x\| \quad \text{and} \quad \mathsf{d}_{\mathsf{G}}(\mathrm{rep}_v, y) \leq (1+\varepsilon)\|y - \mathrm{rep}_v\|.$$

Now, since $\mathrm{rep}_u \mathrm{rep}_v$ is an edge of G, $\mathsf{d}_{\mathsf{G}}(\mathrm{rep}_u, \mathrm{rep}_v) \leq \|\mathrm{rep}_u - \mathrm{rep}_v\|$. Thus, by the inductive hypothesis, the triangle inequality and (3.1), we have that

$$\begin{aligned}\|x - y\| \leq \mathsf{d}_{\mathsf{G}}(x, y) &\leq \mathsf{d}_{\mathsf{G}}(x, \mathrm{rep}_u) + \mathsf{d}_{\mathsf{G}}(\mathrm{rep}_u, \mathrm{rep}_v) + \mathsf{d}_{\mathsf{G}}(\mathrm{rep}_v, y) \\ &\leq (1+\varepsilon)\|\mathrm{rep}_u - x\| + \|\mathrm{rep}_u - \mathrm{rep}_v\| + (1+\varepsilon)\|\mathrm{rep}_v - y\| \\ &\leq 2(1+\varepsilon)\cdot\delta\cdot\|\mathrm{rep}_u - \mathrm{rep}_v\| + \|\mathrm{rep}_u - \mathrm{rep}_v\| \\ &\leq (1 + 2\delta + 2\varepsilon\delta)\|\mathrm{rep}_u - \mathrm{rep}_v\| \leq (1 + 2\delta + 2\varepsilon\delta)(1 + 2\delta)\|x - y\| \\ &\leq (1+\varepsilon)\|x - y\|.\end{aligned}$$

The last step follows by an easy calculation. Indeed, since $c \geq 16$ and $c\delta = \varepsilon \leq 1$, we have that

$$(1 + 2\delta + 2\varepsilon\delta)\,(1 + 2\delta) \leq (1 + 4\delta)(1 + 2\delta) = 1 + 6\delta + 8\delta^2 \leq 1 + 14\delta \leq 1 + \varepsilon,$$

as required. ■

3.2.2. Approximating the minimum spanning tree. For a graph G, let $\mathsf{G}_{\leq r}$ denote the subgraph of G resulting from removing all the edges of weight (strictly) larger than r from G.

Lemma 3.13. *Given a set* P *of n points in* $\mathbb{R}^{\mathsf{d}}$, *one can compute a spanning tree* $\mathcal{T}$ *of* P, *such that* $\omega(\mathcal{T}) \leq (1 + \varepsilon)\omega(\mathcal{M})$, *where* $\mathcal{M}$ *is a minimum spanning tree of* P *and* $\omega(\mathcal{T})$ *is the total weight of the edges of* $\mathcal{T}$. *This takes* $O\left(n \log n + n\varepsilon^{-\mathsf{d}}\right)$ *time. Furthermore, for any* $r \geq 0$ *and a connected component* C *of* $\mathcal{M}_{\leq r}$, *the set* C *is contained in a connected component of* $\mathcal{T}_{\leq (1+\varepsilon)r}$.

Proof. Compute a $(1 + \varepsilon)$-spanner G of P and let $\mathcal{T}$ be a minimum spanning tree of G. We output the tree $\mathcal{T}$ as the approximate minimum spanning tree. Clearly the time to compute $\mathcal{T}$ is $O\left(n \log n + n\varepsilon^{-\mathsf{d}}\right)$, since the MST of a graph with n vertices and m edges can be computed in $O(n \log n + m)$ time.

There remains the task of proving that $\mathcal{T}$ is the required approximation. For any $\mathsf{q}, \mathsf{s} \in$ P, let π_{qs} denote the shortest path between q and s in G. Since G is a $(1 + \varepsilon)$-spanner, we have that $w\left(\pi_{\mathsf{qs}}\right) \leq (1 + \varepsilon)\,\|\mathsf{q} - \mathsf{s}\|$, where $w\left(\pi_{\mathsf{qs}}\right)$ denotes the weight of π_{qs} in G. We have that $\mathsf{G}' = (\mathsf{P}, E)$ is a connected subgraph of G, where

$$E = \bigcup_{\mathsf{qs} \in E(\mathcal{M})} \pi_{uv},$$

where $E(\mathcal{M})$ denotes the set of edges of the graph $\mathcal{M}$. Furthermore,

$$\omega(\mathsf{G}') \leq \sum_{\mathsf{qs} \in \mathcal{M}} w\left(\pi_{\mathsf{qs}}\right) \leq \sum_{\mathsf{qs} \in \mathcal{M}} (1 + \varepsilon)\,\|\mathsf{q} - \mathsf{s}\| = (1 + \varepsilon)\omega(\mathcal{M}),$$

since G is a $(1 + \varepsilon)$-spanner. Since G' is a connected spanning subgraph of G, it follows that $\omega(\mathcal{T}) \leq \omega(\mathsf{G}') \leq (1 + \varepsilon)w(\mathcal{M})$.

The second claim follows by similar argumentation. ■

3.2.3. Approximating the diameter.

Lemma 3.14. *Given a set* P *of n points in* $\mathbb{R}^{\mathsf{d}}$, *one can compute, in* $O\left(n \log n + n\varepsilon^{-\mathsf{d}}\right)$ *time, a pair* $\mathsf{p}, \mathsf{q} \in \mathsf{P}$, *such that* $\|\mathsf{p} - \mathsf{q}\| \geq (1 - \varepsilon)\mathrm{diam}(\mathsf{P})$, *where* $\mathrm{diam}(\mathsf{P})$ *is the diameter of* P.

Proof. Compute a $(4/\varepsilon)$-WSPD $\mathcal{W}$ of P. As before, we assign for each node u of $\mathcal{T}$ an arbitrary representative point that belongs to P_u. This can be done in linear time. Next, for each pair of $\mathcal{W}$, compute the distance of its representative points (for each pair, this takes constant time). Let $\{x, y\}$ be the pair in $\mathcal{W}$ such that the distance between its two representative points is maximal, and return rep_x and rep_y as the two points realizing the approximation. Overall, this takes $O\left(n \log n + n\varepsilon^{-\mathsf{d}}\right)$ time, since $|\mathcal{W}| = O(n/\varepsilon^{\mathsf{d}})$.

To see why it works, consider the pair $\mathsf{q}, \mathsf{s} \in \mathsf{P}$ realizing the diameter of P, and let $\{u, v\} \in \mathcal{W}$ be the pair in the WSPD that contain the two points, respectively (i.e., $\mathsf{q} \in \mathsf{P}_u$ and $\mathsf{s} \in \mathsf{P}_v$). We have that

$$\begin{aligned}\left\|\mathrm{rep}_u - \mathrm{rep}_v\right\| \geq \mathbf{d}(u, v) &\geq \|\mathsf{q} - \mathsf{s}\| - \mathrm{diam}(\mathsf{P}_u) - \mathrm{diam}(\mathsf{P}_v) \\ &\geq (1 - 2(\varepsilon/2))\,\|\mathsf{q} - \mathsf{s}\| = (1 - \varepsilon)\mathrm{diam}(\mathsf{P}),\end{aligned}$$

since, by Corollary 3.4, $\max(\mathrm{diam}(\mathsf{P}_u), \mathrm{diam}(\mathsf{P}_v)) \le 2(\varepsilon/4)\|\mathsf{q} - \mathsf{s}\|$. Namely, the distance of the two points output by the algorithm is at least $(1 - \varepsilon)\mathrm{diam}(\mathsf{P})$. ■

3.2.4. Closest pair. Let P be a set of points in $\mathbb{R}^d$. We would like to compute the ***closest pair***, namely, the two points closest to each other in P.

We need the following observation.

Lemma 3.15. *Let $\mathcal{W}$ be an ε^{-1}-WSPD of* P, *for $\varepsilon \le 1/2$. There exists a pair $\{u, v\} \in \mathcal{W}$, such that (i) $|\mathsf{P}_u| = |\mathsf{P}_v| = 1$ and (ii) $\|\mathrm{rep}_u - \mathrm{rep}_v\|$ is the length of the closest pair, where $\mathsf{P}_u = \{\mathrm{rep}_u\}$ and $\mathsf{P}_v = \{\mathrm{rep}_v\}$.*

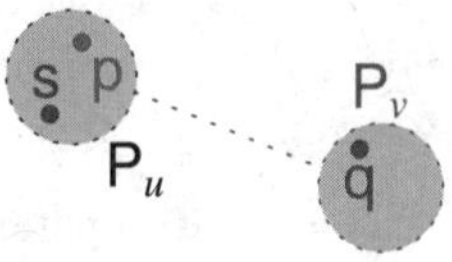

Proof. Consider the pair of closest points p and q in P, and consider the pair $\{u, v\} \in \mathcal{W}$, such that $\mathsf{p} \in \mathsf{P}_u$ and $\mathsf{q} \in \mathsf{P}_v$. If P_u contains an additional point $\mathsf{s} \in \mathsf{P}_u$, then we have that

$$\|\mathsf{p} - \mathsf{s}\| \le \mathrm{diam}(\mathsf{P}_u) \le \varepsilon \cdot \mathbf{d}(u, v) \le \varepsilon \|\mathsf{p} - \mathsf{q}\| < \|\mathsf{p} - \mathsf{q}\|,$$

by Theorem 3.10 and since $\varepsilon = 1/2$. Thus, $\|\mathsf{p} - \mathsf{s}\| < \|\mathsf{p} - \mathsf{q}\|$, a contradiction to the choice of p and q as the closest pair. Thus, $|\mathsf{P}_u| = |\mathsf{P}_v| = 1$ and $\mathrm{rep}_u = \mathsf{q}$ and $\mathrm{rep}_v = \mathsf{s}$. ■

Algorithm. Compute an ε^{-1}-WSPD $\mathcal{W}$ of P, for $\varepsilon = 1/2$. Next, scan all the pairs of $\mathcal{W}$, and compute for all the pairs $\{u, v\}$ which connect singletons (i.e., $|\mathsf{P}_u| = |\mathsf{P}_v| = 1$) the distance between their representatives rep_u and rep_v. The algorithm returns the closest pair of points encountered.

Theorem 3.16. *Given a set* P *of n points in $\mathbb{R}^d$, one can compute the closest pair of points of* P *in $O(n \log n)$ time.*

We remind the reader that we already saw a linear (expected) time algorithm for this problem in Section 1.2. However, this is a deterministic algorithm, and it can be applied in more abstract settings where a small WSPD still exists, while the previous algorithm would not work.

3.2.5. All nearest neighbors. Given a set P of n points in $\mathbb{R}^d$, we would like to compute for each point $\mathsf{q} \in \mathsf{P}$ its ***nearest neighbor*** in P (formally, this is the closest point in $\mathsf{P} \setminus \{\mathsf{q}\}$ to q).

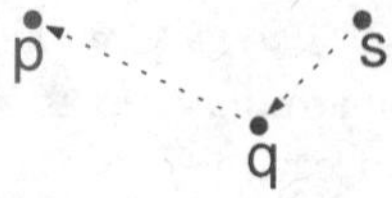

This is harder than it might seem at first, since this is *not* a symmetrical relationship. Indeed, in the figure on the right, q is the nearest neighbor to p, but s is the nearest neighbor to q.

3.2.5.1. *The bounded spread case.*

Algorithm. Assume P is contained in the unit square and $\mathrm{diam}(\mathsf{P}) \ge 1/4$. Furthermore, let $\Phi = \Phi(\mathsf{P})$ denote the spread of P. Compute an ε^{-1}-WSPD $\mathcal{W}$ of P, for $\varepsilon = 1/4$.

Scan all the pairs $\{u, v\}$ with a singleton as one of their sides (i.e., $|P_u| = 1$), and for each such singleton $\mathsf{P}_u = \{\mathsf{p}\}$, record for p the closest point to it in the set P_v. Maintain for each point the closest point to it that was encountered.

We claim that in the end of this process, for every point p in P its recorded nearest point is its nearest neighbor in $\mathsf{P} \setminus \{\mathsf{p}\}$.

Analysis. The analysis of this algorithm is slightly tedious, but it reveals some additional interesting properties of WSPD. We start with a claim that shows that we will indeed find the nearest neighbor for each point.

Lemma 3.17. *Let* p *be any point of* P, *and let* q *be the nearest neighbor to* p *in the set* $\mathsf{P} \setminus \{\mathsf{p}\}$. *Then, there exists a pair* $\{u, v\} \in \mathcal{W}$, *such that* $\mathsf{P}_u = \{\mathsf{p}\}$ *and* $\mathsf{q} \in \mathsf{P}_v$.

PROOF. Consider the pair $\{u, v\} \in \mathcal{W}$ such that $\{\mathsf{p}, \mathsf{q}\} \in \mathsf{P}_u \otimes \mathsf{P}_v$, where $\mathsf{p} \in \mathsf{P}_u$ and $\mathsf{q} \in \mathsf{P}_v$. Now, $\mathrm{diam}(\mathsf{P}_v) \leq \varepsilon \mathbf{d}(\mathsf{P}_u, \mathsf{P}_v) \leq \varepsilon \|\mathsf{p} - \mathsf{q}\| \leq \|\mathsf{p} - \mathsf{q}\|/4$. Namely, if P_v contained any other point except p, then q would not be the nearest neighbor to p. ■

Thus, the above lemma implies that the algorithm will find the nearest neighbor for each point p of P, as the appropriate pair containing only p on one side would be considered by the algorithm. Thus remains the task of bounding the running time.

A pair of nodes $\{x, y\}$ of $\mathcal{T}$ is a ***generator*** of a pair $\{u, v\} \in \mathcal{W}$ if $\{u, v\}$ was computed inside a recursive call **algWSPD**(x, y).

Lemma 3.18. *Let* $\mathcal{W}$ *be an* ε^{-1}*-WSPD of a point set* P *generated by* **algWSPD**. *Consider a pair* $\{u, v\} \in \mathcal{W}$. *Then* $\Delta(\overline{\mathsf{p}}(v)) \geq (\varepsilon/2)\mathbf{d}(u, v)$ *and* $\Delta(\overline{\mathsf{p}}(u)) \geq (\varepsilon/2)\mathbf{d}(u, v)$, *where* $\mathbf{d}(u, v) = \mathbf{d}(\square_u, \square_v)$ *is the distance between the cell of u and the cell of v.*

PROOF. Assume, for the sake of contradiction, that $\Delta(v') < (\varepsilon/2)\ell$, where $\ell = \mathbf{d}(u, v)$ and $v' = \overline{\mathsf{p}}(v)$. By Lemma 3.8, we have that

$$\Delta(u) \leq \Delta(v') < \varepsilon\frac{\ell}{2}.$$

But then

$$\mathbf{d}(u, v') \geq \ell - \Delta(v') \geq \ell - \varepsilon\frac{\ell}{2} \geq \frac{\ell}{2}.$$

Thus,

$$\max(\Delta(u), \Delta(v')) < \varepsilon\frac{\ell}{2} \leq \varepsilon\,\mathbf{d}(u, v').$$

Namely, u and v' are well separated, and as such $\{u, v'\}$ cannot be a generator of $\{u, v\}$. Indeed, if $\{u, v'\}$ was considered by the algorithm, then it would have added it to the WSPD and never created the pair $\{u, v\}$.

So, the other possibility is that $\{u', v\}$ is the generator of $\{u, v\}$, where $u' = \overline{\mathsf{p}}(u)$. But then $\Delta(u') \leq \Delta(v') < \varepsilon\ell/2$, by Lemma 3.8. Using the same argumentation as above, we have that $\{u', v\}$ is a well-separated pair and as such it cannot be a generator of $\{u, v\}$.

But this implies that $\{u, v\}$ cannot be generated by **algWSPD**, since either $\{u, v'\}$ or $\{u', v\}$ must be a generator of $\{u, v\}$, a contradiction. ■

Claim 3.19. *For two pairs* $\{u, v\}, \{u', v'\} \in \mathcal{W}$ *such that* $\square_u \subseteq \square_{u'}$, *the interiors of* $\square_v$ *and* $\square_{v'}$ *are disjoint.*

PROOF. Since $\square_u \subseteq \square_{u'}$, it follows that u' is an ancestor of u.

If v' is an ancestor of v, then **algWSPD** returned the pair $\{u', v'\}$ and it would have never generated the pair $\{u, v\}$.

If v is an ancestor of v' (see the figure on the right), then

$$\begin{aligned}
\Delta(u) &< \Delta(u') && (u' \text{ is an ancestor of } u)\\
&\leq \Delta(\overline{\mathsf{p}}(v')) && (\text{by Lemma 3.8 applied to } \{u', v'\})\\
&\leq \Delta(v) && (v \text{ is an ancestor of } v')\\
&\leq \Delta(\overline{\mathsf{p}}(u)) && (\text{by Lemma 3.8 applied to } \{u, v\})\\
&\leq \Delta(u') && (u' \text{ is an ancestor of } u).
\end{aligned}$$

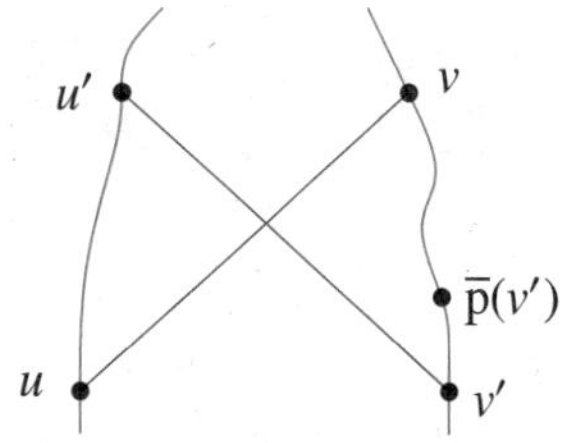

Namely, $\Delta(v) = \Delta(u')$. But then the pair $\{u', v\}$ is a generator of both $\{u, v\}$ and $\{u', v'\}$. To see that, observe that **algWSPD** always tries to consider pairs that have the same diameter (it always splits the bigger side of a pair). As such, for the algorithm to generate the pair $\{u, v\}$, it must have had a generator $\{u'', v\}$, such that $\mathrm{diam}(u'') \geq \mathrm{diam}(v)$. But the lowest ancestor of u that has this property is u'.

Now, when **algWSPD** considered the pair $\{u', v\}$, it split one of its sides (by calling on its children). In either case, it either did not create the pair $\{u, v\}$ or it did not create the pair $\{u', v'\}$, a contradiction.

The case $v = v'$ is handled in a similar fashion. ■

Lemma 3.20. *Let* P *be a set of n points in* $\mathbb{R}^{\mathsf{d}}$, *let* $\mathcal{W}$ *be an* ε^{-1}-*WSPD of* P, *let* $\ell > 0$ *be a distance, and let L be the set of pairs* $\{u, v\} \in \mathcal{W}$ *such that* $\ell \leq \mathbf{d}(u, v) \leq 2\ell$. *Then, for any point* $\mathsf{p} \in \mathsf{P}$, *the number of pairs in L containing* p *is* $O(1/\varepsilon^{\mathsf{d}})$.

PROOF. Let u be the leaf of the quadtree $\mathcal{T}$ (that is used in computing $\mathcal{W}$) storing the point p, and let π be the path between u and the root of $\mathcal{T}$. We claim that L contains at most $O(1/\varepsilon^{\mathsf{d}})$ pairs with nodes that appear along π. Let

$$X = \left\{ v \;\middle|\; v \in \pi, \{u, v\} \in L \right\}.$$

The cells of X are interior disjoint by Claim 3.19, and they contain all the pairs in W that covers p.

So, let r be the largest power of two which is smaller than (say) $\varepsilon\ell / \left(4\sqrt{\mathsf{d}}\right)$. Clearly, there are $O(1/\varepsilon^{\mathsf{d}})$ cells of G_r within a distance at most 2ℓ from $\Box_u$. We account for the nodes $v \in X$, as follows:

(i) If $\Delta(v) \geq r\sqrt{\mathsf{d}}$, then $\Box_v$ contains a cell of G_r, and there are at most $O(1/\varepsilon^{\mathsf{d}})$ such cells.

(ii) If $\Delta(v) < r\sqrt{\mathsf{d}}$ and $\Delta(\overline{\mathsf{p}}(v)) \geq r\sqrt{\mathsf{d}}$, then:

- (a) If $\overline{\mathsf{p}}(v)$ is a compressed node, then $\overline{\mathsf{p}}(v)$ contains a cell of G_r and it has only v as a single child. As such, there are at most $O(1/\varepsilon^{\mathsf{d}})$ such charges.
- (b) Otherwise, $\overline{\mathsf{p}}(v)$ is not compressed, but then $\mathrm{diam}(\Box_v) = \mathrm{diam}\left(\Box_{\overline{\mathsf{p}}(v)}\right)/2$. As such, $\Box_v$ contains a cell of $\mathsf{G}_{r/2}$ within a distance at most 2ℓ from $\Box_u$, and there are $O(1/\varepsilon^{\mathsf{d}})$ such cells.

(iii) The case $\Delta(\overline{\mathsf{p}}(v)) < r\sqrt{\mathsf{d}}$ is impossible. Indeed, by Lemma 3.8, we have $\Delta(\overline{\mathsf{p}}(v)) < r\sqrt{\mathsf{d}} \leq \varepsilon\ell/4 = (\varepsilon/4)\mathbf{d}(u, v)$, a contradiction to Lemma 3.18.

We conclude that there are at most $O(1/\varepsilon^{\mathsf{d}})$ pairs that include p in L. ■

Lemma 3.21. *Let* P *be a set of n points in the plane. Then one can solve the all nearest neighbors problem, in* $O(n(\log n + \log \Phi(P)))$ *time, where* Φ *is the spread of* P.

PROOF. The algorithm is described above. There only remains the task of analyzing the running time. For a number $i \in \{0, -1, \ldots, -\lfloor \lg \Phi \rfloor - 4\}$, consider the set of pairs L_i, such that $\{u, v\} \in L_i$, if and only if $\{u, v\} \in \mathcal{W}$, and $2^{i-1} \leq \mathbf{d}(u, v) \leq 2^i$. Here, $\mathcal{W}$ is a $1/\varepsilon$-WSPD of P, where $\varepsilon = 1/4$. A point $\mathsf{p} \in \mathsf{P}$ can be scanned at most $O(1/\varepsilon^{\mathsf{d}}) = O(1)$ times because of pairs in L_i by Lemma 3.20. As such, a point gets scanned at most $O(\log \Phi)$ times overall, which implies the running time bound. ■

3.2.5.2. *All nearest neighbors – the unbounded spread case.* To handle the unbounded case, we need to use some additional geometric properties.

Lemma 3.22. *Let u be a node in the compressed quadtree of* P, *and partition the space around* rep_u *into cones of angle* $\leq \pi/3$. *Let* ψ *be such a cone, and let* Q *be the set of all* *points in* P *which are within a distance* $\geq 4\mathrm{diam}(\mathsf{P}_u)$ *from* rep_u *and all lie inside* ψ*. Let* q *be the closest point in* Q *to* rep_u*. Then,* q *is the only point in* Q *whose nearest neighbor might be in* P_u*.*

Proof. Let $\mathsf{p} = \mathrm{rep}_u$ and consider any point $\mathsf{s} \in \mathsf{Q}$.

Since $\|\mathsf{s} - \mathsf{p}\| \geq \|\mathsf{q} - \mathsf{p}\|$, it follows that $\alpha = \angle \mathsf{sqp} \geq \angle \mathsf{qsp} = \gamma$. Now, $\alpha + \gamma = \pi - \beta$, where $\beta = \angle \mathsf{spq}$. But $\beta \leq \pi/3$, and as such

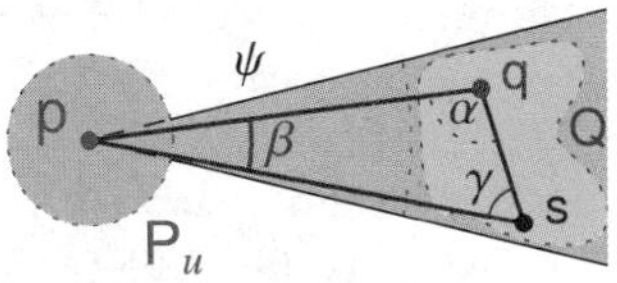

$$2\alpha \geq \alpha + \gamma = \pi - \beta \geq 3\beta - \beta = 2\beta.$$

Namely, α is the largest angle in the triangle $\triangle \mathsf{pqs}$, which implies $\|\mathsf{s} - \mathsf{p}\| \geq \|\mathsf{s} - \mathsf{q}\|$. Namely, q is closer to s than p, and as such p cannot serve as the nearest neighbor to s in P.

It is now straightforward (but tedious) to show that, in fact, for any $\mathsf{t} \in \mathsf{P}_u$, we have $\|\mathsf{s} - \mathsf{t}\| \geq \|\mathsf{s} - \mathsf{q}\|$, which implies the claim.[2] ■

Lemma 3.22 implies that we can do a top-down traversal of the compressed quadtree of P, after computing an ε^{-1}-WSPD $\mathcal{W}$ of P, for $\varepsilon = 1/16$. For every node u, we maintain a (constant size) set R_u of candidate points such that P_u *might* contain their nearest neighbor.

So, assume we had computed $\mathsf{R}_{\overline{\mathsf{p}}(u)}$, and consider the set

$$X(u) = \mathsf{R}_{\overline{\mathsf{p}}(u)} \cup \bigcup_{\{u,v\} \in \mathcal{W}, |\mathsf{P}_v| = 1} \mathsf{P}_v.$$

(Note that we do not have to consider pairs with $|\mathsf{P}_v| > 1$, since no point in P_v can have its nearest neighbor in P_u in such a scenario.) Clearly, we can compute $X(u)$ in linear time in the number of pairs in $\mathcal{W}$ involved with u. Now, we build a "grid" of cones around rep_u and throw the points of $X(u)$ into this grid. For each such cone, we keep only the closest point p' to rep_u (because the other points in this cone would use p' as nearest neighbor before using any point of P_u). Let R_u be the set of these closest points. Since the number of cones is $O(1)$, it follows that $|\mathsf{R}_u| = O(1)$.

We continue this top-down traversal till R_u is computed for all the nodes in the tree.

Now, for every vertex u, if P_u contains only a single point p, then we compute for any point $\mathsf{q} \in \mathsf{R}_u$ its distance to p, and if p is a better candidate to be a nearest neighbor, then we set p as the (current) nearest neighbor to q.

Correctness. Clearly, the resulting running time (ignoring the computation of the WSPD) is linear in the number of pairs of the WSPD and the size of the compressed quadtree. If p is the nearest neighbor to q, then there must be a WSPD pair $\{u, v\}$ such that $\mathsf{P}_v = \{\mathsf{q}\}$ and $\mathsf{p} \in \mathsf{P}_u$. But then the algorithm would add q to the set R_u, and it would be in R_z, for all descendants z of u in the quadtree, such that $\mathsf{p} \in \mathsf{P}_z$. In particular, if y is the leaf of the quadtree storing p, then $\mathsf{q} \in \mathsf{R}_y$, which implies that the algorithm computes correctly the nearest neighbor to q.

This implies the correctness of the algorithm.

Theorem 3.23. *Given a set* P *of* n *points in* $\mathbb{R}^d$*, one can solve the all nearest neighbors problem in* $O(n \log n)$ *time.*

---

[2] Here are the details for readers of little faith. By the law of sines, we have $\frac{\|\mathsf{p}-\mathsf{s}\|}{\sin\alpha} = \frac{\|\mathsf{q}-\mathsf{s}\|}{\sin\beta}$. As such, $\|\mathsf{q} - \mathsf{s}\| = \|\mathsf{p} - \mathsf{s}\| \frac{\sin\beta}{\sin\alpha}$. Now, if $\alpha \leq \pi - 3\beta$, then $\|\mathsf{q} - \mathsf{s}\| = \|\mathsf{p} - \mathsf{s}\| \frac{\sin\beta}{\sin\alpha} \leq \|\mathsf{p} - \mathsf{s}\| \frac{\sin\beta}{\sin(3\beta)} \leq \frac{\|\mathsf{p}-\mathsf{s}\|}{2} < \|\mathsf{p} - \mathsf{s}\| - \Delta(u)$, since $\|\mathsf{p} - \mathsf{s}\| \geq 4\Delta(u)$. This implies that no point of P_u can be the nearest neighbor of s.

If $\alpha \geq \pi - 3\beta$, then the maximum length of qs is achieved when $\gamma = 2\beta$. The law of sines then implies that $\|\mathsf{q} - \mathsf{s}\| = \|\mathsf{p} - \mathsf{q}\| \frac{\sin\beta}{\sin(2\beta)} \leq \frac{3}{4}\|\mathsf{p} - \mathsf{q}\| \leq \frac{3}{4}\|\mathsf{p} - \mathsf{s}\| < \|\mathsf{p} - \mathsf{s}\| - \Delta(u)$, which again implies the claim.

3.3. Semi-separated pair decomposition (SSPD)

Here we present an interesting relaxation of WSPD that has the advantage of having a low total weight.

Definition 3.24. Given a pair decomposition $\mathcal{W} = \left\{\{A_1, B_1\}, \ldots, \{A_s, B_s\}\right\}$ of a point set P, its ***weight*** is $\omega(\mathcal{W}) = \sum_{i=1}^{s}(|A_i| + |B_i|)$.

It is easy to verify that, in the worst case, a WSPD of a set of n points in the plane might have weight $O(n^2)$; see Exercise 3.1. The notation of SSPD circumvents this by requiring a weaker notion of separation.

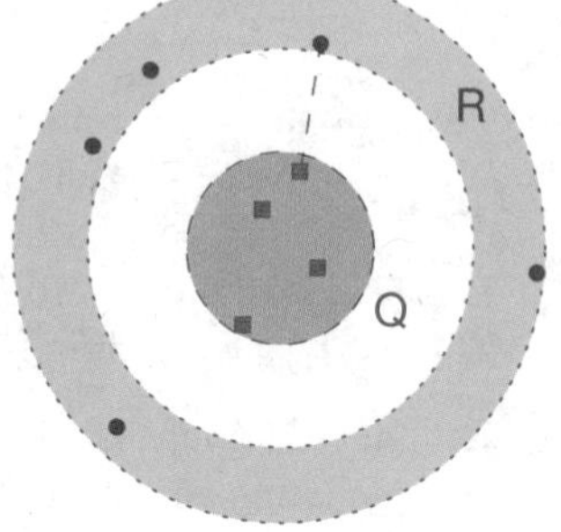

Definition 3.25. Two sets of points Q and R are $(1/\varepsilon)$-***semi-separated*** if

$$\min(\mathrm{diam}(Q), \mathrm{diam}(R)) \le \varepsilon \cdot \mathbf{d}(Q, R),$$

where $\mathbf{d}(Q, R) = \min_{s\in Q, t\in R} \|s - t\|$.

See the figure on the right for an example.

See Definition 3.2p29 for the original notion of sets being well separated; in particular, in the well-separated case we demanded that the separation be large relative to the diameter of both sets, while here it is sufficient that the separation be large for the smaller of the two sets. The following is the analog of the WSPD (Definition 3.3).

Definition 3.26 (SSPD)**.** For a point set P, a ***semi-separated pair decomposition*** (***SSPD***) of P with parameter $1/\varepsilon$, denoted by ε^{-1}-SSPD, is a pair decomposition of P formed by a set of pairs $\mathcal{W}$ such that all the pairs are $1/\varepsilon$-semi-separated.

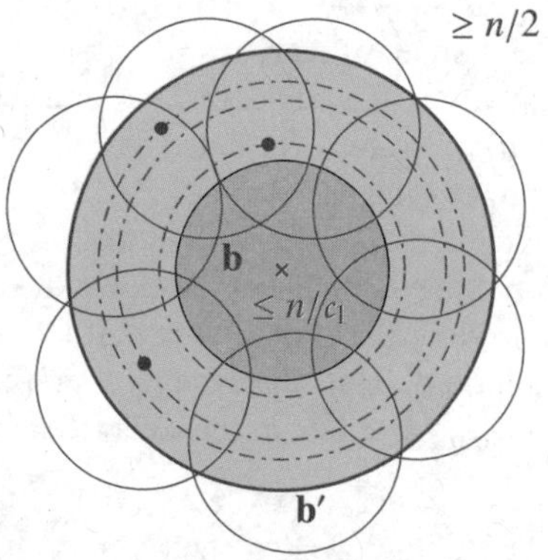

Observation 3.27. *An* ε^{-1}*-WSPD of* P *is an* ε^{-1}*-SSPD of* P.

3.3.1. Construction. We use the property that in low dimensions there always exists a good separating ring that breaks the point set into two reasonably large subsets. Indeed, consider the smallest ball $\mathbf{b} = \mathbf{b}(p, r)$ that contains n/c_1 points of P, where c_1 is a sufficiently large constant. Let $\mathbf{b}'$ be the scaling of this ball by a factor of two. By a standard packing argument, the ring $\mathbf{b}' \setminus \mathbf{b}$ can be covered with $c = O(1)$ copies of $\mathbf{b}$, none of which can contain more than n/c_1 points of P. It follows that by picking $c_1 = 3c$, we are guaranteed that at least half the points of P are outside $\mathbf{b}'$. Now, the ring can be split into $n/2$ empty rings (by taking a sphere that passes through each point inside the ring). One of them would be of thickness at least r/n, and it would separate the inner n/c points of P from the outer $n/2$ points of P. Doing this efficiently requires trading off some constants, and it requires some tedious details, as described in the following lemma.

Lemma 3.28. *Let* P *be a set of* n *points in* $\mathbb{R}^d$*, let* $t > 0$ *be a parameter, and let* c *be a sufficiently large constant. Then one can compute, in linear time, a ball* $\mathbf{b} = \mathbf{b}(p, r)$*, such that (i)* $|\mathbf{b} \cap P| \ge n/c$*, (ii)* $|\mathbf{b}(p, r(1 + 1/t)) \cap P| \le n/2t + |\mathbf{b} \cap P|$*, and (iii)* $|P \setminus \mathbf{b}(p, 2r)| \ge n/2$.

PROOF. Let $\mathbf{b} = \mathbf{b}(\mathsf{p}, \alpha)$ be the disk computed, in $O(n)$ time, by Lemma 1.16$_{p6}$, for $k = n/c$. This ball contains n/c points of P and $\alpha \leq 2r_{\mathrm{opt}}(\mathsf{P}, k)$, where $r_{\mathrm{opt}}(\mathsf{P}, k)$ is the radius of the smallest ball containing k points of P. Observe that the ball $\mathbf{b}_{8\alpha} = \mathbf{b}(\mathsf{p}, 8\alpha)$ can be covered by $M = O(1)$ balls of radius $\alpha/2$. Each of these balls contains at most n/c points, by the construction of $\mathbf{b}$. As such, if $c > 2M$, we have that $|\mathsf{P} \cap \mathbf{b}_{8\alpha}| \leq M(n/c) \leq n/2$ and $|\mathsf{P} \setminus \mathbf{b}_{8\alpha}| \geq n/2$.

We will set $r \in [\alpha, e\alpha]$ in such a way that property (ii) will hold for it. Indeed, set $r_i = \alpha(1 + 1/t)^i$, for $i = 0, \ldots, t$, and consider the rings

$$\mathcal{R}_i = \mathbf{b}(\mathsf{p}, r_i) \setminus \mathbf{b}(\mathsf{p}, r_{i-1}),$$

for $i = 1, \ldots, t$. We have that $r_t = \alpha(1 + 1/t)^t \leq \alpha \exp(t/t) = \alpha e$, since $1 + x \leq e^x$ for all $x \geq 0$. Now, all these (interior disjoint) rings are contained inside $\mathbf{b}_{4\alpha}$. It follows that one of these rings, say the ith ring $\mathcal{R}_i$, contains at most $(n/2)/t$ of the points of P (since $\mathbf{b}(\mathsf{p}, 8\alpha)$ contains at most half of the points of P). For $r = r_{i-1} \leq 4r$ the ball $\mathbf{b} = \mathbf{b}(\mathsf{p}, r)$ has the required properties, as $\mathbf{b}(\mathsf{p}, 2r) \subseteq \mathbf{b}(\mathsf{p}, 8\alpha)$. ∎

We also need the following easy property.

Lemma 3.29. *Let P be a set of n points in $\mathbb{R}^{\mathsf{d}}$, with spread $\Phi = \Phi(\mathsf{P})$, and let $\varepsilon > 0$ be a parameter. Then, one can compute $(1/\varepsilon)$-WSPD (and thus a $(1/\varepsilon)$-SSPD) for P of total weight $O(n\varepsilon^{-\mathsf{d}} \log \Phi)$. Furthermore, any point of P participates in at most $O\left(\varepsilon^{-\mathsf{d}} \log \Phi\right)$ pairs.*

PROOF. Build a regular (i.e., not compressed) quadtree for P, and observe that its depth is $O(\log \Phi)$. Now, construct a WSPD for P using this quadtree. Consider a pair of nodes (u, v) in this WSPD, and observe that the sidelength of u and v is the same up to a factor of two (since we used a non-compressed quadtree). As such, every node participates in $O(1/\varepsilon^{\mathsf{d}})$ pairs in the WSPD. We conclude that each point participates in $O\left(\varepsilon^{-\mathsf{d}} \log \Phi\right)$ pairs, which implies that the total weight of this WSPD is as claimed. ∎

Theorem 3.30. *Let P be a set of points in $\mathbb{R}^{\mathsf{d}}$, and let $\varepsilon > 0$ be a parameter. Then, one can compute a $(1/\varepsilon)$-SSPD for P of total weight $O\left(n\varepsilon^{-\mathsf{d}} \log^2 n\right)$. The number of pairs in the SSPD is $O\left(n\varepsilon^{-\mathsf{d}} \log n\right)$, and the computation time is $O\left(n \log^2 n + n\varepsilon^{-\mathsf{d}} \log n\right)$.*

PROOF. Using Lemma 3.28, with $t = n$, we compute a ball $\mathbf{b}(\mathsf{p}, r)$ that contains at least n/c points of P and such that $\mathcal{R} = \mathbf{b}(\mathsf{p}, (1 + 1/t)r) \setminus \mathbf{b}(\mathsf{p}, r)$ contains no point of P. Let $\mathsf{P}_{\mathrm{in}} = \mathsf{P} \cap \mathbf{b}$, $\mathsf{P}_{\mathrm{far}} = \mathsf{P} \cap \mathbf{b}(\mathsf{p}, 2r/\varepsilon)$, and $\mathsf{P}_{\mathrm{out}} = \mathsf{P} \setminus \left(\mathsf{P}_{\mathrm{in}} \cup \mathsf{P}_{\mathrm{far}}\right)$. Clearly, $\left\{\mathsf{P}_{\mathrm{in}}, \mathsf{P}_{\mathrm{far}}\right\}$ is a $(1/\varepsilon)$-semi-separated set, which we add to our SSPD. Let $\ell = \min_{\mathsf{p} \in \mathsf{P}_{\mathrm{in}}, \mathsf{q} \in \mathsf{P}_{\mathrm{out}}} \|\mathsf{p} - \mathsf{q}\|$. Observe that ℓ is larger than the thickness of the empty ring $\mathcal{R}$; that is, $\ell \geq r/n$.

We would like to compute the SSPD for all pairs in $X = \mathsf{P}_{\mathrm{in}} \otimes \mathsf{P}_{\mathrm{out}}$. The observation is that none of these pairs are of distance smaller than ℓ, and the diameter of the point set $\mathsf{Q} = \mathsf{P}_{\mathrm{in}} \cup \mathsf{P}_{\mathrm{out}}$ is $\mathrm{diam}(\mathsf{Q}) \leq 4\ell n/\varepsilon$. Thus, we can snap the point set Q to a grid of sidelength $\varepsilon\ell/10$. The resulting point set Q' has spread $O(n/\varepsilon^2)$. Next, compute a $2/\varepsilon$-SSPD for the snapped point set Q', using Lemma 3.29. Clearly, the computed SSPD when extended back to the point set Q would cover all the pairs of X, and it would provide a $(1/\varepsilon)$-SSPD for these pairs. By Lemma 3.29, every point of Q would participate in at most $O\left(\varepsilon^{-\mathsf{d}} \log(n/\varepsilon)\right) = O\left(\varepsilon^{-\mathsf{d}} \log n\right)$ pairs.

To complete the construction, we need to construct a $(1/\varepsilon)$-SSPD for the pairs in $\mathsf{P}_{\mathrm{in}} \otimes \mathsf{P}_{\mathrm{in}}$ and in $(\mathsf{P}_{\mathrm{out}} \cup \mathsf{P}_{\mathrm{far}}) \otimes (\mathsf{P}_{\mathrm{out}} \cup \mathsf{P}_{\mathrm{far}})$. This we do by continuing the construction recursively on the point sets P_{in} and $\mathsf{P}_{\mathrm{out}} \cup \mathsf{P}_{\mathrm{far}}$.

In the resulting WSPD, every point participates in at most

$$T(n) = 1 + O\left(\varepsilon^{-\mathsf{d}} \log n\right) + \max(T(n_1), T(n_2))$$

pairs of the resulting SSPD, where $n_1 = \left|\mathsf{P}_{\mathrm{in}}\right|$ and $n_2 = \left|\mathsf{P}_{\mathrm{out}} \cup \mathsf{P}_{\mathrm{far}}\right|$. Since $n_1 + n_2 = n$ and $n_1, n_2 \geq n/c$, where c is some constant, it follows that $T(n) = O\left(\varepsilon^{-\mathsf{d}} \log^2 n\right)$. Namely, every point of P participates in at most $O\left(\varepsilon^{-\mathsf{d}} \log^2 n\right)$ pairs in the resulting SSPD. As such, the total weight of the SSPD is $O\left(n\varepsilon^{-\mathsf{d}} \log^2 n\right)$.

In each level of the recursion, we create at most $O(n/\varepsilon^2)$ pairs. As such, the bound on the number of pairs created follows. The bound on the running time follows by similar argumentation. ■

3.3.2. Lower bound. The result of Theorem 3.30 can be improved so that the total weight of the SSPD is $O\left(n\varepsilon^{-\mathsf{d}} \log n\right)$; see the bibliographical notes in the next section. Interestingly, it turns out that any pair decomposition (without any required separation property) has to be of total weight $\Omega(n \log n)$.

Lemma 3.31. *Let* P *be a set of* n *points, and let* $\mathcal{W} = \left\{\{A_1, B_1\}, \ldots, \{A_s, B_s\}\right\}$ *be a pair decomposition of* P *(see Definition* 3.1_{p29}*). Then,* $\omega(\mathcal{W}) = \sum_i (|A_i| + |B_i|) = \Omega(n \log n)$.

PROOF. Scan the pairs $\{A_i, B_i\}$ one by one, and for each such pair, randomly set Y_i to be either A_i or B_i with equal probability. Let $\mathsf{R} = \mathsf{P} \setminus \left(\bigcup_i Y_i\right)$.

Observe that R is either empty or contain a single point. Indeed, if there are two distinct points $\mathsf{p}, \mathsf{q} \in \mathsf{R}$, then there exists an index j such that $\{\mathsf{p}, \mathsf{q}\} \in A_j \otimes B_j$. In particular, $\left|\{\mathsf{p}, \mathsf{q}\} \cap A_j\right| = 1$ and $\left|\{\mathsf{p}, \mathsf{q}\} \cap B_j\right| = 1$. As such $\left|\{\mathsf{p}, \mathsf{q}\} \cap Y_j\right| = 1$, and this implies that R cannot contain both points, as R does not contain any of the points of Y_j.

Now, let $\mathsf{P} = \{\mathsf{p}_1, \ldots, \mathsf{p}_n\}$ and let x_i be the number of pairs that contain p_i (in either side of the pair), for $i = 1, \ldots, n$. Now, the quantity we need to bound is $\sum_i x_i$ since $\sum_{i=1}^{s} (|A_i| + |B_i|) = \sum_{i=1}^{n} x_i$. Observe that the probability of p_i to be in R is *exactly* $1/2^{x_i}$ since this is the probability that for all the x_i pairs that contain p_i we eliminated the side that does not contain p_i. As such, define an indicator variable Z_i such that $Z_i = 1$ if and only if $\mathsf{p}_i \in \mathsf{R}$. Observe that $\sum_i Z_i = |\mathsf{R}| \leq 1$. As such, by linearity of expectations, we have that $\sum_i \mathbf{E}[Z_i] = \mathbf{E}[\sum_i Z_i] \leq 1$. But $\mathbf{E}[Z_i = 1] = \mathbf{Pr}[\mathsf{p}_i \in \mathsf{R}] = 1/2^{x_i}$. Namely, we have that

$$\sum_{i=1}^{n} \frac{1}{2^{x_i}} = \sum_i \mathbf{Pr}[\mathsf{p}_i \in \mathsf{R}] = \sum_i \mathbf{E}[Z_i] \leq 1.$$

Let $u_1, \ldots, u_n$ be a set of n positive integers that minimizes $\sum_i u_i$, such that $\sum_{i=1}^{n} 2^{-u_i} \leq 1$. Observe that $\sum_i u_i \leq \sum_i x_i$.

Now, if there are i and j such that $u_i > u_j + 1$, then we have that

$$\frac{1}{2^{u_i - 1}} + \frac{1}{2^{u_j + 1}} \leq \frac{2}{2^{u_j + 1}} = \frac{1}{2^{u_j}} \leq \frac{1}{2^{u_i}} + \frac{1}{2^{u_j}}.$$

Namely, setting $u_i - 1$ and $u_j + 1$ as the new values of u_i and u_j, respectively, does not change the quantity $u_1 + \cdots + u_n$, while preserving the inequality $\sum_{i=1}^{n} 2^{-u_i} \leq 1$. We repeat this fix-up process till we have $|u_i - u_j| \leq 1$ for all i and j. So, let t be the number such that $t \leq u_i \leq t + 1$, for all i. We have that

$$\sum_{i=1}^{n} 1/2^{t+1} \leq \sum_{i=1}^{n} 1/2^{u_i} \leq 1 \implies n/2^{t+1} \leq 1 \implies n \leq 2^{t+1} \implies t \geq \lg(n/2).$$

We conclude that $\sum_i x_i \geq \sum_i u_i \geq nt \geq n \lg(n/2)$, as claimed. ■

3.4. Bibliographical notes

Well-separated pair decomposition was defined by Callahan and Kosaraju [**CK95**]. They defined a different space decomposition tree, known as the ***fair split tree***. Here, one computes the axis parallel bounding box of the point set, always splitting along the longest edge by a perpendicular plane in the middle (or near the middle). This splits the point set into two sets, for which we construct the fair split tree recursively. Implementing this in $O(n \log n)$ time requires some cleverness. See [**CK95**] for details.

Our presentation of the WSPD (very roughly) follows [**HM06**]. The (easy) observation that a WSPD can be generated directly from a compressed quadtree (thus avoiding the fair split tree) is from there.

Callahan and Kosaraju [**CK95**] were inspired by the work of Vaidya [**Vai86**] on the all nearest neighbors problem (i.e., compute for each point in P its nearest neighbor in P). He defined the fair split tree and showed how to compute the all nearest neighbors in $O(n \log n)$ time. However, the first to give an $O(n \log n)$ time algorithm for the all nearest neighbors problem was Clarkson [**Cla83**] using similar ideas (this was part of his PhD thesis).

Diameter. The algorithm for computing the diameter in Section 3.2.3 can be improved by not constructing pairs that cannot improve the (current) diameter and by constructing the underlying tree on the fly together with the diameter. This yields a simple algorithm that works quite well in practice; see [**Har01a**].

All nearest neighbors. Section 3.2.5 is a simplification of the algorithm for the all k-nearest neighbors problem. Here, one can compute for every point its k-nearest neighbors in $O(n \log n + nk)$ time. See [**CK95**] for details.

The all nearest neighbors algorithm for the bounded spread case (Section 3.2.5.1) is from [**HM06**]. Note that unlike the unbounded case, this algorithm only uses packing arguments for its correctness. Surprisingly, the usage of the Euclidean nature of the underlying space (as done in Section 3.2.5.2) seems to be crucial in getting a faster algorithm for this problem. In particular, for the case of metric spaces of low doubling dimension (that do have a small WSPD), solving this problem requires $\Omega(n^2)$ time in the worst case.

Dynamic maintenance. WSPD can be maintained in polylogarithmic time under insertions and deletions. This is quite surprising when one considers that, in the worst case, a point might participate in a linear number of pairs, and a node in the quadtree might participate in a linear number of pairs. This is described in detail in Callahan's thesis [**Cal95**]. Interestingly, using randomization, maintaining the WSPD can be considerably simplified; see the work by Fischer and Har-Peled [**FH05**].

High dimension. In high dimensions, as the uniform metric demonstrates (i.e., n points, all of them within a distance of 1 from each other), the WSPD can have quadratic complexity. This metric is easily realizable as the vertices of a simplex in $\mathbb{R}^{n-1}$. On the other hand, doubling metrics have near linear size WSPD. Since WSPDs by themselves are so powerful, it is tempting to try to define the dimension of a point set by the size of the WSPD it has. This seems like an interesting direction for future research, as currently little is known about it (to the best of the author's knowledge).

Semi-separated pair decomposition. The notion of semi-separated pair decomposition was introduced by Varadarajan [**Var98**] who used it to speed up the matching algorithm for points in the plane. His SSPD construction was of total weight $O(n \log^4 n)$ (for a constant ε). This was improved to $O(n \log n)$ by [**AdBFG09**].

SSPDs were used to construct spanners that can survive even if a large fraction of the graph disappears (for example, all the nodes inside the arbitrary convex region disappear) [**AdBFG09**]. In [**AdBF$^+$09**], SSPDs were used for computing additively weighted spanners. Further work on SSPDs can be found in [**ACFS09**]. The simpler construction shown here is due to Abam and Har-Peled [**AH10**] and it also works for metrics with low doubling dimension.

The elegant proof of the lower bound on the size of the SSPD is from [**BS07**].

Euclidean minimum spanning tree. Let P be a set of points in $\mathbb{R}^{\mathsf{d}}$ for which we want to compute its minimum spanning tree (MST). It is easy to verify that if an edge pq is not a Delaunay edge of P, then its diametrical ball must contain some other point s in its interior, but then ps and qs are shorter than pq. This in turn implies that pq cannot be an edge of the MST. As such, the edges of the MST are subset of the edges of the Delaunay triangulation of P. Since the Delaunay triangulation can be computed in $O(n \log n)$ time in the plane, this implies that the MST can be computed in $O(n \log n)$ time in the plane. In $\mathsf{d} \geq 3$ dimensions, the exact MST can be computed in $O\left(n^{2-2/(\lceil \mathsf{d}/2 \rceil+1)+\varepsilon'}\right)$ time, where $\varepsilon' > 0$ is an arbitrary small constant [**AESW91**]. The computation of the MST is closely related to the bichromatic closest pair problem [**KLN99**].

There is a lot of work on the MST in Euclidean settings, from estimating its weight in sublinear time [**CRT05**], to doing this in the streaming model under insertions and deletions [**FIS05**], to implementation of a variant of the approximation algorithm described here [**NZ01**]. This list is by no means exhaustive.

3.5. Exercises

Exercise 3.1 (The unbearable weight of the WSPD). Show that there exists a set of n points (even on the line), such that any ε^{-1}-WSPD for it has weight $\Omega(n^2)$, for $\varepsilon = 1/4$.

Exercise 3.2 (WSPD structure). Let $\varepsilon > 0$ be a sufficiently small constant. For any n sufficiently large, show an example of a point set P of n points, such that its $(1/\varepsilon)$-WSPD (as computed by **algWSPD**) has the property that a single set participates in $\Omega(n)$ sets.[③]

Exercise 3.3 (Number of resolutions that matter). Let P be an n-point set in $\mathbb{R}^{\mathsf{d}}$, and consider the set

$$U = \left\{ i \;\middle|\; 2^i \leq \|\mathsf{p} - \mathsf{q}\| \leq 2^{i+1}, \text{ for } \mathsf{p}, \mathsf{q} \in \mathsf{P} \right\}.$$

Prove that $|U| = O(n)$ (the constant depends on d). Namely, there are only n different resolutions that "matter".

(The claim is a special case of a more general claim; see Exercise 26.3$_{p333}$.)

Exercise 3.4 (WSPD and sum of distances). Let P be a set of n points in $\mathbb{R}^{\mathsf{d}}$. The ***sponginess***[④] of P is $X = \sum_{\{\mathsf{p},\mathsf{q}\} \subseteq \mathsf{P}} \|\mathsf{p} - \mathsf{q}\|$. Provide an efficient algorithm for approximating X. Namely, given P and a parameter $\varepsilon > 0$, it outputs a number Y such that $X \leq Y \leq (1+\varepsilon)X$.

[③]Note that there is always a WSPD construction such that each node participates in a "small" number of pairs.

[④]This is also known as the sum of pairwise distances in the literature, for reasons that the author cannot fathom.

(The interested reader can also verify that computing (exactly) the sum of all ***squared*** distances (i.e., $\sum_{\{p,q\}\subseteq P} \|p - q\|^2$) is considerably easier.)

Exercise 3.5 (SSPD with fewer pairs)**.** Let P be a set of n points in $\mathbb{R}^d$. The result of Theorem 3.30 can be improved so that the number of pairs in the $(1/\varepsilon)$-SSPD $\mathcal{W}$ generated is $O(n/\varepsilon^d)$. To this end, consider a pair separating points of P_{in} from P_{far} as a ***long pair***, and a pair separating P_{in} from P_{out} as a ***short pair***.

(A) Prove that the number of long pairs generated by the construction of Theorem 3.30 is $O(n)$.

(B) For an appropriately small enough constant c, consider a (c/ε)-WSPD $\mathcal{W}'$ of P, and show how to find for each short pair of $\mathcal{W}$ the pair in $\mathcal{W}'$ that looks like it.

(C) Show how to reduce the number of short pairs in the SSPD by merging certain pairs, so that the resulting pairs are still $O(1/\varepsilon)$-separated.

(D) Describe how to reduce the number of pairs $\mathcal{W}$, so that the resulting decomposition is an $O(1/\varepsilon)$-SSPD with $O(n/\varepsilon^d)$ pairs and having the same weight as $\mathcal{W}$.

CHAPTER 4

Clustering – Definitions and Basic Algorithms

In this chapter, we will initiate our discussion of ***clustering***. Clustering is one of the most fundamental computational tasks but, frustratingly, one of the fuzziest. It can be stated informally as: "Given data, find an interesting structure in the data. Go!"

The fuzziness arises naturally from the requirement that the clustering should be "interesting", which is not a well-defined concept and depends on human perception and hence is impossible to quantify clearly. Similarly, the meaning of "structure" is also open to debate. Nevertheless, clustering is inherent to many computational tasks like learning, searching, and data-mining.

Empirical study of clustering concentrates on trying various measures for the clustering and trying out various algorithms and heuristics to compute these clusterings. See the bibliographical notes in this chapter for some relevant references.

Here we will concentrate on some well-defined clustering tasks, including k-center clustering, k-median clustering, and k-means clustering, and some basic algorithms for these problems.

4.1. Preliminaries

A clustering problem is usually defined by a set of items, and a distance function between the items in this set. While these items might be points in $\mathbb{R}^d$ and the distance function just the regular Euclidean distance, it is sometime beneficial to consider the more abstract setting of a general metric space.

4.1.1. Metric spaces.

Definition 4.1. A ***metric space*** is a pair $(\mathcal{X}, \mathbf{d})$ where $\mathcal{X}$ is a set and $\mathbf{d} : \mathcal{X} \times \mathcal{X} \rightarrow [0, \infty)$ is a ***metric*** satisfying the following axioms: (i) $\mathbf{d}_{\mathcal{M}}(x, y) = 0$ if and only if $x = y$, (ii) $\mathbf{d}_{\mathcal{M}}(x, y) = \mathbf{d}_{\mathcal{M}}(y, x)$, and (iii) $\mathbf{d}_{\mathcal{M}}(x, y) + \mathbf{d}_{\mathcal{M}}(y, z) \geq \mathbf{d}_{\mathcal{M}}(x, z)$ (triangle inequality).

For example, $\mathbb{R}^2$ with the regular Euclidean distance is a metric space. In the following, we assume that we are given ***black-box access*** to $\mathbf{d}_{\mathcal{M}}$. Namely, given two points $\mathsf{p}, \mathsf{q} \in \mathcal{X}$, we assume that $\mathbf{d}_{\mathcal{M}}(\mathsf{p}, \mathsf{q})$ can be computed in constant time.

Another standard example for a finite metric space is a graph G with non-negative weights $\omega(\cdot)$ defined on its edges. Let $\mathbf{d}_{\mathsf{G}}(x, y)$ denote the shortest path (under the given weights) between any $x, y \in V(\mathsf{G})$. It is easy to verify that $\mathbf{d}_{\mathsf{G}}(\cdot, \cdot)$ is a metric. In fact, any ***finite metric*** (i.e., a metric defined over a finite set) can be represented by such a weighted graph.

The ***L_p-norm*** defines the distance between two points $\mathsf{p}, \mathsf{q} \in \mathbb{R}^d$ as

$$\left\| \mathsf{p} - \mathsf{q} \right\|_p = \left(\sum_{i=1}^{\mathsf{d}} |\mathsf{p}_i - \mathsf{q}_i|^p \right)^{1/p},$$

for $p \geq 1$. The ***L_2-norm*** is the regular Euclidean distance.

The L_1-***norm***, also known as the ***Manhattan distance*** or ***taxicab distance***, is

$$\left\|p - q\right\|_1 = \sum_{i=1}^{d} |p_i - q_i| .$$

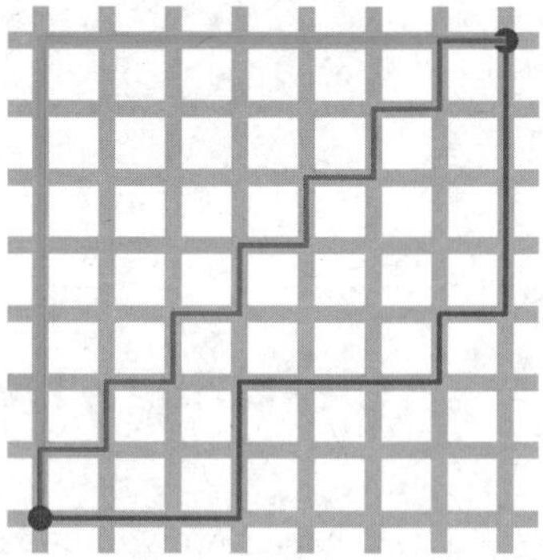

The L_1-norm distance between two points is the minimum path length that is axis parallel and connects the two points. For a uniform grid, it is the minimum number of grid edges (i.e., blocks in Manhattan) one has to travel between two grid points. In particular, the shortest path between two points is no longer unique; see the picture on the right. Of course, in the L_2-norm the shortest path between two points is the segment connecting the two points.

The L_∞-***norm*** is

$$\left\|p - q\right\|_\infty = \lim_{p\to\infty} \left\|p - q\right\|_p = \max_{i=1}^{d} |p_i - q_i| .$$

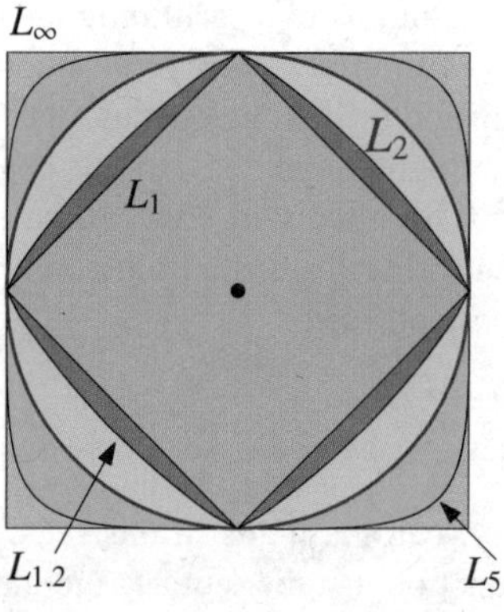

The triangle inequality holds for the L_p-norm, for any $p \geq 1$ (it is called the ***Minkowski inequality*** in this case). In particular, L_p is a metric for $p \geq 1$. Specifically, $\mathbb{R}^d$ with any L_p-norm (i.e., $p \geq 1$) is another example of a metric space.

It is useful to consider the different unit balls of L_p for different value of p; see the figure on the right. The figure implies (and one can prove it formally) that for any point $p \in \mathbb{R}^d$, we have that $\|p\|_p \leq \|p\|_q$ if $p > q$.

Lemma 4.2. *For any* $p \in \mathbb{R}^d$, *we have that* $\|p\|_1 / \sqrt{d} \leq \|p\|_2 \leq \|p\|_1$.

Proof. Indeed, let $p = (p_1, \ldots, p_d)$, and assume that $p_i \geq 0$, for all i. It is easy to verify that for a constant α, the function $f(x) = x^2 + (\alpha - x)^2$ is minimized when $x = \alpha/2$. As such, setting $\alpha = \|p\|_1 = \sum_{i=1}^{d} |p_i|$, we have, by symmetry and by the above observation on $f(x)$, that $\sum_{i=1}^{d} p_i^2$ is minimized under the condition $\|p\|_1 = \alpha$, when all the coordinates of p are equal. As such, we have that $\|p\|_2 \geq \sqrt{d(\alpha/d)^2} = \|p\|_1 / \sqrt{d}$, implying the claim. ■

4.1.2. The clustering problem. There is a metric space $(\mathcal{X}, \mathbf{d})$ and the input is a set of n points $P \subseteq \mathcal{X}$. Given a set of centers $\mathbf{C}$, every point of P is assigned to its nearest neighbor in $\mathbf{C}$. All the points of P that are assigned to a center $\overline{c}$ form the ***cluster*** of $\overline{c}$, denoted by

$$\text{cluster}(\mathbf{C}, \overline{c}) = \left\{ p \in P \;\middle|\; \mathbf{d}_{\mathcal{M}}(p, \overline{c}) = \mathbf{d}(p, \mathbf{C}) \right\}, \tag{4.1}$$

where

$$\mathbf{d}(p, \mathbf{C}) = \min_{\overline{c} \in \mathbf{C}} \mathbf{d}_{\mathcal{M}}(p, \overline{c})$$

denotes the ***distance*** of p to the set $\mathbf{C}$. Namely, the center set $\mathbf{C}$ partition P into clusters. This specific scheme of partitioning points by assigning them to their closest center (in a given set of centers) is known as a ***Voronoi partition***.

In particular, let $\mathsf{P} = \{\mathsf{p}_1, \ldots, \mathsf{p}_n\}$, and consider the n-dimensional point

$$\mathsf{P}_\mathbf{C} = \Big(\mathbf{d}(\mathsf{p}_1, \mathbf{C}), \mathbf{d}(\mathsf{p}_2, \mathbf{C}), \ldots, \mathbf{d}(\mathsf{p}_n, \mathbf{C})\Big).$$

The ith coordinate of the point $\mathsf{P}_\mathbf{C}$ is the distance (i.e., cost of assigning) of p_i to its closest center in $\mathbf{C}$.

4.2. On k-center clustering

In the ***k-center clustering problem***, a set $\mathsf{P} \subseteq \mathcal{X}$ of n points is provided together with a parameter k. We would like to find a set of k points, $\mathbf{C} \subseteq \mathsf{P}$, such that the maximum distance of a point in P to its closest point in $\mathbf{C}$ is minimized.

As a concrete example, consider the set of points to be a set of cities. Distances between points represent the time it takes to travel between the corresponding cities. We would like to build k hospitals and minimize the maximum time it takes a patient to arrive at her closest hospital. Naturally, we want to build the hospitals in the cities and not in the middle of nowhere.[①]

Formally, given a set of centers $\mathbf{C}$, the k-center clustering ***price*** of P by $\mathbf{C}$ is denoted by

$$\|\mathsf{P}_\mathbf{C}\|_\infty = \max_{\mathsf{p} \in \mathsf{P}} \mathbf{d}(\mathsf{p}, \mathbf{C}).$$

Note that every point in a cluster is within a distance at most $\|\mathsf{P}_\mathbf{C}\|_\infty$ from its respective center.

Formally, the ***k-center problem*** is to find a set $\mathbf{C}$ of k points, such that $\|\mathsf{P}_\mathbf{C}\|_\infty$ is minimized; namely,

$$\mathrm{opt}_\infty(\mathsf{P}, k) = \min_{\mathbf{C} \subseteq \mathsf{P}, |\mathbf{C}| = k} \|\mathsf{P}_\mathbf{C}\|_\infty.$$

We will denote the set of centers realizing the optimal clustering by C_{opt}. A more explicit definition (and somewhat more confusing) of the k-center clustering is to compute the set $\mathbf{C}$ of size k realizing $\min_{\mathbf{C} \subseteq \mathsf{P}} \max_{\mathsf{p} \in \mathsf{P}} \min_{\overline{\mathsf{c}} \in \mathbf{C}} \mathbf{d}_\mathcal{M}(\mathsf{p}, \overline{\mathsf{c}})$.

It is known that k-center clustering is NP-HARD, and it is in fact hard to approximate within a factor of 1.86, even for a point set in the plane (under the Euclidean distance). Surprisingly, there is a simple and elegant algorithm that achieves a 2-approximation.

Discrete vs. continuous clustering. If the input is a point set in $\mathbb{R}^\mathsf{d}$, the centers of the clustering are not necessarily restricted to be a subset of the input point, as they might be placed anywhere in $\mathbb{R}^\mathsf{d}$. Allowing this flexibility might further reduce the price of the clustering (by a constant factor). The variant where one is restricted to use the input points as centers is the ***discrete clustering*** problem. The version where centers might be placed anywhere in the given metric space is the ***continuous clustering*** version.

4.2.1. The greedy clustering algorithm.

The algorithm **GreedyKCenter** starts by picking an arbitrary point, $\overline{\mathsf{c}}_1$, and setting $\mathbf{C}_1 = \{\overline{\mathsf{c}}_1\}$. Next, we compute for every point $\mathsf{p} \in \mathsf{P}$ its distance $d_1[\mathsf{p}]$ from $\overline{\mathsf{c}}_1$. Now, consider the point worst served by $\mathbf{C}_1$; this is the point realizing $r_1 = \max_{\mathsf{p} \in \mathsf{P}} d_1[\mathsf{p}]$. Let $\overline{\mathsf{c}}_2$ denote this point, and add it to the set $\mathbf{C}_1$, resulting in the set $\mathbf{C}_2$.

Specifically, in the ith iteration, we compute for each point $\mathsf{p} \in \mathsf{P}$ the quantity $d_{i-1}[\mathsf{p}] = \min_{\overline{\mathsf{c}} \in \mathbf{C}_{i-1}} \mathbf{d}_\mathcal{M}(\mathsf{p}, \overline{\mathsf{c}})$. We also compute the radius of the clustering

$$r_{i-1} = \left\|\mathsf{P}_{\mathbf{C}_{i-1}}\right\|_\infty = \max_{\mathsf{p} \in \mathsf{P}} d_{i-1}[\mathsf{p}] = \max_{\mathsf{p} \in \mathsf{P}} \mathbf{d}(\mathsf{p}, \mathbf{C}_{i-1}) \tag{4.2}$$

[①] Although, there are recorded cases in history of building universities in the middle of nowhere.

and the bottleneck point $\bar{c}_i$ that realizes it. Next, we add $\bar{c}_i$ to $\mathbf{C}_{i-1}$ to form the new set $\mathbf{C}_i$. We repeat this process k times.

Namely, the algorithm repeatedly picks the point furthest away from the current set of centers and adds it to this set.

To make this algorithm slightly faster, observe that

$$d_i[\mathsf{p}] = \mathbf{d}(\mathsf{p}, \mathbf{C}_i) = \min(\mathbf{d}(\mathsf{p}, \mathbf{C}_{i-1}), \mathbf{d}_{\mathcal{M}}(\mathsf{p}, \bar{c}_i)) = \min(d_{i-1}[\mathsf{p}], \mathbf{d}_{\mathcal{M}}(\mathsf{p}, \bar{c}_i)).$$

In particular, if we maintain for each point $\mathsf{p} \in \mathsf{P}$ a single variable $d[\mathsf{p}]$ with its current distance to its closest center in the current center set, then the above formula boils down to

$$d[\mathsf{p}] \leftarrow \min(d[\mathsf{p}], \mathbf{d}_{\mathcal{M}}(\mathsf{p}, \bar{c}_i)).$$

Namely, the above algorithm can be implemented using $O(n)$ space, where $n = |\mathsf{P}|$. The ith iteration of choosing the ith center takes $O(n)$ time. Thus, overall, this approximation algorithm takes $O(nk)$ time.

A ***ball*** of radius r around a point $\mathsf{p} \in \mathsf{P}$ is the set of points in P with distance at most r from p; namely, $\mathbf{b}(\mathsf{p}, r) = \left\{ \mathsf{q} \in \mathsf{P} \;\middle|\; \mathbf{d}_{\mathcal{M}}(\mathsf{p}, \mathsf{q}) \le r \right\}$. Thus, the k-center problem can be interpreted as the problem of covering the points of P using k balls of minimum (maximum) radius.

Theorem 4.3. *Given a set of n points P in a metric space $(\mathcal{X}, \mathbf{d})$, the algorithm* **GreedyK-Center** *computes a set $\mathbf{K}$ of k centers, such that $\mathbf{K}$ is a 2-approximation to the optimal k-center clustering of P; namely, $\|\mathsf{P}_{\mathbf{K}}\|_\infty \le 2\mathrm{opt}_\infty$, where $\mathrm{opt}_\infty = \mathrm{opt}_\infty(\mathsf{P}, k)$ is the price of the optimal clustering. The algorithm takes $O(nk)$ time.*

Proof. The running time follows by the above description, so we concern ourselves only with the approximation quality.

By definition, we have $r_k = \|\mathsf{P}_{\mathbf{K}}\|_\infty$, and let $\bar{c}_{k+1}$ be the point in P realizing $r_k = \max_{\mathsf{p} \in \mathsf{P}} \mathbf{d}(\mathsf{p}, \mathbf{K})$. Let $\mathbf{C} = \mathbf{K} \cup \{\bar{c}_{k+1}\}$. Observe that by the definition of r_i (see (4.2)), we have that $r_1 \ge r_2 \ge \ldots \ge r_k$. Furthermore, for $i < j \le k+1$ we have that

$$\mathbf{d}_{\mathcal{M}}\left(\bar{c}_i, \bar{c}_j\right) \ge \mathbf{d}_{\mathcal{M}}\left(\bar{c}_j, \mathbf{C}_{j-1}\right) = r_{j-1} \ge r_k.$$

Namely, the distance between any pair of points in $\mathbf{C}$ is at least r_k. Now, assume for the sake of contradiction that $r_k > 2\mathrm{opt}_\infty(\mathsf{P}, k)$. Consider the optimal solution that covers P with k balls of radius opt_∞. By the triangle inequality, any two points inside such a ball are within a distance at most $2\mathrm{opt}_\infty$ from each other. Thus, none of these balls can cover two points of $\mathbf{C} \subseteq \mathsf{P}$, since the minimum distance between members of $\mathbf{C}$ is $> 2\mathrm{opt}_\infty$. As such, the optimal cover by k balls of radius opt_∞ cannot cover $\mathbf{C}$ (and thus P), as $|\mathbf{C}| = k + 1$, a contradiction. ■

In the spirit of never trusting a claim that has only a single proof, we provide an alternative proof.[2]

Alternative proof. If every cluster of C_{opt} contains exactly one point of $\mathbf{K}$, then the claim follows. Indeed, consider any point $\mathsf{p} \in \mathsf{P}$, and let $\bar{c}$ be the center it belongs to in C_{opt}. Also, let $\bar{g}$ be the center of $\mathbf{K}$ that is in $\mathrm{cluster}\left(C_{\mathrm{opt}}, \bar{c}\right)$. We have that $\mathbf{d}_{\mathcal{M}}(\mathsf{p}, \bar{c}) = \mathbf{d}\left(\mathsf{p}, C_{\mathrm{opt}}\right) \le \mathrm{opt}_\infty = \mathrm{opt}_\infty(\mathsf{P}, k)$. Similarly, observe that $\mathbf{d}_{\mathcal{M}}(\bar{g}, \bar{c}) = \mathbf{d}\left(\bar{g}, C_{\mathrm{opt}}\right) \le \mathrm{opt}_\infty$. As such, by the triangle inequality, we have that $\mathbf{d}_{\mathcal{M}}(\mathsf{p}, \bar{g}) \le \mathbf{d}_{\mathcal{M}}(\mathsf{p}, \bar{c}) + \mathbf{d}_{\mathcal{M}}(\bar{c}, \bar{g}) \le 2\mathrm{opt}_\infty$.

---

[2] Mark Twain is credited with saying that "I don't give a damn for a man that can only spell a word one way." However, there seems to be some doubt if he really said that, which brings us to the conclusion of never trusting a quote if it is credited only to a single person.

By the pigeon hole principle, the only other possibility is that there are at least two centers $\overline{g}$ and $\overline{h}$ of $\mathbf{K}$ that are both in cluster$\big(C_{opt}, \overline{c}\big)$, for some $\overline{c} \in C_{opt}$. Assume, without loss of generality, that $\overline{h}$ was added later than $\overline{g}$ to the center set $\mathbf{K}$ by the algorithm **GreedyKCenter**, say in the ith iteration. But then, since **GreedyKCenter** always chooses the point furthest away from the current set of centers, we have that

$$\|\mathsf{P}_{\mathbf{K}}\|_\infty \leq \left\|\mathsf{P}_{\mathbf{C}_{i-1}}\right\|_\infty = \mathbf{d}\big(\overline{h}, \mathbf{C}_{i-1}\big) \leq \mathbf{d}_{\mathcal{M}}\big(\overline{h}, \overline{g}\big) \leq \mathbf{d}_{\mathcal{M}}\big(\overline{h}, \overline{c}\big) + \mathbf{d}_{\mathcal{M}}(\overline{c}, \overline{g}) \leq 2\mathrm{opt}_\infty. \qquad \blacksquare$$

4.2.2. The greedy permutation. There is an interesting phenomena associated with **GreedyKCenter**. If we run it till it exhausts all the points of P (i.e., $k = n$), then this algorithm generates a permutation of P; that is, $\langle \mathsf{P} \rangle = \langle \overline{c}_1, \overline{c}_2, \ldots, \overline{c}_n \rangle$. We will refer to $\langle \mathsf{P} \rangle$ as the ***greedy permutation*** of P. There is also an associated sequence of radii $\langle r_1, r_2, \ldots, r_n \rangle$, where all the points of P are within a distance at most r_i from the points of $\mathbf{C}_i = \langle \overline{c}_1, \ldots, \overline{c}_i \rangle$.

Definition 4.4. A set $S \subseteq \mathsf{P}$ is an ***r-packing*** for P if the following two properties hold.
 (i) ***Covering property***: All the points of P are within a distance at most r from the points of S.
 (ii) ***Separation property***: For any pair of points $\mathsf{p}, \mathsf{q} \in S$, we have that $\mathbf{d}_{\mathcal{M}}(\mathsf{p}, \mathsf{q}) \geq r$.
(One can relax the separation property by requiring that the points of S be at a distance $\Omega(r)$ apart.)

Intuitively, an r-packing of a point set P is a compact representation of P in the resolution r. Surprisingly, the greedy permutation of P provides us with such a representation for all resolutions.

Theorem 4.5. *Let P be a set of n points in a finite metric space, and let its greedy permutation be $\langle \overline{c}_1, \overline{c}_2, \ldots, \overline{c}_n \rangle$ with the associated sequence of radii $\langle r_1, r_2, \ldots, r_n \rangle$. For any i, we have that $\mathbf{C}_i = \langle \overline{c}_1, \ldots, \overline{c}_i \rangle$ is an r_i-packing of P.*

Proof. Note that by construction $r_{k-1} = \mathbf{d}\big(\overline{c}_k, \mathbf{C}_{k-1}\big)$, for all $k = 2, \ldots, n$. As such, for $j < k \leq i \leq n$, we have that $\mathbf{d}_{\mathcal{M}}\big(\overline{c}_j, \overline{c}_k\big) \geq \mathbf{d}\big(\overline{c}_k, \mathbf{C}_{k-1}\big) = r_{k-1} \geq r_i$, since $r_1, r_2, \ldots, r_n$ is a monotonically non-increasing sequence. This implies the required separation property.

The covering property follows by definition; see $(4.2)_{\text{p49}}$. $\blacksquare$

4.3. On k-median clustering

In the ***k-median clustering problem***, a set $\mathsf{P} \subseteq \mathcal{X}$ is provided together with a parameter k. We would like to find a set of k points, $\mathbf{C} \subseteq \mathsf{P}$, such that the sum of the distances of points of P to their closest point in $\mathbf{C}$ is minimized.

Formally, given a set of centers $\mathbf{C}$, the k-median clustering ***price*** of clustering P by $\mathbf{C}$ is denoted by

$$\|\mathsf{P}_{\mathbf{C}}\|_1 = \sum_{\mathsf{p} \in \mathsf{P}} \mathbf{d}(\mathsf{p}, \mathbf{C}).$$

Formally, the ***k-median problem*** is to find a set $\mathbf{C}$ of k points, such that $\|\mathsf{P}_{\mathbf{C}}\|_1$ is minimized; namely,

$$\mathrm{opt}_1(\mathsf{P}, k) = \min_{\mathbf{C} \subseteq \mathsf{P}, |\mathbf{C}| = k} \|\mathsf{P}_{\mathbf{C}}\|_1.$$

We will denote the set of centers realizing the optimal clustering by C_{opt}.

There is a simple and elegant constant factor approximation algorithm for k-median clustering using ***local search*** (its analysis however is painful).

A note on notation. Consider the set $U = \left\{ \mathsf{P_C} \,\middle|\, \mathbf{C} \in \mathsf{P}^k \right\}$. Clearly, we have that $\mathrm{opt}_\infty(\mathsf{P}, k) = \min_{\mathsf{q} \in U} \|\mathsf{q}\|_\infty$ and $\mathrm{opt}_1(\mathsf{P}, k) = \min_{\mathsf{q} \in U} \|\mathsf{q}\|_1$.

Namely, k-center clustering under this interpretation is just finding the point minimizing the L_∞-norm in a set U of points in n dimensions. Similarly, the k-median problem is to find the point minimizing the L_1-norm in the set U.

Claim 4.6. *For any point set* P *of* n *points and a parameter* k*, we have that* $\mathrm{opt}_\infty(\mathsf{P}, k) \leq \mathrm{opt}_1(\mathsf{P}, k) \leq n\, \mathrm{opt}_\infty(\mathsf{P}, k)$.

Proof. For any point $\mathsf{p} \in \mathbb{R}^n$, we have that $\|\mathsf{p}\|_\infty = \max_{i=1}^n |\mathsf{p}_i| \leq \sum_{i=1}^n |\mathsf{p}_i| = \|\mathsf{p}\|_1$ and $\|\mathsf{p}\|_1 = \sum_{i=1}^n |\mathsf{p}_i| \leq \sum_{i=1}^n \max_{j=1}^n |\mathsf{p}_j| \leq n \|\mathsf{p}\|_\infty$.

Let $\mathbf{C}$ be the set of k points realizing $\mathrm{opt}_1(\mathsf{P}, k)$; that is, $\mathrm{opt}_1(\mathsf{P}, k) = \|\mathsf{P_C}\|_1$. We have that $\mathrm{opt}_\infty(\mathsf{P}, k) \leq \|\mathsf{P_C}\|_\infty \leq \|\mathsf{P_C}\|_1 = \mathrm{opt}_1(\mathsf{P}, k)$. Similarly, if $\mathbf{K}$ is the set realizing $\mathrm{opt}_\infty(\mathsf{P}, k)$, then $\mathrm{opt}_1(\mathsf{P}, k) = \|\mathsf{P_C}\|_1 \leq \|\mathsf{P_K}\|_1 \leq n \|\mathsf{P_K}\|_\infty = n \cdot \mathrm{opt}_\infty(\mathsf{P}, k)$. ■

4.3.1. Approximation algorithm – local search. We are given a set P of n points and a parameter k. In the following, let C_{opt} denote the set of centers realizing the optimal solution, and let $\mathrm{opt}_1 = \mathrm{opt}_1(\mathsf{P}, k)$.

4.3.1.1. *The algorithm.*

A $2n$-approximation. The algorithm starts by computing a set of k centers L using Theorem 4.3. Claim 4.6 implies that

$$\begin{aligned} &\|\mathsf{P}_L\|_1 / 2n \leq \|\mathsf{P}_L\|_\infty / 2 \leq \mathrm{opt}_\infty(\mathsf{P}, k) \leq \mathrm{opt}_1 \leq \|\mathsf{P}_L\|_1 \\ \Longrightarrow \quad &\mathrm{opt}_1 \leq \|\mathsf{P}_L\|_1 \leq 2n\, \mathrm{opt}_1. \end{aligned} \tag{4.3}$$

Namely, L is a $2n$-approximation to the optimal solution.

Improving it. Let $0 < \tau < 1$ be a parameter to be determined shortly. The local search algorithm **algLocalSearchKMed** initially sets the current set of centers L_{curr} to be L, the set of centers computed above. Next, at each iteration it checks if the current solution L_{curr} can be improved by replacing one of the centers in it by a center from the outside. We will refer to such an operation as a ***swap***. There are at most $|\mathsf{P}|\,|L_{\mathrm{curr}}| = nk$ choices to consider, as we pick a center $\overline{\mathsf{c}} \in L_{\mathrm{curr}}$ to throw away and a new center to replace it by $\overline{\mathsf{o}} \in (\mathsf{P} \setminus L_{\mathrm{curr}})$. We consider the new candidate set of centers $\mathbf{K} \leftarrow (L_{\mathrm{curr}} \setminus \{\overline{\mathsf{c}}\}) \cup \{\overline{\mathsf{o}}\}$. If $\|\mathsf{P_K}\|_1 \leq (1 - \tau) \left\|\mathsf{P}_{L_{\mathrm{curr}}}\right\|_1$, then the algorithm sets $L_{\mathrm{curr}} \leftarrow \mathbf{K}$. The algorithm continues iterating in this fashion over all possible swaps.

The algorithm **algLocalSearchKMed** stops when there is no swap that would improve the current solution by a factor of (at least) $(1 - \tau)$. The final content of the set L_{curr} is the required constant factor approximation.

4.3.1.2. *Running time.* An iteration requires checking $O(nk)$ swaps (i.e., $n - k$ candidates to be swapped in and k candidates to be swapped out). Computing the price of every such swap, done naively, requires computing the distance of every point to its nearest center, and that takes $O(nk)$ time per swap. As such, overall, each iteration takes $O\left((nk)^2\right)$ time.

Since $1/(1 - \tau) \geq 1 + \tau$, the running time of the algorithm is

$$O\left((nk)^2 \log_{1/(1-\tau)} \frac{\|\mathsf{P}_L\|_1}{\mathrm{opt}_1} \right) = O\left((nk)^2 \log_{1+\tau} 2n \right) = O\left((nk)^2 \frac{\log n}{\tau} \right),$$

by (4.3) and Lemma 28.10$_{\mathrm{p348}}$. Thus, if τ is polynomially small, then the running time would be polynomial.

4.3.2. Proof of quality of approximation. We claim that the above algorithm provides a constant factor approximation for the optimal k-median clustering.

4.3.2.1. *Definitions and intuition.* Intuitively, since the local search got stuck in a locally optimal solution, it cannot be too far from the true optimal solution.

For the sake of simplicity of exposition, let us assume (for now) that the solution returned by the algorithm cannot be improved (at all) by any swap, and let L be this set of centers. For a center $\overline{c} \in L$ and a point $\overline{o} \in \mathsf{P} \setminus L$, let $L - \overline{c} + \overline{o} = (L \setminus \{\overline{c}\}) \cup \{\overline{o}\}$ denote the set of centers resulting from applying the swap $\overline{c} \to \overline{o}$ to L. We are assuming that there is no beneficial swap; that is,

$$\forall \overline{c} \in L, \overline{o} \in \mathsf{P} \setminus L \qquad 0 \leq \Delta(\overline{c}, \overline{o}) = \nu_1(L - \overline{c} + \overline{o}) - \nu_1(L), \tag{4.4}$$

where $\nu_1(X) = \|\mathsf{P}_X\|_1$.

Equation (4.4) provides us with a large family of inequalities that all hold together. Each inequality is represented by a swap $\overline{c} \to \overline{o}$. We would like to combine these inequalities such that they will imply that $5 \left\|\mathsf{P}_{C_{\mathrm{opt}}}\right\|_1 \geq \|\mathsf{P}_L\|_1$, namely, that the local search algorithm provides a constant factor approximation to optimal clustering. This idea seems to be somewhat mysterious (or even impossible), but hopefully it will become clearer shortly.

From local clustering to local clustering complying with the optimal clustering. The first hurdle in the analysis is that a cluster of the optimal solution $\mathrm{cluster}(C_{\mathrm{opt}}, \overline{o})$, for $\overline{o} \in C_{\mathrm{opt}}$, might intersect a large number of clusters in the local clustering (i.e., clusters of the form $\mathrm{cluster}(L, \overline{c})$ for $\overline{c} \in L$).

Fortunately, one can modify the assignment of points to clusters in the locally optimal clustering so that the resulting clustering of P complies with the optimal partition and the price of the clustering increases only moderately; that is, every cluster in the optimal clustering would be contained in a single cluster of the modified local solution. In particular, now an optimal cluster would intersect only a single cluster in the modified local solution.

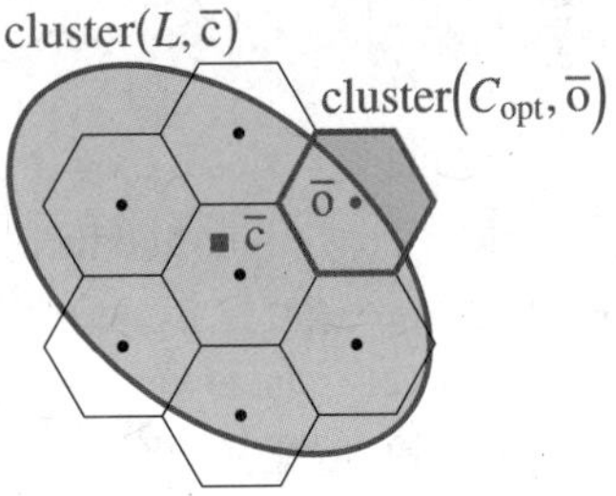

Furthermore, this modified local solution Π is not much more expensive. Now, in this modified partition there are many beneficial swaps (by making it into the optimal clustering). But these swaps cannot be too profitable, since then they would have been profitable for the original local solution. This would imply that the local solution cannot be too expensive. The picture on the right depicts a local cluster and the optimal clusters in its vicinity such that their centers are contained inside it.

In the following, we denote by $\mathsf{nn}(\mathsf{p}, X)$ the nearest neighbor to p in the set X.

For a point $\mathsf{p} \in \mathsf{P}$, let $\overline{o}_{\mathsf{p}} = \mathsf{nn}(\mathsf{p}, C_{\mathrm{opt}})$ be its optimal center, and let $\alpha(\mathsf{p}) = \mathsf{nn}(\overline{o}_{\mathsf{p}}, L)$ be the center p should use if p follows its optimal center's assignment. Let Π be the modified partition of P by the function $\alpha(\cdot)$.

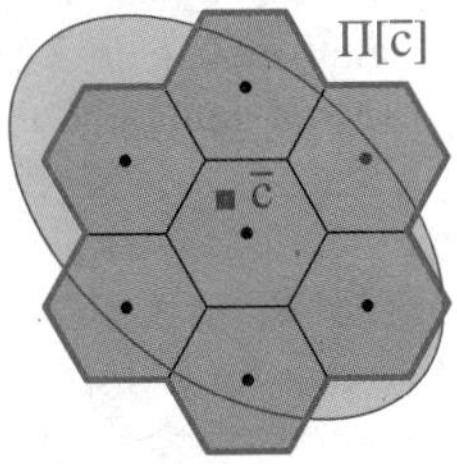

That is, for $\overline{c} \in L$, its cluster in Π, denoted by $\Pi[\overline{c}]$, is the set of all points $\mathsf{p} \in \mathsf{P}$ such that $\alpha(\mathsf{p}) = \overline{c}$.

Now, for any center $\overline{o} \in C_{\mathrm{opt}}$, let $\mathsf{nn}(\overline{o}, L)$ be its nearest neighbor in L, and observe that $\mathrm{cluster}(C_{\mathrm{opt}}, \overline{o}) \subseteq \Pi[\mathsf{nn}(\overline{o}, L)]$ (see $(4.1)_{p48}$). The picture on the right shows the resulting modified cluster for the above example.

Let δ_p denote the price of this reassignment for the point p; that is, $\delta_p = \mathbf{d}_{\mathcal{M}}(p, \alpha(p)) - \mathbf{d}(p, L)$. Note that if p does not get reassigned, then $\delta_p = 0$ and otherwise $\delta_p \ge 0$, since $\alpha(p) \in L$ and $\mathbf{d}(p, L) = \min_{\bar{c} \in L} \mathbf{d}_{\mathcal{M}}(p, \bar{c})$.

Lemma 4.7. *The increase in cost from moving from the clustering induced by L to the clustering of* Π *is bounded by* $2\left\|P_{C_{opt}}\right\|_1$. *That is,* $\sum_{p \in P} \delta_p \le 2\left\|P_{C_{opt}}\right\|_1$.

Proof. For a point $p \in P$, let $\bar{c} = nn(p, L)$ be its local center, let $\bar{o} = nn\left(p, C_{opt}\right)$ be its optimal center, and let $\alpha(p) = nn(\bar{o}, L)$ be its new assigned center in Π. Observe that $\mathbf{d}_{\mathcal{M}}(\bar{o}, \alpha(p)) = \mathbf{d}_{\mathcal{M}}(\bar{o}, nn(\bar{o}, L)) \le \mathbf{d}_{\mathcal{M}}(\bar{o}, \bar{c})$.

As such, by the triangle inequality, we have that

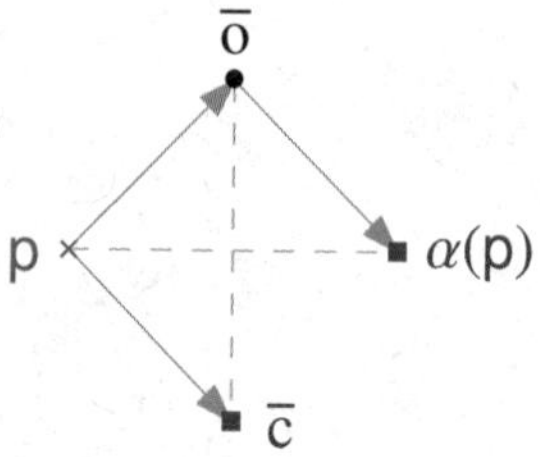

$$\begin{aligned}\mathbf{d}_{\mathcal{M}}(p, \alpha(p)) &\le \mathbf{d}_{\mathcal{M}}(p, \bar{o}) + \mathbf{d}_{\mathcal{M}}(\bar{o}, \alpha(p)) \le \mathbf{d}_{\mathcal{M}}(p, \bar{o}) + \mathbf{d}_{\mathcal{M}}(\bar{o}, \bar{c}) \\ &\le \mathbf{d}_{\mathcal{M}}(p, \bar{o}) + (\mathbf{d}_{\mathcal{M}}(\bar{o}, p) + \mathbf{d}_{\mathcal{M}}(p, \bar{c})) \\ &= 2\mathbf{d}_{\mathcal{M}}(p, \bar{o}) + \mathbf{d}_{\mathcal{M}}(p, \bar{c}).\end{aligned}$$

Finally, $\delta_p = \mathbf{d}_{\mathcal{M}}(p, \alpha(p)) - \mathbf{d}(p, L) \le 2\mathbf{d}_{\mathcal{M}}(p, \bar{o}) + \mathbf{d}_{\mathcal{M}}(p, \bar{c}) - \mathbf{d}_{\mathcal{M}}(p, \bar{c}) = 2\mathbf{d}_{\mathcal{M}}(p, \bar{o}) = 2\mathbf{d}\left(p, C_{opt}\right)$. As such, $\sum_{p \in P} \delta_p \le \sum_{p \in P} 2\mathbf{d}\left(p, C_{opt}\right) = 2\left\|P_{C_{opt}}\right\|_1$. ■

Drifters, anchors, and tyrants. A center of L that does not serve any center of C_{opt} (i.e., its cluster in Π is empty) is a ***drifter***. Formally, we map each center of C_{opt} to its nearest neighbor in L, and for a center $\bar{c} \in L$ its ***degree***, denoted by $\deg(\bar{c})$, is the number of points of C_{opt} mapped to it by this nearest neighbor mapping.

As such, a center $\bar{c} \in L$ is a ***drifter*** if $\deg(\bar{c}) = 0$, an ***anchor*** if $\deg(\bar{c}) = 1$, and a ***tyrant*** if $\deg(\bar{c}) > 1$. Observe that if $\bar{c}$ is a drifter, then $\Pi[\bar{c}] = \emptyset$.

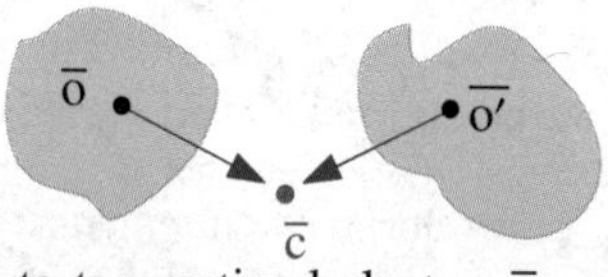

The reader should not take these names too seriously, but observe that centers that are tyrants cannot easily move around and are bad candidates for swaps. Indeed, consider the situation depicted in the figure on the right. Here the center $\bar{c}$ serves points of P that belong to two optimal clusters $\bar{o}$ and $\overline{o'}$, such that $\bar{c} = nn(\bar{o}, L) = nn\left(\overline{o'}, L\right)$. If we swap $\bar{c} \to \bar{o}$, then the points in the cluster $\text{cluster}\left(C_{opt}, \overline{o'}\right)$ might find themselves very far from any center in $L - \bar{c} + \bar{o}$. Similarly, the points of $\text{cluster}\left(C_{opt}, \bar{o}\right)$ might be in trouble if we swap $\bar{c} \to \overline{o'}$.

Intuitively, since we shifted our thinking from the local solution to the partition Π, a drifter center is not being used by the clustering, and we can reassign it so that it decreases the price of the clustering.

That is, since moving from the local clustering of L to Π is relatively cheap, we can free a drifter $\bar{c}$ from all its clients in the local partition. Formally, the ***ransom*** of a drifter center $\bar{c}$ is $\text{ransom}(\bar{c}) = \sum_{p \in \text{cluster}(L, \bar{c})} \delta_p$. This is the price of reassigning all the points that are currently served by the drifter $\bar{c}$ to the center in L serving their optimal center. Once this ransom is paid, $\bar{c}$ serves nobody and can be moved with no further charge.

More generally, the ***ransom*** of any center $\bar{c} \in L$ is

$$\text{ransom}(\bar{c}) = \sum_{p \in \text{cluster}(L, \bar{c}) \setminus \Pi[\bar{c}]} \delta_p.$$

Note that for a drifter $\Pi[\bar{c}] = \emptyset$ and $\text{cluster}(L, \bar{c}) = \text{cluster}(L, \bar{c}) \setminus \Pi[\bar{c}]$, and in general, the points of $\text{cluster}(L, \bar{c}) \setminus \Pi[\bar{c}]$ are exactly the points of $\text{cluster}(L, \bar{c})$ being reassigned.

Hence, ransom($\overline{c}$) is the increase in cost of reassigning the points of cluster $(L, \overline{c})$ when moving from the local clustering of L to the clustering of Π.

Observe that, by Lemma 4.7, we have that

$$\sum_{\overline{c} \in L} \mathrm{ransom}(\overline{c}) \leq 2 \left\| \mathsf{P}_{C_{\mathrm{opt}}} \right\|_1 . \tag{4.5}$$

For $\overline{o} \in C_{\mathrm{opt}}$, the ***optimal price*** and ***local price*** of cluster $\left(C_{\mathrm{opt}}, \overline{o}\right)$ are

$$\mathrm{opt}\,(\overline{o}) = \sum_{\mathsf{p} \in \mathrm{cluster}\left(C_{\mathrm{opt}}, \overline{o}\right)} \mathbf{d}\left(\mathsf{p}, C_{\mathrm{opt}}\right) \qquad \text{and} \qquad \mathrm{local}(\overline{o}) = \sum_{\mathsf{p} \in \mathrm{cluster}\left(C_{\mathrm{opt}}, \overline{o}\right)} \mathbf{d}(\mathsf{p}, L),$$

respectively.

Lemma 4.8. *If $\overline{c} \in L$ is a drifter and $\overline{o}$ is any center of C_{opt}, then* $\mathrm{local}(\overline{o}) \leq \mathrm{ransom}(\overline{c}) + \mathrm{opt}(\overline{o})$.

Proof. Since $\overline{c} \in L$ is a drifter, we can swap it with any center in $\overline{o} \in C_{\mathrm{opt}}$. Since L is a locally optimal solution, we have that the change in the cost caused by the swap $\overline{c} \rightarrow \overline{o}$ is

$$\begin{aligned} 0 \leq \Delta(\overline{c}, \overline{o}) \leq \mathrm{ransom}(\overline{c}) - \mathrm{local}(\overline{o}) + \mathrm{opt}(\overline{o}) \\ \Longrightarrow \qquad \mathrm{local}(\overline{o}) \leq \mathrm{ransom}(\overline{c}) + \mathrm{opt}(\overline{o}) . \end{aligned} \tag{4.6}$$

Indeed, $\overline{c}$ pays its ransom so that all the clients using it are now assigned to some other centers of L. Now, all the points of cluster $\left(C_{\mathrm{opt}}, \overline{o}\right)$ instead of paying $\mathrm{local}(\overline{o})$ are now paying (at most) $\mathrm{opt}(\overline{o})$. (We might pay less for a point $\mathsf{p} \in \mathrm{cluster}\left(C_{\mathrm{opt}}, \overline{o}\right)$ if it is closer to $L - \overline{c} + \overline{o}$ than to $\overline{o}$.) ■

Equation (4.6) provides us with a glimmer of hope that we can bound the price of the local clustering. We next argue that if there are many tyrants, then there must also be many drifters. In particular, with these drifters we can bound the price of the local clustering cost of the optimal clusters assigned to tyrants. Also, we argue that an anchor and its associated optimal center define a natural swap which is relatively cheap. Putting all of these together will imply the desired claim.

There are many drifters. Let S_{opt} (resp. $\mathrm{A}_{\mathrm{opt}}$) be the set of all the centers of C_{opt} that are assigned to tyrants (resp. anchors) by $\mathsf{nn}(\cdot, L)$. Observe that $S_{\mathrm{opt}} \cup \mathrm{A}_{\mathrm{opt}} = C_{\mathrm{opt}}$. Let $\mathcal{D}$ be the set of drifters in L.

Observe that every tyrant has at least two followers in C_{opt}; that is, $\left|S_{\mathrm{opt}}\right| \geq 2\#_{\mathrm{tyrants}}$. Also, $k = \left|C_{\mathrm{opt}}\right| = |L|$ and $\#_{\mathrm{anchors}} = \left|\mathrm{A}_{\mathrm{opt}}\right|$. As such, we have that

$$\begin{aligned} &\#_{\mathrm{tyrants}} + \#_{\mathrm{anchors}} + \#_{\mathrm{drifters}} = |L| = \left|C_{\mathrm{opt}}\right| = \left|S_{\mathrm{opt}}\right| + \left|\mathrm{A}_{\mathrm{opt}}\right| \\ &\Longrightarrow \quad \#_{\mathrm{drifters}} = \left|S_{\mathrm{opt}}\right| + \left|\mathrm{A}_{\mathrm{opt}}\right| - \#_{\mathrm{anchors}} - \#_{\mathrm{tyrants}} = \left|S_{\mathrm{opt}}\right| - \#_{\mathrm{tyrants}} \geq \left|S_{\mathrm{opt}}\right| / 2. \end{aligned} \tag{4.7}$$

Namely, $2\#_{\mathrm{drifters}} \geq \left|S_{\mathrm{opt}}\right|$.

Lemma 4.9. *We have that* $\displaystyle\sum_{\overline{o} \in S_{\mathrm{opt}}} \mathrm{local}(\overline{o}) \leq 2 \sum_{\overline{c} \in \mathcal{D}} \mathrm{ransom}(\overline{c}) + \sum_{\overline{o} \in S_{\mathrm{opt}}} \mathrm{opt}(\overline{o})$.

Proof. If $\left|S_{\mathrm{opt}}\right| = 0$, then the statement holds trivially.

So assume $\left|S_{\mathrm{opt}}\right| > 0$ and let $\overline{c}$ be the drifter with the lowest $\mathrm{ransom}(\overline{c})$. For any $\overline{o} \in S_{\mathrm{opt}}$, we have that $\mathrm{local}(\overline{o}) \leq \mathrm{ransom}(\overline{c}) + \mathrm{opt}(\overline{o})$, by (4.6). Summing over all such centers, we have that

$$\sum_{\overline{o} \in S_{\mathrm{opt}}} \mathrm{local}(\overline{o}) \leq \left|S_{\mathrm{opt}}\right| \mathrm{ransom}(\overline{c}) + \sum_{\overline{o} \in S_{\mathrm{opt}}} \mathrm{opt}(\overline{o}),$$

which is definitely smaller than the stated bound, since $\left|S_{\mathrm{opt}}\right| \leq 2\left|\mathcal{D}\right|$, by (4.7). ∎

Lemma 4.10. *We have that* $\sum_{\overline{o} \in \mathsf{A}_{\mathrm{opt}}} \mathrm{local}(\overline{o}) \leq \sum_{\overline{o} \in \mathsf{A}_{\mathrm{opt}}} \mathrm{ransom}(\mathsf{nn}(\overline{o}, L)) + \sum_{\overline{o} \in \mathsf{A}_{\mathrm{opt}}} \mathrm{opt}(\overline{o})$.

Proof. For a center $\overline{o} \in \mathsf{A}_{\mathrm{opt}}$, its anchor is $\overline{c} = \mathsf{nn}(\overline{o}, L)$. Consider the swap $\overline{c} \to \overline{o}$, and the increase in clustering cost as we move from L to $L - \overline{c} + \overline{o}$.

We claim that $\mathrm{local}(\overline{o}) \leq \mathrm{ransom}(\overline{c}) + \mathrm{opt}(\overline{o})$ (i.e., (4.6)) holds in this setting. The points for which their clustering is negatively affected (i.e., their clustering price might increase) by the swap are in the set $\mathrm{cluster}(L, \overline{c}) \cup \mathrm{cluster}\left(C_{\mathrm{opt}}, \overline{o}\right)$, and we split this set into two disjoint sets $X = \mathrm{cluster}(L, \overline{c}) \setminus \mathrm{cluster}\left(C_{\mathrm{opt}}, \overline{o}\right)$ and $Y = \mathrm{cluster}\left(C_{\mathrm{opt}}, \overline{o}\right)$.

The increase in price by reassigning the points of X to some other center in L is exactly the ransom of $\overline{c}$. Now, the points of Y might get reassigned to $\overline{o}$, and the change in price of the points of Y can now be bounded by $-\mathrm{local}(\overline{o}) + \mathrm{opt}(\overline{o})$, as was argued in the proof of Lemma 4.8.

Note that it might be that points outside $X \cup Y$ get reassigned to $\overline{o}$ in the clustering induced by $L - \overline{c} + \overline{o}$. However, such reassignment only further reduce the price of the swap. As such, we have that $0 \leq \Delta(\overline{c}, \overline{o}) \leq \mathrm{ransom}(\overline{c}) - \mathrm{local}(\overline{o}) + \mathrm{opt}(\overline{o})$. As such, summing up the inequality $\mathrm{local}(\overline{o}) \leq \mathrm{ransom}(\overline{c}) + \mathrm{opt}(\overline{o})$ over all the centers in $\mathsf{A}_{\mathrm{opt}}$ implies the claim. ∎

Lemma 4.11. *Let L be the set of k centers computed by the local search algorithm. We have that* $\left\|\mathsf{P}_L\right\|_1 \leq 5\mathrm{opt}_1(\mathsf{P}, k)$.

Proof. From the above two lemmas, we have that

$$\begin{aligned}
\left\|\mathsf{P}_L\right\|_1 &= \sum_{\overline{o} \in C_{\mathrm{opt}}} \mathrm{local}(\overline{o}) = \sum_{\overline{o} \in S_{\mathrm{opt}}} \mathrm{local}(\overline{o}) + \sum_{\overline{o} \in \mathsf{A}_{\mathrm{opt}}} \mathrm{local}(\overline{o}) \\
&\leq 2 \sum_{\overline{c} \in \mathcal{D}} \mathrm{ransom}(\overline{c}) + \sum_{\overline{o} \in S_{\mathrm{opt}}} \mathrm{opt}(\overline{o}) + \sum_{\overline{o} \in \mathsf{A}_{\mathrm{opt}}} \mathrm{ransom}(\mathsf{nn}(\overline{o}, L)) + \sum_{\overline{o} \in \mathsf{A}_{\mathrm{opt}}} \mathrm{opt}(\overline{o}) \\
&\leq 2 \sum_{\overline{c} \in L} \mathrm{ransom}(\overline{c}) + \sum_{\overline{o} \in C_{\mathrm{opt}}} \mathrm{opt}(\overline{o}) \leq 4 \left\|\mathsf{P}_{C_{\mathrm{opt}}}\right\|_1 + \left\|\mathsf{P}_{C_{\mathrm{opt}}}\right\|_1 = 5\mathrm{opt}_1(\mathsf{P}, k),
\end{aligned}$$

by (4.5). ∎

4.3.2.2. *Removing the strict improvement assumption.* In the above proof, we assumed that the current local minimum cannot be improved by a swap. Of course, this might not hold for the **algLocalSearchKMed** solution, since the algorithm allows a swap only if it makes "significant" progress. In particular, (4.4) is in fact

$$\forall \overline{c} \in L, \overline{o} \in \mathsf{P} \setminus L, \qquad -\tau \left\|\mathsf{P}_L\right\|_1 \leq \left\|\mathsf{P}_{L-\overline{c}+\overline{o}}\right\|_1 - \left\|\mathsf{P}_L\right\|_1. \tag{4.8}$$

To adapt the proof to use this modified inequality, observe that the proof worked by adding up k inequalities defined by (4.4) and getting the inequality $0 \leq 5\left\|\mathsf{P}_{C_{\mathrm{opt}}}\right\|_1 - \left\|\mathsf{P}_L\right\|_1$. Repeating the same argumentation on the modified inequalities, which is tedious but straightforward, yields

$$-\tau k \left\|\mathsf{P}_L\right\|_1 \leq 5\left\|\mathsf{P}_{C_{\mathrm{opt}}}\right\|_1 - \left\|\mathsf{P}_L\right\|_1.$$

This implies $\|\mathsf{P}_L\|_1 \leq 5 \left\|\mathsf{P}_{C_{\mathrm{opt}}}\right\|_1 /(1 - \tau k)$. For arbitrary $0 < \varepsilon < 1$, setting $\tau = \varepsilon/10k$, we have that $\|\mathsf{P}_L\|_1 \leq 5(1 + \varepsilon/5)\mathrm{opt}_1$, since $1/(1 - \tau k) \leq 1 + 2\tau k = 1 + \varepsilon/5$, for $\tau \leq 1/10k$. We summarize:

Theorem 4.12. *Let* P *be a set of n points in a metric space. For* $0 < \varepsilon < 1$, *one can compute a* $(5 + \varepsilon)$*-approximation to the optimal k-median clustering of* P. *The running time of the algorithm is* $O\left(n^2 k^3 \frac{\log n}{\varepsilon}\right)$.

4.4. On k-means clustering

In the ***k-means clustering problem***, a set $\mathsf{P} \subseteq \mathcal{X}$ is provided together with a parameter k. We would like to find a set of k points $\mathbf{C} \subseteq \mathsf{P}$, such that the sum of squared distances of all the points of P to their closest point in **C** is minimized.

Formally, given a set of centers **C**, the k-center clustering ***price*** of clustering P by **C** is denoted by

$$\|\mathsf{P}_{\mathbf{C}}\|_2^2 = \sum_{\mathsf{p} \in \mathsf{P}} (\mathbf{d}_{\mathcal{M}}(\mathsf{p}, \mathbf{C}))^2 ,$$

and the ***k-means problem*** is to find a set **C** of k points, such that $\|\mathsf{P}_{\mathbf{C}}\|_2^2$ is minimized; namely,

$$\mathrm{opt}_2(\mathsf{P}, k) = \min_{\mathbf{C}, |\mathbf{C}|=k} \|\mathsf{P}_{\mathbf{C}}\|_2^2 .$$

Local search also works for k-means and yields a constant factor approximation. We leave the proof of the following theorem to Exercise 4.4.

Theorem 4.13. *Let* P *be a set of n points in a metric space. For* $0 < \varepsilon < 1$, *one can compute a* $(25 + \varepsilon)$*-approximation to the optimal k-means clustering of* P. *The running time of the algorithm is* $O\left(n^2 k^3 \frac{\log n}{\varepsilon}\right)$.

4.5. Bibliographical notes

In this chapter we introduced the problem of clustering and showed some algorithms that achieve constant factor approximations. A lot more is known about these problems including faster and better clustering algorithms, but to discuss them, we need more advanced tools than what we currently have at hand.

Clustering is widely researched. Unfortunately, a large fraction of the work on this topic relies on heuristics or experimental studies. The inherent problem seems to be the lack of a universal definition of what is a good clustering. This depends on the application at hand, which is rarely clearly defined. In particular, no clustering algorithm can achieve all desired properties together; see the work by Kleinberg [**Kle02**] (although it is unclear if all these desired properties are indeed natural or even really desired).

k-center clustering. The algorithm **GreedyKCenter** is by Gonzalez [**Gon85**], but it was probably known before, as the notion of r-packing is much older. The hardness of approximating k-center clustering was shown by Feder and Greene [**FG88**].

k-median/means clustering. The analysis of the local search algorithm is due to Arya et al. [**AGK⁺01**]. Our presentation however follows the simpler proof of Gupta and Tangwongsan [**GT08**]. The extension to k-means is due to Kanungo et al. [**KMN⁺04**]. The extension is not completely trivial since the triangle inequality no longer holds. However, some approximate version of the triangle inequality does hold. Instead of performing a single swap, one can decide to do p swaps simultaneously. Thus, the running time deteriorates since there are more possibilities to check. This improves the approximation constant for the k-median (resp., k-means) to $(3 + 2/p)$ (resp. $(3 + 2/p)^2$). Unfortunately, this is (essentially) tight in the worst case. See [**AGK⁺01, KMN⁺04**] for details.

The k-median and k-means clustering are more interesting in Euclidean settings where there is considerably more structure, and one can compute a $(1+\varepsilon)$-approximation in linear time for fixed ε and k and d; see [**HM04**].

Since k-median and k-means clustering can be used to solve the dominating set in a graph, this implies that both clustering problems are NP-HARD to solve exactly.

One can also compute a permutation similar to the greedy permutation (for k-center clustering) for k-median clustering. See the work by Mettu and Plaxton [**MP03**].

Handling outliers. The problem of handling outliers is still not well understood. See the work of Charikar et al. [**CKMN01**] for some relevant results. In particular, for k-center clustering they get a constant factor approximation, and Exercise 4.3 is taken from there. For k-median clustering they present a constant factor approximation using a linear programming relaxation that also approximates the number of outliers. Recently, Chen [**Che08**] provided a constant factor approximation algorithm by extending the work of Charikar et al. The problem of finding a simple algorithm with simple analysis for k-median clustering with outliers is still open, as Chen's work is quite involved.

Open Problem 4.14. Get a *simple* constant factor k-median clustering algorithm that runs in polynomial time and uses exactly m outliers. Alternatively, solve this problem in the case where P is a set of n points in the plane. (The emphasize here is that the analysis of the algorithm should be simple.)

Bi-criteria approximation. All clustering algorithms tend to become considerably easier if one allows trade-off in the number of clusters. In particular, one can compute a constant factor approximation to the optimal k-median/means clustering using $O(k)$ centers in $O(nk)$ time. The algorithm succeeds with constant probability. See the work by Indyk [**Ind99**] and Chen [**Che06**] and references therein.

Facility location. All the problems mentioned here fall into the family of facility location problems. There are numerous variants. The more specific ***facility location*** problem is a variant of k-median clustering where the number of clusters is not specified, but instead one has to pay to open a facility in a certain location. Local search also works for this variant.

Local search. As mentioned above, ***local search*** also works for k-means clustering [**AGK⁺01**]. A collection of some basic problems for which local search works is described in the book by Kleinberg and Tardos [**KT06**]. Local search is a widely used heuristic for attacking NP-HARD problems. The idea is usually to start from a solution and try to locally improve it. Here, one defines a neighborhood of the current solution, and one tries to move to the best solution in this neighborhood. In this sense, local search can be thought of as a hill-climbing/EM (expectation maximization) algorithm. Problems for which local search was used include vertex cover, traveling salesperson, and satisfiability, and probably many other problems.

Provable cases where local search generates a guaranteed solution are less common and include facility location, k-median clustering [**AGK⁺01**], weighted max cut, k-means [**KMN⁺04**], the metric labeling problem with the truncated linear metric [**GT00**], and image segmentation [**BVZ01**]. See [**KT06**] for more references and a nice discussion of the connection of local search to the ***Metropolis algorithm*** and ***simulated annealing***.

4.6. Exercises

Exercise 4.1 (Another algorithm for k-center clustering)**.** Consider the algorithm that, given a point set P and a parameter k, initially picks an arbitrary set $\mathbf{C} \subseteq \mathsf{P}$ of k points. Next, it computes the closest pair of points $\overline{c}, \overline{f} \in \mathbf{C}$ and the point s realizing $\|\mathsf{P}_{\mathbf{C}}\|_\infty$. If $\mathbf{d}(s, \mathbf{C}) > \mathbf{d}_{\mathcal{M}}(\overline{c}, \overline{f})$, then the algorithm sets $\mathbf{C} \leftarrow \mathbf{C} - \overline{c} + s$ and repeats this process till the condition no longer holds.

(A) Prove that this algorithm outputs a k-center clustering of radius $\leq 2\mathrm{opt}_\infty(\mathsf{P}, k)$.
(B) What is the running time of this algorithm?
(C) If one is willing to trade off the approximation quality of this algorithm, it can be made faster. In particular, suggest a variant of this algorithm that in $O(k)$ iterations computes an $O(1)$-approximation to the optimal k-center clustering.

Exercise 4.2 (Handling outliers)**.** Given a point set P, we would like to perform a k-median clustering of it, where we are allowed to ignore m of the points. These m points are ***outliers*** which we would like to ignore since they represent irrelevant data. Unfortunately, we do not know the m outliers in advance. It is natural to conjecture that one can perform a local search for the optimal solution. Here one maintains a set of k centers and a set of m outliers. At every point in time the algorithm moves one of the centers or the outliers if it improves the solution.

Show that local search does not work for this problem; namely, the approximation factor is not a constant.

Exercise 4.3 (Handling outliers for k-center clustering)**.** Given P, k, and m, present a polynomial time algorithm that computes a constant factor approximation to the optimal k-center clustering of P with m outliers. (Hint: Assume first that you know the radius of the optimal solution.)

Exercise 4.4 (Local search for k-means clustering)**.** Prove Theorem 4.13.

CHAPTER 5

On Complexity, Sampling, and ε-Nets and ε-Samples

In this chapter we will try to quantify the notion of geometric complexity. It is intuitively clear that a ● (i.e., disk) is a simpler shape than an ⬬ (i.e., ellipse), which is in turn simpler than a ☻ (i.e., smiley). This becomes even more important when we consider several such shapes and how they interact with each other. As these examples might demonstrate, this notion of complexity is somewhat elusive.

To this end, we show that one can capture the structure of a distribution/point set by a small subset. The size here would depend on the complexity of the shapes/ranges we care about, but surprisingly it would be independent of the size of the point set.

5.1. VC dimension

Definition 5.1. A ***range space*** S is a pair $(\mathsf{X}, \mathcal{R})$, where X is a ***ground set*** (finite or infinite) and $\mathcal{R}$ is a (finite or infinite) family of subsets of X. The elements of X are ***points*** and the elements of $\mathcal{R}$ are ***ranges***.

Our interest is in the size/weight of the ranges in the range space. For technical reasons, it will be easier to consider a finite subset x as the underlining ground set.

Definition 5.2. Let $\mathsf{S} = (\mathsf{X}, \mathcal{R})$ be a range space, and let x be a finite (fixed) subset of X. For a range $\mathbf{r} \in \mathcal{R}$, its ***measure*** is the quantity

$$\overline{m}(\mathbf{r}) = \frac{|\mathbf{r} \cap \mathsf{x}|}{|\mathsf{x}|}.$$

While x is finite, it might be very large. As such, we are interested in getting a good estimate to $\overline{m}(\mathbf{r})$ by using a more compact set to represent the range space.

Definition 5.3. Let $\mathsf{S} = (\mathsf{X}, \mathcal{R})$ be a range space. For a subset N (which might be a multiset) of x, its ***estimate*** of the measure of $\overline{m}(\mathbf{r})$, for $\mathbf{r} \in \mathcal{R}$, is the quantity

$$\overline{s}(\mathbf{r}) = \frac{|\mathbf{r} \cap N|}{|N|}.$$

The main purpose of this chapter is to come up with methods to generate a sample N, such that $\overline{m}(\mathbf{r}) \approx \overline{s}(\mathbf{r})$, for all the ranges $\mathbf{r} \in \mathcal{R}$.

It is easy to see that in the worst case, no sample can capture the measure of all ranges. Indeed, given a sample N, consider the range $\mathsf{x} \setminus N$ that is being completely missed by N. As such, we need to concentrate on range spaces that are "low dimensional", where not all subsets are allowable ranges. The notion of VC dimension (named after Vapnik and Chervonenkis [**VC71**]) is one way to limit the complexity of a range space.

Definition 5.4. Let $\mathsf{S} = (\mathsf{X}, \mathcal{R})$ be a range space. For $Y \subseteq \mathsf{X}$, let

$$\mathcal{R}_{|Y} = \left\{ \mathbf{r} \cap Y \,\middle|\, \mathbf{r} \in \mathcal{R} \right\} \tag{5.1}$$

denote the ***projection*** of $\mathcal{R}$ on Y. The range space S projected to Y is $\mathsf{S}_{|Y} = (Y, \mathcal{R}_{|Y})$.

If $\mathcal{R}_{|Y}$ contains all subsets of Y (i.e., if Y is finite, we have $\left|\mathcal{R}_{|Y}\right| = 2^{|Y|}$), then Y is ***shattered*** by $\mathcal{R}$ (or equivalently Y is shattered by S).

The ***Vapnik-Chervonenkis*** dimension (or VC ***dimension***) of S, denoted by $\dim_{\mathsf{VC}}(S)$, is the maximum cardinality of a shattered subset of X. If there are arbitrarily large shattered subsets, then $\dim_{\mathsf{VC}}(S) = \infty$.

5.1.1. Examples.

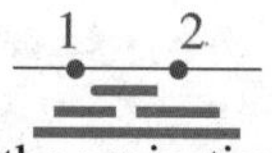

Intervals. Consider the set X to be the real line, and consider $\mathcal{R}$ to be the set of all intervals on the real line. Consider the set $Y = \{1, 2\}$. Clearly, one can find four intervals that contain all possible subsets of Y. Formally, the projection $\mathcal{R}_{|Y} = \left\{ \{\,\}, \{1\}, \{2\}, \{1, 2\} \right\}$. The intervals realizing each of these subsets are depicted on the right.

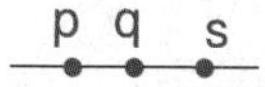

However, this is false for a set of three points $B = \{\mathsf{p}, \mathsf{q}, \mathsf{s}\}$, since there is no interval that can contain the two extreme points p and s without also containing q. Namely, the subset $\{\mathsf{p}, \mathsf{s}\}$ is not realizable for intervals, implying that the largest shattered set by the range space (real line, intervals) is of size two. We conclude that the VC dimension of this space is two.

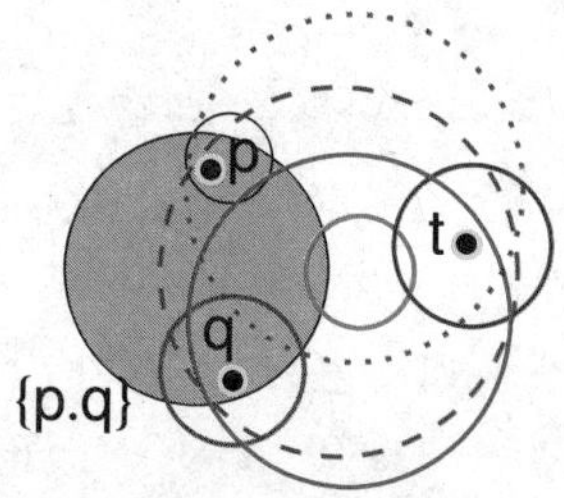

Disks. Let $\mathsf{X} = \mathbb{R}^2$, and let $\mathcal{R}$ be the set of disks in the plane. Clearly, for any three points in the plane (in general position), denoted by p, q, and s, one can find eight disks that realize all possible 2^3 different subsets. See the figure on the right.

But can disks shatter a set with four points? Consider such a set P of four points. If the convex hull of P has only three points on its boundary, then the subset X having only those three vertices (i.e., it does not include the middle point) is impossible, by convexity. Namely, there is no disk that contains only the points of X without the middle point.

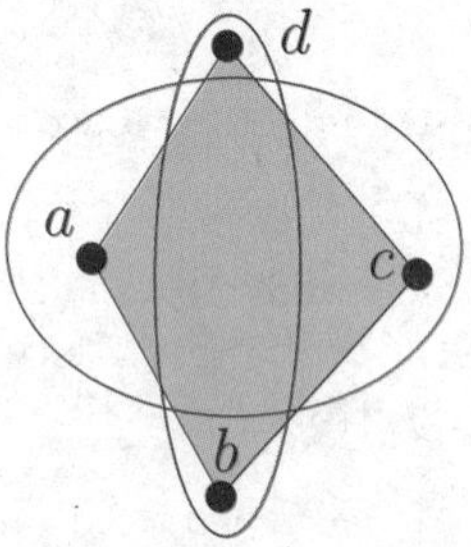

Alternatively, if all four points are vertices of the convex hull and they are a, b, c, d along the boundary of the convex hull, either the set $\{a, c\}$ or the set $\{b, d\}$ is not realizable. Indeed, if both options are realizable, then consider the two disks D_1 and D_2 that realize those assignments. Clearly, ∂D_1 and ∂D_2 must intersect in four points, but this is not possible, since two circles have at most two intersection points. See the figure on the left. Hence the VC dimension of this range space is 3.

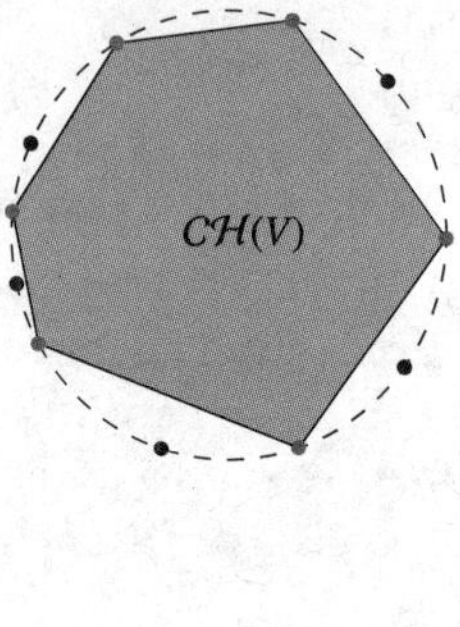

Convex sets. Consider the range space $\mathsf{S} = (\mathbb{R}^2, \mathcal{R})$, where $\mathcal{R}$ is the set of all (closed) convex sets in the plane. We claim that $\dim_{\mathsf{VC}}(\mathsf{S}) = \infty$. Indeed, consider a set U of n points $p_1, \ldots, p_n$ all lying on the boundary of the unit circle in the plane. Let V be any subset of U, and consider the convex hull $\mathcal{CH}(V)$. Clearly, $\mathcal{CH}(V) \in \mathcal{R}$, and furthermore, $\mathcal{CH}(V) \cap U = V$. Namely, any subset of U is realizable by S. Thus, S can shatter sets of arbitrary size, and its VC dimension is unbounded.

Complement. Consider the range space $\mathsf{S} = (\mathsf{X}, \mathcal{R})$ with $\delta = \dim_{\mathsf{VC}}(\mathsf{S})$. Next, consider the complement space, $\overline{\mathsf{S}} = \left(\mathsf{X}, \overline{\mathcal{R}}\right)$, where

$$\overline{\mathcal{R}} = \left\{ \mathsf{X} \setminus \mathbf{r} \;\middle|\; \mathbf{r} \in \mathcal{R} \right\};$$

namely, the ranges of $\overline{\mathsf{S}}$ are the complement of the ranges in S. What is the VC dimension of $\overline{\mathsf{S}}$? Well, a set $B \subseteq \mathsf{X}$ is shattered by $\overline{\mathsf{S}}$ if and only if it is shattered by S. Indeed, if S shatters B, then for any $Z \subseteq B$, we have that $(B \setminus Z) \in \mathcal{R}_{|B}$, which implies that $Z = B \setminus (B \setminus Z) \in \overline{\mathcal{R}}_{|B}$. Namely, $\overline{\mathcal{R}}_{|B}$ contains all the subsets of B, and $\overline{\mathsf{S}}$ shatters B. Thus, $\dim_{\mathsf{VC}}\left(\overline{\mathsf{S}}\right) = \dim_{\mathsf{VC}}(\mathsf{S})$.

Lemma 5.5. *For a range space* $\mathsf{S} = (\mathsf{X}, \mathcal{R})$ *we have that* $\dim_{\mathsf{VC}}(\mathsf{S}) = \dim_{\mathsf{VC}}\left(\overline{\mathsf{S}}\right)$, *where* $\overline{\mathsf{S}}$ *is the complement range space.*

5.1.1.1. *Halfspaces.* Let $\mathsf{S} = (\mathsf{X}, \mathcal{R})$, where $\mathsf{X} = \mathbb{R}^{\mathsf{d}}$ and $\mathcal{R}$ is the set of all (closed) halfspaces in $\mathbb{R}^{\mathsf{d}}$. We need the following technical claim.

Claim 5.6. *Let* $\mathsf{P} = \{\mathsf{p}_1, \ldots, \mathsf{p}_{\mathsf{d}+2}\}$ *be a set of* $\mathsf{d} + 2$ *points in* $\mathbb{R}^{\mathsf{d}}$. *There are real numbers* $\beta_1, \ldots, \beta_{\mathsf{d}+2}$, *not all of them zero, such that* $\sum_i \beta_i \mathsf{p}_i = 0$ *and* $\sum_i \beta_i = 0$.

Proof. Indeed, set $\mathsf{q}_i = (\mathsf{p}_i, 1)$, for $i = 1, \ldots, d+2$. Now, the points $\mathsf{q}_1, \ldots, \mathsf{q}_{\mathsf{d}+2} \in \mathbb{R}^{\mathsf{d}+1}$ are linearly dependent, and there are coefficients $\beta_1, \ldots, \beta_{\mathsf{d}+2}$, not all of them zero, such that $\sum_{i=1}^{\mathsf{d}+2} \beta_i \mathsf{q}_i = 0$. Considering only the first d coordinates of these points implies that $\sum_{i=1}^{\mathsf{d}+2} \beta_i \mathsf{p}_i = 0$. Similarly, by considering only the $(d+1)$st coordinate of these points, we have that $\sum_{i=1}^{\mathsf{d}+2} \beta_i = 0$. ■

To see what the VC dimension of halfspaces in $\mathbb{R}^{\mathsf{d}}$ is, we need the following result of Radon. (For a reminder of the formal definition of convex hulls, see Definition 28.1_{p347}.)

Theorem 5.7 (Radon's theorem). *Let* $\mathsf{P} = \{\mathsf{p}_1, \ldots, \mathsf{p}_{\mathsf{d}+2}\}$ *be a set of* $\mathsf{d} + 2$ *points in* $\mathbb{R}^{\mathsf{d}}$. *Then, there exist two disjoint subsets C and D of* P, *such that* $\mathcal{CH}(C) \cap \mathcal{CH}(D) \neq \emptyset$ *and* $C \cup D = \mathsf{P}$.

Proof. By Claim 5.6 there are real numbers $\beta_1, \ldots, \beta_{\mathsf{d}+2}$, not all of them zero, such that $\sum_i \beta_i \mathsf{p}_i = 0$ and $\sum_i \beta_i = 0$.

Assume, for the sake of simplicity of exposition, that $\beta_1,` \ldots, \beta_k \geq 0$ and $\beta_{k+1}, \ldots, \beta_{\mathsf{d}+2} < 0$. Furthermore, let $\mu = \sum_{i=1}^{k} \beta_i = -\sum_{i=k+1}^{\mathsf{d}+2} \beta_i$. We have that

$$\sum_{i=1}^{k} \beta_i \mathsf{p}_i = -\sum_{i=k+1}^{\mathsf{d}+2} \beta_i \mathsf{p}_i.$$

In particular, $v = \sum_{i=1}^{k} (\beta_i/\mu)\mathsf{p}_i$ is a point in $\mathcal{CH}(\{\mathsf{p}_1, \ldots, \mathsf{p}_k\})$. Furthermore, for the same point v we have $v = \sum_{i=k+1}^{\mathsf{d}+2} -(\beta_i/\mu)\mathsf{p}_i \in \mathcal{CH}(\{\mathsf{p}_{k+1}, \ldots, \mathsf{p}_{\mathsf{d}+2}\})$. We conclude that v is in the intersection of the two convex hulls, as required. ■

The following is a trivial observation, and yet we provide a proof to demonstrate it is true.

Lemma 5.8. *Let* $\mathsf{P} \subseteq \mathbb{R}^{\mathsf{d}}$ *be a finite set, let* s *be any point in* $\mathcal{CH}(\mathsf{P})$, *and let* h^+ *be a halfspace of* $\mathbb{R}^{\mathsf{d}}$ *containing* s. *Then there exists a point of* P *contained inside* h^+.

PROOF. The halfspace h^+ can be written as $h^+ = \left\{ \mathsf{t} \in \mathbb{R}^{\mathsf{d}} \,\middle|\, \langle \mathsf{t}, v \rangle \leq c \right\}$. Now $\mathsf{s} \in \mathcal{CH}(\mathsf{P}) \cap h^+$, and as such there are numbers $\alpha_1, \ldots, \alpha_m \geq 0$ and points $\mathsf{p}_1, \ldots, \mathsf{p}_m \in \mathsf{P}$, such that $\sum_i \alpha_i = 1$ and $\sum_i \alpha_i \mathsf{p}_i = \mathsf{s}$. By the linearity of the dot product, we have that

$$\mathsf{s} \in h^+ \implies \langle \mathsf{s}, v \rangle \leq c \implies \left\langle \sum_{i=1}^{m} \alpha_i \mathsf{p}_i, v \right\rangle \leq c \implies \beta = \sum_{i=1}^{m} \alpha_i \langle \mathsf{p}_i, v \rangle \leq c.$$

Setting $\beta_i = \langle \mathsf{p}_i, v \rangle$, for $i = 1, \ldots, m$, the above implies that β is a weighted average of $\beta_1, \ldots, \beta_m$. In particular, there must be a β_i that is no larger than the average. That is $\beta_i \leq c$. This implies that $\langle \mathsf{p}_i, v \rangle \leq c$. Namely, $\mathsf{p}_i \in h^+$ as claimed. ∎

Let S be the range space having $\mathbb{R}^{\mathsf{d}}$ as the ground set and all the close halfspaces as ranges. Radon's theorem implies that if a set Q of $\mathsf{d}+2$ points is being shattered by S, then we can partition this set Q into two disjoint sets Y and Z such that $\mathcal{CH}(Y) \cap \mathcal{CH}(Z) \neq \emptyset$. In particular, let s be a point in $\mathcal{CH}(Y) \cap \mathcal{CH}(Z)$. If a halfspace h^+ contains all the points of Y, then $\mathcal{CH}(Y) \subseteq h^+$, since a halfspace is a convex set. Thus, any halfspace h^+ containing all the points of Y will contain the point $\mathsf{s} \in \mathcal{CH}(Y)$. But $\mathsf{s} \in \mathcal{CH}(Z) \cap h^+$, and this implies that a point of Z must lie in h^+, by Lemma 5.8. Namely, the subset $Y \subseteq \mathsf{Q}$ cannot be realized by a halfspace, which implies that Q cannot be shattered. Thus $\dim_{\mathsf{VC}}(S) < d+2$. It is also easy to verify that the regular simplex with $\mathsf{d}+1$ vertices is shattered by S. Thus, $\dim_{\mathsf{VC}}(S) = \mathsf{d}+1$.

5.2. Shattering dimension and the dual shattering dimension

The main property of a range space with bounded VC dimension is that the number of ranges for a set of n elements grows polynomially in n (with the power being the dimension) instead of exponentially. Formally, let the ***growth function*** be

$$\mathcal{G}_\delta(n) = \sum_{i=0}^{\delta} \binom{n}{i} \leq \sum_{i=0}^{\delta} \frac{n^i}{i!} \leq n^\delta, \tag{5.2}$$

for $\delta > 1$ (the cases where $\delta = 0$ or $\delta = 1$ are not interesting and we will just ignore them). Note that for all $n, \delta \geq 1$, we have $\mathcal{G}_\delta(n) = \mathcal{G}_\delta(n-1) + \mathcal{G}_{\delta-1}(n-1)$[①].

Lemma 5.9 (Sauer's lemma). *If $(\mathsf{X}, \mathcal{R})$ is a range space of* VC *dimension δ with $|\mathsf{X}| = n$, then $|\mathcal{R}| \leq \mathcal{G}_\delta(n)$.*

PROOF. The claim trivially holds for $\delta = 0$ or $n = 0$.

Let x be any element of X, and consider the sets

$$\mathcal{R}_x = \left\{ \mathbf{r} \setminus \{x\} \,\middle|\, \mathbf{r} \cup \{x\} \in \mathcal{R} \text{ and } \mathbf{r} \setminus \{x\} \in \mathcal{R} \right\} \quad \text{and} \quad \mathcal{R} \setminus x = \left\{ \mathbf{r} \setminus \{x\} \,\middle|\, \mathbf{r} \in \mathcal{R} \right\}.$$

Observe that $|\mathcal{R}| = |\mathcal{R}_x| + |\mathcal{R} \setminus x|$. Indeed, we charge the elements of $\mathcal{R}$ to their corresponding element in $\mathcal{R} \setminus x$. The only bad case is when there is a range $\mathbf{r}$ such that both $\mathbf{r} \cup \{x\} \in \mathcal{R}$ and $\mathbf{r} \setminus \{x\} \in \mathcal{R}$, because then these two distinct ranges get mapped to the same range in $\mathcal{R} \setminus x$. But such ranges contribute exactly one element to $\mathcal{R}_x$. Similarly, every element of $\mathcal{R}_x$ corresponds to two such "twin" ranges in $\mathcal{R}$.

[①]Here is a cute (and standard) counting argument: $\mathcal{G}_\delta(n)$ is just the number of different subsets of size at most δ out of n elements. Now, we either decide to not include the first element in these subsets (i.e., $\mathcal{G}_\delta(n-1)$) or, alternatively, we include the first element in these subsets, but then there are only $\delta - 1$ elements left to pick (i.e., $\mathcal{G}_{\delta-1}(n-1)$).

Observe that $(X\setminus\{x\}, \mathcal{R}_x)$ has VC dimension $\delta-1$, as the largest set that can be shattered is of size $\delta-1$. Indeed, any set $B \subset \mathsf{X}\setminus\{x\}$ shattered by $\mathcal{R}_x$ implies that $B\cup\{x\}$ is shattered in $\mathcal{R}$.

Thus, we have

$$|\mathcal{R}| = |\mathcal{R}_x| + |\mathcal{R}\setminus x| \le \mathcal{G}_{\delta-1}(n-1) + \mathcal{G}_\delta(n-1) = \mathcal{G}_\delta(n),$$

by induction. ■

Interestingly, Lemma 5.9 is tight. See Exercise 5.4.

Next, we show pretty tight bounds on $\mathcal{G}_\delta(n)$. The proof is technical and not very interesting, and it is delegated to Section 5.6.

Lemma 5.10. *For $n \ge 2\delta$ and $\delta \ge 1$, we have $\left(\frac{n}{\delta}\right)^\delta \le \mathcal{G}_\delta(n) \le 2\left(\frac{ne}{\delta}\right)^\delta$, where $\mathcal{G}_\delta(n) = \sum_{i=0}^{\delta}\binom{n}{i}$.*

Definition 5.11 (Shatter function)**.** Given a range space $\mathsf{S} = (\mathsf{X}, \mathcal{R})$, its ***shatter function*** $\pi_{\mathsf{S}}(m)$ is the maximum number of sets that might be created by S when restricted to subsets of size m. Formally,

$$\pi_{\mathsf{S}}(m) = \max_{\substack{B\subset \mathsf{X}\\ |B|=m}} \left|\mathcal{R}_{|B}\right|;$$

see (5.1).

The ***shattering dimension*** of S is the smallest d such that $\pi_{\mathsf{S}}(m) = O(m^{\mathsf{d}})$, for all m.

By applying Lemma 5.9 to a finite subset of X, we get:

Corollary 5.12. *If $\mathsf{S} = (\mathsf{X}, \mathcal{R})$ is a range space of VC dimension δ, then for every finite subset B of X, we have $\left|\mathcal{R}_{|B}\right| \le \pi_{\mathsf{S}}(|B|) \le \mathcal{G}_\delta(|B|)$. That is, the VC dimension of a range space always bounds its shattering dimension.*

Proof. Let $n = |B|$, and observe that $\left|\mathcal{R}_{|B}\right| \le \mathcal{G}_\delta(n) \le n^\delta$, by (5.2). As such, $\left|\mathcal{R}_{|B}\right| \le n^\delta$, and, by definition, the shattering dimension of S is at most δ; namely, the shattering dimension is bounded by the VC dimension. ■

Our arch-nemesis in the following is the function $x/\ln x$. The following lemma states some properties of this function, and its proof is delegated to Exercise 5.2.

Lemma 5.13. *For the function $f(x) = x/\ln x$ the following hold.*
(A) $f(x)$ is monotonically increasing for $x \ge e$.
(B) $f(x) \ge e$, for $x > 1$.
(C) For $u \ge \sqrt{e}$, if $f(x) \le u$, then $x \le 2u\ln u$.
(D) For $u \ge \sqrt{e}$, if $x > 2u\ln u$, then $f(x) > u$.
(E) For $u \ge e$, if $f(x) \ge u$, then $x \ge u\ln u$.

The next lemma introduces a standard argument which is useful in bounding the VC dimension of a range space by its shattering dimension. It is easy to see that the bound is tight in the worst case.

Lemma 5.14. *If $\mathsf{S} = (\mathsf{X}, \mathcal{R})$ is a range space with shattering dimension d, then its VC dimension is bounded by $O(\mathsf{d}\log\mathsf{d})$.*

Proof. Let $N \subseteq \mathsf{X}$ be the largest set shattered by S, and let δ denote its cardinality. We have that $2^\delta = \left|\mathcal{R}_{|N}\right| \le \pi_{\mathsf{S}}(|N|) \le c\delta^{\mathsf{d}}$, where c is a fixed constant. As such, we have that $\delta \le \lg c + \mathsf{d} \lg \delta$, which in turn implies that $\dfrac{\delta - \lg c}{\lg \delta} \le \mathsf{d}$.[2] Assuming $\delta \ge \max(2, 2\lg c)$, we have that

$$\frac{\delta}{2\lg\delta} \le \mathsf{d} \implies \frac{\delta}{\ln\delta} \le \frac{2\mathsf{d}}{\ln 2} \le 6\mathsf{d} \implies \delta \le 2(6\mathsf{d})\ln(6\mathsf{d}),$$

by Lemma 5.13(C). ■

Disks revisited. To see why the shattering dimension is more convenient to work with than the VC dimension, consider the range space $\mathsf{S} = (\mathsf{X}, \mathcal{R})$, where $\mathsf{X} = \mathbb{R}^2$ and $\mathcal{R}$ is the set of disks in the plane. We know that the VC dimension of S is 3 (see Section 5.1.1).

We next use a standard continuous deformation argument to argue that the shattering dimension of this range space is also 3.

Lemma 5.15. *Consider the range space $\mathsf{S} = (\mathsf{X}, \mathcal{R})$, where $\mathsf{X} = \mathbb{R}^2$ and $\mathcal{R}$ is the set of disks in the plane. The shattering dimension of S is 3.*

Proof. Consider any set P of n points in the plane, and consider the set $\mathcal{F} = \mathcal{R}_{|\mathsf{P}}$. We claim that $|\mathcal{F}| \le 4n^3$.

The set $\mathcal{F}$ contains only n sets with a single point in them and only $\binom{n}{2}$ sets with two points in them. So, fix $\mathsf{Q} \in \mathcal{F}$ such that $|\mathsf{Q}| \ge 3$.

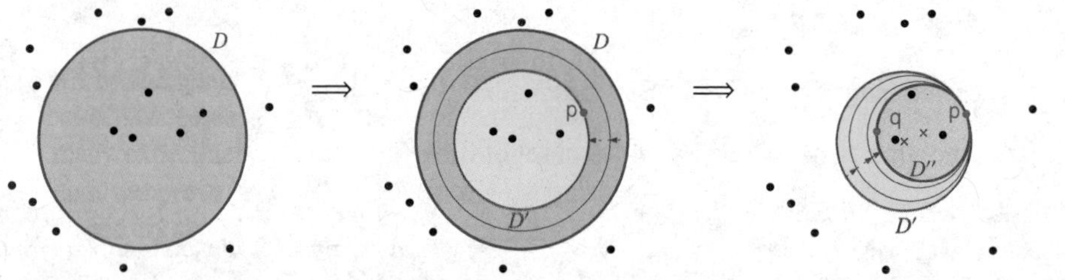

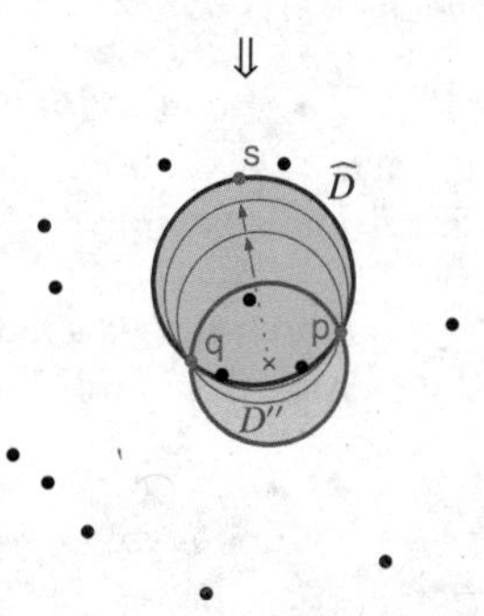

There is a disk D that realizes this subset; that is, $\mathsf{P} \cap D = \mathsf{Q}$. For the sake of simplicity of exposition, assume that P is in general position. Shrink D till its boundary passes through a point p of P.

Now, continue shrinking the new disk D' in such a way that its boundary passes through the point p (this can be done by moving the center of D' towards p). Continue in this continuous deformation till the new boundary hits another point q of P. Let D'' denote this disk.

Next, we continuously deform D'' so that it has both $\mathsf{p} \in \mathsf{Q}$ and $\mathsf{q} \in \mathsf{Q}$ on its boundary. This can be done by moving the center of D'' along the bisector linear between p and q. Stop as soon as the boundary of the disk hits a third point $\mathsf{s} \in \mathsf{P}$. (We have freedom in choosing in which direction to move the center. As such, move in the direction that causes the disk boundary to hit a new point s.) Let $\widehat{D}$ be the resulting disk. The boundary of $\widehat{D}$ is the unique circle passing through p, q, and s. Furthermore, observe that

$$D \cap \left(\mathsf{P} \setminus \{\mathsf{s}\}\right) = \widehat{D} \cap \left(\mathsf{P} \setminus \{\mathsf{s}\}\right).$$

[2] We remind the reader that $\lg = \log_2$.

That is, we can specify the point set $\mathsf{P} \cap D$ by specifying the three points $\mathsf{p}, \mathsf{q}, \mathsf{s}$ (and thus specifying the disk $\widehat{D}$) and the status of the three special points; that is, we specify for each point $\mathsf{p}, \mathsf{q}, \mathsf{s}$ whether or not it is inside the generated subset.

As such, there are at most $8\binom{n}{3}$ different subsets in $\mathcal{F}$ containing more than three points, as each such subset maps to a "canonical" disk, there are at most $\binom{n}{3}$ different such disks, and each such disk defines at most eight different subsets.

Similar argumentation implies that there are at most $4\binom{n}{2}$ subsets that are defined by a pair of points that realizes the diameter of the resulting disk. Overall, we have that

$$|\mathcal{F}| = 1 + n + 4\binom{n}{2} + 8\binom{n}{3} \leq 4n^3,$$

since there is one empty set in $\mathcal{F}$, n sets of size 1, and the rest of the sets are counted as described above. ■

The proof of Lemma 5.15 might not seem like a great simplification over the same bound we got by arguing about the VC dimension. However, the above argumentation gives us a very powerful tool – the shattering dimension of a range space defined by a family of shapes is always bounded by the number of points that determine a shape in the family.

Thus, the shattering dimension of, say, arbitrarily oriented rectangles in the plane is bounded by (and in this case, equal to) five, since such a rectangle is uniquely determined by five points. To see that, observe that if a rectangle has only four points on its boundary, then there is one degree of freedom left, since we can rotate the rectangle "around" these points; see the figure on the right.

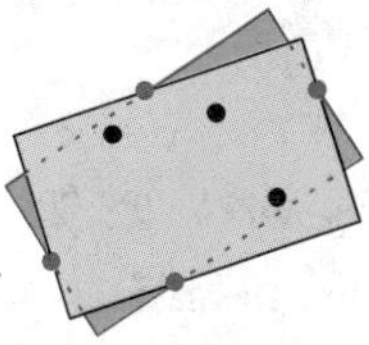

5.2.1. The dual shattering dimension. Given a range space $\mathsf{S} = (\mathsf{X}, \mathcal{R})$, consider a point $\mathsf{p} \in \mathsf{X}$. There is a set of ranges of $\mathcal{R}$ associated with p, namely, the set of all ranges of $\mathcal{R}$ that contains p which we denote by

$$\mathcal{R}_{\mathsf{p}} = \left\{ \mathbf{r} \;\middle|\; \mathbf{r} \in \mathcal{R}, \text{ the range } \mathbf{r} \text{ contains } \mathsf{p} \right\}.$$

This gives rise to a natural dual range space to S.

Definition 5.16. The ***dual range space*** to a range space $\mathsf{S} = (\mathsf{X}, \mathcal{R})$ is the space $\mathsf{S}^\star = (\mathcal{R}, \mathsf{X}^\star)$, where $\mathsf{X}^\star = \left\{ \mathcal{R}_{\mathsf{p}} \;\middle|\; \mathsf{p} \in \mathsf{X} \right\}$.

Naturally, the dual range space to $\mathsf{S}^\star$ is the original S, which is thus sometimes referred to as the ***primal range space***. (In other words, the dual to the dual is the primal.) The easiest way to see this, is to think about it as an abstract set system realized as an incidence matrix, where each point is a column and a set is a row in the matrix having 1 in an entry if and only if it contains the corresponding point; see Figure 5.1. Now, it is easy to verify that the dual range space is the transposed matrix.

To understand what the dual space is, consider X to be the plane and $\mathcal{R}$ to be a set of m disks. Then, in the dual range space $\mathsf{S}^\star = (\mathcal{R}, \mathsf{X}^\star)$, every point p in the plane has a set associated with it in $\mathsf{X}^\star$, which is the set of disks of $\mathcal{R}$ that contains p. In particular, if we consider the arrangement formed by the m disks of $\mathcal{R}$, then all the points lying inside a single face of this arrangement correspond to the same set of $\mathsf{X}^\star$. The number of ranges in $\mathsf{X}^\star$ is bounded by the complexity of the arrangement of these disks, which is $O(m^2)$; see Figure 5.1.

(A)

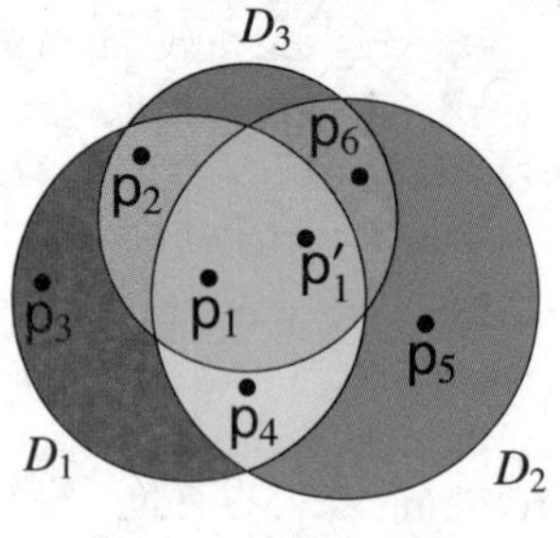

(B)

	p_1	p'_1	p_2	p_3	p_4	p_5	p_6
D_1	1	1	1	1	1	0	0
D_2	1	1	0	0	1	1	1
D_3	1	1	1	0	0	0	1

(C)

	D_1	D_2	D_3
p_1	1	1	1
p'_1	1	1	1
p_2	1	0	1
p_3	1	0	0
p_4	1	1	0
p_5	0	1	0
p_6	0	1	1

FIGURE 5.1. (A) $\mathcal{R}_{p_1} = \mathcal{R}_{p'_1}$. (B) Writing the set system as an incidence matrix where a point is a column and a set is a row. For example, D_2 contains p_4, and as such the column of p_4 has a 1 in the row corresponding to D_2. (C) The dual set system is represented by a matrix which is the transpose of the original incidence matrix.

Let the ***dual shatter function*** of the range space S be $\pi^\star_S(m) = \pi_{S^\star}(m)$, where $S^\star$ is the dual range space to S.

Definition 5.17. The ***dual shattering dimension*** of S is the shattering dimension of the dual range space $S^\star$.

Note that the dual shattering dimension might be smaller than the shattering dimension and hence also smaller than the VC dimension of the range space. Indeed, in the case of disks in the plane, the dual shattering dimension is just 2, while the VC dimension and the shattering dimension of this range space is 3. Note, also, that in geometric settings bounding the dual shattering dimension is relatively easy, as all you have to do is bound the complexity of the arrangement of m ranges of this space.

The following lemma shows a connection between the VC dimension of a space and its dual. The interested reader[3] might find the proof amusing.

Lemma 5.18. *Consider a range space* $S = (X, \mathcal{R})$ *with* VC *dimension* δ. *The dual range space* $S^\star = (\mathcal{R}, X^\star)$ *has* VC *dimension bounded by* $2^{\delta+1}$.

PROOF. Assume that $S^\star$ shatters a set $\mathcal{F} = \{\mathbf{r}_1, \ldots, \mathbf{r}_k\} \subseteq \mathcal{R}$ of k ranges. Then, there is a set $P \subseteq X$ of $m = 2^k$ points that shatters $\mathcal{F}$. Formally, for every subset $V \subseteq \mathcal{F}$, there exists a point $p \in P$, such that $\mathcal{F}_p = V$.

So, consider the matrix $\mathbf{M}$ (of dimensions $k \times 2^k$) having the points $p_1, \ldots, p_{2^k}$ of P as the columns, and every row is a set of $\mathcal{F}$, where the entry in the matrix corresponding to a point $p \in P$ and a range $\mathbf{r} \in \mathcal{F}$ is 1 if and only if $p \in \mathbf{r}$ and zero otherwise. Since P shatters $\mathcal{F}$, we know that this matrix has all possible 2^k binary vectors as columns.

[3] The author is quite aware that the interest of the reader in this issue might not be the result of free choice. Nevertheless, one might draw some comfort from the realization that the existence of the interested reader is as much an illusion as the existence of free choice. Both are convenient to assume, and both are probably false. Or maybe not.

Next, let $\kappa' = 2^{\lfloor \lg k \rfloor} \le k$, and consider the matrix $\mathbf{M}'$ of size $\kappa' \times \lg \kappa'$, where the ith row is the binary representation of the number $i-1$ (formally, the jth entry in the ith row is 1 if the jth bit in the binary representation of $i-1$ is 1), where $i = 1, \ldots, \kappa'$. See the figure on the right. Clearly, the $\lg \kappa'$ columns of $\mathbf{M}'$ are all different, and we can find $\lg \kappa'$ columns of $\mathbf{M}$ that are identical to the columns of $\mathbf{M}'$ (in the first κ' entries starting from the top of the columns).

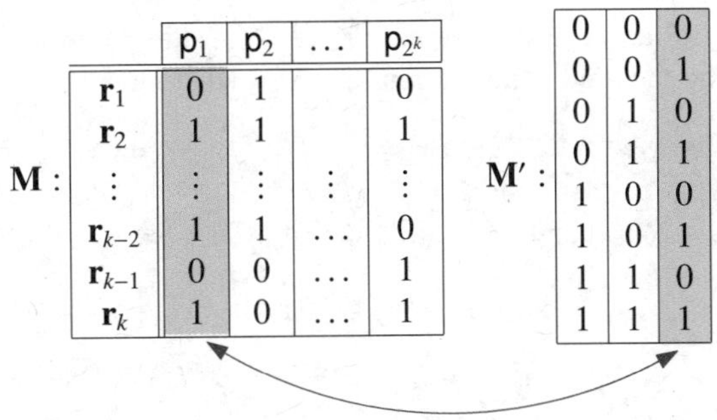

Each such column corresponds to a point $\mathsf{p} \in \mathsf{P}$, and let $\mathsf{Q} \subset \mathsf{P}$ be this set of $\lg \kappa'$ points. Note that for any subset $Z \subseteq \mathsf{Q}$, there is a row t in $\mathbf{M}'$ that encodes this subset. Consider the corresponding row in $\mathbf{M}$; that is, the range $\mathbf{r}_t \in \mathcal{F}$. Since $\mathbf{M}$ and $\mathbf{M}'$ are identical (in the relevant $\lg \kappa'$ columns of $\mathbf{M}$) on the first κ', we have that $\mathbf{r}_t \cap \mathsf{Q} = Z$. Namely, the set of ranges $\mathcal{F}$ shatters Q. But since the original range space has VC dimension δ, it follows that $|\mathsf{Q}| \le \delta$. Namely, $|\mathsf{Q}| = \lg \kappa' = \lfloor \lg k \rfloor \le \delta$, which implies that $\lg k \le \delta + 1$, which in turn implies that $k \le 2^{\delta+1}$. ∎

Lemma 5.19. *If a range space* $\mathsf{S} = (\mathsf{X}, \mathcal{R})$ *has dual shattering dimension* δ, *then its* VC *dimension is bounded by* $\delta^{O(\delta)}$.

PROOF. The shattering dimension of the dual range space $\mathsf{S}^\star$ is bounded by δ, and as such, by Lemma 5.14, its VC dimension is bounded by $\delta' = O(\delta \log \delta)$. Since the dual range space to $\mathsf{S}^\star$ is S, we have by Lemma 5.18 that the VC dimension of S is bounded by $2^{\delta'+1} = \delta^{O(\delta)}$. ∎

The bound of Lemma 5.19 might not be pretty, but it is sufficient in a lot of cases to bound the VC dimension when the shapes involved are simple.

Example 5.20. Consider the range space $\mathsf{S} = \left(\mathbb{R}^2, \mathcal{R}\right)$, where $\mathcal{R}$ is a set of shapes in the plane, so that the boundary of any pair of them intersects at most s times. Then, the VC dimension of S is $O(1)$. Indeed, the dual shattering dimension of S is $O(1)$, since the complexity of the arrangement of n such shapes is $O(sn^2)$. As such, by Lemma 5.19, the VC dimension of S is $O(1)$.

5.2.1.1. *Mixing range spaces.*

Lemma 5.21. *Let* $\mathsf{S} = (\mathsf{X}, \mathcal{R})$ *and* $\mathsf{T} = (\mathsf{X}, \mathcal{R}')$ *be two range spaces of* VC *dimension* δ *and* δ', *respectively, where* $\delta, \delta' > 1$. *Let* $\widehat{\mathcal{R}} = \left\{ \mathbf{r} \cup \mathbf{r}' \;\middle|\; \mathbf{r} \in \mathcal{R}, \mathbf{r}' \in \mathcal{R}' \right\}$. *Then, for the range space* $\widehat{\mathsf{S}} = \left(\mathsf{X}, \widehat{\mathcal{R}}\right)$, *we have that* $\dim_{\mathsf{VC}}\left(\widehat{\mathsf{S}}\right) = O(\delta + \delta')$.

PROOF. As a warm-up exercise, we prove a somewhat weaker bound here of $O((\delta + \delta') \log(\delta + \delta'))$. The stronger bound follows from Theorem 5.22 below. Let B be a set of n points in X that are shattered by $\widehat{\mathsf{S}}$. There are at most $\mathcal{G}_\delta(n)$ and $\mathcal{G}_{\delta'}(n)$ different ranges of B in the range sets $\mathcal{R}_{|B}$ and $\mathcal{R}'_{|B}$, respectively, by Lemma 5.9. Every subset C of B realized by $\widehat{r} \in \widehat{\mathcal{R}}$ is a union of two subsets $B \cap \mathbf{r}$ and $B \cap \mathbf{r}'$, where $\mathbf{r} \in R$ and $\mathbf{r}' \in \mathcal{R}'$, respectively. Thus, the number of different subsets of B realized by $\widehat{\mathsf{S}}$ is bounded by $\mathcal{G}_\delta(n)\mathcal{G}_{\delta'}(n)$. Thus, $2^n \le n^\delta n^{\delta'}$, for $\delta, \delta' > 1$. We conclude that $n \le (\delta + \delta') \lg n$, which implies that $n = O((\delta + \delta') \log(\delta + \delta'))$, by Lemma 5.13(C). ∎

Interestingly, one can prove a considerably more general result with tighter bounds. The required computations are somewhat more painful.

Theorem 5.22. *Let* $\mathsf{S}_1 = \left(\mathsf{X}, \mathcal{R}^1\right), \ldots, \mathsf{S}_k = \left(\mathsf{X}, \mathcal{R}^k\right)$ *be range spaces with* VC *dimension* $\delta_1, \ldots, \delta_k$, *respectively. Next, let* $f(\mathbf{r}_1, \ldots, \mathbf{r}_k)$ *be a function that maps any k-tuple of sets* $\mathbf{r}_1 \in \mathcal{R}^1, \ldots, \mathbf{r}_k \in \mathcal{R}^k$ *into a subset of* X. *Consider the range set*

$$\mathcal{R}' = \left\{ f(\mathbf{r}_1, \ldots, \mathbf{r}_k) \;\middle|\; \mathbf{r}_1 \in \mathcal{R}_1, \ldots, \mathbf{r}_k \in \mathcal{R}_k \right\}$$

and the associated range space $\mathsf{T} = (\mathsf{X}, \mathcal{R}')$. *Then, the* VC *dimension of* T *is bounded by* $O(k\delta \lg k)$, *where* $\delta = \max_i \delta_i$.

PROOF. Assume a set $Y \subseteq \mathsf{X}$ of size t is being shattered by $\mathcal{R}'$, and observe that

$$\left|\mathcal{R}'_{|Y}\right| \leq \left|\left\{ (\mathbf{r}_1, \ldots, \mathbf{r}_k) \;\middle|\; \mathbf{r}_1 \in \mathcal{R}^1_{|Y}, \ldots, \mathbf{r}_k \in \mathcal{R}^k_{|Y} \right\}\right| \leq \left|\mathcal{R}^1_{|Y}\right| \cdots \left|\mathcal{R}^k_{|Y}\right| \leq \mathcal{G}_{\delta_1}(t) \cdot \mathcal{G}_{\delta_2}(t) \cdots \mathcal{G}_{\delta_k}(t)$$

$$\leq \left(\mathcal{G}_\delta(t)\right)^k \leq \left(2\left(\frac{te}{\delta}\right)^\delta\right)^k,$$

by Lemma 5.9 and Lemma 5.10. On the other hand, since Y is being shattered by $\mathcal{R}'$, this implies that $\left|\mathcal{R}'_{|Y}\right| = 2^t$. Thus, we have the inequality $2^t \leq \left(2(te/\delta)^\delta\right)^k$, which implies $t \leq k(1 + \delta \lg(te/\delta))$. Assume that $t \geq e$ and $\delta \lg(te/\delta) \geq 1$ since otherwise the claim is trivial, and observe that $t \leq k(1 + \delta \lg(te/\delta)) \leq 3k\delta \lg(t/\delta)$. Setting $x = t/\delta$, we have

$$\frac{t}{\delta} \leq 3k \frac{\ln(t/\delta)}{\ln 2} \leq 6k \ln \frac{t}{\delta} \implies \frac{x}{\ln x} \leq 6k \implies x \leq 2 \cdot 6k \ln(6k) \implies x \leq 12k \ln(6k),$$

by Lemma 5.13(C). We conclude that $t \leq 12\delta k \ln(6k)$, as claimed. ∎

Corollary 5.23. *Let* $\mathsf{S} = (\mathsf{X}, \mathcal{R})$ *and* $\mathsf{T} = (\mathsf{X}, \mathcal{R}')$ *be two range spaces of* VC *dimension* δ *and* δ', *respectively, where* $\delta, \delta' > 1$. *Let* $\widehat{\mathcal{R}} = \left\{ \mathbf{r} \cap \mathbf{r}' \;\middle|\; \mathbf{r} \in \mathcal{R}, \mathbf{r}' \in \mathcal{R}' \right\}$. *Then, for the range space* $\widehat{\mathsf{S}} = (\mathsf{X}, \widehat{\mathcal{R}})$, *we have that* $\dim_{\mathsf{VC}}(\widehat{\mathsf{S}}) = O(\delta + \delta')$.

Corollary 5.24. *Any finite sequence of combining range spaces with finite* VC *dimension (by intersecting, complementing, or taking their union) results in a range space with a finite* VC *dimension.*

5.3. On ε-nets and ε-sampling

5.3.1. ε-nets and ε-samples.

Definition 5.25 (ε-sample)**.** Let $\mathsf{S} = (\mathsf{X}, \mathcal{R})$ be a range space, and let x be a finite subset of X. For $0 \leq \varepsilon \leq 1$, a subset $C \subseteq \mathsf{x}$ is an ***ε-sample*** for x if for any range $\mathbf{r} \in \mathcal{R}$, we have

$$\left| \overline{m}(\mathbf{r}) - \overline{s}(\mathbf{r}) \right| \leq \varepsilon,$$

where $\overline{m}(\mathbf{r}) = |\mathsf{x} \cap \mathbf{r}| / |\mathsf{x}|$ is the measure of $\mathbf{r}$ (see Definition 5.2) and $\overline{s}(\mathbf{r}) = |C \cap \mathbf{r}| / |C|$ is the estimate of $\mathbf{r}$ (see Definition 5.3). (Here C might be a multi-set, and as such $|C \cap \mathbf{r}|$ is counted with multiplicity.)

As such, an ε-sample is a subset of the ground set x that "captures" the range space up to an error of ε. Specifically, to estimate the fraction of the ground set covered by a range $\mathbf{r}$, it is sufficient to count the points of C that fall inside $\mathbf{r}$.

If X is a finite set, we will abuse notation slightly and refer to C as an ***ε-sample*** for S.

To see the usage of such a sample, consider $\mathsf{x} = \mathsf{X}$ to be, say, the population of a country (i.e., an element of X is a citizen). A range in $\mathcal{R}$ is the set of all people in the country that answer yes to a question (i.e., would you vote for party Y?, would you buy a bridge from me?, questions like that). An ε-sample of this range space enables us to estimate reliably (up to an error of ε) the answers for all these questions, by just asking the people in the sample.

The natural question of course is how to find such a subset of small (or minimal) size.

Theorem 5.26 (ε-sample theorem, **[VC71]**). *There is a positive constant c such that if $(\mathsf{X}, \mathcal{R})$ is any range space with* VC *dimension at most δ, $\mathsf{x} \subseteq \mathsf{X}$ is a finite subset and $\varepsilon, \varphi > 0$, then a random subset $C \subseteq \mathsf{x}$ of cardinality*

$$s = \frac{c}{\varepsilon^2}\left(\delta \log \frac{\delta}{\varepsilon} + \log \frac{1}{\varphi}\right)$$

is an ε-sample for x *with probability at least $1 - \varphi$.*

(In the above theorem, if $s > |\mathsf{x}|$, then we can just take all of x to be the ε-sample.)

For a strengthened version of the above theorem with slightly better bounds, see Theorem 7.13_{p107}.

Sometimes it is sufficient to have (hopefully smaller) samples with a weaker property – if a range is "heavy", then there is an element in our sample that is in this range.

Definition 5.27 (ε-net). A set $N \subseteq \mathsf{x}$ is an ***ε-net*** for x if for any range $\mathbf{r} \in \mathcal{R}$, if $\overline{m}(\mathbf{r}) \geq \varepsilon$ (i.e., $|\mathbf{r} \cap \mathsf{x}| \geq \varepsilon |\mathsf{x}|$), then $\mathbf{r}$ contains at least one point of N (i.e., $\mathbf{r} \cap N \neq \emptyset$).

Theorem 5.28 (ε-net theorem, **[HW87]**). *Let $(\mathsf{X}, \mathcal{R})$ be a range space of* VC *dimension δ, let x be a finite subset of X, and suppose that $0 < \varepsilon \leq 1$ and $\varphi < 1$. Let N be a set obtained by m random independent draws from x, where*

$$m \geq \max\left(\frac{4}{\varepsilon} \lg \frac{4}{\varphi}, \frac{8\delta}{\varepsilon} \lg \frac{16}{\varepsilon}\right). \tag{5.3}$$

Then N is an ε-net for x *with probability at least $1 - \varphi$.*

(We remind the reader that $\lg = \log_2$.)

The proofs of the above theorems are somewhat involved and we first turn our attention to some applications before presenting the proofs.

Remark 5.29. The above two theorems also hold for spaces with shattering dimension at most δ, in which case the sample size is slightly larger. Specifically, for Theorem 5.28, the sample size needed is $O\left(\frac{1}{\varepsilon} \lg \frac{1}{\varphi} + \frac{\delta}{\varepsilon} \lg \frac{\delta}{\varepsilon}\right)$.

5.3.2. Some applications. We mention two (easy) applications of these theorems, which (hopefully) demonstrate their power.

5.3.2.1. *Range searching.* So, consider a (very large) set of points P in the plane. We would like to be able to quickly decide how many points are included inside a query rectangle. Let us assume that we allow ourselves 1% error. What Theorem 5.26 tells us is that there is a subset of *constant size* (that depends only on ε) that can be used to perform this estimation, and it works for *all* query rectangles (we used here the fact that rectangles in the plane have finite VC dimension). In fact, a random sample of this size works with constant probability.

5.3.2.2. *Learning a concept.* Assume that we have a function f defined in the plane that returns '1' inside an (unknown) disk D_{unknown} and '0' outside it. There is some distribution $\mathcal{D}$ defined over the plane, and we pick points from this distribution. Furthermore, we can compute the function for these labels (i.e., we can compute f for certain values, but it is expensive). For a mystery value $\varepsilon > 0$, to be explained shortly, Theorem 5.28 tells us to pick (roughly) $O((1/\varepsilon)\log(1/\varepsilon))$ random points in a sample R from this distribution and to compute the labels for the samples. This is demonstrated in the figure on the right, where black dots are the sample points for which $f(\cdot)$ returned 1.

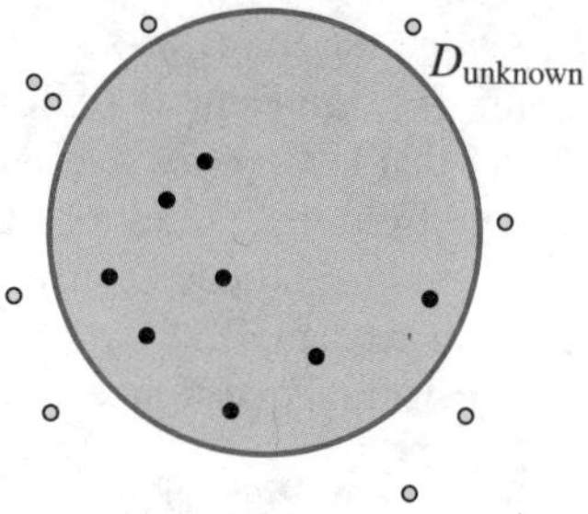

So, now we have positive examples and negative examples. We would like to find a hypothesis that agrees with all the samples we have and that hopefully is close to the true unknown disk underlying the function f. To this end, compute the smallest disk D that contains the sample labeled by '1' and does not contain any of the '0' points, and let $g : \mathbb{R}^2 \rightarrow \{0, 1\}$ be the function g that returns '1' inside the disk and '0' otherwise. We claim that g classifies correctly all but an ε-fraction of the points (i.e., the probability of misclassifying a point picked according to the given distribution is smaller than ε); that is, $\mathbf{Pr}_{\mathsf{p}\in\mathcal{D}}[f(\mathsf{p}) \neq g(\mathsf{p})] \leq \varepsilon$.

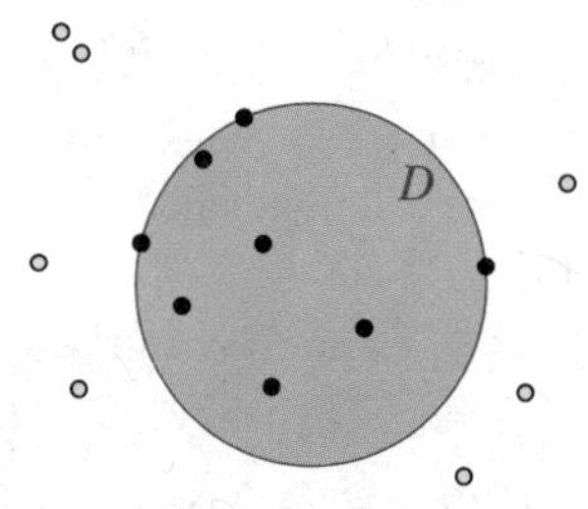

Geometrically, the region where g and f disagree is all the points in the symmetric difference between the two disks. That is, $\mathcal{E} = D \oplus D_{\text{unknown}}$; see the figure on the right.

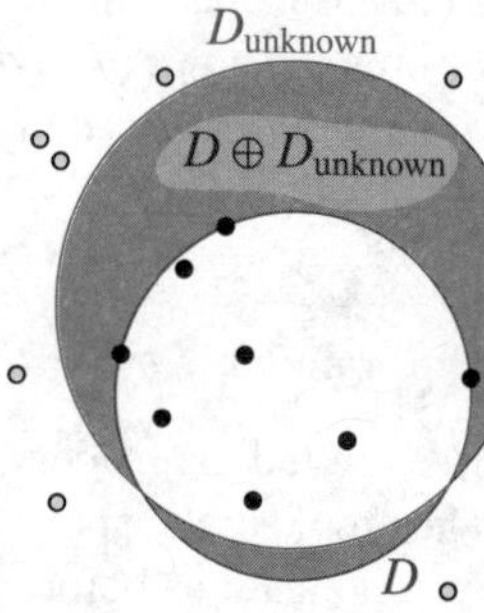

Thus, consider the range space S having the plane as the ground set and the symmetric difference between any two disks as its ranges. By Corollary 5.24, this range space has finite VC dimension. Now, consider the (unknown) disk D' that induces f and the region $\mathbf{r} = D_{\text{unknown}} \oplus D$. Clearly, the learned classifier g returns incorrect answers only for points picked inside $\mathbf{r}$.

Thus, the probability of a mistake in the classification is the measure of $\mathbf{r}$ under the distribution $\mathcal{D}$. So, if $\mathbf{Pr}_{\mathcal{D}}\left[\mathbf{r}\right] > \varepsilon$ (i.e., the probability that a sample point falls inside $\mathbf{r}$), then by the ε-net theorem (i.e., Theorem 5.28) the set R is an ε-net for S (ignore for the time being the possibility that the random sample fails to be an ε-net) and as such, R contains a point q inside $\mathbf{r}$. But, it is not possible for g (which classifies correctly all the sampled points of R) to make a mistake on q, a contradiction, because by construction, the range $\mathbf{r}$ is where g misclassifies points. We conclude that $\mathbf{Pr}_{\mathcal{D}}\left[\mathbf{r}\right] \leq \varepsilon$, as desired.

Little lies. The careful reader might be tearing his or her hair out because of the above description. First, Theorem 5.28 might fail, and the above conclusion might not hold. This is of course true, and in real applications one might use a much larger sample to guarantee that the probability of failure is so small that it can be practically ignored. A more serious issue is that Theorem 5.28 is defined only for finite sets. Nowhere does it speak about a continuous distribution. Intuitively, one can approximate a continuous distribution to an arbitrary precision using a huge sample and apply the theorem to this sample as our ground set. A formal proof is more tedious and requires extending the proof of Theorem 5.28 to continuous distributions. This is straightforward and we will ignore this topic altogether.

5.3.2.3. *A naive proof of the ε-sample theorem.* To demonstrate why the ε-sample/net theorems are interesting, let us try to prove the ε-sample theorem in the natural naive way. Thus, consider a finite range space $\mathsf{S} = (\mathsf{x}, \mathcal{R})$ with shattering dimension δ. Also, consider a range $\mathbf{r}$ that contains, say, a p fraction of the points of x, where $p \geq \varepsilon$. Consider a random sample R of r points from x, picked with replacement.

Let p_i be the ith sample point, and let X_i be an indicator variable which is one if and only if $\mathsf{p}_i \in \mathbf{r}$. Clearly, $(\sum_i X_i)/r$ is an estimate for $p = |\mathbf{r} \cap \mathsf{x}| / |\mathsf{x}|$. We would like this estimate to be within $\pm\varepsilon$ of p and with confidence $\geq 1 - \varphi$.

As such, the sample failed if $\left|\sum_{i=1}^{r} X_i - pr\right| \geq \varepsilon r = (\varepsilon/p)pr$. Set $\phi = \varepsilon/p$ and $\mu = \mathbf{E}[\sum_i X_i] = pr$. Using Chernoff's inequality (Theorem 27.17$_{\text{p340}}$ and Theorem 27.18$_{\text{p341}}$), we have

$$\mathbf{Pr}\left[\left|\sum_{i=1}^{r} X_i - pr\right| \geq (\varepsilon/p)pr\right] = \mathbf{Pr}\left[\left|\sum_{i=1}^{r} X_i - \mu\right| \geq \phi\mu\right] \leq \exp\left(-\mu\phi^2/2\right) + \exp\left(-\mu\phi^2/4\right)$$

$$\leq 2\exp\left(-\mu\phi^2/4\right) = 2\exp\left(-\frac{\varepsilon^2}{4p}r\right) \leq \varphi,$$

for $r \geq \left\lceil \frac{4}{\varepsilon^2} \ln \frac{2}{\varphi} \right\rceil \geq \left\lceil \frac{4p}{\varepsilon^2} \ln \frac{2}{\varphi} \right\rceil$.

Viola! We proved the ε-sample theorem. Well, not quite. We proved that the sample works correctly for a single range. Namely, we proved that for a specific range $\mathbf{r} \in \mathcal{R}$, we have that $\mathbf{Pr}\left[\left|\overline{m}(\mathbf{r}) - \overline{s}(\mathbf{r})\right| > \varepsilon\right] \leq \varphi$. However, we need to prove that $\forall \mathbf{r} \in \mathcal{R}$, $\mathbf{Pr}\left[\left|\overline{m}(\mathbf{r}) - \overline{s}(\mathbf{r})\right| > \varepsilon\right] \leq \varphi$.

Now, naively, we can overcome this by using a union bound on the bad probability. Indeed, if there are k different ranges under consideration, then we can use a sample that is large enough such that the probability of it to fail for each range is at most φ/k. In particular, let $\mathcal{E}_i$ be the bad event that the sample fails for the ith range. We have that $\mathbf{Pr}[\mathcal{E}_i] \leq \varphi/k$, which implies that

$$\mathbf{Pr}\left[\text{sample fails for any range}\right] \leq \mathbf{Pr}\left[\bigcup_{i=1}^{k} \mathcal{E}_i\right] \leq \sum_{i=1}^{k} \mathbf{Pr}\left[\mathcal{E}_i\right] \leq k(\varphi/k) \leq \varphi,$$

by the union bound; that is, the sample works for all ranges with good probability.

However, the number of ranges that we need to prove the theorem for is $\pi_{\mathsf{S}}(|\mathsf{x}|)$ (see Definition 5.11). In particular, if we plug in confidence $\varphi/\pi_{\mathsf{S}}(|\mathsf{x}|)$ to the above analysis and use the union bound, we get that for

$$r \geq \left\lceil \frac{4}{\varepsilon^2} \ln \frac{\pi_{\mathsf{S}}(|\mathsf{x}|)}{\varphi} \right\rceil$$

the sample estimates correctly (up to $\pm\varepsilon$) the size of all ranges with confidence $\geq 1 - \varphi$. Bounding $\pi_{\mathsf{S}}(|\mathsf{x}|)$ by $O\left(|\mathsf{x}|^{\delta}\right)$ (using (5.2)$_{\text{p64}}$ for a space with VC dimension δ), we can bound the required size of r by $O\left(\delta\varepsilon^{-2}\log(|\mathsf{x}|/\varphi)\right)$. We summarize the result.

Lemma 5.30. *Let $(\mathsf{x}, \mathcal{R})$ be a finite range space with VC dimension at most δ, and let $\varepsilon, \varphi > 0$ be parameters. Then a random subset $C \subseteq \mathsf{x}$ of cardinality $O\left(\delta\varepsilon^{-2}\log(|\mathsf{x}|/\varphi)\right)$ is an ε-sample for x with probability at least $1 - \varphi$.*

Namely, the "naive" argumentation gives us a sample bound which depends on the underlying size of the ground set. However, the sample size in the ε-sample theorem (Theorem 5.26) is independent of the size of the ground set. This is the magical property of the ε-sample theorem[4].

Interestingly, using a chaining argument on Lemma 5.30, one can prove the ε-sample theorem for the finite case; see Exercise 5.3. We provide a similar proof when using discrepancy, in Section 5.4. However, the original proof uses a clever double sampling idea that is both interesting and insightful that makes the proof work for the infinite case also.

5.3.3. A quicky proof of the ε-net theorem (Theorem 5.28). Here we provide a sketchy proof of Theorem 5.28, which conveys the main ideas. The full proof in all its glory and details is provided in Section 5.5.

Let $N = (x_1, \ldots, x_m)$ be the sample obtained by m independent samples from X (observe that N might contain the same element several times, and as such it is a multi-set). Let $\mathcal{E}_1$ be the probability that N fails to be an ε-net. Namely, for $n = |\mathsf{X}|$, let

$$\mathcal{E}_1 = \left\{ \exists \mathbf{r} \in \mathcal{R} \;\middle|\; |\mathbf{r} \cap \mathsf{X}| \geq \varepsilon n \text{ and } \mathbf{r} \cap N = \emptyset \right\}.$$

To complete the proof, we must show that $\mathbf{Pr}[\mathcal{E}_1] \leq \varphi$.

Let $T = (y_1, \ldots, y_m)$ be another random sample generated in a similar fashion to N. It might be that N fails for a certain range $\mathbf{r}$, but then since T is an independent sample, we still expect that $|\mathbf{r} \cap T| = \varepsilon m$. In particular, the probability that $\mathbf{Pr}\left[|\mathbf{r} \cap T| \geq \frac{\varepsilon m}{2}\right]$ is a large constant close to 1, regardless of how N performs. Indeed, if m is sufficiently large, we expect the random variable $|\mathbf{r} \cap T|$ to concentrate around εm, and one can argue this formally using Chernoff's inequality. Namely, intuitively, for a heavy range $\mathbf{r}$ we have that

$$\mathbf{Pr}[\mathbf{r} \cap N = \emptyset] \approx \mathbf{Pr}\left[\mathbf{r} \cap N = \emptyset \text{ and } \left(|\mathbf{r} \cap T| \geq \frac{\varepsilon m}{2}\right)\right].$$

Inspired by this, let $\mathcal{E}_2$ be the event that N fails for some range $\mathbf{r}$ but T "works" for $\mathbf{r}$; formally

$$\mathcal{E}_2 = \left\{ \exists \mathbf{r} \in \mathcal{R} \;\middle|\; |\mathbf{r} \cap \mathsf{X}| \geq \varepsilon n, \ \mathbf{r} \cap N = \emptyset \text{ and } |\mathbf{r} \cap T| \geq \frac{\varepsilon m}{2} \right\}.$$

Intuitively, since $\mathbf{E}[|\mathbf{r} \cap T|] \geq \varepsilon m$, then for the range $\mathbf{r}$ that N fails for, we have with "good" probability that $|\mathbf{r} \cap T| \geq \varepsilon m/2$. Namely, $\mathbf{Pr}[\mathcal{E}_1] \approx \mathbf{Pr}[\mathcal{E}_2]$.

Next, let

$$\mathcal{E}_2' = \left\{ \exists \mathbf{r} \in R \;\middle|\; \mathbf{r} \cap N = \emptyset \text{ and } |\mathbf{r} \cap T| \geq \frac{\varepsilon m}{2} \right\}.$$

Clearly, $\mathcal{E}_2 \subseteq \mathcal{E}_2'$ and as such $\mathbf{Pr}[\mathcal{E}_2] \leq \mathbf{Pr}\left[\mathcal{E}_2'\right]$. Now, fix $Z = N \cup T$, and observe that $|Z| = 2m$. Next, fix a range $\mathbf{r}$, and observe that the bad probability of $\mathcal{E}_2'$ is maximized if $|\mathbf{r} \cap Z| = \varepsilon m/2$. Now, the probability that all the elements of $\mathbf{r} \cap Z$ fall only into the second half of the sample is at most $2^{-\varepsilon m/2}$ as a careful calculation shows. Now, there are at most $\left|Z_{|R}\right| \leq \mathcal{G}_d(2m)$ different ranges that one has to consider. As such, $\mathbf{Pr}[\mathcal{E}_1] \approx \mathbf{Pr}[\mathcal{E}_2] \leq \mathbf{Pr}\left[\mathcal{E}_2'\right] \leq \mathcal{G}_d(2m)2^{-\varepsilon m/2}$ and this is smaller than φ, as a careful calculation shows by just plugging the value of m into the right-hand side; see $(5.3)_{p71}$. ■

[4]The notion of magic is used here in the sense of Arthur C. Clarke's statement that "any sufficiently advanced technology is indistinguishable from magic."

5.4. Discrepancy

The proof of the ε-sample/net theorem is somewhat complicated. It turns out that one can get a somewhat similar result by attacking the problem from the other direction; namely, let us assume that we would like to take a truly large sample of a finite range space $\mathsf{S} = (\mathsf{X}, \mathcal{R})$ defined over n elements with m ranges. We would like this sample to be as representative as possible as far as S is concerned. In fact, let us decide that we would like to pick exactly half of the points of X in our sample (assume that $n = |\mathsf{X}|$ is even).

To this end, let us color half of the points of X by -1 (i.e., black) and the other half by 1 (i.e., white). If for every range, $\mathbf{r} \in \mathcal{R}$, the number of black points inside it is equal to the number of white points, then doubling the number of black points inside a range gives us the exact number of points inside the range. Of course, such a perfect coloring is unachievable in almost all situations. To see this, consider the complete graph K_3 – clearly, in any coloring (by two colors) of its vertices, there must be an edge with two endpoints having the same color (i.e., the edges are the ranges).

Formally, let $\chi : \mathsf{X} \to \{-1, 1\}$ be a coloring. The ***discrepancy*** of χ over a range $\mathbf{r}$ is the amount of imbalance in the coloring inside χ. Namely,

$$|\chi(\mathbf{r})| = \left| \sum_{\mathsf{p} \in \mathbf{r}} \chi(\mathsf{p}) \right|.$$

The overall ***discrepancy*** of χ is $\mathrm{disc}(\chi) = \max_{\mathbf{r} \in \mathcal{R}} |\chi(\mathbf{r})|$. The ***discrepancy*** of a (finite) range space $\mathsf{S} = (\mathsf{X}, \mathcal{R})$ is the discrepancy of the best possible coloring; namely,

$$\mathrm{disc}(\mathsf{S}) = \min_{\chi : \mathsf{X} \to \{-1,+1\}} \mathrm{disc}(\chi).$$

The natural question is, of course, how to compute the coloring χ of minimum discrepancy. This seems like a very challenging question, but when you do not know what to do, you might as well do something random. So, let us pick a random coloring χ of X. To this end, let Π be an arbitrary partition of X into pairs (i.e., a perfect matching). For a pair $\{\mathsf{p}, \mathsf{q}\} \in \Pi$, we will either color $\chi(\mathsf{p}) = -1$ and $\chi(\mathsf{q}) = 1$ or the other way around; namely, $\chi(\mathsf{p}) = 1$ and $\chi(\mathsf{q}) = -1$. We will decide how to color this pair using a single coin flip. Thus, our coloring would be induced by making such a decision for every pair of Π, and let χ be the resulting coloring. We will refer to χ as ***compatible*** with the partition Π if, for all $\{\mathsf{p}, \mathsf{q}\} \in \Pi$, we have that $\chi(\{\mathsf{p}, \mathsf{q}\}) = 0$; namely,

$$\forall \{\mathsf{p}, \mathsf{q}\} \in \Pi \quad \big(\chi(\mathsf{p}) = +1 \text{ and } \chi(\mathsf{q}) = -1\big) \quad \text{or} \quad \big(\chi(\mathsf{p}) = -1 \text{ and } \chi(\mathsf{q}) = +1\big).$$

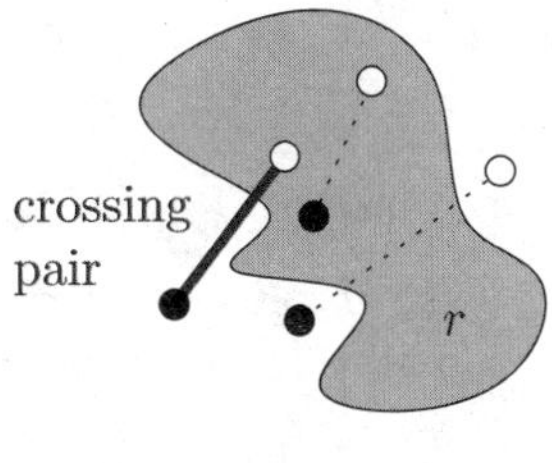

Consider a range $\mathbf{r}$ and a coloring χ compatible with Π. If a pair $\{\mathsf{p}, \mathsf{q}\} \in \Pi$ falls completely inside $\mathbf{r}$ or completely outside $\mathbf{r}$, then it does not contribute anything to the discrepancy of $\mathbf{r}$. Thus, the only pairs that contribute to the discrepancy of $\mathbf{r}$ are the ones that ***cross*** it. Namely, $\{\mathsf{p}, \mathsf{q}\} \cap \mathbf{r} \neq \emptyset$ and $\{\mathsf{p}, \mathsf{q}\} \cap (\mathsf{X} \setminus \mathbf{r}) \neq \emptyset$.

As such, let $\#_{\mathbf{r}}$ denote the ***crossing number*** of $\mathbf{r}$, that is, the number of pairs that cross $\mathbf{r}$. Next, let $X_i \in \{-1, +1\}$ be the indicator variable which is the contribution of the ith crossing pair to the discrepancy of $\mathbf{r}$. For $\Delta_{\mathbf{r}} = \sqrt{2\#_{\mathbf{r}} \ln(4m)}$, we have by Chernoff's inequality (Theorem 27.13$_{p338}$), that

$$\mathbf{Pr}\Big[|\chi(\mathbf{r})| \geq \Delta_{\mathbf{r}} \Big] = \mathbf{Pr}\Big[\chi(\mathbf{r}) \geq \Delta_{\mathbf{r}}\Big] + \mathbf{Pr}\Big[\chi(\mathbf{r}) \leq -\Delta_{\mathbf{r}}\Big] = 2\,\mathbf{Pr}\Big[\sum_i X_i \geq \Delta_{\mathbf{r}}\Big]$$

$$\leq 2\exp\left(-\frac{\Delta_{\mathbf{r}}^2}{2\#_{\mathbf{r}}}\right) = \frac{1}{2m}.$$

Since there are m ranges in $\mathcal{R}$, it follows that with good probability (i.e., at least half) for all $\mathbf{r} \in \mathcal{R}$ the discrepancy of $\mathbf{r}$ is at most $\Delta_{\mathbf{r}}$.

Theorem 5.31. *Let* $\mathsf{S} = (\mathsf{X}, \mathcal{R})$ *be a range space defined over* $n = |\mathsf{X}|$ *elements with* $m = |\mathcal{R}|$ *ranges. Consider any partition* Π *of the elements of* X *into pairs. Then, with probability* $\geq 1/2$, *for any range* $\mathbf{r} \in \mathcal{R}$, *a random coloring* $\chi : \mathsf{X} \to \{-1, +1\}$ *that is compatible with the partition* Π *has discrepancy at most*

$$|\chi(\mathbf{r})| < \Delta_{\mathbf{r}} = \sqrt{2\#_{\mathbf{r}} \ln(4m)},$$

where $\#_{\mathbf{r}}$ *denotes the number of pairs of* Π *that cross* $\mathbf{r}$. *In particular, since* $\#_{\mathbf{r}} \leq |\mathbf{r}|$, *we have* $|\chi(\mathbf{r})| \leq \sqrt{2\,|\mathbf{r}| \ln(4m)}$.

Observe that for every range $\mathbf{r}$ we have that $\#_{\mathbf{r}} \leq n/2$, since $2\#_{\mathbf{r}} \leq |\mathsf{X}|$. As such, we have:

Corollary 5.32. *Let* $\mathsf{S} = (\mathsf{X}, \mathcal{R})$ *be a range space defined over n elements with m ranges. Let* Π *be an arbitrary partition of* X *into pairs. Then a random coloring which is compatible with* Π *has* $\mathrm{disc}(\chi) < \sqrt{n \ln(4m)}$, *with probability* $\geq 1/2$.

One can easily amplify the probability of success of the coloring by increasing the threshold. In particular, for any constant $c \geq 1$, one has that

$$\forall \mathbf{r} \in \mathcal{R} \quad |\chi(\mathbf{r})| \leq \sqrt{2c\,\#_{\mathbf{r}} \ln(4m)},$$

with probability $\geq 1 - \dfrac{2}{(4m)^c}$.

5.4.1. Building ε-sample via discrepancy. Let $\mathsf{S} = (\mathsf{X}, \mathcal{R})$ be a range space with shattering dimension δ. Let $\mathsf{P} \subseteq \mathsf{X}$ be a set of n points, and consider the induced range space $\mathsf{S}_{|\mathsf{P}} = (\mathsf{P}, \mathcal{R}_{|\mathsf{P}})$; see Definition 5.4$_{p62}$. Here, by the definition of shattering dimension, we have that $m = \left|\mathcal{R}_{|\mathsf{P}}\right| = O\left(n^\delta\right)$. Without loss of generality, we assume that n is a power of 2. Consider a coloring χ of P with discrepancy bounded by Corollary 5.32. In particular, let Q be the points of P colored by, say, -1. We know that $|\mathsf{Q}| = n/2$, and for any range $\mathbf{r} \in \mathcal{R}$, we have that

$$\chi(\mathbf{r}) = \Big|\, |(\mathsf{P} \setminus \mathsf{Q}) \cap \mathbf{r}| - |\mathsf{Q} \cap \mathbf{r}| \,\Big| < \sqrt{n \ln(4m)} = \sqrt{n \ln O(n^\delta)} \leq c\sqrt{n \ln(n^\delta)},$$

for some absolute constant c. Observe that $|(\mathsf{P} \setminus \mathsf{Q}) \cap \mathbf{r}| = |\mathsf{P} \cap \mathbf{r}| - |\mathsf{Q} \cap \mathbf{r}|$. In particular, we have that for any range $\mathbf{r}$,

$$\Big|\, |\mathsf{P} \cap \mathbf{r}| - 2\,|\mathsf{Q} \cap \mathbf{r}| \,\Big| \leq c\sqrt{n \ln(n^\delta)}. \tag{5.4}$$

Dividing both sides by $n = |\mathsf{P}| = 2\,|\mathsf{Q}|$, we have that

$$\left| \frac{|\mathsf{P} \cap \mathbf{r}|}{|\mathsf{P}|} - \frac{|\mathsf{Q} \cap \mathbf{r}|}{|\mathsf{Q}|} \right| \leq \tau(n) \quad \text{for} \quad \tau(n) = c\sqrt{\frac{\delta \ln n}{n}}. \tag{5.5}$$

Namely, a coloring with discrepancy bounded by Corollary 5.32 yields a $\tau(n)$-sample. Intuitively, if n is very large, then Q provides a good approximation to P. However, we want an ε-sample for a prespecified $\varepsilon > 0$. Conceptually, ε is a fixed constant while $\tau(n)$ is considerably smaller. Namely, Q is a sample which is too tight for our purposes (and thus too big). As such, we will coarsen (and shrink) Q till we get the desired ε-sample by repeated application of Corollary 5.32. Specifically, we can "chain" together several approximations generated by Corollary 5.32. This is sometime refereed to as the ***sketch*** property of samples. Informally, as testified by the following lemma, a sketch of a sketch is a sketch[5].

Lemma 5.33. *Let $\mathsf{Q} \subseteq \mathsf{P}$ be a ρ-sample for P (in some underlying range space S), and let $\mathsf{R} \subseteq \mathsf{Q}$ be a ρ'-sample for Q. Then R is a $(\rho + \rho')$-sample for P.*

Proof. By definition, we have that, for every $\mathbf{r} \in \mathcal{R}$,

$$\left| \frac{|\mathbf{r} \cap \mathsf{P}|}{|\mathsf{P}|} - \frac{|\mathbf{r} \cap \mathsf{Q}|}{|\mathsf{Q}|} \right| \leq \rho \quad \text{and} \quad \left| \frac{|\mathbf{r} \cap \mathsf{Q}|}{|\mathsf{Q}|} - \frac{|\mathbf{r} \cap \mathsf{R}|}{|\mathsf{R}|} \right| \leq \rho'.$$

By adding the two inequalities together, we get

$$\left| \frac{|\mathbf{r} \cap \mathsf{P}|}{|\mathsf{P}|} - \frac{|\mathbf{r} \cap \mathsf{R}|}{|\mathsf{R}|} \right| = \left| \frac{|\mathbf{r} \cap \mathsf{P}|}{|\mathsf{P}|} - \frac{|\mathbf{r} \cap \mathsf{Q}|}{|\mathsf{Q}|} + \frac{|\mathbf{r} \cap \mathsf{Q}|}{|\mathsf{Q}|} - \frac{|\mathbf{r} \cap \mathsf{R}|}{|\mathsf{R}|} \right| \leq \rho + \rho'.$$ ■

Thus, let $\mathsf{P}_0 = \mathsf{P}$ and $\mathsf{P}_1 = \mathsf{Q}$. Now, in the ith iteration, we will compute a coloring χ_{i-1} of P_{i-1} with low discrepancy, as guaranteed by Corollary 5.32, and let P_i be the points of P_{i-1} colored white by χ_{i-1}. Let $\delta_i = \tau(n_{i-1})$, where $n_{i-1} = |\mathsf{P}_{i-1}| = n/2^{i-1}$. By Lemma 5.33, we have that P_k is a $(\sum_{i=1}^k \delta_i)$-sample for P. Since we would like the smallest set in the sequence $\mathsf{P}_1, \mathsf{P}_2, \ldots$ that is still an ε-sample, we would like to find the maximal k, such that $(\sum_{i=1}^k \delta_i) \leq \varepsilon$. Plugging in the value of δ_i and $\tau(\cdot)$, see (5.5), it is sufficient for our purposes that

$$\sum_{i=1}^{k} \delta_i = \sum_{i=1}^{k} \tau(n_{i-1}) = \sum_{i=1}^{k} c\sqrt{\frac{\delta \ln(n/2^{i-1})}{n/2^{i-1}}} \leq c_1 \sqrt{\frac{\delta \ln(n/2^{k-1})}{n/2^{k-1}}} = c_1 \sqrt{\frac{\delta \ln n_{k-1}}{n_{k-1}}} \leq \varepsilon,$$

since the above series behaves like a geometric series, and as such its total sum is proportional to its largest element[6], where c_1 is a sufficiently large constant. This holds for

$$c_1 \sqrt{\frac{\delta \ln n_{k-1}}{n_{k-1}}} \leq \varepsilon \iff c_1^2 \frac{\delta \ln n_{k-1}}{n_{k-1}} \leq \varepsilon^2 \iff \frac{c_1^2 \delta}{\varepsilon^2} \leq \frac{n_{k-1}}{\ln n_{k-1}}.$$

The last inequality holds for $n_{k-1} \geq 2\frac{c_1^2\delta}{\varepsilon^2} \ln \frac{c_1^2\delta}{\varepsilon^2}$, by Lemma 5.13(D). In particular, taking the largest k for which this holds results in a set P_k of size $O\big((\delta/\varepsilon^2)\ln(\delta/\varepsilon)\big)$ which is an ε-sample for P.

Theorem 5.34 (ε-sample via discrepancy). *For a range space $(X, \mathcal{R})$ with shattering dimension at most δ and $B \subseteq X$ a finite subset and $\varepsilon > 0$, there exists a subset $C \subseteq B$, of cardinality $O\big((\delta/\varepsilon^2)\ln(\delta/\varepsilon)\big)$, such that C is an ε-sample for B.*

Note that it is not obvious how to turn Theorem 5.34 into an efficient construction algorithm of such an ε-sample. Nevertheless, this theorem can be turned into a relatively efficient deterministic algorithm using conditional probabilities. In particular, there is a

[5] Try saying this quickly 100 times.

[6] Formally, one needs to show that the ratio between two consecutive elements in the series is larger than some constant, say 1.1. This is easy but tedious, but the well-motivated reader (of little faith) might want to do this calculation.

deterministic $O\left(n^{\delta+1}\right)$ time algorithm for computing an ε-sample for a range space of VC dimension δ and with n points in its ground set using the above approach (see the bibliographical notes in Section 5.7 for details). Inherently, however, it is a far cry from the simplicity of Theorem 5.26 that just requires us to take a random sample. Interestingly, there are cases where using discrepancy leads to smaller ε-samples; again see bibliographical notes for details.

5.4.1.1. *Faster deterministic construction of ε-samples.* One can speed up the deterministic construction mentioned above by using a sketch-and-merge approach. To this end, we need the following ***merge*** property of ε-samples. (The proof of the following lemma is quite easy. Nevertheless, we provide the proof in excruciating detail for the sake of completeness.)

Lemma 5.35. *Consider the sets* $\mathsf{R} \subseteq \mathsf{P}$ *and* $\mathsf{R}' \subseteq \mathsf{P}'$. *Assume that* P *and* P' *are disjoint,* $|\mathsf{P}| = |\mathsf{P}'|$, *and* $|\mathsf{R}| = |\mathsf{R}'|$. *Then, if* R *is an ε-sample of* P *and* R' *is an ε-sample of* P', *then* $\mathsf{R} \cup \mathsf{R}'$ *is an ε-sample of* $\mathsf{P} \cup \mathsf{P}'$.

PROOF. We have for any range $\mathbf{r}$ that

$$\begin{aligned}
\left| \frac{|\mathbf{r} \cap (\mathsf{P} \cup \mathsf{P}')|}{|\mathsf{P} \cup \mathsf{P}'|} - \frac{|\mathbf{r} \cap (\mathsf{R} \cup \mathsf{R}')|}{|\mathsf{R} \cup \mathsf{R}'|} \right| &= \left| \frac{|\mathbf{r} \cap \mathsf{P}|}{|\mathsf{P} \cup \mathsf{P}'|} + \frac{|\mathbf{r} \cap \mathsf{P}'|}{|\mathsf{P} \cup \mathsf{P}'|} - \frac{|\mathbf{r} \cap \mathsf{R}|}{|\mathsf{R} \cup \mathsf{R}'|} - \frac{|\mathbf{r} \cap \mathsf{R}'|}{|\mathsf{R} \cup \mathsf{R}'|} \right| \\
&= \left| \frac{|\mathbf{r} \cap \mathsf{P}|}{2\,|\mathsf{P}|} + \frac{|\mathbf{r} \cap \mathsf{P}'|}{2\,|\mathsf{P}'|} - \frac{|\mathbf{r} \cap \mathsf{R}|}{2\,|\mathsf{R}|} - \frac{|\mathbf{r} \cap \mathsf{R}'|}{2\,|\mathsf{R}'|} \right| \\
&= \frac{1}{2} \left| \left(\frac{|\mathbf{r} \cap \mathsf{P}|}{|\mathsf{P}|} - \frac{|\mathbf{r} \cap \mathsf{R}|}{|\mathsf{R}|} \right) + \left(\frac{|\mathbf{r} \cap \mathsf{P}'|}{|\mathsf{P}'|} - \frac{|\mathbf{r} \cap \mathsf{R}'|}{|\mathsf{R}'|} \right) \right| \\
&\le \frac{1}{2} \left| \frac{|\mathbf{r} \cap \mathsf{P}|}{|\mathsf{P}|} - \frac{|\mathbf{r} \cap \mathsf{R}|}{|\mathsf{R}|} \right| + \frac{1}{2} \left| \frac{|\mathbf{r} \cap \mathsf{P}'|}{|\mathsf{P}'|} - \frac{|\mathbf{r} \cap \mathsf{R}'|}{|\mathsf{R}'|} \right| \\
&\le \frac{\varepsilon}{2} + \frac{\varepsilon}{2} = \varepsilon.
\end{aligned}$$

■

Interestingly, by breaking the given ground sets into sets of equal size and building a balanced binary tree over these sets, one can speed up the deterministic algorithm for building ε-samples. The idea is to compute the sample bottom-up, where at every node we merge the samples provided by the children (i.e., using Lemma 5.35), and then we sketch the resulting set using Lemma 5.33. By carefully fine-tuning this construction, one can get an algorithm for computing ε-samples in time which is near linear in n (assuming ε and δ are small constants). We delegate the details of this construction to Exercise 5.6.

This algorithmic idea is quite useful and we will refer to it as ***sketch-and-merge***.

5.4.2. Building ε-net via discrepancy. We are given range space $(\mathsf{X}, \mathcal{R})$ with shattering dimension d and $\varepsilon > 0$ and the target is to compute an ε-net for this range space.

We need to be slightly more careful if we want to use discrepancy to build ε-nets, and we will use Theorem 5.31 instead of Corollary 5.32 in the analysis.

The construction is as before – we set $\mathsf{P}_0 = \mathsf{P}$, and P_i is all the points colored +1 in the coloring of P_{i-1} by Theorem 5.31. We repeat this till we get a set that is the required net.

To analyze this construction (and decide when it should stop), let $\mathbf{r}$ be a range in a given range space $(\mathsf{X}, \mathcal{R})$ with shattering dimension d, and let

$$\nu_i = |\mathsf{P}_i \cap \mathbf{r}|$$

denote the size of the range $\mathbf{r}$ in the ith set P_i and let $n_i = |\mathsf{P}_i|$, for $i \geq 0$. Observer that the number of points in $\mathbf{r}$ colored by $+1$ and -1 when coloring P_{i-1} is

$$\alpha_i = |\mathsf{P}_i \cap \mathbf{r}| = \nu_i \quad \text{and} \quad \beta_i = |\mathsf{P}_{i-1} \cap \mathbf{r}| - |\mathsf{P}_i \cap \mathbf{r}| = \nu_{i-1} - \nu_i,$$

respectively. As such, setting $m_i = \left|\mathcal{R}_{|\mathsf{P}_i}\right| = O\left(n_i^d\right)$, we have, by Theorem 5.31, that the discrepancy of $\mathbf{r}$ in this coloring of P_{i-1} is

$$|\alpha_i - \beta_i| = |\nu_i - 2\nu_{i-1}| \leq \sqrt{2\nu_{i-1} \ln 4m_{i-1}} \leq c\sqrt{d\nu_{i-1} \ln n_{i-1}}$$

for some constant c, since the crossing number $\#_{\mathbf{r}}$ of a range $\mathbf{r} \cap \mathsf{P}_{i-1}$ is always bounded by its size. This is equivalent to

$$\left|2^{i-1}\nu_{i-1} - 2^i \nu_i\right| \leq c 2^{i-1} \sqrt{d\nu_{i-1} \ln n_{i-1}}. \tag{5.6}$$

We need the following technical claim that states that the size of ν_k behaves as we expect; as long as the set P_k is large enough, the size of ν_k is roughly $\nu_0/2^k$.

Claim 5.36. *There is a constant c_4 (independent of d), such that for all k with $\nu_0/2^k \geq c_4 d \ln n_k$, $(\nu_0/2^k)/2 \leq \nu_k \leq 2(\nu_0/2^k)$.*

Proof. The proof is by induction. For $k = 0$ the claim trivially holds. Assume that it holds for $i < k$. Adding up the inequalities of (5.6), for $i = 1, \ldots, k$, we have that

$$\left|\nu_0 - 2^k \nu_k\right| \leq \sum_{i=1}^{k} c 2^{i-1} \sqrt{d\nu_{i-1} \ln n_{i-1}} \leq \sum_{i=1}^{k} c 2^{i-1} \sqrt{2d \frac{\nu_0}{2^{i-1}} \ln n_{i-1}} \leq c_3 2^k \sqrt{d \frac{\nu_0}{2^k} \ln n_k},$$

for some constant c_3 since this summation behaves like an increasing geometric series and the last term dominates the summation. Thus,

$$\frac{\nu_0}{2^k} - c_3 \sqrt{d \frac{\nu_0}{2^k} \ln n_k} \leq \nu_k \leq \frac{\nu_0}{2^k} + c_3 \sqrt{d \frac{\nu_0}{2^k} \ln n_k}.$$

By assumption, we have that $\sqrt{\dfrac{\nu_0}{c_4 2^k}} \geq \sqrt{d \ln n_k}$. This implies that

$$\nu_k \leq \frac{\nu_0}{2^k} + c_3 \sqrt{\frac{\nu_0}{2^k} \cdot \frac{\nu_0}{c_4 2^k}} = \frac{\nu_0}{2^k}\left(1 + \frac{c_3}{\sqrt{c_4}}\right) \leq 2\frac{\nu_0}{2^k},$$

by selecting $c_4 \geq 4c_3^2$. Similarly, we have

$$\nu_k \geq \frac{\nu_0}{2^k}\left(1 - \frac{c_3 \sqrt{d \ln n_k}}{\sqrt{\nu_0/2^k}}\right) \geq \frac{\nu_0}{2^k}\left(1 - \frac{c_3 \sqrt{\nu_0/c_4 2^k}}{\sqrt{\nu_0/2^k}}\right) = \frac{\nu_0}{2^k}\left(1 - \frac{c_3}{\sqrt{c_4}}\right) \geq \frac{\nu_0}{2^k}/2.$$

■

So consider a "heavy" range $\mathbf{r}$ that contains at least $\nu_0 \geq \varepsilon n$ points of P. To show that P_k is an ε-net, we need to show that $\mathsf{P}_k \cap \mathbf{r} \neq \emptyset$. To apply Claim 5.36, we need a k such that $\varepsilon n/2^k \geq c_4 d \ln n_{k-1}$, or equivalently, such that

$$\frac{2n_k}{\ln(2n_k)} \geq \frac{2c_4 d}{\varepsilon},$$

which holds for $n_k = \Omega\left(\frac{d}{\varepsilon} \ln \frac{d}{\varepsilon}\right)$, by Lemma 5.13(D). But then, by Claim 5.36, we have that

$$\nu_k = |\mathsf{P}_k \cap \mathbf{r}| \geq \frac{|\mathsf{P} \cap \mathbf{r}|}{2 \cdot 2^k} \geq \frac{1}{2} \cdot \frac{\varepsilon n}{2^k} = \frac{\varepsilon}{2} n_k = \Omega\left(d \ln \frac{d}{\varepsilon}\right) > 0.$$

We conclude that the set P_k, which is of size $\Omega\left(\frac{d}{\varepsilon} \ln \frac{d}{\varepsilon}\right)$, is an ε-net for P.

Theorem 5.37 (ε-net via discrepancy). *For any range space $(\mathsf{X}, \mathcal{R})$ with shattering dimension at most d, a finite subset $B \subseteq \mathsf{X}$, and $\varepsilon > 0$, there exists a subset $C \subseteq B$, of cardinality $O((d/\varepsilon)\ln(d/\varepsilon))$, such that C is an ε-net for B.*

5.5. Proof of the ε-net theorem

In this section, we finally prove Theorem 5.28.

Let $(\mathsf{X}, \mathcal{R})$ be a range space of VC dimension δ, and let x be a subset of X of cardinality n. Suppose that m satisfies $(5.3)_{p71}$. Let $N = (x_1, \dots, x_m)$ be the sample obtained by m independent samples from x (the elements of N are not necessarily distinct, and we treat N as an ordered set). Let $\mathcal{E}_1$ be the probability that N fails to be an ε-net. Namely,

$$\mathcal{E}_1 = \left\{ \exists \mathbf{r} \in \mathcal{R} \;\middle|\; |\mathbf{r} \cap \mathsf{x}| \geq \varepsilon n \text{ and } \mathbf{r} \cap N = \emptyset \right\}.$$

(Namely, there exists a "heavy" range $\mathbf{r}$ that does not contain any point of N.) To complete the proof, we must show that $\mathbf{Pr}[\mathcal{E}_1] \leq \varphi$. Let $T = (y_1, \dots, y_m)$ be another random sample generated in a similar fashion to N. Let $\mathcal{E}_2$ be the event that N fails but T "works"; formally

$$\mathcal{E}_2 = \left\{ \exists \mathbf{r} \in R \;\middle|\; |\mathbf{r} \cap \mathsf{x}| \geq \varepsilon n,\ \mathbf{r} \cap N = \emptyset, \text{ and } |\mathbf{r} \cap T| \geq \frac{\varepsilon m}{2} \right\}.$$

Intuitively, since $\mathbf{E}[|\mathbf{r} \cap T|] \geq \varepsilon m$, we have that for the range $\mathbf{r}$ that N fails for, it follows with "good" probability that $|\mathbf{r} \cap T| \geq \varepsilon m/2$. Namely, $\mathcal{E}_1$ and $\mathcal{E}_2$ have more or less the same probability.

Claim 5.38. $\mathbf{Pr}[\mathcal{E}_2] \leq \mathbf{Pr}[\mathcal{E}_1] \leq 2\,\mathbf{Pr}[\mathcal{E}_2]$.

Proof. Clearly, $\mathcal{E}_2 \subseteq \mathcal{E}_1$, and thus $\mathbf{Pr}[\mathcal{E}_2] \leq \mathbf{Pr}[\mathcal{E}_1]$. As for the other part, note that by the definition of conditional probability, we have

$$\mathbf{Pr}\left[\mathcal{E}_2 \;\middle|\; \mathcal{E}_1\right] = \mathbf{Pr}[\mathcal{E}_2 \cap \mathcal{E}_1] \,/\, \mathbf{Pr}[\mathcal{E}_1] = \mathbf{Pr}[\mathcal{E}_2] \,/\, \mathbf{Pr}[\mathcal{E}_1].$$

It is thus enough to show that $\mathbf{Pr}\left[\mathcal{E}_2 \;\middle|\; \mathcal{E}_1\right] \geq 1/2$.

Assume that $\mathcal{E}_1$ occurs. There is $\mathbf{r} \in R$, such that $|\mathbf{r} \cap \mathsf{x}| > \varepsilon n$ and $\mathbf{r} \cap N = \emptyset$. The required probability is at least the probability that for this specific $\mathbf{r}$, we have $|\mathbf{r} \cap T| \geq \frac{\varepsilon n}{2}$. However, $X = |\mathbf{r} \cap T|$ is a binomial variable with expectation $\mathbf{E}[X] = pm$, and variance $\mathbf{V}[X] = p(1-p)m \leq pm$, where $p = |\mathbf{r} \cap \mathsf{x}|/n \geq \varepsilon$. Thus, by Chebychev's inequality (Theorem 27.3_{p335}),

$$\begin{aligned}
\mathbf{Pr}\left[X < \frac{\varepsilon m}{2}\right] &\leq \mathbf{Pr}\left[X < \frac{pm}{2}\right] \leq \mathbf{Pr}\left[|X - pm| > \frac{pm}{2}\right] \\
&= \mathbf{Pr}\left[|X - pm| > \frac{\sqrt{pm}}{2}\sqrt{pm}\right] \leq \mathbf{Pr}\left[|X - \mathbf{E}[X]| > \frac{\sqrt{pm}}{2}\sqrt{\mathbf{V}[X]}\right] \\
&\leq \left(\frac{2}{\sqrt{pm}}\right)^2 \leq \frac{1}{2},
\end{aligned}$$

since $m \geq 8/\varepsilon \geq 8/p$; see $(5.3)_{p71}$. Thus, for $\mathbf{r} \in \mathcal{E}_1$, we have

$$\frac{\mathbf{Pr}[\mathcal{E}_2]}{\mathbf{Pr}[\mathcal{E}_1]} \geq \mathbf{Pr}\left[|\mathbf{r} \cap T| \geq \tfrac{\varepsilon m}{2}\right] = 1 - \mathbf{Pr}\left[|\mathbf{r} \cap T| < \frac{\varepsilon m}{2}\right] \geq \frac{1}{2}.$$

■

Claim 5.38 implies that to bound the probability of $\mathcal{E}_1$, it is enough to bound the probability of $\mathcal{E}_2$. Let

$$\mathcal{E}_2' = \left\{ \exists \mathbf{r} \in R \;\middle|\; \mathbf{r} \cap N = \emptyset, |\mathbf{r} \cap T| \geq \frac{\varepsilon m}{2} \right\}.$$

Clearly, $\mathcal{E}_2 \subseteq \mathcal{E}_2'$. Thus, bounding the probability of $\mathcal{E}_2'$ is enough to prove Theorem 5.28. Note, however, that a shocking thing happened! We no longer have x participating in our event. Namely, we turned bounding an event that depends on a global quantity (i.e., the ground set x) into bounding a quantity that depends only on a local quantity/experiment (involving only N and T). This is the crucial idea in this proof.

Claim 5.39. $\mathbf{Pr}\left[\mathcal{E}_2\right] \le \mathbf{Pr}\left[\mathcal{E}_2'\right] \le \mathcal{G}_d(2m)2^{-\varepsilon m/2}$.

Proof. We imagine that we sample the elements of $N \cup T$ together, by picking $Z = (z_1, \ldots, z_{2m})$ independently from x. Next, we randomly decide the m elements of Z that go into N, and the remaining elements go into T. Clearly,

$$\mathbf{Pr}\left[\mathcal{E}_2'\right] = \sum_{z \in \mathsf{x}^{2m}} \mathbf{Pr}\left[\mathcal{E}_2' \cap (Z = z)\right] = \sum_{z \in \mathsf{x}^{2m}} \frac{\mathbf{Pr}\left[\mathcal{E}_2' \cap (Z = z)\right]}{\mathbf{Pr}[Z = z]} \cdot \mathbf{Pr}\left[Z = z\right]$$

$$= \sum_{z} \mathbf{Pr}\left[\mathcal{E}_2' \,\middle|\, Z = z\right] \mathbf{Pr}\left[Z = z\right] = \mathbf{E}\left[\mathbf{Pr}\left[\mathcal{E}_2' \,\middle|\, Z = z\right]\right].$$

Thus, from this point on, we fix the set Z, and we bound $\mathbf{Pr}\left[\mathcal{E}_2' \,\middle|\, Z\right]$. Note that $\mathbf{Pr}[\mathcal{E}_2']$ is a weighted average of $\mathbf{Pr}[\mathcal{E}_2' | Z = z]$, and as such a bound on this quantity would imply the same bound on $\mathbf{Pr}[\mathcal{E}_2']$.

It is now enough to consider the ranges in the projection space $(Z, \mathcal{R}_{|Z})$ (which has VC dimension δ). By Lemma 5.9, we have $\left|\mathcal{R}_{|Z}\right| \le \mathcal{G}_\delta(2m)$.

Let us fix any $\mathbf{r} \in \mathcal{R}_{|Z}$, and consider the event

$$\mathcal{E}_{\mathbf{r}} = \left\{ \mathbf{r} \cap N = \emptyset \text{ and } |\mathbf{r} \cap T| > \frac{\varepsilon m}{2} \right\}.$$

We claim that $\mathbf{Pr}[\mathcal{E}_{\mathbf{r}}] \le 2^{-\varepsilon m/2}$. Observe that if $k = |\mathbf{r} \cap (N \cup T)| \le \varepsilon m/2$, then the event is empty, and this claim trivially holds. Otherwise, $\mathbf{Pr}[\mathcal{E}_{\mathbf{r}}] = \mathbf{Pr}[\mathbf{r} \cap N = \emptyset]$. To bound this probability, observe that we have the $2m$ elements of Z, and we can choose any m of them to be N, as long as none of them is one of the k "forbidden" elements of $\mathbf{r} \cap (N \cup T)$. The probability of that is $\binom{2m-k}{m} / \binom{2m}{m}$. We thus have

$$\mathbf{Pr}\left[\mathcal{E}_{\mathbf{r}}\right] \le \mathbf{Pr}\left[\mathbf{r} \cap N = \emptyset\right] = \frac{\binom{2m-k}{m}}{\binom{2m}{m}} = \frac{(2m-k)(2m-k-1)\cdots(m-k+1)}{2m(2m-1)\cdots(m+1)}$$

$$= \frac{m(m-1)\cdots(m-k+1)}{2m(2m-1)\cdots(2m-k+1)} \le 2^{-k} \le 2^{-\varepsilon m/2}.$$

Thus,

$$\mathbf{Pr}\left[\mathcal{E}_2' \,\middle|\, Z\right] = \mathbf{Pr}\left[\bigcup_{\mathbf{r} \in \mathcal{R}_{|Z}} \mathcal{E}_{\mathbf{r}}\right] \le \sum_{\mathbf{r} \in \mathcal{R}_{|Z}} \mathbf{Pr}[\mathcal{E}_{\mathbf{r}}] \le \left|\mathcal{R}_{|Z}\right| 2^{-\varepsilon m/2} \le \mathcal{G}_\delta(2m) 2^{-\varepsilon m/2},$$

implying that $\mathbf{Pr}\left[\mathcal{E}_2'\right] \le \mathcal{G}_\delta(2m)2^{-\varepsilon m/2}$. ■

Proof of Theorem 5.28. By Claim 5.38 and Claim 5.39, we have that $\mathbf{Pr}[\mathcal{E}_1] \le 2\mathcal{G}_\delta(2m)2^{-\varepsilon m/2}$. It thus remains to verify that if m satisfies (5.3), then $2\mathcal{G}_\delta(2m)2^{-\varepsilon m/2} \le \varphi$.

Indeed, we know that $2m \ge 8\delta$ (by $(5.3)_{p71}$) and by Lemma 5.10, $\mathcal{G}_\delta(2m) \le 2(2em/\delta)^\delta$, for $\delta \ge 1$. Thus, it is sufficient to show that the inequality $4(2em/\delta)^\delta 2^{-\varepsilon m/2} \le \varphi$ holds. By rearranging and taking lg of both sides, we have that this is equivalent to

$$2^{\varepsilon m/2} \geq \frac{4}{\varphi}\left(\frac{2em}{\delta}\right)^{\delta} \qquad \Longrightarrow \qquad \frac{\varepsilon m}{2} \geq \delta \lg \frac{2em}{\delta} + \lg \frac{4}{\varphi}.$$

By our choice of m (see (5.3)), we have that $\varepsilon m/4 \geq \lg(4/\varphi)$. Thus, we need to show that

$$\frac{\varepsilon m}{4} \geq \delta \lg \frac{2em}{\delta}.$$

We verify this inequality for $m = \dfrac{8\delta}{\varepsilon} \lg \dfrac{16}{\varepsilon}$ (this would also hold for bigger values, as can be easily verified). Indeed

$$2\delta \lg \frac{16}{\varepsilon} \geq \delta \lg\left(\frac{16e}{\varepsilon} \lg \frac{16}{\varepsilon}\right).$$

This is equivalent to $\left(\dfrac{16}{\varepsilon}\right)^2 \geq \dfrac{16e}{\varepsilon} \lg \dfrac{16}{\varepsilon}$, which is equivalent to $\dfrac{16}{e\,\varepsilon} \geq \lg \dfrac{16}{\varepsilon}$, which is certainly true for $0 < \varepsilon \leq 1$.

This completes the proof of the theorem. ■

5.6. A better bound on the growth function

In this section, we prove Lemma 5.10$_{\text{p65}}$. Since the proof is straightforward but tedious, the reader can safely skip reading this section.

Lemma 5.40. *For any positive integer n, the following hold.*

(i) $(1 + 1/n)^n \leq e$. *(ii)* $(1 - 1/n)^{n-1} \geq e^{-1}$.

(iii) $n! \geq (n/e)^n$. *(iv) For any* $k \leq n$, *we have* $\left(\dfrac{n}{k}\right)^k \leq \dbinom{n}{k} \leq \left(\dfrac{ne}{k}\right)^k$.

Proof. (i) Indeed, $1+1/n \leq \exp(1/n)$, since $1+x \leq e^x$, for $x \geq 0$. As such $(1+1/n)^n \leq \exp(n(1/n)) = e$.

(ii) Rewriting the inequality, we have that we need to prove $\left(\frac{n-1}{n}\right)^{n-1} \geq \frac{1}{e}$. This is equivalent to proving $e \geq \left(\frac{n}{n-1}\right)^{n-1} = \left(1 + \frac{1}{n-1}\right)^{n-1}$, which is our friend from (i).

(iii) Indeed,

$$\frac{n^n}{n!} \leq \sum_{i=0}^{\infty} \frac{n^i}{i!} = e^n,$$

by the Taylor expansion of $e^x = \sum_{i=0}^{\infty} \frac{x^i}{i!}$. This implies that $(n/e)^n \leq n!$, as required.

(iv) Indeed, for any $k \leq n$, we have $\frac{n}{k} \leq \frac{n-1}{k-1}$, as can be easily verified. As such, $\frac{n}{k} \leq \frac{n-i}{k-i}$, for $1 \leq i \leq k-1$. As such,

$$\left(\frac{n}{k}\right)^k \leq \frac{n}{k} \cdot \frac{n-1}{k-1} \cdots \frac{n-k+1}{1} = \binom{n}{k}.$$

As for the other direction, by (iii), we have $\dbinom{n}{k} \leq \dfrac{n^k}{k!} \leq \dfrac{n^k}{\left(\frac{k}{e}\right)^k} = \left(\dfrac{ne}{k}\right)^k$. ■

Lemma 5.10 restated. *For* $n \geq 2\delta$ *and* $\delta \geq 1$, *we have* $\left(\dfrac{n}{\delta}\right)^{\delta} \leq \mathcal{G}_\delta(n) \leq 2\left(\dfrac{ne}{\delta}\right)^{\delta}$, *where* $\mathcal{G}_\delta(n) = \displaystyle\sum_{i=0}^{\delta} \binom{n}{i}$.

PROOF. Note that by Lemma 5.40(iv), we have $\mathcal{G}_\delta(n) = \sum_{i=0}^{\delta}\binom{n}{i} \le 1 + \sum_{i=1}^{\delta}\left(\frac{ne}{i}\right)^i$. This series behaves like a geometric series with constant larger than 2, since

$$\left(\frac{ne}{i}\right)^i \Big/ \left(\frac{ne}{i-1}\right)^{i-1} = \frac{ne}{i}\left(\frac{i-1}{i}\right)^{i-1} = \frac{ne}{i}\left(1-\frac{1}{i}\right)^{i-1} \ge \frac{ne}{i}\frac{1}{e} = \frac{n}{i} \ge \frac{n}{\delta} \ge 2,$$

by Lemma 5.40. As such, this series is bounded by twice the largest element in the series, implying the claim. ■

5.7. Bibliographical notes

The exposition of the ε-net and ε-sample theorems is roughly based on Alon and Spencer [**AS00**] and Komlós et al. [**KPW92**]. In fact, Komlós et al. proved a somewhat stronger bound; that is, a random sample of size $(\delta/\varepsilon)\ln(1/\varepsilon)$ is an ε-net with constant probability. For a proof that shows that in general ε-nets cannot be much smaller in the worst case, see [**PA95**]. The original proof of the ε-net theorem is due to Haussler and Welzl [**HW87**]. The proof of the ε-sample theorem is due to Vapnik and Chervonenkis [**VC71**]. The bound in Theorem 5.26 can be improved to $O\left(\frac{\delta}{\varepsilon^2} + \frac{1}{\varepsilon^2}\log\frac{1}{\varphi}\right)$ [**AB99**].

An alternative proof of the ε-net theorem proceeds by first computing an $(\varepsilon/4)$-sample of sufficient size, using the ε-sample theorem (Theorem 5.26_{p71}), and then computing and $\varepsilon/4$-net for this sample using a direct sample of the right size. It is easy to verify the resulting set is an ε-net. Furthermore, using the "naive" argument (see Section 5.3.2.3) then implies that this holds with the right probability, thus implying the ε-net theorem (the resulting constants might be slightly worse). Exercise 5.3 deploys similar ideas.

The beautiful alternative proof of both theorems via the usage of discrepancy is due to Chazelle and Matoušek [**CM96**]. The discrepancy method is a beautiful topic which is quite deep mathematically, and we have just skimmed the thin layer of melted water on top of the tip of the iceberg[7]. Two nice books on the topic are the books by Chazelle [**Cha01**] and Matoušek [**Mat99**]. The book by Chazelle [**Cha01**] is currently available online for free from Chazelle's webpage.

We will revisit discrepancy since in some geometric cases it yields better results than the ε-sample theorem. In particular, the random coloring of Theorem 5.31 can be derandomized using conditional probabilities. One can then use it to get an ε-sample/net by applying it repeatedly. A faster algorithm results from a careful implementation of the sketch-and-merge approach. The disappointing feature of all the deterministic constructions of ε-samples/nets is that their running time is exponential in the dimension δ, since the number of ranges is usually exponential in δ.

A similar result to the one derived by Haussler and Welzl [**HW87**], using a more geometric approach, was done independently by Clarkson at the same time [**Cla87**], exposing the fact that VC dimension is not necessary if we are interested only in geometric applications. This was later refined by Clarkson [**Cla88**], leading to a general technique that, in geometric settings, yields stronger results than the ε-net theorem. This technique has numerous applications in discrete and computational geometry and leads to several "proofs from the book" in discrete geometry.

Exercise 5.5 is from Anthony and Bartlett [**AB99**].

---

[7] The iceberg is melting because of global warming; so sorry, climate change.

5.7.1. Variants and extensions. A natural application of the ε-sample theorem is to use it to estimate the weights of ranges. In particular, given a finite range space $(\mathsf{X}, \mathcal{R})$, we would like to build a data-structure such that we can decide quickly, given a query range $\mathbf{r}$, what the number of points of X inside $\mathbf{r}$ is. We could always use a sample of size (roughly) $O(\varepsilon^{-2})$ to get an estimate of the weight of a range, using the ε-sample theorem. The error of the estimate of the size $|\mathbf{r} \cap \mathsf{X}|$ is $\leq \varepsilon n$, where $n = |\mathsf{X}|$; namely, the error is additive. The natural question is whether one can get a multiplicative estimate ρ, such that $|\mathbf{r} \cap \mathsf{X}| \leq \rho \leq (1+\varepsilon)\, |\mathbf{r} \cap \mathsf{X}|$, where $|\mathbf{r} \cap \mathsf{X}|$.

In particular, a subset $A \subset \mathsf{X}$ is a (relative) (ε, p)-***sample*** if for each $\mathbf{r} \in \mathcal{R}$ of weight $\geq pn$,

$$\left| \frac{|\mathbf{r} \cap A|}{|A|} - \frac{|\mathbf{r} \cap \mathsf{X}|}{|\mathsf{X}|} \right| \leq \varepsilon \frac{|\mathbf{r} \cap \mathsf{X}|}{|\mathsf{X}|}.$$

Of course, one can simply generate an εp-sample of size (roughly) $O(1/(\varepsilon p)^2)$ by the ε-sample theorem. This is not very interesting when $p = 1/\sqrt{n}$. Interestingly, the dependency on p can be improved.

Theorem 5.41 ([**LLS01**]). *Let $(\mathsf{X}, \mathcal{R})$ be a range space with shattering dimension d, where $|\mathsf{X}| = n$, and let $0 < \varepsilon < 1$ and $0 < p < 1$ be given parameters. Then, consider a random sample $A \subseteq \mathsf{X}$ of size $\dfrac{c}{\varepsilon^2 p}\left(d \log \dfrac{1}{p} + \log \dfrac{1}{\varphi}\right)$, where c is a constant. Then, it holds that for each range $\mathbf{r} \in \mathcal{R}$ of at least pn points, we have*

$$\left| \frac{|\mathbf{r} \cap A|}{|A|} - \frac{|\mathbf{r} \cap \mathsf{X}|}{|\mathsf{X}|} \right| \leq \varepsilon \frac{|\mathbf{r} \cap \mathsf{X}|}{|\mathsf{X}|}.$$

In other words, A is a (p, ε)-sample for $(\mathsf{X}, \mathcal{R})$. The probability of success is $\geq 1 - \varphi$.

A similar result is achievable by using discrepancy; see Exercise 5.7.

5.8. Exercises

Exercise 5.1 (Compute clustering radius). Let C and P be two given sets of points in the plane, such that $k = |C|$ and $n = |P|$. Let $r = \max_{p \in \mathsf{P}} \min_{c \in C} \|c - p\|$ be the ***covering radius*** of P by C (i.e., if we place a disk of radius r centered at each point of C, all those disks cover the points of P).

(A) Give an $O(n + k \log n)$ expected time algorithm that outputs a number α, such that $r \leq \alpha \leq 10r$.

(B) For $\varepsilon > 0$ a prescribed parameter, give an $O(n + k\varepsilon^{-2} \log n)$ expected time algorithm that outputs a number α, such that $r \leq \alpha \leq (1+\varepsilon)r$.

Exercise 5.2 (Some calculus required). Prove Lemma 5.13.

Exercise 5.3 (A direct proof of the ε-sample theorem). For the case that the given range space is finite, one can prove the ε-sample theorem (Theorem 5.26_{p71}) directly. So, we are given a range space $\mathsf{S} = (\mathsf{x}, \mathcal{R})$ with VC dimension δ, where x is a finite set.

(A) Show that there exists an ε-sample of S of size $O\left(\delta\varepsilon^{-2} \log \frac{\log|\mathsf{x}|}{\varepsilon}\right)$ by extracting an $\varepsilon/3$-sample from an $\varepsilon/9$-sample of the original space (i.e., apply Lemma 5.30 twice and use Lemma 5.33).

(B) Show that for any k, there exists an ε-sample of S of size $O\left(\delta\varepsilon^{-2} \log \frac{\log^{(k)}|\mathsf{x}|}{\varepsilon}\right)$.

(C) Show that there exists an ε-sample of S of size $O\left(\delta\varepsilon^{-2} \log \frac{1}{\varepsilon}\right)$.

Exercise 5.4 (Sauer's lemma is tight)**.** Show that Sauer's lemma (Lemma 5.9) is tight. Specifically, provide a finite range space that has the number of ranges as claimed by Lemma 5.9.

Exercise 5.5 (Flip and flop)**.** (A) Let $b_1, \ldots, b_{2m}$ be m binary bits. Let Ψ be the set of all permutations of $1, \ldots, 2m$, such that for any $\sigma \in \Psi$, we have $\sigma(i) = i$ or $\sigma(i) = m + i$, for $1 \le i \le m$, and similarly, $\sigma(m + i) = i$ or $\sigma(m + i) = m + i$. Namely, $\sigma \in \Psi$ either leaves the pair $i, i + m$ in their positions or it exchanges them, for $1 \le i \le m$. As such $|\Psi| = 2^m$.

Prove that for a random $\sigma \in \Psi$, we have

$$\mathbf{Pr}\left[\left|\frac{\sum_{i=1}^{m} b_{\sigma(i)}}{m} - \frac{\sum_{i=1}^{m} b_{\sigma(i+m)}}{m}\right| \ge \varepsilon\right] \le 2e^{-\varepsilon^2 m/2}.$$

(B) Let Ψ' be the set of all permutations of $1, \ldots, 2m$. Prove that for a random $\sigma \in \Psi'$, we have

$$\mathbf{Pr}\left[\left|\frac{\sum_{i=1}^{m} b_{\sigma(i)}}{m} - \frac{\sum_{i=1}^{m} b_{\sigma(i+m)}}{m}\right| \ge \varepsilon\right] \le 2e^{-C\varepsilon^2 m/2},$$

where C is an appropriate constant. [**Hint:** Use (A), but be careful.]

(C) Prove Theorem 5.26 using (B).

Exercise 5.6 (Sketch and merge)**.** Assume that you are given a deterministic algorithm that can compute the discrepancy of Theorem 5.31 in $O(nm)$ time, where n is the size of the ground set and m is the number of induced ranges. We are assuming that the VC dimension δ of the given range space is small and that the algorithm input is only the ground set X (i.e., the algorithm can figure out on its own what the relevant ranges are).

(A) For a prespecified $\varepsilon > 0$, using the ideas described in Section 5.4.1.1, show how to compute a small ε-sample of X quickly. The running time of your algorithm should be (roughly) $O\left(n/\varepsilon^{O(\delta)} \text{polylog}\right)$. What is the exact bound on the running time of your algorithm?

(B) One can slightly improve the running of the above algorithm by more aggressively sketching the sets used. That is, one can add additional sketch layers in the tree. Show how by using such an approach one can improve the running time of the above algorithm by a logarithmic factor.

Exercise 5.7 (Building relative approximations)**.** Prove the following theorem using discrepancy.

Theorem 5.42. *Let $(\mathsf{X}, \mathcal{R})$ be a range space with shattering dimension δ, where $|\mathsf{X}| = n$, and let $0 < \varepsilon < 1$ and $0 < p < 1$ be given parameters. Then one can construct a set $N \subseteq \mathsf{X}$ of size $O\left(\frac{\delta}{\varepsilon^2 p} \ln \frac{\delta}{\varepsilon p}\right)$, such that, for each range $\mathbf{r} \in \mathcal{R}$ of at least pn points, we have*

$$\left|\frac{|\mathbf{r} \cap N|}{|N|} - \frac{|\mathbf{r} \cap \mathsf{X}|}{|\mathsf{X}|}\right| \le \varepsilon \frac{|\mathbf{r} \cap \mathsf{X}|}{|\mathsf{X}|}.$$

In other words, N is a relative (p, ε)-approximation for $(\mathsf{X}, \mathcal{R})$.

CHAPTER 6

Approximation via Reweighting

In this chapter, we will introduce a powerful technique for "structure" approximation. The basic idea is to perform a search by assigning elements weights and picking the elements according to their weights. The element's weight indicates its importance. By repeatedly picking elements according to their weights and updating the weights of objects that are being neglected (i.e., they are more important than the current weights indicate), we end up with a structure that has some desired properties.

We will demonstrate this technique for two problems. In the first problem, we will compute a spanning tree of points that has low stabbing number. In the second problem, we will show how the set cover problem can be approximated efficiently in geometric settings, yielding a better bound than the general approximation algorithm for this problem.

6.1. Preliminaries

In this section, we describe how to implement efficiently some low-level operations required by the algorithms described in this chapter. The reader uninterested in such low-level details can safely skip to Section 6.2.

The following algorithms assign and manipulate weights associated with various entities. There are two main technical problems: (i) elements might have weights that are very large (i.e., require n bits) and (ii) we need to perform (efficiently) random sampling from a set of elements that are weighted using such numbers.

6.1.1. Handling large weights. In the following, some of the weights we have to handle are of the form 2^i, where $i \geq 0$ is some integer. Such numbers can easily be represented efficiently by storing only the index i. In our algorithm i is polynomial in n.

We will also need to handle numbers that are the sum of m such numbers, where m is usually a polynomial of the input size. Naturally, if we care only about polynomial running time, we could just use exact arithmetic to handle these numbers. Alternatively, we can approximately compute such numbers. So, consider such a number $x = \sum_{i=1}^{m} 2^{b_i}$, where $b_1, \ldots, b_m$ are non-negative integers. If the powers are sufficiently small (say all the b_is are at most 16382), we can just use floating-point numbers directly (here, you would have to use the `long double` floating-point number type). Otherwise, one needs to implement a variant of floating-point numbers that can handle large powers. This can easily be done by explicitly storing the exponent part of the floating-point number. Naturally, we need the implementation of only a few specific operations on these numbers, and as such the implementation does not need to be exhaustive. It is easy to verify that one can implement all the required operations so that they take constant time.

The error introduced by this extended floating-point representation is pretty small in our applications. Indeed, all the operations the algorithms below do are addition, multiplication, and division (but no subtraction). As such, the error caused by the representation is too minuscule for us to worry about, and we will ignore it.

6.1.2. Random sampling from a weighted set. We need to do r independent draws from a weighted set N having (say n) elements, where each weight is a large real number (as described above). In particular, for an element $x \in N$, let $\omega(x)$ denote its weight. Assume that we are given a function that can randomly and uniformly pick an integer number in a range $[1, M]$, where M is a parameter given to the function.

One way to do this sampling is to compute the element x of N having the maximum weight; that is, $\omega(x) = \max_{y \in N} \omega(y)$. Observe that all the elements of weight $\leq \omega(x)/n^{10}$ have weight which is so tiny that they can be ignored. Thus, normalize all the weights by dividing them by $2^{\lfloor \lg \omega(x)/n^{10} \rfloor}$, and remove all the elements with weights smaller than 1. For an element $z \in N$, let $\widehat{\omega}(z)$ denote its normalized weight. Clearly, all the normalized weights are integers in the range $1, \ldots, 2n^{10}$.

Thus, we now have to pick elements from a set with (relatively small) integer weights. Place the elements in an array, and compute the prefix sum array of their weights. That is, $\alpha_k = \sum_{i=1}^{k} \widehat{\omega}(z_i)$, for $k = 0, \ldots, n$ (as such, $\alpha_0 = 0$). Next, pick a random (integer) number γ uniformly in the range $[1, \alpha_n]$, and using a binary search, find the j, such that $\alpha_{j-1} < \gamma \leq \alpha_j$. This picks the element z_j to be in the random sample. This requires $O(n)$ preprocessing, but then a single random sample can be done in $O(\log n)$ time. We need to perform r independent samples. Thus, this takes $O(n + r \log n)$ time overall.

Corollary 6.1. *Given a set N of n weighted elements, one can preprocess it in linear time, such that one can randomly pick an element of N uniformly at random (according to the weights of the elements of N) in $O(\log n)$ time. In particular, picking r elements (with replacement) from N can be done in $O(n + r \log n)$ time.*

6.2. Computing a spanning tree with low crossing number

The ***crossing number*** of a set of segments in the plane is the maximum number of segments that can be stabbed by a single line. For a tree $\mathcal{T}$ drawn in the plane, it is the maximum number of intersections of a line with $\mathcal{T}$.

Let P be a set of n points in the plane. We would like to compute a tree $\mathcal{T}$ that spans the points of P such that every line in the plane crosses the edges of the tree at most $O\left(\sqrt{n}\right)$ times. If the points are the $\sqrt{n} \times \sqrt{n}$ grid, this easily holds if we pick any spanning tree that connects only points that are adjacent on the grid. It is not hard to show that one cannot do better (see Exercise 6.2).

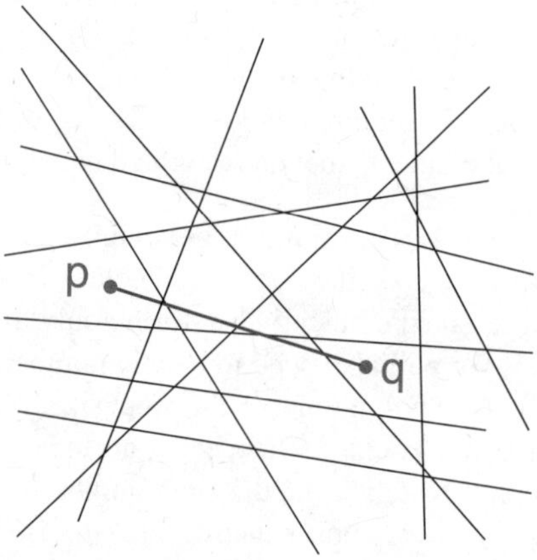

FIGURE 6.1. Here $\mathbf{d}_{\times}(\mathsf{p}, \mathsf{q}) = 5$.

Definition 6.2. Given a weighted set of lines L, the ***crossing distance*** $\mathbf{d}_{\times}(\mathsf{p}, \mathsf{q})$ between two points p and q is the minimum weight of the lines one has to cross (i.e., cut) to get from p to q. The function $\mathbf{d}_{\times}(\mathsf{p}, \mathsf{q})$ complies with the triangle inequality, and it is a ***pseudo-metric*** (since distinct points can have distance zero between them).

Because of the triangle inequality, the crossing distance $\mathbf{d}_{\times}(\mathsf{p}, \mathsf{q})$ is equal to the total weight of the lines intersecting the close segment pq; see Figure 6.1.

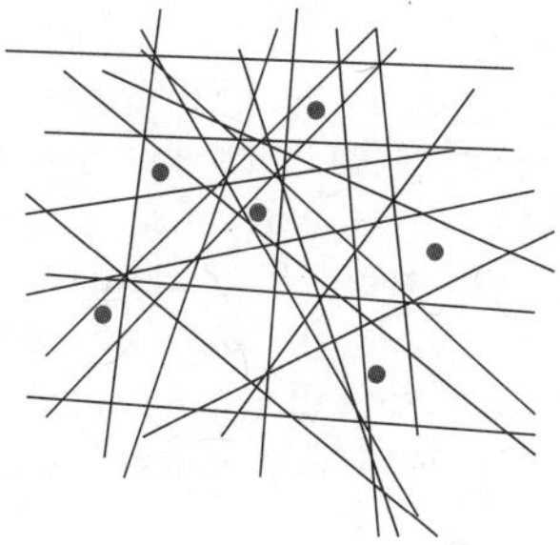

6.2.1. The algorithm. So, consider the set of all separating lines $\widehat{\mathsf{L}}$ of P. Formally, we will consider two lines ℓ and ℓ' to be equivalent if the closed halfplane above ℓ contains the same set of points as the closed halfplane above ℓ' (we assume no two points in P have the same x coordinate, and as such we can ignore vertical lines). For each equivalent class of $\widehat{\mathsf{L}}$ pick one representative line into a set L; see the figure on the right. Clearly, given such a line, we can translate and rotate it till it passes through two points of P or till it is horizontal and passes through one point of P. If it passes through two points, we need to specify for these two points whether they belong to the set defined. As such, there are at most $4\binom{n}{2} + n + 1 = O(n^2)$ different lines in L.

Idea. The algorithm would build the spanning tree by adding the edges one by one. So, consider a candidate edge (i.e., segment) qt, and consider a line $\ell \in \mathsf{L}$. If ℓ already intersects k edges of E_i and k is large (i.e., $\Omega(\sqrt{n})$), then we would like to discourage the usage of qt in the spanning tree. The problem is that the desirability of an edge is determined by the lines that intersect the edge. Furthermore, a line that is already crossing many edges might be close to the limit of what it is allowed to cross, and as such this line would "prefer" not to cross any more edges. Intuitively, as the load (i.e., number of edges it crosses) of a line gets higher, edges that cross this line becomes less desirable.

Algorithm. The input is the set P of n points in the plane. Let $E_0 = \emptyset$, and for $i > 0$ let E_i be the set of edges added by the algorithm by the end of the ith iteration. The weight $\omega(\ell)$ of a line $\ell \in \mathsf{L}$ is initialized to 1. The ***weight*** of the line at the beginning of the ith iteration would be denoted by $\omega_{i-1}(\ell) = 2^{n_{i-1}(\ell)}$, where

$$n_{i-1}(\ell) = \left| \left\{ s \in E_{i-1} \;\middle|\; s \cap \ell \neq \emptyset \right\} \right|$$

is the number of segments of E_{i-1} that intersect ℓ. The ***weight*** of a segment s, in the beginning of the ith iteration, is

$$\omega_{i-1}(s) = \sum_{\ell \in \mathsf{L}, \ell \cap s \neq \emptyset} \omega_{i-1}(\ell).$$

Specifically, the weight of $\omega_{i-1}(s)$ is the total weight of the lines intersecting s. Clearly, the heavier a segment is the less desirable it is to be used for the spanning tree. Motivated by this, we would always pick an edge qt such that $\mathsf{q}, \mathsf{t} \in \mathsf{P}$ belong to two different connected components of the forest induced by E_{i-1}, and its weight is minimal among all such edges. We repeat this process till we end up with a spanning tree of P. To simplify the implementation of the algorithm, when adding s to the set of edges in the forest, we also remove one its endpoints from P (i.e., every connected component of this forest has a single representative point). Thus, the algorithm terminates when P has a single point in it.

6.2.2. Analysis.

6.2.2.1. *Proof of correctness.* We claim that the resulting spanning tree has the required properties. The algorithm performs $n - 1$ iterations and as such, the largest weight used is $\leq 2^n$, and such integer numbers can be manipulated in polynomial time. Thus, overall the running time of the algorithm is polynomial.

For a point $\mathsf{q} \in \mathbb{R}^2$ and a set of lines L, consider the set of all the vertices of the arrangement of $\mathcal{A} = \mathcal{A}(\mathsf{L})$ that are in crossing distance at most r from q. We will refer to this set of vertices of the arrangement, denoted by $\mathbf{b}_{\rtimes}(\mathsf{q}, r)$, as the ***ball*** of radius r under the crossing metric. Such a ball under the crossing distance is depicted on the right. It is star shaped but not necessarily convex.

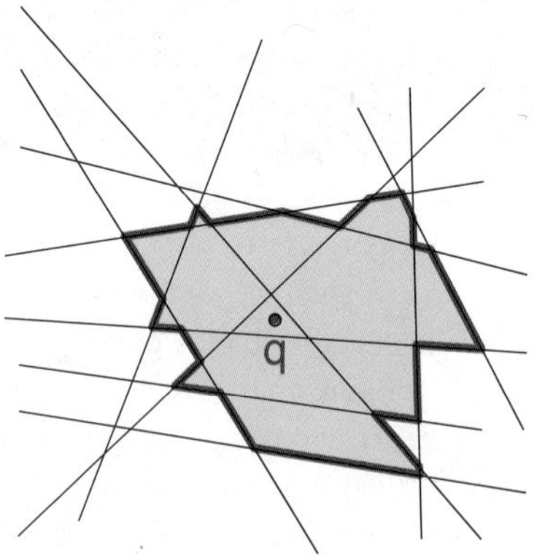

Lemma 6.3. *Let L be a set of n lines in the plane, and let $\mathsf{q} \in \mathbb{R}^2$ be a point (not lying on any of the lines of L). Then, for any $r \leq n/2$, we have that $|\mathbf{b}_{\rtimes}(\mathsf{q}, r)| \geq r^2/8$.*

Proof. Indeed, one can shoot a ray ζ from q that intersects at least $n/2$ lines of L. Let $\ell_1, \ldots, \ell_{r/2}$ be the first $r/2$ lines hit by the ray ζ, and let $\mathsf{t}_1, \ldots, \mathsf{t}_{r/2}$ be the respective intersection points between these lines and ζ. Now, mark all the intersection points of the arrangement $\mathcal{A}(\mathsf{L})$ along the line ℓ_i that are in distance at most $r/2$ from t_i, for $i = 1, \ldots, r/2$.

The picture on the right depict one such vertex being collected in this way. Clearly, we overall marked at least $(r/2)(r/2)/2$ vertices of the arrangement, since we marked (at least) $r/2$ vertices along each of the lines $\ell_1, \ldots, \ell_{r/2}$. Furthermore, each vertex can be counted in this way at most twice. Now, observe that all these vertices are in distance at most $r/2 + r/2$ from q because of the triangle inequality, implying the claim. ■

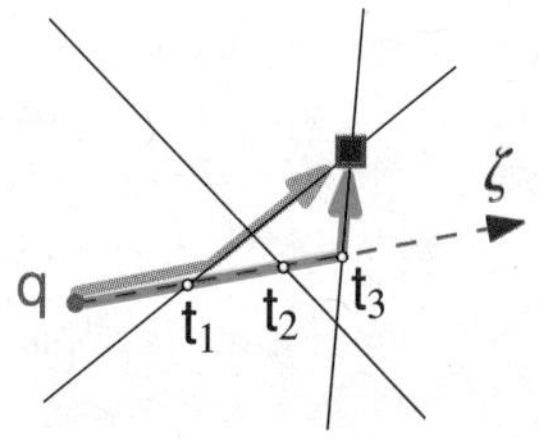

Lemma 6.4. *Let P be a set of n points in the plane, and let L be a set of lines in the plane with total weight W. One can always find a pair of points q and t in P, such that the total weight of the segment $s = \mathsf{qt}$ (i.e., the total weight of the lines of L intersecting s) is at most $4W/\sqrt{n} + 3 \leq cW/\sqrt{n}$, for some constant c.*

Proof. First, since the weights considered are always integers, we can consider all the weights to be 1 by replacing a line ℓ of weight $\omega(\ell)$ by $\omega(\ell)$ copies of it. Perturb slightly the lines, so that there is no pair of them which is parallel.

Next, consider the set of vertices $X(r) = \bigcup_{\mathsf{p} \in \mathsf{P}} \mathbf{b}_{\rtimes}(\mathsf{p}, r)$. Clearly, as long the balls of $X(r)$ are disjoint, the number of vertices of the arrangement $\mathcal{A}$ included in $X(r)$ is at least $nr^2/8$, by Lemma 6.3. In particular, the overall number of vertices in the arrangement is $\binom{W}{2}$, and as such it must be true that two balls of $X(r)$ are not disjoint when $nr^2/8 > \binom{W}{2} = W(W-1)/2$. Namely, this happens when $r^2 > 4W^2/n$. This happens when $r > 2W/\sqrt{n}$. As such, for $r = \lceil 2W/\sqrt{n} \rceil + 1$ there must be a vertex v in the arrangement $\mathcal{A}$ and two points $\mathsf{q}, \mathsf{t} \in \mathsf{P}$, such that $\mathbf{d}_{\rtimes}(\mathsf{q}, v) \leq r$ and $\mathbf{d}_{\rtimes}(v, \mathsf{t}) \leq r$, and by the triangle inequality, we have that

$$\mathbf{d}_{\rtimes}(\mathsf{q}, \mathsf{t}) \leq \mathbf{d}_{\rtimes}(\mathsf{q}, v) + \mathbf{d}_{\rtimes}(v, \mathsf{t}) \leq 2r \leq 4W/\sqrt{n} + 3.$$

Namely, q and t are within the required crossing distance. ■

Claim 6.5. *Any line in the plane crosses at most $O(\sqrt{n})$ edges of the resulting spanning tree $\mathcal{T}$.*

Proof. Let W_i denote the total weight of the lines in L in the end of the ith iteration. We have that $W_0 = |\mathsf{L}| \leq 6\binom{n}{2}$, and since there are $n_i = n - i + 1$ points in P in the beginning of the ith iteration, it follows, by Lemma 6.4, that the algorithm found a segment s_i of weight at most $cW_{i-1}/\sqrt{n_i}$. We double the weight of all the lines that intersect s_i. Thus,

$$W_i \le W_{i-1} + cW_{i-1}/\sqrt{n_i} \le \left(1 + \frac{c}{\sqrt{n_i}}\right) W_{i-1} \le \prod_{k=1}^{i} \left(1 + \frac{c}{\sqrt{n_k}}\right) W_0$$

$$\le W_0 \prod_{k=1}^{i} \exp\left(\frac{c}{\sqrt{n_k}}\right) = W_0 \exp\left(\sum_{k=1}^{i} \frac{c}{\sqrt{n-k+1}}\right),$$

since $1 + x \le e^x$, for all $x \ge 0$. In particular, we have that

$$W_n \le W_0 \exp\left(\sum_{k=1}^{n} \frac{c}{\sqrt{n-k+1}}\right) \le 6\binom{n}{2} \exp\left(\sum_{k=1}^{n} \frac{c}{\sqrt{k}}\right) \le 3n^2 \exp\left(4c\sqrt{n}\right),$$

since $\sum_{k=1}^{n} 1/\sqrt{k} \le 1 + \int_{x=1}^{n} (1/\sqrt{x})dx = 1 + \left[2\sqrt{x}\right]_{x=1}^{x=n+1} \le 4\sqrt{n}$. As for the other direction, consider the heaviest line $\hbar$ in L in the end of the execution of the algorithm. If it crosses Δ segments $\mathcal{T}$, then its weight is 2^Δ, and as such

$$2^\Delta = \omega(\hbar) \le W_n \le 3n^2 \exp\left(4c\sqrt{n}\right).$$

It follows that $\Delta = O\left(\log n + \sqrt{n}\right)$, as required. Namely, any line in the plane crosses at most $O\left(\sqrt{n}\right)$ edges of $\mathcal{T}$. ■

Theorem 6.6. *Given a set* P *of* n *points in the plane, one can compute a spanning tree* $\mathcal{T}$ *of* P *such that each line crosses at most* $O\left(\sqrt{n}\right)$ *edges of* $\mathcal{T}$. *The running time is polynomial in* n.

This result also holds in higher dimensions. The proof is left as an exercise (see Exercise 6.1).

Theorem 6.7. *Given a set* P *of* n *points in* $\mathbb{R}^d$, *one can compute a spanning tree* $\mathcal{T}$ *of* P *such that each line crosses at most* $O(n^{1-1/d})$ *edges of* $\mathcal{T}$. *The running time is polynomial in* n.

6.2.3. An application – better discrepancy. Before extending this result to more abstract settings, let us quickly outline why this spanning tree of low crossing number leads to better discrepancy and a smaller ε-sample for halfplanes.

Indeed, let us turn $\mathcal{T}$ into a cycle by drawing a tour walking around the edges of $\mathcal{T}$; formally, we double every edge of $\mathcal{T}$ and observe that the resulting graph is Eulerian, and we extract the Eulerian tour from this graph. Clearly, this cycle C has twice the crossing number of the tree.

Next, consider a curve γ in the plane and a segment s with the same endpoints, and observe that a line that intersects s must also intersect γ (but not vice versa!). As such, if we shortcut parts of C by replacing them by straight segments, we are only decreasing the crossing number of C. To this end, start traversing C from some arbitrary point $\mathsf{p}_0 \in \mathsf{P}$, and start "walking" along C. Whenever we are at point $\mathsf{p} \in \mathsf{P}$, along the cycle C, we go directly from there (i.e., shortcut) to the next point visited by C that was not visited yet. Let C' be the resulting cycle. Clearly, it visits all the points of P and it has a crossing number which is at most twice the crossing number of $\mathcal{T}$. Now, assuming that $n = |\mathsf{P}|$ is even, pick all the even edges of C' so they form a prefect matching $\mathcal{M}$ of P. We have:

Lemma 6.8. *One can compute a perfect matching* $\mathcal{M}$ *of a set of* n *points in the plane, such that every line crosses at most* $O\left(\sqrt{n}\right)$ *edges of the matching.*

Now, going back to the discrepancy question, we remind the reader that we would like to color the points by $\{-1, 1\}$ such that for any halfplane the 'balance' of the coloring is as close to perfect as possible. To this end, we use the matching of the above lemma and plug it into Theorem 5.31_{p76}. Since any line ℓ crosses at most $\#_\ell = O(\sqrt{n})$ edges of $\mathcal{M}$, we get the following result.

Theorem 6.9. *Let* P *be a set of n points in the plane. One can compute a coloring* χ *of* P *by* $\{-1, 1\}$*, such that for all halfplanes h, it holds that* $|\chi(h)| = O\left(n^{1/4}\sqrt{\ln n}\right)$.

In words, the discrepancy of n points in relation to halfplanes is $O\left(n^{1/4}\sqrt{\ln n}\right)$. This also implies that one can construct a better ε-sample in this case. But before dwelling on this, let us prove a more general version of the spanning tree lemma.

6.2.4. Spanning tree for space with bounded shattering dimension. Let $\mathsf{S} = (\mathsf{X}, \mathcal{R})$ be a range space with shattering dimension δ and dual shattering dimension $\delta^\star$ (see Definition 5.17_{p68}). Let $\mathsf{P} \subseteq \mathsf{X}$ be a set of n points. Consider a spanning tree $\mathcal{T}$ of P. The tree $\mathcal{T}$ is defined by $n-1$ edges $e_i = \{\mathsf{p}_i, \mathsf{q}_i\} \subseteq \mathsf{P}$. An edge $\{\mathsf{p}, \mathsf{q}\}$ ***crosses*** a range $\mathbf{r} \in \mathcal{R}$ if $|\{\mathsf{p}, \mathsf{q}\} \cap \mathbf{r}| = 1$. Our purpose is to build a spanning tree such that every range of $\mathcal{R}$ crosses a small number of edges of $\mathcal{T}$. The reader can verify that this abstract setting corresponds to the more concrete problem described above.

We will concentrate on the restricted space $\mathsf{S}_{|\mathsf{P}} = (\mathsf{P}, \mathcal{R}_{|\mathsf{P}})$. It is easy to verify that $\mathsf{S}_{|\mathsf{P}}$ has shuttering dimension $\leq \delta$ and dual shattering dimension bounded by $\delta^\star$. Observe that $m = \left|\mathcal{R}_{|\mathsf{P}}\right| \leq O\left(n^\delta\right)$. Let $\mathcal{F}$ be a weighted subset of $\mathcal{R}_{|\mathsf{P}}$.

6.2.4.1. *The algorithm.* The algorithm to compute a spanning tree with low crossing number would work as before: Initialize the weight of all the ranges of $\mathcal{R}_{|\mathsf{P}}$ to 1 and the spanning tree to be empty.

Now, at each iteration, the algorithm computes a pair $\{\mathsf{p}, \mathsf{q}\} \subseteq \mathsf{P}$ such that the total weight of the ranges of $\mathcal{F}$ it crosses is minimized. Next, the algorithm doubles the weight of these ranges, adds the edge $\{\mathsf{p}, \mathsf{q}\}$ to the spanning tree, and deletes, say, q from P. The algorithm repeats this process till there remains only a single point in P. Clearly, we have computed a spanning tree.

6.2.4.2. *Analysis.* We need to bound the crossing number of the generated tree. As before, the analysis boils down to proving the existence of two "close" points.

Lemma 6.10. *Let* $\mathsf{S} = (\mathsf{X}, \mathcal{R})$ *be a range space with dual shattering dimension* $\delta^\star$. *Let* $\mathsf{P} \subseteq X$ *be a set of n points, and let* $\mathcal{F}$ *be a weighted set of ranges from* $\mathcal{R}_{|\mathsf{P}}$*, with total weight W. Then, there is a pair of points* $\mathsf{e} = \{\mathsf{p}, \mathsf{q}\} \subseteq \mathsf{P}$*, such that the total weight of the ranges crossed by* e *is at most* $O\left(W\delta^\star n^{-1/\delta^\star} \log n\right)$.

PROOF. Let ε be a parameter to be specified shortly. For an edge $\{u, v\} \subseteq \mathsf{P}$, consider the set of ranges that its crosses:

$$C(u, v) = \left\{ \mathbf{r} \;\middle|\; (u \in \mathbf{r} \text{ and } v \notin \mathbf{r}) \quad \text{or} \quad (u \notin \mathbf{r} \text{ and } v \in \mathbf{r}), \text{ for } \mathbf{r} \in \mathcal{F} \right\}.$$

Observe that the range space $\mathsf{U} = (\mathsf{P}, \mathcal{F})$ has dual shattering dimension $\delta^\star$. Next, consider the dual range space $\mathsf{U}^\star = (\mathcal{F}, \mathsf{P}^\star)$; see Definition 5.16_{p67}. This range space has (primal) shattering dimension bounded by $\delta^\star$, by assumption. Consider the new range space

$$\mathsf{T}^\star = \left(\mathcal{F}, \left\{ \mathcal{F}_\mathsf{p} \oplus \mathcal{F}_\mathsf{q} \;\middle|\; \mathcal{F}_\mathsf{p}, \mathcal{F}_\mathsf{q} \in \mathsf{P}^\star \right\} \right),$$

where $\oplus$ is the symmetric difference of the two sets; namely, $\mathbf{r} \oplus \mathbf{r}' = (\mathbf{r} \setminus \mathbf{r}') \cup (\mathbf{r}' \setminus \mathbf{r})$, and $\mathcal{F}_{\mathsf{p}} = \left\{ \mathbf{r} \,\middle|\, \mathbf{r} \in \mathcal{F} \text{ and } \mathsf{p} \in \mathbf{r} \right\}$. By arguing as in Corollary 5.24_{p70}, we have that $\mathsf{T}^\star$ has a shattering dimension at most $2\delta^\star$. Furthermore, the projected range space $\mathsf{T}^\star$ has $C(u, v)$ as one of its ranges, for any $u, v \in \mathsf{P}$.

So consider a random sample N of size $O((\delta^\star/\varepsilon)\log(\delta^\star/\varepsilon))$ from $\mathcal{F}$ (note that $\mathcal{F}$ is weighted and the random sampling is done accordingly). By the ε-net theorem (see Theorem 5.28_{p71} and Remark 5.29_{p71}), we know that with constant probability, this is an ε-net of $\mathsf{T}^\star$. Namely, a set $C(u, v)$ which does not contain any range of N has weight at most εW (indeed, if the weight of $C(u, v)$ exceeds εW, then it must contain an element of the ε-net N).

On the other hand, the projection of the range space $\mathsf{U}^\star = (\mathcal{F}, \mathsf{P}^\star)$ to N is the range space $\mathsf{U}^\star_{|N} = \left(N, \mathsf{P}^\star_{|N}\right)$. This range space has shattering dimension at most $\delta^\star$, and as such

$$\mu(\varepsilon) = \left|\mathsf{P}^\star_{|N}\right| = O\!\left(|N|^{\delta^\star}\right) \le \left(c\frac{\delta^\star}{\varepsilon}\log\frac{\delta^\star}{\varepsilon}\right)^{\delta^\star}$$

ranges[1], where c is an appropriate constant. In particular, let us pick ε as small as possible, such that $\mu(\varepsilon) < n = |\mathsf{P}|$. We are then guaranteed that there are two points $\mathsf{p}, \mathsf{q} \in \mathsf{P}$ such that $N_{\mathsf{p}} = N_{\mathsf{q}}$.

Let us regroup. We found two points p and q that do not cross any range of N. But N is an ε-net for $\mathsf{T}^\star$ (with constant probability, so assume it is a net). This implies that $C(\mathsf{p}, \mathsf{q})$ has total weight at most εW. Namely, the total weight of the ranges crossing the edge $\{\mathsf{p}, \mathsf{q}\}$ is at most εW. As such, we found a "light" edge. Note that we can verify the weight of $\{\mathsf{p}, \mathsf{q}\}$ by computing explicitly all the ranges of $\mathcal{F}$ that cross it. If it is not light, we can repeat this algorithm till we succeed.

We are left with the task of picking ε. Naturally, we would like ε to be as small as possible. Now, for $\varepsilon = c_1\delta^\star n^{-1/\delta^\star}\log(n)$ it holds that $\mu(\varepsilon) < n$ if c_1 is a sufficiently large constant, as can be easily verified. ■

To complete the argument, we need to bound the total weight of the ranges in the end of this process. For some constant c_2, it is bounded by

$$U \le \left|\mathcal{R}_{|\mathsf{P}}\right| \prod_{i=1}^{n}\left(1 + c_1\frac{\log i}{i^{1/\delta^\star}}\right) \le c_2 n^\delta \prod_{i=1}^{n}\left(1 + c_1\frac{\log i}{i^{1/\delta^\star}}\right) \le c_2 n^\delta \exp\left(\sum_i c_1\frac{\log i}{i^{1/\delta^\star}}\right)$$
$$\le c_2 n^\delta \exp\left(O\left(c_1 n^{1-1/\delta^\star}\log n\right)\right).$$

Now, the crossing number of the resulting tree $\mathcal{T}$, for any range $\mathbf{r} \in \mathcal{R}$, is bounded by $\lg(U) = O\left(\delta\log n + n^{1-1/\delta^\star}\log n\right)$. We thus conclude:

Theorem 6.11. *Given a range space $\mathsf{S} = (\mathsf{X}, \mathcal{R})$ with shattering dimension δ and dual shattering dimension $\delta^\star$ and a set $\mathsf{P} \subseteq \mathsf{X}$ of n points, one can compute, in polynomial time (assuming that δ and $\delta^\star$ are constants), a spanning tree $\mathcal{T}$ of P, such that any range of $\mathcal{R}$ crosses at most $O\left(\delta\log n + n^{1-1/\delta^\star}\log n\right)$ edges of $\mathcal{T}$.*

Plugging this into the discrepancy machinery, we get the following result.

Theorem 6.12. *Given a range space $\mathsf{S} = (\mathsf{X}, \mathcal{R})$ with shattering dimension δ and dual shattering dimension $\delta^\star$ and a set $\mathsf{P} \subseteq \mathsf{X}$ of n points, one can compute, in polynomial time*

[1] If the ranges of $\mathcal{R}$ are geometric shapes, it means that the arrangement formed by the shapes of N has at most $\mu(\varepsilon)$ faces.

(assuming that δ and $\delta^\star$ are constants), a coloring of P *by* $\{-1, 1\}$, *such that for any range* $\mathbf{r} \in \mathcal{R}$, *we have that* $|\chi(\mathbf{r})| = O\left(\delta n^{1/2-1/2\delta^\star} \log n\right)$.

Proof. Indeed, compute a spanning tree with low crossing number using the algorithm of Theorem 6.11. Next, extract a matching from it with low crossing number, and use this matching in computing a good discrepancy; see Theorem 5.31p76. ■

Now, we can extract a small ε-sample from such a range space by using the construction of an ε-sample via discrepancy. We get the following result.

Theorem 6.13. *Given a range space* $\mathsf{S} = (\mathsf{X}, \mathcal{R})$ *with shattering dimension* δ *and dual shattering dimension* $\delta^\star$, *a set* $\mathsf{P} \subseteq \mathsf{X}$ *of* n *points, and* $\varepsilon > 0$, *one can compute, in polynomial time (assuming that* δ *and* $\delta^\star$ *are constants), an* ε*-sample for the pointset* P *of size* $O\left(\left((\delta/\varepsilon)\log(\delta/\varepsilon)\right)^{2-2/(\delta^\star+1)}\right)$.

Proof. We plug the improved bound of Theorem 6.12 into Theorem 5.34p77. Clearly, $\tau(n)$ (which is how good of a sample the points colored by $+1$ are in one round of coloring) in our case is

$$\tau(n) = \frac{\operatorname{disc}(\mathsf{S}_{|\mathsf{P}})}{|n|} = O\left(\delta n^{-(\delta^\star+1)/2\delta^\star} \log n\right).$$

We repeat this halving process, and the required ε-sample is the smallest $m = n/2^i$, such that $c_3\tau(m) < \varepsilon$, for some constant c_3. Namely, it is sufficient that, for some constant c_4, we have that

$$c_4 \delta m^{-(\delta^\star+1)/2\delta^\star} \log m < \varepsilon.$$

Now, since $2\delta^\star/(\delta^\star + 1) = 2 - 2/(\delta^\star + 1)$, it is now easy to verify that this holds for $m = O\left(\left(\frac{\delta}{\varepsilon}\log\frac{\delta}{\varepsilon}\right)^{2-2/(\delta^\star+1)}\right)$. ■

Theorem 6.13 is a surprising result. It implies that one can construct ε-samples of size with subquadratic dependency of ε. Note that a regular random sample cannot provide such a guarantee even for a single range! Namely, we have broken the glass ceiling of what can be achieved by random sampling.

6.3. Geometric set cover

Let $\mathsf{S} = (\mathsf{X}, \mathcal{R})$ be a range space with bounded VC dimension. For example, let X be a set of n points in the plane, and let $\mathcal{R}$ be a set of m allowable disks. The question is to find the minimal number of disks of $\mathcal{R}$ one needs to cover the points of X.

This is an instance of set cover, and it is in general NP-Hard, and the general version is NP-Hard to approximate within a factor of $\Omega(\log n)$. There is an easy greedy algorithm which repeatedly picks the set (i.e., disk) that covers the largest number of points not covered yet. It is easy to show that his algorithm would cover the points using $O(k \log n)$ sets, where k is the number of sets used by the optimal solution.

The algorithm. Interestingly, one can do much better if the set system S has bounded dual shattering dimension $\delta^\star$. Indeed, let us assign weight 1 to each range of $\mathcal{R}$ and pick a random subset R of size $O((\delta^\star/\varepsilon)\log(\delta^\star/\varepsilon))$ from $\mathcal{R}$, where $\varepsilon = 1/4k$ (the sample is done according to the weights). If the sets of R cover all the points of X, then we are done. Otherwise, consider a point $\mathsf{p} \in \mathsf{X}$ which is not covered by R. If the total weight of $\mathcal{R}_\mathsf{p}$ (i.e., the set of ranges covering p) is smaller than $\varepsilon W(\mathcal{R})$, then we double the weight of all the ranges in $\mathcal{R}_\mathsf{p}$, where $W(\mathcal{R}) = \sum_{\mathbf{r}\in\mathcal{R}} \omega(\mathbf{r})$. In any case, even if doubling is not carried out, we repeat this process till it succeeds.

Details and intuition. In the above algorithm, if a random sample fails (i.e., there is an uncovered point), then one of the ranges that covers p must be in the optimal solution. In particular, by increasing the weight of the ranges covering p, we improve the probability that p would be covered in the next iteration. Furthermore, with good probability, the sample is an ε-net, and as such the algorithm doubles the weight of a "few" ranges. One of these few ranges must be in the optimal solution. As such, the weight of the optimal solution grows exponentially in a faster rate than the total weight of the universe, implying that at some point the algorithm must terminate, as the weight of the optimal solution exceeds the total weight of all the ranges, which is of course impossible.

6.3.1. Proof of correctness. Clearly, if the algorithm terminates, then it found the desired cover. As such, in the following, we bound the number of iterations performed by the algorithm. As before, let $W_0 = m$ be the initial weight of the ranges, and W_i would be the weight in the end of the ith iteration. We consider an iteration to be ***successful*** if the doubling stage is being performed in the iteration. Since an iteration is successful if the sample is an ε-net, and by Theorem $5.28_{p71}$the probability for that is at least, say, $1/2$, it follows that we need to bound only the number of successful iterations (indeed, it would be easy to verify that with high probability, using the Chernoff inequality, the number of successful iterations is at least a quarter of all iterations performed).

As before, we know that $W_i \le (1+\varepsilon)W_{i-1} = (1+\varepsilon)^i m \le m\exp(\varepsilon i)$. On the other hand, in each iteration the algorithm "hits" at least one of the ranges in the optimal solution. Let $t_i(j)$ be the number of times the weight of the jth range in the optimal solution was doubled, for $j = 1, \ldots, k$, where k is the size of the optimal solution. Clearly, the weight of the universe in the ith iteration is at least

$$\sum_{j=1}^{k} 2^{t_i(j)}.$$

But this quantity is minimized when $t_i(1), \ldots, t_i(k)$ are as equal to each other as possible. (Indeed, $2^a + 2^b \ge 2 \cdot 2^{\lfloor (a+b)/2 \rfloor}$, for any integers $a, b \ge 0$.) As such, we have that

$$k2^{\lfloor i/k \rfloor} \le \sum_{j=1}^{k} 2^{t_i(j)} \le W_i \le m\exp(\varepsilon i).$$

So, consider $i = tk$, for t an integer. We have that $k2^t \le m\exp(\varepsilon tk) = m\exp(t/4)$, since $\varepsilon = 1/4k$. Namely,

$$\lg k + t \le \lg m + \frac{t}{4}\lg e \le \lg m + \frac{t}{2},$$

as $\lg e \le 1.45$. Namely, $t \le 2\lg(m/k)$ and $i = tk \le 2k\lg(m/k)$. We conclude that the algorithm performs at most $M = 2k\lg(m/k)$ successful iterations.

Number of iterations. Every iteration succeeds with probability larger than, say, $1/2$. As such, in expectation the algorithm performs at most $2M$ iterations till performing M successful iterations and then terminating. Using Chernoff's inequality, it is easy to verify that the number of iterations is $\le 4M = O(k\log m/k)$ with high probability (the high probability here is in terms of m/k).

Searching for the right value of k. Note that we assumed that k is known to us in advance. This can be easily overcome by doing an exponential search for the right value of k. Given a guess $\widehat{k}$ to the value of k, we will run the algorithm with k set to $\widehat{k}$. If the algorithm exceeds $c\log(m/k)$ iterations without terminating, then the guess is too small and the algorithm will next try $\widehat{2k}$, repeating this till successful.

Running time. Here, n is the number of points and m is the number of ranges. It is easy to verify that with careful implementation the sampling stage can be carried out in linear time. The size of the resulting cover is $O((\delta^\star/\varepsilon)\log(\delta^\star/\varepsilon)) = O(\delta^\star k \log(\delta^\star k))$. Checking if all the points are covered by the random sample takes $O(n\delta^\star k \log \delta^\star k)$ time, assuming we can in constant time determine if a point is inside a range. Computing the total weight of the ranges covering p takes $O(m)$ time. Thus, each iteration takes $O(m + n\delta^\star k \log \delta^\star k)$ time. We thus get the following.

Theorem 6.14. *Given a finite range space* $\mathsf{S} = (\mathsf{X}, \mathcal{R})$ *with n points and m ranges, and given that* S *has a dual shattering dimension* $\delta^\star$*, then one can compute a cover of* X*, using the ranges of* $\mathcal{R}$*, such that the cover uses* $O(\delta^\star k \log(\delta^\star k))$ *sets, where k is the size of the smallest set cover of* X *by ranges of* $\mathcal{R}$*.*

The expected running time of the algorithm is $O((m + n\delta^\star k \log(\delta^\star k)) \log(m/k) \log n)$ *time, assuming that in constant time we can decide if a point is inside a range. (The bound on the running time also holds with high probability.)*

6.3.2. Application – guarding an art gallery. Let $\mathcal{P}$ be a simple polygon (i.e., without holes) with n sides in the plane. We would like to find a minimal number of guards that see the whole polygon. A guard is a point that can see in all directions, but, unlike superman, it cannot see through the walls of the polygon. There is a beautiful standard result in computational geometry that shows that such a polygon can always be guarded by $\lfloor n/3 \rfloor$ guards placed on the vertices of the polygon. However, this might be arbitrarily bad compared to the optimal solution using k guards.

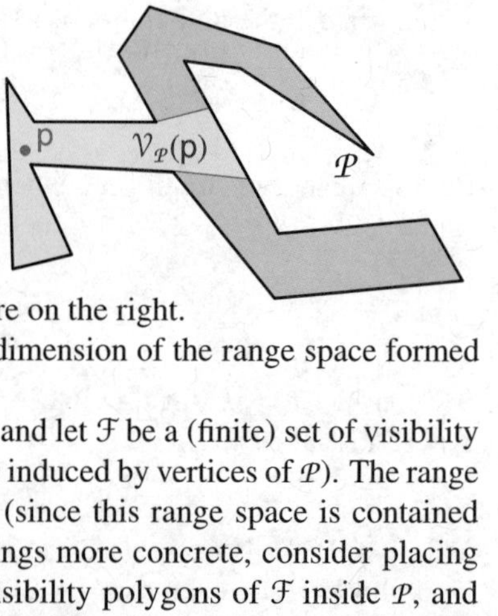

So, consider the range space $\mathsf{S} = (\mathcal{P}, \mathcal{R})$, where $\mathcal{R}$ is the set of all the possible visibility polygons inside $\mathcal{P}$. We remind the reader that for a point $\mathsf{p} \in \mathcal{P}$, the ***visibility polygon*** of p in $\mathcal{P}$ is the polygon

$$\mathcal{V}_\mathcal{P}(\mathsf{p}) = \left\{ \mathsf{q} \;\middle|\; \mathsf{q} \in \mathcal{P}, \mathsf{pq} \subseteq \mathcal{P} \right\},$$

where we consider $\mathcal{P}$ to be a closed set[②]; see the figure on the right.

We prove below (see Lemma 6.16) that the VC dimension of the range space formed by visibility polygons inside a polygon is a constant.

So, consider a simple polygon $\mathcal{P}$ with n vertices, and let $\mathcal{F}$ be a (finite) set of visibility polygons that covers $\mathcal{P}$ (say, all the visibility polygons induced by vertices of $\mathcal{P}$). The range space induced by $\mathcal{P}$ and $\mathcal{F}$ has finite VC dimension (since this range space is contained inside the range space of Lemma 6.16). To make things more concrete, consider placing a point inside each face of the arrangement of the visibility polygons of $\mathcal{F}$ inside $\mathcal{P}$, and let U denote the resulting set of points. Next, consider the projection of this range space into U; namely, consider the range space $\mathsf{S} = \left(U, \left\{ U \cap \mathcal{V} \;\middle|\; \mathcal{V} \in \mathcal{F} \right\}\right)$. Clearly, a set cover of minimal size for S corresponds to a minimal number of visibility polygons of $\mathcal{F}$ that cover $\mathcal{P}$. Now, S has polynomial size in $\mathcal{P}$ and $\mathcal{F}$ and it can clearly be computed efficiently. We can now apply the algorithm of Theorem 6.14 to it and get a "small" set cover. We conclude:

Theorem 6.15. *Given a simple polygon* $\mathcal{P}$ *of size n and a set of visibility polygons* $\mathcal{F}$ *(of polynomial size) that cover* $\mathcal{P}$*, then one can compute, in polynomial time, a cover of* $\mathcal{P}$

[②] As such, under our definition of visibility, one can see through a reflex corner of a polygon.

using $O(k_{\mathrm{opt}} \log k_{\mathrm{opt}})$ polygons of $\mathcal{F}$, where k_{opt} is the smallest number of polygons of $\mathcal{F}$ that completely cover $\mathcal{P}$.

Lemma 6.16. *The* VC *dimension of the range space formed by all visibility polygons inside a polygon $\mathcal{P}$ (i.e.,* S *above) is a constant.*

Proof. Let U be a set of k points shattered by the set of visibility polygons inside $\mathcal{P}$. Let $\mathcal{F}$ be the set of k visibility polygons induced by the points of U. Clearly, by the shattering property, for every subset S of U, there exists a point $\mathsf{p} \in \mathcal{P}$ such that the visibility polygon of p contains exactly S and no other points of U. Conversely, in the arrangement $\mathcal{A}(\mathcal{F})$, the point p is contained in a face that is covered by the visibility polygons of S and is outside all the visibility polygons of $U \setminus S$.

Unfortunately, we cannot just bound the complexity of $\mathcal{A}(\mathcal{F})$, and use it to bound the VC dimension of the given range space, since its complexity involves n (it is $O(nk^2)$ in the worst case).

Instead, consider R to be a set of at most k points inside $\mathcal{P}$ and let $T_{\mathrm{in}}(\mathsf{R}, \mathcal{V})$ be the maximum number of different subsets of R one can see from a point located inside a polygon $\mathcal{V} \subseteq \mathcal{P}$. Similarly, let $T_{\mathrm{out}}(\mathsf{R}, \mathcal{V})$ be the maximum number of different subsets of R that one can see from a point located inside a polygon $\mathcal{V} \subseteq \mathcal{P}$, where all the points of R are outside $\mathcal{V}$ and there is a diagonal s such that all the points are on one side of s and $\mathcal{V}$ is on the other side.

We show below, in Lemma 6.17, that $T_{\mathrm{out}}\big(\mathsf{R}, \mathcal{V}\big) = O\big(|\mathsf{R}|^6\big)$.

Now, consider a triangulation of $\mathcal{P}$. There must be a triangle in this triangulation, such that if we remove it, then every one of the remaining pieces $\mathcal{V}_1, \mathcal{V}_2, \mathcal{V}_3$ contains at most $2k/3$ points of U. Let $\triangle$ be this triangle, and let $U_i = \mathcal{V}_i \cap \mathsf{R}$, for $i = 1, 2, 3$.

Clearly, the complexity of the visibility polygons of $\mathcal{F}$ inside $\triangle$ is $O(k^2)$. Indeed, every visibility polygon intersects $\triangle$ in a region that is a convex polygon with constant complexity, and there are k such polygons. Furthermore, inside $\mathcal{V}_i$ one can see only $T_{\mathrm{out}}(U \setminus \mathcal{V}_i, \mathcal{V}_i) = O\big(k^6\big)$ different subsets of the points of U outside $\mathcal{V}_i$, for $i = 1, 2, 3$. Thus, the total number of different subsets of U one can see is bounded by

$$
\begin{aligned}
T_{\mathrm{in}}(U, \mathcal{P}) &\le O(k^2) + \sum_{i=1}^{3} T_{\mathrm{out}}\big(U \setminus \mathcal{V}_i, \mathcal{V}_i\big) \cdot T_{\mathrm{in}}\big(U \cap \mathcal{V}_i, \mathcal{V}_i\big) \\
&= O(k^2) + O(k^6) \cdot 3 \cdot T_{\mathrm{in}}(2k/3) = k^{O(\log k)},
\end{aligned}
$$

by Lemma 6.17, where $T_{\mathrm{in}}(2k/3)$ is the maximum $T_{\mathrm{in}}(X, \mathcal{P})$ over all subsets of $2k/3$ points in $\mathcal{P}$. However, for U to be shattered, we need that $T_{\mathrm{in}}(k) \ge 2^k$. Namely, we have that $2^k \le k^{O(\log k)}$, and this implies that $k \le O(\log^2 k)$. We conclude that k is a constant, as desired. ■

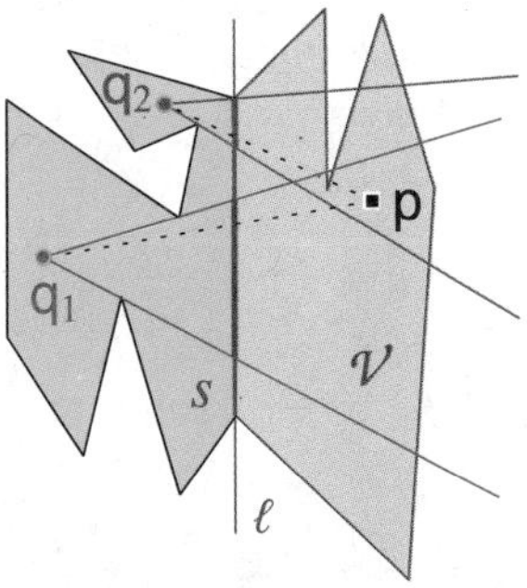

Lemma 6.17. $T_{\mathrm{out}}(k, \mathcal{V}) = O\big(k^6\big)$.

Proof. Let R be the set of k points under consideration, which are on one side of the diagonal s and for which $\mathcal{V}$ is on the other side. Let ℓ be the line containing s. Each of the points of R sees a subinterval of s (inside $\mathcal{P}$), and this interval induces a wedge in the plane on the other side of s (we ignore for the time being the boundary of $\mathcal{P}$). Also, consider all the lines passing through pairs of points of R. Together, this set of $O(k^2)$ lines, rays, and segments partition the halfplane (that is induced by ℓ) into $O(k^4)$ different faces.

Any point inside the same face of this arrangement sees the same subset of points of R through the segment s in the same radial order (ignoring the boundaries of $\mathcal{V}$).

So, consider a point p inside such a face f, and let $q_1, \dots, q_m$ be the (clockwise ordered) list of points of R that p sees. To visualize this list, connect p to each one of these points by a segment. Now, as we introduce the boundary of $\mathcal{V}$, some of these segments are no longer realizable since they intersect the boundary $\partial\mathcal{V} \setminus s$. Observe that the set of visible points from p must be a consecutive subsequence of the above list; namely, p can see inside $\mathcal{P}$ the points $q_L, \dots, q_R$, for some $L \leq R$. We conclude that since there are $O\left(k^2\right)$ choices for the indices L and R, it follows that inside f there are at most $O(k^2)$ different subsets of R that are realizable inside $\mathcal{V}$. Now, there are $O\left(k^4\right)$ such faces in the arrangement, which implies the bound. ■

6.4. Geometric set cover via linear programming

In the following, we are going to assume one can solve LP (even in high dimensions) in polynomial time. For a discussion of this assumption, see the bibliographical notes in this chapter. LP is a powerful tool, and it is natural to ask what can be solved/approximated if one is allowed to use an LP solver as a polynomial time black box. We show here that geometric set cover can be solved efficiently using LP.

A standard approach to computing an approximate solution to an NP-HARD problem is to solve a linear programming relaxation (LP) of the problem and round its fractional solution to an integral solution to the original problem.

In our case, assume we are given a finite range space $\mathsf{S} = (\mathsf{P}, \mathcal{F})$ with VC dimension δ and dual VC dimension $\delta^\star$, where $\mathsf{P} = \{\mathsf{p}_1, \dots, \mathsf{p}_n\}$ and $\mathcal{F} = \{\mathbf{r}_1, \dots, \mathbf{r}_m\}$. The natural integer program (IP) for solving the set cover problem for this instance has a variable y_i for every range $\mathbf{r}_i \in \mathcal{F}$, where $y_i = 1$ if $\mathbf{r}_i$ is in the suggested cover and zero otherwise. As such, the resulting IP is

$$
\begin{aligned}
&\min && \sum_{i=1}^{m} y_i && \\
(6.1)\qquad &\text{subject to} && \sum_{i:\mathsf{p}_j \in \mathbf{r}_i} y_i \geq 1 && \forall \mathsf{p}_j \in \mathsf{P}, \\
& && y_i \in \{0, 1\} && i = 1, \dots, m.
\end{aligned}
$$

Consider the optimal solution (for the above IP) and the values it assigns to its variables, and let $\mathrm{opt} = \sum_{i=1}^{m} y_i$ denote the ***value*** of the (integral) optimal solution.

The natural LP relaxation, for this IP, replaces the integer variable $y_i \in \{0, 1\}$ by a continuous variable x_i, for $i = 1, \dots, m$. The resulting LP is

$$
\begin{aligned}
&\min && \sum_{i=1}^{m} x_i && \\
(6.2)\qquad &\text{subject to} && \sum_{i:\mathsf{p}_j \in \mathbf{r}_i} x_i \geq 1 && \forall \mathsf{p}_j \in \mathsf{P}, \\
& && x_i \in [0, 1] && i = 1, \dots, m.
\end{aligned}
$$

Let $\mathsf{f} = \mathsf{f}(\mathcal{I}) = \sum_{i=1}^{m} x_i$ denote the ***value*** of an optimum solution to the above LP. Clearly, the optimal (integral) solution to the IP is a valid solution of the above LP. As such, we have $\mathrm{opt} \geq \mathsf{f}$. We will refer to the values assigned to the variables x_i for some particular optimal solution to the LP as the ***fractional solution***.

In the following, we will refer to the value of x_i in the solution as the ***weight*** of the range $\mathbf{r}_i$.

Consider a fractional solution to the LP given by x_i assigned to ranges $\mathbf{r}_i \in \mathcal{F}$, with total value $\mathsf{f} = \sum_i x_i$. The above LP can be rewritten as

$$\begin{aligned} &\min \quad \mathsf{f} \\ &\text{subject to} \quad \mathsf{f} = \sum_{i=1}^{m} x_i \\ &\qquad \sum_{i:\mathsf{p}_j \in \mathbf{r}_i} x_i \geq 1 && \forall \mathsf{p}_j \in \mathsf{P}, \\ &\qquad x_i \geq 0 && i = 1, \ldots, m. \end{aligned}$$

(Note that the restriction that $x_i \leq 1$ is redundant since any optimal solution would have this property anyway. As such, we can just drop this inequality.) Dividing this LP by f, we get the following equivalent LP:

$$\begin{aligned} &\min \quad \mathsf{f} \\ &\text{s.t.} \quad \sum_{i=1}^{m} x_i/\mathsf{f} = 1 \\ &\qquad \sum_{i:\mathsf{p}_j \in \mathbf{r}_i} x_i/\mathsf{f} \geq 1/\mathsf{f} && \forall \mathsf{p}_j \in \mathsf{P}, \\ &\qquad x_i \geq 0 && i = 1, \ldots, m \end{aligned} \quad \Leftrightarrow \quad \begin{aligned} &\max \quad \varepsilon \\ &\text{s.t.} \quad \sum_{i=1}^{m} z_i = 1 \\ &\qquad \sum_{i:\mathsf{p}_j \in \mathbf{r}_i} z_i \geq \varepsilon && \forall \mathsf{p}_j \in \mathsf{P}, \\ &\qquad z_i \geq 0 && i = 1, \ldots, m. \end{aligned}$$

The last step followed by rewriting $\varepsilon = 1/\mathsf{f}$ and $z_i = x_i/\mathsf{f}$ and observing that minimizing f is equivalent to maximizing $\varepsilon = 1/\mathsf{f}$.

Observe that solving the original LP or any of the LP listed above is equivalent (i.e., given an optimal solution to one, we can derive an optimal solution to any other). So, consider the optimal solution for the last LP listed above. We assign every range $\mathbf{r}_i$ a number z_i which lies between 0 and 1. Furthermore, the total sum of the z_is is 1. It is thus natural to think about z_is as probabilities; that is, consider choosing every range $\mathbf{r}_i$ with probability z_i.

So, what is the above LP doing? It is trying to assign weights (i.e., probabilities) to the ranges of $\mathcal{F}$ such that any point has weight (i.e., the total weight of the ranges containing it) as large as possible (i.e., larger than ε, and the LP is maximizing ε).

In particular, consider the dual range space $\mathsf{S}^\star = (\mathcal{F}, \mathsf{P}^\star)$, with the weights assigned to the ranges according to the LP solution. Now, every range $\mathcal{F}_\mathsf{p}$ has weight at least ε and the total weight of the ranges of $\mathcal{F}$ is 1. As such, an ε-net N for this range space will contain at least one element in every range that has weight at least ε. Namely, since all the ranges have weight at least ε, an ε-net will contain an element in every range of the range space.

Namely, an ε-net $N \subseteq \mathcal{F}$ generated by a random sample for this dual range space (according to the weights of the ranges assigned to them by the LP) will cover all the points of P. Now, in this case, a random sample of size $O((\delta^\star/\varepsilon)\log(1/\varepsilon))$ is an ε-net with constant probability by Theorem 5.28_{p71}.

Thus, we got a cover for the given set cover problem of size

$$O\Big((\delta^\star/\varepsilon)\log(1/\varepsilon)\Big) = O\Big(\delta^\star \mathsf{f} \log \mathsf{f}\Big) = O\Big(\delta^\star \mathrm{opt} \log \mathrm{opt}\Big),$$

since $\varepsilon = 1/\mathsf{f}$ and $\mathsf{f} \leq \mathrm{opt}$, where opt is the optimal minimum size cover.

The algorithm. Given the set cover instance, we write the required LP as described above and solve the LP using some LP solver. Next, we consider the values of the variables (in the computed solution of the LP) as weights assigned to the relevant ranges. The solution of the LP also specifies the value of ε.

We then compute an ε-net of the dual range space, using these weights, and output this ε-net as the required cover. We thus get the following.

Theorem 6.18. *Given a finite range space* $(\mathsf{P}, \mathcal{F})$ *with dual* VC *dimension* $\delta^\star$, *one can compute, in polynomial time, a set cover of* P *by* $O(\delta^\star \mathrm{opt} \log \mathrm{opt})$ *ranges, where* opt *is the optimal minimum size cover of* P *by ranges of* $\mathcal{R}$.

Computing this cover requires solving a single instance of LP *and computing a single* ε*-net of the appropriately weighted dual space.*

This algorithm gives us a new interpretation of the previous algorithm (i.e., the one that used reweighting). It is easy to verify that the reweighting algorithm works if we redouble the weight of any range that is light. The reweighting algorithm uses ε-nets repeatedly only as a way to find such light ranges to double their weights. As such, one can interpret the reweighting algorithm as an implicit procedure that solves the above LP (using reweighting).

6.5. Bibliographical notes

The stabbing tree result is due to Welzl [**Wel92**], which is in turn a simplification of the work of Chazelle and Welzl [**CW89**]. The running time of computing the good spanning tree can be improved to $O(n^{1+\varepsilon})$ [**Wel92**].

The extension of the spanning tree result to range spaces with low VC dimension is due to Matoušek et al. [**MWW93**].

A natural question is whether one can compute a spanning tree which is weight sensitive. Namely, compute a spanning tree of n points in the plane, such that if a line has only k points on one side of it, it would cross at most $O(\sqrt{k})$ edges of the tree. A slightly weaker bound is currently known; see Exercise 6.3, which is from [**HS11**]. This leads to small ε-samples which work for ranges which are heavy enough.

Another natural question is whether, given a set of lines and points in the plane, one can compute a spanning tree with an overall small number of crossings. Surprisingly, one can compute a $(1 + \varepsilon)$-approximation to the best tree in near linear time; see [**HI00**]. This result also extends to higher dimensions.

Section 6.3 is due to Clarkson [**Cla93**]. This technique was used to approximate terrains [**AD97**] and covering polytopes [**Cla93**].

The algorithm we described for set cover falls into a general method of multiplicative weights update. Algorithms of this family include Littlestone's Winnow algorithm [**Lit88**], AdaBoost [**FS97**], and many more. The basic idea can be tracked back to the 1950s. See [**AHK06**] for a nice survey of this method.

The idea of using LP to solve set cover is by now quite old. The idea of using ε-nets for solving set cover is due to Long [**Lon01**]. The idea of using LP to help solve problems in low-dimensional computational geometry still seems to be under utilized. For some relevant results, see [**CCH09, CC09, CH09**].

6.6. Exercises

Exercise 6.1 (Spanning tree with low crossing number in $\mathbb{R}^d$). Prove Theorem 6.7.

Exercise 6.2 (Tight example for spanning tree with low crossing number)**.** Show that in the worst case, one can pick a point set P of n points in the plane, such that for any spanning tree $\mathcal{T}$ of P, there exists a line ℓ, such that ℓ crosses $\Omega(\sqrt{n})$ edges of $\mathcal{T}$.

Exercise 6.3 (Spanning tree with relative crossing number)**.** Let P be a set of n points in the plane. For a line ℓ, let w_ℓ^+ (resp., w_ℓ^-) be the number of points of P lying above (resp., below or on) ℓ, and define the *weight* of ℓ, denoted by $\omega\ell$, to be $\min(w_\ell^+, w_\ell^-)$.

Show that one can construct a spanning tree $\mathcal{T}$ for P such that any line ℓ crosses at most $O\left(\sqrt{\omega\ell}\log(n/\omega\ell)\right)$ edges of $\mathcal{T}$.

CHAPTER 7

Yet Even More on Sampling

In this chapter, we will extend the sampling results shown in Chapter 5. In the process, we will prove stronger bounds on sampling and provide more general tools for using them. This chapter contains more advanced material and the casual reader will benefit from skipping it.

7.1. Introduction

7.1.1. A not quite new notion of dimension. We are interested in how much information can be extracted by random sampling of a certain size for a range space of VC dimension δ. To make things more interesting, we will extend the notion of VC dimension to real functions.

Definition 7.1. Let X be a ground set, and consider a set of functions $\mathcal{F}$ from X to the interval $\left[0, 1\right]$. For a set $N = \{\mathsf{p}_1, \ldots, \mathsf{p}_d\} \subseteq \mathsf{X}$, a set of real values

$$V = \left\{ v_1, \ldots, v_d \;\middle|\; v_i \in \left[0, 1\right] \right\},$$

and a function $f \in \mathcal{F}$, consider the subset

$$N_f = \left\{ \mathsf{p}_i \in N \;\middle|\; f(\mathsf{p}_i) \geq v_i \right\}$$

of N induced by f. We will refer to N_f as the ***induced subset*** of N formed by f and V.

We remind the reader that any subset $Y \subseteq \mathsf{X}$ can be described using its characteristic function (aka indicator function) $\mathbf{1}_Y : \mathsf{X} \to \{0, 1\}$, where $x \in Y$ if and only if $\mathbf{1}_Y(x) = 1$. A ***set system*** $(\mathsf{X}, \mathcal{R})$ is a pair, where $\mathcal{R}$ is a set of subsets of X. As such, given a set system $(\mathsf{X}, \mathcal{R})$, we can always interpret $\mathcal{R}$ as a set of characteristic functions of these sets. All the induced subsets would just be $\mathcal{R}$ if we set the threshold values $v_1, \ldots, v_d$ to all be 1. Thus, replacing the sets of $\mathcal{R}$ by real functions is a natural extension of the notion of a set system.

We next extend the notion of shattering.

Definition 7.2. Let X be a ground set, and consider a set of functions $\mathcal{F}$ from X to the interval $\left[0, 1\right]$. A subset $N = \{\mathsf{p}_1, \ldots, \mathsf{p}_d\} \subseteq \mathsf{X}$ with associated values $V = \{v_1, \ldots, v_d\}$ is ***shattered*** by $(\mathsf{X}, \mathcal{F})$ (or by $\mathcal{F}$) if

$$P_N = \left\{ N_f \;\middle|\; f \in \mathcal{F} \right\}$$

contains all possible (i.e., 2^d) subsets of N.

In particular, we now have a natural extension of the notion of dimension.

Definition 7.3. Given $\mathsf{S} = (\mathsf{X}, \mathcal{F})$ as above, its ***pseudo-dimension***, denoted by $\mathrm{pDim}(\mathcal{F})$, is the size of the largest subset of X that is shattered by $\mathcal{F}$.

Definition 7.4. Given a function $f \in \mathcal{F}$ and a probability distribution $\mathcal{D}$ on X, consider the average value of f on X. The ***measure*** of f is

$$\overline{m}(f) = \sum_{\mathsf{p} \in \mathsf{X}} \mathbf{Pr}\Big[\mathsf{p}\Big] f(\mathsf{p}) \quad \text{and} \quad \overline{m}(f) = \int_{\mathsf{p} \in \mathsf{X}} f(\mathsf{p})\, d\mathcal{D}$$

if the ground set is finite or infinite, respectively.

Intuitively, if $f = \mathbf{1}_Y$ is an indicator function of a set $Y \subseteq \mathsf{X}$, then this is no more than the measure of Y in X; namely, if the underlying distribution is uniform, then $\overline{m}(\mathbf{1}_Y) = |Y| / |\mathsf{X}|$. Again, we are assuming here that X is finite, but the discussion can easily be extended to infinite domains, by replacing summation by integration. As usual, we are going to ignore this tedious (and somewhat insignificant) technicality.

Definition 7.5. Similarly, for an ordered sample $N = \langle \mathsf{p}_1, \ldots, \mathsf{p}_m \rangle$ of m points from X and a function $f \in \mathcal{F}$, let

$$\overline{s} = \overline{s}_N(f) = \frac{1}{m} \sum_{i=1}^{m} f(\mathsf{p}_i)$$

denote the ***estimate*** of f by N. The quantity $\overline{s}_N(f)$ can be interpreted as approximation to $\overline{m}(f)$ by the sample N.

Remark 7.6. Before the reader gets confused and impressed by the claptrap of pseudo-dimension, observe that we can easily construct a range space with the same VC dimension, so that we can work directly on this "alternative" range space. Indeed, let $\mathsf{X}' = \mathsf{X} \times [0, 1]$ and $\mathcal{G} = \left\{ \widehat{\mathbf{r}}(f) \;\middle|\; f \in \mathcal{F} \right\}$, where

$$\widehat{\mathbf{r}}(f) = \left\{ (x, y) \;\middle|\; x \in \mathsf{X}, y \in [0, 1], \text{ and } y \leq f(x) \right\}. \tag{7.1}$$

Now, consider the range space $\mathsf{T} = (\mathsf{X}', \mathcal{G})$.

A point $\mathsf{p} \in N$ with associated value v is in $\overline{s}_N(f)$ if and only if $f(\mathsf{p}) \geq v$. Namely, the point (p, v) lies below the graph of f. This is equivalent to $(\mathsf{p}, v) \in \widehat{\mathbf{r}}(f)$. As such, $\mathsf{S} = (\mathsf{X}, \mathcal{F})$ has pseudo-dimension δ if and only if T has VC dimension δ. Indeed, if S shatters the set $N = \{\mathsf{p}_1, \ldots, \mathsf{p}_d\} \subseteq \mathsf{X}$ and its associated real values $V = \left\{ v_1, \ldots, v_d \;\middle|\; v_i \in \left[0, 1\right] \right\}$, then T shatters the set of points $\left\{ (\mathsf{p}_1, v_1), \ldots, (\mathsf{p}_d, v_d) \right\}$.

Observe, also, that a distribution $\mathcal{D}$ over X induces a natural distribution on X'. Indeed, to pick a point randomly in the set $\mathsf{X} \times [0, 1]$, pick randomly a point $x \in \mathsf{X}$ according to the given distribution, and then pick uniformly $y \in [0, 1]$. The resulting pair (x, y) is in X', and this process defines a natural distribution on the elements of X'. In fact, given $f \in \mathcal{F}$, we have that $\overline{m}(f)$ is equal to the measure of $\widehat{\mathbf{r}}(f)$ in the ground set X'. Indeed, being slightly informal, we have that

$$\begin{aligned} \overline{m}(\widehat{\mathbf{r}}(f)) &= \int_{(x,y) \in \mathsf{X}'} \mathbf{Pr}[(x, y)]\, \mathbf{1}_{\widehat{\mathbf{r}}(f)}\big((x, y)\big)\, dx\, dy = \int_{x \in \mathsf{X}} \int_{y \in [0,1]} \mathbf{1}_{\widehat{\mathbf{r}}(f)}\big((x, y)\big)\, dy\, d\mathcal{D} \\ &= \int_{x \in \mathsf{X}} f(x)\, d\mathcal{D} = \overline{m}(f). \end{aligned}$$

Thus, we immediately get the ε-net and ε-sample theorems for the pseudo-dimension settings (since we proved these theorems for the VC dimension case).

7.1.2. The result. For a parameter $\nu > 0$, consider the distance function between two real numbers $r \geq 0$ and $s \geq 0$ defined as

$$d_\nu(r, s) = \frac{|r - s|}{r + s + \nu}.$$

It is easy to verify that (i) $0 \leq d_\nu(r, s) < 1$, (ii) $d_\nu(r, s) \leq d_\nu(u, v)$, for $u \leq r \leq s \leq v$, and (iii) $d_\nu(x, y) > 0$ if $x \neq y$.

We will elaborate on the properties of this distance function in Section 7.1.3 below.

Our purpose in this chapter is to prove the following theorem.

Theorem 7.7. *Let $\alpha, \nu, \varphi > 0$ be parameters, let $\mathsf{S} = (\mathsf{X}, \mathcal{F})$ be a range space, and let $\mathcal{F}$ be a set of functions from X to $[0, 1]$, such that the pseudo-dimension of S is δ. For a random sample N (with repetition) from X of size*

$$O\left(\frac{1}{\alpha^2 \nu}\left(\delta \log \frac{1}{\nu} + \log \frac{1}{\varphi}\right)\right),$$

we have

$$\forall f \in \mathcal{F} \quad d_\nu\left(\overline{m}(f), \overline{s}_N(f)\right) < \alpha$$

with probability $\geq 1 - \varphi$.

Before providing a proof of this theorem, we first demonstrate how this theorem implies the ε-net/sampling theorems, and some other results. But first, we must get a better handle on the distance function $d_\nu(\cdot, \cdot)$.

7.1.3. The curious distance function between real numbers. For technical reasons, it is convenient to work with the distance function described above between numbers in describing and deriving this result.

The proof of the following lemma is tedious and is left to Exercise 7.2.

Lemma 7.8. *The triangle inequality holds for $d_\nu(\cdot, \cdot)$. Namely, for any $x, y, z \geq 0$ and $\nu > 0$, we have $d_\nu(x, y) + d_\nu(y, z) \geq d_\nu(x, z)$.*

Lemma 7.9. *The function $d_\nu(\cdot, \cdot)$ is a metric.*

Proof. Indeed, for any $x > 0$, we have $d_\nu(x, x) = 0$. Also, for any $x, y > 0$, we have $d_\nu(x, y) = d_\nu(y, x)$. Finally, the triangle inequality is implied by Lemma 7.8. ∎

Lemma 7.10. *Let $\alpha, \nu, \overline{m}, \overline{s}$ be non-negative numbers. Then $d_\nu(\overline{m}, \overline{s}) < \alpha$ if and only if $\overline{s} \in \mathcal{I} = (u_l, u_r)$, where*

$$u_l = \left(1 - \frac{2\alpha}{1 + \alpha}\right)\overline{m} - \frac{\alpha\nu}{(1 + \alpha)} \quad \text{and} \quad u_r = \left(1 + \frac{2\alpha}{1 - \alpha}\right)\overline{m} + \frac{\alpha\nu}{(1 + \alpha)}.$$

Proof. We have $d_\nu\left(\overline{m}, \overline{s}\right) = \dfrac{|\overline{m} - \overline{s}|}{\overline{m} + \overline{s} + \nu}$. If $\overline{m} > \overline{s}$, then this implies that

$$d_\nu\left(\overline{m}, \overline{s}\right) < \alpha \iff \overline{m} - \overline{s} < \alpha(\overline{m} + \overline{s} + \nu) \iff (1 - \alpha)\overline{m} - \alpha\nu < (1 + \alpha)\overline{s}$$

$$\iff \frac{(1 - \alpha)}{(1 + \alpha)}\overline{m} - \frac{\alpha\nu}{(1 + \alpha)} < \overline{s} \iff u_l = \left(1 - \frac{2\alpha}{1 + \alpha}\right)\overline{m} - \frac{\alpha\nu}{(1 + \alpha)} < \overline{s}.$$

Since $\overline{m} \in \mathcal{I}$, this implies that $u_l < \overline{s} < \overline{m} \leq u_r$; namely, $\overline{s} \in \mathcal{I}$.

TABLE 7.1. The results hold for any range $\mathbf{r}$ in the given range space that has VC dimension δ, where $\overline{m} = \overline{m}(\mathbf{r})$ is the measure of $\mathbf{r}$ and $\overline{s} = \overline{s}(\mathbf{r})$ is its estimate in the random sample. The samples have the required property (for all the ranges in the range space) with constant probability.

Name	Property $\forall \mathbf{r} \in \mathcal{F}$	Sample size
ε-net **[HW87]** Theorem 7.12	$\overline{m} = \overline{m}(r)$, $\overline{s} = \overline{s}(r)$ $\overline{m} \geq \varepsilon \;\Rightarrow\; \overline{s} > 0$	$O\left(\frac{\delta}{\varepsilon} \log \frac{1}{\varepsilon}\right)$
ε-sample **[VC71]** Theorem 7.13	$\lvert\overline{m} - \overline{s}\rvert \leq \varepsilon$	$O\left(\frac{\delta}{\varepsilon^2}\right)$
Sensitive ε-approximation **[Brö95, BCM99]** Theorem 7.15	$\lvert\overline{m} - \overline{s}\rvert \leq \frac{\varepsilon}{2}\left(\sqrt{\overline{m}} + \varepsilon\right)$	$O\left(\frac{\delta}{\varepsilon^2} \log \frac{1}{\varepsilon}\right)$
Relative (ε, p)-approximation **[CKMS06]** Theorem 7.17	$\overline{m} \leq p \;\Rightarrow\; \overline{s} \leq (1+\varepsilon)p$ $\overline{m} \geq p \;\Rightarrow$ $(1-\varepsilon)\overline{m} \leq \overline{s} \leq (1+\varepsilon)\overline{m}$	$O\left(\frac{\delta}{\varepsilon^2 p} \log \frac{1}{p}\right)$

If $\overline{m} < \overline{s}$, then this implies that

$$d_\nu\left(\overline{m}, \overline{s}\right) < \alpha \iff \overline{s} - \overline{m} < \alpha(\overline{m} + \overline{s} + \nu) \iff (1-\alpha)\,\overline{s} < \alpha\nu + (1+\alpha)\,\overline{m}$$
$$\iff \overline{s} < \frac{(1+\alpha)}{(1-\alpha)}\overline{m} + \frac{\alpha\nu}{(1+\alpha)} \iff \overline{s} < u_r = \left(1 + \frac{2\alpha}{1-\alpha}\right)\overline{m} + \frac{\alpha\nu}{(1+\alpha)}.$$

Again, this implies that $u_l \leq \overline{m} \leq \overline{s} < u_r$, implying that $\overline{s} \in \mathcal{I}$. ■

Corollary 7.11. *For any real numbers $\nu, \alpha, \overline{m}, \overline{s} > 0$, we have:*
(i) If $\lvert\overline{s} - \overline{m}\rvert \leq \Delta = (2\overline{m} + \nu)\dfrac{\alpha}{1+\alpha}$, *then* $d_\nu(\overline{m}, \overline{s}) < \alpha$.

(ii) If $d_\nu(\overline{m}, \overline{s}) < \alpha$, *then* $\lvert\overline{s} - \overline{m}\rvert \leq \Delta' = \dfrac{2\alpha}{1-\alpha}\overline{m} + \dfrac{\alpha\nu}{1+\alpha}$.

7.2. Applications

The following assumes that the reader is familiar and comfortable with ε-nets and ε-samples (see Chapter 5). The results implied by Theorem 7.7 are summarized in Table 7.1.

We are given a range space $\mathsf{S} = (\mathsf{X}, \mathcal{F})$ of VC dimension δ, where X is a point set and $\mathcal{F}$ is a set of ranges of X. In our settings, we will usually consider a finite subset $\mathsf{x} \subseteq \mathsf{X}$ and we will be interested in the range space induced by S on x. In particular, let N be a sample of X. For a range $\mathbf{r} \in \mathcal{F}$, let

$$\overline{m} = \overline{m}_{\mathsf{x}}(\mathbf{r}) = \frac{\lvert\mathbf{r} \cap \mathsf{x}\rvert}{\lvert\mathsf{x}\rvert} \quad \text{and} \quad \overline{s} = \overline{s}_N(\mathbf{r}) = \frac{\lvert\mathbf{r} \cap N\rvert}{\lvert N\rvert}.$$

Intuitively, $\overline{m}$ is the total weight of $\mathbf{r}$ in x, while $\overline{s}$ is the sample estimate for $\mathbf{r}$.

7.2.1. Getting the ε-net and ε-approximation theorems.

7.2.1.1. *ε-net theorem.* We remind the reader that $N \subseteq \mathsf{x}$ is an ***ε-net*** for a finite range space $(\mathsf{x}, \mathcal{F})$ if for every $\mathbf{r} \in \mathcal{F}$, such that $\overline{m}_{\mathsf{x}}(\mathbf{r}) \geq \varepsilon$, we have that $N \cap \mathbf{r} \neq \emptyset$.

The following is a restatement of Theorem 5.28_{p71}.

Theorem 7.12 (**[HW87]**, ε-net theorem). *Let $\varphi, \varepsilon > 0$ be parameters and let $\mathsf{S} = (\mathsf{X}, \mathcal{F})$ be a range space with* VC *dimension δ. Let $\mathsf{x} \subset \mathsf{X}$ be a finite subset. Then, a sample of size*

$$O\left(\frac{1}{\varepsilon}\left(\delta \log \frac{1}{\varepsilon} + \log \frac{1}{\varphi}\right)\right)$$

from x *is an ε-net for* x *with probability $\geq 1 - \varphi$.*

Proof. Let $\alpha = 1/4$, $\nu = \varepsilon$, and apply Theorem 7.7. The sample size is

$$O\left(\frac{1}{\alpha^2 \nu}\left(\delta \log \frac{1}{\nu} + \log \frac{1}{\varphi}\right)\right) = O\left(\frac{1}{\varepsilon}\left(\delta \log \frac{1}{\varepsilon} + \log \frac{1}{\varphi}\right)\right).$$

Now, let $\mathbf{r} \in \mathcal{F}$ be a range such that $\overline{m} = \overline{m}(\mathbf{r}) \geq \varepsilon$. We have that $d_\nu(\overline{m}, \overline{s}) < \alpha = 1/4$, where $\overline{s} = \overline{s}(\mathbf{r})$ (and this holds with probably $\geq 1 - \varphi$ for all ranges). Then, by Corollary 7.11(ii), we have that $\overline{s} \in (\overline{m} - \Delta', \overline{m} + \Delta')$, where

$$\Delta' = \frac{2\alpha}{1-\alpha}\overline{m} + \frac{\alpha\nu}{1+\alpha} = \frac{2}{3}\overline{m} + \frac{\varepsilon}{5}.$$

For $\overline{m} \geq \varepsilon$, we have $\overline{s} \geq \overline{m} - \Delta' = \overline{m}/3 - \varepsilon/5 \geq \varepsilon/15 > 0$, implying the claim. ■

7.2.1.2. *ε-sample theorem.* We remind the reader that $N \subseteq \mathsf{x}$ is an ***ε-sample*** for a finite range space $(\mathsf{x}, \mathcal{F})$ if for every $\mathbf{r} \in \mathcal{F}$ we have that $|\overline{m}_{\mathsf{x}}(\mathbf{r}) - \overline{s}_N(\mathbf{r})| \leq \varepsilon$.

The following is a slight strengthening of the ε-sample theorem (Theorem 5.26_{p71}).

Theorem 7.13 (ε-sample theorem). *Let $\varphi, \varepsilon > 0$ be parameters and let $\mathsf{S} = (\mathsf{X}, \mathcal{F})$ be a range space with* VC *dimension δ. Let $\mathsf{x} \subset \mathsf{X}$ be a finite subset. A sample of size*

$$O\left(\frac{1}{\varepsilon^2}\left(\delta + \log \frac{1}{\varphi}\right)\right)$$

from x *is an ε-sample for $\mathsf{S} = (\mathsf{x}, \mathcal{F})$ with probability $\geq 1 - \varphi$.*

Proof. Set $\alpha = \varepsilon/4$ and $\nu = 1/4$. We have, by Theorem 7.7, that for any $\mathbf{r} \in \mathcal{F}$,

$$d_\nu\big(\overline{m}(\mathbf{r}), \overline{s}(\mathbf{r})\big) < \alpha \quad \Longrightarrow \quad \big|\overline{m}(\mathbf{r}) - \overline{s}(\mathbf{r})\big| \leq \frac{\varepsilon}{4}\big(\overline{m}(\mathbf{r}) + \overline{s}(\mathbf{r}) + \nu\big) \leq \varepsilon,$$

implying the claim. ■

7.2.2. Sensitive ε-approximation. A related concept to the above notions of net / sample was introduced by Brönnimann et al. **[BCM99]**.

Definition 7.14. A sample $N \subseteq \mathsf{x}$ is a ***sensitive ε-approximation*** if

$$\forall \mathbf{r} \in \mathcal{F} \quad \big|\overline{m}(\mathbf{r}) - \overline{s}(\mathbf{r})\big| \leq \frac{\varepsilon}{2}\Big(\sqrt{\overline{m}(\mathbf{r})} + \varepsilon\Big).$$

Observe that a set N which is a sensitive ε-approximation is, simultaneously, both an ε^2-net and an ε-sample.

The following theorem shows the existence of a sensitive ε-approximation. Note that the bound on its size is (slightly) better than the bound shown by **[Brö95, BCM99]**.

Theorem 7.15. *A sample N from* x *of size*

$$O\left(\frac{1}{\varepsilon^2}\left(\delta \log \frac{1}{\varepsilon} + \log \frac{1}{\varphi}\right)\right)$$

is a sensitive ε-approximation, with probability $\geq 1 - \varphi$.

PROOF. Let $\nu_i = i\varepsilon^2/c$ and $\alpha_i = \sqrt{1/4i}$, for $i = 1, \ldots, M = \lceil c/\varepsilon^2 \rceil$, where c is some sufficiently large constant. As such, for $i = 1, \ldots, M$, we have that $\alpha_i^2 \nu_i = \varepsilon^2/(4c)$. Consider a single random sample N of size

$$U = O\left(\frac{1}{\varepsilon^2}\left(\delta \log \frac{1}{\varepsilon} + \log \frac{M}{\varphi}\right)\right) = O\left(\frac{1}{\varepsilon^2}\left(\delta \log \frac{1}{\varepsilon} + \log \frac{1}{\varphi}\right)\right).$$

For a fixed i, this sample complies with Theorem 7.7, with parameters ν_i and α_i, with probability at least $1 - \varphi/M$, since $O\left(\frac{1}{\alpha_i^2 \nu_i}\left(\delta \log \frac{1}{\nu_i} + \log \frac{M}{\varphi}\right)\right) = O(U)$.

In particular, Theorem 7.7 holds for N, with probability at least φ, for parameters α_i and ν_i, for all $i = 1, \ldots, M$. Indeed, the probability that it fails for any value of i is bounded by $M(\varphi/M) = \varphi$.

Next, consider a range $\mathbf{r} \in \mathcal{F}$, such that $\overline{m} = \overline{m}(\mathbf{r}) \in [(i-1)\varepsilon^2/800, i\varepsilon^2/800]$ and $\overline{s} = \overline{s}(\mathbf{r})$.

If $i > 1$, we have that $\nu_i/2 \le \overline{m} \le \nu_i$, and as such

$$\alpha_i \overline{m} \le \alpha_i \nu_i = \sqrt{\alpha_i^2 \nu_i \nu_i} \le \sqrt{\alpha_i^2 \nu_i}\,\sqrt{2\overline{m}} = \sqrt{\frac{\varepsilon^2}{4c}}\,\sqrt{2\overline{m}} \le \frac{\varepsilon\sqrt{\overline{m}}}{20}, \tag{7.2}$$

by making c sufficiently large. Now, we have that $d_{\nu_i}(\overline{s}, \overline{m}) \le \alpha_i$, which implies, by Corollary 7.11, that

$$\left|\overline{s} - \overline{m}\right| \le \Delta' = \frac{2\alpha_i}{1-\alpha_i}\overline{m} + \frac{\alpha_i \nu_i}{1+\alpha_i} \le 4\alpha_i \overline{m} + \alpha_i \nu_i \le 6\alpha_i \overline{m} \le \varepsilon\sqrt{\overline{m}} \le \frac{\varepsilon}{2}\left(\sqrt{\overline{m}(\mathbf{r})} + \varepsilon\right),$$

since $\alpha_i \le 1/2$. For $i = 1$, we have $\overline{m} \le \nu_1$, and

$$\left|\overline{s} - \overline{m}\right| \le \Delta' = \frac{2\alpha_1}{1-\alpha_1}\overline{m} + \frac{\alpha_1 \nu_1}{1+\alpha_1} \le 6\alpha_1 \nu_1 = 6\frac{\varepsilon^2}{4c} \le \frac{\varepsilon^2}{2} \le \frac{\varepsilon}{2}\left(\sqrt{\overline{m}(\mathbf{r})} + \varepsilon\right),$$

by picking $c \ge 3$. ■

Looking on the bounds of sensitive ε-approximation as compared to ε-sample, it's natural to ask whether its size can be improved, but observe that since such a sample is also an ε^2-net and it is known that $\Omega\big((\delta/\varepsilon^2)\log(1/\varepsilon)\big)$ is a lower bound on the size of such a net [**KPW92**], this implies that such an improvement is impossible.

7.2.3. Relative ε-approximation.

Definition 7.16. A subset $N \subset \mathsf{X}$ is a ***relative*** (p, ε)***-approximation*** if for each $\mathbf{r} \in \mathcal{F}$ we have:

(i) If $\overline{m}(\mathbf{r}) \ge p$, then $(1-\varepsilon)\overline{m}(\mathbf{r}) \le \overline{s}(\mathbf{r}) \le (1+\varepsilon)\overline{m}(\mathbf{r})$.
(ii) If $\overline{m}(\mathbf{r}) \le p$, then $\overline{s}(\mathbf{r}) \le (1+\varepsilon)p$.

The concept was introduced by [**CKMS06**], except that property (ii) was not required. However, property (ii) is just an easy (but useful) "monotonicity" property that holds for all the constructions that the author is aware of.

There are relative approximations of size (roughly) $1/(\varepsilon^2 p)$. As such, relative approximation is interesting in the case where $p \ll \varepsilon$. Then, we can approximate ranges of weight larger than p with a sample that has only linear dependency on $1/p$. Otherwise, we would have to use the regular p-sample, and there the required sample is of size (roughly) $1/p^2$.

Theorem 7.17. *A sample N of size $O\left(\frac{1}{\varepsilon^2 p}\left(\delta \log \frac{1}{p} + \log \frac{1}{\varphi}\right)\right)$ is a relative (p, ε)-approximation with probability $\ge 1 - \varphi$.*

Proof. Set $\nu = p/2$ and $\alpha = \varepsilon/9$, and apply Theorem 7.7. We get that, for any range $\mathbf{r} \in \mathcal{F}$, such that $\overline{m} = \overline{m}(\mathbf{r})$ and $s = \overline{s}(\mathbf{r})$, by Corollary 7.11(ii),

$$\left|\overline{s} - \overline{m}\right| \le \Delta' = \frac{2\alpha}{1-\alpha}\overline{m} + \frac{\alpha\nu}{1+\alpha} = \frac{2\varepsilon/9}{1-\varepsilon/9}\overline{m} + \frac{p\varepsilon}{18(1+\varepsilon/9)} \le \frac{\varepsilon}{4}\overline{m} + \frac{p\varepsilon}{4},$$

where $\overline{s} = \overline{s}(\mathbf{r})$. For $\overline{m} \ge p$ this implies that $|\overline{s} - \overline{m}| \le \varepsilon\overline{m}$. Similarly, if $\overline{m} \le p$, then $\overline{s} \le \overline{m} + p\varepsilon/2 \le (1+\varepsilon)p$. ■

In fact, one can slightly strengthen the concept by making it "sensitive".

Theorem 7.18. *A sample N of size $O\left(\frac{1}{\varepsilon^2 p}\left(\delta \log \frac{1}{p} + \log \frac{1}{\varphi}\right)\right)$ is a relative $\left(ip, \varepsilon/\sqrt{i}\right)$-approximation with probability $\ge 1 - \varphi$, for all $i \ge 0$.*

Namely, for any range $\mathbf{r} \in \mathcal{F}$, such that $\overline{m}(\mathbf{r}) \ge ip$, we have

$$\left(1 - \frac{\varepsilon}{\sqrt{i}}\right)\overline{m}(\mathbf{r}) \le \overline{s}(\mathbf{r}) \le \left(1 + \frac{\varepsilon}{\sqrt{i}}\right)\overline{m}(\mathbf{r}). \tag{7.3}$$

Proof. Set $p_i = ip$ and $\varepsilon_i = \varepsilon/\sqrt{i}$, for $i = 1, \ldots, 1/p$. Now, apply Theorem 7.17, and observe that all the samples needed are asymptotically of the same size; that is,

$$O\left(\frac{1}{\varepsilon_i^2 p_i}\left(\delta \log \frac{1}{p_i} + \log \frac{1}{p\varphi}\right)\right) = O\left(\frac{1}{\varepsilon^2 p}\left(\delta \log \frac{1}{ip} + \log \frac{1}{p\varphi}\right)\right) = O\left(\frac{1}{\varepsilon^2 p}\left(\delta \log \frac{1}{p} + \log \frac{1}{\varphi}\right)\right).$$

As such, one can use the same sample to get this guarantee for all i, and the probability of this sample to fail for any i is at most $(1/p)p\varphi = \varphi$, as desired. ■

Interestingly, sensitive approximations imply relative approximations.

Lemma 7.19. *Let $\varepsilon, p > 0$ be parameters, and let $\varepsilon' = \varepsilon\sqrt{p}$. Then, if N is a sensitive ε'-approximation to the set system $(\mathsf{X}, \mathcal{F})$, then it is also a relative (ε, p)-approximation.*

Proof. We know that $\forall \mathbf{r} \in \mathcal{F}$, $\left|\overline{m}(\mathbf{r}) - \overline{s}(\mathbf{r})\right| \le \frac{\varepsilon'}{2}\left(\sqrt{\overline{m}(\mathbf{r})} + \varepsilon'\right)$. As such, for $\mathbf{r} \in \mathcal{F}$, if $\overline{m}(\mathbf{r}) = \alpha p$ and $\alpha \ge 1$, then we have

$$\left|\overline{m}(\mathbf{r}) - \overline{s}(\mathbf{r})\right| \le \frac{\varepsilon\sqrt{p}}{2}(\sqrt{\alpha p} + \varepsilon\sqrt{p}) = \frac{\varepsilon^2 p}{2} + \frac{\varepsilon}{2}\sqrt{\alpha}p \le \left(\frac{\varepsilon^2}{2} + \frac{\varepsilon}{2}\right)\alpha p \le \varepsilon\overline{m}(\mathbf{r}),$$

since $\varepsilon < 1$. This implies that N is a relative (ε, p)-approximation. ■

7.3. Proof of Theorem 7.7

7.3.1. Why the sample works for a single function. Let us start by proving the claim for a single range. That is, we want to show that a random sample estimates a single range correctly. In this case, even this is not trivial.

We remind the reader that the ***variance*** of a random variable X is defined to be the quantity $\mathbf{V}[X] = \mathbf{E}\left[(X - \mathbf{E}[X])^2\right] = \mathbf{E}\left[X^2\right] - \mathbf{E}[X]^2$.

Lemma 7.20. *Let X be a random variable in the range $I = [0, M]$, with expectation μ. Then $\mathbf{V}[X] \le \mu(M - \mu)$.*

Proof. For simplicity of exposition assume X is discrete. We then have that

$$\begin{aligned}\mathbf{V}\left[X\right] &= \mathbf{E}\left[X^2\right] - \mu^2 = \sum_x x^2 \,\mathbf{Pr}\left[X = x\right] - \mu^2 \le M \sum_x x \,\mathbf{Pr}\left[X = x\right] - \mu^2 \\ &= M\mu - \mu^2 = \mu(M - \mu).\end{aligned}$$

■

Lemma 7.21. *Let $X_1, \ldots, X_\xi \in [0, M]$ be ξ random independent variables each with expectation μ. Let $Y = \sum_{i=1}^{\xi} X_i/\xi$. Then $\mathbf{E}[Y] = \mu$ and $\mathbf{V}[Y] \le \mu(M-\mu)/\xi$.*

Proof. For any constant c, we have $\mathbf{V}[cX] = c^2\,\mathbf{V}[X]$, and for two independent variables X and Y, we have $\mathbf{V}[X+Y] = \mathbf{V}[X] + \mathbf{V}[Y]$. As such, by Lemma 7.20, we have

$$\mathbf{V}[Y] = \frac{\mathbf{V}[X_1] + \ldots + \mathbf{V}\left[X_\xi\right]}{\xi^2} \le \frac{\xi\mu(M-\mu)}{\xi^2} \le \frac{\mu(M-\mu)}{\xi}.$$

■

Lemma 7.22. *Let $f : \mathsf{X} \to [0,1]$ be a function, and let N be a random sample (with repetition) from X of size ξ. Let $\overline{s} = \overline{s}_N(f)$ and $\overline{m} = \overline{m}(f)$. Then $\mathbf{E}[\overline{s}] = \overline{m}$ and $\mathbf{V}[\overline{s}] \le \overline{m}(1-\overline{m})/\xi$.*

Proof. Let X_i be the value of f on the ith sample of N, for $i = 1, \ldots, \xi$. Clearly, $\mathbf{E}[X_i] = \overline{m}(f)$. Observing that $\overline{s} = (\sum_{i=1}^{\xi} X_i)/\xi$ and using Lemma 7.21 implies the result. ■

Lemma 7.23. *Let α, ν be parameters, let X be a ground set, and let f be a function from X to $\left[0,1\right]$. Then, for a random sample N (with repetition) from X of size $\xi = O\left(\frac{1}{\alpha^2\nu}\right)$, we have, with probability $\ge 3/4$, that $d_\nu(\overline{m}, \overline{s}) < \alpha/2$, where $\overline{m} = \overline{m}(f)$ and $\overline{s} = \overline{s}_N(f)$.*

Proof. Corollary 7.11 implies that if

$$\left|\overline{s} - \overline{m}\right| \le \Delta = (2\overline{m} + \nu)\frac{\alpha/2}{1+\alpha/2},$$

then $d_\nu(\overline{m}, \overline{s}) < \alpha/2$. As such, we need to bound the probability that $|\overline{s} - \overline{m}| \le \Delta$. To this end, we will use Chebychev's inequality. Formally, as in the proof of Lemma 7.22, let $\overline{s} = (\sum_{i=1}^{\xi} X_i)/\xi$, where X_i is the value of f on the ith sample point of N. Assume, for the time being, that for $t = 2$, we have $\Delta \ge t\sqrt{\mathbf{V}[\overline{s}]}$. Then, by Chebychev's inequality (Theorem 27.3$_{\text{p335}}$) applied to $\overline{s}$, we have that

$$\mathbf{Pr}\left[d_\nu(\overline{m}, \overline{s}) < \alpha/2\right] \ge \mathbf{Pr}\left[\left|\overline{s} - \overline{m}\right| \le \Delta\right] \ge \mathbf{Pr}\left[\left|\overline{s} - \overline{m}\right| \le t\sqrt{\mathbf{V}[\overline{s}]}\right] \ge 1 - \frac{1}{t^2} = \frac{3}{4},$$

since $\mathbf{E}[\overline{s}] = \overline{m}$.

Now, consider the quantity $t\sqrt{\mathbf{V}[\overline{s}]}$ (for $t = 2$) and observe that by Lemma 7.22, the variance of $\mathbf{V}[\overline{s}]$ (which is bounded by $\overline{m}(1-\overline{m})/\xi$) decreases as ξ increases, since $0 \le \overline{m} \le 1$. As such, we would like to pick ξ as small as possible such that $t\sqrt{\mathbf{V}[\overline{s}]} \le 2\sqrt{\overline{m}(1-\overline{m})/\xi} \le \Delta$. The last inequality can be rewritten as

$$2\sqrt{\frac{\overline{m}(1-\overline{m})}{\xi}} \le \Delta = \frac{\alpha/2}{1+\alpha/2}(2\overline{m}+\nu) \iff 2 \le (2\overline{m}+\nu)\frac{\alpha/2}{1+\alpha/2}\sqrt{\frac{\xi}{\overline{m}(1-\overline{m})}}.$$

However, the last inequality holds for $\xi \ge 16/(\alpha^2\nu)$, since

$$2 \le \frac{\sqrt{\alpha^2\nu\xi}}{2} \le \frac{\alpha/2}{2}\sqrt{\frac{4\xi\overline{m}\nu}{\overline{m}(1-\overline{m})}} \le \frac{\alpha/2}{1+\alpha/2}\sqrt{\frac{\xi(2\overline{m}+\nu)^2}{\overline{m}(1-\overline{m})}}$$

$$\le (2\overline{m}+\nu)\frac{\alpha/2}{1+\alpha/2}\sqrt{\frac{\xi}{\overline{m}(1-\overline{m})}}.$$

We conclude that for $\xi \ge 16/(\alpha^2\nu)$ we have $\mathbf{Pr}[d_\nu(\overline{m}, \overline{s}) < \alpha/2] \ge 3/4$, as desired. ■

7.3.2. Reduction to double sampling. In the above we proved that for a single function in $\mathcal{F}$ is estimated correctly with reasonable probability. To prove Theorem 7.7, we need to extend this to all the functions in $\mathcal{F}$. Specifically, we need to bound the probability that there exists a function $f \in \mathcal{F}$ such that the random sample S (of size m) fails to estimate it correctly; see Theorem 7.7.

The problem is, of course, that $\mathcal{F}$ is an infinite family. What we do in the following is to reduce bounding the probability of a bad event over an infinite family into a finite event that we can bound the probability of using a direct combinatorial argument.

To this end, for $f \in \mathcal{F}$, let $\mathcal{E}_{1,f}$ denote the event that the sample failed for the function f; specifically, we have

$$\mathcal{E}_{1,f} \equiv d_\nu\left(\overline{m}(f), \overline{s}_S(f)\right) > \alpha.$$

The basic events in the probability space are all the possible values of the sample S; as such $\mathcal{E}_{1,f}$ is the set of values of S for which f is not being estimated correctly. Giving in to the urge to be formal, we have that $\mathcal{E}_{1,f} = \left\{ S \,\middle|\, S \in \mathsf{X}^m \text{ and } d_\nu\left(\overline{m}(f), \overline{s}_S(f)\right) > \alpha \right\}$. For the sake of simplicity of exposition, we will try to be slightly less formal, but it is a good idea to keep track of the underlying probability space that we are arguing about.

As such, the bad event that S fails for any function of $\mathcal{F}$ is the event

$$\mathcal{E}_1 = \bigcup_{f \in \mathcal{F}} \mathcal{E}_{1,f},$$

and our purpose here is to bound $\mathbf{Pr}[\mathcal{E}_1]$.

Let T be a second sample of size m, and let

$$\mathcal{E}_{T \text{ good est. of } f} \equiv d_\nu\left(\overline{m}, \overline{s}_T(f)\right) < \frac{\alpha}{2}$$

be the event that f is being estimated correctly by T. Next, let

$$\mathcal{E}_{2,f} \equiv \mathcal{E}_{1,f} \cap \mathcal{E}_{T \text{ good est. of } f} = d_\nu\left(\overline{m}, \overline{s}_S(f)\right) > \alpha \quad \text{and} \quad d_\nu\left(\overline{m}, \overline{s}_T(f)\right) < \frac{\alpha}{2}.$$

By Lemma 7.23, for a fixed function f, we know that $\mathbf{Pr}\left[\mathcal{E}_{T \text{ good est. of } f}\right] \geq 1/2$ if m is sufficiently large.

Consider the event $\mathcal{E}_2 = \bigcup_{f \in \mathcal{F}} \mathcal{E}_{2,f}$. Intuitively, we expect that $\mathcal{E}_{2,f} \approx \mathcal{E}_{1,f}$ and thus $\mathcal{E}_2 \approx \mathcal{E}_1$, and as such, we can bound $\mathbf{Pr}[\mathcal{E}_2]$ instead of bounding $\mathbf{Pr}[\mathcal{E}_1]$. Clearly, $\mathcal{E}_2 \subseteq \mathcal{E}_1$, and

$$\mathbf{Pr}\left[\mathcal{E}_2 \,\middle|\, \mathcal{E}_1\right] = \frac{\mathbf{Pr}[\mathcal{E}_2 \cap \mathcal{E}_1]}{\mathbf{Pr}[\mathcal{E}_1]} = \frac{\mathbf{Pr}[\mathcal{E}_2]}{\mathbf{Pr}[\mathcal{E}_1]}.$$

However, if $\mathcal{E}_1$ happens (i.e., $S \in \mathcal{E}_1$), then there is at least one function $g^S \in \mathcal{F}$ that S fails for; fix this function (conceptually, think about fixing both S and g^S). Now, S and T are independent, and the event $\mathcal{E}_{1,f}$ depends only on S and the event $\mathcal{E}_{T \text{ good est. of } f}$ depends only on T; namely, the events $S \in \mathcal{E}_{1,f}$ and $T \in \mathcal{E}_{T \text{ good est. of } f}$ are independent, for any f. Now, by Lemma 7.23, we have that

$$\begin{aligned}
\mathbf{Pr}\left[\mathcal{E}_2 \,\middle|\, \mathcal{E}_1\right] &= \mathbf{Pr}\left[(S,T) \in \mathcal{E}_2 \,\middle|\, S \in \mathcal{E}_1\right] \geq \mathbf{Pr}\left[(S,T) \in \mathcal{E}_{2,g^S} \,\middle|\, S \in \mathcal{E}_1\right] \\
&= \mathbf{Pr}\left[\left(S \in \mathcal{E}_{1,g^S}\right) \cap \left(T \in \mathcal{E}_{T \text{ good est. of } g^S}\right) \,\middle|\, S \in \mathcal{E}_1\right] \\
&= \mathbf{Pr}\left[T \in \mathcal{E}_{T \text{ good est. of } g^S} \,\middle|\, S \in \mathcal{E}_1\right] \\
&= \mathbf{Pr}\left[\mathcal{E}_{T \text{ good est. of } g^S}\right] \geq \frac{3}{4} \geq \frac{1}{2}.
\end{aligned}$$

This implies that $\mathbf{Pr}[\mathcal{E}_1] \le 2\,\mathbf{Pr}[\mathcal{E}_2]$. Now, by the triangle inequality applied to $d_\nu(\cdot,\cdot)$ (see Lemma 7.8), we have that if $\mathcal{E}_{2,f}$ happens, then

$$\frac{\alpha}{2} + d_\nu\Big(\overline{s}_T(f)\,,\overline{s}_S(f)\Big) > d_\nu\Big(\overline{m},\overline{s}_T(f)\Big) + d_\nu\Big(\overline{s}_T(f)\,,\overline{s}_S(f)\Big) \ge d_\nu\Big(\overline{m},\overline{s}_S(f)\Big) > \alpha.$$

This implies that the event

$$\mathcal{E}'_{2,f} \equiv d_\nu\Big(\overline{s}_T(f)\,,\overline{s}_S(f)\Big) > \frac{\alpha}{2}$$

happens, and let $\mathcal{E}'_2 = \bigcup_f \mathcal{E}'_{2,f}$. We have that if $\mathcal{E}_2$ happens, then $\mathcal{E}'_2$ happens; that is, $\mathcal{E}_2 \subseteq \mathcal{E}'_2$. We conclude that

$$\mathbf{Pr}[\mathcal{E}_1] \le 2\,\mathbf{Pr}[\mathcal{E}_2] \le 2\,\mathbf{Pr}[\mathcal{E}'_2]\,.$$

Namely, we reduced the task of bounding the probability that there exists a bad function for the given sample into a new event. The new event is to decide if there is a function such that the two estimates for the two samples disagree considerably.

As such, from this point on, we are interested in bounding the probability of the event $\mathcal{E}'_2$. Again, giving in to our urge to be formal, the event $\mathcal{E}'_2$ can be stated as

$$\mathcal{E}'_2 = \left\{(S,T)\;\middle|\;\exists f \in \mathcal{F} \text{ such that } d_\nu\Big(\overline{s}_T(f)\,,\overline{s}_S(f)\Big) > \frac{\alpha}{2}\right\}. \tag{7.4}$$

This does not look like a finite event and it is not. But we can bound the probability to be in this set by bounding a finite event. The reason is that we can fix the content of $S \cup T$ and limit the probability argument into what falls into S and what falls into T.

7.3.3. Double sampling. As in the proof of the ε-net theorem, we will bound the probability of failure of the sample, by using a double sampling argument. So, consider a sample $N = \langle \mathsf{p}_1, \ldots, \mathsf{p}_{2m}\rangle$ of size $2m$. Let Γ_{2m} be the set of permutations over $\{1,\ldots,2m\}$, such that for $\sigma \in \Gamma_{2m}$, we have for $1 \le i \le m$ that either $\sigma(i) = i$ and $\sigma(m+i) = m+i$ or $\sigma(i) = m+i$ and $\sigma(m+i) = i$. Namely, a permutation $\sigma \in \Gamma_{2m}$ either flips the pair $(i, m+i)$ or preserve it, for every i.

Now, if we take any random sequence of independent samples and remix it in any way, without looking at the values of the elements of the sequence, what we get is a random sequence.[①] In fact, this random sequence has the same distribution as the original random sequence.

As such, take a random permutation $\sigma \in \Gamma_{2m}$ and mix the random sample N to get a new sequence

$$N_\sigma = \big\langle \mathsf{p}_{\sigma(1)}, \mathsf{p}_{\sigma(2)}, \ldots, \mathsf{p}_{\sigma(2m)}\big\rangle\,.$$

Clearly, N and N_σ have *exactly* the same distribution. As such, bounding the probability for high disagreement between the first half of the sample of N and the second half can be done on N_σ. In particular, for $f \in \mathcal{F}$, let

$$\mu_1(f) = \frac{1}{m}\sum_{i=1}^{m} f(\mathsf{p}_{\sigma(i)}) \quad\text{and}\quad \mu_2(f) = \frac{1}{m}\sum_{i=m+1}^{2m} f(\mathsf{p}_{\sigma(i)})\,. \tag{7.5}$$

We will be interested in bounding the quantity $d_\nu(\mu_1(f)\,,\mu_2(f))$.

[①] Naturally, we can "lose" randomness by looking at the values. For example, think about sorting the sequence.

7.3.4. Some more definitions. Naturally, we can think about $(f(\mathsf{p}_1), \ldots, f(\mathsf{p}_{2m}))$ as a point in $[0,1]^{2m}$. As such, the ordered sample $N = \langle \mathsf{p}_1, \ldots, \mathsf{p}_{2m} \rangle$ induces a manifold in $\mathbb{R}^{2m}$, by considering the value of any function of $\mathcal{F}$ on the points of N. Formally, consider the set

$$\mathcal{M} = \left\{ f(N) = (f(\mathsf{p}_1), f(\mathsf{p}_2), \ldots, f(\mathsf{p}_{2m})) \;\middle|\; f \in \mathcal{F} \right\}.$$

The set $\mathcal{M}$ has pseudo-dimension δ. The reader can think of $\mathcal{M}$ as being a low-dimensional manifold. From this point on, we will use the notation $f = f(N)$ and $f_i = f(\mathsf{p}_i)$, for $i = 1, \ldots, 2m$.

We remind the reader that the ℓ_1-norm of a point $p \in \mathbb{R}^{2m}$ is $\|p\|_1 = \sum_{i=1}^{2m} |p_i|$. The ***measure*** of $f \in \mathbb{R}^{2m}$ is the quantity

$$\overline{m}(f) = \frac{1}{2m} \sum_{i=1}^{2m} f_i.$$

The ***measure distance*** between $f, g \in \mathcal{M}$ is

$$d_{\overline{m}}(f, g) = \overline{m}(|f - g|) = \frac{1}{2m} \|f - g\|_1 . \tag{7.6}$$

Note that since the triangle inequality holds for the $\|\cdot\|_1$, it also holds for $d_{\overline{m}}(\cdot)$ (which is thus a metric).

In the following discussion, fix $\alpha > 0$, and for any $f \in \mathcal{M}$, we would like to prove that $d_\nu(\mu_1(f), \mu_2(f)) \leq \alpha$. In particular, let $\mathcal{E}(f)$ denote the maximum value of ν, such that this inequality holds. That is, we have

$$\begin{aligned}
d_{\mathcal{E}(f)}\big(\mu_1(f), \mu_2(f)\big) = \frac{|\mu_1(f) - \mu_2(f)|}{\mu_1(f) + \mu_2(f) + \mathcal{E}(f)} &= \alpha \\
\iff \left|\mu_1(f) - \mu_2(f)\right| &= \alpha\left(\mu_1(f) + \mu_2(f) + \mathcal{E}(f)\right) \\
\iff \left|\mu_1(f) - \mu_2(f)\right| &= \alpha\left(\frac{f_\Sigma}{m} + \mathcal{E}(f)\right),
\end{aligned}$$

where $f_\Sigma = \sum_{i=1}^{2m} f_i$. As such, we define the ***error*** of f to be

$$\mathcal{E}(f) = \frac{\left|\mu_1(f) - \mu_2(f)\right|}{\alpha} - \frac{f_\Sigma}{m}. \tag{7.7}$$

This quantity monotonically increases as $|\mu_1(f) - \mu_2(f)|$ increases, and as such it measures very roughly how similar the two estimates of f are to each other.[②]

We would like to argue that, with good probability, this quantity is small for all $f \in \mathcal{M}$.

Observation 7.24. *For $g \in \mathcal{M}$, we have that $d_\nu\big(\mu_1(g), \mu_2(g)\big) \leq \alpha$ if and only if $\mathcal{E}(g) \leq \nu$.*

[②]Ultimately, the form of $\mathcal{E}(f)$ rises from its usage in the following proofs. The reader might consider calling $\mathcal{E}(f)$ – the error of f – a mistake. Calling it the distortion of f or the perversion of f might have been better. Maybe so, but words are just labels, weapons of fortune in conveying ideas. The reader is hopefully willing to suffer the arrows of mislabeling to get the underlying ideas. It is the nature of writing that authors rarely discuss such decisions, mainly because they are ad hoc and lead to pointless tangential discussions as demonstrated by this footnote.

7.3.5. Bounding the error. The following lemma bounds the probability that a single function g in our family (now, a function is just a vector with $2m$ coordinates, since we care about its values only at specific $2m$ points) is being incorrectly estimated. The estimation here is a bit weird – we randomly scramble the coordinates g (see (7.5)), and then compute how much the two parts of g diverge (see (7.7)).

Lemma 7.25. *Let $g \in [0,1]^{2m}$, and let $\sigma \in \Gamma_{2m}$ be a random permutation (chosen uniformly). Then, for any $\nu > 0$, we have that $\mathbf{Pr}[\mathcal{E}(g) \geq \nu] \leq 2\exp\bigl(-\alpha^2 \nu m\bigr)$.*

PROOF. By Observation 7.24, the bad event (i.e., $\mathcal{E}(g) \geq \nu$) is equivalent to the condition $d_\nu\bigl(\mu_1(g), \mu_2(g)\bigr) > \alpha$; that is,

$$d_\nu\bigl(\mu_1(g), \mu_2(g)\bigr) = \frac{|\mu_1(g) - \mu_2(g)|}{\mu_1(g) + \mu_2(g) + \nu} > \alpha \quad \Rightarrow \quad \bigl|\mu_1(g) - \mu_2(g)\bigr| > \alpha\bigl(\mu_1(g) + \mu_2(g) + \nu\bigr).$$

Since $\mu_1(g) = (1/m)\sum_{i=1}^{m} g_{\sigma(i)}$ and $\mu_2(g) = (1/m)\sum_{i=m+1}^{2m} g_{\sigma(i)}$, by multiplying both sides by m, we get

$$\left|\sum_{i=1}^{m}\bigl(g_{\sigma(i)} - g_{\sigma(m+i)}\bigr)\right| > \alpha\left(\sum_{i=1}^{2m} g_i + \nu m\right) = \alpha\bigl(g_\Sigma + \nu m\bigr).$$

To bound the probability of this bad event, let us set $X_i = g_{\sigma(i)} - g_{\sigma(m+i)}$. Clearly, since $\sigma \in \Gamma_{2m}$ either keeps this pair or flips it (with equal probability), it follows that X_i is equal to either $g_i - g_{m+i} \in [-1,1]$ or to $-(g_i - g_{m+i})$. In particular, $\mathbf{E}[X_i] = 0$, and $\mathbf{E}[\sum_i X_i] = 0$. By Hoeffding's inequality (Theorem 27.22$_{\text{p344}}$), we have that the probability of the bad event is bounded by

$$\tau = \mathbf{Pr}\left[\left|\sum_{i=1}^{m} X_i\right| > \alpha\,(g_\Sigma + m\,\nu)\right] \leq 2\exp\left(-\frac{2\bigl(\alpha(g_\Sigma + m\,\nu)\bigr)^2}{\sum_{i=1}^{m} 4(g_i - g_{m+i})^2}\right).$$

Observe that $g_i - g_{m+i} \in [-1,1]$, and as such $(g_i - g_{m+i})^2 \leq |g_i - g_{m+i}| \leq g_i + g_{m+i}$. In particular, $\sum_{i=1}^{m}(g_i - g_{m+i})^2 \leq \sum_{i=1}^{2m} g_i = g_\Sigma$. As such,

$$\begin{aligned}\tau &\leq 2\exp\left(-\frac{(\alpha(g_\Sigma + \nu m))^2}{2g_\Sigma}\right) \leq 2\exp\left(-\alpha^2\,\frac{(g_\Sigma)^2 + 2g_\Sigma \nu m + (\nu m)^2}{2g_\Sigma}\right)\\ &\leq 2\exp\bigl(-\alpha^2\,\nu m\bigr),\end{aligned}$$

as claimed. ∎

The following lemma testifies that a triangle-type inequality holds for the error.

Lemma 7.26. *Consider $f, g \in [-1,1]^{2m}$. Then we have $\mathcal{E}(f+g) \leq \mathcal{E}(f) + \mathcal{E}(g)$.*

PROOF. Since $\mu_1(\cdot)$ and $\mu_2(\cdot)$ are linear functions, we have

$$\begin{aligned}\mathcal{E}(f+g) &= \frac{|\mu_1(f+g) - \mu_2(f+g)|}{\alpha} - \frac{(f+g)_\Sigma}{m}\\ &= \frac{|\mu_1(f) - \mu_2(f) + \mu_1(g) - \mu_2(g)|}{\alpha} - \frac{f_\Sigma}{m} - \frac{g_\Sigma}{m}\\ &\leq \frac{|\mu_1(f) - \mu_2(f)|}{\alpha} - \frac{f_\Sigma}{m} + \frac{|\mu_1(g) - \mu_2(g)|}{\alpha} - \frac{g_\Sigma}{m} = \mathcal{E}(f) + \mathcal{E}(g).\end{aligned}$$

∎

One of the key new ingredients in the proof of Theorem 7.7 is the idea that if two functions $f, h \in [0,1]^{2m}$ are close to each other (in the ℓ_1-norm), then they have similar error. To this end, consider the function $g = f - h$, and observe that it must have a small

ℓ_1-norm. As such, we can strengthen Lemma 7.25 for g if the vector g is short (i.e., its ℓ_1-norm is small). Intuitively, what we are proving is that if h has a small error and $f - h$ is "short", then, with high probability, $f = h + g = h + (f - h)$ would also have small error, since the error complies with the triangle inequality.

Lemma 7.27. *Let $\alpha, v, c > 0$ be some constants, where $c \leq 2/3$. Also, let $g \in [-1, 1]^{2m}$, such that $\|g\|_1 \leq cvm$, where $\|g\|_1 = \sum_{i=1}^{2m} |g_i|$ is the ℓ_1-norm of g. Then, for a random permutation $\sigma \in \Gamma_{2m}$ (chosen uniformly), we have that* $\mathbf{Pr}[\mathcal{E}(g) > v] \leq 2\exp\left(-\frac{\alpha^2 vm}{36c}\right)$.

Proof. Arguing as in Lemma 7.25, the desired probability τ is bounded by

$$\tau \leq 2\exp\left(-\frac{2\left(\alpha(g_\Sigma + \nu m)\right)^2}{4\sum_{i=1}^{m}(g_i - g_{m+i})^2}\right).$$

Now, observe that

$$\sum_{i=1}^{m}(g_i - g_{m+i})^2 \leq 2\sum_{i=1}^{m}|g_i - g_{m+i}| \leq 2\sum_{i=1}^{2m}|g_i| \leq 2cvm,$$

since $g_i \in [-1, 1]$. Furthermore, the quantity $(g_\Sigma + \nu m)^2$ is minimized, under the condition $\sum_{i=1}^{2m}|g_i| \leq cvm$, when $g_\Sigma = -cvm$, which implies that

$$\left(g_\Sigma + \nu m\right)^2 \geq ((1-c)vm)^2 \geq \frac{v^2 m^2}{9},$$

since $c \leq 2/3$. This implies that

$$\tau \leq 2\exp\left(-2\alpha^2 \frac{(g_\Sigma + \nu m)^2}{4\sum_{i=1}^{m}(g_i - g_{m+i})^2}\right) \leq 2\exp\left(-\alpha^2 \frac{v^2 m^2/9}{4cvm}\right) \leq 2\exp\left(-\frac{\alpha^2 vm}{36c}\right).$$ ■

7.3.6. On the number of distinct functions. Let $\mathsf{S} = (\mathsf{X}, \mathcal{F})$ be a range space with ***pseudo-dimension*** δ. We are interested in how many "truly" distinct concepts S really has.

Definition 7.28. A subset $P \subseteq \mathcal{F}$ is an ***ε-packing*** of $\mathcal{F}$ if for any pair of functions $f, g \in P$, we have that

$$d_{\overline{m}}(f, g) = \overline{m}(|f - g|) \geq \varepsilon.$$

Furthermore, for any function $h \in \mathcal{F}$, there exists a function $h' \in P$, such that $d_{\overline{m}}(h, h') \leq \varepsilon$.

In the following we bound the size of such an ε-packing P.

For two sets X and Y, we denote $X \oplus Y = (X \cup Y) \setminus (X \cap Y)$. For two functions f and g, the symmetric difference $\widehat{\mathbf{r}}(f) \oplus \widehat{\mathbf{r}}(g)$ is the region lying between the two functions; see the figure on the right.

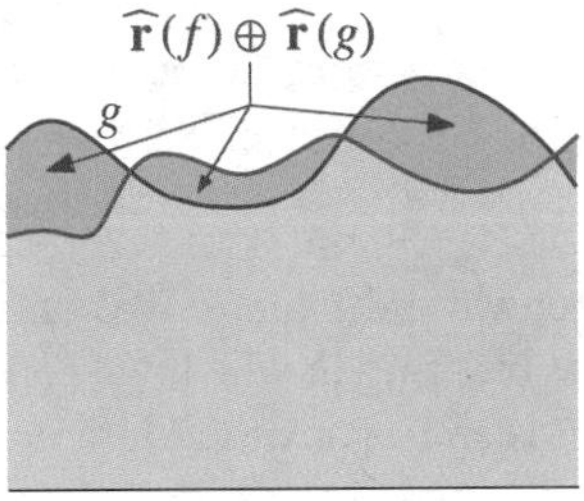

Lemma 7.29. *Let $\mathsf{S} = (\mathsf{X}, \mathcal{F})$ be a range space such that its* **pseudo-dimension** *is δ. Let $\varepsilon > 0$ be a parameter, and let $P \subseteq \mathcal{F}$ be an ε-packing of S. Then $|P| = 1/\varepsilon^{O(\delta)}$.*

Proof. As suggested in Remark 7.6, it would be easier to work in the induced range space

$$\mathsf{T} = (\mathsf{X}', \mathcal{G}),$$

where $\mathcal{G} = \left\{ \widehat{\mathbf{r}}(f) \;\middle|\; f \in \mathcal{F} \right\}$; see $(7.1)_{p104}$. The key observation here is that the VC dimension of T is δ.

Now, for $f \in \mathcal{F}$, we have that $\overline{m}(f) = \overline{m}(\widehat{\mathbf{r}}(f))$, where here $\overline{m}(\widehat{\mathbf{r}}(f))$ is just the measure of the set $\widehat{\mathbf{r}}(f)$. For any $f, g \in P$, we have

$$\varepsilon \le \overline{m}(|f - g|) = \overline{m}\left(\widehat{\mathbf{r}}(f) \oplus \widehat{\mathbf{r}}(g)\right).$$

So, let us consider the range set

$$\mathcal{H} = \left\{ \mathbf{r} \oplus \mathbf{r}' \;\middle|\; \mathbf{r}, \mathbf{r}' \in \mathcal{G} \right\}$$

and the resulting range space $\mathsf{U} = (\mathsf{X}', \mathcal{H})$. Each range in it is formed by combining two ranges of T, and by Theorem 5.22, the VC dimension of U is $O(\delta)$.

By Theorem 5.28_{p71} (which we proved independently of the discussion in this chapter), the range space U has an ε-net $N \subseteq \mathsf{X}'$ of size $k = O((\delta/\varepsilon)\log(1/\varepsilon))$; specifically, a random sample of this size is the desired net with constant probability, and as such a net exists.

Now, consider two distinct functions $f, g \in P$. We have that $\overline{m}(|f - g|) \ge \varepsilon$, and the above discussion implies that $\overline{m}(\widehat{\mathbf{r}}(f) \oplus \widehat{\mathbf{r}}(g)) \ge \varepsilon$. As such, since $\mathbf{r} = \widehat{\mathbf{r}}(f) \oplus \widehat{\mathbf{r}}(g)$ is a range of U and N is an ε-net of U, it follows that $\mathbf{r} \cap N \neq \emptyset$. Namely, $\widehat{\mathbf{r}}(f) \cap N \neq \widehat{\mathbf{r}}(g) \cap N$. As such, $\widehat{\mathbf{r}}(f)$ and $\widehat{\mathbf{r}}(g)$ are distinct ranges, when we project U to N. We conclude that for

$$F' = P_{|N} = \left\{ \widehat{\mathbf{r}}(f) \cap N \;\middle|\; f \in P \right\},$$

we have that

$$|P| = |F'| \le \mathcal{G}_{O(\delta)}(|N|) = |N|^{O(\delta)} = k^{O(\delta)} = 1/\varepsilon^{O(\delta)},$$

by Lemma 5.9. ■

7.3.7. Chaining. We need to show that the error $\mathcal{E}(f) \le \nu$ for all $f \in \mathcal{M}$. The problem is that $\mathcal{M}$ is potentially an infinite set, and our tools can handle only a finite set of points (i.e., Lemma 7.25). To overcome this problem, we define a canonical set of packings of $\mathcal{M}$. So, let

$$\varepsilon_j = \frac{\nu}{2^{2j+4}},$$

for $j \ge 1$. Let $P_{-1} = \{h\}$, for any arbitrary point on $h \in \mathcal{M}$.

Next, for $j \ge 0$, let P_j be an ε_j-packing of $\mathcal{M}$. Here we are using the measure distance between points, which is $d_{\overline{m}}(f, g) = (1/2m)\sum_{i=1}^{2m} |f_i - g_i|$; see $(7.6)_{p113}$. As such, for any $f, g \in P_j$, it follows that $d_{\overline{m}}(f, g) \ge \varepsilon_j$.

One can build the packing P_j, for $j \ge 1$, by starting from the set P_{j-1}. If there is a point $f \in \mathcal{M}$, such that

$$\mathrm{dist}(f, P_j) = \min_{g \in P_j} \overline{m}(f - g) \ge \varepsilon_j,$$

then we add f to P_j. We continue in this fashion till no such point can be found. Note that the ordering in which we consider the points of $\mathcal{M}$ does not matter. Clearly, the resulting set is an ε_j-packing. More importantly, we have the property that

$$P_1 \subseteq P_2 \subseteq \cdots \subseteq P_j \subseteq \cdots.$$

In particular, let $\mathcal{M}^* = \bigcup_{j=0}^{\infty} P_j$. The set $\mathcal{M}^*$ is dense in $\mathcal{M}$, and as such it is sufficient to prove our claim for points on $\mathcal{M}^*$ (because the error function $\mathcal{E}(\cdot)$ is continuous). To see the density claim, observe that for any $\mathsf{p} \in \mathcal{M}$, there exists a sequence $\mathsf{p}_j \in P_j$, such that $d_{\overline{m}}\left(\mathsf{p}, \mathsf{p}_j\right) \le \varepsilon_j$. As such, $\lim_{j\to\infty} \mathsf{p}_j = \mathsf{p}$.

For a point $f \in \mathcal{M}^*$, let $\widehat{f_j}$ be its nearest neighbor in P_j. Formally,

$$\widehat{f_j} = \arg\min_{g \in P_j} \overline{m}(|f - g|) .$$

Since P_j is an ε_j-packing, we have that for any $f \in \mathcal{M}^*$, it follows that $d_{\overline{m}}\left(f, \widehat{f_j}\right) \le \varepsilon_j$. As such, observe that

$$d_{\overline{m}}\left(\widehat{f_j}, \widehat{f_{j+1}}\right) \le d_{\overline{m}}\left(\widehat{f_j}, f\right) + d_{\overline{m}}\left(f, \widehat{f_{j+1}}\right) \le \varepsilon_j + \varepsilon_{j+1} \le 2\varepsilon_j. \tag{7.8}$$

Also $f = \widehat{f_1} + \sum_{j=1}^{\infty} \left(\widehat{f_{j+1}} - \widehat{f_j}\right)$. As such, since the triangle inequality holds for the error function (see Lemma 7.26), we have

$$\mathcal{E}(f) \le \mathcal{E}\left(\widehat{f_1}\right) + \sum_{j=1}^{\infty} \mathcal{E}\left(\widehat{f_{j+1}} - \widehat{f_j}\right).$$

For $j \ge 1$, let

$$\nu_j = \nu \frac{\sqrt{j+1}}{3 \cdot 2^j}. \tag{7.9}$$

Also, assume that, for all f and j, we have

$$\mathcal{E}\left(\widehat{f_1}\right) \le \nu_1 \quad \text{and} \quad \mathcal{E}\left(\widehat{f_{j+1}} - \widehat{f_j}\right) \le \nu_j. \tag{7.10}$$

Then, we have that

$$\mathcal{E}(f) \le \nu_1 + \sum_{j=1}^{\infty} \nu_j = \nu \left(\frac{\sqrt{2}}{3 \cdot 2} + \sum_{j=1}^{\infty} \frac{\sqrt{j+1}}{3 \cdot 2^j} \right) \le \nu,$$

as can be easily verified.

Now, if the above holds for all $f \in \mathcal{M}^*$, then this implies the desired result (that is, Theorem 7.7). Indeed, Observation 7.24 implies that this would imply that for all $f \in \mathcal{M}$, we have that $d_\nu\left(\mu_1(g), \mu_2(g)\right) \le \alpha/2$ (for the sake of simplicity of exposition, we are fidgeting with the constants a bit here). This implies that $\mathcal{E}_2'$ does not happen (see $(7.4)_{\text{p112}}$), which implies that sample estimates all functions correctly, which is what we needed prove.

Thus, to complete the proof, we need to bound the probabilities that the assumptions of (7.10) do not hold.

Lemma 7.29 implies the following result.

Corollary 7.30. *For $j \ge 1$, we have $\left|P_j\right| \le 1/\varepsilon_j^{c\delta}$, where c is some constant.*

Lemma 7.31. *If m is of size as specified by Theorem 7.7, then* $\mathbf{Pr}\left[\exists f \in P_1 \; \mathcal{E}(f) \ge \nu_1\right] \le \frac{\varphi}{2}$.

Proof. By Corollary 7.30, the size of P_1 is bounded by $1/\varepsilon_1^{c\delta} = \left(2^6/\nu\right)^{c\delta} = (64/\nu)^{c\delta}$. We remind the reader that $\nu_1 = \nu\frac{\sqrt{2}}{6}$ (see (7.9)) and $m = O\left(\frac{1}{\alpha^2 \nu}\left(\delta \log \frac{1}{\nu} + \log \frac{1}{\varphi}\right)\right)$; see

Theorem 7.7. As such, by Lemma 7.25, for some constant c_1, we have

$$\begin{aligned}\mathbf{Pr}\Big[\exists f \in P_1 \;\; \mathcal{E}(f) \ge \nu_1\Big] &\le 2\,|P_1|\exp\!\left(-\alpha^2 \nu_1 m\right)\\ &\le 2\left(\frac{64}{\nu}\right)^{c\delta} \exp\!\left(-c_1 \alpha^2 \nu \frac{1}{\alpha^2 \nu}\left(\delta \log \frac{1}{\nu} + \log \frac{1}{\varphi}\right)\right)\\ &\le 2\left(\frac{64}{\nu}\right)^{c\delta} \exp\!\left(-c_1 \delta \log \frac{1}{\nu} - c_1 \log \frac{1}{\varphi}\right) = 2\left(\frac{64}{\nu}\right)^{c\delta} \nu^{c_1\delta} \varphi^{c_1} \le \frac{\varphi}{2},\end{aligned}$$

by making m (and thus c_1) sufficiently large. ■

Lemma 7.32. *If m is of size as specified by Theorem 7.7, then for any $j \ge 1$ and $f \in \mathcal{M}$, we have* $\mathbf{Pr}\Big[\exists f \in \mathcal{M} \;\; \mathcal{E}\big(\widehat{f}_{j+1} - \widehat{f}_j\big) \ge \nu_j\Big] \le \dfrac{\varphi}{2^{j+1}}$.

PROOF. Fix j, and consider the set

$$X = \left\{ \widehat{f}_{j+1} - \widehat{f}_j \;\middle|\; f \in \mathcal{M} \right\}.$$

For any $x \in X$, we have that $\overline{m}(|x|) \le 2\varepsilon_j$, by (7.8). As such, by definition (see $(7.6)_{\text{p113}}$), we have

$$\forall x \in X \;\; \|x\|_1 \le 4\varepsilon_j m = \frac{4}{2^{2j+4}} \nu m,$$

which implies that we can apply Lemma 7.27 to its vectors, with $c = c_j = 2^{-2j-3}/4c$.

Furthermore, an element of X is formed by the difference of a point of P_j and P_{j+1}, and as such

$$|X| \le \left|P_j\right| \cdot \left|P_{j+1}\right| \le \frac{1}{\varepsilon_j^{c\delta} \cdot \varepsilon_{j+1}^{c\delta}} \le \varepsilon_{j+1}^{-2c\delta} \le \left(\frac{2^{2j+4}}{\nu}\right)^{2c\delta},$$

by Corollary 7.30. We remind the reader that $\nu_j = \nu \dfrac{\sqrt{j+1}}{3 \cdot 2^j}$ (see (7.9)) and

$$m = O\!\left(\frac{1}{\alpha^2 \nu}\left(\delta \log \frac{1}{\nu} + \log \frac{1}{\varphi}\right)\right); \tag{7.11}$$

see Theorem 7.7. As such, by Lemma 7.27, we have

$$\begin{aligned}\mathbf{Pr}\Big[\exists x \in X \;\; \mathcal{E}(x) \ge \nu_j\Big] &\le 2\,|X| \exp\!\left(-\frac{\alpha^2 \nu_j m}{9 c_j}\right) \le 2\left(\frac{2^{2j+4}}{\nu}\right)^{2c\delta} \exp\!\left(-\frac{4\alpha^2\left(\nu \frac{\sqrt{j+1}}{3 \cdot 2^j}\right) m}{9 \cdot 2^{-2j-3}}\right)\\ &= 2\left(\frac{2^{2j+4}}{\nu}\right)^{2c\delta} \exp\!\left(-\frac{32 \cdot 2^j \sqrt{j+1}}{27} \alpha^2 \nu m\right).\end{aligned}$$

Now, by (7.11), we observe that there exists a constant c_2, such that

$$\begin{aligned}\mathbf{Pr}\Big[\exists x \in X \;\; \mathcal{E}(x) \ge \nu_j\Big] &\le 2\left(\frac{2^{2j+4}}{\nu}\right)^{2c\delta} \exp\!\left(-2^j \sqrt{j+1}\,\alpha^2 \nu \left(\frac{c_2}{\alpha^2 \nu}\left(\delta \log \frac{1}{\nu} + \log \frac{1}{\varphi}\right)\right)\right)\\ &\le \exp\!\left(1 + 2c\delta(2j+4) + 2c\delta \ln \frac{1}{\nu} - 2^j \sqrt{j+1}\left(c_2\left(\delta \log \frac{1}{\nu} + \log \frac{1}{\varphi}\right)\right)\right)\\ &\le \exp\!\big(-(j+1) - \ln \varphi\big) \le \frac{\varphi}{2^{j+1}},\end{aligned}$$

by making m (and thus c_2) sufficiently large. ■

Lemma 7.31 and Lemma 7.32 imply that the probability that our assumptions of $(7.10)_{p117}$ fail is at most

$$\frac{\varphi}{2} + \sum_{j=1}^{\infty} \frac{\varphi}{2^{j+1}} \leq \varphi,$$

implying that Theorem 7.7 holds. ■

7.4. Bibliographical notes

Theorem 7.7 is from Li et al. [**LLS01**], and we *very* loosely followed their presentation. Their work is the pinnacle of a long sequence of papers by Pollard [**Pol86**] and Haussler [**Hau92**]. The presentation in Section 7.2 follows Har-Peled and Sharir [**HS11**].

The author came up with the proof of Lemma 7.29, although better bounds are known; see Haussler [**Hau95**].

Another take on the proof of Theorem 7.7. The key step in the proof Theorem 7.7 was showing that the manifold $\mathcal{M}$ embeds into a short interval, when projected into a specifically chosen random vector. Lemma 7.29 tells us that the manifold $\mathcal{M}$ is low dimensional. From this point on, the proof now uses the chaining technique due to Kolmogorov. A somewhat similar argument was used recently to show that manifolds with low doubling dimension embed with low distortion [**AHY07**] (see also [**IN07**]). This interpretation of spaces of low VC dimension (so sorry, pseudo-dimension) as being low-dimensional in the sense of the low-dimension manifold is quite interesting.

7.5. Exercises

Exercise 7.1 (Playing with $d_\nu(\cdot,\cdot)$)**.** Prove the following properties of $d_\nu(\cdot,\cdot)$:
(A) For all $r, s \geq 0$, we have $0 \leq d_\nu(r, s) < 1$.
(B) For all non-negative $r \leq s \leq t$, we have $d_\nu(r, s) \leq d_\nu(r, t)$ and $d_\nu(s, t) \leq d_\nu(r, t)$.
(C) For $0 \leq r, s \leq M$, we have $\frac{|r-s|}{2M+\nu} \leq d_\nu(r, s) \leq d_\nu(r, t) \leq \frac{|r-s|}{\nu}$.

Exercise 7.2 (Distance between numbers and the triangle)**.** Prove Lemma 7.8. Namely, for any $r, s \geq 0$ and $\nu > 0$, consider the function

$$d_\nu(r, s) = \frac{|r - s|}{r + s + \nu}.$$

Prove that the triangle inequality holds for $d_\nu(\cdot,\cdot)$. Namely, we have $d_\nu(x, y) + d_\nu(y, z) > d_\nu(x, z)$.

(Warning: This exercise is tedious.)

CHAPTER 8

Sampling and the Moments Technique

8.1. Vertical decomposition

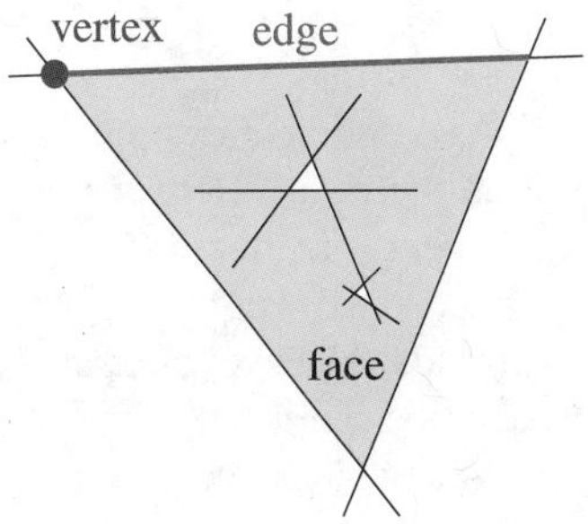

Given a set S of n segments in the plane, its ***arrangement***, denoted by $\mathcal{A}(\mathsf{S})$, is the decomposition of the plane into faces, edges, and vertices. The ***vertices*** of $\mathcal{A}(\mathsf{S})$ are the endpoints and the intersection points of the segments of S, the ***edges*** are the maximal connected portions of the segments not containing any vertex, and the ***faces*** are the connected components of the complement of the union of the segments of S. These definitions are depicted on the right.

We will be interested in computing the arrangement $\mathcal{A}(\mathsf{S})$ and a representation of it that makes it easy to manipulate. In particular, we would like to be able to quickly resolve questions of the type (i) are two points in the same face?, (ii) can one traverse from one point to the other without crossing any segment?, etc. The naive representation of each face as polygons (potentially with holes) is not conducive to carrying out such tasks, since a polygon might be arbitrarily complicated. Instead, we will prefer to break the arrangement into smaller canonical tiles.

To this end, a ***vertical trapezoid*** is a quadrangle with two vertical sides. The breaking of the faces into such trapezoids is the ***vertical decomposition*** of the arrangement $\mathcal{A}(\mathsf{S})$.

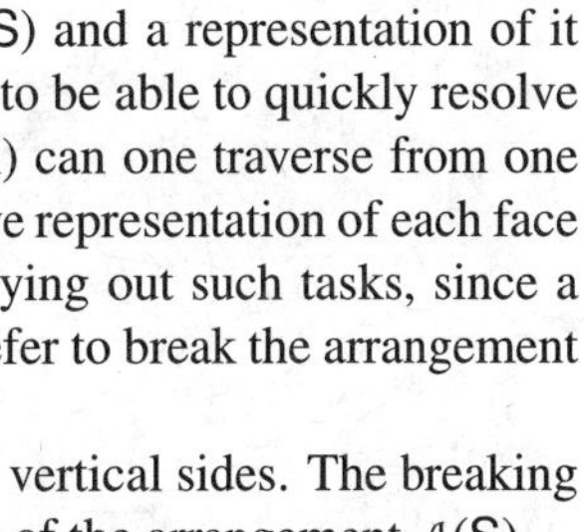

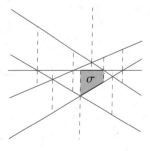

Formally, for a subset $\mathsf{R} \subseteq \mathsf{S}$, let $\mathcal{A}^{|}(\mathsf{R})$ denote the ***vertical decomposition*** of the plane formed by the arrangement $\mathcal{A}(\mathsf{R})$ of the segments of R. This is the partition of the plane into interior disjoint vertical trapezoids formed by erecting vertical walls through each vertex of $\mathcal{A}^{|}(\mathsf{R})$. Formally, a ***vertex*** of $\mathcal{A}^{|}(\mathsf{R})$ is either an endpoint of a segment of R or an intersection point of two of its segments. From each such vertex we shoot up (similarly, down) a vertical ray till it hits a segment of R or it continues all the way to infinity. See the figure on the right.

Note that a vertical trapezoid is defined by at most four segments: two segments defining its ceiling and floor and two segments defining the two intersection points that induce the two vertical walls on its boundary. Of course, a vertical trapezoid might be degenerate and thus be defined by fewer segments (i.e., an unbounded vertical trapezoid or a triangle with a vertical segment as one of its sides).

Vertical decomposition breaks the faces of the arrangement that might be arbitrarily complicated into entities (i.e., vertical trapezoids) of constant complexity. This makes handling arrangements much easier computationally.

In the following, we assume that the n segments of S have k pairwise intersection points overall, and we want to compute the arrangement $\mathcal{A} = \mathcal{A}(\mathsf{S})$; namely, compute the edges, vertices, and faces of $\mathcal{A}(\mathsf{S})$. One possible way is the following: Compute a random permutation of the segments of S: $\mathsf{S} = \langle s_1, \ldots, s_n \rangle$. Let $\mathsf{S}_i = \langle s_1, \ldots, s_i \rangle$ be the prefix of length i of S. Compute $\mathcal{A}^{|}(\mathsf{S}_i)$ from $\mathcal{A}^{|}(\mathsf{S}_{i-1})$, for $i = 1, \ldots, n$. Clearly, $\mathcal{A}^{|}(\mathsf{S}) = \mathcal{A}^{|}(\mathsf{S}_n)$, and we can extract $\mathcal{A}(\mathsf{S})$ from it. Namely, in the ith iteration, we insert the segment s_i into the arrangement $\mathcal{A}^{|}(\mathsf{S}_{i-1})$.

This technique of building the arrangement by inserting the segments one by one is called ***randomized incremental construction***.

8.1.1. Randomized incremental construction (RIC). Imagine that we had computed the arrangement $\mathcal{B}_{i-1} = \mathcal{A}^{|}(\mathsf{S}_{i-1})$. In the ith iteration we compute $\mathcal{B}_i$ by inserting s_i into the arrangement $\mathcal{B}_{i-1}$. This involves splitting some trapezoids (and merging some others).

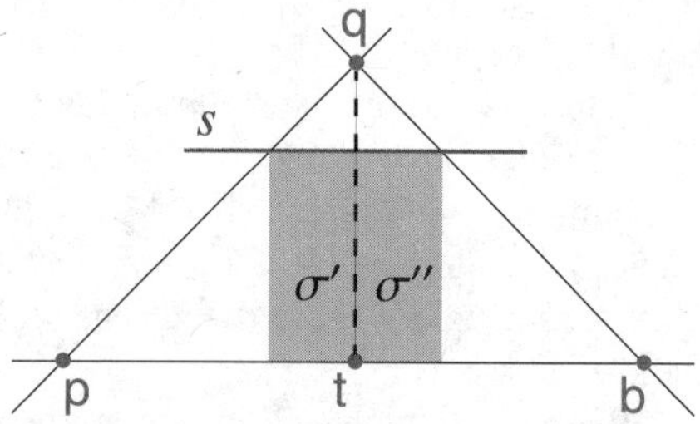

As a concrete example, consider the figure on the right. Here we insert s in the arrangement. To this end we split the "vertical trapezoids" $\triangle \mathsf{pqt}$ and $\triangle \mathsf{bqt}$, each into three trapezoids. The two trapezoids σ' and σ'' now need to be merged together to form the new trapezoid which appears in the vertical decomposition of the new arrangement. (Note that the figure does not show all the trapezoids in the vertical decomposition.)

To facilitate this, we need to compute the trapezoids of $\mathcal{B}_{i-1}$ that intersect s_i. This is done by maintaining a ***conflict graph***. Each trapezoid $\sigma \in \mathcal{A}^{|}(\mathsf{S}_{i-1})$ maintains a ***conflict list*** $\mathrm{cl}(\sigma)$ of *all* the segments of S that intersect its interior. In particular, the conflict list of σ cannot contain any segment of S_{i-1}, and as such it contains only the segments of $\mathsf{S} \setminus \mathsf{S}_{i-1}$ that intersect its interior. We also maintain a similar structure for each segment, listing all the trapezoids of $\mathcal{A}^{|}(\mathsf{S}_{i-1})$ that it currently intersects (in its interior). We maintain those lists with cross pointers, so that given an entry (σ, s) in the conflict list of σ, we can find the entry (s, σ) in the conflict list of s in constant time.

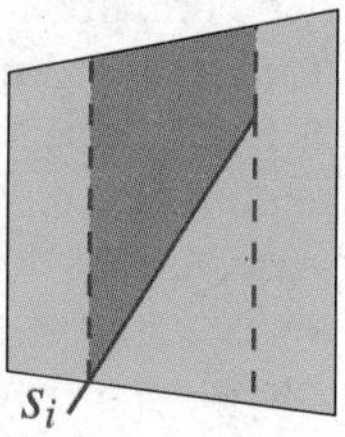

Thus, given s_i, we know what trapezoids need to be split (i.e., all the trapezoids in $\mathrm{cl}(s_i)$). Splitting a trapezoid σ by a segment s_i is the operation of computing a set of (at most) four trapezoids that cover σ and have s_i on their boundary. We compute those new trapezoids, and next we need to compute the conflict lists of the new trapezoids. This can be easily done by taking the conflict list of a trapezoid $\sigma \in \mathrm{cl}(s_i)$ and distributing its segments among the $O(1)$ new trapezoids that cover σ. Using careful implementation, this requires a linear time in the size of the conflict list of σ.

Note that only trapezoids that intersect s_i in their interior get split. Also, we need to update the conflict lists for the segments (that were not inserted yet).

We next sketch the low-level details involved in maintaining these conflict lists. For a segment s that intersects the interior of a trapezoid σ, we maintain the pair (s, σ). For every trapezoid σ, in the current vertical decomposition, we maintain a doubly linked list of all such pairs that contain σ. Similarly, for each segment s we maintain the doubly linked list of all such pairs that contain s. Finally, each such pair contains two pointers to the location in the two respective lists where the pair is being stored.

It is now straightforward to verify that using this data-structure we can implement the required operations in linear time in the size of the relevant conflict lists.

In the above description, we ignored the need to merge adjacent trapezoids if they have identical floor and ceiling – this can be done by a somewhat straightforward and tedious implementation of the vertical decomposition data-structure, by providing pointers between adjacent vertical trapezoids and maintaining the conflict list sorted (or by using hashing) so that merge operations can be done quickly. In any case, this can be done in linear time in the input/output size involved, as can be verified.

8.1.1.1. *Analysis.*

Claim 8.1. *The (amortized) running time of constructing $\mathcal{B}_i$ from $\mathcal{B}_{i-1}$ is proportional to the size of the conflict lists of the vertical trapezoids in $\mathcal{B}_i \setminus \mathcal{B}_{i-1}$ (and the number of such new trapezoids).*

Proof. Observe that we can charge all the work involved in the ith iteration to either the conflict lists of the newly created trapezoids or the deleted conflict lists. Clearly, the running time of the algorithm in the ith iteration is linear in the total size of these conflict lists. Observe that every conflict gets charged twice – when it is being created and when it is being deleted. As such, the (amortized) running time in the ith iteration is proportional to the total length of the newly created conflict lists. ■

Thus, to bound the running time of the algorithm, it is enough to bound the expected size of the destroyed conflict lists in ith iteration (and sum this bound on the n iterations carried out by the algorithm). Or alternatively, bound the expected size of the conflict lists created in the ith iteration.

Lemma 8.2. *Let S be a set of n segments (in general position*[①]*) with k intersection points. Let S_i be the first i segments in a random permutation of S. The expected size of $\mathcal{B}_i = \mathcal{A}^{|}(\mathsf{S}_i)$, denoted by $\tau(i)$ (i.e., the number of trapezoids in $\mathcal{B}_i$), is $O\left(i + k(i/n)^2\right)$.*

Proof. Consider[②] an intersection point $\mathsf{p} = s \cap s'$, where $s, s' \in \mathsf{S}$. The probability that p is present in $\mathcal{A}^{|}(\mathsf{S}_i)$ is equivalent to the probability that both s and s' are in S_i. This probability is

$$\alpha = \frac{\binom{n-2}{i-2}}{\binom{n}{i}} = \frac{(n-2)!}{(i-2)!\,(n-i)!} \cdot \frac{i!\,(n-i)!}{n!} = \frac{i(i-1)}{n(n-1)}.$$

For each intersection point p in $\mathcal{A}(\mathsf{S})$ define an indicator variable X_{p}, which is 1 if the two segments defining p are in the random sample S_i and 0 otherwise. We have that $\mathbf{E}\left[X_{\mathsf{p}}\right] = \alpha$, and as such, by linearity of expectation, the expected number of intersection points in the arrangement $\mathcal{A}(\mathsf{S}_i)$ is

$$\mathbf{E}\left[\sum_{\mathsf{p}\in V} X_{\mathsf{p}}\right] = \sum_{\mathsf{p}\in V} \mathbf{E}\left[X_{\mathsf{p}}\right] = \sum_{\mathsf{p}\in V} \alpha = k\alpha,$$

[①] In this case, no two intersection points of input segments are the same, no two intersection points (or vertices) have the same x-coordinate, no two segments lie on the same line, etc. Making the geometric algorithm work correctly for all degenerate inputs is a huge task that can usually be handled by tedious and careful implementation. Thus, we will always assume general position of the input. In other words, in theory all geometric inputs are inherently good, while in practice they are all evil (as anybody who tried to implement geometric algorithms can testify). The reader is encouraged not to use this to draw any conclusions on the human condition.

[②] The proof is provided in excruciating detail to get the reader used to this kind of argumentation. I would apologize for this pain, but it is a minor trifle, not to be mentioned, when compared to the other offenses in this book.

where V is the set of k intersection points of $\mathcal{A}(\mathsf{S})$. Also, every endpoint of a segment of S_i contributed its two endpoints to the arrangement $\mathcal{A}(\mathsf{S}_i)$. Thus, we have that the expected number of vertices in $\mathcal{A}(\mathsf{S}_i)$ is

$$2i + \frac{i(i-1)}{n(n-1)}k.$$

Now, the number of trapezoids in $\mathcal{A}^{|}(\mathsf{S}_i)$ is proportional to the number of vertices of $\mathcal{A}(\mathsf{S}_i)$, which implies the claim. ■

8.1.2. Backward analysis. In the following, we would like to consider the total amount of work involved in the ith iteration of the algorithm. The way to analyze these iterations is (conceptually) to run the algorithm for the first i iterations and then run "backward" the last iteration.

So, imagine that the overall size of the conflict lists of the trapezoids of $\mathcal{B}_i$ is W_i and the total size of the conflict lists created only in the ith iteration is C_i.

We are interested in bounding the expected size of C_i, since this is (essentially) the amount of work done by the algorithm in this iteration. Observe that the structure of $\mathcal{B}_i$ is defined independently of the permutation S_i and depends only on the (unordered) set $\mathsf{S}_i = \{s_1, \ldots, s_i\}$. So, fix S_i. What is the probability that s_i is a specific segment s of S_i? Clearly, this is $1/i$ since this is the probability of s being the last element in a permutation of the i elements of S_i (i.e., we consider a random permutation of S_i).

Now, consider a trapezoid $\sigma \in \mathcal{B}_i$. If σ was created in the ith iteration, then s_i must be one of the (at most four) segments that define it. Indeed, if s_i is not one of the segments that define σ, then σ existed in the vertical decomposition before s_i was inserted. Since $\mathcal{B}_i$ is independent of the internal ordering of S_i, it follows that $\mathbf{Pr}[\sigma \in (\mathcal{B}_i \setminus \mathcal{B}_{i-1})] \le 4/i$. In particular, the overall size of the conflict lists in the end of the ith iteration is

$$W_i = \sum_{\sigma \in \mathcal{B}_i} |\operatorname{cl}(\sigma)|.$$

As such, the expected overall size of the conflict lists created in the ith iteration is

$$\mathbf{E}\Big[C_i \,\Big|\, \mathcal{B}_i\Big] \le \sum_{\sigma \in \mathcal{B}_i} \frac{4}{i} \Big|\operatorname{cl}(\sigma)\Big| \le \frac{4}{i} W_i.$$

By Lemma 8.2, the expected size of $\mathcal{B}_i$ is $O(i + ki^2/n^2)$. Let us guess (for the time being) that on average the size of the conflict list of a trapezoid of $\mathcal{B}_i$ is about $O(n/i)$. In particular, assume that we know that

$$\mathbf{E}\Big[W_i\Big] = O\left(\left(i + \frac{i^2}{n^2}k\right)\frac{n}{i}\right) = O\left(n + k\frac{i}{n}\right),$$

by Lemma 8.2, implying

$$\mathbf{E}\Big[C_i\Big] = \mathbf{E}\Big[\mathbf{E}\Big[C_i \,\Big|\, \mathcal{B}_i\Big]\Big] \le \mathbf{E}\left[\frac{4}{i} W_i\right] = \frac{4}{i}\,\mathbf{E}[W_i] = O\left(\frac{4}{i}\left(n + \frac{ki}{n}\right)\right) = O\left(\frac{n}{i} + \frac{k}{n}\right), \tag{8.1}$$

using Lemma 27.8. In particular, the expected (amortized) amount of work in the ith iteration is proportional to $\mathbf{E}[C_i]$. Thus, the overall expected running time of the algorithm is

$$\mathbf{E}\left[\sum_{i=1}^{n} C_i\right] = \sum_{i=1}^{n} O\left(\frac{n}{i} + \frac{k}{n}\right) = O(n \log n + k).$$

Theorem 8.3. *Given a set S of n segments in the plane with k intersections, one can compute the vertical decomposition of $\mathcal{A}(\mathsf{S})$ in expected $O(n \log n + k)$ time.*

Intuition and discussion. What remains to be seen is how we came up with the guess that the average size of a conflict list of a trapezoid of $\mathcal{B}_i$ is about $O(n/i)$. Note that using ε-nets implies that the bound $O((n/i)\log i)$ holds with constant probability (see Theorem 5.28$_{p71}$) for all trapezoids in this arrangement. As such, this result is only slightly surprising. To prove this, we present in the next section a "strengthening" of ε-nets to geometric settings.

To get some intuition on how we came up with this guess, consider a set P of n points on the line and a random sample R of i points from P. Let $\widehat{I}$ be the partition of the real line into (maximal) open intervals by the endpoints of R, such that these intervals do not contain points of R in their interior.

Consider an interval (i.e., a one-dimensional trapezoid) of $\widehat{I}$. It is intuitively clear that this interval (in expectation) would contain $O(n/i)$ points. Indeed, fix a point x on the real line, and imagine that we pick each point with probability i/n to be in the random sample. The random variable which is the number of points of P we have to scan starting from x and going to the right of x till we "hit" a point that is in the random sample behaves like a geometric variable with probability i/n, and as such its expected value is n/i. The same argument works if we scan P to the left of x. We conclude that the number of points of P in the interval of $\widehat{I}$ that contains x but does not contain any point of R is $O(n/i)$ in expectation.

Of course, the vertical decomposition case is more involved, as each vertical trapezoid is defined by four input segments. Furthermore, the number of possible vertical trapezoids is larger. Instead of proving the required result for this special case, we will prove a more general result which can be applied in a lot of other settings.

8.2. General settings

8.2.1. Notation. Let S be a set of objects. For a subset $\mathsf{R} \subseteq \mathsf{S}$, we define a collection of 'regions' called $\mathcal{F}(\mathsf{R})$. For the case of vertical decomposition of segments (i.e., Theorem 8.3), the objects are segments, the regions are trapezoids, and $\mathcal{F}(\mathsf{R})$ is the set of vertical trapezoids in $\mathcal{A}^{|}(\mathsf{R})$. Let

$$\mathcal{T} = \mathcal{T}(\mathsf{S}) = \bigcup_{\mathsf{R} \subseteq \mathsf{S}} \mathcal{F}(\mathsf{R})$$

denote the set of ***all possible regions*** defined by subsets of S.

In the vertical trapezoids case, the set $\mathcal{T}$ is the set of all vertical trapezoids that can be defined by any subset of the given input segments.

We associate two subsets $D(\sigma), K(\sigma) \subseteq \mathsf{S}$ with each region $\sigma \in \mathcal{T}$.

The ***defining set*** $D(\sigma)$ of σ is the subset of S defining the region σ (the precise requirements from this set are specified in the axioms below). We assume that for every $\sigma \in \mathcal{T}$, $|D(\sigma)| \leq d$ for a (small) constant d. The constant d is sometime referred to as the ***combinatorial dimension***. In the case of Theorem 8.3, each trapezoid σ is defined by at most four segments (or lines) of S that define the region covered by the trapezoid σ, and this set of segments is $D(\sigma)$. See Figure 8.1.

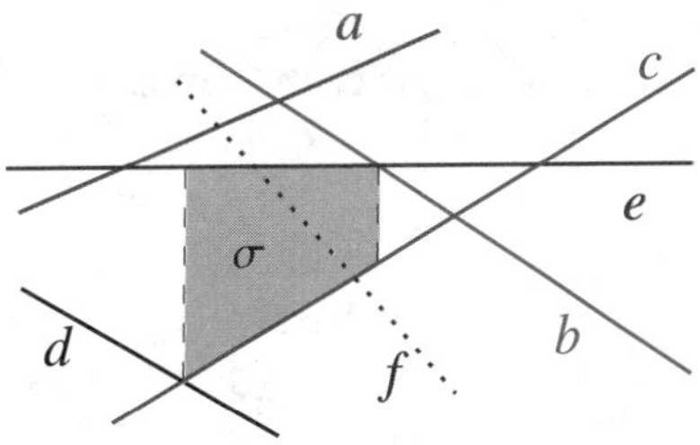

FIGURE 8.1. $D(\sigma) = \{b, c, d, e\}$ and $K(\sigma) = \{f\}$.

The ***stopping set*** $K(\sigma)$ of σ is the set of objects of S such that including any object of $K(\sigma)$ in R prevents σ from appearing in $\mathcal{F}(\mathsf{R})$. In many applications $K(\sigma)$ is just the set of objects intersecting the cell σ; this is also the case in Theorem 8.3, where $K(\sigma)$ is the set of segments of S intersecting the interior of the trapezoid σ (see Figure 8.1). Thus, the stopping set of a region σ, in many cases, is just the conflict list of this region, when it is being created by an RIC algorithm. The ***weight*** of σ is $\omega(\sigma) = |K(\sigma)|$.

Axioms. Let $\mathsf{S}, \mathcal{F}(\mathsf{R}), D(\sigma)$, and $K(\sigma)$ be such that for any subset $\mathsf{R} \subseteq \mathsf{S}$, the set $\mathcal{F}(\mathsf{R})$ satisfies the following axioms:

(i) For any $\sigma \in \mathcal{F}(\mathsf{R})$, we have $D(\sigma) \subseteq \mathsf{R}$ and $\mathsf{R} \cap K(\sigma) = \emptyset$.
(ii) If $D(\sigma) \subseteq \mathsf{R}$ and $K(\sigma) \cap \mathsf{R} = \emptyset$, then $\sigma \in \mathcal{F}(\mathsf{R})$.

8.2.2. Analysis. In the following, S is a set of n objects complying with axioms (i) and (ii).

8.2.2.1. *On the probability of a region to be created.* Inherently, to analyze a randomized algorithm using this framework, we will be interested in the probability that a certain region would be created. Thus, let

$$\rho_{r,n}\left(d, k\right)$$

denote the probability that a region $\sigma \in \mathcal{T}$ appears in $\mathcal{F}(\mathsf{R})$, where its defining set is of size d, its stopping set is of size k, R is a random sample of size r from a set S, and $n = |\mathsf{S}|$.

The sampling model. For describing algorithms it is usually easier to work with samples created by picking a subset of a certain size (without repetition) from the original set of objects. Usually, in the algorithmic applications this would be done by randomly permuting the objects and interpreting a prefix of this permutation as a random sample. Insisting on analyzing this framework in the "right" sampling model creates some non-trivial technical pain.

Lemma 8.4. *We have that* $\rho_{r,n}\left(d, k\right) \approx \left(1 - \frac{r}{n}\right)^k \left(\frac{r}{n}\right)^d$. *Formally,*

$$\frac{1}{2^{2d}} \left(1 - 4 \cdot \frac{r}{n}\right)^k \left(\frac{r}{n}\right)^d \le \rho_{r,n}\left(d, k\right) \le 2^{2d} \left(1 - \frac{1}{2} \cdot \frac{r}{n}\right)^k \left(\frac{r}{n}\right)^d. \tag{8.2}$$

PROOF. Let σ be the region under consideration that is defined by d objects and having k stoppers (i.e., $k = K(\sigma)$). We are interested in the probability of σ being created when taking a sample of size r (without repetition) from a set S of n objects. Clearly, this probability is $\rho_{r,n}\left(d, k\right) = \binom{n-d-k}{r-d} / \binom{n}{r}$, as we have to pick the d defining objects into the random sample and avoid picking any of the k stoppers. A tedious but careful calculation, delegated to Section 8.4, implies (8.2).

Instead, here is an elegant argument for why this estimate is correct in a slightly different sampling model. We pick every element of S into the sample R with probability r/n, and this is done independently for each object. In expectation, the random sample is of size r, and clearly the probability that σ is created is the probability that we pick its d defining objects (that is, $(r/n)^d$) multiplied by the probability that we did not pick any of its k stoppers (that is, $(1 - r/n)^k$). ■

Remark 8.5. The bounds of (8.2) hold only when r, d, and k are in certain (reasonable) ranges. For the sake of simplicity of exposition we ignore this minor issue. With care, all our arguments work when one pays careful attention to this minor technicality.

8.2.2.2. *On exponential decay.* For any natural number r and a number $t > 0$, consider R to be a random sample of size r from S without repetition. We will refer to a region $\sigma \in \mathcal{F}(\mathsf{R})$ as being ***t*-heavy** if $\omega(\sigma) \geq t(n/r)$. Let $\mathcal{F}_{\geq t}(\mathsf{R})$ denote all the t-heavy regions of $\mathcal{F}(\mathsf{R})$.[③] We intuitively expect the size of this set to drop fast as t increases. Indeed, Lemma 8.4 tells us that a trapezoid of weight $t(n/r)$ has probability

$$\rho_{r,n}\left(d, t(n/r)\right) \approx \left(1 - \frac{r}{n}\right)^{t(n/r)} \left(\frac{r}{n}\right)^{d} \approx \exp(-t) \cdot \left(\frac{r}{n}\right)^{d} \approx \exp(-t+1) \cdot \left(1 - \frac{r}{n}\right)^{n/r} \left(\frac{r}{n}\right)^{d}$$
$$\approx \exp(-t+1) \cdot \rho_{r,n}\left(d, n/r\right)$$

to be created, since $(1 - r/n)^{n/r} \approx 1/e$. Namely, a t-heavy region has exponentially lower probability to be created than a region of weight n/r. We next formalize this argument.

Lemma 8.6. *Let $r \leq n$ and let t be parameters, such that $1 \leq t \leq r/d$. Furthermore, let R be a sample of size r, and let R' be a sample of size $r' = \lfloor r/t \rfloor$, both from S. Let $\sigma \in \mathcal{T}$ be a trapezoid with weight $\omega(\sigma) \geq t(n/r)$. Then,* $\mathbf{Pr}[\sigma \in \mathcal{F}(\mathsf{R})] = O\left(\exp\left(-\frac{t}{2}\right) t^d\, \mathbf{Pr}[\sigma \in \mathcal{F}(\mathsf{R}')]\right)$.

PROOF. For the sake of simplicity of exposition, assume that $k = \omega(\sigma) = t(n/r)$. By Lemma 8.4 (i.e., (8.2)) we have

$$\frac{\mathbf{Pr}[\sigma \in \mathcal{F}(\mathsf{R})]}{\mathbf{Pr}[\sigma \in \mathcal{F}(\mathsf{R}')]} = \frac{\rho_{r,n}\left(d,k\right)}{\rho_{r',n}\left(d,k\right)} \leq \frac{2^{2d}\left(1 - \frac{1}{2}\cdot\frac{r}{n}\right)^{k}\left(\frac{r}{n}\right)^{d}}{\frac{1}{2^{2d}}\left(1 - 4\frac{r'}{n}\right)^{k}\left(\frac{r'}{n}\right)^{d}}$$
$$\leq 2^{4d}\exp\left(-\frac{kr}{2n}\right)\left(1 + 8\frac{r'}{n}\right)^{k}\left(\frac{r}{r'}\right)^{d} \leq 2^{4d}\exp\left(8\frac{kr'}{n} - \frac{kr}{2n}\right)\left(\frac{r}{r'}\right)^{d}$$
$$= 2^{4d}\exp\left(8\frac{tn\lfloor r/t\rfloor}{nr} - \frac{tnr}{2nr}\right)\left(\frac{r}{\lfloor r/t\rfloor}\right)^{d} = O\left(\exp(-t/2)\,t^d\right),$$

since $1/(1-x) \leq 1 + 2x$ for $x \leq 1/2$ and $1 + y \leq \exp(y)$, for all y. (The constant in the above $O(\cdot)$ depends exponentially on d.) ■

Let

$$\mathbf{E}f(r) = \mathbf{E}\left[\left|\mathcal{F}(\mathsf{R})\right|\right] \quad \text{and} \quad \mathbf{E}f_{\geq t}(r) = \mathbf{E}\left[\left|\mathcal{F}_{\geq t}(\mathsf{R})\right|\right],$$

where the expectation is over random subsets $\mathsf{R} \subseteq \mathsf{S}$ of size r. Note that $\mathbf{E}f(r) = \mathbf{E}f_{\geq 0}(r)$ is the expected number of regions created by a random sample of size r. In words, $\mathbf{E}f_{\geq t}(r)$ is the expected number of regions in a structure created by a sample of r random objects, such that these regions have weight which is t times larger than the "expected" weight (i.e., n/r). In the following, we assume that $\mathbf{E}f(r)$ is a monotone increasing function.

Lemma 8.7 (The exponential decay lemma). *Given a set S of n objects and parameters $r \leq n$ and $1 \leq t \leq r/d$, where $d = \max_{\sigma\in\mathcal{T}(\mathsf{S})}|D(\sigma)|$, if axioms (i) and (ii) above hold for any subset of S, then*

$$\mathbf{E}f_{\geq t}(r) = O\left(t^d \exp(-t/2)\ \mathbf{E}f(r)\right). \tag{8.3}$$

PROOF. Let R be a random sample of size r from S and let R' be a random sample of size $r' = \lfloor r/t \rfloor$ from S. Let $H = \bigcup_{X\subseteq\mathsf{S},|X|=r}\mathcal{F}_{\geq t}(X)$ denote the set of all t-heavy regions that might be created by a sample of size r. In the following, the expectation is taken over the content of the random samples R and R'.

[③]These are the regions that are at least t times overweight. Speak about an obesity problem.

For a region σ, let X_σ be the indicator variable that is 1 if and only if $\sigma \in \mathcal{F}(\mathsf{R})$. By linearity of expectation and since $\mathbf{E}[X_\sigma] = \mathbf{Pr}[\sigma \in \mathcal{F}(\mathsf{R})]$, we have

$$\begin{aligned}\mathbf{E}f_{\geq t}(r) &= \mathbf{E}\Big[\,|\mathcal{F}_{\geq t}(\mathsf{R})|\,\Big] = \mathbf{E}\Big[\sum_{\sigma\in H} X_\sigma\Big] = \sum_{\sigma\in H}\mathbf{E}\big[X_\sigma\big] = \sum_{\sigma\in H}\mathbf{Pr}\big[\sigma \in \mathcal{F}(\mathsf{R})\big]\\ &= O\Big(t^d \exp(-t/2)\sum_{\sigma\in H}\mathbf{Pr}[\sigma\in\mathcal{F}(\mathsf{R}')]\Big) = O\Big(t^d\exp(-t/2)\sum_{\sigma\in\mathcal{T}}\mathbf{Pr}[\sigma\in\mathcal{F}(\mathsf{R}')]\Big)\\ &= O\big(t^d\exp(-t/2)\,\mathbf{E}f(r')\big) = O\big(t^d\exp(-t/2)\,\mathbf{E}f(r)\big),\end{aligned}$$

by Lemma 8.6 and since $\mathbf{E}f(r)$ is a monotone increasing function. ■

8.2.2.3. *Bounding the moments.* Consider a different randomized algorithm that in a first round samples r objects, $\mathsf{R} \subseteq \mathsf{S}$ (say, segments), computes the arrangement induced by these r objects (i.e., $\mathcal{A}^{|}(\mathsf{R})$), and then inside each region σ it computes the arrangement of the $\omega(\sigma)$ objects intersecting the interior of this region, using an algorithm that takes $O\big((\omega(\sigma))^c\big)$ time, where $c > 0$ is some fixed constant. The overall expected running time of this algorithm is

$$\mathbf{E}\Bigg[\sum_{\sigma\in\mathcal{F}(\mathsf{R})}(\omega(\sigma))^c\Bigg].$$

We are now able to bound this quantity.

Theorem 8.8 (Bounded moments theorem)**.** *Let* $\mathsf{R} \subseteq \mathsf{S}$ *be a random subset of size* r. *Let* $\mathbf{E}f(r) = \mathbf{E}[|\mathcal{F}(\mathsf{R})|]$ *and let* $c \geq 1$ *be an arbitrary constant. Then,*

$$\mathbf{E}\Bigg[\sum_{\sigma\in\mathcal{F}(\mathsf{R})}(\omega(\sigma))^c\Bigg] = O\bigg(\mathbf{E}f(r)\Big(\frac{n}{r}\Big)^c\bigg).$$

PROOF. Let $\mathsf{R} \subseteq \mathsf{S}$ be a random sample of size r. Observe that all the regions with weight in the range $\left[(t-1)\frac{n}{r},\, t\cdot\frac{n}{r}\right)$ are in the set $\mathcal{F}_{\geq t-1}(\mathsf{R}) \setminus \mathcal{F}_{\geq t}(\mathsf{R})$. As such, we have by Lemma 8.7 that

$$\begin{aligned}\mathbf{E}\Bigg[\sum_{\sigma\in\mathcal{F}(\mathsf{R})}\omega(\sigma)^c\Bigg] &\leq \mathbf{E}\Bigg[\sum_{t\geq 1}\Big(t\frac{n}{r}\Big)^c\Big(|\mathcal{F}_{\geq t-1}(\mathsf{R})| - |\mathcal{F}_{\geq t}(\mathsf{R})|\Big)\Bigg] \leq \mathbf{E}\Bigg[\sum_{t\geq 1}\Big(t\frac{n}{r}\Big)^c\,|\mathcal{F}_{\geq t-1}(\mathsf{R})|\Bigg]\\ &\leq \Big(\frac{n}{r}\Big)^c\sum_{t\geq 0}(t+1)^c\cdot\mathbf{E}\Big[\,|\mathcal{F}_{\geq t}(\mathsf{R})|\,\Big]\\ &= \Big(\frac{n}{r}\Big)^c\sum_{t\geq 0}(t+1)^c\,\mathbf{E}f_{\geq t}(r) = \Big(\frac{n}{r}\Big)^c\sum_{t\geq 0}O\big((t+1)^{c+d}\exp(-t/2)\,\mathbf{E}f(r)\big)\\ &= O\Bigg(\mathbf{E}f(r)\Big(\frac{n}{r}\Big)^c\sum_{t\geq 0}(t+1)^{c+d}\exp(-t/2)\Bigg) = O\bigg(\mathbf{E}f(r)\Big(\frac{n}{r}\Big)^c\bigg),\end{aligned}$$

since c and d are both constants. ■

8.3. Applications

8.3.1. Analyzing the RIC algorithm for vertical decomposition. We remind the reader that the input of the algorithm of Section 8.1.2 is a set S of n segments with k intersections, and it uses randomized incremental construction to compute the vertical decomposition of the arrangement $\mathcal{A}(\mathsf{S})$.

Lemma 8.2 shows that the number of vertical trapezoids in the randomized incremental construction is in expectation $\mathbf{E}f(i) = O\left(i + k\,(i/n)^2\right)$. Thus, by Theorem 8.8 (used with $c = 1$), we have that the total expected size of the conflict lists of the vertical decomposition computed in the ith step is

$$\mathbf{E}\left[W_i\right] = \mathbf{E}\left[\sum_{\sigma \in \mathcal{B}_i} \omega(\sigma)\right] = O\left(\mathbf{E}f(i)\,\frac{n}{i}\right) = O\left(n + k\frac{i}{n}\right).$$

This is the missing piece in the analysis of Section 8.1.2. Indeed, the amortized work in the ith step of the algorithm is $O(W_i/i)$ (see $(8.1)_{\mathrm{p124}}$), and as such, the expected running time of this algorithm is

$$\mathbf{E}\left[O\left(\sum_{i=1}^{n} \frac{W_i}{i}\right)\right] = O\left(\sum_{i=1}^{n} \frac{1}{i}\left(n + k\frac{i}{n}\right)\right) = O(n \log n + k).$$

This implies Theorem 8.3.

8.3.2. Cuttings. Let S be a set of n lines in the plane, and let r be an arbitrary parameter. A $(1/r)$-***cutting*** of S is a partition of the plane into constant complexity regions such that each region intersects at most n/r lines of S. It is natural to try to minimize the number of regions in the cutting, as cuttings are a natural tool for performing "divide and conquer".

Consider the range space having S as its ground set and vertical trapezoids as its ranges (i.e., given a vertical trapezoid σ, its corresponding range is the set of all lines of S that intersect the interior of σ). This range space has a VC dimension which is a constant as can be easily verified. Let $X \subseteq \mathsf{S}$ be an ε-net for this range space, for $\varepsilon = 1/r$. By Theorem 5.28_{p71} (ε-net theorem), there exists such an ε-net X of this range space, of size $O((1/\varepsilon)\log(1/\varepsilon)) = O(r \log r)$. In fact, Theorem 5.28_{p71} states that an appropriate random sample is an ε-net with non-zero probability, which implies, by the probabilistic method, that such a net (of this size) exists.

Lemma 8.9. *There exists a $(1/r)$-cutting of a set of lines S in the plane of size $O\left((r \log r)^2\right)$.*

Proof. Consider the vertical decomposition $\mathcal{A}^{|}(X)$, where X is as above. We claim that this collection of trapezoids is the desired cutting.

The bound on the size is immediate, as the complexity of $\mathcal{A}^{|}(X)$ is $O\left(|X|^2\right)$ and $|X| = O(r \log r)$.

As for correctness, consider a vertical trapezoid σ in the arrangement $\mathcal{A}^{|}(X)$. It does not intersect any of the lines of X in its interior, since it is a trapezoid in the vertical decomposition $\mathcal{A}^{|}(X)$. Now, if σ intersected more than n/r lines of S in its interior, where $n = |\mathsf{S}|$, then it must be that the interior of σ intersects one of the lines of X, since X is an ε-net for S, a contradiction.

It follows that σ intersects at most $\varepsilon n = n/r$ lines of S in its interior. ■

Claim 8.10. *Any $(1/r)$-cutting in the plane of n lines contains at least $\Omega\left(r^2\right)$ regions.*

Proof. An arrangement of n lines (in general position) has $M = \binom{n}{2}$ intersections. However, the number of intersections of the lines intersecting a single region in the cutting is at most $m = \binom{n/r}{2}$. This implies that any cutting must be of size at least $M/m = \Omega\left(n^2/(n/r)^2\right) = \Omega\left(r^2\right)$. ■

We can get cuttings of size matching the above lower bound using the moments technique.

Theorem 8.11. *Let* S *be a set of n lines in the plane, and let r be a parameter. One can compute a* $(1/r)$*-cutting of* S *of size* $O(r^2)$.

Proof. Let $\mathsf{R} \subseteq \mathsf{S}$ be a random sample of size r, and consider its vertical decomposition $\mathcal{A}^{|}(\mathsf{R})$. If a vertical trapezoid $\sigma \in \mathcal{A}^{|}(\mathsf{R})$ intersects at most n/r lines of S, then we can add it to the output cutting. The other possibility is that a σ intersects $t(n/r)$ lines of S, for some $t > 1$, and let $\mathrm{cl}(\sigma) \subset \mathsf{S}$ be the conflict list of σ (i.e., the list of lines of S that intersect the interior of σ). Clearly, a $(1/t)$-cutting for the set $\mathrm{cl}(\sigma)$ forms a vertical decomposition (clipped inside σ) such that each trapezoid in this cutting intersects at most n/r lines of S. Thus, we compute such a cutting inside each such "heavy" trapezoid using the algorithm (implicit in the proof) of Lemma 8.9, and these subtrapezoids to the resulting cutting. Clearly, the size of the resulting cutting inside σ is $O\left(t^2 \log^2 t\right) = O\left(t^4\right)$. The resulting two-level partition is clearly the required cutting. By Theorem 8.8, the expected size of the cutting is

$$O\left(\mathbf{E}f(r) + \mathbf{E}\left[\sum_{\sigma\in\mathcal{F}(\mathsf{R})} \left(2\frac{\omega(\sigma)}{n/r}\right)^4\right]\right) = O\left(\mathbf{E}f(r) + \left(\frac{r}{n}\right)^4 \mathbf{E}\left[\sum_{\sigma\in\mathcal{F}(\mathsf{R})} (\omega(\sigma))^4\right]\right)$$

$$= O\left(\mathbf{E}f(r) + \left(\frac{r}{n}\right)^4 \cdot \mathbf{E}f(r)\left(\frac{n}{r}\right)^4\right) = O\left(\mathbf{E}f(r)\right) = O\left(r^2\right),$$

since $\mathbf{E}f(r)$ is proportional to the complexity of $\mathcal{A}(\mathsf{R})$ which is $O(r^2)$. ■

8.4. Bounds on the probability of a region to be created

Here we prove Lemma 8.4$_{\text{p126}}$ in the "right" sampling model. The casual reader is encouraged to skip this section, as it contains mostly tedious (and not very insightful) calculations.

Let S be a given set of n objects. Let $\rho_{r,n}\left(d,k\right)$ be the probability that a region $\sigma \in \mathcal{T}$ whose defining set is of size d and whose stopping set is of size k appears in $\mathcal{F}(\mathsf{R})$, where R is a random sample from S of size r (without repetition).

Lemma 8.12. *We have* $\rho_{r,n}\left(d,k\right) = \dfrac{\binom{n-d-k}{r-d}}{\binom{n}{r}} = \dfrac{\binom{n-d-k}{r-d}}{\binom{n}{r-d}} \cdot \dfrac{\binom{r}{d}}{\binom{n-(r-d)}{d}} = \dfrac{\binom{n-d-k}{r-d}}{\binom{n-d}{r-d}} \cdot \dfrac{\binom{r}{d}}{\binom{n}{d}}.$

Proof. So, consider a region σ with d defining objects in $D(\sigma)$ and k detractors in $K(\sigma)$. We have to pick the d defining objects of $D(\sigma)$ to be in the random sample R of size r but avoid picking any of the k objects of $K(\sigma)$ to be in R.

The second part follows since $\binom{n}{r} = \binom{n}{r-d}\binom{n-(r-d)}{d} \Big/ \binom{r}{d}$. Indeed, for the right-hand side first pick a sample of size $r-d$ and then a sample of size d from the remaining objects. Merging the two random samples, we get a random sample of size r. However, since we do not care if an object is in the first sample or second sample, we observe that every such random sample is being counted $\binom{r}{d}$ times.

The third part is easier, as it follows from $\binom{n}{r-d}\binom{n-(r-d)}{d} = \binom{n}{d}\binom{n-d}{r-d}$. The two sides count the different ways to pick two subsets from a set of size n, the first one of size d and the second one of size $r-d$. ■

Lemma 8.13. *For $M \geq m \geq t \geq 0$, we have* $\left(\frac{m-t}{M-t}\right)^t \leq \frac{\binom{m}{t}}{\binom{M}{t}} \leq \left(\frac{m}{M}\right)^t$.

Proof. We have that $\alpha = \frac{\binom{m}{t}}{\binom{M}{t}} = \frac{m!}{(m-t)!t!} \frac{(M-t)!t!}{M!} = \frac{m}{M} \cdot \frac{m-1}{M-1} \cdots \frac{m-t+1}{M-t+1}$. Now, since $M \geq m$, we have that $\frac{m-i}{M-i} \leq \frac{m}{M}$, for all $i \geq 0$. As such, the maximum (resp. minimum) fraction on the right-hand size is m/M (resp. $\frac{m-t+1}{M-t+1}$). As such, we have $\left(\frac{m-t}{M-t}\right)^t \leq \left(\frac{m-t+1}{M-t+1}\right)^t \leq \alpha \leq (m/M)^t$. ∎

Lemma 8.14. *Let $0 \leq X, Y \leq N$. We have that* $\left(1 - \frac{X}{N}\right)^Y \leq \left(1 - \frac{Y}{2N}\right)^X$.

Proof. Since $1 - \alpha \leq \exp(-\alpha) \leq (1 - \alpha/2)$, for $0 \leq \alpha \leq 1$, it follows that

$$\left(1 - \frac{X}{N}\right)^Y \leq \exp\left(-\frac{XY}{N}\right) = \left(\exp\left(-\frac{Y}{n}\right)\right)^X \leq \left(1 - \frac{Y}{2n}\right)^X.$$

∎

Lemma 8.15. *For $2d \leq r \leq n/8$ and $k \leq n/2$, we have that*

$$\frac{1}{2^{2d}}\left(1 - 4 \cdot \frac{r}{n}\right)^k \left(\frac{r}{n}\right)^d \leq \rho_{r,n}(d,k) \leq 2^{2d}\left(1 - \frac{1}{2} \cdot \frac{r}{n}\right)^k \left(\frac{r}{n}\right)^d.$$

Proof. By Lemma 8.12, Lemma 8.13, and Lemma 8.14 we have

$$\rho_{r,n}(d,k) = \frac{\binom{n-d-k}{r-d}}{\binom{n-d}{r-d}} \cdot \frac{\binom{r}{d}}{\binom{n}{d}} \leq \left(\frac{n-d-k}{n-d}\right)^{r-d} \left(\frac{r}{n}\right)^d \leq \left(1 - \frac{k}{n}\right)^{r-d} \left(\frac{r}{n}\right)^d \leq 2^d \left(1 - \frac{k}{n}\right)^r \left(\frac{r}{n}\right)^d$$
$$\leq 2^d \left(1 - \frac{r}{2n}\right)^k \left(\frac{r}{n}\right)^d,$$

since $k \leq n/2$. As for the other direction, by similar argumentation, we have

$$\rho_{r,n}(d,k) = \frac{\binom{n-d-k}{r-d}}{\binom{n}{r-d}} \cdot \frac{\binom{r}{d}}{\binom{n-(r-d)}{d}} \geq \left(\frac{n-d-k-(r-d)}{n-(r-d)}\right)^{r-d} \left(\frac{r-d}{n-(r-d)-d}\right)^d$$
$$= \left(1 - \frac{d+k}{n-(r-d)}\right)^{r-d} \left(\frac{r-d}{n-r}\right)^d \geq \left(1 - \frac{d+k}{n/2}\right)^r \left(\frac{r/2}{n}\right)^d$$
$$\geq \frac{1}{2^d}\left(1 - \frac{4r}{n}\right)^{d+k} \left(\frac{r}{n}\right)^d \geq \frac{1}{2^{2d}}\left(1 - \frac{4r}{n}\right)^k \left(\frac{r}{n}\right)^d,$$

by Lemma 8.14 (setting $N = n/4$, $X = r$, and $Y = d + k$) and since $r \geq 2d$ and $4r/n \leq 1/2$. ∎

8.5. Bibliographical notes

The technique described in this chapter is generally attributed to the work by Clarkson and Shor [**CS89**], which is historically inaccurate as the technique was developed by Clarkson [**Cla88**]. Instead of mildly confusing the matter by referring to it as the Clarkson technique, we decided to make sure to really confuse the reader and refer to it as the ***moments technique***. The Clarkson technique [**Cla88**] is in fact more general and implies a connection between the number of "heavy" regions and "light" regions. The general framework can be traced back to the earlier paper [**Cla87**]. This implies several beautiful results, some of which we cover later in the book.

For the full details of the algorithm of Section 8.1, the interested reader is refereed to the books [**dBCvKO08, BY98**]. Interestingly, in some cases the merging stage can be skipped; see [**Har00a**].

Agarwal et al. [**AMS98**] presented a slightly stronger variant than the original version of Clarkson [**Cla88**] that allows a region to disappear even if none of the members of its stopping set are in the random sample. This stronger setting is used in computing the vertical decomposition of a single face in an arrangement (instead of the whole arrangement). Here an insertion of a faraway segment of the random sample might cut off a portion of the face of interest. In particular, in the settings of Agarwal et al. Axiom (ii) is replaced by the following:

(ii) If $\sigma \in \mathcal{F}(\mathsf{R})$ and R' is a subset of R with $D(\sigma) \subseteq \mathsf{R}'$, then $\sigma \in \mathcal{F}(\mathsf{R}')$.

Interestingly, Clarkson [**Cla88**] did not prove Theorem 8.8 using the exponential decay lemma but gave a direct proof. In fact, his proof implicitly contains the exponential decay lemma. We chose the current exposition since it is more modular and provides a better intuition of what is really going on and is hopefully slightly simpler. In particular, Lemma 8.4 is inspired by the work of Sharir [**Sha03**].

The exponential decay lemma (Lemma 8.7) was proved by Chazelle and Friedman [**CF90**]. The work of Agarwal et al. [**AMS98**] is a further extension of this result. Another analysis was provided by Clarkson et al. [**CMS93**].

Another way to reach similar results is using the technique of Mulmuley [**Mul94a**], which relies on a direct analysis on 'stoppers' and 'triggers'. This technique is somewhat less convenient to use but is applicable to some settings where the moments technique does not apply directly. Also, his concept of the omega function might explain why randomized incremental algorithms perform better in practice than their worst case analysis [**Mul94b**].

Backwards analysis in geometric settings was first used by Chew [**Che86**] and was formalized by Seidel [**Sei93**]. It is similar to the "leave one out" argument used in statistics for cross validation. The basic idea was probably known to the Greeks (or Russians or French) at some point in time.

(Naturally, our summary of the development is cursory at best and not necessarily accurate, and all possible disclaimers apply. A good summary is provided in the introduction of [**Sei93**].)

Sampling model. As a rule of thumb all the different sampling approaches are similar and yield similar results. For example, we used such an alternative sampling approach in the "proof" of Lemma 8.4. It is a good idea to use whichever sampling scheme is the easiest to analyze in figuring out what's going on. Of course, a formal proof requires analyzing the algorithm in the sampling model its uses.

Lazy randomized incremental construction. If one wants to compute a single face that contains a marking point in an arrangement of curves, then the problem in using randomized incremental construction is that as you add curves, the region of interest shrinks, and regions that were maintained should be ignored. One option is to perform flooding in the vertical decomposition to figure out what trapezoids are still reachable from the marking point and maintaining only these trapezoids in the conflict graph. Doing it in each iteration is way too expensive, but luckily one can use a lazy strategy that performs this cleanup only a logarithmic number of times (i.e., you perform a cleanup in an iteration if the iteration number is, say, a power of 2). This strategy complicates the analysis a bit; see [**dBDS95**] for more details on this ***lazy randomized incremental construction*** technique. An alternative technique was suggested by the author for the (more restricted) case of planar arrangements; see [**Har00b**]. The idea is to compute only what the algorithm really needs to compute the output, by computing the vertical decomposition in an exploratory online fashion. The details are unfortunately overwhelming although the algorithm seems to perform quite well in practice.

Cuttings. The concept of cuttings was introduced by Clarkson. The first optimal size cuttings were constructed by Chazelle and Friedman [**CF90**], who proved the exponential decay lemma to this end. Our elegant proof follows the presentation by de Berg and Schwarzkopf [**dBS95**]. The problem with this approach is that the constant involved in the cutting size is awful[④]. Matoušek [**Mat98**] showed that there $(1/r)$-cuttings with $8r^2 + 6r + 4$ trapezoids, by using level approximation. A different approach was taken by the author [**Har00a**], who showed how to get cuttings which seem to be quite small (i.e., constant-wise) in practice. The basic idea is to do randomized incremental construction but at each iteration greedily add all the trapezoids with conflict list small enough to the cutting being output. One can prove that this algorithm also generates $O(r^2)$ cuttings, but the details are not trivial as the framework described in this chapter is not applicable for analyzing this algorithm.

Cuttings also can be computed in higher dimensions for hyperplanes. In the plane, cuttings can also be computed for well-behaved curves; see [**SA95**].

Another fascinating concept is ***shallow cuttings***. These are cuttings covering only portions of the arrangement that are in the "bottom" of the arrangement. Matoušek came up with the concept [**Mat92b**]. See [**AES99, CCH09**] for extensions and applications of shallow cuttings.

Even more on randomized algorithms in geometry. We have only scratched the surface of this fascinating topic, which is one of the cornerstones of "modern" computational geometry. The interested reader should have a look at the books by Mulmuley [**Mul94a**], Sharir and Agarwal [**SA95**], Matoušek [**Mat02**], and Boissonnat and Yvinec [**BY98**].

8.6. Exercises

Exercise 8.1 (Convex hulls incrementally). Let P be a set of n points in the plane.

(A) Describe a randomized incremental algorithm for computing the convex hull $\mathcal{CH}(\mathsf{P})$. Bound the expected running time of your algorithm.

(B) Assume that for any subset of P, its convex hull has complexity t (i.e., the convex hull of the subset has t edges). What is the expected running time of your algorithm in this case? If your algorithm is not faster for this case (for example, think about the case where $t = O(\log n)$), describe a variant of your algorithm which is faster for this case.

Exercise 8.2 (Compressed quadtree made incremental). Given a set P of n points in $\mathbb{R}^{\mathsf{d}}$, describe a randomized incremental algorithm for building a compressed quadtree for P that works in expected $O(\mathsf{d}n \log n)$ time. Prove the bound on the running time of your algorithm.

[④]This is why all computations related to cuttings should be done on a waiter's bill pad. As Douglas Adams put it: "On a waiter's bill pad, reality and unreality collide on such a fundamental level that each becomes the other and anything is possible, within certain parameters."

CHAPTER 9

Depth Estimation via Sampling

In this chapter, we introduce a "trivial" yet powerful idea. Given a set S of objects and a point p that is contained in some of the objects, let p's ***weight*** be the number of objects that contain it. We can estimate the depth/weight of p by counting the number of objects that contain it in a random sample of the objects. In fact, by considering points induced by the sample, we can bound the number of "light" vertices induced by S (i.e., vertices that are not contained in many objects). This idea can be extended to bounding the number of "light" configurations induced by a set of objects.

This approach leads to a sequence of short, beautiful, elegant, and correct[①] proofs of several hallmark results in discrete geometry.

While the results in this chapter are not directly related to approximation algorithms, the insights and general approach will be useful for us later, or so one hopes.

9.1. The at most k-levels

Let L be a set of n lines in the plane. A point $p \in \bigcup_{\ell \in L} \ell$ is of ***level*** k if there are k lines of L strictly below it. The ***k-level*** is the closure of the set of points of level k. Namely, the k-level is an x-monotone curve along the lines of L.

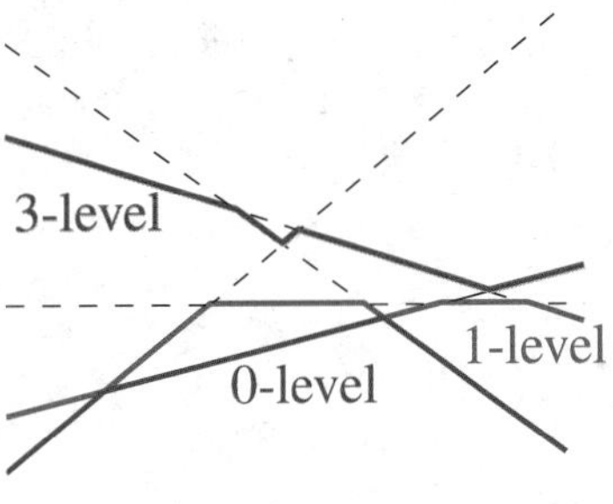

The 0-level is the boundary of the "bottom" face of the arrangement of L (i.e., the face containing the negative y-axis). It is easy to verify that the 0-level has at most $n - 1$ vertices, as each line might contribute at most one segment to the 0-level (which is an unbounded convex polygon).

It is natural to ask what the number of vertices at the k-level is (i.e., what the combinatorial complexity of the polygonal chain forming the k-level is). This is a surprisingly hard question, but the same question on the complexity of the at most k-level is considerably easier.

Theorem 9.1. *The number of vertices of level at most k in an arrangement of n lines in the plane is O(nk).*

Proof. Pick a random sample R of L, by picking each line to be in the sample with probability $1/k$. Observe that

$$\mathbf{E}\Big[|R|\Big] = \frac{n}{k}.$$

Let $\mathbb{L}_{\le k} = \mathbb{L}_{\le k}(L)$ be the set of all vertices of $\mathcal{A}(L)$ of level at most k, for $k > 1$. For a vertex $p \in \mathbb{L}_{\le k}$, let X_p be an indicator variable which is 1 if p is a vertex of the 0-level of

---

[①]The saying goes that "hard theorems have short, elegant, and incorrect proofs". This chapter can maybe serve as a counterexample to this claim.

$\mathcal{A}(\mathsf{R})$. The probability that p is in the 0-level of $\mathcal{A}(\mathsf{R})$ is the probability that none of the j lines below it are picked to be in the sample, and the two lines that define it do get selected to be in the sample. Namely,

$$\mathbf{Pr}\left[X_{\mathsf{p}} = 1\right] = \left(1 - \frac{1}{k}\right)^j \left(\frac{1}{k}\right)^2 \geq \left(1 - \frac{1}{k}\right)^k \frac{1}{k^2} \geq \exp\left(-2\frac{k}{k}\right)\frac{1}{k^2} = \frac{1}{e^2 k^2}$$

since $j \leq k$ and $1 - x \geq e^{-2x}$, for $0 < x \leq 1/2$.

On the other hand, the number of vertices on the 0-level of R is at most $|\mathsf{R}| - 1$. As such,

$$\sum_{\mathsf{p} \in \mathbb{L}_{\leq k}} X_{\mathsf{p}} \leq |\mathsf{R}| - 1.$$

Moreover this, of course, also holds in expectation, implying

$$\mathbf{E}\left[\sum_{\mathsf{p} \in \mathbb{L}_{\leq k}} X_{\mathsf{p}}\right] \leq \mathbf{E}[|\mathsf{R}| - 1] \leq \frac{n}{k}.$$

On the other hand, by linearity of expectation, we have

$$\mathbf{E}\left[\sum_{\mathsf{p} \in \mathbb{L}_{\leq k}} X_{\mathsf{p}}\right] = \sum_{\mathsf{p} \in \mathbb{L}_{\leq k}} \mathbf{E}\left[X_{\mathsf{p}}\right] \geq \frac{|\mathbb{L}_{\leq k}|}{e^2 k^2}.$$

Putting these two inequalities together, we get that $\dfrac{|\mathbb{L}_{\leq k}|}{e^2 k^2} \leq \dfrac{n}{k}$. Namely, $|\mathbb{L}_{\leq k}| \leq e^2 nk$. ∎

The connection to depth is simple. Every line defines a halfplane (i.e., the region above the line). A vertex of depth at most k is contained in at most k halfplanes. The above proof (intuitively) first observed that there are at most n/k vertices of the random sample of zero depth (i.e., 0-level of R) and then showed that every such vertex has probability (roughly) $1/k^2$ to have depth zero in the random sample. It thus follows that if the number of vertices of level at most k is μ, then $\mu/k^2 \leq n/k$; namely, $\mu = O(nk)$.

9.2. The crossing lemma

In the following, for a graph G we denote by $n = |V(\mathsf{G})|$ and $m = |E(\mathsf{G})|$ the number of vertices and edges of G, respectively.

A graph G is ***planar*** if it can be drawn in the plane so that none of its edges are crossing.

We state the following well-known fact.

Theorem 9.2 (Euler's formula). *For a connected planar graph* G, *we have* $f - m + n = 2$, *where* f, m, *and* n *are the number of faces, edges, and vertices in a planar drawing of* G.

Lemma 9.3. *If* G *is a planar graph, then* $m \leq 3n - 6$.

Proof. We assume that the number of edges of G is maximal (i.e., no edges can be added without introducing a crossing). If it is not maximal, then add edges till it becomes maximal. This implies that G is a triangulation (i.e., every face is a triangle). Then, every face is adjacent to three edges, and as such $2m = 3f$. By Euler's formula, we have $f - m + n = (2/3)m - m + n = 2$. Namely, $-m + 3n = 6$. Alternatively, $m = 3n - 6$. However, if m is not maximal, this equality deteriorates to the required inequality. ∎

For example, the above inequality implies that the complete graph over five vertices (i.e., K_5) is not planar. Indeed, it has $m = \binom{5}{2} = 10$ edges and $n = 5$ vertices, but if it were planar, the above inequality would imply that $10 = m \leq 3n - 6 = 9$, which is of course false. (The reader can amuse himself or herself by trying to prove that $K_{3,3}$, the bipartite complete graph with three vertices on each side, is not planar.)

Kuratowski's celebrated theorem states that a graph is planar if and only if it does not contain either K_5 or $K_{3,3}$ induced inside it (formally, it does not have K_5 or $K_{3,3}$ as a minor).

For a graph G, we define the crossing number of G, denoted as $c(\mathsf{G})$, as the minimal number of edge crossings in any drawing of G in the plane. For a planar graph $c(\mathsf{G})$ is zero, and it is "larger" for "less planar" graphs.

Claim 9.4. *For a graph* G, *we have* $c(\mathsf{G}) \geq m - 3n + 6$.

Proof. If $m - 3n + 6 \leq 0 \leq c(\mathsf{G})$, then the claim holds trivially. Otherwise, the graph G is not planar by Lemma 9.3. Draw G in such a way that $c(\mathsf{G})$ is realized and assume, for the sake of contradiction, that $c(\mathsf{G}) < m - 3n + 6$. Let H be the graph resulting from G, by removing one of the edges from each pair of edges of G that intersects in the drawing. We have $m(\mathsf{H}) \geq m(\mathsf{G}) - c(\mathsf{G})$. But H is planar (since its drawing has no crossings), and by Lemma 9.3, we have $m(\mathsf{H}) \leq 3n(\mathsf{H}) - 6$, or equivalently, $m(\mathsf{G}) - c(\mathsf{G}) \leq 3n - 6$. Namely, $m - 3n + 6 \leq c(\mathsf{G})$, which contradicts our assumption. ∎

Lemma 9.5 (Crossing lemma)**.** *For a graph* G, *such that* $m \geq 6n$, *we have* $c(\mathsf{G}) = \Omega(m^3/n^2)$.

Proof. We consider a specific drawing D of G in the plane that has $c(G)$ crossings. Next, let U be a random subset of $V(\mathsf{G})$ selected by choosing each vertex to be in the sample with probability $p > 0$.

Let $\mathsf{H} = \mathsf{G}_U = (U, E')$ be the induced subgraph of G over U. Here, only edges of G with both their endpoints in U "survive" in H; that is, $E' = \left\{ uv \;\middle|\; uv \in E(\mathsf{G}) \text{ and } u, v \in U \right\}$.

Thus, the probability of a vertex v to survive in H is p. The probability of an edge of G to survive in H is p^2, and the probability of a crossing (in this specific drawing D) to survive in the induced drawing D_H (of H) is p^4. Let X_v and X_e denote the (random variables which are the) number of vertices and edges surviving in H, respectively. Similarly, let X_c be the number of crossings surviving in D_H. By Claim 9.4, we have

$$X_c \geq c(\mathsf{H}) \geq X_e - 3X_v + 6.$$

In particular, this holds in the expectation, and by linearity of expectation, we have that

$$\mathbf{E}[X_c] \geq \mathbf{E}[X_e] - 3\,\mathbf{E}[X_v] + 6.$$

Now, by linearity of expectation, we have that $\mathbf{E}[X_c] = c(\mathsf{G})p^4$, $\mathbf{E}[X_e] = mp^2$, and $\mathbf{E}[X_v] = np$, where m and n are the number of edges and vertices of G, respectively. This implies that

$$c(\mathsf{G})p^4 \geq mp^2 - 3np + 6.$$

In particular, $c(\mathsf{G}) \geq m/p^2 - 3n/p^3 + 6/p^4 \geq m/p^2 - 3n/p^3$. In particular, setting $p = 6n/m \leq 1$, we have that

$$c(\mathsf{G}) \geq \frac{m}{p^2} - \frac{3n}{p^3} = \frac{m^3}{6^2 n^2} - \frac{m^3}{2 \cdot 6^2 n^2} = \frac{m^3}{72n^2}.$$

∎

Surprisingly, despite its simplicity, Lemma 9.5 is a very strong tool, as the following results testify.

9.2.1. On the number of incidences. Let P be a set of n distinct points in the plane, and let L be a set of m distinct lines in the plane (note that all the lines might pass through a common point, as we do not assume general position here). Let $I(\mathsf{P}, \mathsf{L})$ denote the number of point/line pairs (p, ℓ), where $\mathsf{p} \in \mathsf{P}, \ell \in \mathsf{L}$, and $\mathsf{p} \in \ell$. The number $I(\mathsf{P}, \mathsf{L})$ is the number of ***incidences*** between lines of L and points of P. Let $I(n, m) = \max_{|\mathsf{P}|=n, |\mathsf{L}|=m} I(\mathsf{P}, \mathsf{L})$.

The following "easy" result has a long history and required major effort to prove before the following elegant proof was discovered[②].

Lemma 9.6. *The maximum number of incidences between n points and m lines is* $I(n, m) = O\left(n^{2/3}m^{2/3} + n\right)$.

Proof. Let P and L be the set of n points and the set of m lines, respectively, realizing $I(m, n)$. Let G be a graph over the points of P (we assume that P contains an additional point at infinity). We connect two points if they lie consecutively on a common line of L, and we also connect the first and last points on each line with the point at infinity.

Clearly, $m(\mathsf{G}) = I + m$ and $n(\mathsf{G}) = n + 1$, where $I = I(m, n)$. Now, we can interpret the arrangement of lines $\mathcal{A}(\mathsf{L})$ as a drawing of G, where a crossing of two edges of G is just a vertex of $\mathcal{A}(\mathsf{L})$. We conclude that $c(\mathsf{G}) \leq m^2$, since m^2 is a trivial bound on the number of vertices of $\mathcal{A}(\mathsf{L})$. On the other hand, by Lemma 9.5, if $I + m = m(\mathsf{G}) \geq 6n(\mathsf{G}) = 6(n + 1)$, we have $c(\mathsf{G}) \geq c_1 m(\mathsf{G})^3 / n(\mathsf{G})^2$, where c_1 is some constant. Thus,

$$c_1 \frac{(I + m)^3}{(n + 1)^2} = c_1 \frac{m(\mathsf{G})^3}{n(\mathsf{G})^2} \leq c(\mathsf{G}) \leq m^2.$$

Thus, we have $I = O\left(m^{2/3}n^{2/3} - m\right) = O\left(m^{2/3}n^{2/3}\right)$.

The other possibility, that $m(\mathsf{G}) < 6n(\mathsf{G})$, implies that $I + m \leq 6(n + 1)$. That is, $I = O(n)$. Putting the two inequalities together, we have that $I = O\left(n^{2/3}m^{2/3} + n\right)$. ■

Lemma 9.7. $I(n, n) = \Omega\left(n^{4/3}\right)$.

Proof. For a positive integer k, let $[\![k]\!] = \{1, \ldots, k\}$. For the sake of simplicity of exposition, assume that $N = n^{1/3}/2$ is an integer.

Consider the integer point set $\mathsf{P} = [\![N]\!] \times [\![8N^2]\!]$ and the set of lines

$$\mathsf{L} = \left\{ y = ax + b \;\middle|\; a \in [\![2N]\!] \text{ and } b \in [\![4N^2]\!] \right\}.$$

Clearly, $|\mathsf{P}| = n$ and $|\mathsf{L}| = n$.

Now, for any $x \in [\![N]\!]$, $a \in [\![2N]\!]$, and $b \in [\![4N^2]\!]$, we have that

$$y = ax + b \leq 2N \cdot N + 4N^2 \leq 8N^2.$$

Namely, any line in L is incident to a point along each vertical line of the grid of P. As such, every line of L is incident to N points of P. Namely, the total number of incidences between the points of P and L is $|\mathsf{L}| N = n \cdot n^{1/3}/2 = n^{4/3}/2$, which implies the claim. ■

9.2.2. On the number of *k*-sets. Let P be a set of n points in the plane in general position (i.e., no three points are collinear). A pair of points $\mathsf{p}, \mathsf{q} \in \mathsf{P}$ form a ***k-set*** if there are exactly k points in the (closed) halfplane below the line passing through p and q. Consider the graph $\mathsf{G} = (\mathsf{P}, E)$ that has an edge for every k-set. We will be interested in bounding the size of E as a function of n. Observe that via duality we have that the number of k-sets is exactly the complexity of the k-level in the dual arrangement $\mathcal{A} = \mathcal{A}(\mathsf{P}^\star)$; that

[②]Or invented – I have no dog in this fight.

is, $\mathsf{P}^\star$ is a set of lines, and every k-set of P corresponds to a vertex on the k-level of $\mathcal{A}$. See Chapter 25 for more information.

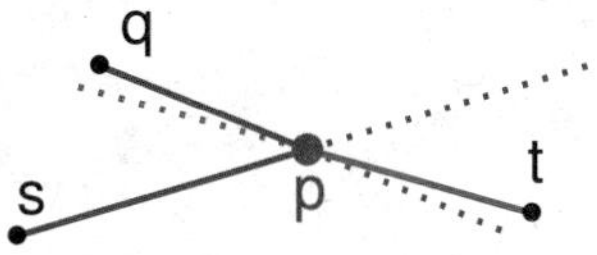

Lemma 9.8 (Antipodality). *Let* qp *and* sp *be two k-set edges of* G, *with* q *and* s *to the left of* p. *Then there exists a point* $\mathsf{t} \in \mathsf{P}$ *to the right of* p *such that* pt *is a k-set, and* $\mathrm{line}(\mathsf{p},\mathsf{t})$ *lies between* $\mathrm{line}(\mathsf{q},\mathsf{p})$ *and* $\mathrm{line}(\mathsf{s},\mathsf{p})$.

Proof. Let $f(\alpha)$ be the number of points below or on the line passing through p and having slope α, where α is a real number. Rotating this line counterclockwise around p corresponds to increasing α. In the following, let $f_+(\alpha)$ (resp., $f_-(\alpha)$) denote the value of $f(\cdot)$ just to the right (resp., left) of α; formally, $f_-(\alpha) = \lim_{x\to\alpha, x<\alpha} f(x)$ and $f_+(\alpha) = \lim_{x\to\alpha, x>\alpha} f(x)$.

Any point swept over by this line which is to the right of p increases f, and any point swept over to the left of p decreases f by 1.

Let α_{q} and α_{s} be the slope of the lines containing qp and sp, respectively. Assume, for the sake of simplicity of exposition, that $\alpha_{\mathsf{q}} < \alpha_{\mathsf{s}}$. Clearly, $f\left(\alpha_{\mathsf{q}}\right) = f(\alpha_{\mathsf{s}}) = k$ and $f_+\left(\alpha_{\mathsf{q}}\right) = k-1$. Let y be the smallest value such that $y > \alpha_{\mathsf{q}}$ and $f(y) = k$. Such a y exists since $\alpha_{\mathsf{q}} < \alpha_{\mathsf{s}}$, $f_+\left(\alpha_{\mathsf{q}}\right) = k-1$, $f(\alpha_{\mathsf{s}}) = k$, and there are no three points that are collinear.

We have that $f_-(y) = k-1$, which implies that the line passing through p with slope $f(y)$ has a point $\mathsf{t} \in \mathsf{P}$ on it and t is to the right of p. Clearly, if we continue sweeping, the line would sweep over sp, which implies the claim. ■

Lemma 9.8 also holds by symmetry in the other direction: Between any two edges to the right of p, there is an antipodal edge on the other side.

Lemma 9.9. *Let* p *be a point of* P, *and let* q *be a point to its left, such that* $\mathsf{qp} \in E(\mathsf{G})$ *and it has the largest slope among all such edges. Furthermore, assume that there are* $k-1$ *points of* P *to the right of* p. *Then, there exists a point* $\mathsf{s} \in \mathsf{P}$, *such that* $\mathsf{ps} \in E(\mathsf{G})$ *and* ps *has larger slope than* qp.

Proof. Let α be the slope of qp, and observe that $f(\alpha) = k$ and $f_+(\alpha) = k-1$ and $f(\infty) \geq k$. Namely, there exists $y > \alpha$ such that $f(y) = k$. We conclude that there is a k-set adjacent to p on the right, with slope larger than α. ■

So, imagine that we are at an edge $\mathsf{e} = \mathsf{qp} \in E(\mathsf{G})$, where q is to the left of p. We rotate a line around p (counterclockwise) till we encounter an edge $\mathsf{e}' = \mathsf{ps} \in E(\mathsf{G})$, where s is a point to the right of p. We can now walk from e to e' and continue walking in this way, forming a chain of edges in G. Note that by Lemma 9.8, no two such chains can be "merged" into using the same edge. Furthermore, by Lemma 9.9, such a chain can end only in the last $k-1$ points of P (in their ordering along the x-axis). Namely, we decomposed the edges of G into $k-1$ edge disjoint convex chains (the chains are convex since we rotate counterclockwise as we walk along a chain). See Figure 9.1 for an example.

Lemma 9.10. *The edges of* G *can be decomposed into* $k-1$ *convex chains* $\mathsf{C}_1, \ldots, \mathsf{C}_{k-1}$.

Similarly, the edges of G *can be decomposed into* $m = n-k+1$ *concave chains* $\mathsf{D}_1, \ldots, \mathsf{D}_m$.

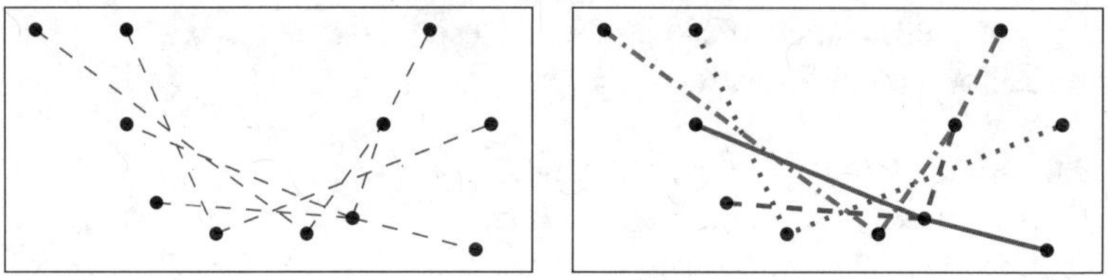

FIGURE 9.1. An example of 5-sets and their decomposition into four convex chains.

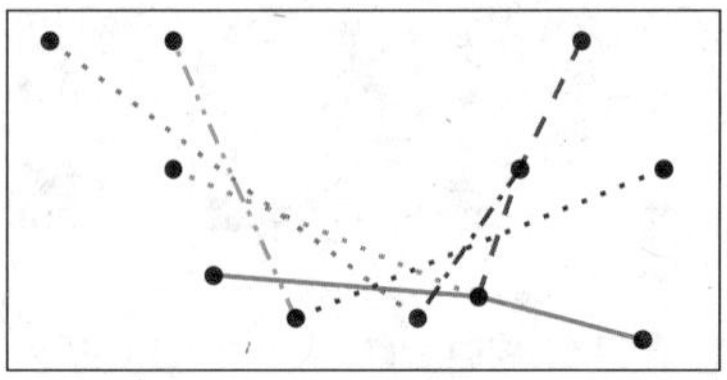

PROOF. The first part of the claim is proved above. As for the second claim, rotate the plane by 180°. Every k-set is now an $(n - k + 2)$-set, and by the above argumentation, the edges of G can be decomposed into $n-k+1$ convex chains, which are concave in the original orientation. (The figure on the left shows these concave chains for $k = 5$ for the example used above.) ■

Theorem 9.11. *The number of k-sets defined by a set of n points in the plane is* $O\left(nk^{1/3}\right)$.

PROOF. The graph G has $n = |\mathsf{P}|$ vertices, and let $m = |E(\mathsf{G})|$ be the number of k-sets. By Lemma 9.10, any crossing of two edges of G is an intersection point of one convex chain of $\mathsf{C}_1, \ldots, \mathsf{C}_{k-1}$ with a concave chain of $\mathsf{D}_1, \ldots, \mathsf{D}_{n-k+1}$. Since a convex chain and a concave chain can have at most two intersections, we conclude that there are at most $2(k-1)(n-k+1)$ crossings in G. By the crossing lemma (Lemma 9.5), there are at least $\Omega\left(m^3/n^2\right)$ crossings. Putting these two inequalities together, we conclude $m^3/n^2 = O(nk)$, which implies $m = O\left(nk^{1/3}\right)$. ■

9.3. A general bound for the at most k-weight

We now extend the at most k-level technique to the general moments technique setting (see Section 8.2). We quickly restate the abstract settings.

Let S be a set of objects. For a subset $\mathsf{R} \subseteq \mathsf{S}$, we define a collection of 'regions' called $\mathcal{F}(\mathsf{R})$. Let $\mathcal{T} = \mathcal{T}(\mathsf{S}) = \bigcup_{\mathsf{R}\subseteq\mathsf{S}} \mathcal{F}(\mathsf{R})$ denote the set of ***all possible regions*** defined by subsets of S. We associate two subsets $D(\sigma), K(\sigma) \subseteq \mathsf{S}$ with each region $\sigma \in \mathcal{T}$. The ***defining set*** $D(\sigma)$ of σ is a subset of S defining the region σ. We assume that for every $\sigma \in \mathcal{T}$, $|D(\sigma)| \leq d$ for a (small) constant d, which is the ***combinatorial dimension***. The ***conflicting set*** $K(\sigma)$ of σ is the set of objects of S such that including any object of $K(\sigma)$ into R prevents σ from appearing in $\mathcal{F}(\mathsf{R})$. The ***weight*** of σ is $\omega(\sigma) = |K(\sigma)|$.

Let $\mathsf{S}, \mathcal{F}(\mathsf{R}), D(\sigma)$, and $K(\sigma)$ be such that for any subset $\mathsf{R} \subseteq \mathsf{S}$, the set $\mathcal{F}(\mathsf{R})$ satisfies the following axioms: (i) For any $\sigma \in \mathcal{F}(\mathsf{R})$, we have $D(\sigma) \subseteq \mathsf{R}$ and $\mathsf{R} \cap K(\sigma) = \emptyset$. (ii) If $D(\sigma) \subseteq \mathsf{R}$ and $K(\sigma) \cap \mathsf{R} = \emptyset$, then $\sigma \in \mathcal{F}(\mathsf{R})$.

Let $\mathcal{T}_{\leq k}(\mathsf{S})$ be the set of regions of $\mathcal{T}$ with weight at most k. Furthermore, assume that the expected number of regions of zero weight of a sample of size r is (or is at most) $\mathbf{f}_0(r)$. Formally, for a sample of size r from S, we denote $\mathbf{f}_0(r) = \mathbf{E}\left[|\mathcal{F}(\mathsf{R})|\right]$. We have the following theorem.

Theorem 9.12. *Let* S *be a set of n objects as above, with combinatorial dimension d, and let k be a parameter. Let* R *be a random sample created by picking each element of* S *with* *probability* $1/k$. *Then, for some constant c, we have*

$$|\mathcal{T}_{\leq k}(\mathsf{S})| \leq c\,\mathbf{E}\Big[k^d \mathbf{f}_0(|\mathsf{R}|)\Big].$$

PROOF. We reproduce the proof of Theorem 9.1. Every region $\sigma \in \mathcal{T}_{\leq k}$ appears in $\mathcal{F}(\mathsf{R})$ with probability $\geq 1/k^d(1-1/k)^k \geq e^{-2}/k^d$. Observe that every sample of size $|\mathsf{R}|$ has equal probability of being picked to be R. As such, we have that $\mathbf{f}_0(r) = \mathbf{E}\Big[|\mathcal{F}(\mathsf{R})| \;\Big|\; |\mathsf{R}| = r\Big]$.

Now, setting $X_\sigma = 1$ if and only if $\sigma \in \mathcal{F}(\mathsf{R})$, we have that

$$\mathbf{E}\Big[\mathbf{f}_0(|\mathsf{R}|)\Big] = \mathbf{E}\Big[\mathbf{E}\Big[|\mathcal{F}(\mathsf{R})| \;\Big|\; |\mathsf{R}| = k\Big]\Big] = \mathbf{E}\Big[|\mathcal{F}(\mathsf{R})|\Big]$$
$$\geq \mathbf{E}\left[\sum_{\sigma \in \mathcal{T}_{\leq k}} X_\sigma\right] = \sum_{\sigma \in \mathcal{T}_{\leq k}} \mathbf{Pr}[\sigma \in \mathcal{F}(\mathsf{R})] \geq \frac{|\mathcal{T}_{\leq k}|}{k^d e^2}.$$

■

Lemma 9.13. *Let* $\mathbf{f}_0(\cdot)$ *be a monotone increasing function which is well behaved; namely, there exists a constant c, such that* $\mathbf{f}_0(xr) \leq c\mathbf{f}_0(r)$, *for any r and* $1 \leq x \leq 2$. *Let Y be the number of heads in n coin flips where the probability for head is* $1/k$. *Then* $\mathbf{E}[\mathbf{f}_0(Y)] = O(\mathbf{f}_0(n/k))$.

PROOF. The claim[③] follows easily from Chernoff's inequality. Indeed, we have that $\mathbf{E}[Y] = n/k$ and $\mathbf{Pr}[Y \geq t(n/k)] \leq 2^{-t(n/k)}$, for $t > 3$, by the simplified form of the Chernoff inequality; see Theorem 27.17$_{\text{p340}}$. Furthermore, by assumption we have that

$$f\left((t+1)n/k\right) \leq cf\left(\frac{t+1}{2}(n/k)\right) \leq c^{\lceil \lg (t+1) \rceil}\mathbf{f}_0(n/k).$$

Putting these two things together, we have that

$$\mathbf{E}\Big[\mathbf{f}_0(Y)\Big] \leq \sum_i \mathbf{f}_0(i)\,\mathbf{Pr}[Y = i] \leq f\left(10\frac{n}{k}\right) + \sum_{t=10}^{k-1} f\left((t+1)\frac{n}{k}\right)\mathbf{Pr}[Y \geq t(n/k)]$$
$$\leq O\Big(\mathbf{f}_0(n/k)\Big) + \sum_{t=10}^{k} c^{\lceil \lg (t+1) \rceil}\mathbf{f}_0(n/k)\,2^{-t(n/k)} = O\Big(\mathbf{f}_0(n/k)\Big).$$

■

The following is an immediate consequence of Theorem 9.12 and Lemma 9.13.

Theorem 9.14. *Let* S *be a set of n objects with combinatorial dimension d, and let k be a parameter. Assume that the number of regions formed by a set of m objects is bounded by a function* $\mathbf{f}_0(m)$ *and, furthermore,* $\mathbf{f}_0(m)$ *is well behaved in the sense of Lemma 9.13. Then,* $|\mathcal{T}_{\leq k}(\mathsf{S})| = O\Big(k^d \mathbf{f}_0(n/k)\Big)$.

In particular, let $\mathbf{f}_{\leq k}(n) = \max_{|S|=n} |\mathcal{T}_{\leq k}(\mathsf{S})|$ *be the maximum number of regions of weight at most k that can be defined by any set of n objects. We have that* $\mathbf{f}_{\leq k}(n) = O\Big(k^d \mathbf{f}_0(n/k)\Big)$.

Note that if the function $\mathbf{f}_0(\cdot)$ grows polynomially, then Theorem 9.14 applies. It fails if $\mathbf{f}_0(\cdot)$ grows exponentially fast.

[③]Note that this lemma is not implied by Jensen's inequality (see Lemma 27.9$_{\text{p337}}$) since if $\mathbf{f}_0(\cdot)$ is convex, then $\mathbf{E}[\mathbf{f}_0(Y)] \geq \mathbf{f}_0(\mathbf{E}[Y]) = \mathbf{f}_0(n/k)$, which goes in the wrong direction.

9.3.1. Example – k-level in higher dimensions. We need the following fact, which we state without proof.

Theorem 9.15 (The upper bound theorem). *The complexity of the convex hull of n points in* d *dimensions is bounded by* $O\left(n^{\lfloor d/2 \rfloor}\right)$.

Example 9.16 (At most k-sets). Let P be a set of n points in $\mathbb{R}^d$. A region here is a halfspace with d points on its boundary. The set of regions defined by P is just the faces of the convex hull of P. The complexity of the convex hull of n points in d dimensions is $\mathbf{f}_0(n) = O\left(n^{\lfloor d/2 \rfloor}\right)$, by Theorem 9.15. Two halfspaces h, h' would be considered to be combinatorially different if $P \cap h \neq P \cap h'$. As such, the number of combinatorially different halfspaces containing at most k points of P is at most $O\left(k^d \mathbf{f}_0(n/k)\right) = O\left(k^{\lceil d/2 \rceil} n^{\lfloor d/2 \rfloor}\right)$.

9.4. Bibliographical notes

The reader should not mistake the simplicity of the proofs in this chapter with easiness. Almost all the results presented have a long and painful history with earlier proofs being considerably more complicated. In some sense, these results are the limit of mathematical evolution: They are simple, breathtakingly elegant (in some cases), and on their own (without exposure to previous work on the topic), it seems inconceivable that one could come up with them.

At most k-level. The technique for bounding the complexity of the at most k-level (or at most depth k) is generally attributed to Clarkson and Shor [**CS89**] and more precisely it is from [**Cla88**]. Previous work on just the two-dimensional variant include [**GP84, Wel86, AG86**]. Our presentation in Section 9.1 and Section 9.3 follows (more or less) Sharir [**Sha03**]. The connection of this technique to the crossing lemma is from there.

For a proof of the upper bound theorem (Theorem 9.15), see Matoušek [**Mat02**].

The crossing lemma. The crossing lemma is by Ajtai et al. [**ACNS82**] and Leighton [**Lei84**]. The current greatly simplified "proof from the book" is attributed to Sharir. The insight that this lemma has something to do with incidences and similar problems is due to Székely [**Szé97**]. Elekes [**Ele97**] used the crossing lemma to prove surprising lower bounds on sum and product problems (the proofs of these bounds are surprisingly elegant).

The complexity of k-level and number of k-sets. This is considered to be one of the hardest problems in discrete geometry, and there is still a big gap between the best lower bound [**Tót01**] and best upper bound currently known [**Dey98**]. Our presentation in Section 9.2.2 follows suggestions by Micha Sharir and is based on the result of Dey [**Dey98**] (which was in turn inspired by the work of Agarwal et al. [**AACS98**]). This problem has a long history, and the reader is referred to Dey [**Dey98**] for its history.

Incidences. This problem again has a long and painful history. The reader is referred to [**PS04, Szé97**] for details. The elegant lower bound proof of Lemma 9.7 is by Elekes [**Ele02**].

We only skimmed the surface of some problems in discrete geometry and results known in this field related to incidences and k-sets. Good starting points for learning more are the books by Brass et al. [**BMP05**] and Matoušek [**Mat02**].

9.5. Exercises

Exercise 9.1 (Incidence lower bound). Prove that $I(m, n) = \Omega\left(m^{2/3} n^{2/3}\right)$ for any n and m.

Exercise 9.2 (The number of heavy disks). Let P be a set of n points in the plane. A disk D is ***canonical*** if its boundary passes through three points of P. Provide a bound on the number of canonical disks that contain at most k points of P in their interior.

Exercise 9.3 (The number of heavy vertical segments). Let L be a set of n lines in the plane. A vertical segment has weight k if it intersects k segments of L of its interior. Two such vertical segments are distinct if the subsets of segments they intersect are different. Bound the number of distinct vertical segments of weight at most k.

CHAPTER 10

Approximating the Depth via Sampling and Emptiness

In this chapter, we show how to build a data-structure that quickly returns an approximate number of objects containing a query point using "few" emptiness queries. Specifically, we show a reduction from emptiness range searching to approximate range counting.

10.1. From emptiness to approximate range counting

Assume that one can construct a data-structure for a set $\mathsf{S} = \{o_1, \ldots, o_n\}$ of n objects, such that given a query range $\mathbf{r}$, we can check in $Q(n)$ time whether $\mathbf{r}$ intersects any of the objects in S. Such queries are referred to as ***range-searching emptiness queries***. For example, consider the case where S is a set of points and the query is a disk. Then, the query becomes whether or not the query disk is empty of points of S.

So, let $\mu_{\mathbf{r}} = \mathrm{depth}(\mathbf{r}, \mathsf{S})$ denote the ***depth*** of $\mathbf{r}$; namely, it is the number of objects of S intersected by $\mathbf{r}$. In our example, this is the number of points of S in the query disk. Such queries are ***range-searching counting*** queries. Such counting queries are harder (i.e., take longer to handle) than emptiness queries.

In this section, we show how to build a data-structure that quickly returns an approximate number of objects in S intersecting $\mathbf{r}$ using emptiness queries as a subroutine. Namely, we show how to answer approximate range-searching counting queries using polylogarithmic emptiness queries.

Specifically, for a prespecified $\varepsilon > 0$, the new data-structure outputs a number $\alpha_{\mathbf{r}}$ such that $(1 - \varepsilon)\mu_{\mathbf{r}} \le \alpha_{\mathbf{r}} \le \mu_{\mathbf{r}}$.

10.1.1. The decision procedure. Given parameters $z \in [1, n]$ and ε, with $1/2 > \varepsilon > 0$, we construct a data-structure, such that given a query range $\mathbf{r}$, we can decide whether $\mu_{\mathbf{r}} < z$ or $\mu_{\mathbf{r}} \ge z$. The data-structure is allowed to make a mistake if $\mu_{\mathbf{r}} \in \Big[(1 - \varepsilon)z, (1 + \varepsilon)z\Big]$.

The data-structure. Let $\mathsf{R}_1, \ldots, \mathsf{R}_M$ be M independent random samples of S formed by picking every element with probability $1/z$, where

$$M = \nu(\varepsilon) = \left\lceil c_2 \varepsilon^{-2} \log n \right\rceil$$

and c_2 is a sufficiently large absolute constant. Build M separate emptiness-query data-structures $D_1, \ldots, D_M$, for the sets $\mathsf{R}_1, \ldots, \mathsf{R}_M$, respectively, and let

$$\mathcal{D} = \mathcal{D}(z, \varepsilon) = \{D_1, \ldots, D_M\}.$$

Answering a query. Consider a query range $\mathbf{r}$, and let $X_i = 1$ if $\mathbf{r}$ intersects any of the objects of R_i and $X_i = 0$ otherwise, for $i = 1, \ldots, M$. The value of X_i can be determined using a single emptiness query in D_i, for $i = 1, \ldots, M$. Compute $Y_{\mathbf{r}} = \sum_i X_i$.

For a range $\mathbf{r}$ of depth k, the probability that $\mathbf{r}$ intersects one of the objects of R_i is

$$\rho_z(k) = 1 - \left(1 - \frac{1}{z}\right)^k. \tag{10.1}$$

Indeed, consider the k objects $o'_1, \ldots, o'_k$ of S that intersect $\mathbf{r}$. Clearly, the probability of o'_i failing to be in the random sample R_i is exactly $1 - 1/z$. As such, the probability that all these k objects will not be in the sample R_i is exactly $(1-1/z)^k$, and $\rho_z(k)$ is the probability of the complement event.

If a range $\mathbf{r}$ has depth z, then $\Lambda = \mathbf{E}[Y_\mathbf{r}] = \nu(\varepsilon)\rho_z(z)$. Our data-structure returns "depth($\mathbf{r}$, S) $< z$" if $Y_\mathbf{r} < \Lambda$ and "depth($\mathbf{r}$, S) $\geq z$" otherwise.

10.1.1.1. *Correctness.* In the following, we show that, with high probability, the data-structure indeed returns the correct answer if the depth of the query range is outside the "uncertainty" range $[(1-\varepsilon)z, (1+\varepsilon)z]$. For simplicity of exposition, we assume in the following that $z \geq 10$ (the case $z < 10$ follows by a similar argument). Consider a range $\mathbf{r}$ of depth at most $(1-\varepsilon)z$. The data-structure returns the wrong answer if $Y_\mathbf{r} > \Lambda$. We will show that the probability of this event is polynomially small. The other case, where $\mathbf{r}$ has depth at least $(1+\varepsilon)z$ but $Y_\mathbf{r} < \Lambda$, is handled in a similar fashion.

Intuition. Before jumping into the murky proof, let us consider the situation. Every sample R_i is an experiment. The experiment succeeds if the sample contains an object that intersects the query range $\mathbf{r}$. The probability of success is $\gamma = \rho_z(k)$ (see (10.1)), where k is the weight of $\mathbf{r}$. By repeating this experiment a sufficient number of times (i.e., M), we get a "reliable" estimate of $\rho_z(k)$. In particular, if there is a big enough gap between $\rho_z((1-\varepsilon)z)$ and $\rho_z(z)$, then we could decide if the range is "heavy" (i.e., weight exceeding z) or "light" (i.e., weight smaller than $(1-\varepsilon)z$) by estimating the probability γ that $\mathbf{r}$ intersects an object in the random sample.

Indeed, if $\mathbf{r}$ is "light", then $\gamma \leq \rho_z((1-\varepsilon)z)$, and if it is "heavy", then $\gamma \geq \rho_z(z)$. We estimate γ by the quantity $Y_\mathbf{r}/M$, namely, repeating the experiment M times and dividing the number of successes by the number of experiments (i.e., M). Now, we need to determine how many experiments we need to perform till we get a good estimate, which is reliable enough to carry out our nefarious task of distinguishing the light case from the heavy case. Clearly, the bigger the gap is between $\rho_z((1-\varepsilon)z)$ and $\rho_z(z)$ the fewer experiments required. Our proof would first establish that the gap between these two probabilities is $\Omega(\varepsilon)$, and next we will plug this into Chernoff's inequality to figure out how large M has to be for this estimate to be reliable.

To estimate this gap, we need to understand what the function $\rho_z(\cdot)$ looks like. So consider the graph on the right. Here $z = 100$ and $\varepsilon = 0.2$. For the heavy case, where the weight of range exceeds z, the probability of success is $\mathsf{p} = \rho_z(100)$, and probability of failure is $\mathsf{q} = \rho_z(80)$. Since the function behaves like a line around these values, it is kind of visually obvious that the required gap (the vertical green segment in the graph) is $\approx \varepsilon$. Proving this formally is somewhat more tedious.

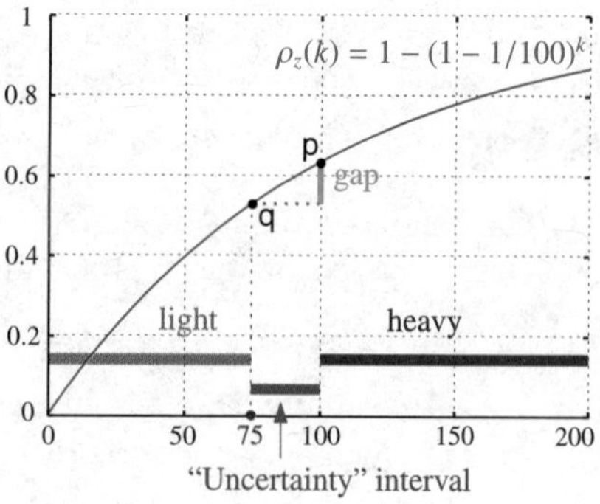

We need the following:

Observation 10.1. *For* $0 \leq x \leq y < 1$, *we have* $\dfrac{1-x}{1-y} = 1 + \dfrac{y-x}{1-y} \geq 1 + y - x$.

Lemma 10.2. *Let* $\Lambda = M\rho_z(z)$. *We have*

$$\alpha = \mathbf{Pr}\Big[Y_\mathbf{r} > \Lambda \,\Big|\, \mathrm{depth}(\mathbf{r}, \mathsf{S}) \leq (1-\varepsilon)z\Big] \leq \frac{1}{n^{c_4}},$$

where $c_4 = c_4(c_2) > 0$ depends only on c_2 and can be made arbitrarily large by a choice of a sufficiently large c_2.

PROOF. The probability α is maximized when $\text{depth}(\mathbf{r}, \mathsf{S}) = (1-\varepsilon)z$. Thus

$$\alpha \le \mathbf{Pr}\left[Y_{\mathbf{r}} > \Lambda \,\middle|\, \text{depth}(\mathbf{r}, \mathsf{S}) = (1-\varepsilon)z\right].$$

Observe that $\mathbf{Pr}[X_i = 1] = \rho_z\big((1-\varepsilon)z\big)$ and

$$\mathbf{E}\left[Y_{\mathbf{r}}\right] = M\rho_z\big((1-\varepsilon)z\big) = M \cdot \left(1 - \left(1 - \frac{1}{z}\right)^{(1-\varepsilon)z}\right) \ge M \cdot \left(1 - e^{-(1-\varepsilon)}\right) \ge \frac{M}{3},$$

since $1 - 1/z \le \exp(-1/z)$ and $\varepsilon \le 1/2$. By definition, $\Lambda = M\rho_z(z)$; therefore, by Observation 10.1, we have

$$\xi = \frac{\Lambda}{\mathbf{E}[Y_{\mathbf{r}}]} = \frac{1 - \left(1 - \frac{1}{z}\right)^z}{1 - \left(1 - \frac{1}{z}\right)^{(1-\varepsilon)z}} \ge 1 + \left(1 - \frac{1}{z}\right)^{(1-\varepsilon)z} - \left(1 - \frac{1}{z}\right)^z$$
$$= 1 + \left(1 - \frac{1}{z}\right)^{(1-\varepsilon)z}\left(1 - \left(1 - \frac{1}{z}\right)^{\varepsilon z}\right).$$

Now, by applying $\exp(-2x) \le 1 - x$ and $1 + x \le \exp(x)$ (see Lemma 28.9$_{\text{p348}}$) repeatedly, we have

$$\xi \ge 1 + \exp\left(-\frac{2}{z}(1-\varepsilon)z\right) \cdot \left(1 - \exp\left(-\frac{1}{z}\varepsilon z\right)\right) = 1 + \frac{1}{e^2}\big(1 - \exp(-\varepsilon)\big)$$
$$\ge 1 + \frac{1}{e^2}\left(1 - \left(1 - \frac{\varepsilon}{2}\right)\right) = 1 + \frac{\varepsilon}{2}.$$

Deploying the Chernoff inequality (Theorem 27.17), we have that if $\mu_{\mathbf{r}} = \text{depth}(\mathbf{r}, \mathsf{S}) = (1-\varepsilon)z$, then

$$\alpha = \mathbf{Pr}[Y_{\mathbf{r}} > \Lambda] \le \mathbf{Pr}\left[Y_{\mathbf{r}} > \xi\,\mathbf{E}[Y_{\mathbf{r}}]\right] \le \mathbf{Pr}\left[Y_{\mathbf{r}} > (1 + \varepsilon/2)\,\mathbf{E}[Y_{\mathbf{r}}]\right]$$
$$\le \exp\left(-\mathbf{E}[Y_{\mathbf{r}}]\frac{1}{4}\left(\frac{\varepsilon}{2}\right)^2\right) \le \exp\left(-\frac{M\varepsilon^2}{c_3}\right) \le \exp\left(-\frac{\varepsilon^2\left\lceil c_2\varepsilon^{-2}\log n\right\rceil}{c_3}\right) \le n^{-c_4},$$

where c_3 is some absolute constant and by setting c_2 to be sufficiently large. ■

This implies the following lemma.

Lemma 10.3. *Given a set S of n objects, a parameter $0 < \varepsilon < 1/2$, and $z \in [0, n]$, one can construct a data-structure $\mathcal{D}$ which, given a range $\mathbf{r}$, returns either* L *or* R*. If it returns* L*, then $\mu_{\mathbf{r}} \le (1+\varepsilon)z$, and if it returns* R*, then $\mu_{\mathbf{r}} \ge (1-\varepsilon)z$. The data-structure might return either answer if $\mu_{\mathbf{r}} \in [(1-\varepsilon)z, (1+\varepsilon)z]$.*

The data-structure $\mathcal{D}$ consists of $M = O(\varepsilon^{-2}\log n)$ emptiness data-structures. The space and preprocessing time needed to build them are $O(S(2n/z)\varepsilon^{-2}\log n)$ where $S(m)$ is the space (and preprocessing time) needed for a single emptiness data-structure storing m objects.

The query time is $O(Q(2n/z)\varepsilon^{-2}\log n)$, where $Q(m)$ is the time needed for a single query in such a structure, respectively. All bounds hold with high probability.

PROOF. The lemma follows immediately from the above discussion. The only missing part is observing that by the Chernoff inequality we have that $|\mathsf{R}_i| \le 2n/z$, and this holds with high probability. ■

10.1.2. Answering the approximate counting query. The path to answering the approximate counting query is now clear. We use Lemma 10.3 to perform a binary search, repeatedly narrowing the range containing the answer. We stop when the size of the range is within our error tolerances.

10.1.3. The data-structure. Lemma 10.3 provides us with a tool for performing a "binary" search for the count value $\mu_{\mathbf{r}}$ of a range $\mathbf{r}$. For small values of i, we just build a separate data-structure of Lemma 10.3 for depth values $i/2$, for $i = 1, \dots, U = O(\varepsilon^{-1})$. For depth i, we use accuracy $1/8i$ (i.e., this is the value of ε when using Lemma 10.3). Using these data-structures, we can decide whether the query range count is at least U or smaller than U. If it is smaller than U, then we can perform a binary search to find its exact value. The result is correct with high probability.

Next, consider the values $v_j = (U/4)(1 + \varepsilon/16)^j$, for $j = U + 1, \dots, W$, where $W = c \log_{1+\varepsilon/16} n = O(\varepsilon^{-1} \log n)$, for an appropriate choice of an absolute constant $c > 0$, so that $v_W = n$. We build a data-structure $\mathcal{D}(v_j)$ for each $z = v_j$, using Lemma 10.3.

Answering a query. Given a range query $\mathbf{r}$, each data-structure in our list returns L or R. Moreover, with high probability, if we were to query all the data-structures, we would get a sequence of Rs, followed by a sequence of Ls. It is easy to verify that the value associated with the last data-structure returning R (rounded to the nearest integer) yields the required approximation. We can use binary search on $\mathcal{D}(v_1), \dots, \mathcal{D}(v_W)$ to locate this changeover value using a total of $O(\log W) = O(\log(\varepsilon^{-1} \log n))$ queries in the structures of $\mathcal{D}_1, \dots, \mathcal{D}_W$. Namely, the overall query time is $O\big(Q(n)\varepsilon^{-2}(\log n)\log(\varepsilon^{-1}\log n)\big)$.

Theorem 10.4. *Given a set* S *of* n *objects, assume that one can construct, using* $S(n)$ *space, in* $T(n)$ *time, a data-structure that answers emptiness queries in* $Q(n)$ *time.*

Then, one can construct, using $O\big(S(n)\varepsilon^{-3}\log^2 n\big)$ *space, in* $O(T(n)\varepsilon^{-3}\log^2 n)$ *time, a data-structure that, given a range* $\mathbf{r}$*, outputs a number* $\alpha_{\mathbf{r}}$*, with* $(1-\varepsilon)\mu_{\mathbf{r}} \le \alpha_{\mathbf{r}} \le \mu_{\mathbf{r}}$*. The query time is* $O\big(\varepsilon^{-2}Q(n)(\log n)\log(\varepsilon^{-1}\log n)\big)$*. The result returned is correct with high probability for any query and the running time bounds hold with high probability.*

The bounds of Theorem 10.4 can be improved; see Section 10.4 for details.

10.2. Application: Halfplane and halfspace range counting

Using the data-structure of Dobkin and Kirkpatrick [**DK85**], one can answer emptiness halfspace range-searching queries in logarithmic time. In this case, we have $S(n) = O(n)$, $T(n) = O(n \log n)$, and $Q(n) = O(\log n)$.

Corollary 10.5. *Given a set* P *of* n *points in two (resp., three) dimensions and a parameter* $\varepsilon > 0$*, one can construct in* $O(n\mathrm{poly}(1/\varepsilon, \log n))$ *time a data-structure, of size* $O(n\mathrm{poly}(1/\varepsilon, \log n))$*, such that given a halfplane (resp., halfspace)* $\mathbf{r}$*, it outputs a number* α*, such that* $(1-\varepsilon)\left|\mathbf{r} \cap \mathsf{P}\right| \le \alpha \le \left|\mathbf{r} \cap \mathsf{P}\right|$ *and the query time is* $O(\mathrm{poly}(1/\varepsilon, \log n))$*. The result returned is correct with high probability for all queries.*

Using the standard lifting of points in $\mathbb{R}^2$ to the paraboloid in $\mathbb{R}^3$ implies a similar result for approximate range counting for disks, as a disk range query in the plane reduces to a halfspace range query in three dimensions.

Corollary 10.6. *Given a set* P *of* n *points in two dimensions and a parameter* ε*, one can construct a data-structure in* $O(n\mathrm{poly}(1/\varepsilon, \log n))$ *time, using* $O(n\mathrm{poly}(1/\varepsilon, \log n))$ *space, such that given a disk* $\mathbf{r}$*, it outputs a number* α*, such that* $(1-\varepsilon)\left|\mathbf{r} \cap \mathsf{P}\right| \le \alpha \le \left|\mathbf{r} \cap \mathsf{P}\right|$ *and* *the query time is* $O(\mathrm{poly}(1/\varepsilon, \log n))$. *The result returned is correct with high probability for all possible queries.*

Depth queries. By computing the union of a set of n pseudo-disks in the plane and preprocessing the union for point-location queries, one can perform "emptiness" queries in this case in logarithmic time. (We are assuming here that we can perform the geometric primitives on the pseudo-disks in constant time.) The space needed is $O(n)$ and it takes $O(n \log n)$ time to construct it. Thus, we get the following result.

Corollary 10.7. *Given a set* S *of* n *pseudo-disks in the plane, one can preprocess them in* $O(n\varepsilon^{-2} \log^2 n)$ *time, using* $O(n\varepsilon^{-2} \log n)$ *space, such that given a query point* q, *one can output a number* α, *such that* $(1 - \varepsilon)\mathrm{depth}(\mathsf{q}, \mathsf{S}) \leq \alpha \leq \mathrm{depth}(\mathsf{q}, \mathsf{S})$ *and the query time is* $O\left(\varepsilon^{-2} \log^2 n\right)$. *The result returned is correct with high probability for all possible queries.*

10.3. Relative approximation via sampling

In the above discussion, we used several random samples of a set S of n objects, and by counting in how many samples a query point lies, we obtained a good estimate of the depth of the point in S. It is natural to ask what can be done if we insist on using a single sample. Intuitively, if we sample each object with probability p in a random sample R, then if the query point $\mathbf{r}$ had depth that is sufficiently large (roughly $1/(p\varepsilon^2)$), then its depth can be estimated reliably by counting the number of objects in R containing $\mathbf{r}$ and multiplying it by $1/p$. The interesting fact is that the deeper $\mathbf{r}$ is the better this estimate is.

Lemma 10.8 (Reliable sampling). *Let* S *be a set of* n *objects,* $0 < \varepsilon < 1/2$, *and let* $\mathbf{r}$ *be a point of depth* $u \geq k$ *in* S. *Let* R *be a random sample of* S, *such that every element is picked to be in the sample with probability*

$$p = \frac{8}{k\varepsilon^2} \ln \frac{1}{\delta}.$$

Let X *be the depth of* $\mathbf{r}$ *in* R. *Then, we have that the estimated depth of* $\mathbf{r}$ *in* R, *that is,* X/p, *lies in the interval* $[(1 - \varepsilon)u, (1 + \varepsilon)u]$. *This estimate succeeds with probability* $\geq 1 - \delta^{u/k} \geq 1 - \delta$.

Proof. We have that $\mu = \mathbf{E}[X] = pu$. As such, by Chernoff's inequality (Theorem 27.17 and Theorem 27.18), we have

$$\begin{aligned}
\mathbf{Pr}\left[X \notin [(1-\varepsilon)\mu, (1+\varepsilon)\mu]\right] &= \mathbf{Pr}\left[X < (1-\varepsilon)\mu\right] + \mathbf{Pr}\left[X > (1+\varepsilon)\mu\right] \\
&\leq \exp\left(-pu\varepsilon^2/2\right) + \exp\left(-pu\varepsilon^2/4\right) \\
&\leq \exp\left(-4\frac{u}{k} \ln \frac{1}{\delta}\right) + \exp\left(-2\frac{u}{k} \ln \frac{1}{\delta}\right) \leq \delta^{u/k},
\end{aligned}$$

since $u \geq k$. ∎

Note that if the depth of $\mathbf{r}$ in S is (say) $u \leq 10k$, then the depth of $\mathbf{r}$ in the sample is (with the stated probabilities)

$$\mathrm{depth}(\mathbf{r}, \mathsf{R}) \leq (1 + \varepsilon)pu = O\left(\frac{1}{\varepsilon^2} \ln \frac{1}{\delta}\right),$$

which is (relatively) a small number. Namely, via sampling, we turned the task of estimating the depth of heavy ranges into the task of estimating the depth of a shallow range. To see why this is true, observe that we can perform a binary (exponential) search for the depth of $\mathbf{r}$ by a sequence of coarser to finer samples.

10.4. Bibliographical notes

The presentation here follows the work by Aronov and Har-Peled [**AH08**]. The basic idea is folklore and predates this paper, but the formal connection between approximate counting to emptiness is from this paper. One can improve the efficiency of this reduction by being more careful; see [**AH08**] for details. Follow-ups to this work include [**KS06, AC09, AHS07, KRS11**]. For more information on relative approximations see the work by Har-Peled and Sharir [**HS11**].

10.5. Exercises

Exercise 10.1 (Heavy disk under restrictions)**.** Let R and B be sets of red and blue points in the plane, respectively, such that $n = |R|+|B|$. A *free disk* is a disk that contains no points of R in its interior, and its *weight* is the maximum number of blue points that it contains.

For a fixed parameter $\varepsilon > 0$, provide an algorithm that computes a free disk of weight $\geq (1 - \varepsilon)k_{\mathrm{opt}}$, where k_{opt} is the weight of the free disk of maximum weight. The running time of your algorithm should be near linear for a fixed ε.

Exercise 10.2 (Relative discrepancy)**.** Let $\mathcal{H}$ be a set of n closed halfplanes. We are also given a function $\chi : \mathcal{H} \to \{-1, +1\}$ that provides a coloring of the halfplanes of $\mathcal{H}$. The *depth* of a point of $\mathsf{p} \in \mathbb{R}^2$, denoted by $\mathtt{depth}(\mathsf{p})$, is the number of halfplanes of $\mathcal{H}$ that contain p. The *discrepancy* of a point $\mathsf{p} \in \mathbb{R}^2$ is $\chi(\mathsf{p}) = \sum_{h\in\mathcal{H},\mathsf{p}\in h} \chi(h)$. The *relative discrepancy* of a point $\mathsf{p} \in \mathbb{R}^2$ is the quantity $\mathrm{rdisc}(\mathsf{p}) = \chi(\mathsf{p})/\mathtt{depth}(\mathsf{p})$.

Provide an algorithm that computes, in near linear time (as a function of n), a point p such that $\mathrm{rdisc}(\mathsf{p}) \geq (\max_{\mathsf{s}\in\mathbb{R}^2} \mathrm{rdisc}(\mathsf{s})) - \varepsilon$, where $\varepsilon > 0$ is a prespecified approximation factor.

(Hint: Show how to handle all the points in the arrangement of depth $O\left(\varepsilon^{-2} \log n\right)$. Then show how to extend this algorithm to the general case.)

CHAPTER 11

Random Partition via Shifting

In this chapter, we investigate a rather simple technique for partitioning a geometric domain. We randomly shift a grid in $\mathbb{R}^d$ and consider each grid cell resulting from this shift. The shifting is done by adding a random vector, chosen uniformly from $[0,1]^d$, so that the new grid has this vector as its origin.

Points that are close together have a good probability of falling into the same cell in the new grid. This idea can be extended to shifting multi-resolution grids over $\mathbb{R}^d$. That is, we randomly shift a quadtree over a region of interest. This yields some simple algorithms for clustering and nearest neighbor search.

11.1. Partition via shifting

11.1.1. Shifted partition of the real line. Consider a real number $\Delta > 0$, and let b be a uniformly distributed number in the interval $[0, \Delta]$. This induces a natural partition of the real line into intervals, by the function

$$h_{\mathsf{b},\Delta}(x) = \left\lfloor \frac{x - \mathsf{b}}{\Delta} \right\rfloor .$$

Hence each interval has size Δ, and the origin is shifted to the right by an amount b.

Remark 11.1. Note that $h_{\mathsf{b},\Delta}(x)$ induces the same partition of the real line as $h_{\mathsf{b}',\Delta}(x)$ (but it is not the same function) if $|\mathsf{b} - \mathsf{b}'| = i\Delta$, where i is an integer. In particular, this implies that we can pick b uniformly from any interval of length Δ and get the same distribution on the partitions of the real line.

Specifically, for our purposes, it is enough if b is distributed uniformly in an interval of the form $[y + i\Delta, y + j\Delta]$, for two integers i and j such that $i < j$ and y is some real number. It is easy to verify that we still get the same distribution on the partitions of the real line.

Lemma 11.2. *For any $x, y \in \mathbb{R}$, we have* $\mathbf{Pr}\left[h_{\mathsf{b},\Delta}(x) \neq h_{\mathsf{b},\Delta}(y)\right] = \min\left(\frac{|x-y|}{\Delta}, 1\right)$.

Proof. Assume $x < y$. Clearly, the claim trivially holds if $x - y > \Delta$, since then x and y are always assigned different values of $h_{\mathsf{b},\Delta}(\cdot)$ and the required probability is 1. Now, imagine we pick b uniformly in the range $[x, x + \Delta]$. Clearly, the probability of the event we are interested in remains the same. But then, $h_{\mathsf{b},\Delta}(x) \neq h_{\mathsf{b},\Delta}(y)$ if and only if $\mathsf{b} \in [x, y]$, which implies the claim. ∎

11.1.2. Shifted partition of space. Let P be a point set in $\mathbb{R}^d$, and consider a point $\mathsf{b} = (\mathsf{b}_1, \ldots, \mathsf{b}_d) \in \mathbb{R}^d$ randomly and uniformly chosen from the hypercube $[0, \Delta]^d$, and consider the grid $\mathsf{G}^d(\mathsf{b}, \Delta)$ that has its origin at b and with sidelength Δ. For a point $\mathsf{p} \in \mathbb{R}^d$ the ID of the grid cell containing it is

$$\mathrm{id}(\mathsf{p}) = h_{\mathsf{b},\Delta}(\mathsf{p}) = \left(h_{\mathsf{b}_1,\Delta}(\mathsf{p}_1), \ldots, h_{\mathsf{b}_d,\Delta}(\mathsf{p}_d)\right).$$

(We used a similar concept when solving the closest pair problem; see Theorem 1.10_{p4}.)

Lemma 11.3. *Given a ball B of radius r in $\mathbb{R}^{\mathsf{d}}$ (or an axis parallel hypercube with side-length $2r$). The probability that B is not contained in a single cell of $\mathsf{G}^{\mathsf{d}}(\mathsf{b}, \Delta)$ is bounded by* $\min(2\mathsf{d}r/\Delta, 1)$, *where $\mathsf{G}^{\mathsf{d}}(\mathsf{b}, \Delta)$ is a randomly shifted grid.*

Proof. Project B into the ith coordinate. It becomes an interval B_i of length $2r$, and $\mathsf{G}^{\mathsf{d}}(\mathsf{b}, \Delta)$ becomes the one-dimensional shifted grid $\mathsf{G}^1(b_i, \Delta)$. Clearly, B is contained in a single cell of $\mathsf{G}^{\mathsf{d}}(\mathsf{b}, \Delta)$ if and only if B_i is contained in a single cell of $\mathsf{G}^1(b_i, \Delta)$, for $i = 1, \ldots, \mathsf{d}$.

Now, B_i is not contained in an interval of $\mathsf{G}^1(\mathsf{b}_i, \Delta)$ if and only if its endpoints are in different cells. Let $\mathcal{E}_i$ denote this event. By Lemma 11.2, the probability of $\mathcal{E}_i$ is $\leq 2r/\Delta$. As such, the probability of B not being contained in a single grid is $\mathbf{Pr}[\bigcup_i \mathcal{E}_i] \leq \sum_i \mathbf{Pr}[\mathcal{E}_i] \leq 2\mathsf{d}r/\Delta$. Since a probability is always bounded by 1, we have that this probability is bounded by $\min(2\mathsf{d}r/\Delta, 1)$. ∎

11.1.2.1. *Application – covering by disks.* Given a set P of n points in the plane, we would like to cover them by a minimal number of unit disks. Here, a ***unit disk*** is a disk of radius 1. Observe that if the cover requires k disks, then we can compute it in $O\left(kn^{2k+1}\right)$ time.

Indeed, consider a unit disk D that covers a subset $\mathsf{Q} \subseteq \mathsf{P}$. We can translate D such that it still covers Q, and either its boundary circle contains two points of Q or the top point of this circle is on an input point.

Observe the following: (I) Every pair of such input points defines two possible disks; see the figure on the right. (II) The same subset covered by a single disk might be coverable by several different such ***canonical*** disks. (III) If a pair of input points is at distance larger than 2, then the canonical disk they define is invalid.

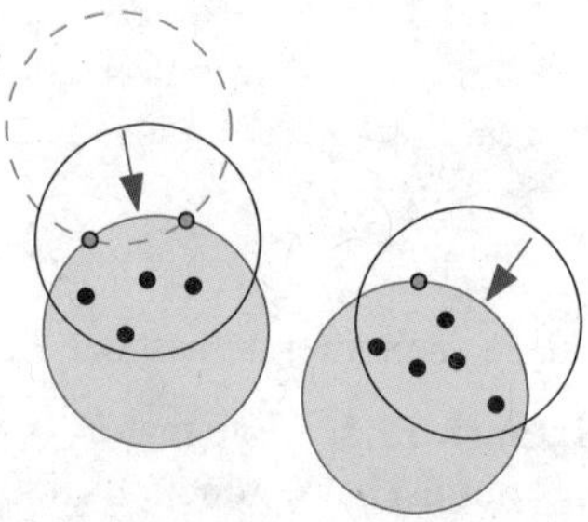

As such, there are

$$2\binom{n}{2} + n \leq n^2$$

such canonical disks. We can assume the cover uses only such disks.

Thus, we exhaustively check all such possible covers formed by k canonical disks. Overall, there are $\leq n^{2k}$ different covers to consider, and each such candidate cover can be verified in $O(nk)$ time. We thus get the following easy result.

Lemma 11.4. *Given a set P of n points in the plane, one can compute, in $O\left(kn^{2k+1}\right)$ time, a cover of P by at most k unit disks, if such a cover exists.*

Proof. We use the above algorithm trying all covers of size i, for $i = 1, \ldots, k$. The algorithm returns the first cover found. Clearly, the running time is dominated by the last iteration of this algorithm. ∎

One can improve the running time of Lemma 11.4 to $n^{O(\sqrt{k})}$; see Exercise 11.2.

The problem with this algorithm is that k might be quite large (say $n/4$). Fortunately, the shifting grid saves the day.

Theorem 11.5. *Given a set P of n points in the plane and a parameter $\varepsilon > 0$, one can compute using a randomized algorithm, in $n^{O(1/\varepsilon^2)}$ time, a cover of P by X unit disks, where $\mathbf{E}[X] \leq (1+\varepsilon)\mathrm{opt}$, where opt is the minimum number of unit disks required to cover P.*

PROOF. Let $\Delta = 12/\varepsilon$, and consider a randomly shifted gird $\mathsf{G}^2(\mathsf{b}, \Delta)$. Compute all the grid cells that contain points of P. This is done by computing for each point $\mathsf{p} \in \mathsf{P}$ its $\mathrm{id}(\mathsf{p})$ and storing it in a hash table. This clearly can be done in linear time.

Now, for each such grid cell $\square$ we now have the points of P falling into this grid cell (they are stored with the same id in the hash table), and let $\mathsf{P}_\square$ denote this set of points.

Observe that any grid cell of $\mathsf{G}^2(\mathsf{b}, \Delta)$ can be covered by $M = (\Delta + 1)^2$ unit disks. Indeed, each unit disk contains a unit square, and clearly a grid cell of sidelength Δ can be covered using at most M unit squares.

Thus, for each such grid cell $\square$, compute the minimum number of unit disks required to cover $\mathsf{P}_\square$. Since this number is at most M, we can compute this minimum cover in $O\left(Mn^{2M+1}\right)$ time, by Lemma 11.4. There are at most n non-empty grid cells, so the total running time of this stage is $O\left(Mn^{2M+2}\right) = n^{O(1/\varepsilon^2)}$. Finally, merge together the covers from each grid cell, and return this as the overall cover.

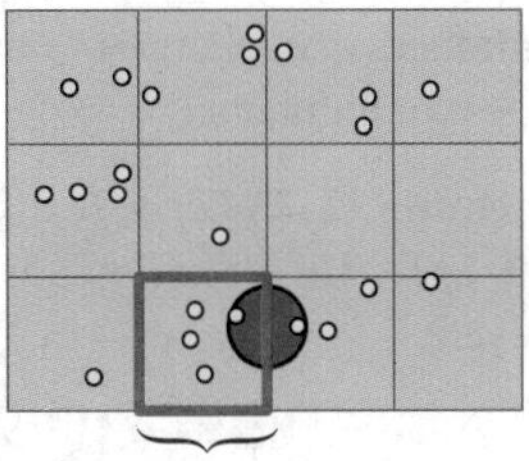

We are left with the task of bounding the expectation of X. So, consider the optimal solution $\mathcal{F} = \left\{D_1, \ldots, D_{\mathrm{opt}}\right\}$. We generate a feasible solution $\mathcal{G}$ from $\mathcal{F}$. Furthermore, the solution $\mathcal{G}$ is one of the possible solutions considered by the algorithm. Specifically, for each grid cell $\square$, let $\mathcal{F}_\square$ denote the set of disks of the optimal solution that intersect $\square$. Consider the multi-set $\mathcal{G} = \bigcup_\square \mathcal{F}_\square$. Clearly, the algorithm returns for each grid cell $\square$ a cover that is of size at most $|\mathcal{F}_\square|$ (indeed, it returns the smallest possible cover for $\mathsf{P}_\square$, and $\mathcal{F}_\square$ is one such possible cover). As such, the cover returned by the algorithm is of size at most $|\mathcal{G}|$.

Clearly, a disk of the optimal solution can intersect at most four cells of the grid, and as such it can appear in $\mathcal{G}$ at most four times. In fact, a disk $D_i \in \mathcal{F}$ will appear in $\mathcal{G}$ more than once if and only if it is not fully contained in a grid cell of $\mathsf{G}^2(\mathsf{b}, \Delta)$. By Lemma 11.3 the probability for that is bounded by $4/\Delta$ (as $r = 1$ and $\mathsf{d} = 2$).

Specifically, let X_i be an indicator variable that is 1 if and only if D_i is not fully contained in a single cell of $\mathsf{G}^2(\mathsf{b}, \Delta)$. We have that

$$\mathbf{E}\left[|\mathcal{G}|\right] \leq \mathbf{E}\left[\mathrm{opt} + \sum_{i=1}^{\mathrm{opt}} 3X_i\right] = \mathrm{opt} + \sum_{i=1}^{\mathrm{opt}} 3\,\mathbf{E}\left[X_i\right] = \mathrm{opt} + \sum_{i=1}^{\mathrm{opt}} 3\,\mathbf{Pr}\left[X_i = 1\right] \leq \mathrm{opt} + \sum_{i=1}^{\mathrm{opt}} 3\frac{4}{\Delta}$$
$$= \left(1 + \frac{12}{\Delta}\right)\mathrm{opt} = (1+\varepsilon)\,\mathrm{opt},$$

since $\Delta = 12/\varepsilon$. As such, in expectation, the solution returned by the algorithm is of size at most $(1+\varepsilon)\mathrm{opt}$. ■

The running time of Theorem 11.5 can be improved to $n^{O(1/\varepsilon)}$; see Exercise 11.2.

11.1.3. Shifting quadtrees.

11.1.3.1. *One-dimensional quadtrees.* Assume that we are given a set P of n numbers contained in the interval $[1/2, 3/4]$. Randomly, and uniformly, choose a number $\mathsf{b} \in [0, 1/2]$, and consider the (one-dimensional) quadtree $\mathcal{T}$ of P, using the interval $\mathsf{b} + [0, 1]$ for the root cell. Now, for two numbers $\alpha, \beta \in \mathsf{P}$ let

$$\mathbb{L}_{\mathsf{b}}(\alpha, \beta) = 1 - \mathrm{bit}_\Delta(\alpha - \mathsf{b}, \beta - \mathsf{b});$$

see Definition 2.7_{p17}.

This is the last ***level*** of the (one-dimensional) shifted canonical grid (i.e., one-dimensional quadtree) that contains α and β in the same interval. Namely, this is the level of the node of $\mathcal{T}$ that is the least common ancestor containing both numbers; that is, the level of $\mathrm{lca}(\alpha,\beta)$ is $\mathbb{L}_{\mathsf{b}}(\alpha,\beta)$. We remind the reader that a node in this quadtree that corresponds to an interval of length 2^{-i} has level $-i$. These definitions are demonstrated in the figure on the right (without shifting).

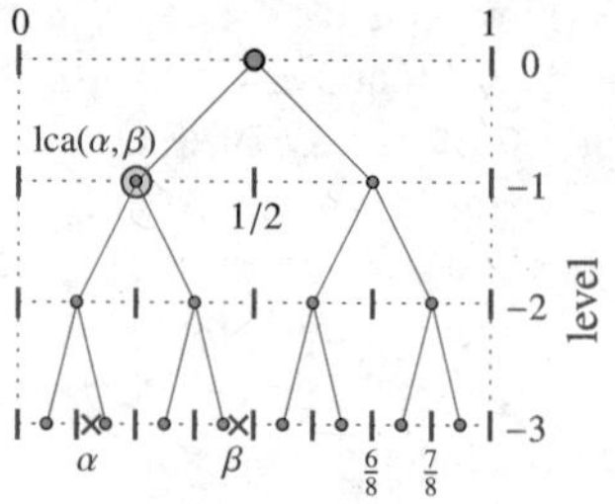

In the following, we will assume that one can compute $\mathbb{L}_{\mathsf{b}}(\alpha,\beta)$ in constant time.

Remark 11.6. Interestingly, the value of $\mathbb{L}_{\mathsf{b}}(\alpha,\beta)$ depends only on α, β, and b but is independent of the other points of P.

The following lemma bounds the probability that the least common ancestor of two numbers is in charge of an interval that is considerably longer than the difference between the two numbers.

Lemma 11.7. *Let $\alpha,\beta \in [1/2, 3/4]$ be two numbers, and consider a random number $\mathsf{b} \in [0, 1/2]$. Then, for any positive integer t, we have that* $\mathbf{Pr}[\mathbb{L}_{\mathsf{b}}(\alpha,\beta) > \lg|\alpha-\beta| + t] \le 4/2^t$, *where* $\lg x = \log_2 x$.

Proof. Let $M = \lfloor \lg|\alpha-\beta| \rfloor$ and consider the shifted partition of the real line by intervals of length $\Delta_{M+i} = 2^{M+i}$ and a shift b. Let X_{M+i} be an indicator variable that is 1 if and only if α and β are in different intervals of this shifted partition. Formally, $X_{M+i} = 1$ if and only if $h_{\mathsf{b},\Delta_{M+i}}(\alpha) \ne h_{\mathsf{b},\Delta_{M+i}}(\beta)$.

Now, if $\mathbb{L}_{\mathsf{b}}(\alpha,\beta) = M + i$, then the highest level such that both α and β lie in the same shifted interval is $M + i$. As such, going one level down, these two numbers are in different intervals, and $h_{\mathsf{b},2^{M+i-1}}(\alpha) \ne h_{\mathsf{b},2^{M+i-1}}(\beta)$. Namely, we have that $X_{M+i-1} = 1$. By Lemma 11.2, we have that $\mathbf{Pr}[X_{M+i} = 1] \le |\alpha-\beta| / \Delta_{M+i}$. As such, the probability we are interested in is

$$\mathbf{Pr}\Big[\mathbb{L}_{\mathsf{b}}(\alpha,\beta) > \lg_2|\alpha-\beta| + t\Big] \le \sum_{i=1+t}^{\infty} \mathbf{Pr}\Big[\mathbb{L}_{\mathsf{b}}(\alpha,\beta) = M + i\Big] \le \sum_{i=t}^{\infty} \mathbf{Pr}\Big[X_{M+i} = 1\Big]$$

$$\le \sum_{i=t}^{\infty} \frac{|\alpha-\beta|}{\Delta_{M+i}} = \sum_{i=t}^{\infty} \frac{|\alpha-\beta|}{2^M \cdot 2^i} \le \sum_{i=t}^{\infty} \frac{|\alpha-\beta|}{(|\alpha-\beta|/2)\, 2^i} \le \sum_{i=t}^{\infty} 2^{1-i} = 2^{2-t}.$$

■

Corollary 11.8. *Let $\alpha,\beta \in \mathsf{P} \subseteq [1/2, 3/4]$ be two numbers, and consider a random number $\mathsf{b} \in [0, 1/2]$. Then, for any parameters $c > 1$, we have* $\mathbf{Pr}[\mathbb{L}_{\mathsf{b}}(\alpha,\beta) > \lg_2|\alpha-\beta| + c\lg n] \le 4/n^c$, *where $n = |\mathsf{P}|$.*

11.1.3.2. *Higher-dimensional quadtrees.* Let P be a set of n points in $[1/2, 3/4]^{\mathsf{d}}$, pick uniformly and randomly a point $\mathsf{b} \in [0, 1/2]^{\mathsf{d}}$, and consider the shifted compressed quadtree $\mathcal{T}$ of P having $\mathsf{b} + [0, 1]^{\mathsf{d}}$ as the root cell.

Consider any two points $\mathsf{p}, \mathsf{q} \in \mathsf{P}$, their $\mathrm{lca}(\mathsf{p}, \mathsf{q})$, and the one-dimensional quadtrees $\mathcal{T}_1, \ldots, \mathcal{T}_{\mathsf{d}}$ built on each of the d coordinates of the point set. Since the quadtree $\mathcal{T}$ is the combination of these one-dimensional quadtrees $\mathcal{T}_1, \ldots, \mathcal{T}_{\mathsf{d}}$, the level where p and q get separated in $\mathcal{T}$ is the first level in any of the quadtrees $\mathcal{T}_1, \ldots, \mathcal{T}_{\mathsf{d}}$ where p and q are separated. In particular, the ***level of the least common ancestor*** (i.e., the level of $\mathrm{lca}(\mathsf{p}, \mathsf{q})$) is

$$\mathbb{L}_{\mathsf{b}}(\mathsf{p}, \mathsf{q}) = \max_{i=1}^{\mathsf{d}} \mathbb{L}_{\mathsf{b}_i}(\mathsf{p}_i, \mathsf{q}_i)\,. \tag{11.1}$$

As in the one-dimensional case (see Remark 11.6) the value of $\mathbb{L}_b(p, q)$ is independent of the other points of P.

Intuitively, $\mathbb{L}_b(p, q)$ is a well-behaved random variable, as testified by the following lemma. Since we do not use this lemma anywhere directly, we leave its proof as an exercise to the reader (see Exercise 11.1).

Lemma 11.9. *For any two fixed points* $p, q \in P$, *the following properties hold.*
(A) For any integer $t > 0$, *we have that* $\mathbf{Pr}\left[\mathbb{L}_b(p, q) > \lg \|p - q\| + t\right] \le 4d/2^t$.
(B) $\mathbf{E}[\mathbb{L}_b(p, q)] \le \lg \|p - q\| + \lg d + 6$.
(C) $\mathbb{L}_b(p, q) \ge \lg \|p - q\| - \lg d - 3$.

11.2. Hierarchical representation of a point set

In the following, it will be convenient to carry out our discussion in a more general setting than in low-dimensional Euclidean space. In particular, we will use the notion of metric space; see Definition 4.1_{p47}. Specifically, we are given a metric space $\mathcal{M}$ with a metric $\mathbf{d}$. We will slightly abuse notation by using $\mathcal{M}$ to refer to the underlying set of points.

11.2.1. Low quality approximation by HST.

We will use the following special type of a metric space.

Definition 11.10. Let P be a set of elements, and let H be a tree having the elements of P as leaves. The tree H defines a ***hierarchically well-separated tree*** (HST) over the points of P if for each vertex $u \in H$ there is associated a label $\Delta(u) \ge 0$, such that $\Delta(u) = 0$ if and only if u is a leaf of H. Furthermore, the labels are such that if a vertex u is a child of a vertex v, then $\Delta(u) \le \Delta(v)$. The distance between two leaves $x, y \in H$ is defined as $\Delta(\mathrm{lca}(x, y))$, where $\mathrm{lca}(x, y)$ is the least common ancestor of x and y in H.

If every internal node of H has exactly two children, we will refer to it as being a ***binary HST*** (BHST).

It is easy to verify that the distances defined by an HST comply with the triangle inequality, and as such the HST defines a metric. The usefulness of an HST is that the metric it defines has a very simple structure, and it can be easily manipulated algorithmically.

Example 11.11. Consider a point set $P \subseteq \mathbb{R}^d$ and a compressed quadtree $\mathcal{T}$ storing P, where for each node $v \in \mathcal{T}$, we set the diameter of $\square_v$ to be its label. It is easy to verify that this is an HST.

For convenience, from now on, we will work with BHSTs, since any HST can be converted into a binary HST in linear time while retaining the underlying distances. We will also associate with every vertex $u \in H$ an arbitrary representative point $\mathrm{rep}_u \in P_u$ (i.e., a point stored in the subtree rooted at u). We also require that $\mathrm{rep}_u \in \left\{\mathrm{rep}_v \;\middle|\; v \text{ is a child of } u\right\}$.

Definition 11.12. A metric space $\mathcal{N}$ is said to ***t-approximate*** the metric $\mathcal{M}$ if they are defined over the same set of points P and $\mathbf{d}_{\mathcal{M}}(u, v) \le \mathbf{d}_{\mathcal{N}}(u, v) \le t \cdot \mathbf{d}_{\mathcal{M}}(u, v)$, for any $u, v \in P$.

It is not hard to see that any n-point metric is $(n-1)$-approximated by some HST.

Lemma 11.13. *Given a weighted connected graph* G *on* n *vertices and* m *edges, it is possible to construct, in* $O(n \log n + m)$ *time, a binary* HST H *that* $(n-1)$*-approximates the shortest path metric of* G.

PROOF. Compute the minimum spanning tree $\mathcal{T}$ of G in $O(n \log n + m)$ time[1].

Sort the $n-1$ edges of $\mathcal{T}$ in non-decreasing order, and add them to the graph one by one, starting with an empty graph on $V(\mathsf{G})$. The HST is built bottom up. At each stage, we have a collection of HSTs, each corresponding to a connected component of the current graph. Each added edge merges two connected components, and we merge the two corresponding HSTs into a single HST by adding a new common root v for the two HSTs and labeling this root with the edge's weight times $|\mathsf{P}_v| - 1$, where P_v is the set of points stored in this subtree. This algorithm is only a slight variation on Kruskal algorithm and hence has the same running time.

Let $\mathtt{H}$ denote the resulting HST. As for the approximation factor, let x, y be any two vertices of G, and let e be the first edge added such that x and y are in the same connected component C, created by merging two connected components C_x and C_y. Observe that e must be the lightest edge in the cut between C_x and the rest of the vertices of G. As such, any path between x and y in G must contain an edge of weight at least $\omega(\mathsf{e})$. Now, e is the heaviest edge in C and

$$\mathbf{d}_{\mathsf{G}}(x, y) \leq (|C| - 1)\, \omega(\mathsf{e}) \leq (n-1)\omega(\mathsf{e}) \leq (n-1)\mathbf{d}_{\mathsf{G}}(x, y),$$

since $\omega(\mathsf{e}) \leq \mathbf{d}_{\mathsf{G}}(x, y)$. Now, $\mathbf{d}_{\mathtt{H}}(x, y) = (|C| - 1)\, \omega(\mathsf{e})$, and the claim follows. ∎

Since any n-point metric $\mathsf{P} \subseteq \mathcal{M}$ can be represented using the complete graph over n vertices (with edge weights $\omega(xy) = \mathbf{d}_{\mathcal{M}}(x, y)$ for all $x, y \in \mathsf{P}$), we get the following.

Corollary 11.14. *For a set* P *of* n *points in a metric space* $\mathcal{M}$, *one can compute, in* $O(n^2)$ *time, an HST* $\mathtt{H}$ *that* $(n-1)$*-approximates the metric* $\mathbf{d}_{\mathcal{M}}$.

One can improve the running time in low-dimensional Euclidean space (the approximation factor deteriorates slightly).

Corollary 11.15. *For a set* P *of* n *points in* $\mathbb{R}^{\mathsf{d}}$, *one can construct, in* $O(n \log n)$ *time (the constant in the* $O(\cdot)$ *depends exponentially on the dimension), a BHST* $\mathtt{H}$ *that* $(2n-2)$*-approximates the distances of points in* P. *That is, for any* $\mathsf{p}, \mathsf{q} \in \mathsf{P}$, *we have* $\mathbf{d}_{\mathtt{H}}(\mathsf{p}, \mathsf{q}) / (2n-2) \leq \|\mathsf{p} - \mathsf{q}\| \leq \mathbf{d}_{\mathtt{H}}(\mathsf{p}, \mathsf{q})$.

PROOF. We remind the reader, that in $\mathbb{R}^{\mathsf{d}}$ one can compute a 2-spanner for P of size $O(n)$, in $O(n \log n)$ time (see Theorem 3.12_{p34}). Let G be this spanner, and we apply Lemma 11.13 to this spanner. Let $\mathtt{H}$ be the resulting HST metric. For any $\mathsf{p}, \mathsf{q} \in \mathsf{P}$, we have $\|\mathsf{p} - \mathsf{q}\| \leq \mathbf{d}_{\mathtt{H}}(\mathsf{p}, \mathsf{q}) \leq (n-1)\mathbf{d}_{\mathsf{G}}(\mathsf{p}, \mathsf{q}) \leq 2(n-1)\, \|\mathsf{p} - \mathsf{q}\|$. ∎

Corollary 11.15 is unique to $\mathbb{R}^{\mathsf{d}}$ since for general metric spaces no HST can be computed in subquadratic time; see Exercise 11.3.

11.2.2. Fast and dirty HST in high dimensions. The above construction of an HST has exponential dependency on the dimension. We next show how one can get an approximate HST of low quality but in polynomial time in the dimension.

Lemma 11.16. *Let* P *be a set of* n *points in* $[1/2, 3/4]^{\mathsf{d}}$, *pick uniformly and randomly a point* $\mathsf{b} \in [0, 1/2]^{\mathsf{d}}$, *and consider the shifted compressed quadtree* $\mathcal{T}$ *of* P *having* $\mathsf{b} + [0, 1]^{\mathsf{d}}$ *as the root cell. Then, for any constant* $c > 1$, *with probability* $\geq 1 - 4\mathsf{d}/n^{c-2}$, *for all* $i = 1, \ldots, \mathsf{d}$ *and for all pairs of points of* P, *we have*

$$\mathbb{L}_{\mathsf{b}_i}(\mathsf{p}_i, \mathsf{q}_i) \leq \lg |\mathsf{p}_i - \mathsf{q}_i| + c \lg n. \tag{11.2}$$

[1]Using, say, Prim's algorithm implemented using a Fibonacci heap.

In particular, this property implies that the compressed quadtree $\mathcal{T}$ *is a* $2\sqrt{\mathsf{d}}n^c$*-approximate* HST *for* P.

Proof. Consider two points $\mathsf{p}, \mathsf{q} \in \mathsf{P}$ and a coordinate i. By Corollary 11.8, we have that

$$\mathbf{Pr}\Big[\mathbb{L}_{\mathsf{b}_i}(\mathsf{p}_i, \mathsf{q}_i) > \lg |\mathsf{p}_i - \mathsf{q}_i| + c \lg n\Big] \le 4/n^c.$$

There are d possible coordinates and $\binom{n}{2}$ possible pairs, and as such, by the union bound, this does not happen for any pair of points of P, for any coordinate, with probability $\ge 1 - \mathsf{d}\binom{n}{2}\frac{4}{n^c} \ge 1 - 2\mathsf{d}/n^{c-2}$.

As such, with probability $\ge 1 - 2\mathsf{d}/n^c$, the level of the lca of any two points $\mathsf{p}, \mathsf{q} \in \mathsf{P}$ is at most

$$U = \mathbb{L}_{\mathsf{b}}(\mathsf{p}, \mathsf{q}) = \max_{i=1}^{\mathsf{d}} \mathbb{L}_{\mathsf{b}_i}(\mathsf{p}_i, \mathsf{q}_i) \le \max(\lg |\mathsf{p}_i - \mathsf{q}_i| + c \lg n) \le \left\lceil \lg \|\mathsf{p} - \mathsf{q}\| + c \lg n \right\rceil,$$

by (11.1). The diameter of a cell at level $U(\mathsf{p}, \mathsf{q})$ is at most $\sqrt{\mathsf{d}}2^U \le 2\sqrt{\mathsf{d}}\,\|\mathsf{p} - \mathsf{q}\|\, n^c$. Namely, $\mathcal{T}$ is a $2\sqrt{\mathsf{d}}n^c$-approximate HST. ■

Verifying *quickly* that $\mathcal{T}$ is an acceptable HST (as far as the quality of approximation goes) is quite challenging in general. Fortunately, one can easily check that (11.2) holds.

Claim 11.17. *One can verify that (11.2) holds for the quadtree* $\mathcal{T}$ *computed by the algorithm of Lemma 11.16 in* $O(\mathsf{d}n \log n)$ *time.*

Proof. If (11.2) fails, then it must be that there are two points $\mathsf{p}, \mathsf{q} \in \mathsf{P}$ and a coordinate j such that $\mathbb{L}_{\mathsf{b}_j}\big(\mathsf{p}_j, \mathsf{q}_j\big) > \lg \big|\mathsf{p}_j - \mathsf{q}_j\big| + t$, for $t = c \lg n$.

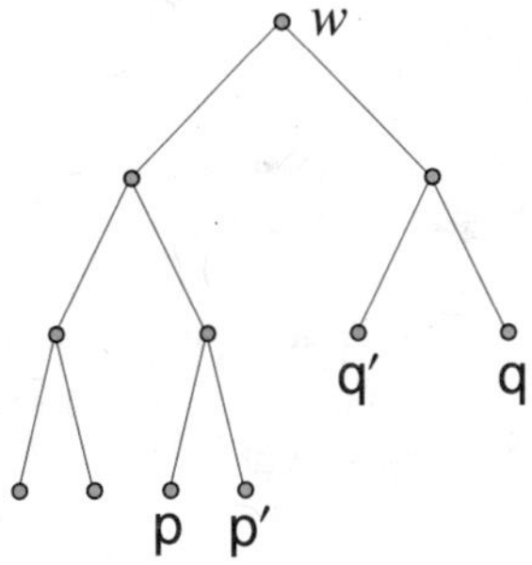

Now, observe that if there are two such bad points for the jth coordinate, then there are two bad points that are consecutive in the order along the jth coordinate. Indeed, consider $w = \mathrm{lca}(\mathsf{p}_j, \mathsf{q}_j)$ in the one-dimensional compressed quadtree $\mathcal{T}_j$ of P on the jth coordinate. Let p′ (resp. q′) be the point with maximal (resp. minimal) value in the jth coordinate that is still in the left (resp. right) subtree of w; see the figure on the right. Clearly, (i) $\mathrm{lca}(\mathsf{p}', \mathsf{q}') = w$, (ii) $\big|\mathsf{p}'_j - \mathsf{q}'_j\big| \le \big|\mathsf{p}_j - \mathsf{q}_j\big|$, and (iii) p′ and q′ are consecutive in the ordering of P according to the values in the jth coordinate. As such, $\mathbb{L}_{\mathsf{b}_j}\big(\mathsf{p}'_j, \mathsf{q}'_j\big) = \mathbb{L}_{\mathsf{b}_j}\big(\mathsf{p}_j, \mathsf{q}_j\big) > \lg \big|\mathsf{p}_j - \mathsf{q}_j\big| + t \ge \lg \big|\mathsf{p}'_j - \mathsf{q}'_j\big| + t$. Namely, (11.2) fails for p′ and q′ on the jth coordinate.

As such, to verify (11.2) on the jth coordinate, we sort the points according to their order on the jth coordinate and verify (11.2) for each consecutive pair. This takes $O(n \log n)$ time per coordinate, since computing $\mathbb{L}_{\mathsf{b}_j}(\cdot, \cdot)$ takes constant time. Doing this for all d coordinates takes $O(\mathsf{d}n \log n)$ time overall. ■

Theorem 11.18. *Given a set* P *of* n *points in* $\mathbb{R}^{\mathsf{d}}$*, for* $\mathsf{d} \le n$*, one can compute a* $2\sqrt{\mathsf{d}}n^5$*-approximate* HST *of* P *in* $O(\mathsf{d}n \log n)$ *expected time.*

Proof. Set $c = 5$, use the algorithm of Lemma 11.16, and verify it, as described in Claim 11.17. If the compressed quadtree fails, we repeat the construction until succeeding. Computing and verifying the compressed quadtree takes $O(\mathsf{d}n \log n)$ time, by Theorem 2.23$_{\text{p24}}$ and Claim 11.17. Now, since the probability of success is $\ge 1 - 4\mathsf{d}/n^{c-2} \ge$

$1 - 1/n$ (assuming $n \geq 3$), it follows that the algorithm would have to perform, in expectation, $1/(1 - 1/n) \leq 2$ iterations till it succeeds. ■

11.2.2.1. *An alternative deterministic construction of* HST. Our construction is based on a recursive decomposition of the point set. In each stage, we split the point set into two subsets. We recursively compute an $n\mathsf{d}$-HST for each point set, and we merge the two trees into a single tree, by creating a new vertex, assigning it an appropriate value, and hang the two subtrees from this node. To carry this out, we try to separate the set into two subsets that are furthest away from each other.

Lemma 11.19. *Let* P *be a set of n points in* $\mathbb{R}^{\mathsf{d}}$. *One can compute an* $n\mathsf{d}$-HST *of* P *in* $O(n\mathsf{d}\log^2 n)$ *time (note that the constant hidden by the O notation does not depend on* d*).*

Proof. Let $R = R(\mathsf{P})$ be the minimum axis parallel box containing P, and let $\nu = \sum_{i=1}^{\mathsf{d}} \|I_i(R)\|$, where $I_i(R)$ is the projection of R to the ith dimension.

Clearly, one find an axis parallel strip H of width $\geq \nu/((n-1)\mathsf{d})$, such that there is at least one point of P on each of its sides and there is no point of P inside H. Indeed, to find this strip, project the point set into the ith dimension, and find the longest interval between two consecutive points. Repeat this process for $i = 1, \ldots, \mathsf{d}$, and use the longest interval encountered. Clearly, the strip H corresponding to this interval is of width $\geq \nu/((n-1)\mathsf{d})$. On the other hand, $\operatorname{diam}(\mathsf{P}) \leq \nu$.

Now recursively continue the construction of two trees $\mathcal{T}^+, \mathcal{T}^-$, for $\mathsf{P}^+, \mathsf{P}^-$, respectively, where $\mathsf{P}^+, \mathsf{P}^-$ is the splitting of P into two sets by H. We hung $\mathcal{T}^+$ and $\mathcal{T}^-$ on the root node v and set $\Delta(v) = \nu$. We claim that the resulting tree $\mathcal{T}$ is an $n\mathsf{d}$-HST. To this end, observe that $\operatorname{diam}(\mathsf{P}) \leq \Delta(v)$, and for a point $\mathsf{p} \in \mathsf{P}^-$ and a point $\mathsf{q} \in \mathsf{P}^+$, we have $\|\mathsf{p} - \mathsf{q}\| \geq \nu/((n-1)\mathsf{d})$, which implies the claim.

To construct this efficiently, we use an efficient search trees to store the points according to their order in each coordinate. Let $\mathcal{D}_1, \ldots, \mathcal{D}_{\mathsf{d}}$ be those trees, where $\mathcal{D}_i$ stores the points of P in ascending order according to the ith axis, for $i = 1, \ldots, \mathsf{d}$. We modify them, such that for every node $v \in \mathcal{D}_i$, we know what the largest empty interval along the ith axis is for the points P_v (i.e., the points stored in the subtree of v in $\mathcal{D}_i$). Thus, finding the largest strip to split along can be done in $O(\mathsf{d}\log n)$ time. Now, we need to split the d trees into two families of d trees. Assume we split according to the first axis. We can split $\mathcal{D}_1$ in $O(\log n)$ time using the splitting operation provided by the search tree (Treaps **[SA96]** for example can do this split in $O(\log n)$ time). Let us assume that this splits P into two sets L and R, where $|L| \leq |R|$.

We still need to split the other $\mathsf{d} - 1$ search trees. This is going to be done by deleting all the points of L from those trees and building $\mathsf{d} - 1$ new search trees for L. This takes $O(|L|\,\mathsf{d}\log n)$ time. We charge this work to the points of L.

Since in every split only the points in the smaller portion of the split get charged, it follows that every point can be charged at most $O(\log n)$ time during this construction algorithm. Thus, the overall construction time is $O(\mathsf{d}n\log^2 n)$. ■

11.3. Low quality ANN search

We are interested in answering ***approximate nearest neighbor*** (***ANN***) queries in $\mathbb{R}^{\mathsf{d}}$. Namely, given a set P of n points in $\mathbb{R}^{\mathsf{d}}$ and a parameter $\varepsilon > 0$, we want to preprocess P, such that given a query point q, we can compute (quickly) a point $\mathsf{p} \in \mathsf{P}$, such that p is a $(1+\varepsilon)$-approximate nearest neighbor to q in P. Formally, $\|\mathsf{q} - \mathsf{p}\| \leq (1+\varepsilon)\mathbf{d}(\mathsf{q}, \mathsf{P})$, where $\mathbf{d}(\mathsf{q}, \mathsf{P}) = \min_{\mathsf{p}\in\mathsf{P}} \|\mathsf{q} - \mathsf{p}\|$.

11.3.1. The data-structure and search procedure. Let P be a set of n points in $\mathbb{R}^d$, contained inside the cube $[1/2, 3/4]^d$. Let b be a random vector in the cube $[0, 1/2]^d$, and consider the compressed shifted quadtree $\mathcal{T}$ having the hypercube $b + [0, 1]^d$ as its root. We choose for each node v of the quadtree a representative point $\mathrm{rep}_v \in P_v$.

Given a query point $q \in [1/2, 3/4]^d$, let v be the lowest node of $\mathcal{T}$ whose region rg_v contains q.

(A) If rep_v is defined, then we return it as the ANN.
(B) If rep_v is not defined, then it must be that v is an empty leaf, so let u be its parent, and return rep_u as q's ANN.

11.3.2. Analysis. Let us consider the above search procedure. When answering a query, there are several possibilities:

(A) If rg_v is a cube (i.e., v is a leaf) and v stores a point $p \in Q$ inside it, then we return p as the ANN.
(B) v is a leaf but there is no point associated with it.
In this case rep_v is not defined, and we return rep_u, where $u = \overline{p}(v)$. Observe that $\|q - \mathrm{rep}_u\| \le 2\mathrm{diam}(\mathrm{rg}_v)$. This case happens if the parent u of v has two children that contain points of P and v is one of the empty children. This situation is depicted in the figure on the right.

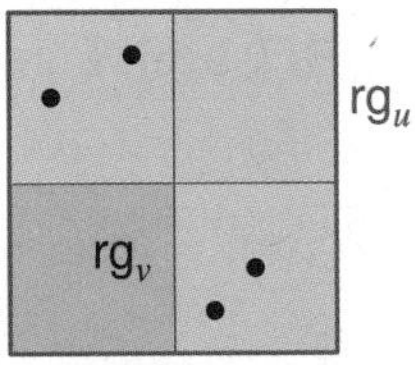

(C) If rg_v is an annulus, then v is a compressed node. In this case, we return rep_v as the ANN. Observe that $\mathbf{d}(q, \mathrm{rep}_v) \le \mathrm{diam}(\mathrm{rg}_v)$.

In all these cases, the distance to the ANN found is at most $2\mathrm{diam}(\Box_v)$.

Lemma 11.20. *For any $\tau > 1$ and a query point* q*, the above data-structure returns a τ-approximation to the distance to the nearest neighbor of* q *in* P*, with probability* $\ge 1 - 4d^{3/2}/\tau$.

PROOF. Let p be q's nearest neighbor in P (i.e., $= \|q - p\| = \mathbf{d}(q, P)$). Let $\mathbf{b}$ be the ball with diameter $\|q - p\|$ that contains q and p (i.e., this is the diametrical ball of qp). Consider the lowest node u in the compressed quadtree that contains $\mathbf{b}$ completely. By construction, the node v fetched by the point-location query for q must be either u or one of its descendants. As such, the ANN returned is of distance $\le 2\mathrm{diam}(\Box_v) \le 2\mathrm{diam}(\Box_u)$.

Let $\ell = \|q - p\|$. By Lemma 11.3, the probability that $\mathbf{b}$ is fully contained inside a single cell in the ith level of $\mathcal{T}$ is at least $1 - d\ell/2^i$ (i is a non-positive integer). In such a case, let $\Box$ be this grid cell, and observe that the distance to the ANN returned is at most $2\mathrm{diam}(\Box) \le 2\sqrt{d}2^i$. As such, the quality of the ANN returned in such a case is bounded by $2\sqrt{d}2^i/\ell$. If we want this ANN to be of quality τ, then we require that

$$2\sqrt{d}2^i/\ell \le \tau \implies 2^i \le \frac{\ell\tau}{2\sqrt{d}} \implies i \le \lg \frac{\ell\tau}{2\sqrt{d}}.$$

In particular, setting $i = \left\lfloor \lg\left(\ell\tau/2\sqrt{d}\right)\right\rfloor$, we get that the returned point is a τ-ANN with probability at least

$$1 - \frac{d\ell}{2^i} \ge 1 - \frac{4d^{3/2}\ell}{\ell\tau} = 1 - \frac{4d^{3/2}}{\tau},$$

implying the claim. ■

One thing we glossed over in the above is how to handle queries outside the square $[1/2, 3/4]^d$. This can be handled by scaling the input point set P to lie inside a hypercube of diameter (say) $1/n^2$ centered at the middle of the domain $[1/2, 3/4]^d$. Let T be the affine transformation (i.e., it is scaling and translation) realizing this. Clearly, the ANN to q in P is the point corresponding to the ANN to $T(\mathsf{q})$ in $T(\mathsf{P})$. In particular, given a query point q, we answer the ANN query on the transformed point set $T(\mathsf{P})$ using the query point $T(\mathsf{q})$.

Now, if the query point $T(\mathsf{q})$ is outside $[1/2, 3/4]^{\mathsf{d}}$, then any point of $T(\mathsf{P})$ is $(1+1/n)$-ANN to the query point as can be easily verified.

Combining Lemma 11.20 with Theorem 2.23_{p24}, we get the following result.

Theorem 11.21. *For a point set $\mathsf{P} \subseteq [1/2, 3/4]^{\mathsf{d}}$, a randomly shifted compressed quadtree $\mathcal{T}$ of* P *can answer ANN queries in $O(\mathsf{d} \log n)$ time. The time to build this data-structure is $O(\mathsf{d}n \log n)$. Furthermore, for any $\tau > 1$, the returned point is a τ-ANN with probability $\geq 1 - 4\mathsf{d}^{3/2}/\tau$.*

Remark 11.22. (A) Theorem 11.21 is usually interesting for τ being polynomially large in n, where the probability of success is quite high.

(B) Note that even for a large τ, this data-structure does not necessarily return the guaranteed quality of approximation for all the points in space. One can prove that this data-structure works with high probability for all the points in the space (but the guarantee of approximation deteriorates, naturally). See Exercise 11.4.

11.4. Bibliographical notes

The approximation algorithm for covering by unit disks is due to Hochbaum and Maas [**HM85**], and our presentation is a variant of their algorithm. Exercise 11.2 is similar in spirit to the work of Agarwal and Procopiuc [**AP02**]. The idea of HSTs was used by Bartal [**Bar98**] to get a similar result for more general settings.

The idea of shifted quadtrees can be derandomized [**Cha98**] and still yield interesting results. This is done by picking the shift carefully and inspecting the resulting $\mathcal{Q}$-order. The idea of doing point-location queries in a compressed quadtree to answer ANN queries is also from [**Cha98**] but it is probably found in earlier work.

The low-quality high-dimensional HST construction of Section 11.2.2.1 is taken from [**Har01b**]. The running time of this lemma can be further improved to $O(\mathsf{d}n \log n)$ by more careful and involved implementation; see [**CK95**] for details.

11.5. Exercises

Exercise 11.1 (My level it is nothing[②]). Prove Lemma 11.9.

Exercise 11.2 (Faster exact cover by unit disks). Let P be a set of n points in the plane that can be covered by k unit disks, and let $\mathcal{F}$ be this set of disks. Furthermore, assume that P cannot be covered by fewer unit disks.

(A) Prove that there exists an axis parallel line that passes through one of the points of P that intersects $O\left(\sqrt{k}\right)$ disks of $\mathcal{F}$.

(B) Provide an $n^{O(\sqrt{k})}$ time algorithm for computing a cover of P by k unit disks.

(C) Given a set P of n points in the plane, show how to $(1+\varepsilon)$-approximate the minimum cover of P by a set of k disks, in $n^{O(1/\varepsilon)}$ time.

Exercise 11.3 (Lower bound on computing an HST). Show, by adversarial argument, that for any $t > 1$, we have the following: Any algorithm computing an HST H for n points in a metric space $\mathcal{M}$ that t-approximates $\mathbf{d}_{\mathcal{M}}$ must in the worst case inspect all $\binom{n}{2}$ distances in the metric. Thus, computing an HST requires quadratic time in the worst case.

[②]Extra credit for getting the "popular" culture reference.

Exercise 11.4 (ANN for all). (Hard) Prove that the data-structure of Theorem 11.21 answers n^{10}-ANN queries for all the points in the space with high probability.

CHAPTER 12

Good Triangulations and Meshing

In this chapter, we show a beautiful application of quadtrees, demonstrating one of their most useful properties: Once we build a quadtree for a point set P, the quadtree captures the "shape" of the point set. We show how, using a quadtree, one can insert points into the given point set so that it looks smooth (i.e., the point density changes slowly).

12.1. Introduction – good triangulations

On simulation and meshing. In a lot of real world applications we are given a model of a physical object, and we would like to simulate some physical process on this object. For example, we are given a model of a solid engine of a rocket, and we would like to simulate how it burns out.

This process involves solving numerically a PDE (partial differential equation) on the given domain, describing the physical process involved. A numerical solver works better if the domain is geometrically simple. As such, we want to break the domain into triangles. The nicer the triangles are (i.e., fat and juicy) the easier it is for the numerical simulation. Naturally, as the simulation progresses, information is being passed between adjacent triangles; as such, the smaller the number of triangles the faster the simulation is.

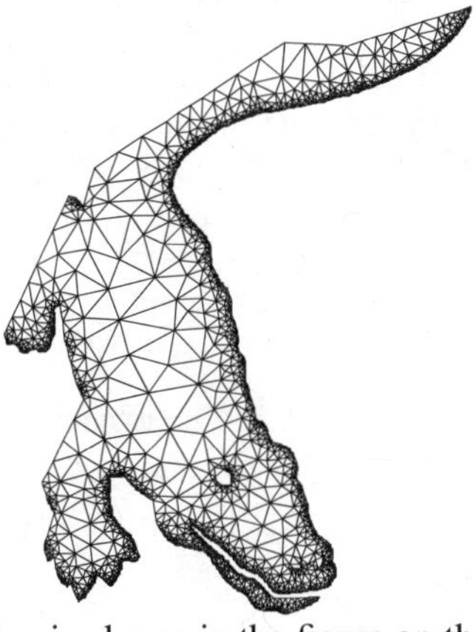

So, given a geometric domain, we would like to compute a triangulation of the domain breaking it into a small number of fat triangles. This process is usually referred to as ***meshing***. An example of such a triangulation in shown in the figure on the right.[1]

To simplify the presentation, we assume that P has diameter $\geq 1/2$ and it is contained in the square $[1/4, 3/4]^2$. This can be achieved by scaling and translation.

Meshes without borders. Instead of meshing an arbitrary domain (bounded by some polygons), we will mesh a point set that "defines" this domain. The point set marks the critical features of the underlying domain and what features need to be preserved and respected by the resulting triangulation. Intuitively, by placing many points on the boundary of the original domain, we guarantee that the resulting triangulation of the point set would respect this boundary. Handling boundaries in meshing is a major headache but conceptually it is enough to solve the problem when we ignore boundary issues. Such algorithms can then be extended to handle input with boundary.

[1] Sadly, the research field of meshing two-dimensional alligators has only recently received the level of attention it deserves.

12.2. Triangulations and fat triangulations

Definition 12.1 (Triangulation)**.** A ***triangulation*** of a domain $\mathcal{D} \subseteq \mathbb{R}^2$ is a partition of the domain into a set of interior disjoint triangles $\mathcal{M}$, such that for any pair of triangles $\triangle, \triangle' \in \mathcal{M}$, we have that either (i) $\triangle$ and $\triangle'$ are disjoint, (ii) $\triangle \cap \triangle'$ is a point which is a common vertex of both triangles or (iii) $\triangle \cap \triangle'$ is a complete edge of both triangles. The ***size*** of the triangulation $\mathcal{M}$, denoted by $|\mathcal{M}|$, is the number of triangles in it.

Observe that a triangulation does not allow a vertex of a triangle in the middle of an edge of an adjacent triangle.

Definition 12.2 (Aspect ratio)**.** The ***aspect ratio*** of a convex body is the ratio between its longest dimension and its shortest dimension. For a triangle $\triangle = \triangle abc$, the aspect ratio $\mathcal{A}_{\text{ratio}}(\triangle)$ is the length of the longest side divided by the height of the triangle when it's on the longest side.

Lemma 12.3. *Let ϕ be the smallest angle of a triangle $\triangle$. We have that $1/\sin\phi \le \mathcal{A}_{\text{ratio}}(\triangle) \le 2/\sin\phi$. In particular, we have $\phi \ge 1/\mathcal{A}_{\text{ratio}}(\triangle)$.*

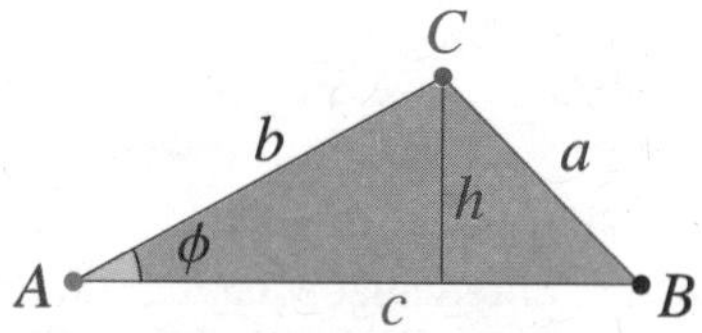

Proof. Consider the triangle $\triangle = \triangle ABC$, depicted on the right. We have $\mathcal{A}_{\text{ratio}}(\triangle) = c/h$. However, $h = b\sin\phi$, and since a is the shortest edge in the triangle (since it is facing the smallest angle), it must be that b is the middle length edge. As such, by the triangle inequality, we have $2b \ge a + b \ge c$. Thus, $\mathcal{A}_{\text{ratio}}(\triangle) = c/h \ge b/h = b/(b\sin\phi) = 1/\sin\phi$. Similarly, $\mathcal{A}_{\text{ratio}}(\triangle) = c/h \le 2b/h = 2b/(b\sin\phi) = 2/\sin\phi$.

As for the second claim, observe that $\sin\phi \le \phi$, for all $\phi \ge 0$. As such, $\phi \ge \sin\phi \ge 1/\mathcal{A}_{\text{ratio}}(\triangle)$. ■

Another natural measure of fatness is edge ratio.

Definition 12.4. The ***edge ratio*** of a triangle $\triangle$, denoted by $E_{\text{ratio}}(\triangle)$, is the ratio between a triangle's longest and shortest edges.

Clearly,

$$\mathcal{A}_{\text{ratio}}(\triangle) > E_{\text{ratio}}(\triangle), \tag{12.1}$$

for any triangle $\triangle$. For a triangulation $\mathcal{M}$, we denote by $\mathcal{A}_{\text{ratio}}(\mathcal{M})$ the maximum aspect ratio of a triangle in $\mathcal{M}$. Similarly, $E_{\text{ratio}}(\mathcal{M})$ denotes the maximum edge ratio of a triangle in $\mathcal{M}$.

Definition 12.5. A triangle is ***α-fat*** if its aspect ratio is bounded by α. We will refer to a triangulation $\mathcal{M}$ as ***α-fat*** if all its triangles are α-fat.

12.2.1. Building well-balanced quadtrees.

Definition 12.6. A ***corner*** of a quadtree cell is one of the four vertices of its square. The *corners* of the quadtree are the points that are corners of its cells. We say that the side of a cell is ***split*** if any of the neighboring boxes sharing it is split (i.e., the neighbor has a corner in the interior of this side). A quadtree is ***balanced*** if any side of a leaf cell contains at most one quadtree corner in its interior. Namely, adjacent leaves are either of the same level or of adjacent levels. (Naturally, a balanced quadtree cannot have compressed nodes.)

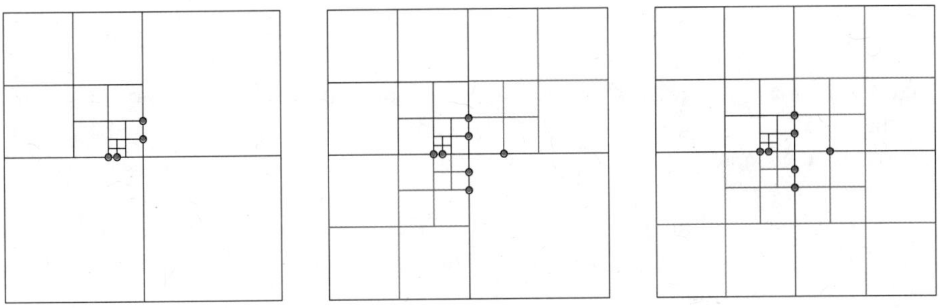

(i) (ii) (iii)

FIGURE 12.1. Rebalancing a quadtree: (i) the marked points are where the quadtree fails to be balanced, (ii) the quadtree resulting from refining nodes till the original unbalancing is fixed (however, new points where the quadtree is not balanced are now created), and (iii) the resulting balanced quadtree after the refining process continues.

Given a quadtree, we can easily rebalance it by adding required nodes to satisfy the balancing requirement in a greedy fashion – if a leaf has a neighbor which is too large, we split the neighbor. An example of a rebalanced quadtree is depicted in Figure 12.1[2].

Definition 12.7. The ***extended cluster*** of a cell c in a quadtree $\mathcal{T}$ is the set of 5×5 neighboring cells of c in the grid containing c. These are all the cells within a distance $< 2\varkappa(\square)$ from c in this grid, where $\varkappa(\square)$ is the sidelength of c.

A quadtree $\mathcal{T}$ over a point set P is ***well-balanced*** if it is balanced and for every leaf node v that contains a (single) point of P, we have the property that all the nodes of the extended cluster of v are in $\mathcal{T}$ and they do not contain any other point of P.

Lemma 12.8. *Let* P *be a set of points in the plane, such that* $\mathrm{diam}(\mathsf{P}) = \Omega(1)$. *Then, one can compute a well-balanced quadtree* $\mathcal{T}$ *of* P, *in* $O(n \log n + m)$ *time, where* m *is the size of the output quadtree.*

PROOF. Compute a compressed quadtree $\mathcal{T}$ of P, in $O(n \log n)$ time, using the algorithm of Theorem 2.23_{p24}. Next, we traverse $\mathcal{T}$ and replace every compressed node of $\mathcal{T}$ by the sequence of quadtree nodes that define it. To guarantee the balance condition, we create a queue of the nodes of $\mathcal{T}$ and store the nodes of $\mathcal{T}$ in a hash table, with their IDs.

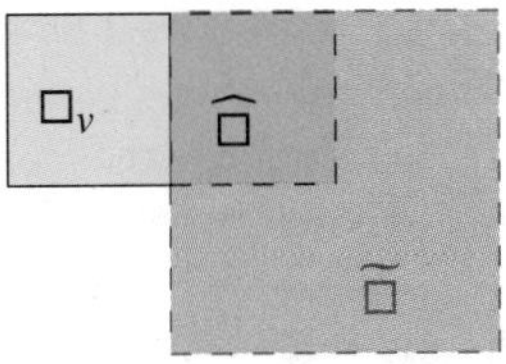

We handle the nodes in the queue one by one. For a node v, we check whether the current adjacent nodes to $\square_v$ are balanced. Specifically, let $\widehat{\square}$ be one of $\square_v$'s neighboring cells in the grid of $\square_v$ (there are four such neighbor cells), and let $\widetilde{\square}$ be the square containing $\widehat{\square}$ in the grid one level up; see the figure on the right. We compute $\mathrm{id}(\widetilde{\square})$ and check if there is a node in $\mathcal{T}$ with the ID $\mathrm{id}(\widetilde{\square})$. If not, then $\widehat{\square}$ and $\widetilde{\square}$ are both missing in the quadtree, and we need to create at least the node $\widetilde{\square}$ to fix the balancing property locally for the node $\square_v$. In this case, we create a node w with region $\widetilde{\square}$ and $\mathrm{id}(\widetilde{\square})$ and recursively retrieve its parent (i.e., if it exists, we retrieve it; otherwise, we create it) and hang w from the parent node. We add all the new nodes to the queue. We repeat the process till the queue is empty.

We charge the work involved in creating new nodes to the output size.

[2]It would be nice if people could be rebalanced as easily as quadtrees.

Now, for every leaf node v of $\mathcal{T}$ which contains a point of P, we check if its extended cluster contains any other point of P, and if so, we split it. Furthermore, for each child u of v we insert all the nodes in the extended cluster of u into the quadtree (if they do not already exist there). We repeat this process till all non-empty leaves have the required separation property. Of course, during this process, we keep the balanced property of the quadtree by adding necessary nodes.

The resulting quadtree is well-balanced, as required. Furthermore, the work can be charged to newly created nodes and as such takes linear time in the output size once the balanced compressed quadtree is computed.

Let t be the length of the closest pair of P, and observe that the algorithm will never create a node with side length smaller than t/c_1, where c_1 is some sufficiently large constant. As such, this algorithm terminates. Overall, the running time of the algorithm is $O(n \log n + m)$, since the work associated with any newly created quadtree node is a constant and can be charged to the output. ■

Remark 12.9. For the sake of simplicity of exposition, the algorithm of Lemma 12.8 generates some nodes that are redundant. Indeed, the algorithm inserts all cluster nodes of a node if its parent cluster contains two points. In particular, it does not output the minimal well-balanced quadtree of the input point set. One can directly argue that the size of the tree generated is a constant factor larger than the minimal well-balanced tree or the given point set. Alternatively, our analysis below implies the same.

Remark 12.10. A technicality we glossed over in describing the algorithm of Lemma 12.8 is how to implement the empty extended cluster queries efficiently. Specifically, given a cell $\square$ and a point p in it, we need to check if there is any other point in its extended cluster. Naturally, if the twenty five squares of the extended cluster exist in the quadtree, we can answer this query using hashing in constant time (in particular, when constructing a node v, a 'contain-points' flag is created that indicates if any point is stored in the subtree of v). However, it might be that such a cell query $\widehat{\square}$ (in the extended cluster of $\square$) does not exist in the tree $\mathcal{T}$.

To this end, we mark every node of the initial balanced quadtree if it contains points in its subtree. This information can be computed bottom-up in the balanced quadtree in linear time.

Next, we compute this flag whenever we create a new node u. We remind the reader that we store the points of P in the leaves of the quadtree. As such, when u is being created, it is formed by splitting a leaf node $\overline{\mathsf{p}}(u)$ of the quadtree, and this leaf node can contain only a single point of P. If $\overline{\mathsf{p}}(u)$ contains a point, we store it in the correct child of $\overline{\mathsf{p}}(u)$ and compute the 'contain-points' flag for all the newly minted children (which will be true only for one child[3]).

Now, during the well-balancing stage, we maintain the balancing property of $\mathcal{T}$ on the fly, and as such, the parent of such a query cell $\widehat{\square}$ must exist in the quadtree, and if not, the algorithm creates it since it needs to be created (so that the balancing condition is maintained). Let u' be the node of this bigger cell. If the flag indicates that u' does not contain points, then we are done. Otherwise, u' must be a leaf (otherwise, the cell of $\widehat{\square}$ would exist in the quadtree), and it contains a single point of P. We can now check in constant time if this point is inside $\widehat{\square}$.

---

[3]This is, the parents' favorite child, who is disliked by all the other children.

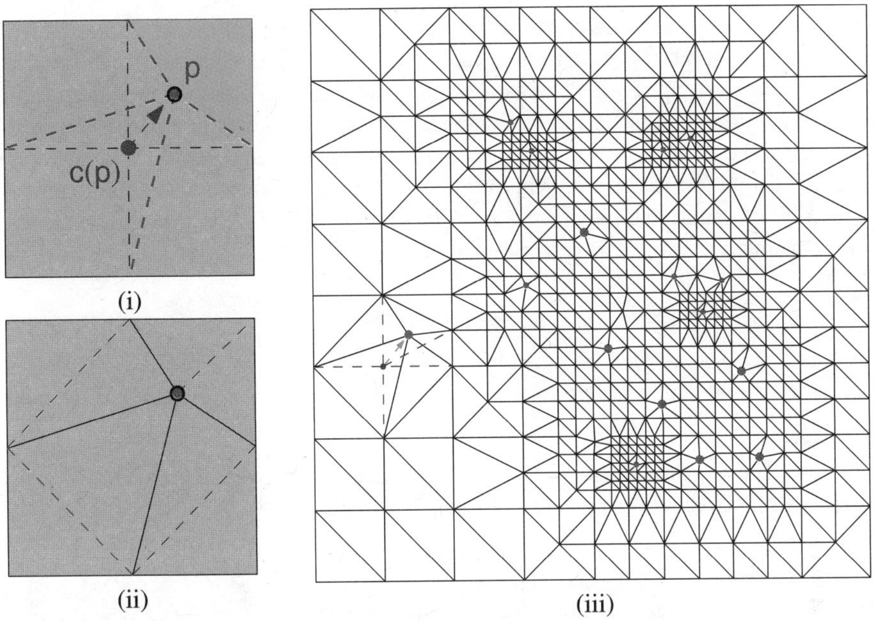

FIGURE 12.2. (i) We warp the square corner into the input point, turning the four squares into quadrangles. (ii) The triangulation of these quadrangles. (iii) A fat triangulation extracted from a quadtree. Note that this figure was generated by a variant of the algorithm we describe that does less refining (i.e., the quadtree used is not quite well-balanced).

12.2.2. The algorithm – extracting a good triangulation. The intuition behind the triangulation algorithm is that the well-balanced quadtree has leaves that have cells of the right size – they are roughly of the size that any triangle in a (small) fat triangulation of P has to be to cover the same region. Thus, we just need to decompose these cells into triangles in such a way that the points of P are vertices of the resulting triangulation.

Equipped with this vision, let us now plunge into the murky details. A well-balanced quadtree $\mathcal{T}$ of P provides for every point a region (i.e., its extended cluster) where it is well protected from other points. It is now possible to turn the partition of the plane induced by the leaves of $\mathcal{T}$ into a triangulation of P.

We "warp" the quadtree framework as follows. For a point $\mathsf{p} \in \mathsf{P}$, let $\mathsf{c}(\mathsf{p})$ be the corner nearest p of the leaf of $\mathcal{T}$ containing p; we replace $\mathsf{c}(\mathsf{p})$ by p. Thus, a 2×2 group of square leaves turns into a square divided into four (fat) quadrangles; see Figure 12.2(i).

Next, we triangulate the resulting planar subdivision. Regular squares can be triangulated by just introducing the diagonal, creating two isosceles right triangles. For cells containing points of P, we warped them into quadrangles, and we triangulate them by choosing the diagonal that gives a better aspect ratio (i.e., we choose the diagonal such that the maximum aspect ratio of the resulting two triangles is minimized); see Figure 12.2(ii).

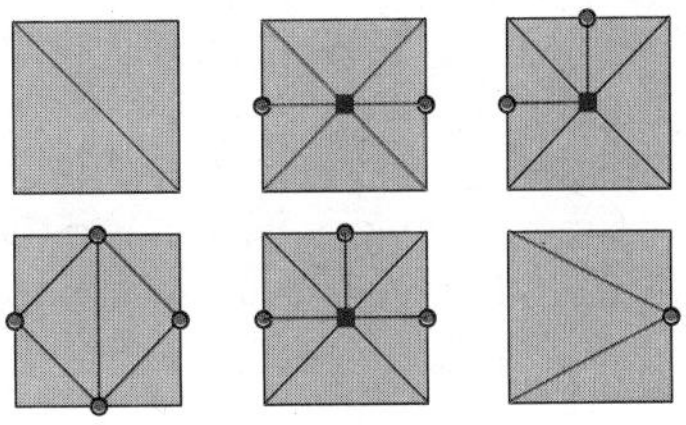

FIGURE 12.3.

We also have to triangulate squares that have one or more split sides. Note that by our well-balanced property (see Definition 12.7), we have that *no* such node can be warped – so it is indeed a square. If there is a single split side, we break the square into three triangles by introducing the two long diagonals. If there are more split sides, we introduce the middle of the square and we split the square into right triangles in the natural way; see Figure 12.3. If all four sides are split, then we do not need to introduce a middle point.

Let $\mathcal{QT}(\mathsf{P})$ denote the resulting ***triangulation***. Figure 12.2(iii) shows a triangulation resulting from a variant of this method.

12.3. Analysis

Lemma 12.11. *The method above gives a triangulation* $\mathcal{QT}(\mathsf{P})$ *with* $\mathcal{A}_{\mathrm{ratio}}(\mathcal{QT}(\mathsf{P})) \leq 4$. *The time to compute this triangulation is* $O(n \log n + m)$, *where* $n = |\mathsf{P}|$ *and* m *is the number of triangles in the triangulation.*

Proof. The right triangles used for the unwarped cells have aspect ratio 2. If an original cell $\square$ with sidelength l is warped, we have two cases.

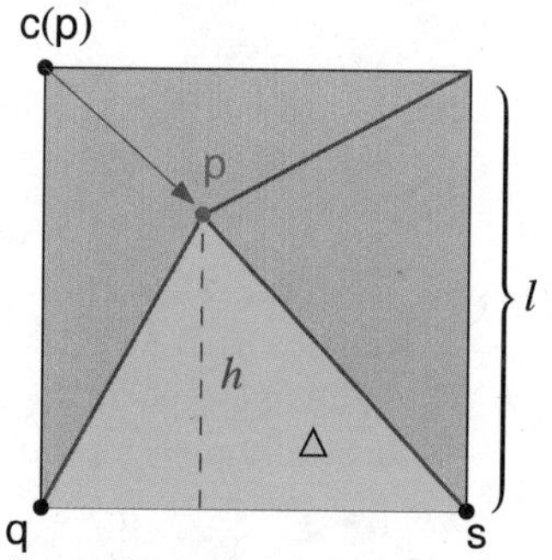

In the first case, the input point of P is inside the square of the original cell. To bound the aspect ratio, we assume that the diagonal touching the warped point is chosen; otherwise, the aspect ratio can only be smaller than what we prove (since we choose the diagonal that minimizes the aspect ratio). Consider one of the two triangles, say $\triangle = \triangle \mathsf{pqs}$, formed by the input point p (which is a warped corner) and the two other cell corners q and s. The maximum length diagonal is formed when the warped point p is at the original location $\mathsf{c}(\mathsf{p})$ and has length $x = \sqrt{2}l$. The minimum area of the triangle $\triangle$ is formed when the point is in the center of the square and has area $\alpha = l^2/4$. We have that $\mathrm{area}(\triangle) = \|\mathsf{q} - \mathsf{s}\| \cdot h/2$, where h is the height of $\triangle$ when on the edge qs. Thus, we have

$$\text{height of the triangle } \triangle \geq 2\frac{\text{min area } \triangle}{\text{max length edge of } \triangle} = \frac{2\alpha}{x}$$

and

$$\mathcal{A}_{\mathrm{ratio}}(\triangle) = \frac{\text{length of longest edge of } \triangle}{\text{height of } \triangle} \leq \frac{x}{2\alpha/x} = \frac{x^2}{2\alpha} = \frac{2l^2}{2l^2/4} = 4.$$

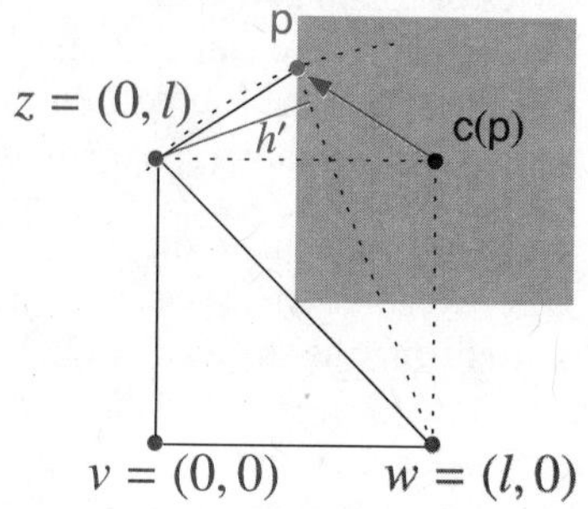

Figure 12.4.

In the second case, the input point is outside the original square. Since the quadtree is well balanced, the new point p is somewhere inside a square of sidelength l centered at $\mathsf{c}(\mathsf{p})$ (since we always move the closest leaf corner to the new point). In this case, we assume that the diagonal not touching the warped point is chosen. This divides the cell into an isosceles right triangle $\triangle zvw$ and the other triangle $\triangle = \triangle z\mathsf{p}w$. If the chosen diagonal is the longest edge of $\triangle$, then one can argue as before, and the aspect ratio is bounded by 4. Otherwise, the longest edge of $\triangle$ is adjacent to the input point p. The altitude h' is minimized when the triangle is isosceles with as sharp an angle as possible; see Figure 12.4.

Using the notation of the figure, we have $\mathsf{p} = (l/2, \sqrt{7}l/2)$. Thus, computing the area of a triangle, using Lemma 28.3,

$$\mu = \text{area}(\triangle z\mathsf{p}w) = \frac{1}{2}\begin{vmatrix} 1 & 0 & l \\ 1 & l & 0 \\ 1 & l/2 & (\sqrt{7}/2)l \end{vmatrix} = \frac{1}{2}\begin{vmatrix} l & -l \\ l/2 & (\sqrt{7}/2-1)l \end{vmatrix} = \frac{\sqrt{7}-1}{4}l^2.$$

We have $h'\sqrt{2}l/2 = \mu$, and thus $h' = \sqrt{2}\mu/l = \frac{\sqrt{7}-1}{2\sqrt{2}}l$. The longest distance p can be from w is $\beta = \sqrt{(1/2)^2 + (3/2)^2}l = (\sqrt{10}/2)l$. Thus, the aspect ratio of the new triangle is bounded by $\beta/h' = \left(\sqrt{10}/2\right)/\frac{\sqrt{7}-1}{2\sqrt{2}} \approx 2.717 \le 4$. ■

12.3.1. On local feature size.

Definition 12.12. The ***local feature size*** of a point p in the plane, with relation to a point set P, denoted by $\text{lfs}_{\mathsf{P}}(\mathsf{p})$, is the radius of the smallest closed disk centered at p containing at least two points of P.

The following lemma testifies that the local feature size function changes slowly.

Lemma 12.13. *The local feature is* 1*-Lipschitz; that is, for any* $\mathsf{p}, \mathsf{q} \in \mathbb{R}^2$, *we have that*

$$\text{lfs}_{\mathsf{P}}(\mathsf{p}) - \|\mathsf{p}-\mathsf{q}\| \le \text{lfs}_{\mathsf{P}}(\mathsf{q}) \le \text{lfs}_{\mathsf{P}}(\mathsf{p}) + \|\mathsf{p}-\mathsf{q}\|.$$

Proof. Let s and s' be the two nearest points to p in P. As such, we have $\text{lfs}_{\mathsf{P}}(\mathsf{p}) = \max(\|\mathsf{p}-\mathsf{s}\|, \|\mathsf{p}-\mathsf{s}'\|)$. Also, by the triangle inequality, we have that $\|\mathsf{q}-\mathsf{s}\| \le \|\mathsf{p}-\mathsf{q}\| + \|\mathsf{p}-\mathsf{s}\| \le \|\mathsf{p}-\mathsf{q}\| + \text{lfs}_{\mathsf{P}}(\mathsf{p})$ and $\|\mathsf{q}-\mathsf{s}'\| \le \|\mathsf{p}-\mathsf{q}\| + \text{lfs}_{\mathsf{P}}(\mathsf{p})$. As such,

$$\text{lfs}_{\mathsf{P}}(\mathsf{q}) \le \max\left(\|\mathsf{q}-\mathsf{s}\|, \left\|\mathsf{q}-\mathsf{s}'\right\|\right) \le \|\mathsf{p}-\mathsf{q}\| + \text{lfs}_{\mathsf{P}}(\mathsf{p}).$$

The other direction follows by a symmetric argument. ■

12.3.2. From triangulations to covering by disks and back. In this section, we provide some intuition about how the local feature size relates to the meshing problem, by investigating a somewhat tangential problem of how to cover the domain with a few disks such that no disk covers too many points of P. In particular, the material here, except for building the intuition for our later proofs, is not strictly needed for our analysis, and the reader looking for a quick enlightenment[4] might skip directly to Section 12.3.3.

Intuitively, the local feature size captures the local size of fat triangles that one has to use, such that a fat triangulation P_{map} would contain P (as vertices). Here, the triangulation would have a set of vertices Q that contains P but might potentially contain many additional points inserted to make such a triangulation possible. Indeed, not all point sets have a fat triangulation if we do not allow inserting additional points.

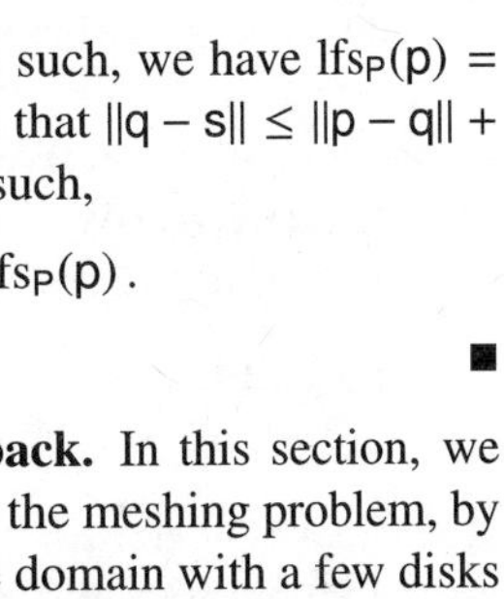

The leap of faith here[5] is to think about a fat triangle in the triangulation P_{map} that we want to construct as being roughly a disk. Indeed, consider an α-fat triangle $\triangle$ and two disks associated with it, that is, the largest disk inscribed inside it with radius r and the smallest disk containing $\triangle$ of radius R; see the figure on the right. It is not hard to verify that $R/r = O(\alpha)$. Now, if we take the inscribed disk and move it a bit locally, it would not contain more than one point of P

[4]And who isn't, in this world, on the lookout for a quick enlightenment on the cheap? Naturally, if you are looking for enlightenment in the middle of a section about meshing, you are looking for a very strange kind of enlightenment. The author can only hope that the text would provide it.

(since the adjacent triangles to $\triangle$ in the triangulation P_{map} have roughly the same size as $\triangle$).

So, for the time being, forget about the triangles in the triangulation, and imagine considering only their inscribed disks. Slightly changing the problem, we would like to cover the domain $\mathcal{D} = [0, 1]^2$ with such disks. Conceptually, the question becomes how many disks we need to cover the domain. The name of the game is that we do not allow a disk to contain more than one point of the original point set P in its interior. Also, we will require that any two disks that intersect have roughly (up to a constant factor) the same radius.

Specifically, we are trying to find a minimum number of disks that cover the domain with these properties. Clearly, a disk in this collection centered at a point p can have radius at most $\mathrm{lfs}_{\mathsf{P}}(\mathsf{p})$.

Let us construct such a covering in a greedy fashion. Let $\mathcal{F}$ be the current set of disks chosen so far (initially, it is empty). As long as there is a point $\mathsf{p} \in \mathcal{D}$ that is not covered by any disk of $\mathcal{F}$, let us add the disk centered at p of radius $\mathrm{lfs}_{\mathsf{P}}(\mathsf{p})/2$ to $\mathcal{F}$. We stop when the domain is fully covered by the disks of $\mathcal{F}$. Since the points of P are distinct and the domain $\mathcal{D} = [0, 1]^2$ is finite, this algorithm indeed stops. See the figure on the right for an example of what the output of this algorithm looks like. We refer to this algorithm as **algCoverDisksGreedy**.

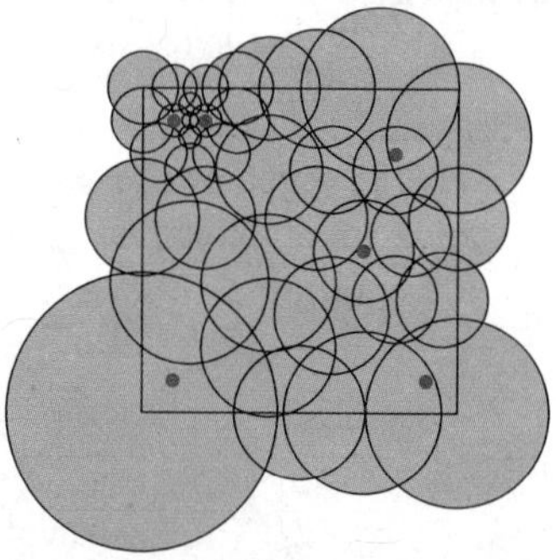

Let us start by proving that the resulting set of disks has indeed the required properties.

Lemma 12.14. *If* $\mathbf{b}(\mathsf{p}, r) \in \mathcal{F}$ *and* $\mathbf{b}(\mathsf{q}, r') \in \mathcal{F}$ *intersect and* $r' \leq r$, *then* $r/3 \leq r' \leq r$ *and* $\|\mathsf{p} - \mathsf{q}\| \geq r'$, *where* $\mathbf{b}(\mathsf{p}, r)$ *denotes the disk of radius* r *centered at* p.

Proof. By construction, $r = \mathrm{lfs}_{\mathsf{P}}(\mathsf{p})/2$ and $r' = \mathrm{lfs}_{\mathsf{P}}(\mathsf{q})/2$. Since the two disks intersect, it follows that $\|\mathsf{p} - \mathsf{q}\| \leq r + r'$. Observe that, by Lemma 12.13, we have that

$$2r' = \mathrm{lfs}_{\mathsf{P}}(\mathsf{q}) \geq \mathrm{lfs}_{\mathsf{P}}(\mathsf{p}) - \|\mathsf{p} - \mathsf{q}\| \geq 2r - r - r' \implies r' \geq r/3.$$

As for the second claim, if $\mathbf{b}(\mathsf{q}, r')$ was added to $\mathcal{F}$ after $\mathbf{b}(\mathsf{p}, r)$, then $\mathsf{q} \notin \mathbf{b}(\mathsf{p}, r)$ (otherwise q would be covered and the ball would have never been created). Namely, $\|\mathsf{p} - \mathsf{q}\| \geq r \geq r'$. Arguing similarly, if $\mathbf{b}(\mathsf{p}, r)$ was added to $\mathcal{F}$ after $\mathbf{b}(\mathsf{q}, r')$, then $\|\mathsf{p} - \mathsf{q}\| \geq r'$. ∎

Claim 12.15. *No point in the plane is covered by more than* $c = O(1)$ *disks of* $\mathcal{F}$.

Proof. The proof follows by a relatively easy packing argument. Indeed, let p be an arbitrary point in the plane that is covered by the disks of $\mathcal{F}$. Now, let X be the set of all disks of $\mathcal{F}$ covering p, and let $r_{\min}$ and $r_{\max}$ be the minimum and maximum radii of the disks of X. All the pairs of disks of X intersect since they contain p. As such, by Lemma 12.14, we have that $r_{\min} \leq r_{\max} \leq 3r_{\min}$, and the distance between any pair of centers of disks of X is at least $r_{\min}$. In particular, the centers of the disks of X are contained inside a disk of radius $r_{\max}$ centered at p.

So, place a disk of radius $r_{\min}/2$ centered at each one of these centers and let $\mathcal{G}$ be the resulting set of disks. The disks of $\mathcal{G}$ are interior disjoint disks and are all contained in a

[5] Naturally, if you do not like the resulting faith, you can always leap back.

disk of radius $r_{max} + r_{min}/2 \le 2r_{max}$ centered at p. As such, by charging these disks to the area they occupy in $\mathbf{b}(\mathsf{p}, 2r_{max})$, we have that

$$|X| = |\mathcal{G}| \le \frac{\text{area}(\mathbf{b}(\mathsf{p}, 2r_{max}))}{\text{area}(\mathbf{b}(r_{min}/2))} = \frac{4\pi r_{max}^2}{\pi r_{min}^2/4} \le \frac{16 r_{max}^2}{(r_{max}/3)^2} = 144.$$

■

So, how many disks do we expect to have in such a cover? Well, if a point p has local feature size α, then the disk of $\mathcal{F}$ centered at p has radius $\alpha/2$, and it has area $\Theta(\alpha^2)$. If the local feature size was the same everywhere, then all the disks we would use would be of area $\pi\alpha^2/4$ and the number of disks needed would be proportional to $\text{area}(\mathcal{D})/\alpha^2$. Since the local feature size changes over the domain, we have to adapt to it in bounding the number of disks used. One possible way to do so is to use an integral. Thus, we claim that the number of disks we have to use is roughly

$$\#\Delta(\mathsf{P}, \mathcal{D}) = \int_{\mathsf{p}\in\mathcal{D}} \frac{1}{(\text{lfs}_\mathsf{P}(\mathsf{p}))^2}\, d\mathsf{p}, \tag{12.2}$$

where $\mathcal{D} = [0,1]^2$ is the domain being covered.

Lemma 12.16. *For a set of points* $\mathsf{P} \subseteq [1/4, 3/4]^2$*, let* $\mathcal{F}$ *be the cover of* $\mathcal{D}$ *computed by* **algCoverDisksGreedy**$(\mathsf{P}, \mathcal{D})$*. Then* $|\mathcal{F}| = \Theta(\#\Delta(\mathsf{P}, \mathcal{D}))$.

Proof. For a ball $\mathbf{b}(\mathsf{p}, r) \in \mathcal{F}$, we have that $r = \text{lfs}_\mathsf{P}(\mathsf{p})/2$. As such, by Lemma 12.13, we have, for any point $\mathsf{t} \in \mathbf{b}(\mathsf{p}, r)$, that $\|\mathsf{p} - \mathsf{t}\| \le r$ and

$$r = \text{lfs}_\mathsf{P}(\mathsf{p}) - r \le \text{lfs}_\mathsf{P}(\mathsf{p}) - \|\mathsf{p}-\mathsf{t}\| \le \text{lfs}_\mathsf{P}(\mathsf{t}) \le \text{lfs}_\mathsf{P}(\mathsf{p}) + \|\mathsf{p}-\mathsf{t}\| \le \text{lfs}_\mathsf{P}(\mathsf{p}) + r = 3r.$$

As such,

$$\#\Delta(\mathsf{P}, \mathcal{D}) = \int_{\mathsf{t}\in\mathcal{D}} \frac{1}{(\text{lfs}_\mathsf{P}(\mathsf{t}))^2}\, dt \le \sum_{\mathbf{b}(\mathsf{p},r)\in\mathcal{F}} \int_{\mathsf{t}\in\mathbf{b}(\mathsf{p},r)} \frac{1}{(\text{lfs}_\mathsf{P}(\mathsf{t}))^2}\, dt \le \sum_{\mathbf{b}(\mathsf{p},r)\in\mathcal{F}} \int_{\mathsf{t}\in\mathbf{b}(\mathsf{p},r)} \frac{1}{r^2}\, dt$$
$$\le \sum_{\mathbf{b}(\mathsf{p},r)\in\mathcal{F}} \frac{\pi r^2}{r^2} \le \pi\,|\mathcal{F}|.$$

As for the other direction, observe that, by construction, every disk of $\mathcal{F}$ has its center in the domain $\mathcal{D}$. As such, at least a quarter of each disk lies inside the domain and contributes to the integral of $\#\Delta(\mathsf{P}, \mathcal{D})$. Also, every point in the plane in the domain is covered by at most c disks of $\mathcal{F}$, by Claim 12.15. As such, we have that

$$\#\Delta(\mathsf{P}, \mathcal{D}) = \int_{\mathsf{t}\in\mathcal{D}} \frac{1}{(\text{lfs}_\mathsf{P}(\mathsf{t}))^2}\, dt \ge \frac{1}{c} \sum_{\mathbf{b}(\mathsf{p},r)\in\mathcal{F}} \int_{\mathsf{t}\in\mathbf{b}(\mathsf{p},r)\cap\mathcal{D}} \frac{1}{(\text{lfs}_\mathsf{P}(\mathsf{t}))^2}\, dt$$

$$\ge \frac{1}{c} \sum_{\mathbf{b}(\mathsf{p},r)\in\mathcal{F}} \int_{\mathsf{t}\in\mathbf{b}(\mathsf{p},r)\cap\mathcal{D}} \frac{1}{(3r)^2}\, dt \ge \frac{1}{c} \sum_{\mathbf{b}(\mathsf{p},r)\in\mathcal{F}} \frac{1}{4} \int_{\mathsf{t}\in\mathbf{b}(\mathsf{p},r)} \frac{1}{9r^2}\, dt$$
$$= \frac{1}{36c} \sum_{\mathbf{b}(\mathsf{p},r)\in\mathcal{F}} \frac{\pi r^2}{r^2} = \frac{\pi}{36c}\,|\mathcal{F}|.$$

■

12.3.3. Upper bound. The problem at hand is to bound the number of triangles in any "minimal" fat triangulation (of $\mathcal{D}$) that has P as a subset of the vertices. Specifically, we would like to bound the number of triangles in the triangulation generated by the algorithm of Section 12.2.2.

Intuitively, a triangle in such a triangulation containing a point p should have diameter roughly $\mathrm{lfs}_{\mathsf{P}}(\mathsf{p})$. Naturally, we can use smaller triangles than the local feature size recommends, but luckily, we do not have to. To this end, we argue that the local feature size at a point p is bounded by the size of the leaf of the quadtree containing p. In fact, the other direction also holds (that is, the local feature size, up to a constant, bounds the diameter of the leaf), but we do not need it for our argument.

First, we will prove that leaves of the well-balanced quadtree of P are not too small. This implies that the size of the triangulation that the algorithm extracts is $O(\#\triangle(\mathsf{P}, \mathcal{D}))$, see (12.2). As for the other direction, we will prove that any good fat triangulation of P must have $\Omega(\#\triangle(\mathsf{P}, \mathcal{D}))$ triangles in it.

Observation 12.17. *Let $\square$ be a cell in some canonical grid. Then, for any two points p and s that lie in the extended cluster of $\square$, we have that $\|\mathsf{p} - \mathsf{s}\| \leq \tau(\square) = 5\sqrt{2}\varkappa(\square)$, where $\varkappa(\square)$ denotes the sidelength of $\square$. See the figure on the right.*

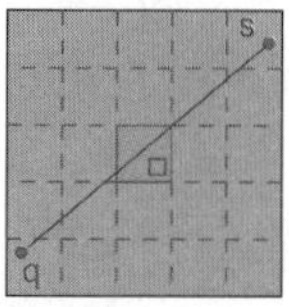

Lemma 12.18. *Let v be a leaf of the well-balanced quadtree of P (computed by the algorithm of Lemma 12.8), and let p be any point in $\square_v$. Then, $\mathrm{lfs}_{\mathsf{P}}(\mathsf{p}) \leq c\,\varkappa(\square_v)$, where $\varkappa(\square_v)$ is the sidelength of the cell $\square_v$ and c is some constant.*

Proof. If the cell $\square_v$ was created because there was an extended cluster that contained two points $\mathsf{q}, \mathsf{s} \in \mathsf{P}$ and the algorithm refined the parent $\overline{\mathsf{p}}(v)$ of v because of this, then q, s are in distance at most $\alpha = 5\sqrt{2}\varkappa\big(\square_{\overline{\mathsf{p}}(v)}\big) = 2 \cdot 5 \cdot \sqrt{2}\varkappa(\square_v)$ from any point of $\square_v \subseteq \square_{\overline{\mathsf{p}}(v)}$, by Observation 12.17. We have

$$\mathrm{lfs}_{\mathsf{P}}(\mathsf{p}) \leq \max(\|\mathsf{p} - \mathsf{q}\|, \|\mathsf{p} - \mathsf{s}\|) \leq \alpha = 10\sqrt{2}\varkappa(\square_v),$$

establishing the claim in this case.

From this point on, the remainder of the proof is a merry-go-round game of blame. A node u where its parent got refined because of balancing considerations ***blames*** the node $w = \text{J'accuse}(u)$ that caused this splitting for its existence. The cell $\square_w$ is adjacent to the cell $\square_{\overline{\mathsf{p}}(u)}$ and $\varkappa(\square_w) \leq \varkappa\big(\square_{\overline{\mathsf{p}}(u)}\big)/4 = \varkappa(\square_u)/2$. Otherwise the parent of $u' = \overline{\mathsf{p}}(u)$ would be balanced with relation to its neighbors and would not get split. Note that w must have been created in the tree before u' got split (and u got created).

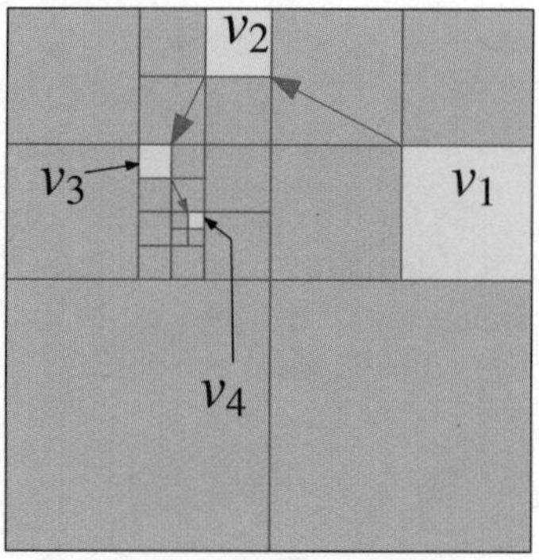

So, let $v_1 = v$, and let v_i be the J'accuse(v_{i-1}), for $i = 1, \ldots, k$; see the figure on the right. Clearly, the blame game must end at some node v_k (because the quadtree is finite, and every time we play the blame game, the cell of the current node shrinks by half at least). The vertex v_k must have been inserted because it participates in some extended cluster of a node u_k that contains some point of P, such that the algorithm inserted all the nodes of the extended cluster of u_k into the quadtree. However, this happens only if the extended cluster of the parent of u_k contains two points of P. It easy to verify that since the cells of v_k and u_k are in the same grid (and belong to the extended clusters of each other), then the extended cluster of the grandparent of v_k contains the extended cluster of the parent of u_k. Namely, the extended cluster of the grandparent of v_k contained two points of P.

As such, we have that

$$\mathrm{lfs}_{\mathsf{P}}(\mathsf{p}) \leq \sum_{i=1}^{k-1} \left(\mathrm{diam}(\square_{v_i}) + \mathbf{d}(\square_{v_i}, \square_{v_{i+1}})\right) + \tau(\overline{\mathrm{p}}(\overline{\mathrm{p}}(v_k))),$$

where $\mathbf{d}(X, Y) = \min_{x \in X, y \in Y} \|x - y\|$ and $\tau(\square) = 5\sqrt{2}\varkappa(\square)$ (see Observation 12.17). By the above, we know that $\mathrm{diam}(v_i) \leq \mathrm{diam}(v_{i-1})/2 \leq \mathrm{diam}(v)/2^{i-1}$, for all i. Using similar argumentation, we have that

$$\mathbf{d}(\square_{v_i}, \square_{v_{i+1}}) \leq \mathrm{diam}(\overline{\mathrm{p}}(v_i)) \leq 2\mathrm{diam}(\square_{v_i}) \leq \mathrm{diam}(\square_v)/2^{i-2}.$$

As such, we have that

$$\mathrm{lfs}_{\mathsf{P}}(\mathsf{p}) \leq \sum_{i=1}^{k-1} \left(\frac{\mathrm{diam}(\square_v)}{2^{i-1}} + \frac{\mathrm{diam}(\square_v)}{2^{i-2}} \right) + 20\sqrt{2}\varkappa(\square_{v_k}) = O(\varkappa(\square_v)),$$

establishing the claim. ∎

Lemma 12.19. *The number of leaves in the well-balanced quadtree of* P *(computed by the algorithm of Lemma 12.8) is* $O(\#\Delta(\mathsf{P}, \mathcal{D}))$, *where* $\mathcal{D} = [0, 1]^2$. *Namely, the size of the triangulation* $\mathcal{QT}(\mathsf{P})$ *is* $O(\#\Delta(\mathsf{P}, \mathcal{D}))$.

PROOF. Consider a leaf v of the well-balance quadtree of P, and observe that by Lemma 12.18, we have that

$$\int_{\mathsf{p} \in \square_v} \frac{1}{(\mathrm{lfs}_{\mathsf{P}}(\mathsf{p}))^2}\, d\mathsf{p} \geq \int_{\mathsf{p} \in \square_v} \frac{1}{(c\,\varkappa(\square_v))^2} d\mathsf{p} = \frac{\mathrm{area}(\square_v)}{(c\,\varkappa(\square_v))^2} = \frac{(\varkappa(\square_v))^2}{(c\,\varkappa(\square_v))^2} = \frac{1}{c^2}.$$

Namely, since the leaves of the well-balanced quadtree are interior disjoint, and each one contributes (at least) $1/c^2$ to $\#\Delta(\mathsf{P}, \mathcal{D})$, it follows that the number of leaves of the quadtree is at most $\#\Delta(\mathsf{P}, \mathcal{D})/(1/c^2) = O(\#\Delta(\mathsf{P}, \mathcal{D}))$; see $(12.2)_{\mathrm{p}171}$. (Note that c is a universal fixed constant that depends only on the dimension; see Claim 12.15). ∎

12.3.4. Lower bound. The other direction of showing that every triangle (in a fat triangulation) contributes at most a constant to $\#\Delta(\mathsf{P}, \mathcal{D})$ requires us to prove that the local feature size anywhere inside such a triangle Δ is $\Omega(\mathrm{diam}(\Delta))$. The key observation is that a triangle in such a triangulation is an island – it has no vertices of the triangulation close to it, thus implying that the local feature size is indeed sufficiently large inside it.

Lemma 12.20. *Let* $\mathcal{M}$ *be an* α*-fat triangulation, and let* p *be a vertex of this triangulation. Consider the shortest edge* $\mathsf{e}_{\mathrm{short}}$ *and the longest edge* $\mathsf{e}_{\mathrm{long}}$ *in the triangulation adjacent to* p. *We have that* $\|\mathsf{e}_{\mathrm{long}}\| / \|\mathsf{e}_{\mathrm{short}}\| \leq c_2$, *where* c_2 *is a constant that depends only on* α.

PROOF. Let $\mathsf{e}_1, \mathsf{e}_2, \ldots, \mathsf{e}_k$ be the edges of $\mathcal{M}$ in (say) clockwise order around p, where $\mathsf{e}_1 = \mathsf{e}_{\mathrm{short}}$ and e_k is adjacent to e_1. Let Δ_i be the triangle formed by the edges e_i and e_{i+1}, for $i = 1, \ldots, k-1$, and let Δ_k be the triangle formed by e_k and e_1. These triangles appear in $\mathcal{M}$, and since this triangulation is α-fat, we have that the aspect ratio of all these triangles is bounded by α (see Definition 12.5). As such, we have that

$$\frac{\|\mathsf{e}_i\|}{\|\mathsf{e}_1\|} = \frac{\|\mathsf{e}_i\|}{\|\mathsf{e}_{i-1}\|} \cdot \frac{\|\mathsf{e}_{i-1}\|}{\|\mathsf{e}_{i-2}\|} \cdots \frac{\|\mathsf{e}_2\|}{\|\mathsf{e}_1\|} \leq \alpha^{i-1}.$$

Similarly, by going around p in the other direction, we have that $\|\mathsf{e}_i\|/\|\mathsf{e}_1\| \leq \alpha^{k-i+1}$.

Now, all the angles of these triangles are at least $\geq 1/\mathcal{A}_{\text{ratio}}(\triangle_i) \geq 1/\alpha$, by Lemma 12.3. As such, $k \leq \lceil 2\pi/(1/\alpha) \rceil = \lceil 2\pi\alpha \rceil$. We conclude that

$$\frac{\|\mathsf{e}_i\|}{\|\mathsf{e}_{\text{short}}\|} \leq \max\left(\alpha^i, \alpha^{k-i+1}\right) \leq \alpha^{\lceil 2\pi\alpha \rceil/2},$$

for all i. In particular, setting $c_2 = \alpha^{\lceil 2\pi\alpha \rceil/2}$ implies the claim. ■

It is useful to know the following concepts even if we will make only light use of them.

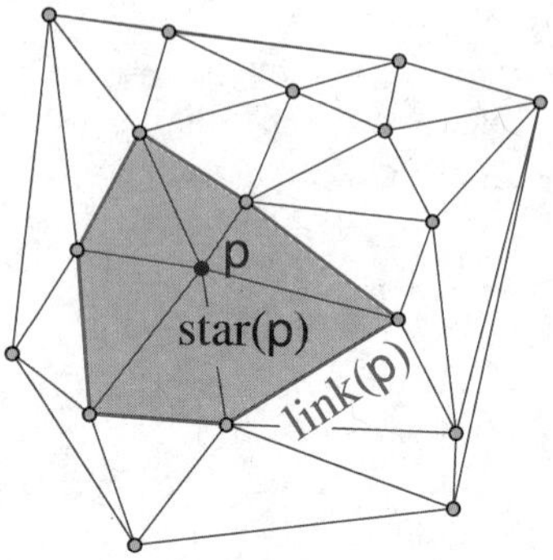

Definition 12.21. In a triangulation $\mathcal{M}$, the ***star*** of a vertex $\mathsf{p} \in \mathcal{M}$ is the union of triangles of $\mathcal{M}$ adjacent to p. The ***link*** of p is the boundary of its star. See the figure on the right.

Lemma 12.22. *Let $\mathcal{M}$ be an α-fat triangulation, and let p be a vertex of this triangulation. Let $\mathsf{e} \in \mathcal{M}$ be the longest edge adjacent to p. Then, the disk of radius $r = c_3 \|\mathsf{e}\|$ centered at p is contained in* star(p), *where c_3 is some constant that depends only on α.*

Proof. By Lemma 12.20, the shortest edge adjacent to p has length at least $\|\mathsf{e}\| / c_2$. Consider a triangle $\triangle = \triangle \mathsf{qps}$ of $\mathcal{M}$ adjacent to p, depicted on the right, and observe that since this triangle is α-fat, we have that angle $1/\alpha \leq \angle \mathsf{pqs} \leq \pi - 2\alpha$, since, by Lemma 12.3, the minimum angle of this triangle is at least $1/\alpha$. As such, the distance of p to the edge qs is at least

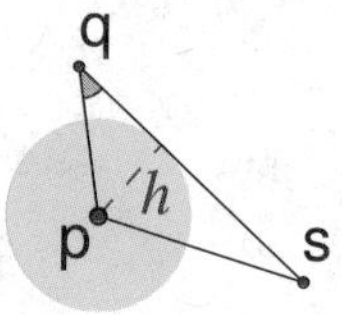

$$h = \|\mathsf{p} - \mathsf{q}\| \sin \angle \mathsf{pqs} \geq \frac{\|\mathsf{e}\|}{c_2} \sin \frac{1}{\alpha} \geq \frac{\|\mathsf{e}\|}{2c_2\alpha},$$

since $\sin x \geq x/2$ for $x \in [0, 1]$. Thus, the claim holds with $c_3 = 1/(2c_2\alpha)$. ■

The following lemma testifies that for a triangle in a fat triangulation there is a buffer zone around it (i.e., a moat) such that no point of the triangulation might lie in it.[6]

Lemma 12.23. *Let $\mathcal{M}$ be an α-fat triangulation, with V being the set of vertices of $\mathcal{M}$. For any triangle $\triangle \in \mathcal{M}$, we have that*

$$\mathbf{d}(V \setminus \triangle, \triangle) \geq c_4 \text{diam}(\triangle),$$

where $\mathbf{d}(X, Y)$ is the minimum distance between any pair of points of X and Y and c_4 is some constant that depends only on α.

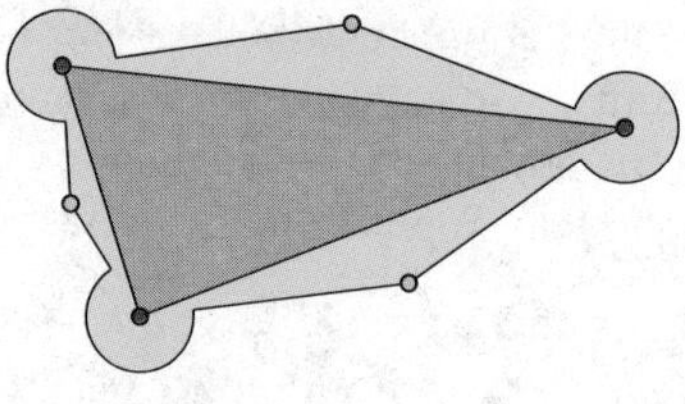

Figure 12.5.

Proof. Let e be the longest edge of $\triangle$, and observe that the shortest edge of $\triangle$ has length $\geq \|\mathsf{e}\| / \alpha$. By Lemma 12.22, one can place disks of radius $\beta = c_3 \|\mathsf{e}\| / \alpha$ at the vertices of $\triangle$, such that these disks are contained inside the respective stars of their vertices.

Now, erect on each edge of $\triangle$ an isosceles triangle having base angle $1/\alpha$ and lying outside $\triangle$. Clearly, these triangles are contained inside the real triangles of the mesh lying

[6]Completely irrelevant, but here is a nonsensical geometric statement using biblical language: "Why do you see the moat that is in your brother's triangle, but don't consider the moat that is in your own triangle?" Answer me that!

on their respective edges since all angles (of triangles) in the mesh are larger than or equal to $1/\alpha$.

Let T denote the union of these three triangles together with the protective disks; see Figure 12.5. The "moat" T forms a region that contains no vertex of the mesh except the vertices of the triangle $\triangle$. It is now easy to verify that the outer boundary of T is within a distance at least

$$\beta \sin \frac{1}{\alpha} \geq \frac{c_3 \|\mathsf{e}\|}{2\alpha^2} = \frac{c_3}{2\alpha^2} \operatorname{diam}(\triangle)$$

from any point in $\triangle$, since $\sin(1/\alpha) \geq 1/2\alpha$ for $\alpha \geq 1$. The claim now follows by setting $c_4 = c_3/2\alpha^2$. ■

Lemma 12.24. *For any α-fat triangulation $\mathcal{M}$ of $\mathcal{D} = [0,1]^2$ that includes the points of* P *as vertices, we have that $|\mathcal{M}| = \Omega(\#\triangle(\mathsf{P}, \mathcal{D}))$; see $(12.2)_{p171}$.*

PROOF. Consider a triangle $\triangle \in \mathcal{M}$, and observe that the length of the edges of $\triangle$ are at least $\operatorname{diam}(\triangle)/\alpha$. By Lemma 12.23, all the other vertices of $\mathcal{M}$ are within a distance at least $c_4 \operatorname{diam}(\triangle)$ from $\triangle$. Thus, for a point $\mathsf{p} \in \triangle$, it might be that the nearest neighbor to p, among the vertices of $\mathcal{M}$, is relatively close (i.e., one of the vertices of $\triangle$), but the second nearest neighbor must be within a distance at least $\min(\operatorname{diam}(\triangle)/2\alpha, c_4 \operatorname{diam}(\triangle)) = c_4 \operatorname{diam}(\triangle)$, as $c_4 \leq 1/2\alpha$ (which can be easily verified). We conclude that

$$\forall \mathsf{p} \in \triangle \quad \operatorname{lfs}_V(\mathsf{p}) \geq c_4 \operatorname{diam}(\triangle), \tag{12.3}$$

where V is the set of vertices of $\mathcal{M}$ (and $\mathsf{P} \subseteq V$).

Observe that the height of a triangle is bounded by its diameter. As such, a triangle $\triangle$ with diameter $\operatorname{diam}(\triangle)$ has $\operatorname{area}(\triangle) \leq (\operatorname{diam}(\triangle))^2/2$. We are now close enough to the end of the tunnel to see the light; indeed, we have that

$$\int_{\mathsf{p}\in\triangle} \frac{1}{(\operatorname{lfs}_{\mathsf{P}}(\mathsf{p}))^2}\, d\mathsf{p} \leq \int_{\mathsf{p}\in\triangle} \frac{1}{(\operatorname{lfs}_V(\mathsf{p}))^2}\, d\mathsf{p} \leq \int_{\mathsf{p}\in\triangle} \frac{1}{(c_4 \operatorname{diam}(\triangle))^2}\, d\mathsf{p}$$
$$= \frac{\operatorname{area}(\triangle)}{(c_4 \operatorname{diam}(\triangle))^2} \leq \frac{(\operatorname{diam}(\triangle))^2}{2(c_4 \operatorname{diam}(\triangle))^2} = \frac{1}{2c_4^2},$$

since $\operatorname{lfs}_{\mathsf{P}}(\mathsf{p}) \geq \operatorname{lfs}_V(\mathsf{p})$ and by (12.3). As such

$$\#\triangle(\mathsf{P}, \mathcal{D}) = \int_{\mathsf{p}\in\mathcal{D}} \frac{1}{(\operatorname{lfs}_{\mathsf{P}}(\mathsf{p}))^2}\, d\mathsf{p} = \sum_{\triangle\in\mathcal{M}} \int_{\mathsf{p}\in\triangle} \frac{1}{(\operatorname{lfs}_{\mathsf{P}}(\mathsf{p}))^2}\, d\mathsf{p} \leq \sum_{\triangle\in\mathcal{M}} \frac{1}{2c_4^2} = \frac{|\mathcal{M}|}{2c_4^2},$$

implying the claim. ■

12.4. The result

Theorem 12.25. *Given a point set* $\mathsf{P} \subseteq [1/4, 3/4]^2$ *with* $\operatorname{diam}(\mathsf{P}) \geq 1/2$*, one can compute a 4-fat triangulation $\mathcal{M}$ of $\mathcal{D} = [0,1]^2$ in $O(n \log n + m)$ time, where $n = |\mathsf{P}|$ and $m = |\mathcal{M}|$. Furthermore, any $O(1)$-fat triangulation of $\mathcal{D}$ having* P *as part of its vertices must be of size $\Omega(m)$.*

PROOF. The aspect ratio of the output and the running time of the algorithm are guaranteed by Lemma 12.11. The upper bound on the size of the triangulation is implied by Lemma 12.19, and the lower bound of the size of such a triangulation is from Lemma 12.24. Putting everything together implies the result. ■

12.5. Bibliographical notes

Balanced quadtree and good triangulations are due to Bern et al. [**BEG94**]. The analysis we present uses ideas from the work of Ruppert [**Rup95**]. While using this approach to analyze [**BEG94**] is well known to people working in the field, I am unaware of a reference in the literature where it is presented in this way.

The problem of generating good triangulations has received considerable attention, as it is central to the problem of generating good meshes, which in turn are important for efficient numerical simulations of physical processes. One of the main techniques used in generating good triangulations is the method of Delaunay refinement. Here, one computes the Delaunay triangulation of the point set and inserts circumscribed centers as new points, for "bad" triangles. Proving that this method converges and generates optimal triangulations is a non-trivial undertaking and is due to Ruppert [**Rup93**]. Extending it to higher dimensions and handling boundary conditions make it even more challenging. However, in practice, the Delaunay refinement method outperforms the (more elegant and simpler to analyze) method of Bern et al. [**BEG94**], which easily extends to higher dimensions. Namely, the Delaunay refinement method generates good meshes with fewer triangles.

Furthermore, Delaunay refinement methods are slower in theory. Getting an algorithm to perform Delaunay refinement in the same time as the algorithm of Bern et al. is still open, although Miller [**Mil04**] obtained an algorithm with only slightly slower running time.

Recently, Alper Üngör came up with a "Delaunay-refinement type" algorithm, which outputs better meshes than the classical Delaunay refinement algorithm [**Üng09**]. Furthermore, by merging the quadtree approach with the Üngör technique, one can get an optimal running time algorithm [**HÜ05**].

CHAPTER 13

Approximating the Euclidean Traveling Salesman Problem (TSP)

In this chapter, we introduce a general technique for approximating the shortest traveling salesperson tour in the plane (i.e., TSP). This technique found wide usage in developing approximation algorithms for various problems. We will present two different variants of the technique. The first is quadtree based and is faster and easier to generalize to higher dimension. The other variant will be presented in the next chapter and is slower but seems to be somewhat stronger in some cases.

13.1. The TSP problem – introduction

Let P be a set of n points in the plane. We would like to compute the shortest tour visiting all the points of P. That is, it is the shortest closed polygonal chain where its vertices are all the points of P. We will refer to such a chain as a ***tour***. The problem of computing the shortest TSP tour is NP-HARD, even in the plane with Euclidean distances. In the general graph setting (where the points are vertices of a complete graph and every edge has an associated weight with it) no approximation is possible. For the metric case, where the weights comply with the triangle inequality, a $3/2$ approximation is known. However, in the low-dimensional Euclidean case (i.e., Euclidean distance between points) a PTAS (polynomial time approximation scheme) is known.

The problem has attracted a vast amount of research in operation research and computer science, due to its simplicity and the ability to easily draw and inspect a solution. In fact, for small instances people can do a decent job in solving such problems manually.

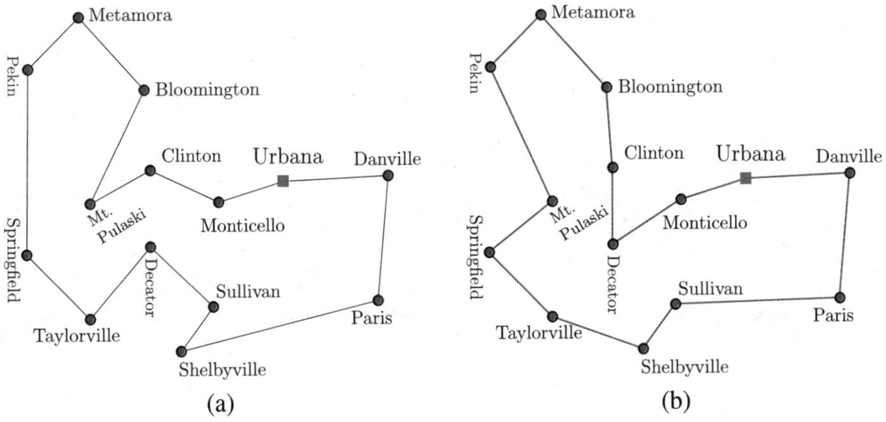

FIGURE 13.1. (a) Lincoln's tour in 1850 in Illinois and (b) the optimal TSP.

A nice historical example of TSP is Lincoln's tour of Illinois in 1850. Lincoln worked as a lawyer traveling with a circuit court. At the time, a circuit court would travel through several cities, where in each city they would stop and handle local cases, spending a few days in each city. The tour that the circuit court (and Lincoln) took in 1850 is depicted in Figure 13.1(a) (they started from Springfield and returned to it in the end of the tour). Clearly, it is very close to being the shortest TSP of these points. In this case, there is a shorter solution, but it is not clear that it was shorter as far as traveling time in 1850; see Figure 13.1.

More modern applications of TSP vary from control systems for telescopes (i.e., the telescope uses the minimum amount of motion while taking pictures of all required spots in the sky during one night) to scheduling of rock group or political tours, and many other applications.

13.2. When the optimal solution is friendly

Here we are interested in the Euclidean variant. We are given a set P of n points in the plane, with the Euclidean distance, and we are looking for the shortest TSP tour.

First try – divide and conquer. A natural approach to this problem would be to try to divide the plane into two parts, say by a line, solve (a variant of) the problem recursively on each side of the line, and then stitch the solutions together. So let ℓ be such a separating vertical line, having at least one point of P on each size of it, and we divide the problem into two subproblems along the line ℓ.

To this end, we need to guess where the optimal TSP crosses the separating line. Let us assume, for the time being, that the TSP crosses this separating line at most m times, where m is some small constant. To solve the left (resp. right) subproblem, we need to specify how the TSP crosses the "iron curtain" formed by ℓ.

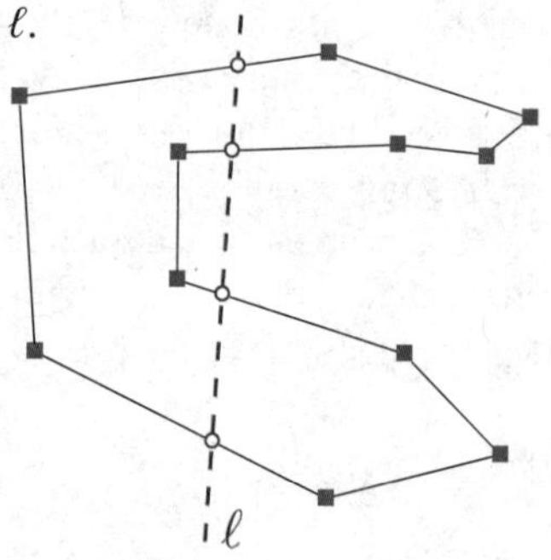

The optimal TSP path is made of segments connecting two points of P. As such, there are $N = \binom{n}{2}$ potential segments that the TSP might use. Choosing the m intersection points of the tour with ℓ is equivalent to choosing the (at most) m segments of the TSP tour crossing the splitting line ℓ. As such, the number of possibilities is

$$\sum_{i=0}^{m} \binom{N}{i} \leq 2\left(\frac{Ne}{m}\right)^m = N^{O(m)} = n^{O(m)},$$

by Lemma 5.10_{p65}.

Note, however, that it is not sufficient to solve the problem on the left and on the right by just knowing the segments crossing the middle line. Indeed, whether or not a solution on the left subproblem is valid depends on the solution on the right subproblem; see Figure 13.2. As such, for every subproblem, we do not only have to specify the points where the tour enters the subproblem (we will refer to these points as ***checkpoints***), but also the connectivity constraints for the checkpoints of each subproblem must be specified.

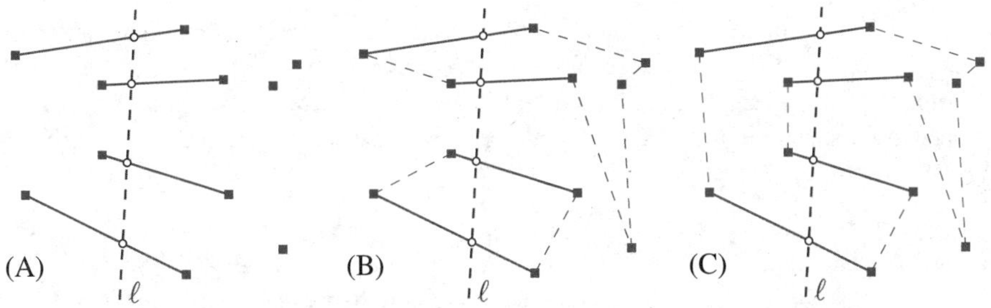

FIGURE 13.2. (A) The guessed middle solution fed into the two subproblems. (B) A solution to the left subproblem resulting in a disconnected tour. (C) A valid solution.

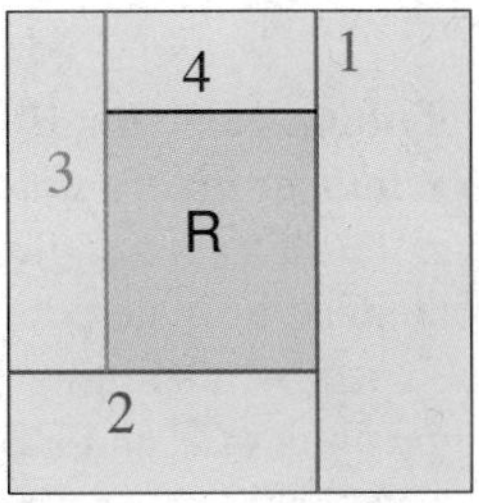

Before going into the exact details of how to do this, imagine continuing recursively in this fashion. Namely, we try to break the problem recursively into smaller and smaller regions. To this end, we will cut each subproblem by (say) a line, alternating between horizontal and vertical lines. A rectangle R generated by a sequence of such cuts is depicted on the right. A subproblem is thus a rectangle R having checkpoints on its boundary, with ordering constraints (i.e., in which order the tour visits these checkpoints), where we need to find a solution of total minimal length connecting the right checkpoints to each other and spanning all the points of P inside R.

Our intention is to solve this problem using dynamic programming. To this end, we would like to minimize the number of subproblems one needs to consider. Note that the cutting lines we use can be ones that pass through the given points. As such, there are $2n$ possible cutting lines, since only horizontal or vertical lines are considered. Every subrectangle is defined by the four cutting lines that form its boundary, and as such there are $(2n)^4 = O(n^4)$ possible rectangles that might appear in a subproblem.

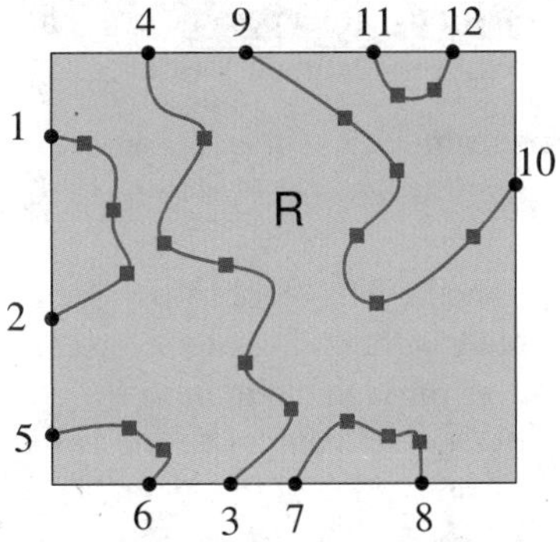

So, consider such a subproblem with a rectangle R with t checkpoints on its boundary. Specifying the order in which the checkpoints are being visited by the tour can be done by numbering these endpoints. Namely, for a specific rectangle with t checkpoints on its boundary, there are $t!$ possible subproblems for which one needs to compute a solution. An example of such a numbering and one possible solution compatible with this numbering are depicted on the right. Here, the point numbered 1 is the first visited by the tour when it enters R. Then the tour leaves the rectangle through the endpoint 2, reenters it using 3, and so on. In the figure, the squares are the points of P inside this rectangle.

One can perform some additional checks on a given instance of a subproblem to verify that it is indeed realizable and should be considered. For example, the total number of times the tour enters/leaves a subrectangle is the same.

Interestingly, the number of possible subproblems one has to consider, for such a rectangle R, is somewhat smaller than $t!$. However, since it is still exponential, it will be easier for us to use the naive bound $t!$.

Problem 13.1. Given a rectangle R and t checkpoints specified on its boundary, show that there are at least $2^{t/4}$ different patterns for a path to enter and leave R using these endpoints. This holds even if we require the path not to self-intersect. Here, a pattern is a matching of the t checkpoints, such that if p_i is matched to p_j, then the path enters the rectangle R at p_i and leaves through p_j (or vice versa).

This implies that if t is large (say $\Omega(n)$), then the running time of the algorithm is inherently exponential.

In the following, we will treat rectangles as being half-opened; that is, an axis parallel rectangle will be a region of the form $[x_1, x_2) \times [y_1, y_2)$. We next define a bounding rectangle for the entire point set, such that there are no points on its boundary.

Definition 13.2. For a point set P in the plane, its ***frame*** is the rectangle $[x_1 - 1, x_2 + 1) \times [y_1 - 1, y_2 + 1)$, where $[x_1, x_2] \times [y_1, y_2]$ is the smallest axis parallel closed rectangle containing P.

Definition 13.3. Let P be a set of n points in the plane, and let R be its frame. An ***RBSP*** (***rectilinear binary space partition***) is a recursive partition of R into rectangles, such that each resulting rectangle contains a single point of P. Formally, an axis parallel rectangle containing a single point is a valid RBSP. As such, an RBSP of R and the point set P is specified by a splitting line ℓ (which is either horizontal or vertical) that passes through some point of P and RBSPs of both $R^+ = \ell^+ \cap R$ and $R^- = \ell^- \cap R$, where ℓ^- and ℓ^+ are the two halfplanes defined by ℓ.[①] Here we required that $R^+ \cap P \neq \emptyset$ and $R^- \cap P \neq \emptyset$.

A rectangle considered during this recursive construction for a given RBSP is a ***subrectangle*** of this RBSP.

Since we always take cutting lines that pass through the given points, there are $2n + 4$ lines that can be used by a subrectangle (the vertical and horizontal lines through the points and the four lines defining the frame of P), which implies that there are at most $O(n^4)$ subrectangles that can participate in an RBSP of the given point set.

Definition 13.4. Let P be a set of n points in the plane, and let π be a tour of P. The tour π is ***t-friendly***, for a parameter $t > 0$, if there exists an RBSP $\mathcal{R}$ of the frame of P, such that π crosses the boundary of every subrectangle of $\mathcal{R}$ at most t times.

Theorem 13.5. *Given a point set* P *and a parameter* $t > 0$, *one can compute the shortest t-friendly tour of* P *in* $n^{O(t)}$ *time.*

PROOF. The algorithm is recursive, and we will use memoization to turn it into a dynamic programming algorithm. Let L be the set of $2n + 4$ vertical and horizontal lines passing through the points of P and the four lines used by the frame of P. Let $\mathcal{F}$ be the set of all $O(n^4)$ rectangles defined by the grid induced by the lines of L.

An instance of the recursive call is defined by (R, U, M), where R is a rectangle of $\mathcal{F}$, U is a set of at most t checkpoints placed on the boundary of R, and M is a permutation of U (this is the order in which the tour visits the checkpoints). As such, there are at most

$$O\left(n^4 \cdot n^{2t} \cdot t!\right) = n^{O(t)}$$

different recursive calls. Indeed, there are $O(n^4)$ possible subrectangles. For a given subrectangle, a checkpoint is the intersection of a segment defined by two input points and

①Formally, one of them is open and the other is closed, such that R^+ and R^- are rectangles, and they are defined in a consistent way with the frames above.

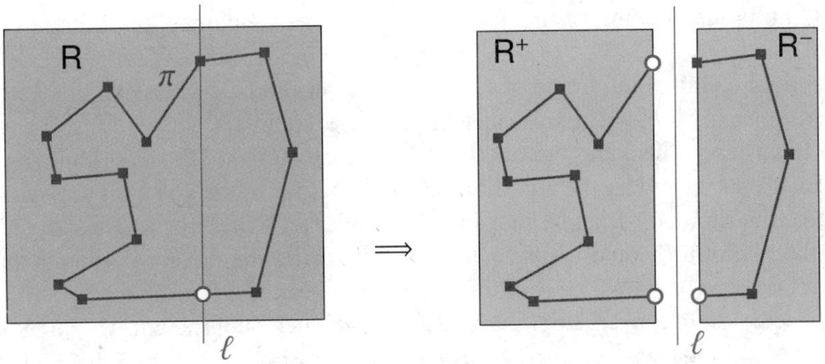

FIGURE 13.3. The top-level split of the RBSP of the optimal solution.

the boundary of the rectangle. Since the boundary of the rectangle might intersect such a segment twice, the number of possible checkpoints for a specified rectangle is $2\binom{n}{2}$. Now, if a given subrectangle has i checkpoints, then there are $\left(2\binom{n}{2}\right)^i$ possible sets of checkpoints that it could have and $i!$ orderings for each set of checkpoints. Since a tour visiting the rectangle can use up to t checkpoints, a subrectangle has at most $\sum_{i=2}^{t}\left(2\binom{n}{2}\right)^i i! \leq 2n^{2t} \cdot t!$ different instances of the subproblems defined for it.

The recursive call (R, U, M) computes the best possible solution to this instance by trying all possibilities to split it. As a base case, if $|\mathsf{P} \cap \mathsf{R}| \leq t$, then we solve this instance using brute force. This takes $t^{O(t)}$ time, by just trying all possible tours restricted to the points inside R and the checkpoints. For each such solution, we can easily verify that it complies with the given constraints.

Otherwise, we try all ways to split it by a line $\ell \in \mathsf{L}$. We require that $\mathsf{R} \cap \ell^+ \cap \mathsf{P} \neq \emptyset$ and $\mathsf{R} \cap \ell^- \cap \mathsf{P} \neq \emptyset$, where ℓ^+ and ℓ^- are the two halfplanes induced by ℓ. For such a line, we compute the two subrectangles $\mathsf{R}^+ = \mathsf{R} \cap \ell^+$ and $\mathsf{R}^- = \mathsf{R} \cap \ell^-$ and the splitting segment $s = \mathsf{R} \cap \ell$. Next, we enumerate all possibilities of the at most t checkpoints on s. For each such set U' of checkpoints, we enumerate all possible subproblems defined for R^+ and R^- using the checkpoints of $U \cup U'$.

For any two such subproblems for R^+ and R^- we verify that they are consistent with the given larger instance for R. In particular, the given instance R has a given permutation on it checkpoints, and the two permutations for the subinstances R^+ and R^- need to be consistent with this permutation and need to be consistent on the shared checkpoints along the splitting line. Note that each one of these permutations has at most t checkpoints in it, and as such the check can be easily done in time $t^{O(t)}$. If so, we recursively compute the shortest solution for these two subproblems. We combine the two recursive solutions into a solution for the given problem for R; that is, we add up the price of these two solutions.

We try all such possible solutions, and we store the cheapest solution found as the solution to this instance of the subproblem.

The work associated with each such subproblem is $n^{O(t)}$ (ignoring the work in the recursive calls). Indeed, we generate all possible cuts (i.e., $O(n)$) and enumerate all possible subsets, up to size t, of checkpoints on the cutting line (i.e., $n^{O(t)}$). Then, we try all possible permutations on the two subproblems and verify that they are consistent with the parent permutation and between the two subproblems (this takes $t^{O(t)}$ time). Finally, if we find such a consistent instance for the two subproblems, we recursively compute their optimal solution.

This implies that the total running time is $n^{O(t)}$, since the number of different subproblems is $n^{O(t)}$.

The initial call uses the frame of P and has no checkpoints (note that this is the only feasible instance in all the recursive calls without checkpoints).

The consistency check throughout the recursive execution of the algorithm guarantees that the solution returned by the algorithm is indeed a tour of the given input points. Now, let π denote the shortest t-friendly tour of P. Consider an RBSP for which π is t-friendly. Clearly, the algorithm would consider this RBSP during the recursive computation and would consider π as one possible solution to the given instance. Indeed, this specific RBSP defines a recursion tree over subproblems. For each such subproblem, we know exactly how the optimal solution behaves in these subproblems and uses their checkpoints; see Figure 13.3 for an example. As such, the algorithm would return the portions of π for each subproblems of this RBSP which glued together yields π.

Specifically, one can use simple induction to argue that inside each such subproblem the algorithm computed the optimal solution (complying with the exact constraints on the checkpoints and how they should be visited as induced by the optimal solution). As such, the solution returned by the algorithm is at least as good as the optimal solution (there might be several equivalent optimal solutions). Namely, it would return π as the computed solution (assuming, of course, that the optimal solution in this case is unique). ∎

It is not too hard to verify that any point set has a tour that is 2-friendly (if we do not require that vertical and horizontal cuts alternate). Intuitively, as t increases, the length of the shortest t-friendly tour decreases. Assume that we could show that the shortest t-friendly tour is of length $\leq (1 + 1/t) \|\pi_{\text{opt}}\|$, where π_{opt} is the shortest TSP tour of the given point set and $\|\pi_{\text{opt}}\|$ is its length. Then, the above algorithm would provide us with a $(1 + \varepsilon)$-approximation algorithm, with running time $n^{O(1/\varepsilon)}$, by setting $t = 1/\varepsilon$.

Surprisingly, showing that such a t-friendly tour exists is not trivial (and it might not exist at all). Instead, we will use similar concepts that use the underlying idea of limited interaction between subproblems to achieve a PTAS for the approximation problem.

13.3. TSP approximation via portals and sliding quadtrees

13.3.1. Portals and sliding – idea and intuition. The basic idea is to do a direct divide and conquer by imposing a grid on the point set. We will compute the optimal solution inside each grid cell and then stitch the solutions together into a global solution. As such, the basic subproblem we will look at is a square. To reduce the interaction between subproblems, we will place m equally spaced points on each boundary of a square and require the solution under consideration to use these ***portals***. Naturally, we will have to snap the solution to these portals.

For efficient implementation, the stitching has to be done in a controlled fashion that involves only a constant number of subsquares being "stitched" together. The most natural way to achieve this is by using a hierarchical grid, that is, a quadtree.

It might be that the boundary of a large square of the grid intersects the optimal solution many times. As such, the snapping to portals might introduce a huge error. To avoid this, we will randomly shift the quadtree/grid. This will distribute the snapping error in a more uniform way in the quadtree. The exact details of this are described below.

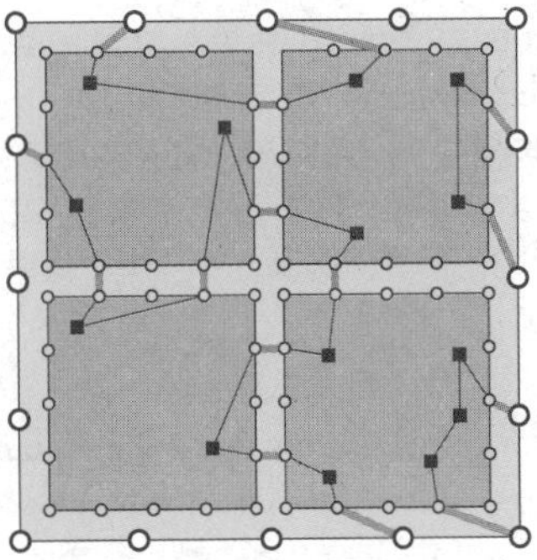

The picture on the right depicts a square and a solution for it using its portals and how this solution propagates to the four subsquares. (The yellow region is of zero width and is enlarged for the sake of clarity.)

13.3.2. The algorithm.

13.3.2.1. *Normalizing the point set.* The first stage is to snap the points to a grid, such that the resulting point set has a bounded polynomial spread. This would guarantee that the quadtree for the point set would have logarithmic depth. Naturally, this would introduce some error and has to be done carefully so that the error is sufficiently small.

So, let P be a set of n points in the plane, and let $1 > \varepsilon > 1/n$ be a prespecified constant. Assume that P is contained in the square $[1/2, 1]^2$ and $\text{diam}(P) \geq 1/4$. This can be guaranteed by scaling and translation, and clearly, given a solution to this instance, we can map it to the original point set.

Note that the optimal solution π_{opt} has to be of length at least $\text{diam}(P) \geq 1/4$ and at most $n\sqrt{2}/2$, as the TSP is made out of n segments, none of which can be longer than the diameter of the square $[1/2, 1]^2$. Let $\mathcal{E} = \lceil 32/\varepsilon \rceil$. Consider the grid

$$G = \left\{ \frac{1}{n\mathcal{E}}(i, j) \;\middle|\; i, j \text{ are integers} \right\}$$

and snap each point of P to its closest point on this grid (if several points get snapped to the same grid point, we treat the snapped points as a single point). Clearly, every point of P is moved by at most a distance of $z = \sqrt{2}/n\mathcal{E}$. As such, solving the problem for the snapped point set can be translated back to the original point set. Indeed, by taking this tour and for each snapped point walking out to its true location and back, we get a tour of the original points which is not much longer. Specifically, this introduces an error of length at most $2z$ per point (as each point needs to move from its true location to its snapped location and back) and overall an error of

$$n \cdot 2z = 2n\frac{\sqrt{2}}{n\mathcal{E}} \leq \frac{4\varepsilon}{32} \leq \frac{\varepsilon}{2}\left\|\pi_{\text{opt}}\right\|, \tag{13.1}$$

since $\left\|\pi_{\text{opt}}\right\| \geq 1/4$. Let Q denote the resulting point set.

13.3.2.2. *The dynamic programming over the quadtree.* We randomly select a point $(x, y) \in [0, 1/2]^2$. Consider the translated canonical grid

$$G^i = (x, y) + G_{2^{-i}}, \tag{13.2}$$

for $i = 0, 1, 2, \ldots$; that is, G^i is a grid with sidelength 2^{-i} with its "origin" at (x, y).

Construct (a regular) quadtree $\mathcal{T}$ over Q using the square $\square = (x, y) + [0, 1]^2 \in G^0$ as the root (note that Q is contained inside $\square$). Since the diameter of Q is larger than $1/4$ and the minimal distance between a pair of points of Q is at least $1/n\mathcal{E}$, the height of the quadtree is going to be at most

$$H \leq 1 + \lceil \log_2 n\mathcal{E} \rceil \leq 6 + \lceil \log_2(n/\varepsilon) \rceil = O(\log n).$$

Along each edge of a quadtree node, we will place

$$m = O(H/\varepsilon) = O\left(\frac{\log n}{\varepsilon}\right)$$

equally spaced points, such that the path must use these ***portals*** whenever entering/exiting the square. We also add (as portals) the four corners of the square. Every portal can be used at most twice. Indeed, if a portal is being used more than twice by a tour, it can be modified to use it at most twice (see Remark 13.7 below for a proof of this). Furthermore, the path is restricted to use at most

$$k = O(1/\varepsilon)$$

portals on each side of the square.

This (uncompressed) quadtree has $O(nH) = O(n \log n)$ nodes. For each such square a subproblem is specified by how the tour interacts with the portals. We consider its four children, generate all the possible ways the tour might interact with the four children, and verify that information is consistent across the subproblems and with the parent constraints, and we recursively solve these subproblems. For each such consistent collection of four subproblems, we add up the costs of the solutions of the subproblems and set it as a candidate price to the parent instance. We try all possible such solutions and return the cheapest one.

One minor technicality is that the portals of the children are not necessarily compatible with the parent portals, even if they lie on the same edge. As such, the recursive program adds to the cost of the subproblem the price of the tour moving from the parent portals to the children portals.

Naturally, if a square in the quadtree has only $O(1/\varepsilon)$ points of Q stored in it, then given a subproblem for this square, the algorithm would solve the problem directly by brute force enumeration. Of course, only solutions that visit all the points in the square would be considered.

Thus, by using memoization, this recursive algorithm becomes a dynamic programming algorithm. The resulting dynamic programming algorithm is similar to the one we saw for the t-friendly case, with the difference here that the portals are fixed, and the recursive partition scheme used is determined by the quadtree and is as such simpler.

We claim that the resulting tour of the points of Q is a $(1 + \varepsilon)$-approximation to the shortest TSP tour of Q. (To get a $(1 + \varepsilon)$-approximation to TSP for the original point set, we need to use a slightly smaller approximation parameter, say $\varepsilon/10$. For the sake of simplicity we will ignore this minor technicality.)

13.3.3. Analysis.

13.3.3.1. *Running time analysis.*

Number of subproblems. Given a square and its portals, a tour needs to list the portals it uses in the order it uses them. The square has $4m + 4$ portals on its boundary, and each portal can be used at most twice. To count this in naive fashion, we duplicate every portal twice and select $i \leq 8k$ of them to be used. We then select one of the $i!$ different orderings for which these portals might be visited by the tour. Overall, the total number of different ways such a tour might interact with a square is bounded by

$$T = \sum_{i=0}^{8k} \binom{2(4m+4)}{i} i! \leq 8k \binom{2(4m+4)}{8k} (8k)! \leq 8k\,(8m+8)^{8k} = \left(\varepsilon^{-1} \log n\right)^{O(1/\varepsilon)},$$

as $k = O(1/\varepsilon)$ and $m = O\left(\varepsilon^{-1} \log n\right)$.

Overall running time. It easy to verify that each subproblem can be solved recursively in $T^{O(1)}$ time, ignoring the time it takes the solve the recursive subproblems. Since the quadtree has $O(n \log n)$ nodes, we conclude that the algorithm has overall running time $n\left(\varepsilon^{-1} \log n\right)^{O(1/\varepsilon)}$.

13.3.3.2. *Quality of approximation.* We need to argue that the optimal solution for Q can be made to comply with the constraints above without increasing its cost too much. Namely, subproblems can connect to one another only by using the portals, such that no more than k portals are used on each side of a subsquare, out of the $4m+4$ portals allocated to each subsquare.

We will first argue that restricting the optimal TSP to cross each side of a square at most k times introduces a small error. This will be done by arguing that one can patch a path so that it crosses every such square edge only a few times. We will argue that this increases the cost of the tour only mildly. Next, we will argue that moving the paths to use the portals incurs only a low cost.

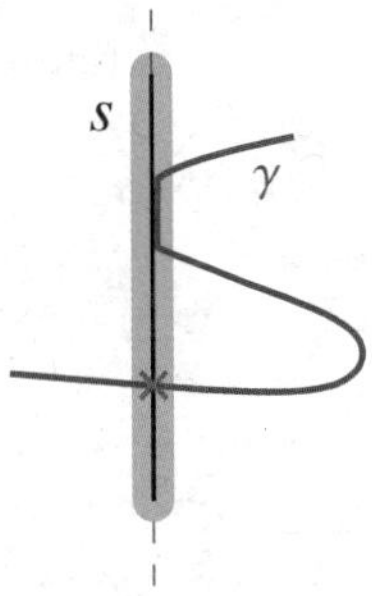

Fit the first – the patching lemma. In the following, we consider a segment as having two sides. Consider a curve π traveling along a segment s, such that it enters s on one side and leaves s on the *same* side. Conceptually, the reader might want to think of s as being a very thin rectangle[②]. We will not consider this to be a crossing of s by π; see the figure on the right for an example of a curve that crosses s once. Formally, a subcurve γ ***crosses*** a segment s if $\gamma \setminus s$ is made out of two non-empty connected components and these two connected components lie on different sides of the line supporting s (of course π might go from one side of s to the other side of s by bypassing it altogether). The number of times a curve crosses a segment s is the maximum number of subcurves it can be broken into, such that each one of these disjoint subcurves crosses s.

Lemma 13.6. *Let π be a closed curve crossing a segment s at least three times. One can replace it by a new curve π' that is identical to π outside s, that crosses s at most twice, and its total length is at most $\|\pi\| + 4\|s\|$.*

PROOF. Conceptually, think about s as being a vertical thin rectangle. Let $\mathsf{p}_1, \ldots, \mathsf{p}_k$ (resp. $\mathsf{q}_1, \ldots, \mathsf{q}_k$) be the k intersection points of π with the left (resp. right) side of s. We build an Eulerian graph having $V = \{\mathsf{p}_1, \ldots, \mathsf{p}_k, \mathsf{q}_1, \ldots, \mathsf{q}_k\}$ as the vertices. Let $s_i = \mathsf{p}_i\mathsf{p}_{i+1}$ and $s'_i = \mathsf{q}_i\mathsf{q}_{i+1}$, for $i = 1, \ldots, k-1$.

Let E_π be the edges formed by taking each connected component of $\pi \setminus s$ as an edge. Consider the multi-set of edges

$$E = \left\{s_1, \ldots, s_{k-1}, s'_1, \ldots, s'_{k-1}\right\} \cup \left\{s_1, s_3, \ldots\right\} \cup \left\{s'_1, s'_3, \ldots\right\} \cup \left\{\mathsf{p}_1\mathsf{q}_1\right\} \cup E_\pi$$

and the connected graph $\mathsf{G} = (V, E)$ they form. If k is odd, then all the vertices in the graph have even degree. As such, it has an Eulerian tour that crosses s exactly once using the zero length bridge $\mathsf{p}_1\mathsf{q}_1$ and is identical to π outside s, and as such it's the required path. (Naturally, this implies that the modified tour crosses s and goes around s to form a closed tour.)

[②]Naturally, the reader might not want to think about this segment at all. Be that as it may, this paragraph deals with this segment and not with the reader's wishes. Hopefully the reader will not become bitter over this injustice.

Otherwise, if k is even, then p_k and q_k are the only odd degree vertices in G. We modify G by adding $\mathsf{p}_k\mathsf{q}_k$ as an edge (which has zero length). Clearly, the resulting graph is now Eulerian, and the same argument goes through. (In this case, however, there are exactly two crossings of s.)

The total length of the edges of this graph (and as such of the resulting curve) is at most

$$\|\pi\| + 2\Big(\text{total length of odd segments along } s\Big) + 2\|s\| \leq \|\pi\| + 4\|s\|,$$

as required. ∎

By being more careful in the patching, the resulting path can be made to be of length at most $\|\pi\| + 3\|s\|$; see Exercise 13.1.

Remark 13.7. Applying Lemma 13.6 to a segment of length zero, we conclude that an optimal solution needs to use a portal at most twice.

Second fiddle – properties of sliding grids. Consider a segment s of length r and a grid G^i. We claim that the expected number of intersections of s with G^i, for $i \geq 1$, is roughly $\|s\|/2^{-i}$, where 2^{-i} is the sidelength of a cell of the grid G^i. Intuitively, this implies that the number of times a grid intersects a segment is a good estimator of its length (up to the scaling by the resolution of the grid). More importantly, it implies that if a segment is very short and the grid is large (i.e., $\|s\| \ll 2^{-i}$), then the probability of a segment intersecting the grid is small and is proportional to its length. We first prove this in one dimension. (A similar argumentation was used in Chapter 11, where we used random shifting for partitioning and clustering.)

Lemma 13.8. *Let s be an interval on the x-axis, and let x be a random number picked in the range $[0, 1/2]$. For $i \geq 1$, consider the periodic point set $T[x] = \left\{ x + j/2^i \;\middle|\; j \text{ is an integer} \right\}$. (The set $T[z]$ can be interpreted as the set of intersections of G^i, centered at z in the x-coordinate, with the x-axis.)*

If $\|s\| \leq 2^{-i}$, then the probability of the segment s containing a point of $T[x]$ is $\|s\|/\mathrm{sidelength}(\mathsf{G}^i) = 2^i\|s\|$.

Furthermore, the expected number of intersections of s with $T[x]$ (i.e., $|s \cap T[x]|$) is exactly $2^i\|s\|$, even if $\|s\| \geq 2^{-i}$.

Proof. Assume that $\|s\| \leq 2^{-i}$ and consider the set

$$U = \left\{ x \;\middle|\; T[x] \cap s \neq \emptyset \right\}.$$

Observe that U is a periodic set. Indeed, if $y = x + j/2^i \in U = T[x] \cap s$, then $y = (x+1/2^i)+(j-1)/2^i \in T[x+1/2^i] \cap s$. Thus $x+1/2^i$ is also in U. As such, $U = \bigcup_{j\in\mathbb{Z}}(s+j/2^i)$. Since U is periodic (with period $1/2^i$), the required probability is

$$\mathbf{Pr}\Big[T[x] \cap s \neq \emptyset\Big] = \frac{\|U \cap [0, 1/2]\|}{\|[0, 1/2]\|} = \frac{\left\|U \cap [0, 1/2^i]\right\|}{\left\|[0, 1/2^i]\right\|} = \frac{\|s\|}{1/2^i} = 2^i\|s\|.$$

As for the second claim, break s into m subsegments $s_1, \ldots, s_m$, such that each one of them is of length at most 2^{-i-1}. Note that each s_j can contain at most one point of $T[x]$, and let X_j be an indicator variable that is 1 if this happens, for $j = 1, \ldots, m$. Clearly, the number of points of $T[x]$ in s is exactly $X_1 + X_2 + \cdots + X_m$. By linearity of expectations, we have that

$$\mathbf{E}\Big[|T[x] \cap s|\Big] = \mathbf{E}\left[\sum_{j=1}^{m} X_j\right] = \sum_{j=1}^{m} \mathbf{E}\big[X_j\big] = \sum_{j=1}^{m} \mathbf{Pr}\big[T[x] \cap s_j \neq \emptyset\big] = \sum_{j=1}^{m} 2^i \left\|s_j\right\| = 2^i \|s\|. \qquad \blacksquare$$

Lemma 13.9. *Let s be a segment in the plane. For $i \geq 1$, the probability that s intersects the edges of the grid G^i is bounded by $\sqrt{2}\|s\|$. Furthermore, the expected total number of intersections of s with the vertical and horizontal lines of G^i is in the range $\left[2^i \|s\|, \sqrt{2} \cdot 2^i \|s\|\right]$.*

Proof. Let $\mathcal{I}_x$ and $\mathcal{I}_y$ be the projections of s into the x- and y-axes, respectively. By Lemma 13.8, the probability that a vertical (resp. horizontal) line of G^i intersects $\mathcal{I}_x$ (resp. $\mathcal{I}_y$) is bounded by $2^i \|\mathcal{I}_x\|$. By Lemma 4.2$_{\text{p48}}$, we have that $\|\mathcal{I}_x\| + \|\mathcal{I}_y\| \leq \sqrt{2}\|s\|$ and

$$\mathbf{Pr}\Big[s \text{ intersects an edge of } \mathsf{G}^i\Big] \leq 2^i \|\mathcal{I}_x\| + 2^i \|\mathcal{I}_y\| \leq \sqrt{2} \cdot 2^i \|s\|.$$

As for the second claim, the expected number of intersections of s with the horizontal (resp. vertical) lines of G^i is, by Lemma 13.8, exactly $2^i \|\mathcal{I}_y\|$ (resp. $2^i \|\mathcal{I}_x\|$). As such, the expected number of intersections of s with the grid lines is exactly $2^i \|\mathcal{I}_x\| + 2^i \|\mathcal{I}_y\|$ (note that the probability that s passes through a vertex of the shifted grid is zero). Now, $\|s\| \leq \|\mathcal{I}_x\| + \|\mathcal{I}_y\| \leq \sqrt{2}\|s\|$, which implies the claim. ■

Remark 13.10. Note that the bounds of Lemma 13.9 hold even if s is a polygonal curve. This follows by breaking it into segments and using linearity of expectation.

We also need the following easy observation.

Claim 13.11. *Let E_i be the union of the open edges of G^i (i.e., it does not include the vertices of G^i). Then, for a point p in the plane, we have that* $\mathbf{Pr}\Big[\mathsf{p} \in E_{i-1} \,\Big|\, \mathsf{p} \in E_i\Big] = 1/2$.

Intuitively, every line of E_i has probability $1/2$ to survive and be a line of E_{i-1}. The formal argument requires a bit more care.

Proof. Assume that p lies on a vertical line of G^i, as the other possibility follows by similar argumentation. Consider the points of G^i in the square $[0, 1/2)^2$. There are $\left((1/2)/2^{-i}\right)^2 = 2^{2(i-1)} > 2$ such points. Each of these points has the same probability to be the shift (x, y) used in generating G^i; see (13.2). For exactly half of these points, if they had been chosen to be the shift, then p would be on an edge of G^{i-1}; see the figure on the right. ■

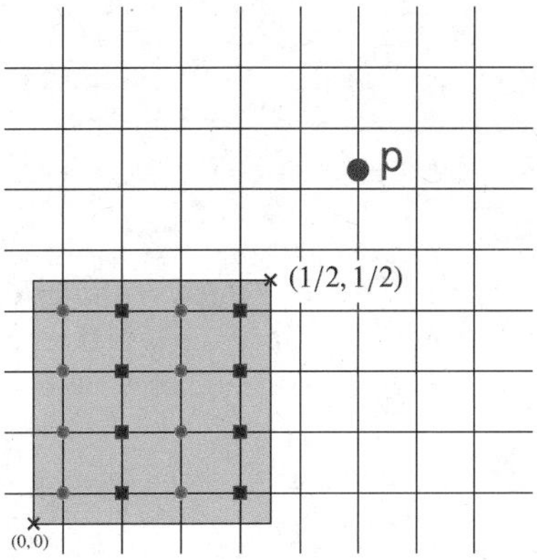

The third wheel – the cost of reducing the number of intersections. We patch the path[3] starting in the lowest canonical grid G^H (i.e., the bottom of the quadtree). Then we fix the resulting path in G^{H-1} (i.e., the next level of the quadtree) and so on, till we reach the canonical grid G^0. Let π_i denote the resulting path after handling the grid G^i. As such, π_{H+1} is just the original optimal TSP π_{opt}, and π_1 is the resulting patched tour (note that G^0 does not intersect the generated path).

[3]Say it quickly 100 times.

The patching is done as follows. In the ith step, for $i = 1, \ldots, H$, we start with π_{H+2-i} and consider the grid G^{H+1-i}. If an edge of the grid intersects π_{H+2-i} more than k times, then we patch it, using Lemma 13.6. We repeat this process over all the edges of the grid, and the resulting tour is π_{H+1-i}. Here k is a parameter of this patching process.

Since we are doing the patching process bottom up, every edge of the grid G^{H+1-i} corresponds to two edges of the finer grid G^{H+2-i}. Each of these two edges intersects the tour at most k times, but together they might intersect this merged edge more than k times, requiring a patch.

A minor technicality is that we assume that π_{opt} does not pass through any vertex of the grid G^H. Note that since we randomly shifted the quadtree, the probability that any grid vertex of G^H would lie on π_{opt} is zero.

Intuitively, this patching process introduces only a low error because when we fix an edge of a grid so that the tour does not intersect it too many times, the number of times the patched tour crosses boundaries of higher-level nodes of the quadtree also goes down. Thus, fix-ups in low levels of the quadtree help us also in higher levels. Similarly, the total number of crossings (of the tour with the grids) drop exponentially as we use larger and larger grids, thus requiring fewer fix-ups. Thus, intuitively, one can think about all the patching happening in the bottom level of the quadtree. The formal argument, below, is slightly more involved and one has to be careful, because we are dealing with expectations.

Claim 13.12. *Let π_{opt} be the optimal TSP of Q. There exists a tour π_1 that is a TSP of Q that crosses every side of a square of the quadtree $\mathcal{T}$ at most k times, and the total length of π_1, in expectation, is $\leq (1 + 8/(k-2)) \left\|\pi_{\mathrm{opt}}\right\|$.*

Proof. The algorithm to compute this modified path is outlined above. We need to analyze how much error this patching process introduced.

Let Y_i denote the number of times the path π_{i+1} crosses the edges of the canonical grid G^i. When the path π_{i+1} crosses an edge of this grid more than k times, we apply the patching operation (i.e., Lemma 13.6), so that it crosses this edge at most twice. Let F_i be the number of patching operations performed at this level, in generating the path π_i from π_{i+1}.

Note that after all the fix-ups are applied to π_{i+1} in the grid G^i, it has at most

$$n_i \leq Y_i - (k-2)F_i$$

crossings with the edges of the grid G^i, for $i \geq 1$. Indeed, every patching of an edge removes at least k crossings and replaces it with at most two crossings. As such, since the probability of a grid line of G^i to be a grid line of G^{i-1} is *exactly* $1/2$, by Claim 13.11, we have that

$$\mathbf{E}\Big[Y_{i-1}\Big] = \mathbf{E}\Big[\mathbf{E}\Big[Y_{i-1}\,\Big|\,n_i\Big]\Big] = \mathbf{E}\left[\frac{n_i}{2}\right] \leq \mathbf{E}\left[\frac{Y_i - (k-2)F_i}{2}\right] = \frac{1}{2}\mathbf{E}[Y_i] - \frac{k-2}{2}\mathbf{E}[F_i]$$

$$\implies \quad \mathbf{E}[F_i] \leq \frac{1}{k-2}(\mathbf{E}[Y_i] - 2\,\mathbf{E}[Y_{i-1}]).$$

The price of each fix-up in the grid G^i is $4/2^i$, by Lemma 13.6. As such, the expected total cost of this patching operation is

$$\mathbf{E}\Big[\mathrm{error}\Big] = \mathbf{E}\left[\sum_{i=1}^{H}\frac{4F_i}{2^i}\right] \leq \mathbf{E}\left[2F_1 + \sum_{i=2}^{H}\frac{4F_i}{2^i}\right] \leq \mathbf{E}[2F_1] + \frac{4}{k-2}\sum_{i=2}^{H}\frac{(\mathbf{E}[Y_i] - 2\,\mathbf{E}[Y_{i-1}])}{2^i}$$

$$\leq \mathbf{E}\left[2\frac{Y_1}{k-2}\right] + \frac{4}{k-2}\left(\frac{\mathbf{E}[Y_H]}{2^H} - \frac{\mathbf{E}[Y_1]}{2}\right) = \frac{4}{k-2}\left(\frac{\mathbf{E}[Y_1]}{2} + \frac{\mathbf{E}[Y_H]}{2^H} - \frac{\mathbf{E}[Y_1]}{2}\right).$$

since $F_1 \leq Y_1/k \leq Y_1/(k-2)$. We conclude that $\mathbf{E}\big[\mathrm{error}\big] \leq \frac{4}{k-2} \cdot \frac{\mathbf{E}[Y_H]}{2^H}$. The number Y_H is the number of times π_{opt} crosses the edges of the bottom grid G^H. By Remark 13.10 and Lemma 13.9 we have that $\mathbf{E}[Y_H] \leq 2 \cdot 2^H \|\pi_{\mathrm{opt}}\|$. As such, we have that

$$\mathbf{E}\big[\mathrm{error}\big] \leq \frac{4}{k-2} \cdot \frac{\mathbf{E}[Y_H]}{2^H} = \frac{4}{k-2} \frac{2 \cdot 2^H \|\pi_{\mathrm{opt}}\|}{2^H} = \frac{8}{k-2} \|\pi_{\mathrm{opt}}\| .$$

■

The fourth wall – bounding the price of snapping the tour to the portals. We finally need to bound the price of snapping the tour to the portals. We remind the reader that every edge e of a square of the quadtree has $m+2$ (two of them are on the endpoints of the edge) equally spaced portals on it. As such, the distance between an intersection of the tour (with the edge) and its closest portal (on the edge) is at most $\|e\|/2(m+1)$. As such, a single snapping operation on e introduces an error of twice the snapping distance by the triangle inequality; that is, $2(\|e\|/2(m+1)) = \|e\|/(m+1)$.

Note that if an edge of a grid has portals from several levels, the snapped tour uses only the portals on this edge that belong to the highest level.

Lemma 13.13. *The error introduced by snapping π_1 so that it uses only portals in all levels of the quadtree is bounded, in expectation, by $\frac{2H}{m+1} \|\pi_{\mathrm{opt}}\|$.*

PROOF. Let Z_i be the number of crossings of π_{opt} with G^i, for $i \geq 1$. Clearly, the tour π_1 has fewer crossings (because of the patching operations) with G^i than π_{opt}. Thus, we have that the expected price of snapping π_1 to the portals of the quadtree (summed over all the levels) is bounded by

$$\sum_{i=1}^{H} Z_i \cdot \frac{1/2^i}{m+1}.$$

As such, in expectation this error is bounded by

$$\mathbf{E}\big[\mathrm{error}_2\big] \leq \mathbf{E}\left[\sum_{i=1}^{H} \frac{Z_i}{2^i(m+1)}\right] \leq \sum_{i=1}^{H} \mathbf{E}\left[\frac{Z_i}{2^i(m+1)}\right] = \sum_{i=1}^{H} \frac{\mathbf{E}[Z_i]}{2^i(m+1)}$$
$$\leq \sum_{i=1}^{H} \frac{2 \cdot 2^i \|\pi_{\mathrm{opt}}\|}{2^i(m+1)} = \frac{2H}{m+1} \|\pi_{\mathrm{opt}}\|$$

by Remark 13.10 and Lemma 13.9. ■

The fifth elephant – putting things together. Normalizing the point set might cause the optimal solution to deteriorate by a factor of $1 + \varepsilon/2$; see (13.1). Forcing the tour to cross every edge of the quadtree at most k times increases its length by a factor of $1 + 8/(k-2)$; see Claim 13.12. Finally, forcing this tour to use the appropriate portals for every level increases its length by a factor of $1 + 2H/(m+1)$; see Lemma 13.13. Putting everything together, the expected length of the optimal solution when forced to be in the form considered by the dynamic programming is at most

$$\left(1 + \frac{\varepsilon}{2}\right)\left(1 + \frac{8}{k-2}\right)\left(1 + \frac{2H}{m+1}\right) \|\pi_{\mathrm{opt}}\| \leq \left(1 + \frac{\varepsilon}{2}\right)\left(1 + \frac{\varepsilon}{10}\right)^2 \|\pi_{\mathrm{opt}}\| \leq (1+\varepsilon) \|\pi_{\mathrm{opt}}\| ,$$

by selecting $k = 90/\varepsilon = O(1/\varepsilon)$ and $m \geq 20H/\varepsilon = \Theta((\log n)/\varepsilon)$. As such, we proved the following theorem.

Theorem 13.14. *Let* P *be a set of n points in the plane, and let* $\varepsilon > 0$ *be a parameter. One can compute, in* $n\left(\varepsilon^{-1}\log n\right)^{O(1/\varepsilon)}$ *time, a path* π *that visits all the points of* P *and whose total length (in expectation) is* $(1+\varepsilon)\left\|\pi_{\text{opt}}\right\|$, *where* π_{opt} *is the shortest* TSP *visiting all the points of* P.

13.4. Bibliographical notes

A beautiful historical survey of the TSP problem is provided in the first chapter of the book by Applegate et al. [**ABCC07**]. This chapter is available online for free, and I highly recommend that the interested reader bless it with his or her attention (i.e., read it).

The approximation algorithm of Section 13.3 is due to Arora [**Aro98**]. An alternative technique by Mitchell [**Mit99**] is described in the next chapter. Arora's technique had a significant impact and it was widely used, from faster algorithms [**RS98**], to k-median clustering algorithms [**ARR98, KR99**].

A nice presentation of the Arora technique and some of the algorithms using it are provided by a survey of Arora [**Aro03**]. Our presentation follows to some extent his presentation. However, we argued directly on the shifted quadtree, while Arora uses instead a clever argument about lines in the grid and the expected error that each one of them introduces. In particular, our direct proof of Claim 13.12 seems to be new, and we believe it might be marginally simpler and more intuitive than the analysis of Arora.

On the number of tours in a subproblem. Concerning the number of different subproblems one has to consider for a given rectangle and t checkpoints (i.e., Problem 13.1), this is equal to the number of matchings of t points in strictly convex position, which is related to the Catalan number. It is known that the number of non-crossing matchings of n points (not necessarily in convex position) in the plane is $2^{\Theta(n)}$; see [**SW06**].

Patching lemma. The patching lemma is not required for the algorithm to work in the planar case. Without it, the running time deteriorates to $n^{O(1/\varepsilon)}$, using the Catalan number argument to carefully bound the number of relevant subproblems. Extending the algorithm to higher dimensions requires the patching lemma.

But is it practical? The clear answer seems to be a resounding no. Finding a PTAS to a problem usually implies an impractical algorithm. It does however imply that a PTAS exists. At this point, one should start hunting for a more practical approach that might work better in practice.

Much research went into solving TSP in practice. Algorithms that work well in practice seem to be based on integer programming coupled with various heuristics. See [**ABCC07**] for more details.

13.5. Exercises

Exercise 13.1 (Better shortcutting). Prove a slightly stronger version of Lemma 13.6. Specifically, given a closed curve π crossing a segment s at least three times, prove that one can replace it by a new curve π' that is identical to π outside s that crosses s at most twice and for which its total length is at most $\|\pi\| + 3\|s\|$.

Exercise 13.2 (TSP with neighborhoods). Given a set of n unit disks $\mathcal{F}$, we are interested in the problem of computing the minimum length TSP that visits all the disks. Here, the tour has to intersect each disk somewhere.

(A) Let P be the set of centers of the disks of $\mathcal{F}$. Show how to get a constant factor approximation for the case that diam(P) ≤ 1.
(B) Extend the above algorithm to the general case.

CHAPTER 14

Approximating the Euclidean TSP Using Bridges

In this chapter, we present an alternative TSP approximation algorithm. It is slower, more complicated, and can be used in fewer cases than the portals-based algorithm already presented in Chapter 13. We present this algorithm for several reasons: (i) it uses clever ideas that should be useful in developing other algorithms, (ii) it is better than the portals algorithm in some very special cases, and (iii) it has a conceptual advantage over the portal technique in the sense that it introduces error into the approximation only if needed. Specifically, for cases where the optimal solution is friendly, this algorithm will output the optimal TSP. Conceptually, it is an extension of the algorithm for the "friendly" case; see Theorem 13.5_{p180}.

That said, a first-time reader might benefit from skipping this chapter. It is targeted at the more experienced reader that is interested in a deeper understanding of this problem.

14.1. Overview

The algorithm for the t-friendly case (see Section 13.1) reveals that TSP can be solved exactly (or approximately) if somehow one can divide and conquer such that each subproblem has limited interaction with its adjacent subproblems and its parent subproblem. Of course, it is not clear that such friendly short solutions to TSP always exist.

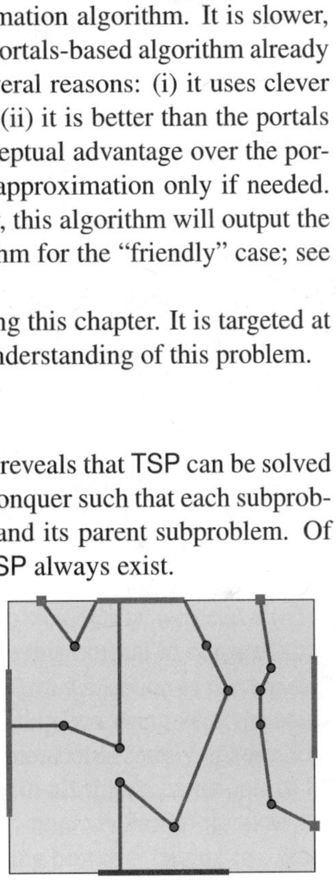
A bridge

Fortunately, one can introduce bridges (functionally, they are more like promenades) into a subrectangle. The idea is that we place an interval (i.e., bridge) on each side of a subrectangle, such that the solution uses either the at most $O(t)$ intersection points along the edge to enter or leave the subproblem or, alternatively, it uses the bridge on this edge, by just connecting to it (i.e., it does whatever is cheapest). See the figure on the right for an example of a subrectangle with its associated bridges and one possible solution. Since bridges hide interaction between subproblems, we get bounded interaction and so a dynamic programming approach will yield a PTAS.

This approach is quite challenging since bridges have to be paid for[①]; that is, we need to be able to argue that there is going to be an RBSP (see Definition 13.3_{p180}) for the optimal solution with additional bridges, such that the modified tour that will be considered by the dynamic programming is not much longer than the optimal solution. Naturally, bridges cannot be built everywhere – they need the right environment to support them.

Our plan of attack is the following: We will first show that "affordable" bridges exist; that is, we will show that converting the optimal solution into a "friendly" instance by adding bridges can be done with only a moderate increase in its length. Then, we will

①Even if they are bridges to nowhere.

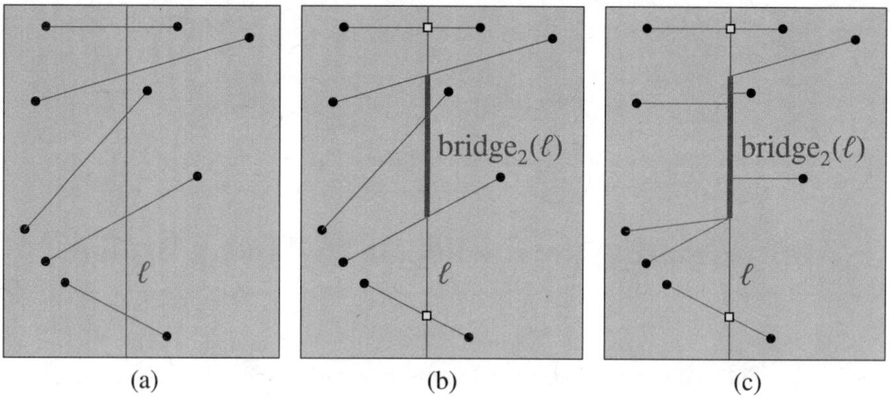

FIGURE 14.1. How a cut and its bridge interact with the optimal solution: (a) an optimal solution and its intersections with the cut ℓ, (b) a bridge and its two associated checkpoints, (c) one possible solution generated by the dynamic programming.

define what canonical rectangles are (imprecisely, they are the subrectangles considered by the dynamic programming). Finally, we will put all the details together to get the required dynamic programming that provides a PTAS for TSP.

14.2. Cuts and bridges

The purpose of this section is to prove that one can introduce interaction reducing cuts (details below) and pay for the error they introduce as we construct an RBSP that will be used in the dynamic programming (see Definition 13.3_{p180} for a reminder of what RBSP is).

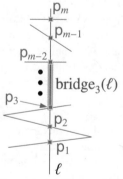

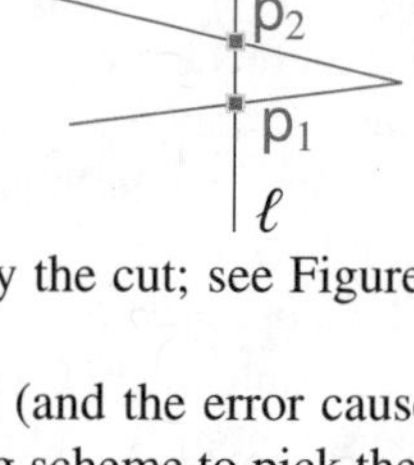

So, let S be a set of segments contained in a rectangle R (some of the segments of S have their endpoints on the boundary of R). For our purposes, think about S as the collection of segments forming the optimal TSP tour inside R. For the sake of simplicity of exposition we assume that the segments are in general position and no segment is horizontal or vertical. A ***cut*** is a vertical or a horizontal line that intersects R. For a cut ℓ, let $p_1, p_2, \ldots, p_m$ be the intersection points of ℓ with the segments of S, sorted along ℓ. The ***t-bridge*** of ℓ is the segment on ℓ connecting p_t with p_{m-t+1}. The t-bridge of ℓ is denoted by $\mathrm{bridge}_t(\ell)$; see the figure on the right. Such a bridge would serve to reduce the interaction between two subproblems separated by the cut; see Figure 14.1 for an example.

In the following, the total length of the bridges introduced (and the error caused by them) is going to be $O(\varepsilon \|S\|)$. We will use a non-trivial charging scheme to pick the right bridges and make sure this property holds.

If a cut ℓ intersects at most $4t$ segments of S, then we will just use this cut directly without introducing a bridge, as was done in the previous chapter (see Section 13.2). We will refer to such a cut as a ***free cut***[②]. If R has a free cut, then we use it. As such, in

[②] A free cut is like a free lunch, except that you cut it instead of eating it.

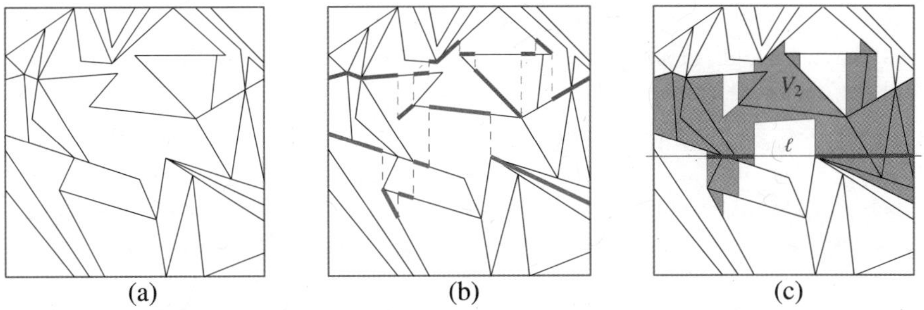

FIGURE 14.2. (a) A set of segments S, (b) the bottom and top 2-levels of S, and (c) the polygon V_2 and its intersection with a horizontal line. The total length of the intersection is the total budget of a bridge using this line.

the following, we concentrate on the case that any cut intersects S in strictly more than $4t$ distinct points.

Intuitively, we would like to find an interesting cut (i.e., one that splits the set of segments in a non-trivial way) that somehow we can pay for. Our scheme is to charge the length of a t-bridge to the segments of S. Specifically, we will break R along the cut and continue recursively to apply such cuts to the subrectangles.

Definition 14.1. A point p on a segment of S is on the ***bottom k-level*** if there is an open vertical downward ray from p that intersects exactly k edges of S. The closure of such points is the ***bottom k-level*** of S; see Figure 14.2. The k-level when we shoot the rays upward is the ***top k-level***. Similarly, when the rays are shot horizontally and to the right (resp. left), the corresponding level is the ***right k-level*** (resp. ***left k-level***).

A point lying on a segment of S is ***vertically t-shallow*** (resp. ***horizontally t-shallow***) if it lies on some top or bottom k-level (resp. right or left k-level) of S, for $k \leq t$.

Now, consider a (say, vertical) cut ℓ that breaks R into two new subrectangles R^+ and R^-. A point p lying on a segment of S is ***horizontally t-exposed*** by ℓ if p is not horizontally t-shallow in R, but it is horizontally t-shallow in R^+ (if p is contained in R^+) or in R^- (if p is contained in R^-). The figure on the right depicts the portions of S being horizontally 1-exposed by the cut ℓ.

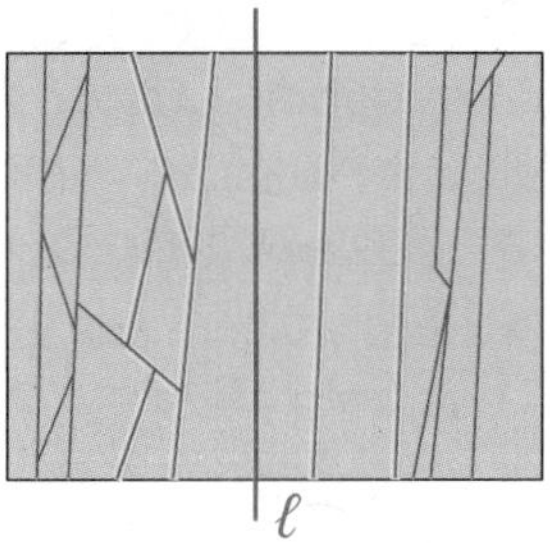

Consider all the fragments (i.e., portions of segments) of S that are t-exposed by ℓ, and let exposed(S, ℓ, R) denote the total length of these fragments.

If $\text{exposed}(S, \ell, R)/t \geq \left\|\text{bridge}_t(\ell)\right\|$, then we charge the length of $\text{bridge}_t(\ell)$ to the points t-exposed by this cut. Such a t-bridge is being paid for by this charging scheme, and it is ***feasible***.

Since a point can be t-exposed only twice, during an RBSP construction (once vertically and once horizontally), it follows that the total charge for S in a recursive RBSP using such cuts is going to be at most $2\left\|S\right\|/t$, where $\left\|S\right\|$ denotes the total length of segments in S.

Putting it slightly differently, if a cut exposes segments of total length exposed(S, ℓ, R), then we have a budget of size roughly $O(\varepsilon \cdot \text{exposed}(\mathsf{S}, \ell, \mathsf{R}))$ to pay for the bridge on the cut (by setting $t = O(1/\varepsilon)$).

14.2.1. Why do feasible bridges exist? Our purpose here is to prove that a feasible cut always exists. We are given a set S of segments and a bounding rectangle R. First, remember that if there is a free cut (i.e., a line intersecting at most $4t$ segments of S), we will use this cut directly without introducing a bridge. As such, assume that there is no free cut.

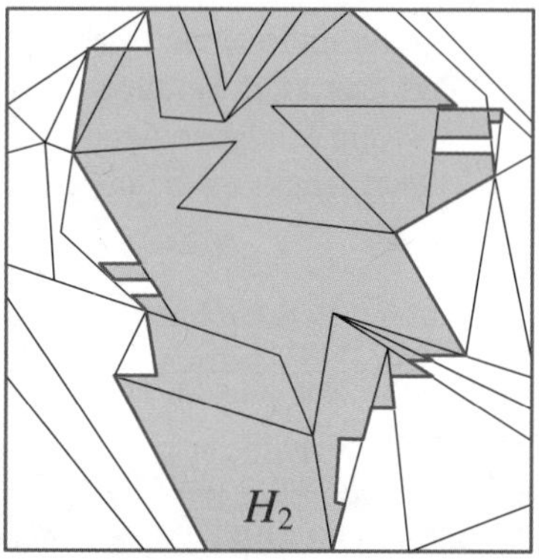

Consider the top t-level $\mathbb{L}^t$ and the bottom t-level $\mathbb{L}_t$ of S. Together with the two vertical sides of R, this forms an x-monotone polygon V_t. See Figure 14.2 for an example. In a similar fashion we define a y-monotone polygon H_t defined with respect to horizontal t-levels; see the figure on the right. Since there are no free cuts, the polygon V_t (resp. H_t) is connected and spans the whole x-range (resp. y-range) of R.

Consider an arbitrary horizontal line ℓ intersecting the polygon V_t, and let $\mathcal{I}$ be a maximum interval contained inside this intersection; see Figure 14.2. All the points of $\mathcal{I}$ are between the bottom t-level and the top t-level of S. As such, when we cut along ℓ, there are at least t layers of segments above or below $\mathcal{I}$ that get t-exposed (since there is no free cut, any vertical line intersects S in at least $4t$ points). As such, the total length of the segments t-exposed by ℓ vertically either below or above $\mathcal{I}$ is at least $t\,\|\mathcal{I}\|$. Applying this argument for all the connected components of the intersection $V_t \cap \ell$ implies the following.

Lemma 14.2. *Let* S *be a set of segments contained in the axis rectangle* R. *Furthermore, assume that there is no free cut (i.e., any horizontal or vertical line intersecting* R *intersects at least* $4t$ *segments of* S*). Then, a horizontal (resp. vertical) cut* ℓ *t-exposes portions of* S *of total length at least* $t\,\|\ell \cap V_t\|$ *(resp.* $t\,\|\ell \cap H_t\|$*).*

Namely, ℓ *can pay for a bridge of length* $\leq \|\ell \cap V_t\|$ *(resp.* $\|\ell \cap H_t\|$*).*

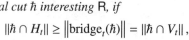

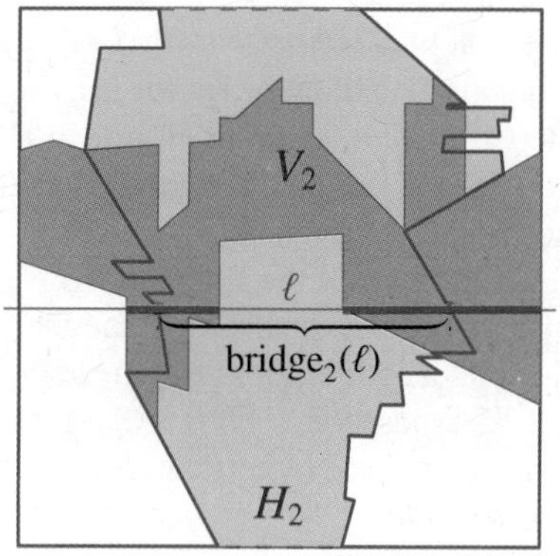

Consider an arbitrary horizontal line ℓ and its bridge. By definition of a t-bridge, it connects the left t-level with the right t-level. But these are exactly the boundary points of H_t on the line ℓ. Namely, the bridge on the line ℓ is the segment $\text{bridge}_t(\ell) = \ell \cap H_t$, and its total length is $\|\ell \cap H_t\|$.

See the figure on the right for an example.

Observation 14.3. *For a horizontal line* ℓ *intersecting* R*, if*

$$\|\ell \cap V_t\| \geq \left\|\text{bridge}_t(\ell)\right\| = \|\ell \cap H_t\|,$$

then, by Lemma 14.2, the cut ℓ *is feasible; see the figure on the right.*

Similarly, for a vertical cut $\hbar$ *interesting* R*, if*

$$\|\hbar \cap H_t\| \geq \left\|\text{bridge}_t(\hbar)\right\| = \|\hbar \cap V_t\|,$$

then the cut $\hbar$ is feasible.

Claim 14.4. *If* area$(V_t) \geq$ area(H_t)*, then there exists a feasible horizontal cut. Similarly, if* area$(H_t) \geq$ area(V_t)*, then there exists a feasible vertical cut.*

PROOF. So, assume that $\mathsf{R} = [a, b] \times [c, d]$. For $\alpha \in [c, d]$, let

$$h(\alpha) = \|\ell_\alpha \cap H_t\| \quad \text{and} \quad v(\alpha) = \|\ell_\alpha \cap V_t\|,$$

where ℓ_α denotes the x-axis parallel line $y = \alpha$.

Clearly, we have that $\operatorname{area}(H_t) = \int_{y=c}^{d} h(y)\,dy$ and $\operatorname{area}(V_t) = \int_{y=c}^{d} v(y)\,dy$. If $\operatorname{area}(V_t) \geq \operatorname{area}(H_t)$, then

$$\int_{y=c}^{d} (v(y) - h(y))\,dy = \operatorname{area}(V_t) - \operatorname{area}(H_t) \geq 0.$$

This implies that there is some $\beta \in [c, d]$, such that for $y = \beta$ we have that $v(\beta) \geq h(\beta)$. By Observation 14.3, since $\left\|\ell_\beta \cap V_t\right\| = v(\beta) \geq h(\beta) = \left\|\ell_\beta \cap H_t\right\| = \operatorname{bridge}_t\!\left(\ell_\beta\right)$, the line $\ell_\beta \equiv (y = \beta)$ induces a feasible horizontal bridge.

The second claim follows by a symmetric argument. ∎

Remark 14.5. Let us push the argument a bit further. The functions $v(t)$ and $h(t)$ are piecewise linear, as can be easily verified. The slope of these functions changes at a value α if the line ℓ_α passes through a vertex of S (this is implied as the vertices of the segments of S are either in R or on its boundary since S is contained in R). As such, we can build the bridge where $v(y) - h(y)$ is maximized, which must be at a cutting line that passes through a vertex of S. We will refer to such a bridge/cut as being ***canonical***[③].

Definition 14.6. Let R be an axis parallel rectangle, and let S be a set of interior disjoint segments. Clipping S to R, let $\mathsf{S}_\mathsf{R} = \left\{ s \cap \mathsf{R} \;\middle|\; s \in \mathsf{S} \right\}$ (if S is contained in R, then $\mathsf{S}_\mathsf{R} = \mathsf{S}$). A ***canonical cut*** is a horizontal or vertical line ℓ that passes through an endpoint of a segment of S_R (we require that the cut splits the rectangle non-trivially; that is, the cut intersects the interior of R).

A ***canonical bridge*** is an interval lying on a canonical cut, having both endpoints on some segments of S_R. A canonical t-bridge that is chargeable to its environment, as above, is a ***feasible canonical bridge***.

The reader might worry that a feasible bridge (as computed by Claim 14.4), canonical or not, might be trivial; that is, it lies on the boundary of R. However, for the horizontal bridge case, observe that, by construction, we have that $h(c) > 0$, $h(d) > 0$, while $v(c) = v(d) = 0$ (since the value of t we are going to use is larger than 3, we are assuming general position and no free cuts). This implies that since no three segments of S can meet in a single point, the boundaries of H_t do not intersect the vertical boundaries of R. Namely, the function $v(y) - h(y)$ is negative for $y = c$ or $y = d$. Therefore, the cut and its feasible bridge (since $v(y) - h(y)$ is not negative for a feasible horizontal cutting line) would not be on the boundary of the rectangle. The same argumentation works for the vertical bridge.

We now summarize what we have just shown.

Lemma 14.7. *Let* S *be a set of segments contained in the axis rectangle* R*, and let t be a parameter. Then, either there exists a free canonical cut of* R *or, alternatively, there exists*

[③]If one wants to use holy bridges, then one has to canonize them....

a feasible canonical cut of R*, such that its t-bridge (i.e., feasible canonical bridge) can be charged to the portions of the segments of* S *that are being t-exposed by this cut.*

In particular, any recursive partition scheme for S *using such cuts would overall introduce bridges of total length at most* $2\,\|S\|\,/t$.

14.2.2. Canonical rectangles. We now have a tool that enables us to recursively cut a subproblem into two, in such a way that the interaction along the cut is bounded. The problem is that the resulting subproblems are no longer nicely defined. Indeed, in the friendly case, a subproblem was defined by an axis parallel rectangle, which was the bounding rectangle of four specific points. As such, the total number of possible subrectangles was $O(n^4)$.

In our case, however, it is not clear how to bound the number of possible subrectangles.

A naive effort to bound the number of possible subrectangles and why it fails. The following is a wishy-washy argument for why the naive approach to bounding the number of subproblems fails and why we have to complicate things so one can bound the number of subproblems. As usual with such arguments, the reader might not find it convincing, as a simple modification might (seem to) make it work. Formally, this wishy-washy argument should be taken for what it's worth, that is, nothing at all (except for maybe building the reader's intuition, which hopefully it does).

Consider an initial rectangle R and a canonical cut line. This canonical cut line might be defined by a point that lies on the intersection of the side of the parent rectangle and a segment that belongs to the optimal solution (there are other possibilities, for example the cut line passes through an endpoint of one of the segments). As such, starting with a rectangle R, we perform a sequence of canonical cuts and get possible subrectangles for which we need to solve the TSP subproblem.

However, as we go down the recursion, a subrectangle might be defined by a long sequence of such operations, which would in turn imply that the number of possible subrectangles is unacceptably large.

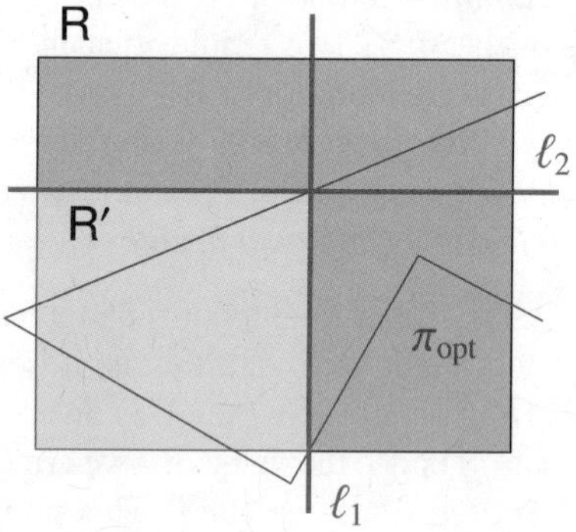

As a concrete example, consider the rectangle R depicted on the right. We start with a canonical cut ℓ_1. Then, we perform another canonical cut with a line ℓ_2. Note that the y-coordinate of ℓ_2 is a function of the location of ℓ_1. Namely, specifying the resulting subrectangle R′ will require us to specify both ℓ_1 and ℓ_2. Clearly, one can continue in such a fashion such that the number of possible subrectangles generated by a sufficiently large sequence of such cuts is exponentially large. As such, a major road block towards a PTAS is somehow reducing the number of subproblems considered.

Reducing the number of subproblems using minimal rectangles. Now, our purpose is to get a dynamic programming algorithm. By modifying the dynamic programming itself, we can avoid the exponential blowup described above.

To minimize the number of subproblems, the idea is to shrink a given rectangle into a "canonical" rectangle and argue that the number of such rectangles is small. Unfortunately, this will increase the complexity of the dynamic programming as well.

So consider a rectangle R and a set of segments S that might have endpoints outside R. The first step will be to trim away segments that have both endpoints outside R (intuitively, such segments [and their lengths] will be handled by a higher level in the recursion and as such can be ignored by the current dynamic program). As such, we assume all the segments of S have at least one endpoint inside R.

Given R, we will start shrinking it by continuously pushing its sides into the interior. Now, an axis parallel rectangle is defined by four real numbers, and as such it has four degrees of freedom (left, right, top, and bottom edges). We start by picking an arbitrary side and moving it inward. There are two cases to consider.

(A) The inward moving side of the rectangle hits an endpoint of a segment of S. In this case, we cannot move the side any further and so we consider this side as ***pinned*** (i.e., this degree of freedom is fixed), and we move on to the next side.

(B) A corner of the inward moving edge passes over a segment s of S. In this case, we fix the corner (but not the corner's position) and require that the final rectangle have this corner lie on s. We then move this corner inwards along s (and move the two adjacent edges with this corner) until we reach an endpoint of a segment of S (which could be an endpoint of s) or another corner passes over a segment of S. In either case, the corner that we moved becomes pinned (note that in this case it fixes at least two degrees of freedom of the rectangle).

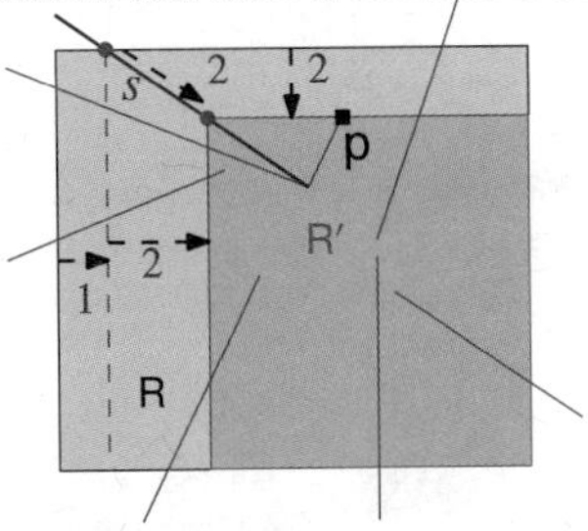

Case (B) above is demonstrated by the figure on the right, where the rectangle R is being shrunk into the rectangle R′. Here, the left edge is being pushed in, followed by this edge hitting a segment of S and then moving the top left corner along this segment, till the top edge (that is moving in tandem with the left edge as the corner slides down along s) hits the endpoint p.

We continue this process until no degrees of freedom are left. Note that in some cases when a corner runs into a segment, it might not be possible to move this corner along this segment, since it might violate other fixed degrees of freedom. In this case, we move on to the next unpinned side. Clearly, the rectangle can lose at most four degrees of freedom before it is completely stuck and cannot move any more.

The beauty of this process is that in the end of it we get a ***minimal rectangle*** min(R) which is "stuck", it is contained in the original rectangle R, and min(R) is uniquely defined by the segments and the endpoints that pin it.

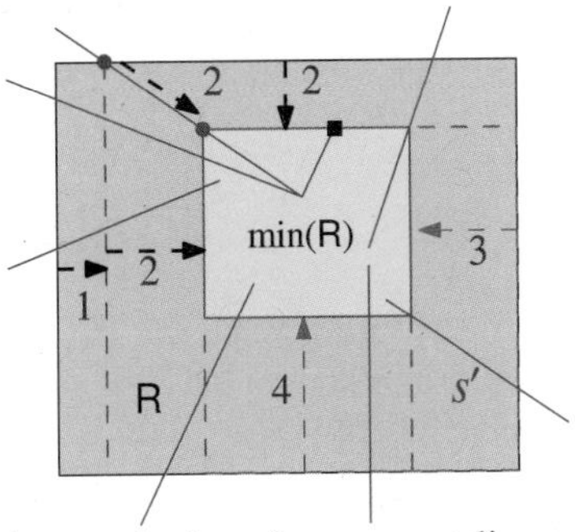

An example is depicted on the right – arrows with the same number indicate that these edges / corners are moving simultaneously during the shrinking processes. Note that in the fourth step the bottom edge cannot move any further, since we require that s' keep intersecting the right edge and since the right edge is pinned by s' and cannot be moved.

It is easy[④] to verify that in the worst case (allowing degeneracies) it takes five segments to determine min(R). In this case the rectangle is pinned by four segments (each corner lies on a segment), and an endpoint of a segment lies on the interior of some edge of min(R). Hence we know that (in any case) the minimal rectangle min(R) is defined by at most five such pinnings, and it is uniquely defined by its pinnings. Each pinning is specified by a segment of S, maybe specifying which endpoint

[④]Case analysis done by the reader and not the author is always easy.

of the segment is involved in the pinning and what kind of pinning it is (i.e., endpoint of a segment on a side of the minimal rectangle, endpoint of a segment on a corner of the minimal rectangle, and corner of the minimal rectangle on a segment).

Note that given this description of the pinning of a minimal rectangle min(R), we can compute the minimal rectangle itself. As such, this canonical description of minimal rectangles uniquely defines all such rectangles.

In particular, in our setting, the segments of S are going to be a subset of all possible segments induced by the set of points P. As such, it follows that there are $\binom{n}{2}$ possible segments that might participate in describing such a minimal rectangle, where $n = |\mathsf{P}|$. For each such segment, there are a constant number of different ways that it could be used in pinning the minimal rectangle.

Observation 14.8. *For any rectangle* R *and a set of segments* S *each having at least one endpoint inside* R*, we have that* $\min(\mathsf{R}) \subseteq \mathsf{R}$.

Furthermore, if S *is a subset of the segments induced by a set of points* P*, then the total number of distinct minimal rectangles that can be defined is* $O\left(\binom{n}{2}^5 2^{O(1)}\right) = O\left(n^{10}\right)$, *where* $n = |\mathsf{P}|$.

Example 14.9. Consider a rectangle R, a set of segments S, and the minimal rectangle min(R). Here, let e be an arbitrary side of R, and let $e_{\min}$ be its respective side in min(R). Intuitively, one might think that the sets $e \cap \mathsf{S} = \left\{ s \in \mathsf{S} \;\middle|\; s \cap e \neq \emptyset \right\}$ and $e_{\min} \cap \mathsf{S}$ are the same. This is not quite correct, but it is true that $e \cap \mathsf{S} \subseteq e_{\min} \cap \mathsf{S}$. The difference might be at most (two) extra segments that pass through the endpoints of $e_{\min}$ that it shares with its neighbor. Furthermore, it might also gain some edges from segments pinned to the shrunk edge $e_{\min}$ (as is the case with 4 in the example below).

See the figure on the right for an example. Here, we have that

$$\mathsf{e}_l \cap \mathsf{S} = \{1, 2\} \subseteq \mathsf{e}'_l \cap \mathsf{S} = \{1, 2, 3\},$$

$$\mathsf{e}_t \cap \mathsf{S} = \{3, 5\} \subseteq \mathsf{e}'_t \cap \mathsf{S} = \{3, 4, 5\},$$

$$\mathsf{e}_r \cap \mathsf{S} = \{6\} \subseteq \mathsf{e}'_r \cap \mathsf{S} = \{5, 6\},$$

$$\mathsf{e}_b \cap \mathsf{S} = \{7, 8\} \subseteq \mathsf{e}'_b \cap \mathsf{S} = \{6, 7, 8\}.$$

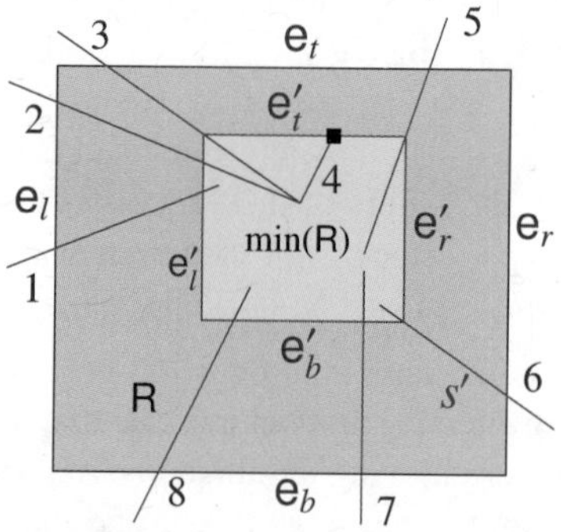

14.3. The dynamic programming

So we are given a set P of n points in the plane in general position (no three are collinear and no two have the same x or y value). Our purpose is to compute a short TSP of the points of P.

Let S be the set of all $\binom{n}{2}$ segments defined by pairs of points of P, and let $t > 0$ be a prespecified parameter. We first compute the set of all possible minimal rectangles $\mathcal{R}_{\min}$ induced by S or subsets of S. This is done by generating all canonical descriptions of minimal rectangles and computing for each such description the minimal rectangle it corresponds to.

Next, we compute all the canonical bridges $\mathcal{B}$ that cut a minimal rectangle (in relation to the set of segments S), and let $\mathcal{B}$ denote the resulting set of bridges. By Remark 14.5, a canonical cut of a rectangle R passes through the intersection of a segment of S with the boundary of R (or the cut line passes through a point of P). As such, there are at most $\phi = 2|\mathsf{S}| + 2|\mathsf{P}| = 2\binom{n}{2} + 2n \le 2n^2$ possible canonical cuts per rectangle. For a specific cut ℓ, there is only a single unique bridge$_t(\ell)$. As such, by Observation 14.8, we have that the number of possible bridges (and this also bounds the number of possible cuts overall) is bounded by

$$|\mathcal{B}| = O(|\mathcal{R}_{\min}| \cdot \phi) = O\left(n^{10} \cdot n^2\right) = O\left(n^{12}\right).$$

We will start by sketching the recursive algorithm and then add details to get to a full description of the algorithm[5].

14.3.1. The basic scheme. The dynamic programming (i.e., recursive algorithm) will work on a minimal rectangle R and will start from the bounding rectangle of P (which is definitely a minimum rectangle defined by S). Given such a minimal rectangle R, the algorithm will guess (try all possibilities for) a cut, with an associated bridge $b \in \mathcal{B}$ that is contained in the interior of R (here b might be an empty bridge). We split the rectangle R by the cut line supporting b. This results in two rectangles R^- and R^+. The problem is that we cannot recursively call on these two rectangles, as R^- and R^+ might not be minimal. As such, the dynamic programming guesses using R (i.e., tries all possibilities) for the two minimal rectangles of R^- and R^+. Let R_g^- and R_g^+ denote these two guesses. The algorithm will then continue recursively to solve the problem on R_g^- and R_g^+. In addition, the algorithm will also try all possible cuts with no bridges on them (i.e., think about them as empty bridges).

14.3.1.1. *A bridge too far.* The recursive algorithm is given a subrectangle R and a "guess" specifying exactly how the optimal solution interacts with R. Naively, each such subrectangle R might have $O(t)$ checkpoints (i.e., either a bridge or an intersection of the optimal solution with the boundary R). An intersection point can be specified by the segment inducing it (such a segment has both endpoints in P). (Note that including such a checkpoint in a subproblem implies that the solution must use the segment inducing the checkpoint in the generated solution.) However, R is a minimal rectangle that was shrunk (in several steps) from bigger rectangles, and these rectangles might contain bridges on their boundaries. As such, the up to four bridges of R might be "hanging" in the air far from R. As such, in a subproblem defined for R, every edge of R would have $O(t)$ checkpoints associated with it (that lie on the edge) and potentially also some bridge either on it or hanging next to it, such that the solution to the subproblem needs to use the checkpoints and the bridges specified.

Such an instance of a subproblem is depicted on the right. Here the parent subproblem has four bridges. After introducing the cut ℓ and its bridge and computing the two minimum rectangles, one gets two minimum rectangles R and R′ for the two subproblems. Here, the rectangle R (resp. R′) has three bridges that it inherited from its parent rectangle (before it was shrunk). In addition, it has another bridge that lies on the last cut that created it. Note that every edge of the subproblem (i.e., the four sides of its rectangle) has the checkpoints used on the edge guessed for it.

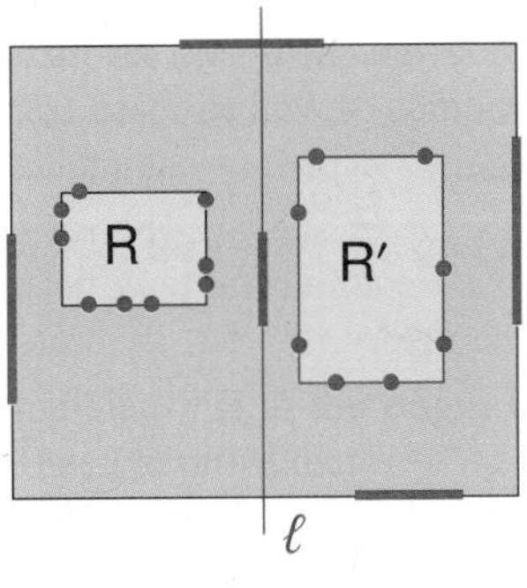

[5]Translation: We will add details till the mess becomes both undecipherable and incomprehensible at the same time. Hopefully, the inner poetical and rhythmical beauty of the text will keep the reader going.

14.3.1.2. *What is specified by the recursive call.* A recursive call would be specified by a minimum rectangle. For each edge of the rectangle, we guess $4t$ checkpoints and a bridge associated with it. We also need to guess the connectivity of the generated solution; see below for details.

Bridge parity. For reasons that (hopefully) will become clear shortly, we will also specify for each bridge whether the number of times the generated solution for the subproblem connects with it is even or odd (this would be done in such a way that a bridge would be visited by the generated solution exactly an even number of times overall). Note also that a single bridge might be used a large number of times by the computed solution.

Specifying connectivity. As usual, we also enumerate the checkpoints and bridges and specify for each of them which other bridge/checkpoint it is connected to. Note that several checkpoints might be connected to the same bridge (in the generated solution). Now, given this fully specified subproblem, we are looking for a set of paths: (i) that connects the checkpoints and the bridges in a way that complies with the specification, (ii) for which the paths visit all the points of P inside the subrectangle R, and (iii) for which all endpoints of the paths lie on the specified bridges and checkpoints (i.e., no path can start in the middle of a subrectangle).

This connectivity information for a subproblem is specified as follows. There are $O(t)$ specified checkpoints and at most four bridges. For every bridge, we specify all the checkpoints and bridges it is connected to. Clearly, per bridge this requires specifying $O(t)$ bits. Now, for every checkpoint we specify the other checkpoint it has to be connected to by the generated solution (if it has to be connected to a bridge, this was already specified). Overall, there are $t^{O(t)}$ different connectivity configurations that can be specify per subproblem.

Bottom of the recursion. Naturally, if the subrectangle has $O(t)$ points of P inside it, we can solve the subproblem directly by brute force by trying all possible paths.

Number of different recursive calls. Clearly, there are

$$O\left(|\mathcal{R}_{\min}| \cdot |\mathcal{B}|^4 \binom{n}{2}^{O(t)} t^{O(t)}\right) = n^{O(t)}$$

different recursive calls. There are $|\mathcal{R}_{\min}|$ minimal rectangles, and every one of them might have four bridges associated with it and $O(t)$ checkpoints on its boundary. We remind the reader that a checkpoint is the intersection of a segment (specified by a pair of points of P) and the boundary of the rectangle.

Accounting for length of suggested solution. So given a subproblem with minimum rectangle R, the recursive program tries all canonical cuts (with or without a bridge) that split the rectangle and all possible pairs of minimum rectangles on both sides of the cut.

Let R^+ and R^- be such a pair of two minimal rectangles. The algorithm tries all possible connectivity information on both minimal rectangles and also on the cutting line. The algorithm tries, by brute force, to find a way to connect the checkpoints and bridges of R to the checkpoints and bridges of R^+ and R^-, while (i) respecting each connectivity constraints and (ii) using only the region between $\mathsf{R} \setminus (\mathsf{R}^+ \cup \mathsf{R}^-)$. In addition, we have to include the cost (computed by the recursive calls) for the subproblems corresponding to R^+ and R^- (with the given/guessed constraints). Let α be the cheapest solution found that connects the checkpoints and bridges of R to the subproblems of R^- and R^+ (plus the cost of these two subproblems).

Here are some low-level tedious technicalities:

(A) The cost α includes the price of the bridge (if it was created) and in fact, we will include the length of the bridge twice (for reasons that will become clear shortly).
(B) If a bridge of R is also used by a subproblem, say R^+, the price of connecting to this bridge in R^+ would be paid by the subproblems (despite the fact that this subproblem pays here for the portion of the solution outside its region). Note that the price of the bridge is being paid when the bridge is being created, and it is only the connections to the bridge that are paid for later when they are being created.
(C) If a bridge is introduced between R^+ and R^-, the recursive algorithm makes sure that the two subproblems connect to the bridge with the same parity; that is, either both connect to the bridge an even number of times or both connect to the bridge an odd number of times. This requirement is enforced by the recursive calls.

It is now tedious but straightforward to verify that the recursive procedure performs at most $n^{O(t)}$ recursive calls. Each recursive call (ignoring the time of the recursive calls it performs) takes $n^{O(t)}$ time. Indeed, this requires guessing the cut line, its bridge, the checkpoints for the two subproblems, and their connectivity information. As shown above, there are $n^{O(t)}$ possibilities to be considered. Since the number of subproblems is $n^{O(t)}$, the overall running time is $n^{O(t)}$.

The algorithm returns the shortest connected graph (with the bridges being paid for twice) that spans all the points of P. We next describe how to convert such a graph into a tour of P of shortest distance.

Generating the tour. We are given a connected collection of segments S computed by the algorithm that spans all the points of P. We want a short tour of it that visits all the points of P. To this end, we will define a natural graph of S and modify it to be Eulerian. An Eulerian graph has a tour visiting all edges of the graph exactly once, and this tour can be computed in linear time. We remind the reader that a graph is ***Eulerian*** if and only if it is connected and the degree of all its vertices is even.

To this end, we will introduce a vertex at each point of P and at each endpoint of S. A maximal connected portion of a segment of S connecting two consecutive such vertices will be interpreted as an edge of the resulting graph G (i.e., a bridge might be fragment into several segments). Clearly, the only place in this graph that we might get vertices of odd degree is on bridges. Observe, however, that by the dynamic programming, we know that the number of edges connected to the bridge on one side of it is of the same parity as the number of edges connected to it on the other side. As such, the total number of edges connected to a bridge overall is even. So consider such a bridge b, and let $V_b = \langle v_1, v_2, \ldots, v_m \rangle$ be all the odd degree vertices on the bridge, sorted in order along it (observe that m is even).

We introduce the edges $v_{2i-1}v_{2i}$ (i.e., or you can think of it as adding a segment) into the graph, for $i = 1, \ldots, m/2$ (note that this edge is already in the graph, and we set it to multiplicity 2). The resulting graph G′ is now a multi-graph (i.e., it might have parallel edges), it is connected, and all its vertices have even degree. As such, an Eulerian tour π of G′ exists and can be computed in linear time.

Clearly, in the worst case, for each bridge, we added an extra copy of all its edges and hence added at most the total length of the bridge. As such, if $\beta(S)$ is the total length of the bridges in S, then the total length of the returned solution is at most

$$\|S\| + \beta(S).$$

We remind the reader that in the dynamic programming, we did charge twice for every bridge. As such, the price of the optimal solution computed by the dynamic programming is a bound on the length of the resulting TSP π.

14.3.2. Bounding the length of the solution. We are left with the task of bounding the length of the solution returned by the dynamic programming. The idea is to take the optimal solution and modify it into a (slightly longer) solution that is considered by the dynamic programming. Clearly, the solution returned by the dynamic programming is not going to be longer than this modified solution, as the dynamic programming returns the shortest solution it considered.

So, let π_{opt} be the optimal solution. Conceptually, we now run the recursive algorithm on π_{opt} to compute the recursive binary partition using cuts (and their bridges). Since we are now working with the real solution (conceptually!), there is no guessing involved. Clearly, we can find a recursive partition of the solution. Modeled as a tree, in this recursive partition, every node is a minimal rectangle of π_{opt}, and there is a feasible or free cut associated with it that created its children, and its two children are the minimal rectangles of the two split parts.

As such, by Lemma 14.7 the total length of the bridges introduces is at most $2 \left\|\pi_{opt}\right\| / t$. Let S be the resulting set of segments from π_{opt} after introducing the bridges.

Clearly, all the segments of π_{opt} are from the set of segments induced by pairs of points of P. As such, the dynamic programming (i.e., recursive algorithm) would consider S as one possible solution (paying twice for the bridges). We conclude that the algorithm would output a TSP of length $\leq \left\|\pi_{opt}\right\| + 2 \left\|\pi_{opt}\right\| / t$.

14.4. The result

For $\varepsilon > 0$, by setting $t = 2/\varepsilon$, the following result is implied.

Theorem 14.10. *For $\varepsilon > 0$ and a set P of n points in the plane, one can compute in $n^{O(1+1/\varepsilon)}$ time a tour visiting all the points of P of length at most $(1 + \varepsilon) \left\|\pi_{opt}\right\|$, where π_{opt} is the shortest tour visiting all the points of P.*

14.5. Bibliographical notes

The approximation algorithm described in this chapter is due to Mitchell [**Mit99**]. Arora's technique (that was described in the previous chapter) is the other approach to solving this problem.

Historically, Arora's technique predates Mitchell's work by a few months/weeks. It is my (not completely substantiated) belief that the Mitchell technique is under-utilized and some interesting results would still follow from it (and that was one motivation to present it here). For example, the PTAS for orienteering in the plane uses the Mitchell technique and rose out of the failure of applying the Arora technique to this case and trying the Mitchell technique in due process [**CH08**].

In fact, the basic insight of Mitchell that one does not have to introduce error unless needed can be incorporated into Arora's technique. This was done for example in the higher-dimensional algorithm for orienteering [**CH08**].

CHAPTER 15

Linear Programming in Low Dimensions

In this chapter, we shortly describe (and analyze) a simple randomized algorithm for linear programming in low dimensions. Next, we show how to extend this algorithm to solve linear programming with violations. Finally, we will show how one can efficiently approximate the number constraints that one needs to violate to make a linear program feasible. This serves as a fruitful ground to demonstrate some of the techniques we visited already.

Our discussion is going to be somewhat intuitive. We will fill in the details and prove the correctness of our algorithms formally in the next chapter.

15.1. Linear programming

Assume we are given a set of n linear inequalities of the form $a_1x_1 + \cdots + a_d x_d \le b$, where $a_1, \ldots, a_d, b$ are constants and $x_1, \ldots, x_d$ are the variables. In the ***linear programming*** (LP) problem, one has to find a ***feasible solution***, that is, a point $(x_1, \ldots, x_d)$ for which all the linear inequalities hold. In the following, we use the shorthand ***LPI*** to stand for ***linear programming instance***. Usually we would like to find a feasible point that maximizes a linear expression (referred to as the ***target function*** of the given LPI) of the form $c_1x_1 + \cdots + c_d x_d$, where $c_1, \ldots, c_d$ are prespecified constants.

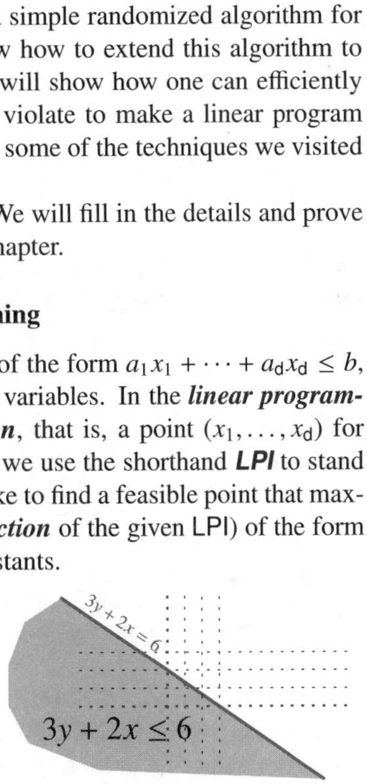

The set of points complying with a linear inequality $a_1x_1 + \cdots + a_d x_d \le b$ is a halfspace of $\mathbb{R}^d$ having the hyperplane $a_1x_1 + \cdots + a_d x_d = b$ as a boundary; see the figure on the right. As such, the feasible region of a LPI is the intersection of n halfspaces; that is, it is a ***polyhedron***. If the polyhedron is bounded, then it is a ***polytope***. The linear target function is no more than specifying a direction, such that we need to find the point inside the polyhedron which is extreme in this direction. If the polyhedron is unbounded in this direction, the optimal solution is ***unbounded***.

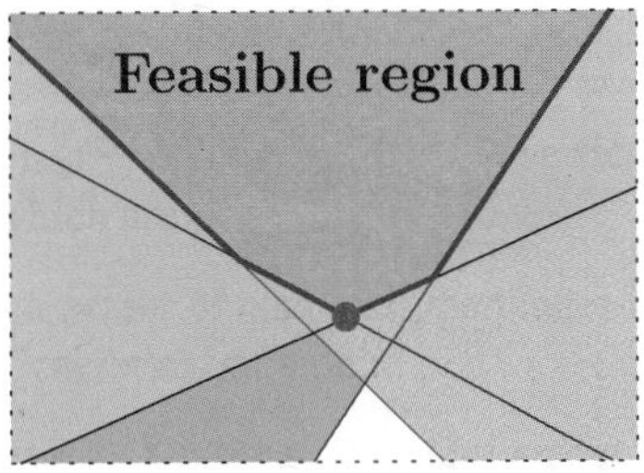

For the sake of simplicity of exposition, it will be easier to think of the direction for which one has to optimize as the negative x_d-axis direction. This can be easily realized by rotating the space such that the required direction is pointing downward. Since the feasible region is the intersection of convex sets (i.e., halfspaces), it is convex. As such, one can imagine the boundary of the feasible region as a vessel (with a convex interior). Next, we release a ball at the top of the vessel, and the ball rolls down (by "gravity" in the direction of the negative x_d-axis) till it reaches the lowest point in the vessel and gets "stuck". This point is the optimal solution to the LPI that we are interested in computing.

In the following, we will assume that the given LPI is in general position. Namely, if we intersect k hyperplanes, induced by k inequalities in the given LPI (the hyperplanes are the result of taking each of this inequalities as an equality), then their intersection is a $(d - k)$-dimensional affine subspace. In particular, the intersection of d of them is a point (referred to as a ***vertex***). Similarly, the intersection of any $d + 1$ of them is empty.

A polyhedron defined by an LPI with n constraints might have $O\left(n^{\lfloor d/2 \rfloor}\right)$ vertices on its boundary (this is known as the upper-bound theorem [**Grü03**]). As we argue below, the optimal solution is a vertex. As such, a naive algorithm would enumerate all relevant vertices (this is a non-trivial undertaking) and return the best possible vertex. Surprisingly, in low dimension, one can do much better and get an algorithm with linear running time.

We are interested in the best vertex of the feasible region, while this polyhedron is defined implicitly as the intersection of halfspaces, and this hints to the quandary that we are in: We are looking for an optimal vertex in a large graph that is defined implicitly. Intuitively, this is why proving the correctness of the algorithms we present here is a non-trivial undertaking (as already mentioned, we will prove correctness in the next chapter).

15.1.1. A solution and how to verify it. Observe that an optimal solution of an LPI is either a vertex or unbounded. Indeed, if the optimal solution p lies in the middle of a segment s, such that s is feasible, then either one of its endpoints provides a better solution (i.e., one of them is lower in the x_d-direction than p) or both endpoints of s have the same target value. But then, we can move the solution to one of the endpoints of s. In particular, if the solution lies on a k-dimensional facet F of the boundary of the feasible polyhedron (i.e., formally F is a set with affine dimension k formed by the intersection of the boundary of the polyhedron with a hyperplane), we can move it so that it lies on a $(k - 1)$-dimensional facet F' of the feasible polyhedron, using the proceeding argumentation. Using it repeatedly, one ends up in a vertex of the polyhedron or in an unbounded solution.

Thus, given an instance of LPI, the LP solver should output one of the following answers.

(A) **Finite.** The optimal solution is finite, and the solver would provide a vertex which realizes the optimal solution.
(B) **Unbounded.** The given LPI has an unbounded solution. In this case, the LP solver would output a ray ζ, such that the ζ lies inside the feasible region and it points down the negative x_d-axis direction.
(C) **Infeasible.** The given LPI does not have any point which complies with all the given inequalities. In this case the solver would output $d + 1$ constraints which are infeasible on their own.

Lemma 15.1. *Given a set of* d *linear inequalities in* $\mathbb{R}^d$, *one can compute the vertex induced by the intersection of their boundaries in* $O\left(d^3\right)$ *time.*

Proof. Write down the system of equalities that the vertex must fulfill. It is a system of d equalities in d variables and it can be solved in $O\left(d^3\right)$ time using Gaussian elimination. ∎

A ***cone*** is the intersection of d constraints, where its apex is the vertex associated with this set of constraints. A set of such d constraints is a ***basis***. An intersection of $d - 1$ of the hyperplanes of a basis forms a line and intersecting this line with the cone of the basis forms a ray. Clipping the same line to the feasible region would yield either a segment, referred to as an ***edge*** of the polyhedron, or a ray (if the feasible region is an unbounded polyhedron). An edge of the polyhedron connects two vertices of the polyhedron.

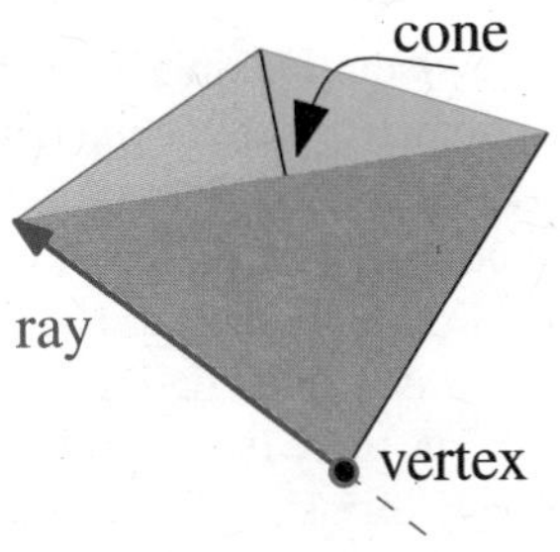

As such, one can think about the boundary of the feasible region as inducing a graph – its vertices and edges are the vertices and edges of the polyhedron, respectively. Since every vertex has d hyperplanes defining it (its basis) and an adjacent edge is defined by d − 1 of these hyperplanes, it follows that each vertex has $\binom{d}{d-1} = d$ edges adjacent to it.

The following lemma tells us when we have an optimal vertex. While it is intuitively clear, its proof requires a systematic understanding of what the feasible region of a linear program looks like, and we delegate it to the next chapter.

Lemma 15.2. *Let L be a given LPI, and let $\mathcal{P}$ denote its feasible region. Let v be a vertex of $\mathcal{P}$, such that all the d rays emanating from v are in the upward x_d-axis direction (i.e., the direction vectors of all these d rays have positive x_d-coordinate). Then v is the lowest (in the x_d-axis direction) point in $\mathcal{P}$ and it is thus the optimal solution to L.*

Interestingly, when we are at a vertex v of the feasible region, it is easy to find the adjacent vertices. Indeed, compute the d rays emanating from v. For such a ray, intersect it with all the constraints of the LPI. The closest intersection point along this ray is the vertex u of the feasible region adjacent to v. Doing this naively takes $O\left(dn + d^{O(1)}\right)$ time.

Lemma 15.2 offers a simple algorithm for computing the optimal solution for an LPI. Start from a feasible vertex of the LPI. As long as this vertex has at least one ray that points downward, follow this ray to an adjacent vertex on the feasible polytope that is lower than the current vertex (i.e., compute the d rays emanating from the current vertex, and follow one of the rays that points downward, till you hit a new vertex). Repeat this till the current vertex has all rays pointing upward, by Lemma 15.2 this is the optimal solution. Up to tedious (and non-trivial) details this is the ***simplex*** algorithm.

We need the following lemma, whose proof is also delegated to the next chapter.

Lemma 15.3. *If L is an LPI in d dimensions which is not feasible, then there exist d + 1 inequalities in L which are infeasible on their own.*

Note that given a set of d+1 inequalities, it is easy to verify (in polynomial time in d) if they are feasible or not. Indeed, compute the $\binom{d+1}{d}$ vertices formed by this set of constraints, and check whether any of these vertices are feasible (for these d + 1 constraints). If all of them are infeasible, then this set of constraints is infeasible.

15.2. Low-dimensional linear programming

15.2.1. An algorithm for a restricted case. There are a lot of tedious details that one has to take care of to make things work with linear programming. As such, we will first describe the algorithm for a special case and then provide the envelope required so that one can use it to solve the general case.

A vertex v is ***acceptable*** if all the d rays associated with it point upward (note that the vertex might not be feasible). The optimal solution (if it is finite) must be located at an acceptable vertex.

Input for the restricted case. The input for the restricted case is an LPI L, which is defined by a set of n linear inequalities in $\mathbb{R}^d$, and a basis $B = \{h_1, \ldots, h_d\}$ of an acceptable vertex.

Let $h_{d+1}, \ldots, h_m$ be a random permutation of the remaining constraints of the LPI L.

We are looking for the lowest point in $\mathbb{R}^{\mathsf{d}}$ which is feasible for L. Our algorithm is randomized incremental. At the ith step, for $i > \mathsf{d}$, it will maintain the optimal solution for the first i constraints. As such, in the ith step, the algorithm checks whether the optimal solution v_{i-1} of the previous iteration is still feasible with the new constraint h_i (namely, the algorithm checks if v_{i-1} is inside the halfspace defined by h_i). If v_{i-1} is still feasible, then it is still the optimal solution, and we set $\mathsf{v}_i \leftarrow \mathsf{v}_{i-1}$.

The more interesting case is when $\mathsf{v}_{i-1} \notin h_i$. First, we check if the basis of v_{i-1} together with h_i forms a set of constraints which is infeasible. If so, the given LPI is infeasible, and we output $B(\mathsf{v}_{i-1}) \cup \{h_i\}$ as the proof of infeasibility.

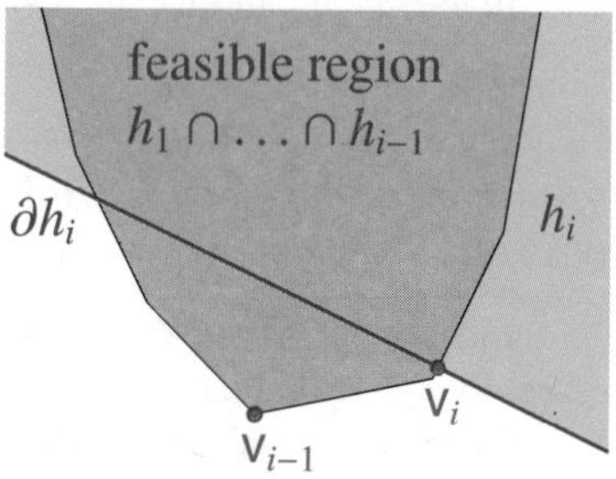

Otherwise, the new optimal solution must lie on the hyperplane associated with h_i. As such, we recursively compute the lowest vertex in the $(\mathsf{d}-1)$-dimensional polyhedron $(\partial h_i) \cap \bigcap_{j=1}^{i-1} h_j$, where ∂h_i denotes the hyperplane which is the boundary of the halfspace h_i. This is a linear program involving $i-1$ constraints, and it involves $\mathsf{d}-1$ variables since the LPI lies on the $(\mathsf{d}-1)$-dimensional hyperplane ∂h_i. The solution found, v_i, is defined by a basis of $\mathsf{d}-1$ constraints in the $(\mathsf{d}-1)$-dimensional subspace ∂h_i, and adding h_i to it results in an acceptable vertex that is feasible in the original d-dimensional space. We continue to the next iteration.

Clearly, the vertex v_n is the required optimal solution.

15.2.1.1. *Running time analysis.* Every set of d constraints is feasible and computing the vertex formed by this constraint takes $O(\mathsf{d}^3)$ time, by Lemma 15.1.

Let X_i be an indicator variable that is 1 if and only if the vertex v_i is recomputed in the ith iteration (by performing a recursive call). This happens only if h_i is one of the d constraints in the basis of v_i. Since there are most d constraints that define the basis and there are at least $i-\mathsf{d}$ constraints that are being randomly ordered (as the first d slots are fixed), we have that the probability that $\mathsf{v}_i \neq \mathsf{v}_{i-1}$ is

$$\alpha_i = \mathbf{Pr}[X_i = 1] \le \min\left(\frac{\mathsf{d}}{i-\mathsf{d}}, 1\right) \le \frac{2\mathsf{d}}{i},$$

for $i \ge \mathsf{d}+1$, as can be easily verified.[①] So, let $T(m,\mathsf{d})$ be the expected time to solve an LPI with m constraints in d dimensions. We have that $T(\mathsf{d},\mathsf{d}) = O(\mathsf{d}^3)$ by the above. Now, in every iteration, we need to check if the current solution lies inside the new constraint, which takes $O(\mathsf{d})$ time per iteration and $O(\mathsf{d}m)$ time overall.

Now, if $X_i = 1$, then we need to update each of the $i-1$ constraints to lie on the hyperplane h_i. The hyperplane h_i defines a linear equality, which we can use to eliminate one of the variables. This takes $O(\mathsf{d}i)$ time, and we have to do the recursive call. The probability that this happens is α_i. As such, we have

$$\begin{aligned} T(m,\mathsf{d}) &= \mathbf{E}\left[O(m\mathsf{d}) + \sum_{i=\mathsf{d}+1}^{m} X_i \left(\mathsf{d}i + T(i-1,\mathsf{d}-1)\right)\right] \\ &= O(m\mathsf{d}) + \sum_{i=\mathsf{d}+1}^{m} \alpha_i(\mathsf{d}i + T(i-1,\mathsf{d}-1)) \end{aligned}$$

---

①Indeed, $\frac{(\mathsf{d})+\mathsf{d}}{(i-\mathsf{d})+\mathsf{d}}$ lies between $\frac{\mathsf{d}}{i-\mathsf{d}}$ and $\frac{\mathsf{d}}{\mathsf{d}} = 1$.

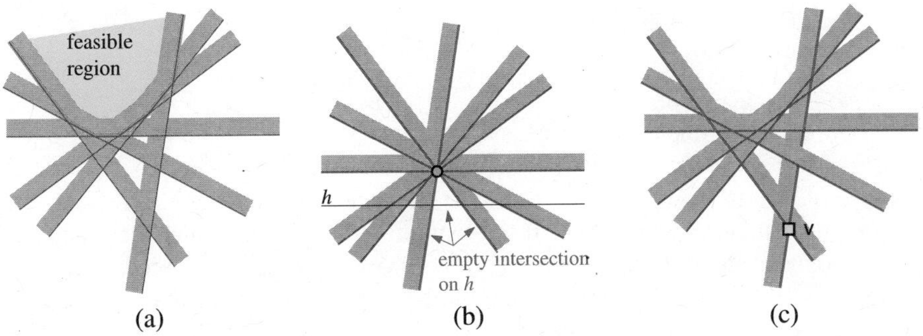

FIGURE 15.1. Demonstrating the algorithm for the general case: (a) given constraints and feasible region, (b) constraints moved to pass through the origin, and (c) the resulting acceptable vertex v.

$$= O(m\mathsf{d}) + \sum_{i=\mathsf{d}+1}^{m} \frac{2\mathsf{d}}{i}(\mathsf{d}i + T(i-1,\mathsf{d}-1))$$

$$= O\left(m\mathsf{d}^2\right) + \sum_{i=\mathsf{d}+1}^{m} \frac{2\mathsf{d}}{i} T(i-1,\mathsf{d}-1).$$

Guessing that $T(m,\mathsf{d}) \le c_{\mathsf{d}} m$, we have that

$$T(m,\mathsf{d}) \le \widehat{c_1} m\mathsf{d}^2 + \sum_{i=\mathsf{d}+1}^{m} \frac{2\mathsf{d}}{i} c_{\mathsf{d}-1}(i-1) \le \widehat{c_1} m\mathsf{d}^2 + \sum_{i=\mathsf{d}+1}^{m} 2\mathsf{d}c_{\mathsf{d}-1} = \left(\widehat{c_1}\mathsf{d}^2 + 2\mathsf{d}c_{\mathsf{d}-1}\right)m,$$

where $\widehat{c_1}$ is some absolute constant. We need that $\widehat{c_1}\mathsf{d}^2 + 2c_{\mathsf{d}-1}\mathsf{d} \le c_{\mathsf{d}}$, which holds for $c_{\mathsf{d}} = O\left((3d)^{\mathsf{d}}\right)$ and $T(m,\mathsf{d}) = O\left((3d)^{\mathsf{d}}\, m\right)$.

Lemma 15.4. *Given an LPI with n constraints in d dimensions and an acceptable vertex for this LPI, then can compute the optimal solution in expected $O\left((3d)^{\mathsf{d}}\, n\right)$ time.*

15.2.2. The algorithm for the general case. Let L be the given LPI, and let L' be the instance formed by translating all the constraints so that they pass through the origin. Next, let h be the hyperplane $x_{\mathsf{d}} = -1$. Consider a solution to the LP L' when restricted to h. This is a $(\mathsf{d}-1)$-dimensional instance of linear programming, and it can be solved recursively.

If the recursive call on $L' \cap h$ returned no solution, then the d constraints that prove that the LP L' is infeasible on h corresponds to a basis in L of a vertex v which is acceptable in the original LPI. Indeed, as we move these d constraints to the origin, their intersection on h is empty (i.e., the "quadrant" that their intersection forms is unbounded only in the upward direction). As such, we can now apply the algorithm of Lemma 15.4 to solve the given LPI. See Figure 15.1.

If there is a solution to $L' \cap h$, then it is a vertex v on h which is feasible. Thus, consider the original set of $\mathsf{d}-1$ constraints in L that corresponds to the basis B of v. Let ℓ be the line formed by the intersection of the hyperplanes of B. It is now easy to verify that the intersection of the feasible region with this line is an unbounded ray, and the algorithm returns this unbounded (downward oriented) ray, as a proof that the LPI is unbounded.

Theorem 15.5. *Given an LP instance with n constraints defined over d variables, it can be solved in expected $O\left((3\mathsf{d})^{\mathsf{d}} n\right)$ time.*

PROOF. The expected running time is

$$S(n, \mathsf{d}) = O(n\mathsf{d}) + S(n, \mathsf{d} - 1) + T(m, \mathsf{d}),$$

where $T(m, \mathsf{d})$ is the time to solve an LP in the restricted case of Section 15.2.1. Indeed, we first solve the problem on the $(\mathsf{d} - 1)$-dimensional subspace $h \equiv x_{\mathsf{d}} = -1$. This takes $O(\mathsf{d}n) + S(n, \mathsf{d} - 1)$ time (we need to rewrite the constraints for the lower-dimensional instance, and that takes $O(\mathsf{d}n)$ time). If the solution on h is feasible, then the original LPI has an unbounded solution, and we return it. Otherwise, we obtained an acceptable vertex, and we can use the special case algorithm on the original LPI. Now, the solution to this recurrence is $O\left((3\mathsf{d})^{\mathsf{d}} n\right)$; see Lemma 15.4. ■

15.3. Linear programming with violations

The problem. Let L be a linear program with d variables and n constraints, and let $k > 0$ be a parameter. We are interested in the optimal solution of L where we are allowed to throw away k constraints (i.e., the solution computed would ***violate*** these k constraints). A naive solution would be to try to throw away all possible subsets of k constraints, solve each one of these instances, and return the best solution found. This would require $O\left(n^{k+1}\right)$ time. Luckily, it turns out that one can do much better if the dimension is small enough.

Note that the given LPI might not be feasible but the solution violating k constraints might be unbounded. For the sake of simplicity of exposition, we will assume here that the optimal solution violating k constraints is bounded (i.e., it is a vertex).

Algorithm. We are given an LPI L with n constraints in d dimensions, a parameter k, and a confidence parameter $\varphi > 0$. The algorithm repeats the following $M = O(k^{\mathsf{d}} \log 1/\varphi)$ times:

(A) Pick, with a certain probability to be specified shortly, each constraint of L be in a random sample R.
(B) Compute the vertex u realizing the optimal solution for R if such a vertex exists.
(C) Count the number of constraints of L that u violates. If this number is at most k, the vertex u is a candidate for the solution.

Finally, the algorithm returns the best candidate vertex computed by step (C) above.

Analysis. The vertex realizing the optimal k-violated solution is a vertex v defined by d constraints (let B denote its basis) and is of depth k. We remind the reader that a point p has ***depth*** k (in L) if it is outside k halfspaces of L (namely, we complement each constraint of L, and p is contained inside k of these complemented hyperplanes). We denote the depth of p by $\mathtt{depth}(\mathsf{p})$. As such, we can use the depth estimation technique we encountered before; see Chapter 9. Specifically, if we pick each constraint of L to be in a new instance of LP with probability $1/k$, then the probability that the new instance R would have all the elements of B in its random sample and would not contain any of the k constraints opposing v is

$$\alpha = \left(\frac{1}{k}\right)^{|B|} \left(1 - \frac{1}{k}\right)^{\mathtt{depth}(\mathsf{v})} \geq \frac{1}{k^{\mathsf{d}}}\left(1 - \frac{1}{k}\right)^{k} \geq \frac{1}{k^{\mathsf{d}}} \exp\left(-\frac{2}{k}k\right) \geq \frac{1}{8k^{\mathsf{d}}},$$

since $1 - x \geq e^{-2x}$, for $0 < x < 1/2$. If this happens, then the optimal solution for R is v, since the basis of v is in R so the solution cannot be any lower and no constraint that violates v is in R.

Next, the algorithm amplified the probability of success by repeating this process $M = 8k^{\mathsf{d}} \ln(1/\varphi)$ times, returning the best solution found. The probability that in all these (independent) iterations we had failed to generate the optimal (violated) solution is at most

$$(1-\alpha)^M \leq \left(1 - \frac{1}{8k^{\mathsf{d}}}\right)^M \leq \exp\left(-\frac{M}{8k^{\mathsf{d}}}\right) = \exp\left(-\ln\left(\frac{1}{\varphi}\right)\right) = \varphi.$$

Clearly, each iteration of the algorithm takes linear time.

Theorem 15.6. *Let L be a linear programming instance with n constraints over d variables, let $k > 0$ be a parameter, and let $\varphi > 0$ be a confidence parameter. Then one can compute the optimal solution to L violating at most k constraints of L, in $O\left(nk^{\mathsf{d}} \log(1/\varphi)\right)$ expected time. The solution returned is correct with probability $\geq 1 - \varphi$.*

15.4. Approximate linear programming with violations

The magic of Theorem 15.6 is that it provides us with a linear programming solver which is robust and can handle a small number of factious constraints. But what happens if the number of violated constraints k is large?[②] As a concrete example, for $k = \sqrt{n}$ and an LPI with n constraints (defined over d variables) the algorithm for computing the optimal solution violating k constraints has running time roughly $O(n^{1+d/2})$. In this case, if one still wants a near linear running time, one can use random sampling to get an approximate solution in near linear time.

Lemma 15.7. *Let L be a linear program with n constraints over d variables, let $k > 0$ and $\varepsilon > 0$ be parameters. Then one can compute a solution to L violating at most $(1+\varepsilon)k$ constraints of L such that its value is better than the optimal solution violating k constraints of L. The expected running time of the algorithm is*

$$O\left(n + n \min\left(\frac{\log^{\mathsf{d}+1} n}{\varepsilon^{2\mathsf{d}}}, \frac{\log^{\mathsf{d}+2} n}{k\,\varepsilon^{2\mathsf{d}+2}}\right)\right).$$

The algorithm succeeds with high probability.

Proof. Let $\rho = O\left(\frac{\mathsf{d}}{k\varepsilon^2} \ln n\right)$ and pick each constraint of L to be in L' with probability ρ. Next, the algorithm computes the optimal solution u in L' violating at most

$$k' = (1 + \varepsilon/3)\rho k$$

constraints and returns this as the required solution.

We need to prove the correctness of this algorithm. To this end, the reliable sampling lemma (Lemma 10.8$_{\text{p149}}$) states that any vertex v of depth u in L has depth in the range

$$\left[(1-\varepsilon/3)u\rho,\ (1+\varepsilon/3)u\rho\right]$$

in L', and this holds with high probability, where $u \geq k$. Specifically, this holds with probability $\geq 1 - 1/n^{O(\mathsf{d})}$. We need this to hold for all the vertices in the original arrangement, and there are $\binom{n}{\mathsf{d}} = O(n^{\mathsf{d}})$ such vertices. Thus, this property holds for all the vertices with high probability.

In particular, let v_{opt} be the optimal solution for L of depth k. With high probability, v_{opt} has depth $\leq (1+\varepsilon/3)\rho k = k'$ in L', which implies that there is at least one vertex of depth $\leq k'$ in L'.

[②] I am sure the reader guessed correctly the consequences of such a despicable scenario: The universe collapses and is replaced by a cucumber.

Now, we argue that any vertex of depth $\leq k'$ in L', with high probability, is a valid approximate solution. So, assume that we have a vertex v of depth β in L and its depth in L' is γ, where $\gamma \leq k'$.

By the reliable sampling lemma, we have that the depth γ of v in L' is in the range

$$\left[(1-\varepsilon/3)\beta\rho, (1+\varepsilon/3)\beta\rho\right].$$

This implies that $(1-\varepsilon/3)\beta\rho \leq k'$. Now, $k' = (1+\varepsilon/3)\rho k$, which implies that

$$(1-\varepsilon/3)\beta\rho \leq (1+\varepsilon/3)\rho k \implies \beta \leq \frac{1+\varepsilon/3}{1-\varepsilon/3}k \leq (1+\varepsilon/3)(1+\varepsilon/2)k \leq (1+\varepsilon)k,$$

since $1/(1-\varepsilon/3) \leq 1+\varepsilon/2$ for $\varepsilon \leq 1$[3].

As for the running time, we are using the algorithm of Theorem 15.6, with $\varphi = 1/n^{O(\mathsf{d})}$. The input size is $O(\min(n, n\rho))$ and the depth threshold is k'. (The bound on the input size holds with high probability. We omit the easy but tedious proof of that using Chernoff's inequality.) As such, ignoring constants in the $O(\cdot)$ that depends exponentially on d, the running time is

$$O\Big(n + \min(n, n\rho)(k')^{\mathsf{d}} \log n\Big) = O\Big(n + \min\big(n, n\rho\big)(\rho k)^{\mathsf{d}} \log n\Big)$$
$$= O\left(n + n\min\left(\frac{\log^{\mathsf{d}+1} n}{\varepsilon^{2\mathsf{d}}}, \frac{\log^{\mathsf{d}+2} n}{k\,\varepsilon^{2\mathsf{d}+2}}\right)\right). \quad \blacksquare$$

Note that the running time of Lemma 15.7 is linear if k is sufficiently large and ε is fixed.

15.5. LP-type problems

Interestingly, the algorithm presented for linear programming can be extended to more abstract settings. Indeed, assume we are given a set of constraints $\mathcal{H}$ and a function w, such that for any subset $G \subset \mathcal{H}$ it returns the value of the optimal solution of the constraint problem when restricted to G. We denote this value by $w(G)$. Our purpose is to compute $w(\mathcal{H})$.

For example, $\mathcal{H}$ is a set of points in $\mathbb{R}^{\mathsf{d}}$, and $w(G)$ is the radius of the smallest ball containing all the points of $G \subseteq \mathcal{H}$. As such, in this case, we would like to compute (the radius of) the smallest enclosing ball for $\mathcal{H}$.

We assume that the following axioms hold:

(1) (**Monotonicity**) For any $F \subseteq G \subseteq \mathcal{H}$, we have

$$w(F) \leq w(G).$$

(2) (**Locality**) For any $F \subseteq G \subseteq \mathcal{H}$, with $-\infty < w(F) = w(G)$, and any $h \in \mathcal{H}$, if

$$w(G) < w(G \cup \{h\}), \quad \text{then} \quad w(F) < w(F \cup \{h\}).$$

If these two axioms hold, we refer to $(\mathcal{H}, w)$ as an ***LP-type*** problem. It is easy to verify that linear programming is an LP-type problem.

Definition 15.8. A ***basis*** is a subset $B \subseteq \mathcal{H}$ such that $w(B) > -\infty$ and $w(B') < w(B)$, for any proper subset B' of B.

As in linear programming, we have to assume that certain ***basic operations*** can be performed quickly:

---

[3]Indeed $(1-\varepsilon/3)(1+\varepsilon/2) = 1 - \varepsilon/3 + \varepsilon/2 - \varepsilon^2/6 \geq 1$.

```
solveLPType(B_0, C)
    // B_0: initial basis, C: set of constraints
    ⟨c'_1, ..., c'_n⟩: random permutation of C \ B_0.
    for i = 1 to n do
        if compTarget(B_{i-1} ∪ {c'_i}) > compTarget(B_{i-1}) then
            T ← compBasis(B_{i-1} ∪ {c'_i})
            B_i ← solveLPType(T, B_0 ∪ {c'_1, ..., c'_i})
        else
            B_i ← B_{i-1}
    return B_n
```

FIGURE 15.2. The algorithm for solving LP-type problems.

(A) (**Violation test**) For a constraint h and a basis B, test whether h is violated by B or not. Namely, test if $w(B \cup \{h\}) > w(B)$. Let **compTarget** $(B \cup \{h\})$ be the provided function that returns $w(B \cup \{h\})$.
(B) (**Basis computation**) For a constraint h and a basis B, compute the basis of $B \cup \{h\}$. Let **compBasis** $(B \cup \{h\})$ be the procedure computing this basis.

We also need to assume (similar to the special case) that we are given an initial basis B_0 from which to start our computation. The ***combinatorial dimension*** of $(\mathcal{H}, w)$ is the maximum size of a basis of $\mathcal{H}$. A variant of the algorithm we presented for linear programming (the special case of Section 15.2.1) works in these settings. Indeed, start with B_0, and add the remaining constraints in a random order. At each step check if the new constraint violates the current solution, and if so, update the basis by performing a recursive call. Intuitively, the recursive call corresponds to solving a subproblem where some members of the basis are fixed. The algorithm is described in Figure 15.2.

15.5.1. Analysis of the algorithm.

Lemma 15.9. *The algorithm* **solveLPType** (B_0, C) *terminates.*

PROOF. Observe that every time that **solveLPType** calls recursively, it is done with a new initial basis T such that $w(T) > w(B_i) \geq w(B_0)$. Furthermore, the recursive call always returns a basis with value $\geq w(T)$. As such, the depth of the recursion is finite, and the algorithm terminates. ■

Lemma 15.10. *In the end of the ith iteration of the loop of* **solveLPType**(B_0, C), *we have that* $w\left(B_0 \cup \left\{c'_1, \dots c'_i\right\}\right) = w(B_i)$, *assuming that all calls to* **solveLPType** *with an initial basis T such that $w(T) > w(B_0)$ are successful.*

PROOF. The claim trivially holds for $i = 0$. So assume the claim holds for $i \leq k$ and $i = k + 1$.

If $w\left(B_0 \cup \left\{c'_0, \dots, c'_i\right\}\right) > w\left(B_0 \cup \left\{c'_0, \dots, c'_{i-1}\right\}\right)$, then, by the locality axiom, we have $w\left(B_{i-1} \cup \left\{c'_i\right\}\right) > w(B_{i-1})$. This implies that c'_i violates B_{i-1}. The algorithm then recursively computes the basis T, which has $w(T) > w(B_{i-1}) \geq w(B_0)$. The algorithm then calls recursively on the set $B_0 \cup \left\{c'_0, \dots, c'_i\right\}$, with an initial basis that has a higher value (i.e., there is progress being made before issuing this recursive call). By assumption, we have that the returned basis B_i is the desired basis of $B_0 \cup \left\{c'_0, \dots, c'_i\right\}$.

If $w\left(B_{i-1} \cup c'_i\right) = w(B_{i-1})$, then, by locality, we have that

$$w\left(B_0 \cup \left\{c'_0, \ldots, c'_i\right\}\right) = w\left(B_0 \cup \left\{c'_0, \ldots, c'_{i-1}\right\}\right) = w(B_{i-1}) = w(B_i),$$

as $B_i = B_{i-1}$, which implies the claim. ∎

By arguing inductively on the value of the initial basis and using the above claim, we get the following.

Lemma 15.11. *The function* **solveLPType**(B_0, C) *returns a basis* $B \subseteq B_0 \cup C$ *such that* $w(B) = w(B_0 \cup C)$.

Lemma 15.12. *If the combinatorial dimension of* $(\mathcal{H}, w)$ *is* d, *then depth of the recursion of* **solveLPType**(B_0, C) *is bounded by* d.

Proof. To bound the depth of the recursion, we argue about the decreasing dimension of the recursive calls being made. Observe that if a constraint c'_i violates the current basis B_{i-1}, then any subset $X \subseteq B_0 \cup \left\{c'_1, \ldots c'_i\right\}$ that has $w(X) > w(B_{i-1})$ must have that $c'_i \in X$. Indeed, otherwise $X \subseteq Y_{i-1} = B_0 \cup \left\{c'_1, \ldots c'_{i-1}\right\}$ and then $w(X) \leq w(Y_{i-1}) = w(B_{i-1})$. As such, if a recursive call is of depth k, then k elements of the basis returned by this recursive call are fixed in advance; that is, there are k constraints in the initial basis (provided to this recursive call) that must appear in any basis output by this call. In particular, since no basis has size larger than d, it follows that the depth of the recursion is at most d. ∎

Theorem 15.13. *Let* $(\mathcal{H}, w)$ *be an instance of an* LP*-type problem with* n *constraints and with combinatorial dimension* d. *Assuming that the basic operations take constant time, we have that* $(\mathcal{H}, w)$ *can be solved by* **solveLPType** *using* $\mathsf{d}^{O(\mathsf{d})}n$ *basic operations (in expectation).*

Proof. If the number of constraints $n = O(\mathsf{d})$, then since the depth of the recursion is at most d by Lemma 15.12, the running time is $\mathsf{d}^{O(\mathsf{d})}$.

The analysis is now similar to the linear programming case. Indeed, the probability that in the ith iteration we need to perform a recursive call is bounded by $\min(1, \mathsf{d}/i)$. As such, if $T(n, i)$ is the expected running time, when the depth of the recursion is i and there are n constraints, then we have the recurrence

$$T(n, i) = O(n) + \sum_{k=1}^{n} \min\left(1, \frac{\mathsf{d}}{k}\right) T(k, i+1).$$

It is now easy to verify that $T(n, 0) = \mathsf{d}^{O(\mathsf{d})}n$, as the depth of recursion is bounded by d. ∎

15.5.2. Examples for LP-type problems.

Smallest enclosing ball. Given a set P of n points in $\mathbb{R}^{\mathsf{d}}$, let $r(\mathsf{P})$ denote the radius of the smallest enclosing ball in $\mathbb{R}^{\mathsf{d}}$. Under general position assumptions, there are at most $\mathsf{d}+1$ points on the boundary of this smallest enclosing ball. We claim that this problem is an LP-type problem. Indeed, the basis in this case is the set of points determining the smallest enclosing ball. The combinatorial dimension is thus $\mathsf{d}+1$. The monotonicity property holds trivially. As for the locality property, assume that we have a set $\mathsf{Q} \subseteq \mathsf{P}$ such that $r(\mathsf{Q}) = r(\mathsf{P})$. As such, P and Q have the same enclosing ball. Now, if we add a point p to Q and the radius of its minimum enclosing ball increases, then the ball enclosing P must also change (and get bigger) when we insert p into P. Thus, this is a LP-type problem, and it can be solved in linear time.

Theorem 15.14. *Given a set* P *of* n *points in* $\mathbb{R}^{\mathsf{d}}$*, one can compute its smallest enclosing ball in (expected) linear time.*

Computing time of first intersection. Let $C(t)$ be a parameterized convex shape in $\mathbb{R}^{\mathsf{d}}$, such that $C(0)$ is empty and $C(t) \subsetneq C(t')$ if $t < t'$. We are given n such shapes $C_1, \ldots, C_n$, and we would like to decide the minimal t for which they all have a common intersection. Assume that given a point p and such a shape C, we can decide (in constant time) the minimum t for which $\mathsf{p} \in C(t)$. Similarly, given (say) $\mathsf{d} + 1$ of these shapes, we can decide, in constant time, the minimum t for which they intersect and this common point of intersection. We would like to find the minimum t for which they all intersect. Let us also assume that these shapes are well behaved in the sense that, for any t, we have $\lim_{\Delta \to 0} \mathrm{Vol}(C(t + \Delta) \setminus C(t)) = 0$ (namely, such a shape cannot "jump" – it grows continuously). It is easy to verify that this is an LP-type problem, and as such it can be solved in linear time.

Note that this problem is an extension of the previous problem. Indeed, if we place a ball of radius t at each point of P, then the problem of deciding the minimal t when all these growing balls have a non-empty intersection is equivalent to finding the minimum radius ball enclosing all points.

15.6. Bibliographical notes

History. Linear programming has a fascinating history. It can be traced back to the early 20th century. It started in earnest in 1939 when L. V. Kantorovich noticed the importance of certain types of linear programming problems for economic planning.

Dantzig, in 1947, invented the simplex method for solving LP problems for the US Air Force planning problems. T. C. Koopmans, in 1947, showed that LP provides the right model for the analysis of classical economic theories. In 1975, both Koopmans and Kantorovich got the Nobel prize for economics. Dantzig probably did not receive it because his work was too mathematical. So it goes. Interestingly, Kantorovich is the only Russian who was awarded the Nobel prize in economics during the cold war.

The simplex algorithm was developed before computers (and computer science) really existed in wide usage, and its standard description is via a careful maintenance of a tableau of the LP, which is easy to handle by hand (this might also explain the unfortunate name "linear programming"). As such, the usual description of the simplex algorithm is somewhat mysterious and counterintuitive (at least for the author). Furthermore, since the universe is not in general position (as we assumed), there are numerous technical difficulties (that we glossed over) in implementing any of these algorithms, and the descriptions of the simplex algorithm usually detail how to handle these cases. See the book by Vanderbei **[Van97]** for an accessible description of this topic.

Linear programming in low dimensions. The first to realize that linear programming can be solved in linear time in low dimensions was Megiddo **[Meg83, Meg84]**. His algorithm was deterministic but considerably more complicated than the randomized algorithm we present. Clarkson **[Cla95]** showed how to use randomization to get a simple algorithm for linear programming with running time $O(\mathsf{d}^2 n + \mathit{noise})$, where the noise is a constant exponential in d. Our presentation follows the paper by Seidel **[Sei91]**. Surprisingly, one can achieve running time with the noise being subexponential in d. This follows by plugging the subexponential algorithms of Kalai **[Kal92]** or Matoušek et al. **[MSW96]** into Clarkson's algorithm **[Cla95]**. The resulting algorithm has expected running time

$O\left(\mathsf{d}^2 n + \exp\left(c\sqrt{\mathsf{d}\log \mathsf{d}}\right)\right)$, for some constant c. See the survey by Goldwasser [**Gol95**] for more details.

More information on Clarkson's algorithm. Clarkson's algorithm contains some interesting new ideas. (The algorithm of Matoušek et al. [**MSW96**] is somewhat similar to the algorithm we presented.)

Observe that if the solution for a random sample R is being violated by a set X of constraints, then X must contain (at least) one constraint which is in the basis of the optimal solution. Thus, by picking R to be of size (roughly) $\sqrt{n}$, we know that it is a $1/\sqrt{n}$-net and there would be at most $\sqrt{n}$ constraints violating the solution of R; see Theorem 5.28_{p71}. Thus, repeating this d times, at each stage solving the problem on the collected constraints from the previous iteration, together with the current random sample, results in a set of $O\left(\mathsf{d}\sqrt{n}\right)$ constraints containing the optimal basis. Now solve recursively the linear program on this (greatly reduced) set of constraints. Namely, we spent $O\left(\mathsf{d}^2 n\right)$ time (d times checking if the n constraints violate a given solution), called recursively d times on "small" subproblems of size (roughly) $O\left(\sqrt{n}\right)$, resulting in a fast algorithm.

An alternative algorithm uses the same observation, by using the reweighting technique. Here each constraint is sampled according to its weight (which is initially 1). By doubling the weight of the violated constraints, one can argue that after a small number of iterations, the sample would contain the required basis, while being small. See Chapter 6 for more details.

Clarkson's final algorithm works by combining these two algorithms together.

Linear programming with violations. The algorithm of Section 15.3 seems to be new, although it is implicit in the work of Matoušek [**Mat95b**], which presents a slightly faster deterministic algorithm. The first paper on this problem (in two dimensions) is due to Everett et al. [**ERvK96**]. This was extended by Matoušek to higher dimensions [**Mat95b**]. His algorithm relies on the idea of computing all $O(k^{\mathsf{d}})$ local maximas in the "k-level" explicitly, by traveling between them. This is done by solving linear programming instances which are "similar". As such, these results can be further improved using techniques for dynamic linear programming that allow insertion and deletions of constraints; see the work by Chan [**Cha96**]. Chan [**Cha05**] showed how to further improve these algorithms for dimensions 2, 3, and 4, although these improvements disappear if k is close to linear.

The idea of approximate linear programming with violations is due to Aronov and Har-Peled [**AH08**], and our presentation follows their results. Using more advanced data-structures, these results can be further improved (as far as the polylog noise is concerned); see the work by Afshani and Chan [**AC09**].

LP-type problems. The notion of LP-type algorithms is mentioned in the work of Sharir and Welzl [**SW92**]. They also showed that deciding if a set of (axis parallel) rectangles can be pierced by three points is an LP-type problem (quite surprising as the problem has no convex programming flavor). Our presentation is influenced by the subsequent work by Matoušek et al. [**MSW96**]. (But our analysis is significantly weaker, and it is inspired by the author's laziness.) Our example of computing the first intersection of growing convex sets is motivated by the work of Amenta [**Ame94**] on the connection between LP-type problems and Helly-type theorems.

Intuitively, any lower-dimensional convex programming problem is a natural candidate to be solved using LP-type techniques.

15.7. Exercises

Exercise 15.1 (Blue/red separation)**.** Let R and B be sets of red and blue points, respectively, in $\mathbb{R}^{\mathsf{d}}$. Describe a linear time algorithm that computes the maximum width slab (i.e., the region enclosed by two parallel hyperplanes) that separates R from B.

Exercise 15.2 (Approximate blue/red separation)**.** Let R and B be sets of red and blue points, respectively, in the high-dimensional Euclidean space $\mathbb{R}^{\mathsf{d}}$. Let γ be the width of the maximum width slab separating R from B, and let $\varepsilon > 0$ be a parameter.

Present an algorithm, with running time $O\left(\mathsf{d}n\left(\frac{\operatorname{diam}(B \cup R)}{\gamma}\right)^c\right)$, that computes a slab separating R from B of width $\geq (1 - \varepsilon)\gamma$, where c is some absolute constant independent of d.

(Hint: Maintain two points x and y that are inside the convex hulls of R and B, respectively. Consider the slab defined by hyperplanes perpendicular to the segment xy and that pass through the points x and y. At each step either this slab is "good enough" or, alternatively, it contains a point of $R \cup B$ deep inside the slab. Consider such a bad point, and update either x or y to define a smaller slab. Bound the speed of convergence of this process.)

CHAPTER 16

Polyhedrons, Polytopes, and Linear Programming

In this chapter, we formally investigate what the feasible region of a linear program looks like and establish the correctness of the algorithm we presented for linear programming in Chapter 15. Linear programming is a case where the geometric intuition is quite clear, but crystallizing it into a formal proof requires quite a bit of work. In particular, we prove in this chapter more than we strictly need, since it supports (and may we dare suggest, with due humbleness and respect to the most esteemed reader, that it even expands[①]) our natural intuition.

Underlining our discussion is the dichotomy between the input to the LP, which is a set of halfspaces, and the entities the LP algorithm works with, which are vertices. In particular, we need to establish that arguing about the feasible region of an LP in terms of (the convex hull of) vertices or, alternatively, as the intersection of halfspaces is equivalent.

16.1. Preliminaries

We had already encountered Radon's theorem (see Theorem 5.7$_{\text{p63}}$), which we restate.

Theorem 16.1 (Radon's theorem). *Let* $\mathsf{P} = \{\mathsf{p}_1, \ldots, \mathsf{p}_{\mathsf{d}+2}\}$ *be a set of* $\mathsf{d}+2$ *points in* $\mathbb{R}^{\mathsf{d}}$*. Then, there exist two disjoint subsets* Q *and* R *of* P*, such that* $\mathcal{CH}(\mathsf{Q}) \cap \mathcal{CH}(\mathsf{R}) \neq \emptyset$*.*

Theorem 16.2 (Helly's theorem). *Let* $\mathcal{F}$ *be a set of* n *convex sets in* $\mathbb{R}^{\mathsf{d}}$*. The intersection of all the sets of* $\mathcal{F}$ *is non-empty if and only if any* $\mathsf{d}+1$ *of them has non-empty intersection.*

Proof. This theorem is the "dual" to Radon's theorem.

If the intersection of all sets in $\mathcal{F}$ is non-empty, then any intersection of $\mathsf{d}+1$ of them is non-empty. As for the other direction, assume for the sake of contradiction that $\mathcal{F}$ is the minimal set of convex sets for which the claim fails. Namely, for $m = |\mathcal{F}| > \mathsf{d}+1$, any subset of $m-1$ sets of $\mathcal{F}$ has non-empty intersection, and yet the intersection of all the sets of $\mathcal{F}$ is empty.

As such, for $X \in \mathcal{F}$, let p_X be a point in the intersection of all sets of $\mathcal{F}$ excluding X. Let $\mathsf{P} = \left\{\mathsf{p}_X \,\middle|\, X \in \mathcal{F}\right\}$. Here $|\mathsf{P}| = |\mathcal{F}| > \mathsf{d}+1$. By Radon's theorem, there is a partition of P into two disjoint sets R and Q such that $\mathcal{CH}(\mathsf{R}) \cap \mathcal{CH}(\mathsf{Q}) \neq \emptyset$. Let s be any point inside this non-empty intersection.

Let $U(\mathsf{R}) = \left\{X \,\middle|\, \mathsf{p}_X \in \mathsf{R}\right\}$ and $U(\mathsf{Q}) = \left\{X \,\middle|\, \mathsf{p}_X \in \mathsf{Q}\right\}$ be the two subsets of $\mathcal{F}$ corresponding to R and Q, respectively. By definition, for $X \in U(\mathsf{R})$, we have that

$$\mathsf{p}_X \in \bigcap_{Y \in \mathcal{F}, Y \neq X} Y \subseteq \bigcap_{Y \in \mathcal{F} \setminus U(\mathsf{R})} Y = \bigcap_{Y \in U(\mathsf{Q})} Y,$$

[①] Hopefully the reader is happy we will be less polite to him/her in the rest of the book, since otherwise the text would be insufferably tedious.

since $U(\mathsf{Q}) \cup U(\mathsf{R}) = \mathcal{F}$ and $U(\mathsf{Q}) \cap U(\mathsf{R}) = \emptyset$. As such, $\mathsf{R} \subseteq \bigcap_{Y \in U(\mathsf{Q})} Y$ and $\mathsf{Q} \subseteq \bigcap_{Y \in U(\mathsf{R})} Y$. Now, by the convexity of the sets of $\mathcal{F}$, we have $\mathcal{CH}(\mathsf{R}) \subseteq \bigcap_{Y \in U(\mathsf{Q})} Y$ and $\mathcal{CH}(\mathsf{Q}) \subseteq \bigcap_{Y \in U(\mathsf{R})} Y$. Namely, we have

$$\mathsf{s} \in \mathcal{CH}(\mathsf{R}) \cap \mathcal{CH}(\mathsf{Q}) \subseteq \left(\bigcap_{Y \in U(\mathsf{Q})} Y \right) \cap \left(\bigcap_{Y \in U(\mathsf{R})} Y \right) = \bigcap_{Y \in \mathcal{F}} Y.$$

Namely, the intersection of all the sets of $\mathcal{F}$ is not empty, a contradiction. ■

Theorem 16.3 (Carathéodory's theorem). *Let X be a convex set in $\mathbb{R}^{\mathsf{d}}$, and let p be some point in the interior of X. Then p is a convex combination of $\mathsf{d} + 1$ points of X.*

Proof. Suppose $\mathsf{p} = \sum_{k=1}^{m} \lambda_i x_i$ is a convex combination of $m > \mathsf{d} + 1$ points, where $\{x_1, \ldots, x_m\} \subseteq X$, $\lambda_1, \ldots, \lambda_m > 0$, and $\sum_i \lambda_i = 1$. We will show that p can be rewritten as a convex combination of $m - 1$ of these points, as long as $m > \mathsf{d} + 1$.

So, consider the following system of equations:

$$\sum_{i=1}^{m} \gamma_i x_i = 0 \quad \text{and} \quad \sum_{i=1}^{m} \gamma_i = 0. \tag{16.1}$$

It has $m > \mathsf{d} + 1$ variables (i.e., $\gamma_1, \ldots, \gamma_m$) but only $\mathsf{d} + 1$ equations (the first equation corresponds to d equalities, one for each dimension). As such, there is a non-trivial solution to this system of equations, and we denote it by $\widehat{\gamma}_1, \ldots, \widehat{\gamma}_m$. Since $\widehat{\gamma}_1 + \cdots + \widehat{\gamma}_m = 0$, some of the $\widehat{\gamma}_i$s are strictly positive, and some of them are strictly negative. Let

$$\tau = \min_{j, \widehat{\gamma}_j > 0} \frac{\lambda_j}{\widehat{\gamma}_j} > 0.$$

Also, assume without loss of generality that $\tau = \lambda_1 / \widehat{\gamma}_1$. Let

$$\widetilde{\lambda}_i = \lambda_i - \tau \widehat{\gamma}_i,$$

for $i = 1, \ldots, m$. Then $\widetilde{\lambda}_1 = \lambda_1 - (\lambda_1 / \widehat{\gamma}_1)\, \widehat{\gamma}_1 = 0$. Furthermore, if $\widehat{\gamma}_i < 0$, then $\widetilde{\lambda}_i = \lambda_i - \tau \widehat{\gamma}_i \geq \lambda_i > 0$. Otherwise, if $\widehat{\gamma}_i > 0$, then

$$\widetilde{\lambda}_i = \lambda_i - \left(\min_{j, \widehat{\gamma}_j > 0} \frac{\lambda_j}{\widehat{\gamma}_j} \right) \widehat{\gamma}_i \geq \lambda_i - \frac{\lambda_i}{\widehat{\gamma}_i} \widehat{\gamma}_i \geq 0.$$

So, $\widetilde{\lambda}_1 = 0$ and $\widetilde{\lambda}_2, \ldots, \widetilde{\lambda}_m \geq 0$. Furthermore,

$$\sum_{i=2}^{m} \widetilde{\lambda}_i = \sum_{i=1}^{m} \widetilde{\lambda}_i = \sum_{i=1}^{m} (\lambda_i - \tau \widehat{\gamma}_i) = \sum_{i=1}^{m} \lambda_i - \tau \sum_{i=1}^{m} \widehat{\gamma}_i = \sum_{i=1}^{m} \lambda_i = 1,$$

since $\widehat{\gamma}_1, \ldots, \widehat{\gamma}_m$ is a solution to (16.1). As such, $\mathsf{q} = \sum_{i=2}^{m} \widetilde{\lambda}_i x_i$ is a convex combination of $m - 1$ points of X. Furthermore, since $\widetilde{\lambda}_1 = 0$, we have

$$\begin{aligned} \mathsf{q} &= \sum_{i=2}^{m} \widetilde{\lambda}_i x_i = \sum_{i=1}^{m} \widetilde{\lambda}_i x_i = \sum_{i=1}^{m} (\lambda_i - \tau \widehat{\gamma}_i)\, x_i = \sum_{i=1}^{m} \lambda_i x_i - \tau \sum_{i=1}^{m} \widehat{\gamma}_i x_i \\ &= \sum_{i=1}^{m} \lambda_i x_i - \tau 0 = \sum_{i=1}^{m} \lambda_i x_i = \mathsf{p}, \end{aligned}$$

since (again) $\widehat{\gamma}_1, \ldots, \widehat{\gamma}_m$ is a solution to (16.1). As such, we found a representation of p as a convex combination of $m - 1$ points, and we can continue this process till $m = \mathsf{d} + 1$, establishing the theorem. ■

16.2. Properties of polyhedrons

An ***$\mathcal{H}$-polyhedron*** (or just polyhedron) is the region formed by the intersection of (a finite number of) halfspaces. Namely, it is the feasible region of an LP.

It will be convenient to consider the LP (i.e., the polyhedron) as being specified in matrix form. Namely, we are given a matrix $\mathsf{M} \in \mathbb{R}^{m \times \mathsf{d}}$ with m rows and d columns, an m-dimensional vector b, and a d-dimensional vector c. Our purpose is to find an $x \in \mathbb{R}^{\mathsf{d}}$, such that $\mathsf{M}x \leq \mathsf{b}$, while the target function $\langle \mathsf{c}, x\rangle$ is minimized. We remind the reader that for two vectors $x, y \in \mathbb{R}^{\mathsf{d}}$, we denote the dot product of x and y by $\langle x, y\rangle = \sum_{i=1}^{\mathsf{d}} x_i y_i$. Namely, the ith row of M corresponds to the ith inequality of the LP, that is, the inequality $\langle \mathsf{M}_i, x\rangle \leq \mathsf{b}_i$, for $i = 1, \ldots, m$, where M_i denotes the ith row of M.

In this form it is easy to combine inequalities together: You multiply several rows by positive constants and add them up (you also need to do this for the relevant constants in b) and get a new inequality. Formally, given a row vector $\mathsf{w} \in \mathbb{R}^m$, such that $\mathsf{w} \geq 0$ (i.e., all entries are non-negative), the resulting inequality is $\mathsf{w}\mathsf{M}x \leq \langle \mathsf{w}, \mathsf{b}\rangle$. Note that such an inequality *must* hold inside the feasible region of the original LP.

16.2.1. Fourier-Motzkin elimination. Let $L = (\mathsf{M}, \mathsf{b})$ be an instance of an LP with d variables and m constraints (we care only about feasibility here, so we ignore the target function). Consider the ith variable x_i in L. If it appears only with a positive coefficient in all the inequalities (i.e., the ith column of M has only positive numbers), then we can set the ith variable to be $-\alpha$, where α is a sufficiently large positive number and all the inequalities where it appears would be immediately feasible. The same holds if all such coefficients are negative (except we then set x_i to be a sufficiently large positive number). Thus, consider the case where the variable x_i appears with both negative coefficients and positive coefficients in the LP. Let us inspect two such inequalities, say the kth and lth inequalities, and assume, for the sake of concreteness, that $\mathsf{M}_{ki} > 0$ and $\mathsf{M}_{li} < 0$. Clearly, we can multiply the lth inequality by a positive number and add it to the kth inequality, so that in the resulting inequality the coefficient of x_i is zero. Formally, we combine the inequalities $\mathsf{M}_k x \leq \mathsf{b}_k$ and $\mathsf{M}_l x \leq \mathsf{b}_l$ into the new inequality

$$\mathrm{elim}(L, i, k, l) \equiv \Big(\big(|\mathsf{M}_{ki}|\, \mathsf{M}_l + |\mathsf{M}_{li}|\, \mathsf{M}_k \big) x \;\leq\; |\mathsf{M}_{ki}|\, \mathsf{b}_l + |\mathsf{M}_{li}|\, \mathsf{b}_k \Big)$$

that is defined over $\mathsf{d} - 1$ variables, as x_i in this inequality has coefficient 0.

Let $L' = \mathrm{elim}(L, i)$ denote the resulting linear program, where we copy all inequalities of the original LP where x_i has zero as a coefficient. In addition, we add all the inequalities $\mathrm{elim}(L, i, k, l)$, where M_{ki} and M_{li} are both non-zero and have different signs. Note that L' might have at most $m^2/4$ inequalities, but since x_i is now eliminated (i.e., all of its appearances are with coefficient zero), the LP L' is defined over $\mathsf{d} - 1$ variables.

Lemma 16.4. *Let L be an instance of* LP *with* d *variables and m inequalities. The linear program $L' = \mathrm{elim}(L, i)$ is feasible if and only if L is feasible, for any $i \in \{1, \ldots, \mathsf{d}\}$.*

PROOF. One direction is easy. If L is feasible, then its solution (omitting the ith variable) is feasible for L'.

The other direction becomes clear once we understand what the elimination really does. So, consider two inequalities in L, such that $\mathsf{M}_{ki} < 0$ and $\mathsf{M}_{ji} > 0$. We can rewrite these inequalities such that they become

(A) $$a_0 + \sum_{\tau \neq i} a_\tau x_\tau \leq x_i$$

(B) $$\text{and} \quad x_i \leq b_0 + \sum_{\tau \neq i} b_\tau x_\tau,$$

respectively.

The eliminated inequality described above is no more than the inequality we get by chaining these inequalities together; that is,

$$a_0 + \sum_{\tau \neq i} a_\tau x_\tau \leq x_i \leq b_0 + \sum_{\tau \neq i} b_\tau x_\tau \implies a_0 + \sum_{\tau \neq i} a_\tau x_\tau \leq b_0 + \sum_{\tau \neq i} b_\tau x_\tau$$
$$\iff \sum_{\tau \neq i} (a_\tau - b_\tau) x_\tau \leq (b_0 - a_0).$$

In particular, for a feasible solution to L', all the left sides of inequalities of type (A) must be smaller than (or equal to) all the right sides of the inequalities of type (B), since we combined all such pairs of inequalities into an inequality in L'.

In particular, given a feasible solution, sol, to L', there exists a value x_i^* that is larger than (or equal to) the left side of all the inequalities of type (A) (for the given feasible solution) and smaller than (or equal to) the right sides of all the inequalities of type (B).

Thus, one can extend sol into a solution for L by setting $x_i = x_i^*$. Indeed, when we substitute the values of $x_1, \ldots, x_{i-1}, x_{i+1}, \ldots, x_{\mathsf{d}}$ in sol into L, the set of all pairs of inequalities of type (A) and type (B) (which are all the inequalities in L in which x_i appears) defines the set of all intervals that x_i must lie in, in order to satisfy L. Each inequality of type (A) induces a left endpoint of such an interval, and each inequality of type (B) induces a right endpoint of such an interval. We create all possible intervals of this type (using these endpoints) when creating L'.

Now, L' is feasible if and only if the intersection of all these intervals is non-empty. Since x_i^* is a feasible value, this implies that all these intervals have non-empty intersection containing x_i^*. Namely, we found a feasible solution to the original LP. ∎

The proof of the following (easy) lemma elaborates upon the argument used in the previous lemma (Lemma 16.4), and the enlighten reader might benefit from skipping it.

Lemma 16.5. *Let $\mathcal{P}$ be an $\mathcal{H}$-polyhedron in $\mathbb{R}^{\mathsf{d}}$. For any $1 \leq i \leq \mathsf{d}$, we have that the $\mathcal{H}$-polyhedron* $\mathrm{elim}(L, i)$ *is the projection of $\mathcal{P}$ to the hyperplane $x_i = 0$.*

Proof. Let L be the instance of an LP with d variables and m inequalities that defines $\mathcal{P}$. For the sake of simplicity of exposition, assume that $i = 1$.

For any point $\mathsf{p} = (0, \mathsf{p}_2, \ldots, \mathsf{p}_{\mathsf{d}}) \in (x_1 = 0)$ consider the associated line ℓ_{p} that passes through p and is parallel to the x_1-axis. Any linear inequality $\langle \mathsf{m}, x \rangle \leq b$ in L that has a non-zero coefficient for x_1 induces a linear inequality on the points of ℓ_{p}. Specifically, this is a one-dimensional inequality (since all the coordinates of the points of ℓ_{p} are fixed except for x_1) that specifies the portion of the line ℓ_{p} that is feasible for this constraint. Formally, the inequality is

$$\sum_{k=2}^{\mathsf{d}} \mathsf{m}_k \mathsf{p}_k + \mathsf{m}_1 x_1 \leq b \implies \begin{cases} \mathsf{m}_1 < 0, & -\sum_{k=2}^{\mathsf{d}} \frac{\mathsf{m}_k}{\mathsf{m}_1} \mathsf{p}_k - b \leq x_1, \\ \mathsf{m}_1 > 0, & x_1 \leq -\sum_{k=2}^{\mathsf{d}} \frac{\mathsf{m}_k}{\mathsf{m}_1} \mathsf{p}_k - b. \end{cases}$$

Namely, for a fixed $\mathsf{p} = (\mathsf{p}_2, \ldots, \mathsf{p}_{\mathsf{d}}) \in \mathbb{R}^{\mathsf{d}-1}$, each linear constraint in L induces a ray that is a subset of ℓ_{p}. The intersection of $\mathcal{P}$ with ℓ_{p} is non-empty if and only if the intersection of all these rays is non-empty. Or putting it differently, assume that the kth and lth constraints in L have non-zero coefficients for x_1 and they are of different signs. Then $\mathrm{elim}(L, 1, k, l)$ is feasible for $(\mathsf{p}_2, \ldots, \mathsf{p}_\mathsf{d})$ if and only if the corresponding rays have non-empty intersection on ℓ_p (here, the intersection is a finite interval). Specifically, the LP $\mathrm{elim}(L, 1)$ is feasible for $(\mathsf{p}_2, \ldots, \mathsf{p}_\mathsf{d})$ if and only if the line ℓ_p intersects $\mathcal{P}$. But $\mathsf{p} = (\mathsf{p}_2, \ldots, \mathsf{p}_\mathsf{d})$ is in the projection of $\mathcal{P}$ into $x_1 = 0$ if and only if the line ℓ_p has non-empty intersection with $\mathcal{P}$. This implies that the feasible region for $\mathrm{elim}(L, 1)$ is the projection of $\mathcal{P}$ to the hyperplane $x_1 = 0$. ■

The above implies, by change of variables, that any projection of an $\mathcal{H}$-polyhedron into a hyperplane is an $\mathcal{H}$-polyhedron. We can repeat this process of projecting down the polyhedron several times, which implies the following lemma.

Lemma 16.6. *The projection of any $\mathcal{H}$-polyhedron into any affine subspace is an $\mathcal{H}$-polyhedron.*

16.2.2. Farakas lemma.

Lemma 16.7 (Farakas lemma)**.** *Let $\mathsf{M} \in \mathbb{R}^{m \times \mathsf{d}}$ and $\mathsf{b} \in \mathbb{R}^m$ specify an LP. Then:*

(i) either there exists a feasible solution $x \in \mathbb{R}^\mathsf{d}$ to the LP, that is, $\mathsf{M}x \leq \mathsf{b}$,

(ii) or there is no feasible solution, and we can prove it by combining the inequalities of the LP into an infeasible inequality.

Formally, there exists a vector $\mathsf{w} \in \mathbb{R}^m$, such that $\mathsf{w} \geq 0$, $\mathsf{wM} = 0$, and $\langle \mathsf{w}, \mathsf{b} \rangle < 0$. Namely, the inequality (that must hold if the LP is feasible)

$$(\mathsf{wM})\, x \leq \langle \mathsf{w}, \mathsf{b} \rangle$$

is infeasible.

Proof. Clearly, the two options cannot hold together, so all we need to show is that if an LP is infeasible, then the second option holds.

Observe that if we apply a sequence of eliminations to an LP, all the resulting inequalities in the new LP are positive combinations of the original inequalities. So, let us apply this process of elimination to each one of the variables in turn.

Consider the resulting inequalities once all the variables are eliminated. Each such inequality would be of the form $0 \leq \kappa$, where κ is some real number. Specifically, there exists a vector $\mathsf{w} \geq 0$ such that $0 = \mathsf{wM} \leq \langle \mathsf{w}, \mathsf{b} \rangle = \kappa$. By Lemma 16.4, if the original LP is not feasible, then this resulting system of inequalities must be infeasible. This implies one of these inequalities must be of the form $0 \leq \kappa$ where $\kappa < 0$. ■

Note that the above elimination process provides us with a naive algorithm for solving an LP (i.e., eliminate all variables, and check if all resulting inequalities are feasible). This algorithm is extremely inefficient, as in the last elimination stage, we might end up with $O\left(m^{2^\mathsf{d}}\right)$ inequalities. This is unacceptably slow for solving LP.

We remind the reader that a linear equality of the form $\sum_i a_i x_i = c$ can be rewritten as the two inequalities $\sum_i a_i x_i \leq c$ and $\sum_i a_i x_i \geq c$. The great success of the Farakas lemma in the marketplace has lead to several sequels to it, and here is one of them.[2]

Lemma 16.8 (Farakas lemma II)**.** *Let $\mathsf{M} \in \mathbb{R}^{m \times \mathsf{d}}$ and $\mathsf{b} \in \mathbb{R}^m$, and consider the linear program $\mathsf{M}x = \mathsf{b}$, $x \geq 0$. Then, either (i) this linear program is feasible or (ii) there is a vector $\mathsf{w} \in \mathbb{R}^m$ such that $\mathsf{wM} \geq 0$ and $\langle \mathsf{w}, \mathsf{b} \rangle < 0$.*

---

[2] As in most sequels, Farakas lemma II is equivalent to the original Farakas lemma. I only hope the reader does not feel cheated.

PROOF. Suppose (ii) holds and there is a feasible solution x to the LP. We have that $\mathsf{M}x = \mathsf{b}$. Multiplying this equality from the left by w, we have that

$$(\mathsf{wM})x = \langle \mathsf{w}, \mathsf{b} \rangle .$$

But (ii) claims that the quantity on the left is non-negative (since $x \geq 0$) and the quantity on the right is negative, a contradiction.

We conclude that (i) and (ii) are mutually exclusive. As such, to complete the proof, we need to show that if (i) does not hold, then (ii) holds; that is, if the given LP is not feasible, then (ii) holds.

The linear program $\mathsf{M}x = \mathsf{b}$, $x \geq 0$, can be rewritten as $\mathsf{M}x \leq \mathsf{b}$, $\mathsf{M}x \geq \mathsf{b}$, $x \geq 0$, which in turn is equivalent to the LP:

$$\begin{array}{c} \mathsf{M}x \leq \mathsf{b} \\ -\mathsf{M}x \leq -\mathsf{b} \\ -x \leq 0 \end{array} \iff \begin{pmatrix} \mathsf{M} \\ -\mathsf{M} \\ -I_{\mathsf{d}} \end{pmatrix} x \leq \begin{pmatrix} \mathsf{b} \\ -\mathsf{b} \\ 0 \end{pmatrix},$$

where I_{d} is the $\mathsf{d} \times \mathsf{d}$ identity matrix. Now, if the original LP does not have a feasible solution, then by the original Farakas lemma (Lemma 16.7), there must be a vector $(\mathsf{w}_1, \mathsf{w}_2, \mathsf{w}_3) \geq 0$ such that

$$(\mathsf{w}_1, \mathsf{w}_2, \mathsf{w}_3) \begin{pmatrix} \mathsf{M} \\ -\mathsf{M} \\ -I_{\mathsf{d}} \end{pmatrix} = 0 \quad \text{and} \quad \left\langle (\mathsf{w}_1, \mathsf{w}_2, \mathsf{w}_3), \begin{pmatrix} \mathsf{b} \\ -\mathsf{b} \\ 0 \end{pmatrix} \right\rangle < 0.$$

Namely, $(\mathsf{w}_1 - \mathsf{w}_2)\,\mathsf{M} - \mathsf{w}_3 = 0$ and $\langle (\mathsf{w}_1 - \mathsf{w}_2), \mathsf{b} \rangle < 0$. But $\mathsf{w}_3 \geq 0$, which implies that

$$(\mathsf{w}_1 - \mathsf{w}_2)\,\mathsf{M} = \mathsf{w}_3 \geq 0.$$

Namely, (ii) holds for $\mathsf{w} = \mathsf{w}_1 - \mathsf{w}_2$. ■

16.2.3. Cones and vertices. For a set of vectors $\mathcal{V} \subseteq \mathbb{R}^{\mathsf{d}}$, let $\mathrm{cone}(\mathcal{V})$ denote the ***cone*** these vectors generate. Formally,

$$\mathrm{cone}(\mathcal{V}) = \left\{ \sum_i t_i \vec{\mathsf{v}}_i \;\middle|\; \vec{\mathsf{v}}_i \in \mathcal{V}, t_i \geq 0 \right\}.$$

Intuitively, a cone is a collection of a contiguous set of rays emanating for the origin. Now, a halfspace ***passes*** through a point p if p is contained in the hyperplane bounding this halfspace. Since 0 is the apex of $\mathrm{cone}(\mathcal{V})$, it is natural to presume that $\mathrm{cone}(\mathcal{V})$ can be generated by a finite intersection of halfspaces, all of them passing through the origin (which is indeed true). This is formally stated in Theorem 16.12 below. To this end, we first state and prove a few useful lemmas.

In the following, let $\mathsf{e}(i, \mathsf{d})$ denote the ith orthonormal vector in $\mathbb{R}^{\mathsf{d}}$; namely,

$$\mathsf{e}(i, \mathsf{d}) = (\overbrace{0, \ldots, 0}^{i-1 \text{ coords}}, 1, \overbrace{0, \ldots, 0}^{\mathsf{d}-i \text{ coords}}).$$

Lemma 16.9. *Let $\mathsf{M} \in \mathbb{R}^{m \times \mathsf{d}}$ be a given matrix, and consider the $\mathcal{H}$-polyhedron $\mathcal{P}$ formed by all points $(x, w) \in \mathbb{R}^{\mathsf{d}+m}$, such that $(x, w) \geq 0$ and $\mathsf{M}x \leq w$ (note that $\mathcal{P}$ is the intersection of halfspaces in $\mathbb{R}^{\mathsf{d}+m}$ as $\mathsf{M}x - w \leq 0$ is a collection of linear inequalities in x and w). Then $\mathcal{P} = \mathrm{cone}(\mathcal{V})$, where $\mathcal{V}$ is a finite set of vectors in $\mathbb{R}^{\mathsf{d}+m}$.*

PROOF. Note that the feasible region, $\mathcal{P}$, is the set of all pairs (x', w'), such that x' is a feasible solution to the linear program $\mathbf{M}x \leq w'$.

Let $E_i = \mathsf{e}(i, \mathsf{d} + m)$, for $i = \mathsf{d} + 1, \ldots, \mathsf{d} + m$. Clearly, the inequality $\mathsf{M}x \le w$ trivially holds if $(x, w) = E_i$, for $i = \mathsf{d} + 1, \ldots, \mathsf{d} + m$ (since $x = 0$ in such a case). Also, let

$$\vec{\mathsf{v}}_i = \big(\mathsf{e}(i, \mathsf{d}), \mathsf{M}\,\mathsf{e}(i, \mathsf{d})\big),$$

for $i = 1, \ldots, \mathsf{d}$. Clearly, the inequality $\mathsf{M}x \le w$ holds for $\vec{\mathsf{v}}_i$, since, trivially, $\mathsf{M}\,\mathsf{e}(i, \mathsf{d}) \le \mathsf{M}\,\mathsf{e}(i, \mathsf{d})$, for $i = 1, \ldots, \mathsf{d}$. Let $\mathcal{V} = \left\{\vec{\mathsf{v}}_1, \ldots, \vec{\mathsf{v}}_{\mathsf{d}}, E_{\mathsf{d}+1}, \ldots, E_{\mathsf{d}+m}\right\}$.

We claim that $\mathrm{cone}(\mathcal{V}) = \mathcal{P}$. One direction is easy. As the inequality $\mathsf{M}x \le w$ holds for all the vectors in $\mathcal{V}$, it also holds for any positive combination of these vectors, implying that $\mathrm{cone}(\mathcal{V}) \subseteq \mathcal{P}$. The other direction is slightly more challenging. Consider an $(x, w) \in \mathbb{R}^{\mathsf{d}+m}$ such that $\mathsf{M}x \le w$. Clearly, $w - \mathsf{M}x \ge 0$. Also, note that $(x, w) = (x, \mathsf{M}x) + (0, w - \mathsf{M}x)$. Now, let $x = (x_1, x_2, \ldots, x_{\mathsf{d}})$, and observe that

$$(x, \mathsf{M}x) = \left(\sum_{i=1}^{\mathsf{d}} x_i \mathsf{e}(i, \mathsf{d}), \mathsf{M} \sum_{i=1}^{\mathsf{d}} x_i \mathsf{e}(i, \mathsf{d})\right) = \left(\sum_{i=1}^{\mathsf{d}} x_i \mathsf{e}(i, \mathsf{d}), \sum_{i=1}^{\mathsf{d}} x_i \mathsf{M}\mathsf{e}(i, \mathsf{d})\right)$$
$$= \sum_{i=1}^{\mathsf{d}} x_i \big(\mathsf{e}(i, \mathsf{d}), \mathsf{M}\mathsf{e}(i, \mathsf{d})\big) = \sum_{i=1}^{\mathsf{d}} x_i \vec{\mathsf{v}}_i.$$

Since $x \ge 0$, this implies that $(x, \mathsf{M}x) \in \mathrm{cone}\big(\{\vec{\mathsf{v}}_1, \ldots, \vec{\mathsf{v}}_{\mathsf{d}}\}\big)$ and since $w - \mathsf{M}x \ge 0$, we have $(0, w - \mathsf{M}x) \in \mathrm{cone}(E_{\mathsf{d}+1}, \ldots, E_{\mathsf{d}+m})$. Thus, we have that $\mathcal{P} \subseteq \mathrm{cone}\big(\{\vec{\mathsf{v}}_1, \ldots, \vec{\mathsf{v}}_{\mathsf{d}}\}\big) + \mathrm{cone}(E_{\mathsf{d}+1}, \ldots, E_{\mathsf{d}+m}) = \mathrm{cone}(\mathcal{V})$. ■

We need the following simple pairing lemma, which we leave as an exercise for the reader (see Exercise 16.1).

Lemma 16.10. *Let $\alpha_1, \ldots, \alpha_n, \beta_1, \ldots, \beta_m$ be positive numbers, such that $\sum_{i=1}^{n} \alpha_i = \sum_{j=1}^{m} \beta_j$, and consider the positive combination $\vec{\mathsf{w}} = \sum_{i=1}^{n} \alpha_i \vec{\mathsf{v}}_i + \sum_{i=j}^{m} \beta_j \vec{\mathsf{u}}_j$, where $\vec{\mathsf{v}}_1, \ldots, \vec{\mathsf{v}}_n$, $\vec{\mathsf{u}}_1, \ldots, \vec{\mathsf{u}}_m$ are vectors (say, in $\mathbb{R}^{\mathsf{d}}$). Then, there are non-negative $\delta_{i,j}$s, such that $\vec{\mathsf{w}} = \sum_{i,j} \delta_{i,j} \left(\vec{\mathsf{v}}_i + \vec{\mathsf{u}}_j\right)$.*

Lemma 16.11. *Let $C = \mathrm{cone}(\mathcal{V})$ be a cone generated by a set of vectors $\mathcal{V}$ in $\mathbb{R}^{\mathsf{d}}$. Consider the region $\mathcal{P} = C \cap h$, where h is a hyperplane that passes through the origin. Then, $\mathcal{P}$ is a cone; namely, there exists a set of vectors $\mathcal{V}'$ such that $\mathcal{P} = \mathrm{cone}(\mathcal{V}')$.*

PROOF. By rigid rotation of the axis system, we can assume that $h \equiv (x_1 = 0)$. Furthermore, by scaling (by multiplying by positive constants), we can assume that the first coordinate of all vectors in $\mathcal{V}$ is either $-1, 0$, or 1 (clearly, scaling of the vectors generating the cone does not affect the cone itself). Hence $\mathcal{V} = \mathcal{V}_{-1} \cup \mathcal{V}_0 \cup \mathcal{V}_1$, where $\mathcal{V}_0$ is the set of vectors in $\mathcal{V}$ with the first coordinate being zero and $\mathcal{V}_{-1}$ (resp. $\mathcal{V}_1$) is the set of vectors in $\mathcal{V}$ with the first coordinate being -1 (resp. 1).

Let $\mathcal{V}_{-1} = \left\{\vec{\mathsf{v}}_1, \ldots, \vec{\mathsf{v}}_n\right\}$, $\mathcal{V}_1 = \left\{\vec{\mathsf{u}}_1, \ldots, \vec{\mathsf{u}}_m\right\}$, and let

$$\mathcal{V}' = \mathcal{V}_0 \cup \left\{\vec{\mathsf{v}}_i + \vec{\mathsf{u}}_j \;\middle|\; i = 1, \ldots, n, j = 1, \ldots, m\right\}.$$

Clearly, all the vectors of $\mathcal{V}'$ have zero in their first coordinate, and as such $\mathcal{V}' \subseteq h$, implying that $\mathrm{cone}(\mathcal{V}') \subseteq h$. Clearly, $\mathrm{cone}(\mathcal{V}') \subseteq C$, which implies that $\mathrm{cone}(\mathcal{V}') \subseteq C \cap h = \mathcal{P}$.

As for the other direction, consider a vector $\vec{\mathsf{w}} \in C \cap h$. Since $\vec{\mathsf{w}}$ is a positive linear combination of vectors in $\mathcal{V}$, it can be rewritten as $\vec{\mathsf{w}} = \vec{\mathsf{w}}_0 + \sum_i \alpha_i \vec{\mathsf{v}}_i + \sum_j \beta_j \vec{\mathsf{u}}_j$, where $\vec{\mathsf{w}}_0 \in \mathrm{cone}(\mathcal{V}_0)$. Since the first coordinate of $\vec{\mathsf{w}}$ is zero, we must have that $\sum_i \alpha_i = \sum_j \beta_j$.

Now, by the above pairing lemma (Lemma 16.10), we have that there are (non-negative) $\delta_{i,j}$s, such that

$$\vec{w} = \vec{w}_0 + \sum_{i,j} \delta_{i,j}\left(\vec{v}_i + \vec{u}_j\right) \in \text{cone}(\mathcal{V}'),$$

since $\vec{v}_i + \vec{u}_j \in \mathcal{V}'$. This implies that $C \cap h \subseteq \text{cone}(\mathcal{V}')$. ■

Theorem 16.12. *A cone C is generated by a finite set $\mathcal{V} \subseteq \mathbb{R}^{\mathsf{d}}$ (that is, $C = \text{cone}(\mathcal{V})$) if and only if there exists a finite set of halfspaces, all passing through the origin, such that their intersection is C.*

In linear programming lingo, a cone C is finitely generated by $\mathcal{V}$ if and only if there exists a matrix $\mathsf{M} \in \mathbb{R}^{m \times \mathsf{d}}$, such that $x \in C$ if and only if $\mathsf{M}x \leq 0$.

Proof. Let $C = \text{cone}(\mathcal{V})$, and observe that a point $x \in \text{cone}(\mathcal{V})$ can be written as (part of) a solution to a linear program. Indeed, let $\mathcal{V} = \left\{\vec{v}_1, \ldots, \vec{v}_m\right\}$, and consider the linear program:

$$\begin{array}{ll} x \in \mathbb{R}^{\mathsf{d}} & \displaystyle\sum_{i=1}^{m} t_i \vec{v}_i = x \\ \forall i & t_i \geq 0. \end{array}$$

Note that $x = (x_1, \ldots, x_{\mathsf{d}})$ defines some of the variables of the LP, and hence this LP is of the form $\mathsf{M}(x_1, \ldots, x_{\mathsf{d}}, t_1, \ldots, t_m) = 0$ and $t_1, \ldots, t_m \geq 0$.

Now, by definition, the feasible region of the above LP is an $\mathcal{H}$-polyhedron $\mathcal{Q} \subseteq \mathbb{R}^{\mathsf{d}+m}$. Clearly, for any $x \in C$, there are $t_1, \ldots, t_m \geq 0$, such that $\sum_i t_i \vec{v}_i = x$. Namely, if $x \in C$, then $(x, t_1, \ldots, t_m) \in \mathcal{Q}$.

Similarly, if $(y, t'_1, \ldots, t'_m) \in \mathcal{Q}$, then $y \in C$. We conclude that the projection of the $\mathcal{H}$-polyhedron $\mathcal{Q}$ into the subspace formed by the first d coordinates is the set C. Now, by Lemma 16.6, the projection of the $\mathcal{H}$-polyhedron $\mathcal{Q}$ is also an $\mathcal{H}$-polyhedron, implying that C is an $\mathcal{H}$-polyhedron.

As for the other direction, assume we are given an $\mathcal{H}$-polyhedron such that for all the halfspaces defining it, their boundary hyperplane passes through the origin. Clearly, such an $\mathcal{H}$-polyhedron can be written as $\mathsf{M}x \leq 0$. Next, consider the $\mathcal{H}$-polyhedron $\mathcal{P}$ formed by the set of points $(x, w) \in \mathbb{R}^{\mathsf{d}+m}$, such that $\mathsf{M}x \leq w$; that is, the feasible region of this $\mathcal{H}$-polyhedron is the set C. By Lemma 16.9, there exists a finite set of vectors $\mathcal{V} \subseteq \mathbb{R}^{\mathsf{d}+m}$ such that $\text{cone}(\mathcal{V}) = \mathcal{P}$. Now,

$$C = \mathcal{P} \cap (w = 0) = \text{cone}(\mathcal{V}) \cap (w_1 = 0) \cap \cdots \cap (w_m = 0).$$

A repeated application of Lemma 16.11 implies that the above set is a cone generated by a (finite) set of vectors, since $w_i = 0$ is a hyperplane passing through the origin, for $i = 1, \ldots, m$. ■

Now that we have the above theorem about the structure of a cone, we would like to understand the general structure of an $\mathcal{H}$-polyhedron, specifically that any $\mathcal{H}$-polyhedron is the combination of a cone and a convex polytope. First we prove the following relevant lemma.

Lemma 16.13. *Let $C = \text{cone}(\mathcal{V})$ be a cone generated by a set of vectors $\mathcal{V}$ in $\mathbb{R}^{\mathsf{d}}$. Consider the region $\mathcal{P} = C \cap h$, where h is any hyperplane in $\mathbb{R}^{\mathsf{d}}$. Then, there exist sets of vectors $\mathcal{U}$ and $\mathcal{W}$ such that $\mathcal{P} = \mathcal{CH}(\mathcal{U}) + \text{cone}(\mathcal{W})$.*

Proof. If h passes through the origin, then we just apply Lemma 16.11. Otherwise, by change of variables (done by rotating and scaling space), we can assume that $h \equiv (x_1 = 1)$ and that we can normalize the vectors of $\mathcal{V}$ so that the first coordinate is either $0, -1$, or 1. Next, also as before, we can break $\mathcal{V}$ into three disjoint sets, such that $\mathcal{V} = \mathcal{V}_{-1} \cup \mathcal{V}_0 \cup \mathcal{V}_1$, where $\mathcal{V}_i$ are the vectors in $\mathcal{V}$ with the first coordinate being i, for $i \in \{-1, 0, 1\}$.

By Lemma 16.11, there exists a set of vectors $\mathcal{W}_0$ such that $\text{cone}(\mathcal{W}_0) = C \cap (x_1 = 0)$. Let $X = \text{cone}(\mathcal{V}_1) \cap (x_1 = 1)$. A point $\mathsf{p} \in X$ is a positive combination of vectors $\mathsf{p} = \sum_i t_i \vec{\mathsf{v}}_i$, where $\vec{\mathsf{v}}_i \in \mathcal{V}_1$ and $t_i \geq 0$ for all i. But the first coordinate of all the points of $\mathcal{V}_1$ is 1, and so is the first coordinate of p. Namely, it must be that $\sum_i t_i = 1$. This implies that $X \subseteq \mathcal{CH}(\mathcal{V}_1)$. Now, by definition, the set $\mathcal{V}_1$ is contained in the hyperplane $x_1 = 1$, and as such this hyperplane also contains $\mathcal{CH}(\mathcal{V}_1)$. Clearly, $\mathcal{CH}(\mathcal{V}_1) \subseteq \text{cone}(\mathcal{V}_1)$, which implies that $\mathcal{CH}(\mathcal{V}_1) \subseteq X$. We conclude that $\mathcal{CH}(\mathcal{V}_1) = X$.

We claim that $\mathcal{P} = \text{cone}(\mathcal{V}) \cap (x_1 = 1)$ is equal to $Y = \mathcal{CH}(\mathcal{V}_1) + \text{cone}(\mathcal{W}_0)$. One direction is easy. As $\mathcal{V}_1, \mathcal{W}_0 \subseteq \text{cone}(\mathcal{V})$, it follows that $Y \subseteq \text{cone}(\mathcal{V})$. Now, a point of Y is the sum of two vectors. One of them has 1 in the first coordinate, and the other has 0. As such, a point of Y lies on the hyperplane $x_1 = 1$. This implies, by the definition of $\mathcal{P}$, that $Y \subseteq \text{cone}(\mathcal{V}) \cap \big(x_1 = 1\big) = \mathcal{P}$.

As for the other direction, consider a point $\mathsf{p} \in \mathcal{P}$. It can be written as

$$\mathsf{p} = \sum_{i, \vec{\mathsf{v}}_i \in \mathcal{V}_1} \alpha_i \vec{\mathsf{v}}_i + \sum_{j, \vec{\mathsf{u}}_j \in \mathcal{V}_0} \beta_j \vec{\mathsf{u}}_j + \sum_{k, \vec{\mathsf{w}}_k \in \mathcal{V}_{-1}} \gamma_k \vec{\mathsf{w}}_k,$$

where $\alpha_i, \beta_j, \gamma_k \geq 0$, for all i, j, k. Now, by considering only the first coordinate of this sum of vectors, we have

$$\sum_i \alpha_i - \sum_k \gamma_k = 1.$$

In particular, α_i can be rewritten as $\alpha_i = a_i + b_i$, where $a_i, b_i \geq 0$, $\sum_i b_i = \sum_k \gamma_k$, and $\sum_i a_i = 1$. As such,

$$\mathsf{p} = \sum_{i, \vec{\mathsf{v}}_i \in \mathcal{V}_1} a_i \vec{\mathsf{v}}_i + \left(\sum_{j, \vec{\mathsf{u}}_j \in \mathcal{V}_0} \beta_j \vec{\mathsf{u}}_j + \sum_{i, \vec{\mathsf{v}}_i \in \mathcal{V}_1} b_i \vec{\mathsf{v}}_i + \sum_{k, \vec{\mathsf{w}}_k \in \mathcal{V}_{-1}} \gamma_k \vec{\mathsf{w}}_k \right).$$

Now, since $\sum_i b_i = \sum_k \gamma_k$, we have $\left(\sum_{i, \vec{\mathsf{v}}_i \in \mathcal{V}_1} b_i \vec{\mathsf{v}}_i + \sum_{k, \vec{\mathsf{w}}_k \in \mathcal{V}_{-1}} \gamma_k \vec{\mathsf{w}}_k \right) \in (x_1 = 0)$. As such, we have

$$\sum_{j, \vec{\mathsf{u}}_j \in \mathcal{V}_0} \beta_j \vec{\mathsf{u}}_j + \sum_{i, \vec{\mathsf{v}}_i \in \mathcal{V}_1} b_i \vec{\mathsf{v}}_i + \sum_{k, \vec{\mathsf{w}}_k \in \mathcal{V}_{-1}} \gamma_k \vec{\mathsf{w}}_k \quad \in \quad \text{cone}(\mathcal{V}) \cap \big(x_1 = 0\big) = \text{cone}(\mathcal{W}_0).$$

Also, $\sum_i a_i = 1$ and, for all i, $a_i \geq 0$, implying that $\sum_{i, \vec{\mathsf{v}}_i \in \mathcal{V}_1} a_i \vec{\mathsf{v}}_i \in \mathcal{CH}(\mathcal{V}_1)$. Thus p is a sum of two points, one of them in $\text{cone}(\mathcal{W}_0)$ and the other in $\mathcal{CH}(\mathcal{V}_1)$. This implies that $\mathsf{p} \in Y$ and thus $\mathcal{P} \subseteq Y$. We conclude that $\mathcal{P} = Y$. ■

Theorem 16.14. *A region $\mathcal{P}$ is an $\mathcal{H}$-polyhedron in $\mathbb{R}^{\mathsf{d}}$ if and only if there exist finite sets $\mathsf{P}, \mathcal{V} \subseteq \mathbb{R}^{\mathsf{d}}$ such that $\mathcal{P} = \mathcal{CH}(\mathsf{P}) + \text{cone}(\mathcal{V})$.*

Proof. Consider the linear inequality $\sum_{i=1}^{\mathsf{d}} a_i x_i \leq b$, which is one of the constraints defining the polyhedron $\mathcal{P}$. We can lift this inequality into an inequality passing through the origin, by introducing an extra variable. The resulting inequality in $\mathbb{R}^{\mathsf{d}+1}$ is $\sum_{i=1}^{\mathsf{d}} a_i x_i - b x_{\mathsf{d}+1} \leq 0$. Clearly, this inequality defines a halfspace that goes through the origin, and furthermore, its intersection with $x_{\mathsf{d}+1} = 1$ is exactly $(\mathcal{P}, 1)$ (i.e., the set of points in the polyhedron $\mathcal{P}$ concatenated with 1 as the last coordinate). Thus, by doing this "lifting" for all the linear constraints defining $\mathcal{P}$ (note that we can use the same variable x_{d+1} for all the constraints), we get a region C defined by the intersection of a finite set of halfspaces passing through the origin. As such, by Theorem 16.12, there exists a set of vectors $\mathcal{V} \subseteq \mathbb{R}^{d+1}$, such that $C = \text{cone}(\mathcal{V})$. Moreover, $C \cap (x_{d+1} = 1) = (\mathcal{P}, 1)$. Now, Lemma 16.13 implies that there exist $\mathsf{P}, \mathcal{W} \subseteq \mathbb{R}^{d+1}$, such that $C \cap (x_{d+1} = 1) = \mathcal{CH}(\mathsf{P}) + \text{cone}(\mathcal{W})$. Dropping the $(d + 1)$st coordinate from these points implies that $\mathcal{P} = \mathcal{CH}(\mathsf{P}') + \text{cone}(\mathcal{W}')$, where P' and $\mathcal{W}'$ are P and $\mathcal{W}$ projected onto the first d coordinates, respectively. (Note that we know that convex hulls and cones remain convex hulls and cones, respectively, under projection.)

As for the other direction, assume that $\mathcal{P} = \mathcal{CH}(\mathsf{P}) + \text{cone}(\mathcal{V})$. Let $\mathsf{P}' = (\mathsf{P}, 1)$ and $\mathcal{V}' = (\mathcal{V}, 0)$ be a lifting of P and $\mathcal{V}$ to $d+1$ dimensions by padding it with an extra coordinate. Now, clearly,

$$(\mathcal{P}, 1) = \mathcal{CH}(\mathsf{P}') + \text{cone}(\mathcal{V}') \subseteq \text{cone}(\mathsf{P}' \cup \mathcal{V}') \cap (x_{d+1} = 1).$$

In fact, the containment also holds in the other direction, since a point $\mathsf{p} \in \text{cone}(\mathsf{P}' \cup \mathcal{V}') \cap (x_{d+1} = 1)$ is made out of a convex combination of the points of P' (since they have 1 in the $(d + 1)$st coordinate) and a positive combination of the points of $\mathcal{V}'$ (that have 0 in the $(d + 1)$st coordinate). Thus, we have that $(\mathcal{P}, 1) = \text{cone}(\mathsf{P}' \cup \mathcal{V}') \cap (x_{d+1} = 1)$. The cone $C' = \text{cone}(\mathsf{P}' \cup \mathcal{V}')$ can be described as a finite intersection of halfspaces (all passing through the origin), by Theorem 16.12. Let L be the equivalent linear program (we ignore the target function). Now, add the constraint $x_{d+1} = 1$ to this linear program. This results in a linear program L' over $\mathbb{R}^{d+1}$, such that its feasible region is $(\mathcal{P}, 1)$. Or, interpreting this LP as having only d variables (since x_{d+1} is fixed and any of its appearances can be replaced by 1) results in an LP with its feasible region being $\mathcal{P}$. Namely, $\mathcal{P}$ is an $\mathcal{H}$-polyhedron. ∎

A ***polytope*** is the convex hull of a finite point set. Theorem 16.14 implies that a polytope is also formed by the intersection of a *finite* set of halfspaces (i.e., $\text{cone}(V) = \emptyset$ since a polytope is bounded). Namely, a polytope is a bounded $\mathcal{H}$-polyhedron.

A linear inequality $\langle a, x\rangle \leq b$ is ***valid*** for a polytope $\mathcal{P}$ if it holds for all $x \in \mathcal{P}$. A ***face*** of $\mathcal{P}$ is a set

$$F = \mathcal{P} \cap (\langle a, x\rangle = b),$$

where $\langle a, x\rangle \leq b$ is a valid inequality for $\mathcal{P}$. The ***dimension*** of F is the affine dimension of the affine space it spans. A ***vertex*** is a face that consists only a single point; that is, a vertex is a 0-dimensional face. Intuitively, vertices are the "corners" of a polytope.

Lemma 16.15. *A vertex p of a polyhedron $\mathcal{P}$ cannot be written as a convex combination of a set X of points, such that $X \subseteq \mathcal{P}$ and $\mathsf{p} \notin X$.*

Proof. Let h be the hyperplane whose intersection with $\mathcal{P}$ is (only) p. Now, since $\mathcal{P}$ is a convex set and since there is no point of X on h itself, it must be that all the points of X must lie on one side of h. As such, any convex combination of the points of X would lie strictly on one side of h. ∎

Claim 16.16. *Let $\mathcal{P}$ be a polytope in $\mathbb{R}^d$. Then, $\mathcal{P}$ is the convex hull of its vertices; namely, $\mathcal{P} = \mathcal{CH}(V(\mathcal{P}))$. Furthermore, if for a set $V \subseteq \mathbb{R}^d$, we have $\mathcal{P} = \mathcal{CH}(V)$, then $V(\mathcal{P}) \subseteq V$.*

Proof. The polytope $\mathcal{P}$ is a bounded intersection of halfspaces. By Theorem 16.14, it can be represented as the sum of a convex hull of a finite point set, with a cone. But the cone here is empty, since $\mathcal{P}$ is bounded. Thus, there exists a finite set V, such that $\mathcal{P} = \mathcal{CH}(V)$. Let p be a point of V. If p can be expressed as a convex combination of the points of $V \setminus \{\mathsf{p}\}$, then $\mathcal{CH}(V \setminus \{\mathsf{p}\}) = \mathcal{CH}(V)$, since any point expressed as a convex combination involving p can be rewritten to exclude p. So, let X be the resulting set after we remove all such superfluous vertices, and observe that $\mathcal{P} = \mathcal{CH}(X)$. We claim that $X = V(\mathcal{P})$.

Indeed, if $\mathsf{p} \in X$, then the following LP (where we ignore the target function) does not have a solution (since we reduced V to X): $\sum_i t_i = 1$, $(t_1, \ldots, t_m) \geq 0$, and $\sum_{i=1}^m t_i \mathsf{p}_i = \mathsf{p}$, where $X \setminus \mathsf{p} = \{\mathsf{p}_1, \ldots, \mathsf{p}_m\}$. In matrix form, this LP is

$$\underbrace{\begin{pmatrix} 1 & 1 & \cdots & 1 \\ \mathsf{p}_1 & \mathsf{p}_2 & \cdots & \mathsf{p}_m \end{pmatrix}}_{\mathsf{M}} \begin{pmatrix} t_1 \\ t_2 \\ \vdots \\ t_m \end{pmatrix} = \begin{pmatrix} 1 \\ \mathsf{p} \end{pmatrix} \quad \text{and} \quad (t_1, \ldots, t_m) \geq 0.$$

By Farakas lemma II (Lemma 16.8), since this LP is not feasible, there exists a vector $\mathsf{w} \in \mathbb{R}^{\mathsf{d}+1}$, such that $\mathsf{w}\mathsf{M} \geq 0$ and $\langle \mathsf{w}, (1, \mathsf{p}) \rangle < 0$. Writing $\mathsf{w} = (\alpha_0, \mathsf{t}_0)$, for $\alpha_0 \in \mathbb{R}$ and $\mathsf{t}_0 \in \mathbb{R}^{\mathsf{d}}$, we can restate these inequalities as

$$\text{for } i = 1, \ldots, m, \quad \alpha_0 + \langle \mathsf{t}_0, \mathsf{p}_i \rangle \geq 0 \quad \text{and} \quad \alpha_0 + \langle \mathsf{t}_0, \mathsf{p} \rangle < 0.$$

In particular, this implies that

$$\langle \mathsf{t}_0, \mathsf{p}_i \rangle > \langle \mathsf{t}_0, \mathsf{p} \rangle, \quad \text{for } i = 1, \ldots, m. \tag{16.2}$$

Now, by definition, p lies on the hyperplane $h \equiv \big(\langle \mathsf{t}_0, x \rangle = \langle \mathsf{t}_0, \mathsf{p} \rangle \big)$, as t_0 and p are fixed (and $x \in \mathbb{R}^{\mathsf{d}}$ is the variable). Furthermore, by (16.2), all the other points of X are strictly on one side of h. In particular, $\mathcal{CH}(X)$ intersects h only at the point p, and $\mathcal{CH}(X) \setminus \{\mathsf{p}\}$ lies strictly on one side of h. Thus, we conclude that p is a vertex of $\mathcal{P} = \mathcal{CH}(X)$, since h defined a valid inequality and $h \cap \mathcal{P} = \mathsf{p}$. Namely, $X \subseteq V(\mathcal{P})$, as p was an arbitrary point of X.

Since $\mathcal{P} = \mathcal{CH}(X)$ and a vertex of $\mathcal{P}$ cannot be written as a convex combination of other points inside $\mathcal{P}$, by Lemma 16.15, it follows that $V(\mathcal{P}) \subseteq X$. Thus, $\mathcal{P} = \mathcal{CH}(V(\mathcal{P}))$. ■

16.3. Vertices of a polytope

Since a vertex is a corner of a polytope, we can cut it off by a hyperplane. This introduces a new face which captures the structure of how the vertex was connected to the rest of the polytope. Formally, consider a polytope $\mathcal{P}$, with $V = V(\mathcal{P})$. Let $\langle \mathsf{w}, x \rangle \leq c$ be a valid inequality for $\mathcal{P}$, such that the intersection of $\mathcal{P}$ with the hyperplane $\langle \mathsf{w}, x \rangle = c$ is a vertex $\mathsf{v} \in V$. Furthermore, for all other $\mathsf{u} \in V$, we have $\langle \mathsf{w}, \mathsf{u} \rangle < c_1 < c$, where c_1 is some constant. Consider the hyperplane $h \equiv \big(\langle \mathsf{w}, x \rangle = c_1 \big)$.

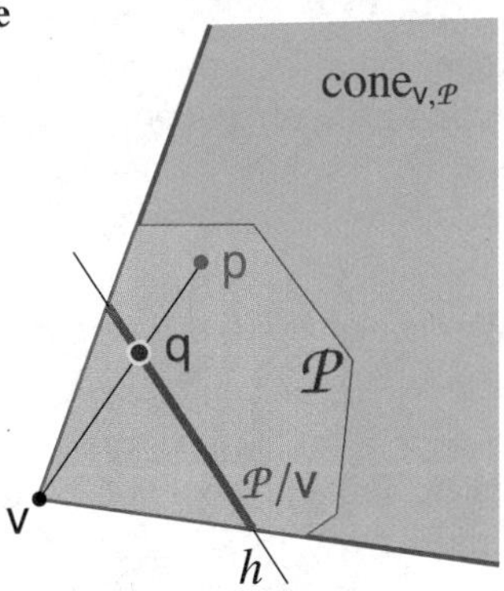

FIGURE 16.1.

The ***vertex figure*** of $\mathcal{P}$ at v is

$$\mathcal{P}/\mathsf{v} = \mathcal{P} \cap h.$$

The set $\mathcal{P}/\mathsf{v}$ depends (of course) on w and c_1, but its structure is independent of these values.

For a vertex v, let $\mathrm{cone}_{\mathsf{v},\mathcal{P}}$ denote the cone spanned locally at v by $\mathcal{P}$; formally,

$$\mathrm{cone}_{\mathsf{v},\mathcal{P}} = \mathsf{v} + \mathrm{cone}\big(\mathcal{P}/\mathsf{v} - \mathsf{v}\big).$$

To visualize $\text{cone}_{v,\mathcal{P}}$, imagine that v is in the origin. Then, this set is no more than the cone formed by the points of $\mathcal{P}$ sufficiently close to v.

Lemma 16.17. *We have* $\mathcal{P} \subseteq \text{cone}_{v,\mathcal{P}}$.

Proof. Consider a point $p \in \mathcal{P}$. We assume that for the hyperplane $h \equiv (\langle w, x\rangle = c_1)$, defining $\mathcal{P}/v$, it is true that $\langle w, p\rangle < c_1$ and $\langle w, v\rangle > c_1$ (otherwise, we can translate h so that it intersects the interior of the segment connecting v to p and then this holds). In particular, consider the point $q = vp \cap (\mathcal{P}/v)$; see Figure 16.1. By the convexity of $\mathcal{P}$, we have $q \in \mathcal{P}$, and by definition $q \in \mathcal{P}/v$. Thus, $q - v \in \text{cone}\big(\mathcal{P}/v - v\big)$ and thus $p - v \in \text{cone}\big(\mathcal{P}/v - v\big)$, since $p - v$ is a scaling of $q - v$. This implies that $p \in \big(v + \text{cone}\big(\mathcal{P}/v - v\big)\big) = \text{cone}_{v,\mathcal{P}}$. ∎

Lemma 16.18. *Let g be a halfspace defined by $\langle w, x\rangle \le c_3$, such that for a vertex v of $\mathcal{P}$, we have $\langle w, v\rangle = c_3$ and g is valid for $\mathcal{P}/v$ (i.e., for all $x \in \mathcal{P}/v$ we have $\langle w, x\rangle \le c_3$). Then g is valid for $\mathcal{P}$.*

Proof. Consider the linear function $f(x) = c_3 - \langle w, x\rangle$.

It is zero for v and non-negative on $\mathcal{P}/v$. Next, consider a line ℓ that passes through v and a point q of $\mathcal{P}/v$. The line ℓ can be written as

$$\ell(t) = \bigcup_{t\in\mathbb{R}} \big((1-t)v + tq\big)$$

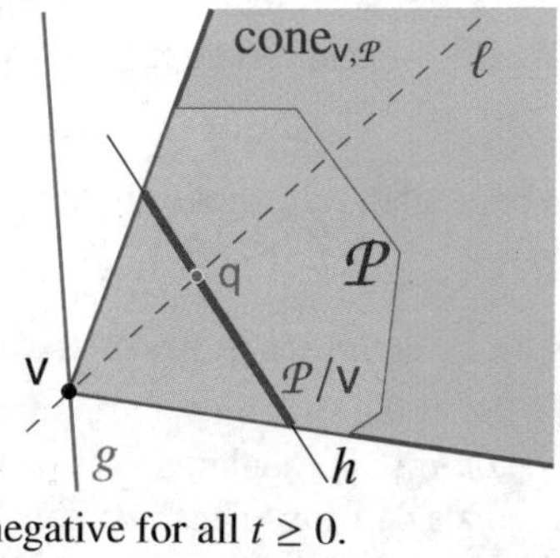

(as such, $\ell(0) = v$ and $\ell(1) = q$). The function $f(\cdot)$ when restricted to ℓ is thus the function $f_\ell(t) = f(\ell(t)) = c_3 - \langle w, (1-t)v + tq\rangle$, which is zero at v (i.e., $t = 0$) and non-negative at q (i.e., $t = 1$). Clearly, the function $f_\ell(t)$ is a one-dimensional linear function, which is thus non-negative for all $t \ge 0$.

Namely $f(\cdot)$ is non-negative on the side of ℓ that intersects $\mathcal{P}/v$; namely, $f(\cdot)$ is positive for any point on ℓ inside $\text{cone}_{v,\mathcal{P}}$. As such, $f(\cdot)$ is non-negative for any point in $\text{cone}_{v,\mathcal{P}}$. This in turn implies, by Lemma 16.17, that $f(\cdot)$ is non-negative for $\mathcal{P}$. ∎

16.3.1. Correspondence between faces of a vertex figure and faces of the polytope. Consider a polyhedron $\mathcal{P}$ and a vertex v of this polyhedron. Let h be the hyperplane defining the vertex figure of v; that is, $\mathcal{P}/v = \mathcal{P} \cap h$.

For a k-dimensional face f of $\mathcal{P}$ that contains v, for $k \ge 1$, consider the corresponding set

$$\pi(f) = f \cap h.$$

We claim that $\pi(f)$ is a face of $\mathcal{P}/v$. Similarly, for a face g of $\mathcal{P}/v$ the corresponding face of $\mathcal{P}$ is

$$\sigma(g) = \text{affine}(\{v\} \cup g) \cap \mathcal{P},$$

where $\text{affine}(\{v\} \cup g)$ is the affine subspace spanned by v and the points of g (see Definition 28.5_{p348}).

In fact, these two mappings are inverses of each other and as such provide us with a complete description of what $\mathcal{P}$ looks like locally around v.

We next formally prove these claims. For the sake of simplicity of exposition, assume that $h \equiv (x_d = 0)$ and furthermore $v_d > 0$, where v_d denotes the dth coordinate of v. This can always be realized by a rigid rotation and translation of space.

For the sake of simplicity of exposition, in the following, we assume that $\mathcal{P}$ is a bounded polytope. (The following claims extend easily to the unbounded case but require some minor tedium which is not very insightful.)

Claim 16.19. *A bounded k-dimensional face* f *of a polyhedron* $\mathcal{P}$ *that contains a specific vertex* v *also contains at least k additional distinct vertices* $v_1, \ldots, v_k$*, such that the k-dimensional affine subspace containing* f *is spanned by* $v, v_1, \ldots, v_k$.

PROOF. By definition, there exists a hyperplane such that $f = \mathcal{P} \cap h$. Now, f is clearly an $\mathcal{H}$-polyhedron, and since it is bounded, it is a polytope. Now, by Claim 16.16, we have $f = \mathcal{CH}(V(f))$. In particular, the minimum-dimensional affine subspace containing f is spanned by $V(f) = \{v'_1, \ldots, v'_m\}$. This implies that $v'_1 - v, \ldots, v'_m - v$ must contain k linearly independent vectors. The original vertices corresponding to these vectors together with v form the required set of spanning points. ■

Lemma 16.20. *Consider a k-dimensional face* f *of* $\mathcal{P}$*, for* $k \geq 1$*, such that* $v \in f$*. Then, the set* $\pi(f)$ *is a non-empty face of* $\mathcal{P}/v$.

PROOF. The face f is defined as $f = \mathcal{P} \cap h_f$ where $h_f \equiv \big(\langle w_2, x\rangle = c_2\big)$ is a hyperplane and $\langle w_2, x\rangle \leq c_2$ is a valid inequality for $\mathcal{P}$.

Now, since f is k-dimensional, by Claim 16.19, f contains another vertex u that is on the other side of h than v (since, by definition, all other vertices of $\mathcal{P}$ are on the other side of the hyperplane h defining the vertex figure of v).

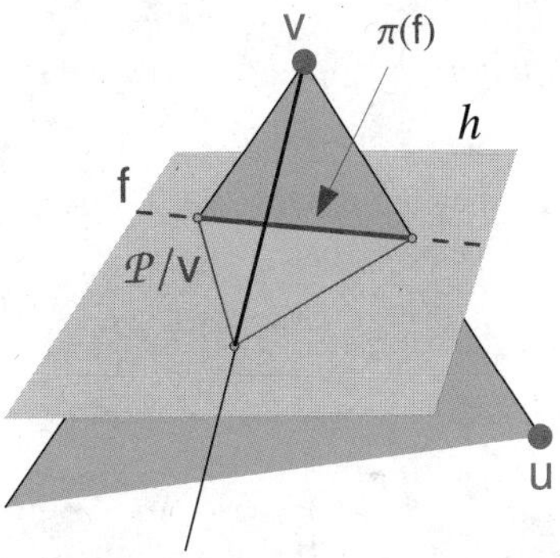

As such, the segment $vu \subseteq f \subseteq h_f$ and $vu \cap h \neq \emptyset$. Thus, $vu \cap (\mathcal{P}/v) \neq \emptyset$.

Furthermore, we have that

$$\pi(f) = f \cap h = \mathcal{P} \cap h_f \cap h = (\mathcal{P} \cap h) \cap h_f = (\mathcal{P}/v) \cap h_f.$$

This implies that $\pi(f)$ is a face of $\mathcal{P}/v$ as the inequality associated with h_f is valid for $\mathcal{P}/v$. Furthermore, since $\emptyset \neq vu \cap (\mathcal{P}/v) \subseteq \pi(f)$, this face is not empty. ■

Lemma 16.21. *For a face* g *of* $\mathcal{P}/v$*, the set* $\sigma(g) = \text{affine}(\{v\} \cup g) \cap \mathcal{P}$ *is a face of* $\mathcal{P}$.

PROOF. Let $g = \mathcal{P}/v \cap h_g$ for $h_g \equiv \big(\langle w_3, x\rangle = c_3\big)$, where $\langle w_3, x\rangle \leq c_3$ is a valid inequality for $\mathcal{P}/v$.

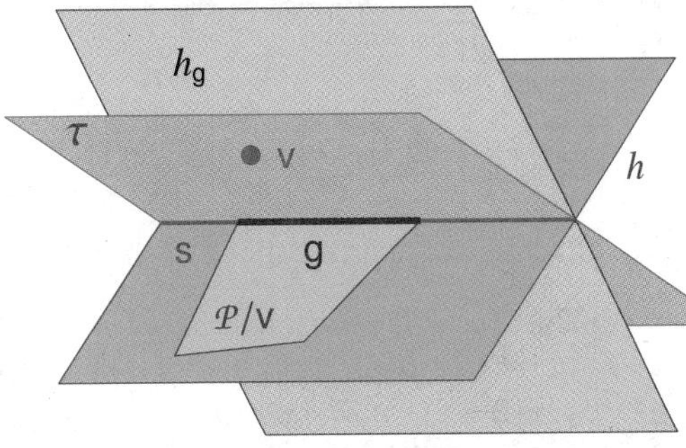

Note that by setting the dth coordinate of w_3 to be a sufficiently large negative number, we can guarantee that $\langle w_3, v\rangle < c_3$ as $v_d > 0$. Indeed, this can be done since all the points of $\mathcal{P}/v \subseteq h \equiv (x_d = 0)$, and as such they are oblivious to the value of the dth coordinate of w_3.

For $\lambda \in [0, 1]$, consider the convex combination of the two inequalities $\langle w_3, x\rangle \leq c_3$ and $x_d \leq 0$; that is,

$$z(\lambda) \equiv \Big\langle (1-\lambda)\, w_3, x\Big\rangle + \lambda x_d \leq (1-\lambda)c_3 + \lambda 0.$$

Geometrically, as we increase λ from 0 to 1, the halfspace $z(\lambda)$ is formed by a hyperplane rotating around the affine subspace $s = h_g \cap h$, where $\partial z(0) = h_g$ and $\partial z(1) = h$.

Since the two original inequalities are valid for $\mathcal{P}/\mathsf{v}$, it follows that $z(\lambda)$ is valid for $\mathcal{P}/\mathsf{v}$, for any $\lambda \in [0,1]$, as $z(\lambda)$ is the positive combination of these two inequalities. On the other hand, $\mathsf{v} \in z(0)$ and $\mathsf{v} \notin z(1)$. It follows that there is a value λ_0 of λ, such that v lies on $\partial z(\lambda_0)$. Since $z(\lambda_0)$ is valid for $\mathcal{P}/\mathsf{v}$, it follows, by Lemma 16.18, that $z(\lambda_0)$ is valid for $\mathcal{P}$. Setting $\tau = \partial z(\lambda_0)$, we have that $\mathsf{f} = \tau \cap \mathcal{P}$ is a face of $\mathcal{P}$ that contains v. Furthermore, f also contains g, as $\mathsf{g} = \mathcal{P}/\mathsf{v} \cap h_\mathsf{g} = \mathcal{P} \cap h \cap h_\mathsf{g} = \mathsf{s} \cap \mathcal{P} \subseteq \tau \cap \mathcal{P} = \mathsf{f}$. Since the hyperplane τ contains both v and g, we have that $\mathrm{affine}(\{\mathsf{v}\} \cup \mathsf{g}) \cap \mathcal{P} \subseteq \mathrm{affine}(\mathsf{f}) \cap \mathcal{P} = \mathsf{f}$.

We next show inclusion in the other direction. Consider any point $\mathsf{p} \in \mathsf{f}$ such that $\mathsf{p}_\mathsf{d} < 0$ (i.e., p is on the other side of the hyperplane $h \equiv (x_\mathsf{d} = 0)$ defining the vertex figure of v). By convexity $\mathsf{pv} \subseteq \mathsf{f} \subseteq h_\mathsf{f}$ and $\mathsf{t} = \mathsf{pv} \cap h \subseteq h_\mathsf{f} \cap h = \mathsf{s}$. Now, $\mathsf{g} = \mathcal{P}/\mathsf{v} \cap h_\mathsf{g} = \mathcal{P} \cap h \cap h_\mathsf{g} = \mathcal{P} \cap \mathsf{s}$, which implies that $\mathsf{t} \in \mathsf{g}$. As such, the line spanned by v and t is contained in $\mathrm{affine}(\{\mathsf{v}\} \cup \mathsf{g})$, which implies that $\mathsf{p} \in \mathrm{affine}(\{\mathsf{v}\} \cup \mathsf{g})$; namely, $\mathsf{p} \in \mathrm{affine}(\{\mathsf{v}\} \cup \mathsf{g}) \cap \mathcal{P}$. We conclude that $\mathsf{f} \subseteq \mathrm{affine}(\{\mathsf{v}\} \cup \mathsf{g}) \cap \mathcal{P}$.

The above implies that $\mathsf{f} = \mathrm{affine}(\{\mathsf{v}\} \cup \mathsf{g}) \cap \mathcal{P}$. ∎

Observation 16.22. *(A) Let* v *be a vertex of* $\mathcal{P}$*, and let* h *be the hyperplane containing the vertex figure* $\mathcal{P}/\mathsf{v}$*. For any set* $\mathsf{g} \subseteq h$*, we have that* $\mathrm{affine}(\{\mathsf{v}\} \cup \mathsf{g}) \cap h = \mathrm{affine}(\mathsf{g}) \cap h$*, since* $\mathsf{v} \notin h$*. (B) For any* $\mathcal{H}$*-polyhedron* $\mathcal{P}$ *and a face* f *of* $\mathcal{P}$*, we have that* $\mathsf{f} = \mathrm{affine}(\mathsf{f}) \cap \mathcal{P}$*.*

Lemma 16.23. *The mapping* π *is a bijection between the k-dimensional faces of* $\mathcal{P}$ *that contain* v *and the* $(k-1)$*-dimensional faces of* $\mathcal{P}/\mathsf{v}$*.*

Proof. The above implies that the mappings π and σ are well defined. We now verify that they are the inverses of each other. Indeed, for a face g of $\mathcal{P}/\mathsf{v}$, by Observation 16.22 (A), we have that

$$\begin{aligned}\pi\Big(\sigma(\mathsf{g})\Big) &= \pi\Big(\mathrm{affine}(\{\mathsf{v}\} \cup \mathsf{g}) \cap \mathcal{P}\Big) = \Big(\mathrm{affine}(\{\mathsf{v}\} \cup \mathsf{g}) \cap \mathcal{P}\Big) \cap h \\ &= \Big(\mathrm{affine}(\{\mathsf{v}\} \cup \mathsf{g}) \cap h\Big) \cap \mathcal{P} \cap h = \mathrm{affine}(\mathsf{g}) \cap (\mathcal{P} \cap h) \\ &= \mathrm{affine}(\mathsf{g}) \cap \mathcal{P}/\mathsf{v} = \mathsf{g},\end{aligned}$$

since $\mathsf{v} \notin h$ and $\mathsf{g} \subseteq h$ and by Observation 16.22(B).

Now, for a face f of $\mathcal{P}$ that contains v, we have $\mathrm{affine}\Big(\{\mathsf{v}\} \cup (\mathsf{f} \cap h)\Big) \subseteq \mathrm{affine}(\mathsf{f})$. The other direction of inclusion also holds, by a similar argumentation to the one used in the proof of Lemma 16.21. Indeed, for any point p of f that is on the other side of h, let $\mathsf{t} = \mathsf{vp} \cap h \in \mathsf{f} \cap h$. We have $\mathsf{p} \in \mathrm{affine}(\{\mathsf{v}, \mathsf{t}\}) \subseteq \mathrm{affine}(\{\mathsf{v}\} \cup (\mathsf{f} \cap h))$. We conclude that $\mathrm{affine}(\{\mathsf{v}\} \cup (\mathsf{f} \cap h)) = \mathrm{affine}(\mathsf{f})$.

This implies that

$$\sigma\Big(\pi(\mathsf{f})\Big) = \sigma\Big(\mathsf{f} \cap h\Big) = \mathrm{affine}\Big(\{\mathsf{v}\} \cup (\mathsf{f} \cap h)\Big) \cap \mathcal{P} = \mathrm{affine}(\mathsf{f}) \cap \mathcal{P} = \mathsf{f}.$$

∎

The following summarizes the above.

Lemma 16.24. *There is a bijection between the k-dimensional faces of* $\mathcal{P}$ *that contain* v *and the* $(k-1)$*-dimensional faces of* $\mathcal{P}/\mathsf{v}$*. Specifically, for a face* f *of* $\mathcal{P}$*, the corresponding face of* $\mathcal{P}/\mathsf{v}$ *is* $\pi(\mathsf{f}) = \mathsf{f} \cap h$*, where* h *is the hyperplane defining* $\mathcal{P}/\mathsf{v}$*. Similarly, for a* $(k-1)$*-dimensional face* g *of* $\mathcal{P}/\mathsf{v}$*, the corresponding face of* $\mathcal{P}$ *is* $\sigma(\mathsf{g}) = \mathrm{affine}(\{\mathsf{v}\} \cup \mathsf{g}) \cap \mathcal{P}$*.*

16.4. Linear programming correctness

We are now ready to prove that the algorithms we presented (see Theorem 15.5_{p207}) for linear programing do indeed work.

We remind the reader that either the LP algorithm outputs a vertex that is locally optimal (and thus globally optimal, as proven below) with the basis generating this vertex as a proof of its local optimality, or it outputs a set of $d + 1$ halfspaces for which their intersection is not feasible.

Surprisingly, all we need to prove is that locally checking if we are in the optimal vertex is sufficient to detect if we have reached the optimal solution. This is sufficient, as the algorithm outputs a basis together with the supposedly optimal vertex. In $\mathbb{R}^d$, the basis is made out of d constraints and it defines a collection of d rays emanating from this vertex (they span the cone of this vertex). We can compute these rays in time polynomial in d from the basis, and one can verify that for all these rays, the vertex achieves the minimal value (along these rays), which implies that this vertex realizes the local optimum solution.

Namely, the LP algorithm provides a solution that is locally optimal (because it has this property that along all its adjacent edges the target function gets worse along these edges), and this can be efficiently verified. Thus, locally, the output vertex is optimal, which, by the following theorem, implies that it is the global optimal solution.

A one-dimensional face is an ***edge*** of a polyhedron. Locally, from a vertex, an edge is a ray emanating out of the vertex.

Theorem 16.25. *Let v be a vertex of an $\mathcal{H}$-polyhedron $\mathcal{P}$ in $\mathbb{R}^d$, and let $f(\cdot)$ be a linear function, such that $f(\cdot)$ is non-decreasing along all the edges leaving v (i.e., v is a "local" minimum of $f(\cdot)$ along the edges adjacent to v). Then v realizes the global minimum of $f(\cdot)$ on $\mathcal{P}$.*

Proof. Assume for the sake of contradiction that this is false, and let x be a point in $\mathcal{P}$, such that $f(x) < f(v)$. Let $\mathcal{P}/v = \mathcal{P} \cap h$, where h is a hyperplane such that v and x are on different sides of h.

By convexity, the segment xv must intersect $\mathcal{P}/v$, and let y be this intersection point. Since $\mathcal{P}/v$ is a convex polytope in $d - 1$ dimensions, y can be written as a convex combination of d of its vertices $u_1, \ldots, u_d$; namely, $y = \sum_i \alpha_i u_i$, where $\alpha_i \geq 0$ and $\sum_i \alpha_i = 1$. By Lemma 16.24, each one of these vertices lies on an edge of $\mathcal{P}$ adjacent to v, and by the local optimality of v, we have $f(u_i) \geq f(v)$. Now, by the linearity of $f(\cdot)$, we have

$$f(y) = f\left(\sum_i \alpha_i u_i\right) = \sum_i \alpha_i f(u_i) \geq f(v).$$

But y is a convex combination of x and v; namely, there exists $\alpha \in (0, 1)$ such that $y = \alpha x + (1 - \alpha)v$. Now, $f(x) < f(v)$. As such,

$$f(y) = f(\alpha x + (1 - \alpha)v) = \alpha f(x) + (1 - \alpha) f(v) < f(v),$$

a contradiction. ∎

The LP algorithm also used the property that if the LP is not feasible, then the algorithm outputs a compact proof of this fact. Formally, if the LP is not feasible, then there exist $d + 1$ constraints such that these constraints, by themselves, are not feasible. But this is an immediate consequence of Helly's theorem (Theorem 16.2_{p217}). Indeed, if every subset of $d + 1$ halfspaces has a non-empty intersection, then by Helly's theorem they all have a non-empty intersection and the LP is feasible.

This implies that for any LP either there is a compact proof of the solution being optimal or, alternatively, there is a compact proof that the LP is not feasible. This implies that the LP algorithm we described will not fail – it would always output one of these two compact proofs.

16.5. Bibliographical notes

There are several good texts on linear programming and polytopes that cover the topics discussed here, including the books by Ziegler [**Zie94**] and Grünbaum [**Grü03**]. Low-dimensional linear programming is covered by Matoušek and Gärtner [**MG06**]. Our presentation is a convex combination of some of these sources.

16.6. Exercises

Exercise 16.1 (Pairing lemma)**.** Prove Lemma 16.10. Specifically, let $\alpha_1, \ldots, \alpha_n, \beta_1, \ldots,$ β_m be positive numbers, such that $\sum_{i=1}^{n} \alpha_i = \sum_{j=1}^{m} \beta_j$, and consider the positive combination $\overrightarrow{\mathsf{w}} = \sum_{i=1}^{n} \alpha_i \overrightarrow{\mathsf{v}}_i + \sum_{i=j}^{m} \beta_j \overrightarrow{\mathsf{u}}_j$, where $\overrightarrow{\mathsf{v}}_1, \ldots, \overrightarrow{\mathsf{v}}_n, \overrightarrow{\mathsf{u}}_1, \ldots, \overrightarrow{\mathsf{u}}_m$ are vectors (say, in $\mathbb{R}^{\mathsf{d}}$). Then, there are non-negative $\delta_{i,j}$s, such that $\overrightarrow{\mathsf{w}} = \sum_{i,j} \delta_{i,j} \left(\overrightarrow{\mathsf{v}}_i + \overrightarrow{\mathsf{u}}_j \right)$.

In fact, at most $n + m - 1$ of the $\delta_{i,j}$s have to be non-zero.

CHAPTER 17

Approximate Nearest Neighbor Search in Low Dimension

17.1. Introduction

Let P be a set of n points in $\mathbb{R}^d$. We would like to preprocess it, such that given a query point q, one can determine the closest point in P to q quickly. Unfortunately, the exact problem seems to require prohibitive preprocessing time. (Namely, it requires computing the Voronoi diagram of P and preprocessing it for point-location queries. This requires (roughly) $O(n^{\lceil d/2 \rceil})$ time.)

Instead, we will specify a parameter $\varepsilon > 0$ and build a data-structure that answers $(1 + \varepsilon)$-***approximate nearest neighbor*** queries.

Definition 17.1. For a set $P \subseteq \mathbb{R}^d$ and a query point q, we denote by $nn(q) = nn(q, P)$ the ***closest point*** (i.e., ***nearest neighbor***) in P to q. We denote by $\mathbf{d}(q, P)$ the distances between q and its closest point in P; that is, $\mathbf{d}(q, P) = \|q - nn(q)\|$.

For a query point q and a set P of n points in $\mathbb{R}^d$, a point $s \in P$ is a $(1+\varepsilon)$-***approximate nearest neighbor*** (or just $(1 + \varepsilon)$-***ANN***) if $\|q - s\| \leq (1 + \varepsilon)\mathbf{d}(q, P)$. Alternatively, for any $t \in P$, we have $\|q - s\| \leq (1 + \varepsilon) \|q - t\|$.

This is yet another instance where solving the bounded spread case is relatively easy. (We remind the reader that the ***spread*** of a point set P, denoted by $\Phi(P)$, is the ratio between the diameter and the distance of the closest pair of P.)

17.2. The bounded spread case

Let P be a set of n points contained inside the unit hypercube in $\mathbb{R}^d$, and let $\mathcal{T}$ be a quadtree of P, where $\text{diam}(P) = \Omega(1)$. We assume that with each (internal) node u of $\mathcal{T}$ there is an associated representative point, rep_u, such that rep_u is one of the points of P stored in the subtree rooted at u.

Let q be a query point and let $\varepsilon > 0$ be a parameter. Here our purpose is to find a $(1 + \varepsilon)$-ANN to q in P.

Idea of the algorithm. The algorithm would maintain a set of nodes of $\mathcal{T}$ that might contain the ANN to the query point. Each such node has a representative point associated with it, and we compute its distance to the query point and maintain the nearest neighbor found so far. At each stage, the search would be refined by replacing a node by its children (and computing the distance from the query point to all the new representatives of these new nodes).

The key observation is that we need to continue searching in a subtree of a node only if the node can contain a point that is significantly closer to the query point than the current best candidate found.

We keep only the "promising" nodes and continue this search till there are no more candidates to check. We claim that we had found the ANN to q, and furthermore the query time is fast. This idea is depicted in Figure 17.2.

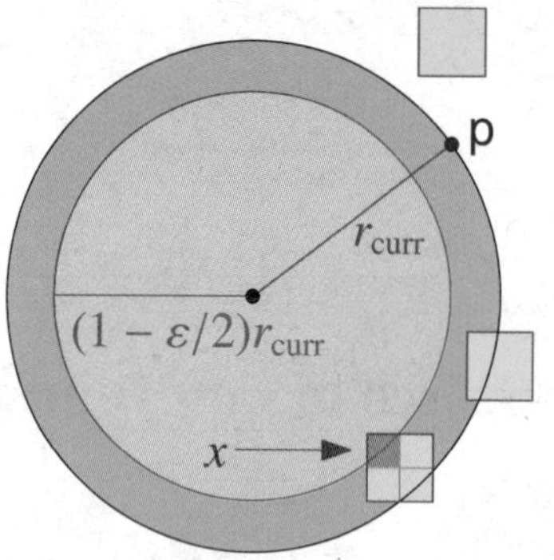

FIGURE 17.1. Out of the three nodes currently in the candidate set, only one of them (i.e., x) has the potential to contain a better ANN to q than the current one (i.e., p).

Formally, let p be the current best candidate found so far, and let its distance from q be r_{curr}. Now, consider a node w in $\mathcal{T}$, and observe that the point set P_w, stored in the subtree of w, might contain a "significantly" nearer neighbor to q only if it contains a point $s \in rg_w \cap P$ such that $\|q - s\| < (1 - \varepsilon/2)r_{curr}$. A conservative lower bound on the distance of any point in rg_w to q is $\|q - rep_w\| - diam(\square_w)$. In particular, if $\|q - rep_w\| - diam(\square_w) > (1-\varepsilon/2)r_{curr}$, then we can abort the search for ANN in the subtree of w.

The algorithm. Let $A_0 = \{root(T)\}$, and let $r_{curr} = \|q - rep_{root(T)}\|$. The value of r_{curr} is the distance to the closest neighbor of q that was found so far by the algorithm.

In the ith iteration, for $i > 0$, the algorithm expands the nodes of A_{i-1} to get A_i. Formally, for $v \in A_{i-1}$, let C_v be the set of children of v in $\mathcal{T}$ and let $\square_v$ denote the cell (i.e., region) that v corresponds to. For every node $w \in C_v$, we compute

$$r_{curr} \leftarrow \min\left(r_{curr}, \|q - rep_w\|\right).$$

The algorithm checks if

$$\|q - rep_w\| - diam(\square_w) < (1 - \varepsilon/2)r_{curr}, \tag{17.1}$$

and if so, it adds w to A_i. The algorithm continues in this expansion process till all the elements of A_{i-1} are considered, and then it moves to the next iteration. The algorithm stops when the generated set A_i is empty. The algorithm returns the point realizing the value of r_{curr} as the ANN.

The set A_i is a set of nodes of depth i in the quadtree that the algorithm visits. Note that all these nodes belong to the canonical grid $G_{2^{-i}}$ of level $-i$, where every canonical square has sidelength 2^{-i}. (Thus, nodes of depth i in the quadtree are of ***level*** $-i$. This is somewhat confusing but it in fact makes the presentation simpler.)

Correctness. Note that the algorithm adds a node w to A_i only if the set P_w might contain points which are closer to q than the (best) current nearest neighbor the algorithm found, where P_w is the set of points stored in the subtree of w. (More precisely, P_w might contain a point which is $1 - \varepsilon/2$ closer to q than any point encountered so far.)

Consider the last node w inspected by the algorithm such that $nn(q) \in P_w$. Since the algorithm decided to throw this node away, we have, by the triangle inequality, that

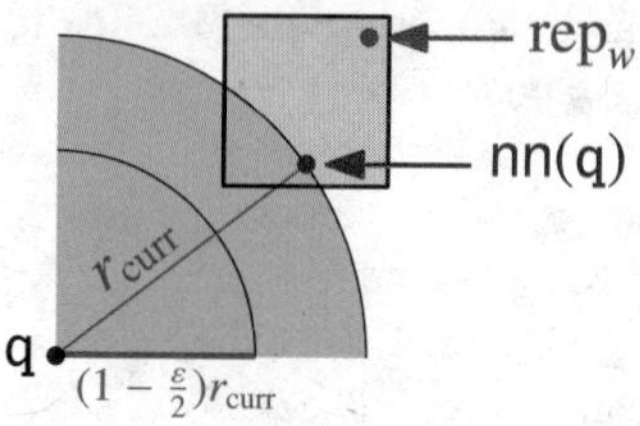

$$\begin{aligned}\|q - nn(q)\| &\geq \|q - rep_w\| - \|rep_w - nn(q)\| \\ &\geq \|q - rep_w\| - diam(\square_w) \geq (1 - \varepsilon/2)r_{curr}.\end{aligned}$$

Thus, $\|q - nn(q)\| / (1 - \varepsilon/2) \geq r_{curr}$. However, $1/(1 - \varepsilon/2) \leq 1 + \varepsilon$, for $1 \geq \varepsilon > 0$, as can be easily verified. Thus, $r_{curr} \leq (1 + \varepsilon)\mathbf{d}(q, P)$, and the algorithm returns $(1 + \varepsilon)$-ANN to q.

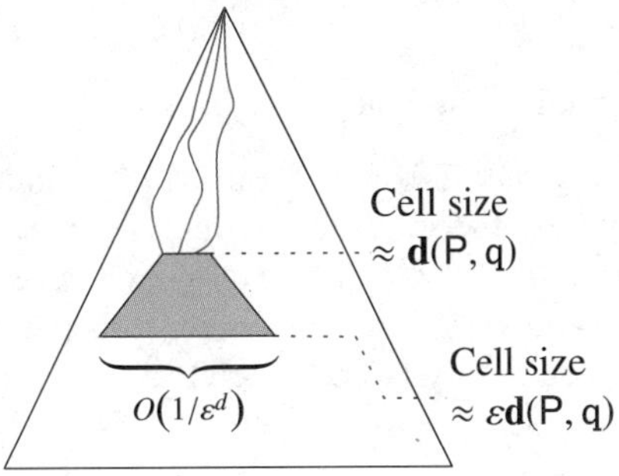

Running time analysis. Before barging into a formal proof of the running time of the above search procedure, it is useful to visualize the execution of the algorithm. It visits the quadtree level by level. As long as the level's grid cells are bigger than the ANN distance $r = \mathbf{d}(q, P)$, the number of nodes visited is a constant (i.e., $|A_i| = O(1)$). (This is not obvious, and at this stage the reader should take this statement with due skepticism.) This number "explodes" only when the cell size become smaller than r, but then the search stops when we reach grid size $O(\varepsilon r)$. In particular, since the number of grid cells visited (in the second stage) grows exponentially with the level, we can use the number of nodes visited in the bottom level (i.e., $O(1/\varepsilon^d)$) to bound the query running time for this part of the query.

Lemma 17.2. *Let* P *be a set of n points contained inside the unit hypercube in* $\mathbb{R}^d$*, and let* $\mathcal{T}$ *be a quadtree of* P*, where* $\mathrm{diam}(P) = \Omega(1)$*. Let* q *be a query point, and let* $\varepsilon > 0$ *be a parameter. A* $(1 + \varepsilon)$*-ANN to* q *can be computed in* $O\left(\varepsilon^{-d} + \log(1/\varpi)\right)$ *time, where* $\varpi = \mathbf{d}(q, P)$.

Proof. The algorithm is described above. We are only left with the task of bounding the query time. Observe that if a node $w \in \mathcal{T}$ is considered by the algorithm and $\mathrm{diam}(\square_w) < (\varepsilon/4)\varpi$, then

$$\left\|q - \mathrm{rep}_w\right\| - \mathrm{diam}(\square_w) \geq \left\|q - \mathrm{rep}_w\right\| - (\varepsilon/4)\varpi \geq r_{curr} - (\varepsilon/4)r_{curr} \geq (1 - \varepsilon/4)r_{curr},$$

which implies that neither w nor any of its children would be inserted into the sets $A_1, \ldots, A_m$, where m is the depth $\mathcal{T}$, by (17.1). Thus, no nodes of depth $\geq h = \lceil -\lg(\varpi\varepsilon/4)\rceil$ are being considered by the algorithm.

Consider the node u of $\mathcal{T}$ of depth i containing $nn(q, P)$. Clearly, the distance between q and rep_u is at most $\ell_i = \varpi + \mathrm{diam}_u = \varpi + \sqrt{d}2^{-i}$. As such, in the end of the ith iteration, we have $r_{curr} \leq \ell_i$, since the algorithm had inspected u.

Thus, the only cells of $G_{2^{-i-1}}$ that might be considered by the algorithm are the ones in distance $\leq \ell_i$ from q. Indeed, in the end of the ith iteration, $r_{curr} \leq \ell_i$. As such, any node of $G_{2^{-i-1}}$ (i.e., nodes considered in the $(i + 1)$st iteration of the algorithm) that is within a distance larger than r_{curr} from q cannot improve the distance to the current nearest neighbor and can just be thrown away if it is in the queue. (We charge the operation of putting a node into the queue to its parent. As such, nodes that get inserted and deleted in the next iteration are paid for by their parents.)

The number of such relevant cells (i.e., cells that the algorithm dequeues and do not get immediately thrown out) is the number of grid cells of $G_{2^{-i-1}}$ that intersect a box of sidelength $2\ell_i$ centered at q; that is,

$$n_i = \left(2\left\lceil\frac{\ell_i}{2^{-i-1}}\right\rceil\right)^d = O\left(\left(1 + \frac{\varpi + \sqrt{d}2^{-i}}{2^{-i-1}}\right)^d\right) = O\left(\left(1 + \frac{\varpi}{2^{-i-1}}\right)^d\right) = O\left(1 + \left(2^i\varpi\right)^d\right),$$

since for any $a, b \geq 0$ we have $(a+b)^{\mathrm{d}} \leq (2\max(a,b))^{\mathrm{d}} \leq 2^{\mathrm{d}}\left(a^{\mathrm{d}} + b^{\mathrm{d}}\right)$. Thus, the total number of nodes visited is

$$\sum_{i=0}^{h} n_i = O\left(\sum_{i=0}^{\lceil -\lg(\varpi\varepsilon/4)\rceil} \left(1 + \left(2^i \varpi\right)^{\mathrm{d}}\right)\right) = O\left(\lg \frac{1}{\varpi\varepsilon} + \left(\frac{\varpi}{\varpi\varepsilon/4}\right)^{\mathrm{d}}\right) = O\left(\log \frac{1}{\varpi} + \frac{1}{\varepsilon^{\mathrm{d}}}\right),$$

and this also bounds the overall query time. ■

One can apply Lemma 17.2 to the case for which the input has spread bounded from above. Indeed, if the distance between the closest pair of points of P is $\mu = \mathcal{CP}(\mathsf{P})$, then the algorithm would never search in cells that have diameter $\leq \mu/8$. Indeed, no such nodes would exist in the quadtree, to begin with, since the parent node of such a node would contain only a single point of the input. As such, we can replace ϖ by μ in the above argumentation.

Lemma 17.3. *Let* P *be a set of* n *points in* $\mathbb{R}^{\mathrm{d}}$*, and let* $\mathcal{T}$ *be a quadtree of* P*, where* $\mathrm{diam}(\mathsf{P}) = \Omega(1)$*. Given a query point* q *and* $1 \geq \varepsilon > 0$*, one can return a* $(1+\varepsilon)$*-ANN to* q *in* $O\left(1/\varepsilon^{\mathrm{d}} + \log \Phi(\mathsf{P})\right)$ *time, where* $\Phi(\mathsf{P})$ *is the spread of* P.

A less trivial task is to adapt the algorithm, so that it uses compressed quadtrees. To this end, the algorithm would still handle the nodes by levels. This requires us to keep a heap of integers in the range $0, -1, \ldots, -\lfloor \lg \Phi(\mathsf{P}) \rfloor$. This can be easily done by maintaining an array of size $O(\log \Phi(\mathsf{P}))$, where each array cell maintains a linked list of all nodes with this level. Clearly, an insertion/deletion into this heap data-structure can be handled in constant time by augmenting it with a hash table. Thus, the above algorithm would work for this case after modifying it to use this "level" heap instead of just the sets A_i.

Theorem 17.4. *Let* P *be a set of* n *points in* $\mathbb{R}^{\mathrm{d}}$*. One can preprocess* P *in* $O(n \log n)$ *time and using linear space, such that given a query point* q *and parameter* $1 \geq \varepsilon > 0$*, one can return a* $(1+\varepsilon)$*-ANN to* q *in* $O\left(1/\varepsilon^{\mathrm{d}} + \log \Phi(\mathsf{P})\right)$ *time. In fact, the query time is* $O(1/\varepsilon^{\mathrm{d}} + \log(\mathrm{diam}(\mathsf{P})/\varpi))$*, where* $\varpi = \mathbf{d}(\mathsf{q}, \mathsf{P})$.

17.3. ANN – the unbounded general case

The snark and the unbounded spread case (or a metaphilosophical pretentious discussion that the reader might want to skip). (The reader might consider this to be a footnote to a footnote, which finds itself inside the text because of lack of space at the bottom of the page.) We have a data-structure that supports insertions, deletions, and approximate nearest neighbor reasonably quickly. The running time for such operations is roughly $O(\log \Phi(\mathsf{P}))$ (ignoring additive terms in $1/\varepsilon$). Since the spread of P in most real world applications is going to be bounded by a constant degree polynomial in n, it seems this is sufficient for our purposes, and we should stop now, while we are ahead in the game. But the nagging question remains: If the spread of P is not bounded by something reasonable, what can be done?

The rule of thumb is that $\Phi(\mathsf{P})$ can be replaced by n (for this problem, but also in a lot of other problems). This usually requires some additional machinery, and sometimes this machinery is quite sophisticated and complicated. At times, the search for the ultimate algorithm that can work for such "strange" inputs looks like the hunting of the snark **[Car76]** – a futile waste of energy looking for some imaginary top-of-the-mountain, which has no practical importance.

Solving the bounded spread case can be acceptable in many situations, and it is the first stop in trying to solve the general case. However, solving the general case provides us with more insight into the problem and in some cases leads to more efficient solutions than the bounded spread case.

With this caveat emptor[①] warning duly given, we plunge ahead into solving the ANN for the unbounded spread case.

Plan of attack. To answer the ANN query in the general case, we will first get a fast rough approximation. Next, using a compressed quadtree, we will find a constant number of relevant nodes and apply Theorem 17.4 to those nodes. This will yield the required approximation. Before solving this problem, we need a minor extension of the compressed quadtree data-structure.

Extending a compressed quadtree to support cell queries. Let $\widehat{\square}$ be a canonical grid cell (we remind the reader that this is a cell of the grid G_{2^i}, for some integer $i \leq 0$). Given a compressed quadtree $\widehat{T}$, we would like to find the *single* node $v \in \widehat{T}$, such that $\mathsf{P} \cap \widehat{\square} = \mathsf{P}_v$. (Note that the node v might be compressed, and the square associated with it might be much larger than $\widehat{\square}$, but its only child w is such that $\square_w \subseteq \widehat{\square} \subseteq \square_v$. However, the annulus $\square_v \subseteq \square_w$ contains no input point.) We will refer to such a query as a ***cell query***.

It is not hard to see that the quadtree data-structure can be modified to support cell queries in logarithmic time (it's a glorified point-location query), and we omit the easy but tedious details. See Exercise 2.4_{p26}.

Lemma 17.5. *One can perform a cell query in a compressed quadtree $\widehat{T}$, in $O(\log n)$ time, where n is the size of $\widehat{T}$. Namely, given a query canonical cell $\widehat{\square}$, one can find, in $O(\log n)$ time, the node $w \in \widehat{T}$ such that $\square_w \subseteq \widehat{\square}$ and $\mathsf{P} \cap \widehat{\square} = \mathsf{P}_w$.*

17.3.1. Putting things together – answering ANN queries. Let P be a set of n points in $\mathbb{R}^{\mathsf{d}}$ contained in the unit hypercube. We build the compressed quadtree $\widehat{T}$ of P, so that it supports cell queries, using Lemma 17.5. We will also need a data-structure that supports very rough ANN queries quickly. We describe one way to build such a data-structure in the next section, and in particular, we will use the following result (see Theorem 17.10). Note that a similar result can be derived by using a shifted quadtree and a simple point-location query; see Theorem 11.21_{p160}. Explicitly, we need the following (provided by Theorem 17.10_{p240}):

> Let P be a set of n points in $\mathbb{R}^{\mathsf{d}}$. One can build a data-structure $\mathcal{T}_R$, in $O(n \log n)$ time, such that given a query point $\mathsf{q} \in \mathbb{R}^{\mathsf{d}}$, one can return a $(1 + 4n)$-ANN of q in P in $O(\log n)$ time.

Given a query point q, using $\mathcal{T}_R$, we compute a point $u \in \mathsf{P}$, such that $\mathbf{d}(\mathsf{q}, \mathsf{P}) \leq \|u - \mathsf{q}\| \leq (1+4n)\mathbf{d}(\mathsf{q}, \mathsf{P})$. Let $R = \|u - \mathsf{q}\|$ and $r = \|u - \mathsf{q}\| / (4n+1)$. Clearly, $r \leq \mathbf{d}(\mathsf{q}, \mathsf{P}) \leq R$. Next, compute $\mathbb{L} = \lceil \lg R \rceil$, and let C be the set of cells of $\mathsf{G}_{2^{\mathbb{L}}}$ that are within a distance $\leq R$ from q; that is, C is the set of grid cells of $\mathsf{G}_{2^{\mathbb{L}}}$ whose union (completely) covers $\mathbf{b}(\mathsf{q}, R)$. Clearly, $\mathsf{nn}(\mathsf{q}, \mathsf{P}) \in \mathbf{b}(\mathsf{q}, R) \subseteq \bigcup_{\square \in C} \square$. Next, for each cell $\square \in C$, we compute the node $v \in \widehat{T}$ such that $\mathsf{P} \cap \square = \mathsf{P}_v$, using a cell query (i.e., Lemma 17.5). (Note that if $\square$ does not contain any point of P, this query would return a leaf or a compressed node whose region contains $\square$, and it might contain at most one point of P.) Let V be the resulting set of nodes of $\widehat{T}$.

For each node of $v \in V$, we now apply the algorithm of Theorem 17.4 to the compressed quadtree rooted at v. We return the nearest neighbor found.

[①]Buyer beware in Latin.

Since $|V| = O(1)$ and $\operatorname{diam}(\mathsf{P}_v) = O(R)$, for all $v \in V$, the query time is

$$\sum_{v \in V} O\left(\frac{1}{\varepsilon^{\mathsf{d}}} + \log \frac{\operatorname{diam}(\mathsf{P}_v)}{r}\right) = O\left(\frac{1}{\varepsilon^{\mathsf{d}}} + \sum_{v \in V} \log \frac{\operatorname{diam}(\mathsf{P}_v)}{r}\right) = O\left(\frac{1}{\varepsilon^{\mathsf{d}}} + \sum_{v \in V} \log \frac{R}{r}\right)$$
$$= O\left(\frac{1}{\varepsilon^{\mathsf{d}}} + \log n\right).$$

As for the correctness, observe that there is a node $w \in V$, such that $\mathsf{nn}(\mathsf{q}, \mathsf{P}) \in \mathsf{P}_w$. As such, when we apply the algorithm of Theorem 17.4 to w, it would return us a $(1+\varepsilon)$-ANN to q.

Theorem 17.6. *Let* P *be a set of* n *points in* $\mathbb{R}^{\mathsf{d}}$. *One can construct a data-structure of linear size, in* $O(n \log n)$ *time, such that given a query point* $\mathsf{q} \in \mathbb{R}^{\mathsf{d}}$ *and a parameter* $1 \geq \varepsilon > 0$, *one can compute a* $(1+\varepsilon)$*-ANN to* q *in* $O(1/\varepsilon^{\mathsf{d}} + \log n)$ *time.*

17.4. Low quality ANN search via the ring separator tree

To perform ANN in the unbounded spread case, all we need is a rough approximation (i.e., polynomial factor in n) to the distance to the nearest neighbor (note that we need only the distance). One way to achieve that was described in Theorem 11.21_{p160}, and we present another alternative construction that uses a more direct argument.

Definition 17.7. A binary tree $\mathcal{T}$ having the points of P as leaves is a ***t-ring tree*** for P if every node $v \in T$ is associated with a ring (hopefully "thick"), such that the ring separates the points into two sets (hopefully both relatively large) and the interior of the ring is empty of any point of P.

For a node v of $\mathcal{T}$, let P_v denote the subset of points of P stored in the subtree of v, and let rep_v be a point stored in v. We require that for any node v of $\mathcal{T}$, there is an associated ball $\mathbf{b}_v = \mathbf{b}(\mathsf{c}_v, r_v)$, such that all the points of $\mathsf{P}^v_{\text{in}} = \mathsf{P}_v \cap \mathbf{b}_v$ are in one child of $\mathcal{T}$. Furthermore, all the other points of P_v are outside the interior of the enlarged ball $\mathbf{b}(\mathsf{c}_v, (1+t)r_v)$ and are stored in the other child of v. (Note that c_v might not be a point of P.)

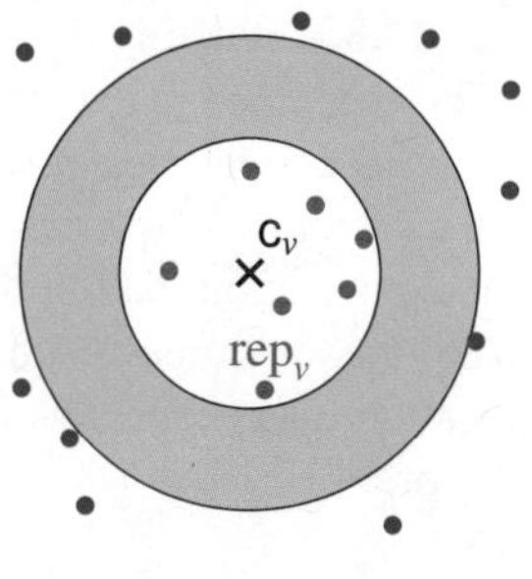

We will also store an arbitrary representative point $\mathrm{rep}_v \in \mathsf{P}^v_{\text{in}}$ in v (rep_v is not necessarily c_v).

To see what the above definition implies, consider a t-ring tree $\mathcal{T}$. For any node $v \in \mathcal{T}$, the interior of the ring associated with v (i.e., $\mathbf{b}(\mathsf{c}_v, (1+t)r_v) \setminus \mathbf{b}(\mathsf{c}_v, r_v)$) is empty of any point of P. Intuitively, the bigger t is, the better $\mathcal{T}$ clusters P. Furthermore, every internal node v of $\mathcal{T}$ has the following quantities associated with it:

(i) r_v: radius of the inner ball of the ring of v,
(ii) c_v: center of the ring of v (the point c_v is not necessarily in P), and
(iii) $\mathrm{rep}_v \in \mathsf{P}^v_{\text{in}}$: a representative from the point set stored in the inner ball of v (note that rep_v might be distinct from c_v).

The ANN search procedure. Let q denote the query point. Initially, set v to be the root of $\mathcal{T}$ and $r_{\text{curr}} \leftarrow \infty$. The algorithm answers the ANN query by traversing down $\mathcal{T}$.

During the traversal, we first compute the distance $l = \left\|\mathsf{q} - \mathrm{rep}_v\right\|$. If this is smaller than r_{curr} (the distance to the current nearest neighbor found), then we update r_{curr} (and store the point realizing the new value of r_{curr}).

If $\|\mathsf{q} - \mathsf{c}_v\| \leq \widehat{r}$, we continue the search recursively in the child containing P^v_{in}, where $\widehat{r} = (1+t/2)r_v$ is the "middle" radius of the ring centered at c_v. Otherwise, we continue the search in the subtree containing $\mathsf{P}^v_{\mathrm{out}}$. The algorithm stops when reaching a leaf of $\mathcal{T}$ and returns the point realizing r_{curr} as the approximate nearest neighbor.

Intuition. If the query point q is outside the outer ball of a node v, it is so far from the points inside the inner ball (i.e., $\mathsf{P}^v_{\mathrm{in}}$), and we can treat all of them as a single point (i.e., rep_v). On the other hand, if the query point q′ is inside the inner ball, then it must have a neighbor nearby (i.e., a point of $\mathsf{P}^v_{\mathrm{in}}$), and all the points of $\mathsf{P}^v_{\mathrm{out}}$ are far enough away that they can be ignored. Naturally, if the query point falls inside the ring, the same argumentation works (with slightly worst constants), using the middle radius as the splitting boundary in the search. See the figure on the right.

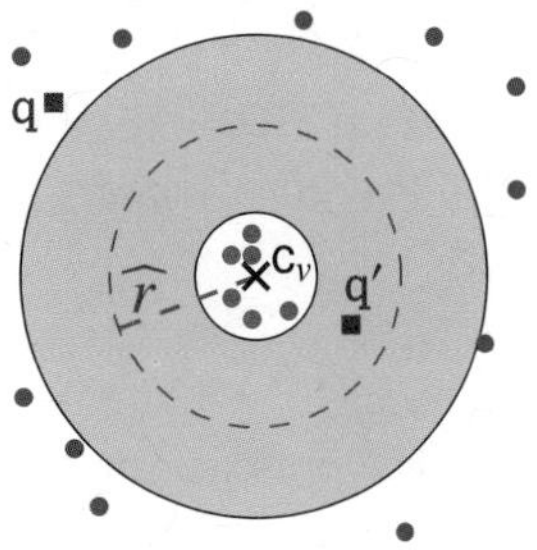

Lemma 17.8. *Given a t-ring tree $\mathcal{T}$, one can answer $(1 + 4/t)$-approximate nearest neighbor queries, in $O(\mathtt{depth}(\mathcal{T}))$ time.*

Proof. Clearly, the query time is $O(\mathtt{depth}(\mathcal{T}))$. As for the quality of approximation, let π denote the generated search path in $\mathcal{T}$ and let $\mathsf{nn}(\mathsf{q})$ denote the nearest neighbor to q in P. Furthermore, let w denote the last node in the search path π, such that $\mathsf{nn}(\mathsf{q}) \in \mathsf{P}_w$. Clearly, if $\mathsf{nn}(\mathsf{q}) \in \mathsf{P}^w_{\mathrm{in}}$ but we continued the search in $\mathsf{P}^w_{\mathrm{out}}$, then q is outside the middle sphere (i.e., $\|\mathsf{q} - \mathsf{c}_w\| \geq (1 + t/2)r_w$) and $\|\mathsf{q} - \mathsf{nn}(\mathsf{q})\| \geq (t/2)r_w$ (since this is the distance between the middle sphere and the inner sphere). Thus,

$$\left\|\mathsf{q} - \mathrm{rep}_w\right\| \leq \|\mathsf{q} - \mathsf{nn}(\mathsf{q})\| + \left\|\mathsf{nn}(\mathsf{q}) - \mathrm{rep}_w\right\| \leq \|\mathsf{q} - \mathsf{nn}(\mathsf{q})\| + 2r_w,$$

since $\mathsf{nn}(\mathsf{q}), \mathrm{rep}_w \in \mathbf{b}_w = \mathbf{b}(\mathsf{c}_w, r_w)$. In particular,

$$\frac{\left\|\mathsf{q} - \mathrm{rep}_w\right\|}{\|\mathsf{q} - \mathsf{nn}(\mathsf{q})\|} \leq \frac{\|\mathsf{q} - \mathsf{nn}(\mathsf{q})\| + 2r_w}{\|\mathsf{q} - \mathsf{nn}(\mathsf{q})\|} \leq 1 + \frac{2r_w}{\|\mathsf{q} - \mathsf{nn}(\mathsf{q})\|} \leq 1 + \frac{2r_w}{(t/2)r_w} = 1 + \frac{4}{t}.$$

Namely, rep_w is a $(1 + 4/t)$-approximate nearest neighbor to q.

Similarly, if $\mathsf{nn}(\mathsf{q}) \in \mathsf{P}^w_{\mathrm{out}}$ but we continued the search in $\mathsf{P}^w_{\mathrm{in}}$, then (i) $\|\mathsf{q} - \mathsf{nn}(\mathsf{q})\| \geq (t/2)r_w$, (ii) $\|\mathsf{q} - \mathsf{c}_w\| \leq (1 + t/2)r_w$, and (iii) $\left\|\mathsf{q} - \mathrm{rep}_w\right\| \leq \|\mathsf{q} - \mathsf{c}_w\| + \left\|\mathsf{c}_w - \mathrm{rep}_w\right\| \leq (t/2 + 2)r_w$. As such, we have that

$$\frac{\left\|\mathsf{q} - \mathrm{rep}_w\right\|}{\|\mathsf{q} - \mathsf{nn}(\mathsf{q})\|} \leq \frac{(t/2 + 2)r_w}{(t/2)r_w} \leq 1 + \frac{4}{t}.$$

Since rep_w is considered as one of the candidates to be nearest neighbor in the search, this implies that the algorithm returns a $(1 + 4/t)$-ANN. ■

In low dimensions, there is always a good separating ring (see Lemma 3.28_{p40}), and we use it in the following construction.

Lemma 17.9. *Given a set* P *of n points in $\mathbb{R}^{\mathsf{d}}$, one can compute a $(1/n)$-ring tree of* P *in $O(n \log n)$ time.*

Proof. The construction is recursive. Setting $t = n$, by Lemma 3.28_{p40}, one can compute, in linear time, a ball $\mathbf{b}(\mathsf{p}, r)$, such that (i) $|\mathbf{b}(\mathsf{p}, r) \cap \mathsf{P}| \geq n/c$, (ii) the ring $\mathbf{b}(\mathsf{p}, r(1 + 1/t)) \setminus \mathbf{b}(\mathsf{p}, r)$ contains no point of P, and (iii) $|\mathsf{P} \setminus \mathbf{b}(\mathsf{p}, r(1 + 1/t))| \geq n/2$.

Let v be the root of the new tree, set $\mathsf{P}^v_{\mathrm{in}}$ to be $\mathsf{P} \cap \mathbf{b}(\mathsf{p}, r)$ and $\mathsf{P}^v_{\mathrm{out}} = \mathsf{P} \setminus \mathsf{P}^v_{\mathrm{in}}$, and store $\mathbf{b}_v = \mathbf{b}(\mathsf{p}, r')$ and $\mathrm{rep}_v = \mathsf{p}$. Continue the construction recursively on those two sets. Observe that $\left|\mathsf{P}^v_{\mathrm{in}}\right|, \left|\mathsf{P}^v_{\mathrm{out}}\right| \geq n/c$, where c is a constant. It follows that the construction time of the algorithm is $T(n) = O(n) + T\left(\left|\mathsf{P}^{v}_{\text{in}}\right|\right) + T\left(\left|\mathsf{P}^{v}_{\text{out}}\right|\right) = O(n \log n)$, and the depth of the resulting tree is $O(\log n)$. ■

Combining the above two lemmas, we get the following result.

Theorem 17.10. *Let* P *be a set of n points in* $\mathbb{R}^{\mathsf{d}}$. *One can preprocess it in* $O(n \log n)$ *time, such that given a query point* $\mathsf{q} \in \mathbb{R}^{\mathsf{d}}$, *one can return a* $(1 + 4n)$*-ANN of* q *in* P *in* $O(\log n)$ *time.*

17.5. Bibliographical notes

The presentation of the ring tree follows the recent work of Har-Peled and Mendel [**HM06**]. Ring trees are probably an old idea. A more elaborate but similar data-structure is described by Indyk and Motwani [**IM98**]. Of course, the property that "thick" ring separators exist is inherently low dimensional, as the regular simplex in n dimensions demonstrates. One option to fix this is to allow the rings to contain points and to replicate the points inside the ring in both subtrees. As such, the size of the resulting tree is not necessarily linear. However, careful implementation yields linear (or small) size; see Exercise 17.1 for more details. This and several additional ideas are used in the construction of the cover tree of Indyk and Motwani [**IM98**].

Section 17.2 is a simplification of the work of Arya et al. [**AMN⁺98**]. Section 17.3 is also inspired to a certain extent by Arya et al., although it is essentially a simplification of Har-Peled and Mendel [**HM06**] data-structure to the case of compressed quadtrees. In particular, we believe that the data-structure presented is conceptually simpler than in previously published work.

There is a huge amount of literature on approximate nearest neighbor search, both in low and high dimensions in the theory, learning and database communities. The reason for this lies in the importance of this problem on special input distributions encountered in practice, different computation models (i.e., I/O-efficient algorithms), search in high dimensions, and practical efficiency.

Liner space. In low dimensions, the seminal work of Arya et al. [**AMN⁺98**], mentioned above, was the first to offer linear size data-structure, with logarithmic query time, such that the approximation quality is specified with the query. The query time of Arya et al. is slightly worse than the running time of Theorem 17.6, since they maintain a heap of cells, always handling the cell closest to the query point. This results in query time $O(\varepsilon^{-d} \log n)$. It can be further improved to $O(1/\varepsilon^{\mathsf{d}} \log(1/\varepsilon) + \log n)$ by observing that this heap has only very few delete-mins and many insertions. This observation is due to Duncan [**Dun99**].

Instead of having a separate ring tree, Arya et al. rebalance the compressed quadtree directly. This results in nodes which correspond to cells that have the shape of an annulus (i.e., the region formed by the difference between two canonical grid cells).

Duncan [**Dun99**] and some other authors offered a data-structure (called the ***BAR-tree***) with similar query time, but it seems to be inferior, in practice, to the work of Arya et al., for the reason that while the regions that the nodes correspond to are convex, they have higher descriptive complexity, and it is harder to compute the distance of the query point to a cell.

Faster query time. One can improve the query time if one is willing to specify ε during the construction of the data-structure, resulting in a trade-off between space and query time. In particular, Clarkson [**Cla94**] showed that one can construct a data-structure of (roughly) size $O(n/\varepsilon^{(\mathsf{d}-1)/2})$ and query time $O(\varepsilon^{-(\mathsf{d}-1)/2}\log n)$. Chan simplified and cleaned up this result [**Cha98**] and also presented some other results.

Details on faster query time. A set of points Q is $\sqrt{\varepsilon}$-*far* from a query point q if the $\|\mathsf{q}-c_{\mathsf{Q}}\| \geq \mathrm{diam}(\mathsf{Q})/\sqrt{\varepsilon}$, where c_{Q} is some point of Q. It is easy to verify that if we partition space around c_{Q} into cones with central angle $O(\sqrt{\varepsilon})$ (this requires $O(1/\varepsilon^{(\mathsf{d}-1)/2})$ cones), then the most extreme point of Q in such a cone ψ, furthest away from c_{Q}, is the $(1+\varepsilon)$-approximate nearest neighbor for any query point inside ψ which is $\sqrt{\varepsilon}$-far. Namely, we precompute the ANN inside each cone if the point is far enough away. Furthermore, by careful implementation (i.e., grid in the angles space), we can decide, in constant time, which cone the query point lies in. Thus, using $O(1/\varepsilon^{(\mathsf{d}-1)/2})$ space, we can answer $(1+\varepsilon)$-ANN queries for q if the query point is $\sqrt{\varepsilon}$-far, in constant time.

Next, construct this data-structure for every set P_v, for $v \in \widehat{T}(\mathsf{P})$, where $\widehat{T}(\mathsf{P})$ is a compressed quadtree for P. This results in a data-structure of size $O(n/\varepsilon^{(\mathsf{d}-1)/2})$. Given a query point q, we use the algorithm of Theorem 17.6 and stop for a node v as soon P_v is $\sqrt{\varepsilon}$-far, and then we use the secondary data-structure for P_v. It is easy to verify that the algorithm would stop as soon as $\mathrm{diam}(\Box_v) = O(\sqrt{\varepsilon}\mathbf{d}(\mathsf{q},\mathsf{P}))$. As such, the number of nodes visited would be $O(\log n + 1/\varepsilon^{\mathsf{d}/2})$, and the bound on the query time is identical.

Note that we omitted the construction time (which requires some additional work to be done efficiently), and our query time is slightly worse than the best known. The interested reader can check out the work by Chan [**Cha98**], which is somewhat more complicated than what is outlined here.

Even faster query time. The first to achieve $O(\log(n/\varepsilon))$ query time (using near linear space) was Har-Peled [**Har01b**], using space roughly $O(n\varepsilon^{-d}\log^2 n)$. This was later simplified and improved by Arya and Malamatos [**AM02**], who presented a data-structure with the same query time and of size $O(n/\varepsilon^{\mathsf{d}})$. These data-structure relies on the notion of computing approximate Voronoi diagrams and performing point-location queries in those diagrams. By extending the notion of approximate Voronoi diagrams, Arya, Malamatos, and Mount [**AMM02**] showed that one can answer $(1+\varepsilon)$-ANN queries in $O(\log(n/\varepsilon))$ time, using $O(n/\varepsilon^{(\mathsf{d}-1)})$ space. On the other end of the spectrum, they showed that one can construct a data-structure of size $O(n)$ and query time $O(\log n + 1/\varepsilon^{(\mathsf{d}-1)/2})$ (note that for this data-structure $\varepsilon > 0$ has to be specified in advance). In particular, the later result breaks a space/query time trade-off that all other results suffer from (i.e., the query time multiplied by the construction time has dependency of $1/\varepsilon^{\mathsf{d}}$ on ε).

Practical considerations. Arya et al. [**AMN⁺98**] implemented their algorithm. For most inputs, it is essentially a *kd*-tree. The code of their library was carefully optimized and is very efficient. In particular, in practice, the author would expect it to beat most of the algorithms mentioned above. The code of their implementation is available online as a library [**AM98**].

Higher dimensions. All our results have exponential dependency on the dimension, in query and preprocessing time (although the space can probably be made subexponential with careful implementation). Getting a subexponential algorithm requires a completely different technique and is discussed in detail later (see Theorem 20.20_{p276}).

Stronger computation models. If one assume that the points have integer coordinates in the range $[1,U]$, then approximate nearest neighbor queries can be answered in (roughly) $O(\log\log U + 1/\varepsilon^{\mathsf{d}})$ time [**AEIS99**] or even $O(\log\log(U/\varepsilon))$ time [**Har01b**]. The algorithm of Har-Peled [**Har01b**] relies on computing a compressed quadtree of height $O(\log(U/\varepsilon))$ and performing a fast point-location query in it. This only requires using the floor function and hashing (note that the algorithm of Theorem 17.6 uses the floor function and hashing during the construction, but it is not used during the query). In fact, if one is allowed to slightly blow up the space (by a factor U^δ, where $\delta > 0$ is an arbitrary constant), the ANN query time can be improved to constant [**HM04**].

By shifting quadtrees and creating $\mathsf{d}+1$ quadtrees, one can argue that the approximate nearest neighbor must lie in the same cell (and this cell is of the "right" size) of the query point in one of those quadtrees. Next, one can map the points into a real number, by using the natural space filling curve associated with each quadtree. This results in $\mathsf{d}+1$ lists of points. One can argue that a constant approximate neighbor must be adjacent to the query point in one of those lists. This can be later improved into $(1+\varepsilon)$-ANN by spreading $1/\varepsilon^{\mathsf{d}}$ points. This simple algorithm is due to Chan [**Cha02**].

The reader might wonder why we bothered with a considerably more involved algorithm. There are several reasons: (i) This algorithm requires the numbers to be integers of limited length (i.e., $O(\log U)$ bits), (ii) it requires shuffling of bits on those integers (i.e., for computing the inverse of the space filling curve) in constant time, and (iii) the assumption is that one can combine d such integers into a single integer and perform exclusive-or on their bits in constant time. The last two assumptions are not reasonable when the input is made out of floating point numbers.

Further research. In low dimensions, the ANN problem seems to be essentially solved both in theory and practice (such proclamations are inherently dangerous and should be taken with a considerable amount of healthy skepticism). Indeed, for $\varepsilon > 1/\log^{1/d} n$, the current data structure of Theorem 17.6 provides logarithmic query time. Thus, ε has to be quite small for the query time to become bad enough that one would wish to speed it up.

The main directions for further research seem to be on this problem in higher dimensions and solving it in other computation models.

Surveys. A survey on the approximate nearest neighbor search problem in high dimensions is by Indyk [**Ind04**]. In low dimensions, there is a survey by Arya and Mount [**AM05**].

17.6. Exercises

Exercise 17.1 (Better ring tree). Let P be a set of n points in $\mathbb{R}^{\mathsf{d}}$. Show how to build a ring tree, of linear size, that can answer $O(\log n)$-ANN queries in $O(\log n)$ time. [**Hint:** Show that there is always a ring containing $O(n/\log n)$ points, such that it is of width w and its interior radius is $O(w \log n)$. Next, build a ring tree, replicating the points in both children of this ring node. Argue that the size of the resulting tree is linear, and prove the claimed bound on the query time and quality of approximation.]

Exercise 17.2 (k-ANN from ANN). In the ***k-approximate nearest neighbor*** problem one has to preprocess a point set P, such that given a query point q, one needs to return k distinct points of P such that their distance from q is at most $(1+\varepsilon)\ell$, where ℓ is the distance from q to its kth nearest neighbor in P.

Show how to build $O(k \log n)$ ANN data-structures, such that one can answer such a k-ANN query using $O(k \log n)$ "regular" ANN queries in these data-structures and the result returned is correct with high probability. (Prove that your data-structure indeed returns k distinct points.)

CHAPTER 18

Approximate Nearest Neighbor via Point-Location

18.1. ANN using point-location among balls

In the following, let P be a set of n points in a metric space $\mathcal{M}$. We remind the reader that a ***ball*** is defined by its center p and a radius r, such that

$$\mathbf{b}(\mathsf{p}, r) = \left\{ \mathsf{q} \in \mathcal{M} \;\middle|\; \mathbf{d}_{\mathcal{M}}(\mathsf{p}, \mathsf{q}) \leq r \right\}.$$

In particular, if $\mathcal{M}$ is a finite metric space, then the ball is the finite set of all the points of $\mathcal{M}$ within a distance at most r from p.

Definition 18.1. For a set of balls $\mathcal{B}$ such that $\bigcup_{\mathbf{b} \in \mathcal{B}} \mathbf{b} = \mathcal{M}$ and a query point $\mathsf{q} \in \mathcal{M}$, the ***target ball*** of q in $\mathcal{B}$, denoted by $\odot_{\mathcal{B}}(\mathsf{q})$, is the smallest ball (i.e., the ball of smallest radius) of $\mathcal{B}$ that contains q (if several equal radius balls contain q, then we resolve this in an arbitrary fashion).

An example of such a set of balls and their induced partition of space is depicted in Figure 18.1.

Our objective is to show that $(1+\varepsilon)$-ANN queries can be reduced to target ball queries, on a "small" set of balls. Let us start from a "silly" result, to get some intuition about the problem.

Observation 18.2. *Consider a set of balls $\mathcal{B}$ centered at the points of P. Let q be a query point, and let $\mathbf{b}(\mathsf{p}, r)$ be the target ball $\odot_{\mathcal{B}}(\mathsf{q})$. Then, the following hold:*
(A) $\mathbf{d}(\mathsf{q}, \mathsf{P}) \leq \mathbf{d}_{\mathcal{M}}(\mathsf{q}, \mathsf{p})$, *where $\mathbf{d}(\mathsf{q}, \mathsf{P})$ is the distance of q to its nearest neighbor $\mathsf{nn}(\mathsf{q}, \mathsf{P})$ in P. This holds since the target ball is centered at a point p of P.*
(B) If there is a ball $\mathbf{b}(\mathsf{s}, r') \in \mathcal{B}$ that contains q and $r' \leq (1+\varepsilon)\mathbf{d}(\mathsf{q}, \mathsf{P})$, then $\mathbf{d}_{\mathcal{M}}(\mathsf{q}, \mathsf{p}) \leq (1+\varepsilon)\mathbf{d}(\mathsf{q}, \mathsf{P})$. Indeed, the target ball $\mathbf{b}(\mathsf{p}, r)$ has radius $r \leq r' \leq (1+\varepsilon)\mathbf{d}(\mathsf{q}, \mathsf{P})$. Now, by definition, $\mathsf{q} \in \mathbf{b}(\mathsf{p}, r)$, which implies that $\mathbf{d}_{\mathcal{M}}(\mathsf{q}, \mathsf{p}) \leq r \leq (1+\varepsilon)\mathbf{d}(\mathsf{q}, \mathsf{P})$.

Definition 18.3. For a set P of n points in a metric space $\mathcal{M}$, let the ***union of balls of radius*** r centered at the points of P be denoted by $\mathcal{U}_{\text{balls}}(\mathsf{P}, r) = \bigcup_{\mathsf{p} \in \mathsf{P}} \mathbf{b}(\mathsf{p}, r)$.

Lemma 18.4. *Let P be a set of points in $\mathcal{M}$. Let $\mathcal{B} = \bigcup_{i=-\infty}^{\infty} \mathcal{U}_{\text{balls}}(\mathsf{P}, (1+\varepsilon)^i)$. (Note that $\mathcal{B}$ is an infinite set.) For a query $\mathsf{q} \in \mathcal{M}$, let $\mathbf{b} = \odot_{\mathcal{B}}(\mathsf{q})$, and let $\mathsf{p} \in \mathsf{P}$ be the center of $\mathbf{b}$. Then, p is $(1+\varepsilon)$-ANN in P to q.*

Proof. Let $\mathsf{s} = \mathsf{nn}(\mathsf{q}, \mathsf{P})$ be the nearest neighbor to q in P, and let $r = \|\mathsf{q} - \mathsf{s}\| = \mathbf{d}(\mathsf{q}, \mathsf{P})$. Let i be the index such that $(1+\varepsilon)^i < r \leq (1+\varepsilon)^{i+1}$. Observe that $\mathsf{q} \in \mathbf{b}\big(\mathsf{s}, (1+\varepsilon)^{i+1}\big) \in \mathcal{B}$. As such, the target ball $\mathbf{b}$ must be of radius $\leq (1+\varepsilon)^{i+1}$. Similarly, the target ball radius cannot be smaller than r, since all such balls in $\mathcal{B}$ do not contain the query point q. We conclude that $(1+\varepsilon)^{i+1} \geq \text{radius}(\mathbf{b}) \geq r > (1+\varepsilon)^i$. It follows that

$$\|\mathsf{q} - \mathsf{p}\| \leq \text{radius}(\mathbf{b}) \leq \text{radius}\Big(\mathbf{b}\big(\mathsf{s}, (1+\varepsilon)^{i+1}\big)\Big) = (1+\varepsilon)^{i+1} < (1+\varepsilon)\mathbf{d}(\mathsf{q}, \mathsf{P}).$$

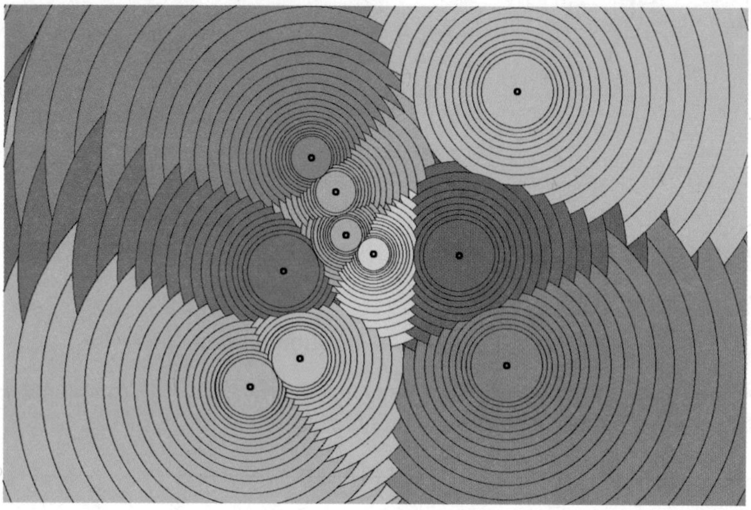

FIGURE 18.1. Example of a collection of disks (i.e., balls), generated by the algorithm of Theorem 18.18, such that a target query among them can resolve approximate nearest neighbor queries for the original points. Since we give preference to smaller balls, the region returning a specific point p as an answer to the target query is an approximation to the region where p is the nearest neighbor among all the points of P.

This implies that p is $(1+\varepsilon)$-ANN to q in P. ∎

18.1.1. A simple quadratic construction. Here is a construction of a set of balls of polynomial size, such that target queries answer $(1+\varepsilon)$-ANN correctly. Think about ANN as a competition where the points of P compete to be the nearest neighbor of the query point. Also, consider two specific points $u, v \in \mathsf{P}$. If q is much closer to u than to v (say $\mathbf{d}_{\mathcal{M}}(\mathsf{q}, u) \leq \mathbf{d}_{\mathcal{M}}(u, v)/4$), then, intuitively, deciding that u is preferable to v is easy. Now, if the query point is sufficiently far from both u and v (say $\mathbf{d}_{\mathcal{M}}(\mathsf{q}, u) \geq 2\mathbf{d}_{\mathcal{M}}(u, v)/\varepsilon$), then they are both equivalent as far as the ANN search is concerned, and the search algorithm can pick either one to continue in the competition for computing the ANN.

Thus, as far as resolving whether u or v is the correct answer, the problematic range is where the query is sufficiently close to u and v but not too close to one of them. Specifically, the algorithm needs to work "harder" to decide whether to return u or v if and only if

$$\mathbf{d}(\mathsf{q}, \{u, v\}) \in \mathcal{I}(u, v) = \left[\frac{1}{8}, \frac{4}{\varepsilon}\right] \mathbf{d}_{\mathcal{M}}(u, v).$$

In particular, consider the set of balls that can resolve the ANN question for these two points; that is,

$$\mathcal{B}(u, v) = \left\{ \mathbf{b}(u, r), \ \mathbf{b}(v, r) \ \middle| \ r = (1+\varepsilon)^i \in \mathcal{I}(u, v) \text{ and } i \in \mathbb{Z} \right\}. \tag{18.1}$$

The set of resulting balls is depicted on the right, and as the following claim testifies, it is pretty small.

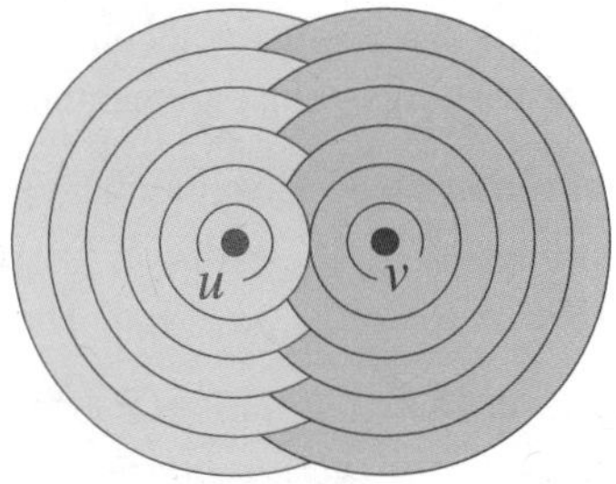

Claim 18.5. *We have* $|\mathcal{B}(u,v)| = O((1/\varepsilon)\log(1/\varepsilon))$.

Proof. Let $R = \frac{2\mathbf{d}_{\mathcal{M}}(u,v)}{\varepsilon} / \frac{\mathbf{d}_{\mathcal{M}}(u,v)}{4} = O(1/\varepsilon)$ be the ratio between the high and low values in $\mathcal{I}(u,v)$. Observe that the number of balls used is $O(\log_{1+\varepsilon} R) = O(\log(R)/\varepsilon) = O\left(\varepsilon^{-1}\log\varepsilon^{-1}\right)$; see Lemma 28.10_{p348}. ■

Construction. For every pair of points $u, v \in \mathsf{P}$, add the balls in the set $\mathcal{B}(u,v)$ to the set of balls $\mathcal{B}$. In addition, we pick an arbitrary point $\mathsf{p} \in \mathsf{P}$ and add the ball $\mathbf{b}(\mathsf{p},\infty)$ to $\mathcal{B}$.

Lemma 18.6. *For a set* P *of* n *points in a metric space, one can compute a set of balls* $\mathcal{B}$ *of size* $O\left(n^2\varepsilon^{-1}\log\varepsilon^{-1}\right)$, *such that one can answer* ANN *queries using target queries on* $\mathcal{B}$. *Specifically, for a query point* q, *let* $\mathbf{b} = \odot_{\mathcal{B}}(\mathsf{q})$, *and let* $\mathsf{p} \in \mathsf{P}$ *be the center of* $\mathbf{b}$. *Then,* p *is* $(1+\varepsilon)$*-ANN in* P *to* q.

Proof. The construction is described above and the bound on the size of $\mathcal{B}$ is straightforward. As for the correctness, let $\mathsf{q}' = \mathsf{nn}(\mathsf{q},\mathsf{P})$ be the nearest neighbor to q in P. There are three potential cases to consider.

(A) If $\mathbf{d}(\mathsf{q},\mathsf{P}) = \mathbf{d}_{\mathcal{M}}(\mathsf{q},\mathsf{q}') \geq 2\mathbf{d}_{\mathcal{M}}(\mathsf{p},\mathsf{q}')/\varepsilon$, then, by the triangle inequality, we have that
$$\mathbf{d}_{\mathcal{M}}(\mathsf{q},\mathsf{p}) \leq \mathbf{d}_{\mathcal{M}}(\mathsf{q},\mathsf{q}') + \mathbf{d}_{\mathcal{M}}(\mathsf{p},\mathsf{q}') \leq (1+\varepsilon/2)\mathbf{d}_{\mathcal{M}}(\mathsf{q},\mathsf{q}') = (1+\varepsilon/2)\mathbf{d}(\mathsf{q},\mathsf{P}).$$

(B) If $\mathbf{d}(\mathsf{q},\mathsf{P}) < \mathbf{d}_{\mathcal{M}}(\mathsf{p},\mathsf{q}')/8$, then, by construction, there is a ball $\mathbf{g}$ of radius at most $\mathbf{d}_{\mathcal{M}}(\mathsf{p},\mathsf{q}')/4$ of $\mathcal{B}(\mathsf{p},\mathsf{q}')$ centered at q' that contains q. By the triangle inequality, the distance of q to p is at least $(3/4)\mathbf{d}_{\mathcal{M}}(\mathsf{p},\mathsf{q}')$. Namely, all the balls of $\mathcal{B}$ centered at p have considerably larger radius than $\mathbf{g} \in \mathcal{B}$. As such, it is impossible that a target query on $\mathcal{B}$ returned a ball centered at p.

(C) Finally, if $\mathbf{d}(\mathsf{q},\mathsf{P}) \in [1/4, 2/\varepsilon]\,\mathbf{d}_{\mathcal{M}}(\mathsf{p},\mathsf{q}')$, then, by the construction of $\mathcal{B}(\mathsf{p},\mathsf{q}') \subseteq \mathcal{B}$, there exists a ball $\mathbf{g}'$ of radius $\leq (1+\varepsilon)\mathbf{d}(\mathsf{q},\mathsf{P})$ centered at q' in $\mathcal{B}$ that contains q. Now, by Observation 18.2(B) the claim follows. ■

Interestingly, in low dimensional Euclidean space, by using a WSPD, we can group together these sets of balls, so that the number of balls needed is only $n/\varepsilon^{O(1)}$; see Section 18.3_{p252}.

Getting the number of balls to be near linear (for the general metric space case) requires us to be more careful, and this is the main task in this chapter.

18.1.2. Handling a range of distances.

Definition 18.7. For a real number $r > 0$, a ***near neighbor data structure***, denoted by $\mathcal{D}_{\text{Near}} = \mathcal{D}_{\text{Near}}(\mathsf{P}, r)$, is a data-structure, such that given a query point q, it can decide whether $\mathbf{d}(\mathsf{q},\mathsf{P}) \leq r$ or $\mathbf{d}(\mathsf{q},\mathsf{P}) > r$. If $\mathbf{d}(\mathsf{q},\mathsf{P}) \leq r$, then it also returns a witness point $\mathsf{p} \in \mathsf{P}$ such that $\mathbf{d}(\mathsf{q},\mathsf{p}) \leq r$.

The data-structure $\mathcal{D}_{\text{Near}}(\mathsf{P}, r)$ can be realized as a set of n balls (of radius r) around the points of P. Performing a target ball query on this set is done by checking, for each ball, if it contains the query point. For the time being, the reader can consider $\mathcal{D}_{\text{Near}}(\mathsf{P}, r)$ to be such a set. (Naturally, if the query point is further away than r from P, then the data-structure would return that there is no ball containing the query.)

One can resolve ANN queries on a range of distances $[a, b]$ by building near neighbor data-structures, with exponential growth of radii in this range.

Definition 18.8. Given a range $[a, b]$, let $\mathcal{N}_i = \mathcal{D}_{\mathrm{Near}}(\mathsf{P}, r_i)$, where $r_i = \min\bigl((1+\varepsilon)^i a,\, b\bigr)$, for $i = 0, \ldots, M$, where $M = \lceil \log_{1+\varepsilon}(b/a) \rceil$. We denote this set of data-structures by $\widehat{\mathcal{I}}(\mathsf{P}, a, b, \varepsilon) = \bigl\{\mathcal{N}_0, \ldots, \mathcal{N}_M\bigr\}$. The set $\widehat{\mathcal{I}}$ is an ***interval near neighbor*** data-structure.

Lemma 18.9. *Given a set* P *of* n *points in a metric space and parameters* $a \le b$ *and* $\varepsilon > 0$, *one can construct an interval near neighbor data-structure* $\widehat{\mathcal{I}} = \widehat{\mathcal{I}}(\mathsf{P}, a, b, \varepsilon)$, *such that the following properties hold for* $\widehat{\mathcal{I}}$: *(A)* $\widehat{\mathcal{I}}$ *is made out of* $O(\varepsilon^{-1}\log(b/a))$ *near neighbor data-structures, and (B)* $\widehat{\mathcal{I}}$ *contains* $O(\varepsilon^{-1} n \log(b/a))$ *balls overall. Furthermore, given a query point* q, *one can decide if either*

(i) $\mathbf{d}(\mathsf{q}, \mathsf{P}) \le a$ *or*

(ii) $\mathbf{d}(\mathsf{q}, \mathsf{P}) > b$, *and otherwise*

(iii) return a number r *and a point* $\mathsf{p} \in \mathsf{P}$, *such that* $\mathbf{d}(\mathsf{q}, \mathsf{P}) \le r = \mathbf{d}(\mathsf{q}, \mathsf{p}) \le (1 + \varepsilon)\mathbf{d}(\mathsf{q}, \mathsf{P})$.

This requires two near neighbor queries if (i) or (ii) holds. Otherwise, the number of near neighbor queries needed is $O(\log(\varepsilon^{-1}\log(b/a)))$.

Proof. The construction of $\widehat{\mathcal{I}}$ is described above. Given a query point q, we first check if $\mathbf{d}(\mathsf{q}, \mathsf{P}) \le a$, by querying $\mathcal{N}_0$, and if so, the algorithm returns "$\mathbf{d}(\mathsf{q}, \mathsf{P}) \le a$". Otherwise, we check if $\mathbf{d}(\mathsf{q}, \mathsf{P}) > b$, by querying $\mathcal{N}_M$, and if so, the algorithm returns "$\mathbf{d}(\mathsf{q}, \mathsf{P}) > b$".

Otherwise, let $X_i = 1$ if $\mathbf{d}(\mathsf{q}, \mathsf{P}) \le r_i$ and zero otherwise, for $i = 0, \ldots, M$. We can determine the value of X_i by performing a query in the data-structure $\mathcal{N}_i$. Clearly, $X_0, X_1, \ldots, X_M$ is a sequence of zeros, followed by a sequence of ones. As such, we can find the i, such that $X_i = 0$ and $X_{i+1} = 1$, by performing a binary search. This would require $O(\log M)$ queries. In this case, we have that $r_i < \mathbf{d}(\mathsf{q}, \mathsf{P}) \le r_{i+1} \le (1 + \varepsilon) r_i$.

Clearly, the ball of radius $r_{i+1} \le (1 + \varepsilon)\mathbf{d}(\mathsf{q}, \mathsf{P})$ centered at $\mathsf{nn}(\mathsf{q}, \mathsf{P})$ is in $\mathcal{N}_{i+1} = \mathcal{D}_{\mathrm{Near}}(\mathsf{P}, r_{i+1}) \subseteq \widehat{\mathcal{I}}$, which by Observation 18.2(B) implies that the returned point is the required ANN.

The number of queries is $O(\log M)$ and $M = \lceil \log_{1+\varepsilon}(b/a) \rceil = O(\log(b/a)/\varepsilon)$, by Lemma 28.10$_{\mathrm{p348}}$. ∎

Corollary 18.10. *Let* P *be a set of* n *points in a metric space* $\mathcal{M}$, *and let* $a < b$ *be real numbers. For a query point* $\mathsf{q} \in \mathcal{M}$, *such that* $\mathbf{d}(\mathsf{q}, \mathsf{P}) \in [a, b]$, *the target query over the set of balls* $\widehat{\mathcal{I}}(\mathsf{P}, a, b, \varepsilon)$ *returns a ball centered at a point which is a* $(1 + \varepsilon)$*-ANN to* q.

Lemma 18.9 implies that we can "cheaply" resolve $(1 + \varepsilon)$-ANN over intervals which are not too long.

Definition 18.11. Any two points p and q of $\mathcal{M}$ are ***in the same connected component*** of $\mathcal{U}_{\mathrm{balls}}(\mathsf{P}, r) = \bigcup_{\mathsf{p} \in \mathsf{P}} \mathbf{b}(\mathsf{p}, r)$ if there exists a sequence of points $\mathsf{s}_1, \ldots, \mathsf{s}_k \in \mathsf{P}$ such that (i) $\mathbf{d}_{\mathcal{M}}(\mathsf{p}, \mathsf{s}_1) \le r$, (ii) $\mathbf{d}_{\mathcal{M}}(\mathsf{s}_i, \mathsf{s}_{i+1}) \le 2r$, for $i = 1, \ldots, k-1$, and (iii) $\mathbf{d}_{\mathcal{M}}(\mathsf{s}_k, \mathsf{p}) \le r$.

Lemma 18.12. *Let* Q *be a set of* m *points in a metric space* $\mathcal{M}$ *and let* $r > 0$ *be a real number, such that* $\mathcal{U}_{\mathrm{balls}}(\mathsf{Q}, r)$ *is a connected set. Then we have the following:*

(A) Any two points $\mathsf{p}, \mathsf{q} \in \mathsf{Q}$ *are within a distance at most* $2r(m-1)$ *from each other.*

(B) For a query point $\mathsf{q} \in \mathcal{M}$, *such that* $\mathbf{d}(\mathsf{q}, \mathsf{Q}) \ge \mathrm{diam}(\mathsf{Q})/\delta$, *we have that any point of* Q *is a* $(1 + \delta)$*-ANN to* q *in the set* Q.

Proof. (A) Since $\mathcal{U}_{\mathrm{balls}}(\mathsf{Q}, r)$ is a connected set, by Definition 18.11, there is a path of length $\le (m-1)2r$ between any two points x and y of Q.

(B) Let $s = nn(q, Q)$ denote the nearest neighbor to q in Q; that is, $\mathbf{d}_{\mathcal{M}}(q, s) = \mathbf{d}(q, Q)$. By assumption, $\operatorname{diam}(Q) \le \delta\, \mathbf{d}(q, Q) = \delta\, \mathbf{d}_{\mathcal{M}}(q, s)$. Now, for any $p \in Q$, we have

$$\begin{aligned}\mathbf{d}_{\mathcal{M}}(q, p) &\le \mathbf{d}_{\mathcal{M}}(q, s) + \mathbf{d}_{\mathcal{M}}(s, p) \le \mathbf{d}_{\mathcal{M}}(q, s) + \operatorname{diam}(Q) \le (1 + \delta)\mathbf{d}_{\mathcal{M}}(q, s)\\ &= (1 + \delta)\mathbf{d}(q, Q).\end{aligned}$$

■

Lemma 18.12 implies that for faraway query points, a cluster of points Q which are close together can be treated as a single point.

18.1.3. The ANN data-structure. Here, we describe how to build a tree to answer ANN queries and analyze the parameters of the construction.

18.1.3.1. *The construction.* We are given a set of points P and a t-approximate BHST H of P; see Definition 11.10_{p155}. For a leaf u of H, we consider rep_u to be the point stored in u. For an internal node u, we require that rep_u is the representative of one of its children. As such, given any subtree $H' \subseteq H$, we denote by $P(H') = \left\{ \mathrm{rep}_u \,\middle|\, u \in H' \right\}$ the set of all the representatives stored in the subtree H′.

We remind the reader that a ***separator*** is a node u of a tree H such that its removal breaks H into connected components such that each one of them contains at most $|V(H)|/2$ vertices. A separator can be computed in linear time; see Lemma 2.13_{p19}.

We are going to build a tree $\mathcal{T}$, such that each node v (of the new tree $\mathcal{T}$) will have an interval near neighbor data-structure $\widehat{\mathcal{I}}_v$ associated with it. As we traverse down the tree $\mathcal{T}$, we will use those data-structures to decide what child to continue the search into.

During the recursive construction, we are given a subtree S of the t-approximate BHST H (initially, of course, $S = H$), and we build an ANN data-structure (i.e., a tree) for the points (i.e., representatives) stored in S. The newly created search tree $\mathcal{T}(S)$ will be rooted at a node $v = v(S)$. For such a node $v = v(S) \in \mathcal{T}$, let n_v denote the number of nodes of S, and let $P^v = P(S) = \bigcup_{u \in S} \mathrm{rep}_u$ denote the set of points for which the subtree rooted at v handles ANN queries. Clearly, we have that $|P^v| \le n_v$ (as the same point might be the representative used for several nodes of S).

So, let $u^v \in V(S)$ be a separator of S, and consider the two subtrees S_L and S_R rooted in the left and right children of u^v, respectively. Now, let S_{out} be the subtree resulting from S by removing S_L and S_R. Also, let $P^v_L = P(S_L)$, $P^v_R = P(S_R)$, and $P^v_{out} = P(S_{out})$.

Given a query point q, we need to decide whether to continue the search in S_L, S_R, or S_{out}. To this end, we build an interval nearest neighbor data-structure and store it in the node v; that is, $\widehat{\mathcal{I}}_v = \widehat{\mathcal{I}}(P^v, r_v, R_v, \varepsilon/4)$, where

$$r_v = \frac{\Delta(u^v)}{4t} \quad \text{and} \quad R_v = \mu\Delta(u^v) \tag{18.2}$$

and $\mu = O(\varepsilon^{-1} \log n)$ is a global parameter. Note that $\Delta(u^v)$ is the label of u^v (at S). This label is an upper bound, which is a t-approximation, to the diameter of the points in the HST stored in the subtree of u^v; see Definition 11.10_{p155}. We now recursively build ANN search trees for S_L, S_R and S_{out}, and hang them on the children of v denoted by v_L, v_R, and v_{out}, respectively.

A second look. Observe that in the above recursive construction, we do not directly partition the point set P^v but rather we recursively break up the HST S storing these points. Note that some points will be stored in the HST not only in a leaf but also as the representative point along a path starting in this leaf and going up the HST. This creates the somewhat counterintuitive situation that while P^v_L and P^v_R are disjoint, the representative point rep_{u^v} (which is in $P^v_L \cup P^v_R$) is also in P^v_{out}. In particular, this is the only point shared by these sets; that is, $\left(P^v_L \cup P^v_R\right) \cap P^v_{out} = \{\text{rep}_{u^v}\}$ (equality holds here, as $\text{rep}_{u^v} \in P^v_{out}$).

18.1.3.2. *Answering a query.* Given a query point q and a node v in the resulting tree $\mathcal{T}$ (initially v would be the root of $\mathcal{T}$), we perform a query into $\widehat{\mathcal{I}}_v$. We then get one of the following three results.

(A) $\mathbf{d}(q, P^v) \le r_v$: Then $q \in \mathcal{U}_{balls}(P^v, r_v)$, and so the data-structure returns a point $p \in P^v$ such that $\mathbf{d}_{\mathcal{M}}(q, p) \le r_v$. If $p \in P^v_L$ (resp. $p \in P^v_R$), then the search continues recursively in v_L (resp. v_R). Otherwise, the search continues recursively in v_{out}.

(B) $\mathbf{d}(q, P^v) \in (r_v, R_v]$: Then the interval near neighbor data-structure finds a $(1 + \varepsilon/4)$-ANN point $s \in P^v$ and returns it as the answer to the query.

(C) $\mathbf{d}(q, P^v) > R_v$: Then the search continues recursively in v_{out}.

18.1.3.3. *Analysis.*

Observation 18.13. *Since* H *is a t-approximation to the metric* $\mathcal{M}$*, we get that the following properties hold, where* $\Delta(u^v)$ *is the label of* u^v *in* H*.*

(A) $P^v_R \cap P^v_L = \emptyset$.

(B) $P^v_R \cap \left(P^v_{out} \setminus \{\text{rep}_{u^v}\}\right) = \emptyset$.

(C) $P^v_L \cap \left(P^v_{out} \setminus \{\text{rep}_{u^v}\}\right) = \emptyset$.

(D) $\mathbf{d}_{\mathcal{M}}\left(P^v_L \cup P^v_R, P^v_{out} \setminus \{\text{rep}_{u^v}\}\right) \ge \Delta(u^v)/t$.

(E) $\mathbf{d}_{\mathcal{M}}\left(P^v_L, P^v_R\right) \ge \Delta(u^v)/t$.

(F) $\text{diam}\left(P^v_L \cup P^v_R\right) \le \Delta(u^v)$.

Lemma 18.14. *The point returned by the above data-structure is a* $(1 + \varepsilon)$*-ANN to the query point in* P.

Proof. Consider a node v on the search path. We have the following possibilities:

(A) $\mathbf{d}(q, P^v) \le r_v$. Then the data-structure continues the search in a subtree that must contain the ANN to q in P^v.

For example, if the algorithm continues to v_L, then $\mathbf{d}\left(q, P^v_L\right) \le r_v$. Furthermore, by Observation 18.13(D) and (E), we have that $\mathbf{d}_{\mathcal{M}}\left(P^v_L, P^v \setminus P^v_L\right) \ge \Delta(u^v)/t$. As such, setting $q_L = \text{nn}\left(q, P^v_L\right)$, we have by the triangle inequality that

$$\mathbf{d}\left(q, P^v \setminus P^v_L\right) \ge \mathbf{d}\left(q_L, P^v \setminus P^v_L\right) - \mathbf{d}_{\mathcal{M}}(q, q_L) \ge \frac{\Delta(u^v)}{t} - r_v > \frac{\Delta(u^v)}{2t} > r_v,$$

as $r_v = \Delta(u^v)/4t$.

Similar arguments work for the other two possibilities.

One subtle issue is that $\text{rep}_{u^v} \in P^v_L \cup P^v_R$. As such, if the search continues into v_{out}, then it must be that the distance between q and rep_{u^v} is considerably larger than r_v. That is, the nearest neighbor to q in P^v_{out} is not rep_{u^v}. Thus, this claim also holds if the search continues into P^v_{out}.

(B) $\mathbf{d}(q, P^v) \in (r_v, R_v]$: Then the interval near neighbor data-structure found a $(1 + \varepsilon/4)$-ANN in P^v and it returns it as the answer to the query. Specifically, a multiplicative error of $1 + \varepsilon/4$ is introduced into the quality of the approximation.

(C) $\mathbf{d}(q, P^v) > R_v$: Then the search continued recursively in v_{out}. Let $s = \text{nn}(q, P^v)$ denote the nearest neighbor to q in P^v.

If $s \in P^v_{out}$, then no error is introduced in this step.

Otherwise, if $s \in P_L^v \cup P_R^v$, then, by Observation 18.13, $\text{diam}\left(P_L^v \cup P_R^v\right) \leq \Delta(u^v)$. Now, $R_v = \mu\Delta(u^v)$, which implies that $\text{rep}_{u^v} \in P_{\text{out}}^v$ is a $(1 + 1/\mu)$-ANN to q, by Lemma 18.12(B). Indeed, any point of $P_L^v \cup P_R^v$ is $(1 + 1/\mu)$-ANN to q, and $\text{rep}_{u^v} \in P_L^v \cup P_R^v$. Namely, this step introduces a multiplicative error of $1 + 1/\mu$.

Now, initially, the given BHST has at most $2n$ nodes (indeed, the points of P are stored in the leaves of H). At a node v, the number of nodes that the subtree of the HST used to construct v is n_v, and the number of nodes in the subtrees sent to the children of v is $\leq n_v/2 + 1$ for each child. As such, the depth of the generated tree $\mathcal{T}$ is bounded by $N = 2 \lg 2n$ (the extra factor of 2 rises because each subproblem is $+1$ larger than half of the parent subproblem). Setting $\mu = 4N/\varepsilon$, we have that the quality of the ANN returned by the search tree is

$$\left(1 + \frac{1}{\mu}\right)^N \left(1 + \frac{\varepsilon}{4}\right) \leq \exp\left(\frac{N}{\mu} + \frac{\varepsilon}{4}\right) = \exp\left(\frac{N}{4N/\varepsilon} + \frac{\varepsilon}{4}\right) = \exp\left(\frac{\varepsilon}{2}\right) \leq 1 + \varepsilon,$$

since $1+x \leq e^x \leq 1+2x$, for $0 \leq x \leq 1/2$. Namely, this data-structure returns a $(1+\varepsilon)$-ANN to the query point. ■

This ANN data-structure is made out of near neighbor data-structures that in turn can be interpreted as sets of balls.

Lemma 18.15. *For $t = n^{O(1)}$, the above ANN data-structure is made out of $O\left(\frac{n}{\varepsilon} \log^2 n\right)$ balls.*

PROOF. Since $\mu = O(\varepsilon^{-1} \log n)$, the number of balls used in an interval near neighbor data-structure at a node v, by Lemma 18.9, is

$$U(n_v) = O\left(\frac{n_v}{\varepsilon} \log \frac{R_v}{r_v}\right) = O\left(\frac{n_v}{\varepsilon} \log \frac{\mu\Delta(u^v)}{\Delta(u^v)/4t}\right) = O\left(\frac{n_v}{\varepsilon} \log\left(\frac{t \log n}{\varepsilon}\right)\right).$$

We get the recurrence $B(n) = U(n) + B(n_L) + B(n_R) + B(n_{\text{out}})$, where $n = n_L + n_R + n_{\text{out}}$, $n_L \leq n/2 + 1$, $n_R \leq n/2 + 1$, and $n_{\text{out}} \leq n/2 + 1$. Clearly, the solution of this recurrence for $B(2n) = O\left(\varepsilon^{-1} n \log n \log(\varepsilon^{-1} t \log n)\right)$.

Now, for $t = n^{O(1)}$ and $1/\varepsilon = O(n)$, we have that $B(2n) = O\left((n/\varepsilon) \log^2 n\right)$. ■

Lemma 18.16. *Assuming $t = n^{O(1)}$, the above ANN data-structure performs $O(\log(n/\varepsilon))$ near neighbor queries.*

PROOF. Consider the search path π in $\mathcal{T}$ for a query point q. Every internal (to the path π) node $v \in V(\mathcal{T})$ on this path corresponds to a situation where $\mathbf{d}(q, P^v) < r_v$ or $\mathbf{d}(q, P^v) > R_v$, and this is decided by two near neighbor queries, by Lemma 18.9. In the final node v of π, the search algorithm resolves the query using

$$O\left(\log \frac{\log(R_v/r_v)}{\varepsilon}\right) = O\left(\log \frac{1}{\varepsilon} + \log \log\left(\frac{t \log n}{\varepsilon}\right)\right) = O\left(\log \frac{1}{\varepsilon} + \log \log n\right)$$

near neighbor queries, by Lemma 18.9, and since $t = n^{O(1)}$. Now, the depth of the tree is $O(\log n)$, and so the claim follows, as $O(\log n + \log(1/\varepsilon) + \log \log n) = O(\log(n/\varepsilon))$. ■

Lemma 18.17. *Using the set of balls $\mathcal{B}$ realizing the above ANN data-structure, one can answer a $(1 + \varepsilon)$-ANN query by performing a single target query on $\mathcal{B}$.*

PROOF. The proof is an easy adaption of the proof of Lemma 18.14, and we provide it for the sake of completeness; in particular, the reader can safely skip this proof.

So, assume the query reached a node v of the ANN data-structure. Let $\mathcal{B}_v$ be the set of balls stored in the subtree of v. We claim that if the algorithm continues the search in a child w of v, then all the balls of $U = \mathcal{B}_v \setminus \mathcal{B}_w$ are not relevant for the target query. That is, the ball returned by the target query on $\mathcal{B}_v$ belongs to the set $\mathcal{B}_w$.

Again, we have the following three possibilities:

(A) $\mathbf{d}(\mathsf{q}, \mathsf{P}^v) \leq r_v$: Then $\mathsf{q} \in \mathcal{U}_{\text{balls}}(\mathsf{P}^v, r_v)$, and hence there exists a ball of radius r_v centered at a point p of P^v in $\mathcal{B}$ that contains q.

In particular, if $\mathsf{p} \in \mathsf{P}_\mathsf{L}$, then $w = v_\mathsf{L}$. Arguing as in the proof of Lemma 18.14, it follows that any ball centered in $\mathsf{P}^v \setminus \mathsf{P}^v_\mathsf{L}$ that covers q has radius strictly larger than r_v, implying that the target query can be restricted to $\mathcal{B}_w$. A similar argument applies if $\mathsf{p} \in \mathsf{P}_\mathsf{R}$.

The above argument is slightly more involved if p is $\mathrm{rep}_{u^v} \in \mathsf{P}_\mathsf{R} \cup \mathsf{P}_\mathsf{L}$. Then, observe that all the points of $\mathsf{P}^v_{\mathsf{out}} \setminus \{\mathrm{rep}_{u^v}\}$ are within a distance at least $\Delta(u^v)/t \geq 4r_v$ from $\mathsf{P}^v_\mathsf{L} \cup \mathsf{P}^v_\mathsf{R}$. As such, no ball (that covers q) centered at these points can be the smallest ball containing q in $\mathcal{B}_v$. If $\mathsf{p} \neq \mathrm{rep}_{u^v}$, then this trivially holds.

The same argumentation as above applies to the case $\mathsf{p} \in \mathsf{P}^v_{\mathsf{out}} \setminus \{\mathrm{rep}_{u^v}\}$, for $w = v_{\mathsf{out}}$.

(B) $\mathbf{d}(\mathsf{q}, \mathsf{P}_v) \in (r_v, R_v]$: Then one of the balls of the interval near neighbor data-structure of v covers q, and hence the point (that is the center of the ball) returned by the target query is an ANN of the same quality as the one returned by a direct query on this interval near neighbor data-structure.

(C) $\mathbf{d}(\mathsf{q}, \mathsf{P}_v) > R_v$: Then the target ball containing p was generated in the subtree of v_{out} and we continue the search there.

As for the quality of the ANN returned, the argument of Lemma 18.14 applies verbatim here, implying the claim. ∎

18.1.3.4. *The result.* We thus get the following result.

Theorem 18.18. *Given a set* P *of* n *points in a metric space* $\mathcal{M}$ *and an HST* H *of* P *that* t*-approximates* $\mathcal{M}$*, for* $t = n^{O(1)}$*, one can construct a data-structure* $\mathcal{D}$ *that answers* $(1+\varepsilon)$*-ANN queries, by performing* $O(\log(n/\varepsilon))$ *near neighbor queries. The data-structure* $\mathcal{D}$ *is realized by a set* $\mathcal{B}$ *of* $O(n\varepsilon^{-1}\log^2 n)$ *balls, and answering* $(1+\varepsilon)$*-ANN queries on* P *can be done by performing a target query on* $\mathcal{B}$.

By being more clever about the construction of the balls using the HST, one can save a polylogarithmic factor. We only state the result, and we leave proving it to Exercise 18.2.

Theorem 18.19. *Given a set* P *of* n *points in a metric space* $\mathcal{M}$ *and an HST* H *of* P *that* t*-approximates* $\mathcal{M}$*, for* $t = n^{O(1)}$*, one can construct a data-structure* $\mathcal{D}$ *that answers* $(1+\varepsilon)$*-ANN queries, by performing* $O(\log(n/\varepsilon))$ *near neighbor queries.*

Interpreting the near neighbor data-structure as a set of balls, the data-structure $\mathcal{D}$ *is realized by a set* $\mathcal{B}$ *of* $O(n\varepsilon^{-1}\log n)$ *balls, and answering* $(1+\varepsilon)$*-ANN queries on* P *can be done by performing a target query on* $\mathcal{B}$.

18.2. ANN using point-location among approximate balls

While Theorem 18.19 might be surprising, it cannot be used immediately to solve ANN, since it reduces the ANN problem into answering near neighbor queries, which also seem to be a hard problem to solve efficiently.

The key observation is that we do not need to use exact balls. We are allowed to slightly deform the balls. This approximate near neighbor problem is considerably easier both in lower and higher dimensions.

Definition 18.20. For a ball $\mathbf{b} = \mathbf{b}(p, r)$, a set $\mathbf{b}_{\approx}$ is a $(1+\varepsilon)$-***approximation to*** $\mathbf{b}$ if $\mathbf{b} \subseteq \mathbf{b}_{\approx} \subseteq \mathbf{b}(p, (1+\varepsilon)r)$.

For a set of balls $\mathcal{B}$, the set $\mathcal{B}_{\approx}$ is a $(1+\varepsilon)$-***approximation*** to $\mathcal{B}$ if for any ball $\mathbf{b} \in \mathcal{B}$ there is a corresponding $(1+\varepsilon)$-approximation $\mathbf{b}_{\approx} \in \mathcal{B}_{\approx}$. For a set $\mathbf{b}_{\approx} \in \mathcal{B}_{\approx}$, let $\mathbf{b} \in \mathcal{B}$ denote the ball corresponding to $\mathbf{b}_{\approx}$, let $r_{\mathbf{b}}$ be the radius of $\mathbf{b}$, and let $p_{\mathbf{b}} \in \mathsf{P}$ denote the center of $\mathbf{b}$.

For a query point $\mathsf{q} \in \mathcal{M}$, the ***target set*** of $\mathcal{B}_{\approx}$ for q is the set $\mathbf{b}_{\approx}$ of $\mathcal{B}_{\approx}$ that contains q and has the smallest radius $r_{\mathbf{b}}$.

Example 18.21. As a concrete example of how algorithmically making the balls approximate helps, consider the problem of deciding, given a point set P in $\mathbb{R}^{\mathsf{d}}$, if a query point is within distance at most 1 of any point of P.

In the exact variant, we need to perform a point-location to decide if we are inside the union of balls of radius 1 centered at the points of P. This problem is solvable but requires quite a bit of non-trivial machinery if one wants a fast query time. In particular, for $\mathsf{P} \subseteq \mathbb{R}^{\mathsf{d}}$, the best algorithm currently known requires $\Omega\left(n^{\lceil \mathsf{d}/2 \rceil}\right)$ space if one wants logarithmic query time. Alternatively, even in three dimensions, the best solution known requires query time $\Omega\left(n^{2/3}\right)$ if one insists on using only near linear space.

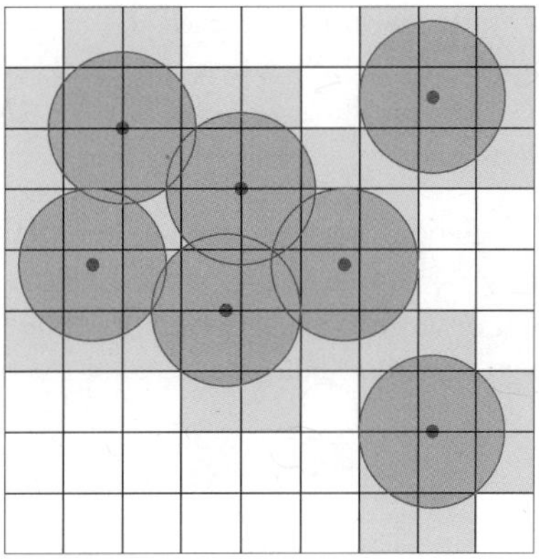

On the other hand, in the approximate variant, we can replace every ball by the grid cells it covers (the sidelength of the grid is roughly ε, the approximation parameter). Thus every ball is replaced by the union of grid cells that intersect it; see the figure on the right. Now, we can throw these grid cells into a hash table, such that deciding if a query point is inside the union of the approximate balls is equivalent to deciding if the query point is inside a marked grid cell, and this can be decided in constant time (note that here all balls are of the same radius, so deciding if one is inside the union of the marked grid cells is sufficient).

Note that the same approach works in any constant dimension d and requires linear space, that is, $O\left(n/\varepsilon^{\mathsf{d}}\right)$ space.

Even in the approximate ball case, the interval near neighbor data-structure still works.

Lemma 18.22. *Let $\mathcal{I}_{\approx} = \mathcal{I}_{\approx}(\mathsf{P}, r, R, \varepsilon/16)$ be a $(1+\varepsilon/16)$-approximation to the data-structure $\widehat{\mathcal{I}} = \widehat{\mathcal{I}}(\mathsf{P}, r, R, \varepsilon/16)$. For a query point, $\mathsf{q} \in \mathcal{M}$, if $\mathcal{I}_{\approx}$ returns a target set that is an approximation to a ball in $\widehat{\mathcal{I}}$, centered at a point $\mathsf{p} \in \mathsf{P}$ of radius α, and $\alpha \in [r, R]$, then p is $(1+\varepsilon/4)$-ANN to q.*

Proof. The data-structure of $\mathcal{I}_{\approx}$ returns p as an ANN to q if and only if there are two consecutive indices, such that q is inside the union of the approximate balls of $\mathcal{N}_{i+1}$ but not inside the union of approximate balls of $\mathcal{N}_i$. As such q is outside the union of the original balls of $\mathcal{N}_i$. Thus,

$$r(1+\varepsilon/16)^i \le \mathbf{d}(\mathsf{q},\mathsf{P}) \le \mathbf{d}(\mathsf{p},\mathsf{q}) \le r(1+\varepsilon/16)^{i+1}(1+\varepsilon/16) \le (1+\varepsilon/16)^2\mathbf{d}(\mathsf{q},\mathsf{P}) \le (1+\varepsilon/4)\mathbf{d}(\mathsf{q},\mathsf{P}).$$

Thus p is indeed $(1+\varepsilon/4)$-ANN. ∎

Lemma 18.23. *Let P be a set of n points in a metric space $\mathcal{M}$, and let $\mathcal{B}$ be a set of balls with centers at P, computed by the algorithm of Theorem 18.19, such that one can answer $(1+\varepsilon/16)$-ANN queries on P, by performing a single target query on $\mathcal{B}$.*

Let $\mathcal{B}_{\approx}$ be a $(1+\varepsilon/16)$-approximation to $\mathcal{B}$. Then, any target query q on $\mathcal{B}_{\approx}$ returns a $(1+\varepsilon)$-ANN to q in P.

Proof. The argument follows the recursive proof of Lemma 18.14. Since the adaption is straightforward, we omit it here. (Interestingly, the "obvious" proof of this lemma is false; see Exercise 18.1.) ∎

By Theorem 11.18$_{\text{p157}}$, for a set P of n points in $\mathbb{R}^{\mathsf{d}}$ one can construct a BHST that t-approximates the distances of P, in $O(\mathsf{d}n\log n)$ time, where $t = 2\sqrt{\mathsf{d}}n^5 \le n^7$, since $\mathsf{d} \le n$. Plugging this into the above, we get the following result.

Theorem 18.24. *Given a set P of n points in $\mathbb{R}^{\mathsf{d}}$, one can compute a set $\mathcal{B}$ of $O(n\varepsilon^{-1}\log n)$ balls, such that answering $(1+\varepsilon)$-ANN queries on P can be done by performing a single target query on $\mathcal{B}$. Computing this set of balls takes $O(\mathsf{d}n\log n + n\varepsilon^{-1}\log n)$ time.*

Furthermore, one can $(1+\varepsilon/16)$-approximate each ball of $\mathcal{B}$ and answer $(1+\varepsilon)$-ANN queries by performing a single target query on the resulting set of approximate balls.

Alternatively, one can answer ANN queries using $O(\log(n/\varepsilon))$ approximate near neighbor queries.

Proof. We only need to bound the construction time. The first bound follows from an analysis similar to the one in Lemma 18.15, and the second bound follows by modifying it and observing that the running time is dominated by the time it takes to build the interval near neighbor data-structure for the root node. ∎

18.3. ANN using point-location among balls in low dimensions

Surprisingly, one can extend the naive construction of Section 18.1.1 to get a linear number of balls in low dimensions. The idea is to use WSPD to group together pairs such that each such group uses only $O(\varepsilon^{-1}\log\varepsilon^{-1})$ balls.

The input is a set P of n points in $\mathbb{R}^{\mathsf{d}}$.

Construction. Using the algorithm of Theorem 3.10$_{\text{p33}}$, construct a (c/ε)-WSPD of P of size $O(n/\varepsilon^{\mathsf{d}})$, where c is a sufficiently large constant, to be determined below. For every pair $\{u,v\} \in \mathcal{W}$, compute the set $\mathcal{B}(\mathrm{rep}_u,\mathrm{rep}_v)$ and add it to the set of balls $\mathcal{B}$, where

$$\mathcal{B}(\mathrm{rep}_u,\mathrm{rep}_v) = \left\{ \mathbf{b}(\mathrm{rep}_u,r),\ \mathbf{b}(\mathrm{rep}_v,r) \;\middle|\; r = (1+\varepsilon/3)^i \in \mathcal{J}(u,v) \text{ and } i \in \mathbb{Z} \right\}$$

$$\text{and}\quad \mathcal{J}(u',v') = \left[\frac{1}{8},\frac{4}{\varepsilon}\right]\cdot \left\|\mathrm{rep}_u - \mathrm{rep}_v\right\|.$$

We also add a ball of infinite radius centered at an arbitrary point of P to the resulting set $\mathcal{B}$.

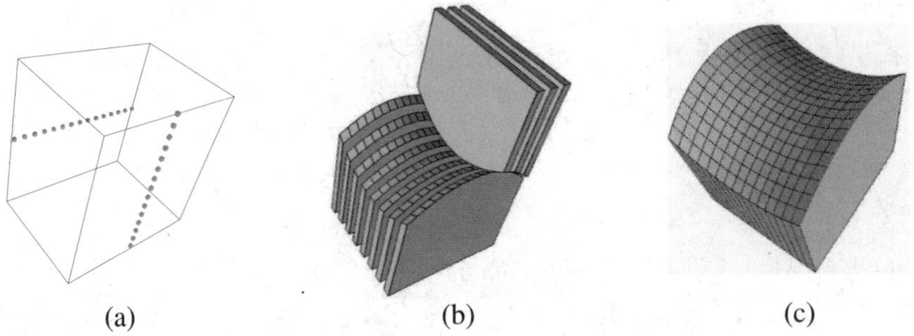

FIGURE 18.2. (a) The point set in 3D inducting a Voronoi diagram of quadratic complexity. (b) Some cells in this Voronoi diagram (the view is rotated). Note that the cells are thin and flat, and every cell from the lower part touches the cells on the upper part. (c) The contact surface between the two parts of the Voronoi diagram has quadratic complexity, and thus the Voronoi diagram itself has quadratic complexity.

Correctness. The correctness follows by modifying the argument in the proof of Lemma 18.6, so that it is being applied to the relevant representatives of the pair of the WSPD covering the two points under consideration. The proof is straightforward but somewhat tedious and is delegated to Exercise 18.3.

Lemma 18.25. *For a set* P *of* n *points in* $\mathbb{R}^d$, *one can compute a set of balls* $\mathcal{B}$, *of size* $O\left(n\varepsilon^{-d-1}\log\varepsilon^{-1}\right)$, *such that one can answer* $(1+\varepsilon)$*-ANN queries using a target query on* $\mathcal{B}$.

It is easy to verify that the same holds even if the balls used are approximate. Thus, we get the following result.

Theorem 18.26. *Given a set* P *of* n *points in* $\mathbb{R}^d$, *one can compute a set* $\mathcal{B}$ *of balls of size* $O(n\varepsilon^{-d-1}\log\varepsilon^{-1})$ *(the constant in the* $O(\cdot)$ *depends exponentially on the dimension), such that answering* $(1+\varepsilon)$*-ANN queries on* P *can be done by performing a single target query on* $\mathcal{B}$. *Computing this set of balls takes* $O(dn\log n + n\varepsilon^{-d-1}\log\varepsilon^{-1})$ *time.*

Furthermore, one can $(1+\varepsilon/16)$*-approximate each ball of* $\mathcal{B}$ *and answer* $(1+\varepsilon)$*-ANN queries by performing a single target query on the resulting set of approximate balls.*

PROOF. The construction and correctness is outlined above. We only bound the number of generated balls. Observe that every pair in the WSPD generates $O(\varepsilon^{-1}\log\varepsilon^{-1})$ balls. There are $O(n\varepsilon^{-d})$ pairs in the WSPD, which implies the claim. ∎

18.4. Approximate Voronoi diagrams

18.4.1. Introduction. A Voronoi diagram of a point set $P \subseteq \mathbb{R}^d$ is a partition of space into regions, such that the ***cell*** of a point $p \in P$ is

$$V(p, P) = \left\{ s \in \mathbb{R}^d \;\middle|\; \|s - p\| \le \|s - p'\| \text{ for all } p' \in P \right\}.$$

Voronoi diagrams are a powerful tool and have numerous applications [**Aur91**].

One problem with Voronoi diagrams is that their descriptive complexity is $O\left(n^{\lceil d/2 \rceil}\right)$ in $\mathbb{R}^d$, in the worst case. See Figure 18.2 for an example in three dimensions. It is a natural question to ask whether one can reduce the complexity to linear (or near linear) by allowing some approximations.

Definition 18.27 (Approximate Voronoi diagram)**.** Given a set P of n points in $\mathbb{R}^{\mathsf{d}}$ and parameter $\varepsilon > 0$, a $(1+\varepsilon)$***-approximated Voronoi diagram*** of P is a partition, $\mathcal{V}$, of $\mathbb{R}^{\mathsf{d}}$ into regions, such that for any region $\varphi \in \mathcal{V}$, there is an associated point $\mathrm{rep}_\varphi \in \mathsf{P}$, such that for any $x \in \varphi$, we have that rep_φ is a $(1+\varepsilon)$-ANN for x; that is,

$$\forall x \in \varphi \quad \left\| x - \mathrm{rep}_\varphi \right\| \le (1+\varepsilon)\mathbf{d}(\mathsf{q}, \mathsf{P}).$$

We will refer to $\mathcal{V}$ as a $(1+\varepsilon)$-***AVD*** of P.

18.4.2. Fast ANN in $\mathbb{R}^{\mathsf{d}}$. In the following, assume P is a set of points contained in the hypercube $[0.5 - \varepsilon/\mathsf{d}, 0.5 + \varepsilon/\mathsf{d}]^{\mathsf{d}}$. This can be easily guaranteed by an affine linear transformation T which preserves relative distances (i.e., computing the ANN for q on P is equivalent to computing the ANN for $T(\mathsf{q})$ on $T(\mathsf{P})$).

In particular, if the query point is outside the unit hypercube $[0,1]^{\mathsf{d}}$, then any point of P is $(1+\varepsilon)$-ANN, and we are done. Thus, we need to answer ANN only for points inside the unit hypercube.

We use the reduction between ANN and target queries among approximate balls. In particular, by Theorem $18.26_{\mathrm{p}253}$, one can compute a set $\mathcal{B}$ of $O\left(n\varepsilon^{-\mathsf{d}-1}\log \varepsilon^{-1}\right)$ balls centered at the points of P. This takes $O\left(n \log n + n\varepsilon^{-\mathsf{d}-1}\log\varepsilon^{-1}\right)$ time. Next, we approximate each ball $\mathbf{b} \in \mathcal{B}$. By Theorem $18.26_{\mathrm{p}253}$, the $(1+\varepsilon)$-ANN query can be answered by performing a target query on the resulting set $\mathcal{B}_\approx$ of approximate balls.

To this end, we approximate a ball $\mathbf{b}$ of radius r of $\mathcal{B}$ by the set of grid cells of G_{2^i} that intersect it, where $\sqrt{\mathsf{d}}2^i \le (\varepsilon/16)r$; namely, $i = \left\lfloor \lg\left(\varepsilon r/\sqrt{\mathsf{d}}\right)\right\rfloor$. In particular, let $\mathbf{b}_\approx$ denote the region corresponding to the grid cells intersected by $\mathbf{b}$. Clearly, $\mathbf{b}_\approx$ is an approximate ball of $\mathbf{b}$. Let $\mathcal{B}_\approx$ be the resulting set of approximate balls for $\mathcal{B}$, and let C' be the associated (multi-)set of grid cells formed by those balls.

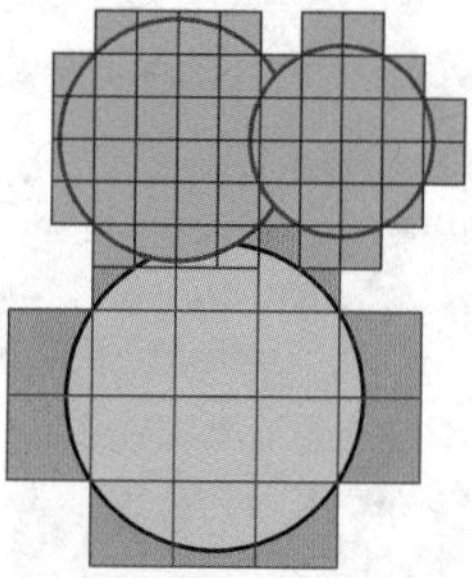

The multi-set C' is a collection of canonical grid cells from different resolutions. We create a set out of C', by picking from all the instances of the same cell $\square \in C'$ the instance associated with the smallest ball of $\mathcal{B}$. Roughly speaking, the center of this smallest ball represents the closest point to this grid cell in P. Let C be the resulting set of canonical grid cells. What the resulting set of grid cell might look like is depicted on the right (note that a point might be covered several times by grid cells of different resolutions).

Thus, by Theorem $18.24_{\mathrm{p}252}$, answering a $(1+\varepsilon)$-ANN query is no more than performing a target query on $\mathcal{B}_\approx$. This, in turn, requires finding the smallest canonical grid cell of C which contains the query point q. Earlier, we showed that one can construct a compressed quadtree $\widehat{T}$ that has all the cells of C appearing as nodes of $\widehat{T}$. The construction of this compressed quadtree takes $O(|C| \log |C|)$ time, by Lemma $2.11_{\mathrm{p}18}$.

By performing a point-location query in $\widehat{T}$, we can compute the smallest grid cell of C that contains the query point in $O(\log |C|)$ time. Specifically, the query returns a leaf v of $\widehat{T}$. We need to find, along the path of v to the root, the smallest ball associated with those nodes; that is, we find the first node u along this path from the leaf that is one of the grid cells of C. The node u remembers the grid cell that it corresponds to and, furthermore, the ball that generates this grid cell. This information can be propagated down the compressed quadtree during the preprocessing stage, by every marked node that corresponds to a cell of C sending the information down the tree. As such, once we have v at hand, one can answer the ANN query in constant time. We conclude with the following.

Theorem 18.28. *Let* P *be a set of n points in* $\mathbb{R}^d$. *One can build a compressed quadtree* $\widehat{T}$, *in* $O(n\varepsilon^{-2d-1} \log n/\varepsilon)$ *time, of size* $O(n\varepsilon^{-2d-1} \log \varepsilon^{-1})$, *such that a* $(1 + \varepsilon)$*-ANN query on* P *can be answered by a single point-location query in* $\widehat{T}$. *Such a point-location query takes* $O(\log(n/\varepsilon))$ *time.*

Proof. The construction is described above. We only need to prove the bounds on the running time. It takes $O(n \log n + n\varepsilon^{-d-1} \log \varepsilon^{-1})$ time to compute $\mathcal{B}$. For every such ball, we generate $O(1/\varepsilon^d)$ canonical grid cells that cover it. Let C be the resulting set of grid cells (after filtering multiple instances of the same grid cell). Naively, the size of C is bounded by $N = O\left(|\mathcal{B}| / \varepsilon^d\right)$. We can compute C in this time, by careful implementation (we omit the straightforward and tedious details).

Constructing $\widehat{T}$ for C takes $O(N \log N) = O(n\varepsilon^{-2d-1} \log(n/\varepsilon))$ time (Lemma 2.11_{p18}). The resulting compressed quadtree is of size $O(N)$. Point-location queries in $\widehat{T}$ take $O(\log N) = O(\log(n/\varepsilon))$ time. Given a query point and the leaf v, such that $q \in \square_v$, we need to find the first ancestor above v that has a point associated with it. This can be done in constant time, by preprocessing $\widehat{T}$, by propagating down the compressed quadtree the nodes that the leaves are associated with. ■

Better construction. The dependency on ε in Theorem 18.28 can be improved by being a bit more careful about the details. We quickly outline the idea: Use a WSPD but only with a constant separation. For each such pair $\{u, v\}$, observe that the construction of $\mathcal{B}(\mathrm{rep}_u, \mathrm{rep}_v)$ generates a set of $O(\varepsilon^{-1} \log \varepsilon^{-1})$ balls. These balls when translated into a grid form an exponential grid around both points. This grid has $O(\varepsilon^{-d} \log \varepsilon^{-1})$ cells in it. For each such cell, explicitly compute its nearest neighbor in $P_u \cup P_v$. It is easy to argue that these sets of cells now answer ANN correctly for all points that are within a distance range $[1/4, 4/\varepsilon] \cdot \left\| \mathrm{rep}_u - \mathrm{rep}_v \right\|$. Clearly, the resulting set of grid cells can be used to construct an AVD. The resulting AVD has complexity $O(n\varepsilon^{-d} \log \varepsilon^{-1})$. Bringing the construction time down requires some additional work.

Theorem 18.29. *Let* P *be a set of n points in* $\mathbb{R}^d$. *One can build a compressed quadtree* $\widehat{T}$, *in* $O\left(n\varepsilon^{-d} \log n/\varepsilon\right)$ *time, of size* $O(n\varepsilon^{-d} \log \varepsilon^{-1})$, *such that a* $(1 + \varepsilon)$*-ANN query on* P *can be answered by a single point-location query in* $\widehat{T}$. *Such a point-location query takes* $O(\log(n/\varepsilon))$ *time.*

18.5. Bibliographical notes

As demonstrated in this chapter, HSTs, despite their simplicity, are a powerful tool in giving a compact (but approximate) representation of metric spaces.

Indyk and Motwani showed that ANN can be reduced to a small number of near neighbor queries (i.e., this is the PLEB data-structure of [**IM98**]). In particular, the elegant argument in Section 18.1.1 is due to Indyk (personal communications). Indyk and Motwani also provided a rather involved reduction showing that a near linear number of balls is sufficient. A simpler reduction was provided by Har-Peled [**Har01b**], and here we presented a variant which is even simpler. Theorem 18.19 and Exercise 18.2 are from [**Har01b**]. The number of balls used in the construction can be further reduced; see Sabharwal et al. [**SSS02**].

For more information on how to answer point-location queries exactly among balls, see the survey by Agarwal and Erickson [**AE98**] on range searching.

The ideas of Section 18.3 are from Arya and Malamatos [**AM02**], although our presentation is somewhat different. The notion of AVD as described here was introduced by Har-Peled [**Har01b**]. In particular, the idea of first building a global set of balls and using a compressed quadtree on the associated set of approximate balls is due to Har-Peled [**Har01b**].

Further work on trade-offs between query time and space for AVDs was done by Arya et al. [**AMM09**]. In particular, Theorem 18.29 is described there but is originally from Arya and Malamatos [**AM02**].

The reader might wonder why we went through the excruciating pain of first approximating the metric by balls and then using it to construct an AVD. The reduction to point-location among approximate balls would prove to be useful when dealing with proximity in high dimensions. Of course, once we have the facts about ANN via point-location among approximate balls, the AVD construction is relatively easy.

Open problems. The argument showing that one can use approximate near neighbor data-structures instead of exact ones (i.e., Lemma 18.23) is tedious and far from being elegant; see Exercise 18.1. A natural question for further research is to try to give a simple concrete condition on the set of balls, such that using approximate near neighbor data-structures still gives the correct result. Currently, it seems that what we need is some kind of separability property. However, it would be nice to give a simpler direct condition.

As mentioned above, Sabharwal et al. [**SSS02**] showed a reduction from ANN to a linear number of balls (ignoring the dependency on ε). However, it seems like a simpler construction should work.

18.6. Exercises

Exercise 18.1 (No direct approximation to the set of balls)**.** Lemma 18.23 is not elegant. A more natural conjecture would be the following:

> **Conjecture 18.30.** *Let* P *be a set of* n *points in the plane, and let* $\mathcal{B}$ *be a set of balls such that one can answer* $(1+\varepsilon/c)$*-ANN queries on* P *by performing a single target query on* $\mathcal{B}$*. Now, let* $\mathcal{B}_{\approx}$ *be a* $(1+\varepsilon/c)$*-approximation to* $\mathcal{B}$*. Then a target query on* $\mathcal{B}_{\approx}$ *answers a* $(1+\varepsilon)$*-ANN query on* P*, for* c *a large enough constant.*

Give a counterexample to the above conjecture.

Exercise 18.2 (Better reduction to PLEB)**.** Prove Theorem 18.19. (This is not easy.)

Exercise 18.3 (PLEB in low dimensions)**.** Prove Lemma 18.25.

CHAPTER 19

Dimension Reduction – The Johnson-Lindenstrauss (JL) Lemma

In this chapter, we will prove that given a set P of n points in $\mathbb{R}^{\mathsf{d}}$, one can reduce the dimension of the points to $k = O(\varepsilon^{-2} \log n)$ such that distances are $1 \pm \varepsilon$ preserved. Surprisingly, this reduction is done by randomly picking a subspace of k dimensions and projecting the points into this random subspace. One way of thinking about this result is that we are "compressing" the input of size $n\mathsf{d}$ (i.e., n points with d coordinates) into size $O(n\varepsilon^{-2} \log n)$, while (approximately) preserving distances.

19.1. The Brunn-Minkowski inequality

For a set $\mathcal{A} \subseteq \mathbb{R}^{\mathsf{d}}$ and a point $\mathsf{p} \in \mathbb{R}^{\mathsf{d}}$, let $\mathcal{A} + \mathsf{p}$ denote the translation of $\mathcal{A}$ by p. Formally, $\mathcal{A} + \mathsf{p} = \left\{ \mathsf{q} + \mathsf{p} \;\middle|\; \mathsf{q} \in \mathcal{A} \right\}$.

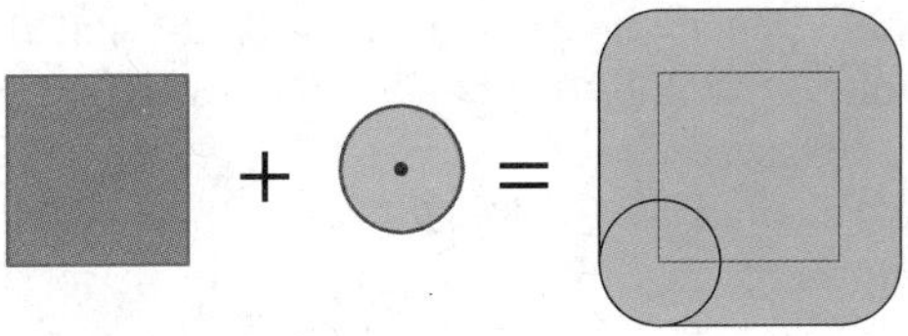

Definition 19.1. For two sets $\mathcal{A}$ and $\mathcal{B}$ in $\mathbb{R}^n$, let $\mathcal{A} + \mathcal{B}$ denote the ***Minkowski sum*** of $\mathcal{A}$ and $\mathcal{B}$. Formally,

$$\mathcal{A} + \mathcal{B} = \left\{ a + b \;\middle|\; a \in \mathcal{A}, b \in \mathcal{B} \right\} = \bigcup_{\mathsf{p} \in \mathcal{A}} (\mathsf{p} + \mathcal{B}).$$

Remark 19.2. It is easy to verify that if $\mathcal{A}', \mathcal{B}'$ are translated copies of $\mathcal{A}, \mathcal{B}$ (that is, $\mathcal{A}' = \mathcal{A} + \mathsf{p}$ and $\mathcal{B}' = \mathcal{B} + \mathsf{q}$, for some points $\mathsf{p}, \mathsf{q} \in \mathbb{R}^{\mathsf{d}}$) respectively, then $\mathcal{A}' + \mathcal{B}'$ is a translated copy of $\mathcal{A} + \mathcal{B}$. In particular, since volume is preserved under translation, we have that $\mathrm{Vol}(\mathcal{A}' + \mathcal{B}') = \mathrm{Vol}((\mathcal{A} + \mathcal{B}) + \mathsf{p} + \mathsf{q}) = \mathrm{Vol}(\mathcal{A} + \mathcal{B})$.

Our purpose here is to prove the following theorem.

Theorem 19.3 (Brunn-Minkowski inequality)**.** *Let $\mathcal{A}$ and $\mathcal{B}$ be two non-empty compact sets in $\mathbb{R}^n$. Then*

$$\mathrm{Vol}(\mathcal{A} + \mathcal{B})^{1/n} \geq \mathrm{Vol}(\mathcal{A})^{1/n} + \mathrm{Vol}(\mathcal{B})^{1/n}.$$

Definition 19.4. A set $\mathcal{A} \subseteq \mathbb{R}^n$ is a ***brick set*** if it is the union of finitely many (close) axis parallel boxes with disjoint interiors.

It is intuitively clear, by limit arguments, that proving Theorem 19.3 for brick sets will imply it for the general case.

Lemma 19.5 (Brunn-Minkowski inequality for brick sets)**.** *Let $\mathcal{A}$ and $\mathcal{B}$ be two non-empty brick sets in $\mathbb{R}^n$. Then*

$$\mathrm{Vol}(\mathcal{A} + \mathcal{B})^{1/n} \geq \mathrm{Vol}(\mathcal{A})^{1/n} + \mathrm{Vol}(\mathcal{B})^{1/n}.$$

PROOF. We prove by induction on the number k of bricks in $\mathcal{A}$ and $\mathcal{B}$. If $k = 2$, then $\mathcal{A}$ and $\mathcal{B}$ are just bricks, with dimensions $\alpha_1, \ldots, \alpha_n$ and $\beta_1, \ldots, \beta_n$, respectively. In this case, the dimensions of $\mathcal{A} + \mathcal{B}$ are $\alpha_1 + \beta_1, \ldots, \alpha_n + \beta_n$, as can be easily verified. Thus, we need to prove that $\left(\prod_{i=1}^n \alpha_i\right)^{1/n} + \left(\prod_{i=1}^n \beta_i\right)^{1/n} \le \left(\prod_{i=1}^n (\alpha_i + \beta_i)\right)^{1/n}$. Dividing the left side by the right side, we have

$$\left(\prod_{i=1}^n \frac{\alpha_i}{\alpha_i + \beta_i}\right)^{1/n} + \left(\prod_{i=1}^n \frac{\beta_i}{\alpha_i + \beta_i}\right)^{1/n} \le \frac{1}{n}\sum_{i=1}^n \frac{\alpha_i}{\alpha_i + \beta_i} + \frac{1}{n}\sum_{i=1}^n \frac{\beta_i}{\alpha_i + \beta_i} = 1,$$

by the generalized arithmetic-geometric mean inequality[1], and the claim follows for this case.

Now let $k > 2$ and suppose that the Brunn-Minkowski inequality holds for any pair of brick sets with fewer than k bricks (together). Let $\mathcal{A}$ and $\mathcal{B}$ be a pair of sets having k bricks together. Assume that $\mathcal{A}$ has at least two (disjoint) bricks. However, this implies that there is an axis parallel hyperplane h that separates the interior of one brick of $\mathcal{A}$ from the interior of another brick of $\mathcal{A}$ (the hyperplane h might intersect other bricks of $\mathcal{A}$). Assume that h is the hyperplane $x_1 = 0$ (this can be achieved by translation and renaming of coordinates).

Let $\overline{\mathcal{A}^+} = \mathcal{A} \cap h^+$ and $\overline{\mathcal{A}^-} = \mathcal{A} \cap h^-$, where h^+ and h^- are the two open halfspaces induced by h. Let $\mathcal{A}^+$ and $\mathcal{A}^-$ be the closure of $\overline{\mathcal{A}^+}$ and $\overline{\mathcal{A}^-}$, respectively. Clearly, $\mathcal{A}^+$ and $\mathcal{A}^-$ are both brick sets with (at least) one fewer brick than $\mathcal{A}$.

Next, observe that the claim is translation invariant (see Remark 19.2), and as such, let us translate $\mathcal{B}$ so that its volume is split by h in the same ratio $\mathcal{A}$'s volume is being split. Denote the two parts of $\mathcal{B}$ by $\mathcal{B}^+$ and $\mathcal{B}^-$, respectively. Let $\rho = \mathrm{Vol}(\mathcal{A}^+)/\mathrm{Vol}(\mathcal{A}) = \mathrm{Vol}(\mathcal{B}^+)/\mathrm{Vol}(\mathcal{B})$ (if $\mathrm{Vol}(\mathcal{A}) = 0$ or $\mathrm{Vol}(\mathcal{B}) = 0$, the claim trivially holds).

Observe that $\mathcal{X}^+ = \mathcal{A}^+ + \mathcal{B}^+ \subseteq \mathcal{A} + \mathcal{B}$, and $\mathcal{X}^+$ lies on one side of h (since $h \equiv (x_1 = 0)$), and similarly $\mathcal{X}^- = \mathcal{A}^- + \mathcal{B}^- \subseteq \mathcal{A} + \mathcal{B}$ and $\mathcal{X}^-$ lies on the other side of h. Thus, by induction and since $\mathcal{A}^+ + \mathcal{B}^+$ and $\mathcal{A}^- + \mathcal{B}^-$ are interior disjoint, we have

$$\begin{aligned}
\mathrm{Vol}(\mathcal{A} + \mathcal{B}) &\ge \mathrm{Vol}(\mathcal{A}^+ + \mathcal{B}^+) + \mathrm{Vol}(\mathcal{A}^- + \mathcal{B}^-) \\
&\ge \left(\mathrm{Vol}(\mathcal{A}^+)^{1/n} + \mathrm{Vol}(\mathcal{B}^+)^{1/n}\right)^n + \left(\mathrm{Vol}(\mathcal{A}^-)^{1/n} + \mathrm{Vol}(\mathcal{B}^-)^{1/n}\right)^n \\
&= \left[\rho^{1/n}\,\mathrm{Vol}(\mathcal{A})^{1/n} + \rho^{1/n}\,\mathrm{Vol}(\mathcal{B})^{1/n}\right]^n \\
&\quad + \left[(1-\rho)^{1/n}\,\mathrm{Vol}(\mathcal{A})^{1/n} + (1-\rho)^{1/n}\,\mathrm{Vol}(\mathcal{B})^{1/n}\right]^n \\
&= \left(\rho + (1-\rho)\right)\left[\mathrm{Vol}(\mathcal{A})^{1/n} + \mathrm{Vol}(\mathcal{B})^{1/n}\right]^n \\
&= \left[\mathrm{Vol}(\mathcal{A})^{1/n} + \mathrm{Vol}(\mathcal{B})^{1/n}\right]^n,
\end{aligned}$$

establishing the claim. ■

PROOF OF THEOREM 19.3. Let $\mathcal{A}_1 \subseteq \mathcal{A}_2 \subseteq \cdots \subseteq \mathcal{A}_i \subseteq \cdots$ be a sequence of finite brick sets, such that $\bigcup_i \mathcal{A}_i = \mathcal{A}$, and similarly let $\mathcal{B}_1 \subseteq \mathcal{B}_2 \subseteq \cdots \subseteq \mathcal{B}_i \subseteq \cdots$ be a sequence of finite brick sets, such that $\bigcup_i \mathcal{B}_i = \mathcal{B}$. By the definition of volume[2], we have that $\lim_{i\to\infty} \mathrm{Vol}(\mathcal{A}_i) = \mathrm{Vol}(\mathcal{A})$ and $\lim_{i\to\infty} \mathrm{Vol}(\mathcal{B}_i) = \mathrm{Vol}(\mathcal{B})$.

---

[1] Here is a proof of the generalized form: Let $x_1, \ldots, x_n$ be n positive real numbers. Consider the quantity $R = x_1 x_2 \cdots x_n$. If we fix the sum of the n numbers to be equal to γ, then R is maximized when all the x_is are equal, as can be easily verified. Thus, $\sqrt[n]{x_1 x_2 \cdots x_n} \le \sqrt[n]{(\gamma/n)^n} = \gamma/n = (x_1 + \cdots + x_n)/n$.

[2] This is the standard definition of volume. The reader unfamiliar with this fanfare can either consult a standard text on calculus (or measure theory) or take it for granted as this is intuitively clear.

We claim that $\lim_{i\to\infty} \mathrm{Vol}(\mathcal{A}_i+\mathcal{B}_i) = \mathrm{Vol}(\mathcal{A}+\mathcal{B})$. Indeed, consider any point $z \in \mathcal{A}+\mathcal{B}$, and let $u \in \mathcal{A}$ and $v \in \mathcal{B}$ be such that $u + v = z$. By definition, there exists an i, such that for all $j > i$ we have $u \in \mathcal{A}_j$, $v \in \mathcal{B}_j$, and as such $z \in \mathcal{A}_j + \mathcal{B}_j$. Thus, $\mathcal{A} + \mathcal{B} \subseteq \bigcup_j(\mathcal{A}_j + \mathcal{B}_j)$ and $\bigcup_j(\mathcal{A}_j + \mathcal{B}_j) \subseteq \bigcup_j(\mathcal{A} + \mathcal{B}) \subseteq \mathcal{A} + \mathcal{B}$; namely, $\bigcup_j(\mathcal{A}_j + \mathcal{B}_j) = \mathcal{A} + \mathcal{B}$.

Furthermore, for any $i > 0$, since $\mathcal{A}_i$ and $\mathcal{B}_i$ are brick sets, we have

$$\mathrm{Vol}(\mathcal{A}_i + \mathcal{B}_i)^{1/n} \geq \mathrm{Vol}(\mathcal{A}_i)^{1/n} + \mathrm{Vol}(\mathcal{B}_i)^{1/n},$$

by Lemma 19.5. Thus,

$$\begin{aligned}\mathrm{Vol}(\mathcal{A} + \mathcal{B})^{1/n} &= \lim_{i\to\infty} \mathrm{Vol}(\mathcal{A}_i + \mathcal{B}_i)^{1/n} \geq \lim_{i\to\infty}\Big(\mathrm{Vol}(\mathcal{A}_i)^{1/n} + \mathrm{Vol}(\mathcal{B}_i)^{1/n}\Big)\\ &= \mathrm{Vol}(\mathcal{A})^{1/n} + \mathrm{Vol}(\mathcal{B})^{1/n}.\end{aligned}$$

■

Theorem 19.6 (Brunn-Minkowski theorem for slice volumes). *Let $\mathcal{P}$ be a convex set in $\mathbb{R}^{n+1}$, and let $\mathcal{A} = \mathcal{P} \cap (x_1 = a)$, $\mathcal{B} = \mathcal{P} \cap (x_1 = b)$, and $\mathcal{C} = \mathcal{P} \cap (x_1 = c)$ be three slices of $\mathcal{A}$, for $a < b < c$. We have* $\mathrm{Vol}(\mathcal{B}) \geq \min(\mathrm{Vol}(\mathcal{A}), \mathrm{Vol}(\mathcal{C}))$.

In fact, consider the function

$$v(t) = (\mathrm{Vol}(\mathcal{P} \cap (x_1 = t)))^{1/n},$$

and let $\mathcal{I} = [t_{\min}, t_{\max}]$ be the interval where the hyperplane $x_1 = t$ intersects $\mathcal{P}$. Then, $v(t)$ is concave on $\mathcal{I}$.

Proof. If a or c is outside $\mathcal{I}$, then $\mathrm{Vol}(\mathcal{A}) = 0$ or $\mathrm{Vol}(\mathcal{C}) = 0$, respectively, and then the claim trivially holds.

Otherwise, let $\alpha = (b - a)/(c - a)$. We have that $b = (1 - \alpha) \cdot a + \alpha \cdot c$, and by the convexity of $\mathcal{P}$, we have $(1 - \alpha)\mathcal{A} + \alpha\mathcal{C} \subseteq \mathcal{B}$. Thus, by Theorem 19.3 we have

$$\begin{aligned}v(b) = \mathrm{Vol}(\mathcal{B})^{1/n} &\geq \mathrm{Vol}((1-\alpha)\mathcal{A} + \alpha\mathcal{C})^{1/n} \geq \mathrm{Vol}((1-\alpha)\mathcal{A})^{1/n} + \mathrm{Vol}(\alpha\mathcal{C})^{1/n}\\ &= \Big((1-\alpha)^n \mathrm{Vol}(\mathcal{A})\Big)^{1/n} + \Big(\alpha^n \mathrm{Vol}(\mathcal{C})\Big)^{1/n}\\ &= (1-\alpha)\mathrm{Vol}(\mathcal{A})^{1/n} + \alpha \mathrm{Vol}(\mathcal{C})^{1/n}\\ &= (1-\alpha)v(a) + \alpha v(c).\end{aligned}$$

Namely, $v(\cdot)$ is concave on $\mathcal{I}$, and in particular $v(b) \geq \min(v(a), v(c))$, which in turn implies that $\mathrm{Vol}(\mathcal{B}) = v(b)^n \geq (\min(v(a), v(c)))^n = \min(\mathrm{Vol}(\mathcal{A}), \mathrm{Vol}(\mathcal{C}))$, as claimed. ■

Corollary 19.7. *For $\mathcal{A}$ and $\mathcal{B}$ compact sets in $\mathbb{R}^n$,* $\mathrm{Vol}((\mathcal{A} + \mathcal{B})/2) \geq \sqrt{\mathrm{Vol}(\mathcal{A})\mathrm{Vol}(\mathcal{B})}$.

Proof. We have that

$$\begin{aligned}\mathrm{Vol}((\mathcal{A} + \mathcal{B})/2)^{1/n} &= \mathrm{Vol}(\mathcal{A}/2 + \mathcal{B}/2)^{1/n} \geq \mathrm{Vol}(\mathcal{A}/2)^{1/n} + \mathrm{Vol}(\mathcal{B}/2)^{1/n}\\ &= (\mathrm{Vol}(\mathcal{A})^{1/n} + \mathrm{Vol}(\mathcal{B})^{1/n})/2 \geq \sqrt{\mathrm{Vol}(\mathcal{A})^{1/n}\mathrm{Vol}(\mathcal{B})^{1/n}}\end{aligned}$$

by Theorem 19.3 and since $(a + b)/2 \geq \sqrt{ab}$ for any $a, b \geq 0$. The claim now follows by raising this inequality to the power n. ■

19.1.1. The isoperimetric inequality. The following is not used anywhere else and is provided because of its mathematical elegance. The skip-able reader can thus employ their special gift and move on to Section 19.2.

The ***isoperimetric inequality*** states that among all convex bodies of a fixed surface area, the ball has the largest volume (in particular, the unit circle is the largest area planar region with perimeter 2π). This problem can be traced back to antiquity; in particular Zenodorus (200–140 BC) wrote a monograph (which was lost) that seemed to have proved the claim in the plane for some special cases. The first formal proof for the planar case was done by Steiner in 1841. Interestingly, the more general claim is an easy consequence of the Brunn-Minkowski inequality.

Let K be a convex body in $\mathbb{R}^n$ and let $\mathbf{b}$ be the n-dimensional ball of radius 1 centered at the origin. Let $\mathsf{S}(X)$ denote the surface area of a compact set $X \subseteq \mathbb{R}^n$. The ***isoperimetric inequality*** states that

$$\left(\frac{\mathrm{Vol}(K)}{\mathrm{Vol}(\mathbf{b})}\right)^{1/n} \leq \left(\frac{\mathsf{S}(K)}{\mathsf{S}(\mathbf{b})}\right)^{1/(n-1)}. \tag{19.1}$$

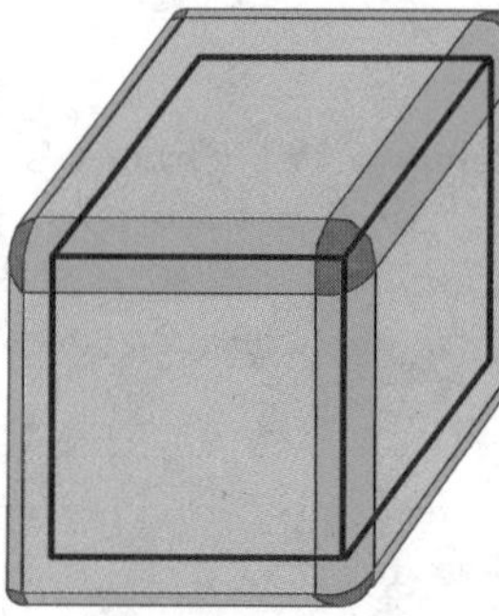

Namely, the left side is the radius of a ball having the same volume as K, and the right side is the radius of a sphere having the same surface area as K. In particular, if we scale K so that its surface area is the same as $\mathbf{b}$, then the above inequality implies that $\mathrm{Vol}(K) \leq \mathrm{Vol}(\mathbf{b})$.

To prove (19.1), observe that $\mathrm{Vol}(\mathbf{b}) = \mathsf{S}(\mathbf{b})/n$[3]. Also, observe that $K + \varepsilon\mathbf{b}$ is the body K together with a small "atmosphere" around it of thickness ε. In particular, the volume of this "atmosphere" is (roughly) $\varepsilon\,\mathsf{S}(K)$ (in fact, Minkowski defined the surface area of a convex body to be the limit stated next).

Formally, we have

$$\mathsf{S}(K) = \lim_{\varepsilon \to 0+} \frac{\mathrm{Vol}(K + \varepsilon\mathbf{b}) - \mathrm{Vol}(K)}{\varepsilon} \geq \lim_{\varepsilon \to 0+} \frac{\left(\mathrm{Vol}(K)^{1/n} + \mathrm{Vol}(\varepsilon\mathbf{b})^{1/n}\right)^n - \mathrm{Vol}(K)}{\varepsilon},$$

by the Brunn-Minkowski inequality. Now $\mathrm{Vol}(\varepsilon\mathbf{b})^{1/n} = \varepsilon\,\mathrm{Vol}(\mathbf{b})^{1/n}$, and as such

$$\begin{aligned}\mathsf{S}(K) &\geq \lim_{\varepsilon \to 0+} \frac{\mathrm{Vol}(K) + \binom{n}{1}\varepsilon\,\mathrm{Vol}(K)^{\frac{n-1}{n}}\,\mathrm{Vol}(\mathbf{b})^{\frac{1}{n}} + \binom{n}{2}\varepsilon^2\langle\cdots\rangle + \cdots + \varepsilon^n\,\mathrm{Vol}(\mathbf{b}) - \mathrm{Vol}(K)}{\varepsilon}\\ &= \lim_{\varepsilon \to 0+} \frac{n\varepsilon\,\mathrm{Vol}(K)^{\frac{n-1}{n}}\,\mathrm{Vol}(\mathbf{b})^{\frac{1}{n}}}{\varepsilon} = n\,\mathrm{Vol}(K)^{\frac{n-1}{n}}\,\mathrm{Vol}(\mathbf{b})^{\frac{1}{n}}.\end{aligned}$$

Dividing both sides by $\mathsf{S}(\mathbf{b}) = n\,\mathrm{Vol}(\mathbf{b})$, we have

$$\frac{\mathsf{S}(K)}{\mathsf{S}(\mathbf{b})} \geq \frac{\mathrm{Vol}(K)^{(n-1)/n}}{\mathrm{Vol}(\mathbf{b})^{(n-1)/n}} \quad\Rightarrow\quad \left(\frac{\mathsf{S}(K)}{\mathsf{S}(\mathbf{b})}\right)^{1/(n-1)} \geq \left(\frac{\mathrm{Vol}(K)}{\mathrm{Vol}(\mathbf{b})}\right)^{1/n},$$

establishing the isoperimetric inequality.

19.2. Measure concentration on the sphere

Let $\mathbb{S}^{(n-1)}$ be the unit sphere in $\mathbb{R}^n$. We assume there is a uniform probability measure defined over $\mathbb{S}^{(n-1)}$, such that its total measure is 1. Surprisingly, most of the mass of this measure is near the equator. Indeed, consider an arbitrary equator π on $\mathbb{S}^{(n-1)}$ (that it, it is the intersection of the sphere with a hyperplane passing through the center of the ball inducing the sphere). Next, consider all the points that are within distance $\approx \ell(n) = c/n^{1/3}$ from π. The question we are interested in is what fraction of the sphere is covered by this strip T (depicted in Figure 19.1).

[3]Indeed, $\mathrm{Vol}(\mathbf{b}) = \int_{r=0}^{1} \mathsf{S}(\mathbf{b})\, r^{n-1}\,dr = \mathsf{S}(\mathbf{b})/n$.

Notice that as the dimension increases, the width $\ell(n)$ of this strip decreases. But surprisingly, despite its width becoming smaller, as the dimension increases, this strip contains a larger and larger fraction of the sphere. In particular, the total fraction of the sphere not covered by this (shrinking!) strip converges to zero.

Furthermore, counterintuitively, this is true for *any* equator. We are going to show that even a stronger result holds: The mass of the sphere is concentrated close to the boundary of any set $A \subseteq \mathbb{S}^{(n-1)}$ such that $\mathbf{Pr}[A] = 1/2$.

Before proving this somewhat surprising theorem, we will first try to get an intuition about the behavior of the hypersphere in high dimensions.

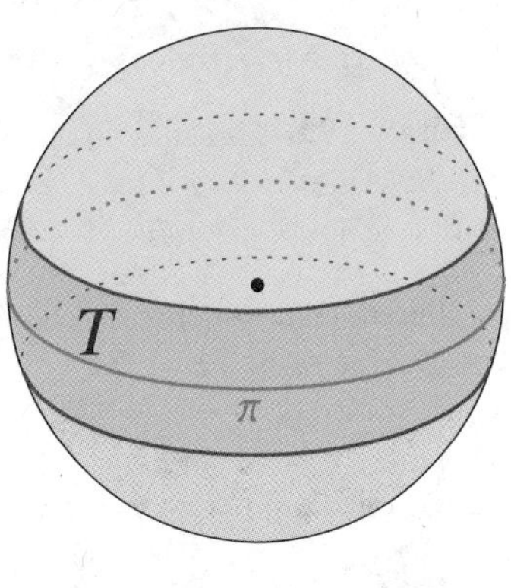

FIGURE 19.1.

19.2.1. The strange and curious life of the hypersphere. Consider the ball of radius r in $\mathbb{R}^n$ (denoted by $r\,\mathbf{b}^n$), where $\mathbf{b}^n$ is the unit radius ball centered at the origin. Clearly, $\mathrm{Vol}(r\,\mathbf{b}^n) = r^n\,\mathrm{Vol}(\mathbf{b}^n)$. Now, even if r is very close to 1, the quantity r^n might be very close to zero if n is sufficiently large. Indeed, if $r = 1 - \delta$, then $r^n = (1-\delta)^n \le \exp(-\delta n)$, which is very small if $\delta \gg 1/n$. (Here, we used $1 - x \le e^x$, for $x \ge 0$.) Namely, for the ball in high dimensions, its mass is concentrated in a very thin shell close to its surface.

The volume of a ball and the surface area of a hypersphere. Let $\mathrm{Vol}(r\mathbf{b}^n)$ denote the volume of the ball of radius r in $\mathbb{R}^n$, and let $\mathsf{S}(r\mathbf{b}^n)$ denote the surface area of its bounding sphere (i.e., the surface area of $r\mathbb{S}^{(n-1)}$). It is known that

$$\mathrm{Vol}(r\mathbf{b}^n) = \frac{\pi^{n/2} r^n}{\Gamma(n/2+1)} \quad \text{and} \quad \mathsf{S}(r\mathbf{b}^n) = \frac{2\pi^{n/2} r^{n-1}}{\Gamma(n/2)},$$

where the gamma function, $\Gamma(\cdot)$, is an extension of the factorial function. Specifically, if n is even, then $\Gamma(n/2+1) = (n/2)!$, and for n odd $\Gamma(n/2+1) = \sqrt{\pi}(n!!)/2^{(n+1)/2}$, where $n!! = 1 \cdot 3 \cdot 5 \cdots n$ is the ***double factorial***. The most surprising implication of these two formulas is that, as n increases, the volume of the unit ball first increases (till dimension 5) and then starts decreasing to zero.

Similarly, the surface area of the unit sphere $\mathbb{S}^{(n-1)}$ in $\mathbb{R}^n$ tends to zero as the dimension increases. To see the above explicitly, compute the volume of the unit ball using an integral of its slice volume, when it is being sliced by a hyperplane perpendicular to the nth coordinate.

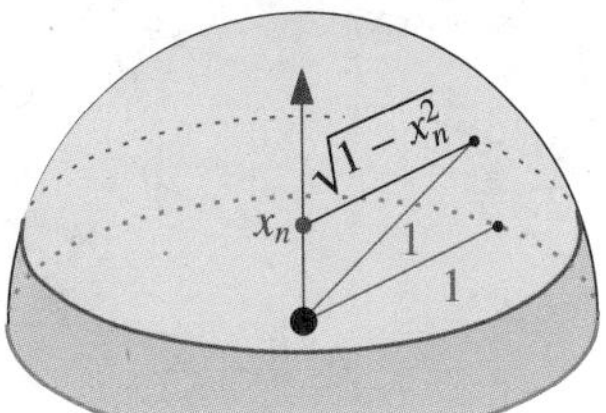

We have (see figure on the right) that

$$\mathrm{Vol}(\mathbf{b}^n) = \int_{x_n=-1}^{1} \mathrm{Vol}\left(\sqrt{1-x_n^2}\,\mathbf{b}^{n-1}\right) dx_n = \mathrm{Vol}\left(\mathbf{b}^{n-1}\right) \int_{x_n=-1}^{1} \left(1-x_n^2\right)^{(n-1)/2} dx_n.$$

Now, the integral on the right side tends to zero as n increases. In fact, for n very large, the term $\left(1-x_n^2\right)^{(n-1)/2}$ is very close to 0 everywhere except for a small interval around 0. This implies that the main contribution of the volume of the ball happens when we consider slices of the ball created by hyperplanes of the form $x_n = \delta$, where δ is small (roughly, for $|\delta| \le 1/\sqrt{n}$).

If one has to visualize what such a ball in high dimensions looks like, it might be best to think about it as a star-like creature: It has very little mass close to the tips of any set of orthogonal directions we pick, and most of its mass (somehow) lies on hyperplanes passing close to its center.[4]

19.2.2. Measure concentration on the sphere.

Theorem 19.8 (Measure concentration on the sphere). *Let $A \subseteq \mathbb{S}^{(n-1)}$ be a measurable set with $\mathbf{Pr}[A] \geq 1/2$, and let A_t denote the set of points of $\mathbb{S}^{(n-1)}$ within a distance at most t from A, where $t \leq 2$. Then $1 - \mathbf{Pr}[A_t] \leq 2\exp\left(-nt^2/2\right)$.*

Proof. We will prove a slightly weaker bound, with $-nt^2/4$ in the exponent. Let $\widehat{A} = T(A)$, where

$$T(X) = \left\{ \alpha x \;\middle|\; x \in X, \alpha \in [0,1] \right\} \subseteq \mathbf{b}^n$$

and $\mathbf{b}^n$ is the unit ball in $\mathbb{R}^n$. We have that $\mathbf{Pr}[A] = \mu\left(\widehat{A}\right)$, where $\mu\left(\widehat{A}\right) = \mathrm{Vol}\left(\widehat{A}\right) / \mathrm{Vol}(\mathbf{b}^n)$[5].

Let $B = \mathbb{S}^{(n-1)} \setminus A_t$ and $\widehat{B} = T(B)$. We have that $\|a - b\| \geq t$ for all $a \in A$ and $b \in B$. By Lemma 19.9 below, the set $\left(\widehat{A} + \widehat{B}\right)/2$ is contained in the ball $r\mathbf{b}^n$ centered at the origin, where $r = 1 - t^2/8$. Observe that $\mu(r\mathbf{b}^n) = \mathrm{Vol}(r\mathbf{b}^n)/\mathrm{Vol}(\mathbf{b}^n) = r^n = \left(1 - t^2/8\right)^n$. As such, applying the Brunn-Minkowski inequality in the form of Corollary 19.7, we have

$$\left(1 - \frac{t^2}{8}\right)^n = \mu(r\mathbf{b}^n) \geq \mu\left(\frac{\widehat{A} + \widehat{B}}{2}\right) \geq \sqrt{\mu\left(\widehat{A}\right)\mu\left(\widehat{B}\right)} = \sqrt{\mathbf{Pr}[A]\,\mathbf{Pr}[B]} \geq \sqrt{\mathbf{Pr}[B]/2}.$$

Thus, $\mathbf{Pr}[B] \leq 2(1 - t^2/8)^{2n} \leq 2\exp(-2nt^2/8)$, since $1 - x \leq \exp(-x)$, for $x \geq 0$. ■

Lemma 19.9. *For any $\widehat{a} \in \widehat{A}$ and $\widehat{b} \in \widehat{B}$, we have* $\left\| \dfrac{\widehat{a} + \widehat{b}}{2} \right\| \leq 1 - \dfrac{t^2}{8}$.

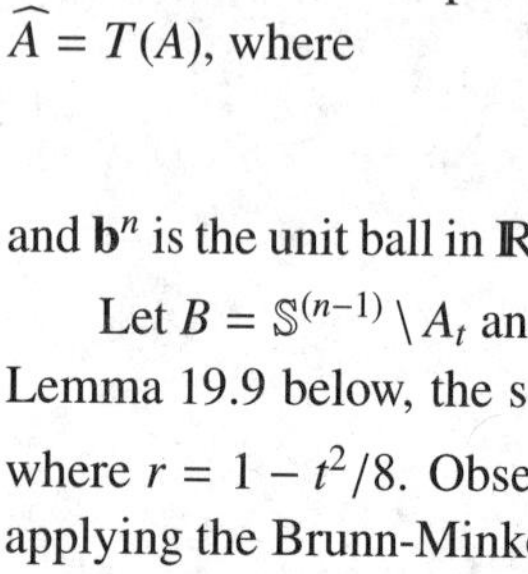

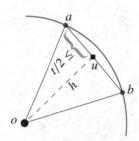

Proof. Let $\widehat{a} = \alpha a$ and $\widehat{b} = \beta b$, where $a \in A$ and $b \in B$. We have

$$\|u\| = \left\| \frac{a + b}{2} \right\| = \sqrt{1^2 - \left\| \frac{a - b}{2} \right\|^2} \leq \sqrt{1 - \frac{t^2}{4}} \leq 1 - \frac{t^2}{8}, \tag{19.2}$$

since $\|a - b\| \geq t$. As for $\widehat{a}$ and $\widehat{b}$, assume that $\alpha \leq \beta$, and observe that the quantity $\left\|\widehat{a} + \widehat{b}\right\|$ is maximized when $\beta = 1$. As such, by the triangle inequality, we have

$$\left\| \frac{\widehat{a} + \widehat{b}}{2} \right\| = \left\| \frac{\alpha a + b}{2} \right\| \leq \left\| \frac{\alpha(a + b)}{2} \right\| + \left\| (1 - \alpha)\frac{b}{2} \right\| \leq \alpha\left(1 - \frac{t^2}{8}\right) + (1 - \alpha)\frac{1}{2} = \tau,$$

by (19.2) and since $\|b\| = 1$. Now, τ is a convex combination of the two numbers $1/2$ and $1 - t^2/8$. In particular, we conclude that $\tau \leq \max(1/2, 1 - t^2/8) \leq 1 - t^2/8$, since $t \leq 2$. ■

---

[4] In short, it looks like a Boojum **[Car76]**.

[5] This is one of these "trivial" claims that might give the reader a pause, so here is a formal proof. Pick a random point p uniformly inside the ball $\mathbf{b}^n$. Let ψ be the probability that $\mathsf{p} \in \widehat{A}$. Clearly, $\mathrm{Vol}\left(\widehat{A}\right) = \psi\,\mathrm{Vol}(\mathbf{b}^n)$. So, consider the normalized point $\mathsf{q} = \mathsf{p}/\|\mathsf{p}\|$. Clearly, $\mathsf{p} \in \widehat{A}$ if and only if $\mathsf{q} \in A$, by the definition of $\widehat{A}$. Thus, $\mu\left(\widehat{A}\right) = \mathrm{Vol}\left(\widehat{A}\right)/\mathrm{Vol}(\mathbf{b}^n) = \psi = \mathbf{Pr}\left[\mathsf{p} \in \widehat{A}\right] = \mathbf{Pr}\left[\mathsf{q} \in A\right] = \mathbf{Pr}[A]$, since q has a uniform distribution on the hypersphere by assumption.

19.3. Concentration of Lipschitz functions

Consider a function $f : \mathbb{S}^{(n-1)} \to \mathbb{R}$, and imagine that we have a probability density function defined over the sphere. Let $\mathbf{Pr}[f \le t] = \mathbf{Pr}\Big[\Big\{x \in S^{n-1} \;\Big|\; f(x) \le t\Big\}\Big]$. We define the ***median*** of f, denoted by $\mathrm{med}(f)$, to be $\sup t$, such that $\mathbf{Pr}[f \le t] \le 1/2$.

We define $\mathbf{Pr}[f < \mathrm{med}(f)] = \sup_{x<\mathrm{med}(f)} \mathbf{Pr}[f \le x]$. The following is obvious but requires a formal proof.

Lemma 19.10. *We have* $\mathbf{Pr}\big[f < \mathrm{med}(f)\big] \le 1/2$ *and* $\mathbf{Pr}\big[f > \mathrm{med}(f)\big] \le 1/2$.

PROOF. Since $\bigcup_{k\ge 1}(-\infty, \mathrm{med}(f) - 1/k] = (-\infty, \mathrm{med}(f))$, we have

$$\mathbf{Pr}\big[f < \mathrm{med}(f)\big] = \sup_{k\ge 1} \mathbf{Pr}\left[f \le \mathrm{med}(f) - \frac{1}{k}\right] \le \sup_{k\ge 1} \frac{1}{2} = \frac{1}{2}.$$

The second claim follows by a symmetric argument. ∎

Definition 19.11 (*c-Lipschitz*). A function $f : A \to B$ is ***c-Lipschitz*** if, for any $x, y \in A$, we have $\|f(x) - f(y)\| \le c\,\|x - y\|$.

Theorem 19.12 (Lévy's lemma). *Let* $f : \mathbb{S}^{(n-1)} \to \mathbb{R}$ *be 1-Lipschitz. Then for all* $t \in [0, 1]$, *we have*

$$\mathbf{Pr}\big[f > \mathrm{med}(f) + t\big] \le 2\exp\big(-t^2 n/2\big) \quad \text{and} \quad \mathbf{Pr}\big[f < \mathrm{med}(f) - t\big] \le 2\exp\big(-t^2 n/2\big).$$

PROOF. We prove only the first inequality; the second follows by symmetry. Let

$$A = \Big\{x \in \mathbb{S}^{(n-1)} \;\Big|\; f(x) \le \mathrm{med}(f)\Big\}.$$

By Lemma 19.10, we have $\mathbf{Pr}[A] \ge 1/2$. Consider a point $x \in A_t$, where A_t is as defined in Theorem 19.8. Let $\mathsf{nn}(x)$ be the nearest point in A to x. We have by definition that $\|x - \mathsf{nn}(x)\| \le t$. As such, since f is 1-Lipschitz and $\mathsf{nn}(x) \in A$, we have that

$$f(x) \le f(\mathsf{nn}(x)) + \|\mathsf{nn}(x) - x\| \le \mathrm{med}(f) + t.$$

Thus, by Theorem 19.8, we get $\mathbf{Pr}[f > \mathrm{med}(f) + t] \le 1 - \mathbf{Pr}[A_t] \le 2\exp\big(-t^2 n/2\big)$. ∎

19.4. The Johnson-Lindenstrauss lemma

Lemma 19.13. *For a unit vector* $x \in \mathbb{S}^{(n-1)}$, *let*

$$f(x) = \sqrt{x_1^2 + x_2^2 + \cdots + x_k^2}$$

be the length of the projection of x into the subspace formed by the first k coordinates. Let x be a vector randomly chosen with uniform distribution from $\mathbb{S}^{(n-1)}$. Then $f(x)$ is sharply concentrated. Namely, there exists $m = m(n, k)$ such that

$$\mathbf{Pr}\big[f(x) \ge m + t\big] \le 2\exp(-t^2 n/2) \quad \text{and} \quad \mathbf{Pr}\big[f(x) \le m - t\big] \le 2\exp(-t^2 n/2),$$

for any $t \in [0, 1]$. Furthermore, for $k \ge 10 \ln n$, we have $m \ge \frac{1}{2}\sqrt{k/n}$.

PROOF. The orthogonal projection $p : \mathbb{R}^n \to \mathbb{R}^k$ given by $p(x_1, \ldots, x_n) = (x_1, \ldots, x_k)$ is 1-Lipschitz (since projections can only shrink distances; see Exercise 19.3). As such, $f(x) = \|p(x)\|$ is 1-Lipschitz, since for any x, y we have

$$|f(x) - f(y)| = \Big|\|p(x)\| - \|p(y)\|\Big| \le \|p(x) - p(y)\| \le \|x - y\|,$$

by the triangle inequality and since p is 1-Lipschitz. Theorem 19.12 (i.e., Lévy's lemma) gives the required tail estimate with $m = \mathrm{med}(f)$.

Thus, we only need to prove the lower bound on m. For a random $x = (x_1, \ldots, x_n) \in \mathbb{S}^{(n-1)}$, we have $\mathbf{E}\big[\|x\|^2\big] = 1$. By linearity of expectations and by symmetry, we have $1 = \mathbf{E}\big[\|x\|^2\big] = \mathbf{E}\big[\sum_{i=1}^n x_i^2\big] = \sum_{i=1}^n \mathbf{E}\big[x_i^2\big] = n\,\mathbf{E}\big[x_j^2\big]$, for any $1 \le j \le n$. Thus, $\mathbf{E}\big[x_j^2\big] = 1/n$, for $j = 1, \ldots, n$. Thus,

$$\mathbf{E}\Big[(f(x))^2\Big] = \mathbf{E}\left[\sum_{i=1}^{k} x_i^2\right] = \sum_{i=1}^{k} \mathbf{E}[x_i] = \frac{k}{n},$$

by linearity of expectation.

We next use that f is concentrated to show that f^2 is also relatively concentrated. For any $t \ge 0$, we have

$$\frac{k}{n} = \mathbf{E}\big[f^2\big] \le \mathbf{Pr}\big[f \le m + t\big](m+t)^2 + \mathbf{Pr}\big[f \ge m + t\big] \cdot 1 \le 1 \cdot (m+t)^2 + 2\exp(-t^2 n/2),$$

since $f(x) \le 1$, for any $x \in \mathbb{S}^{(n-1)}$. Let $t = \sqrt{k/5n}$. Since $k \ge 10 \ln n$, we have that $2\exp(-t^2 n/2) \le 2/n$. We get that

$$\frac{k}{n} \le \left(m + \sqrt{k/5n}\right)^2 + 2/n,$$

implying that $\sqrt{(k-2)/n} \le m + \sqrt{k/5n}$, which in turn implies that $m \ge \sqrt{(k-2)/n} - \sqrt{k/5n} \ge \frac{1}{2}\sqrt{k/n}$. ∎

Next, we would like to argue that given a fixed vector, projecting it down into a random k-dimensional subspace results in a random vector such that its length is highly concentrated. This would imply that we can do dimension reduction and still preserve distances between points that we care about.

To this end, we would like to flip Lemma 19.13 around. Instead of randomly picking a point and projecting it down to the first k-dimensional space, we would like x to be fixed and randomly pick the k-dimensional subspace we project into. However, we need to pick this random k-dimensional space carefully. Indeed, if we rotate this random subspace, by a transformation T, so that it occupies the first k dimensions, then the point $T(x)$ needs to be uniformly distributed on the hypersphere if we want to use Lemma 19.13.

As such, we would like to randomly pick a rotation of $\mathbf{R}^n$. This maps the standard orthonormal basis into a randomly rotated orthonormal space. Taking the subspace spanned by the first k vectors of the rotated basis results in a k-dimensional random subspace. Such a rotation is an orthonormal matrix with determinant 1. We can generate such a matrix, by randomly picking a vector $e_1 \in \mathbb{S}^{(n-1)}$. Next, we set e_1 as the first column of our rotation matrix and generate the other $n-1$ columns, by generating recursively $n-1$ orthonormal vectors in the space orthogonal to e_1.

Remark 19.14 (Generating a random point on the sphere). At this point, the reader might wonder how we pick a point uniformly from the unit hypersphere. The idea is to pick a point from the multi-dimensional normal distribution $N^n(0,1)$ and normalize it to have length 1. Since the multi-dimensional normal distribution has the density function

$$(2\pi)^{-n/2} \exp\Big(-(x_1^2 + x_2^2 + \cdots + x_n^2)/2\Big),$$

which is symmetric (i.e., all the points at distance r from the origin have the same distribution), it follows that this indeed generates a point randomly and uniformly on $\mathbb{S}^{(n-1)}$.

Generating a vector with multi-dimensional normal distribution is no more than picking each coordinate according to the normal distribution; see Lemma 27.11$_{p338}$. Given a source of random numbers according to the uniform distribution, this can be done using $O(1)$ computations per coordinate, using the Box-Muller transformation [**BM58**]. Overall, each random vector can be generated in $O(n)$ time.

Since projecting down the n-dimensional normal distribution to the lower-dimensional space yields a normal distribution, it follows that generating a random projection is no more than randomly picking n vectors according to the multi-dimensional normal distribution $v_1, \ldots, v_n$. Then, we orthonormalize them, using Graham-Schmidt, where $\widehat{v_1} = v_1/\|v_1\|$ and $\widehat{v_i}$ is the normalized vector of $v_i - w_i$, where w_i is the projection of v_i to the space spanned by $v_1, \ldots, v_{i-1}$.

Taking those vectors as columns of a matrix generates a matrix A, with determinant either 1 or -1. We multiply one of the vectors by -1 if the determinant is -1. The resulting matrix is a random rotation matrix.

We can now restate Lemma 19.13 in the setting where the vector is fixed and the projection is into a random subspace.

Lemma 19.15. *Let $x \in \mathbb{S}^{(n-1)}$ be an arbitrary unit vector. Now, consider a random k-dimensional subspace $\mathcal{F}$, and let $f(x)$ be the length of the projection of x into $\mathcal{F}$. Then, there exists $m = m(n,k)$ such that*

$$\mathbf{Pr}\Big[f(x) \geq m + t\Big] \leq 2\exp\Big(-t^2 n/2\Big) \quad \textit{and} \quad \mathbf{Pr}\Big[f(x) \leq m - t\Big] \leq 2\exp\Big(-t^2 n/2\Big),$$

for any $t \in [0,1]$. Furthermore, for $k \geq 10 \ln n$, we have $m \geq \frac{1}{2}\sqrt{k/n}$.

Proof. Let v_i be the ith orthonormal vector having 1 at the ith coordinate and 0 everywhere else. Let $\mathbf{M}$ be a random translation of space generated as described above. Clearly, for arbitrary fixed unit vector x, the vector $\mathbf{M}x$ is distributed uniformly on the sphere. Now, the ith column of the matrix $\mathbf{M}$ is the random vector e_i, and $e_i = \mathbf{M}^T v_i$. As such, we have

$$\langle \mathbf{M}x, v_i \rangle = (\mathbf{M}x)^T v_i = x^T \mathbf{M}^T v_i = x^T e_i = \langle x, e_i \rangle .$$

In particular, treating $\mathbf{M}x$ as a random vector and projecting it on the first k coordinates, we have that

$$f(x) = \sqrt{\sum_{i=1}^{k} \langle \mathbf{M}x, v_i \rangle^2} = \sqrt{\sum_{i=1}^{k} \langle x, e_i \rangle^2}.$$

But $e_1, \ldots, e_k$ is just an orthonormal basis of a random k-dimensional subspace. As such, the expression on the right is the length of the projection of x into a k-dimensional random subspace. As such, the length of the projection of x into a random k-dimensional subspace has exactly the same distribution as the length of the projection of a random vector into the first k coordinates. The claim now follows by Lemma 19.13. ■

Definition 19.16. The mapping $f : \mathbb{R}^n \to \mathbb{R}^k$ is called ***K-bi-Lipschitz*** for a subset $X \subseteq \mathbb{R}^n$ if there exists a constant $c > 0$ such that

$$cK^{-1} \cdot \|\mathsf{p} - \mathsf{q}\| \leq \|f(\mathsf{p}) - f(\mathsf{q})\| \leq c \cdot \|\mathsf{p} - \mathsf{q}\|,$$

for all $\mathsf{p}, \mathsf{q} \in X$.

The least K for which f is K-bi-Lipschitz is called the ***distortion*** of f and is denoted $\mathrm{dist}(f)$. We will refer to f as a ***K-embedding*** of X.

Remark 19.17. Let $X \subseteq \mathbb{R}^m$ be a set of n points, where m potentially might be much larger than n. Observe that in this case, since we only care about the inter-point distances of points in X, we can consider X to be a set of points lying in the affine subspace $\mathcal{F}$ spanned by the points of X. Note that this subspace has dimension $n-1$. As such, each point of X can be interpreted as an $(n-1)$-dimensional point in $\mathcal{F}$. Namely, we can assume, for our purposes, that the set of n points in Euclidean space we care about lies in $\mathbb{R}^n$ (i.e., $\mathbb{R}^{n-1}$).

Note that if $m < n$, we can always pad all the coordinates of the points of X by zeros, such that the resulting point set lies in $\mathbb{R}^n$.

Theorem 19.18 (Johnson-Lindenstrauss lemma / JL lemma)**.** *Let X be an n-point set in a Euclidean space, and let $\varepsilon \in (0,1]$ be given. Then there exists a $(1+\varepsilon)$-embedding of X into $\mathbb{R}^k$, where $k = O(\varepsilon^{-2}\log n)$.*

Proof. By Remark 19.17, we can assume that $X \subseteq \mathbb{R}^n$. Let $k = 200\varepsilon^{-2}\ln n$. Assume $k < n$, and let $\mathcal{F}$ be a random k-dimensional linear subspace of $\mathbb{R}^n$. Let $P_{\mathcal{F}} : \mathbb{R}^n \to \mathcal{F}$ be the orthogonal projection operator of $\mathbb{R}^n$ into $\mathcal{F}$. Let m be the number around which $\|P_{\mathcal{F}}(x)\|$ is concentrated, for $x \in \mathbb{S}^{(n-1)}$, as in Lemma 19.15.

Fix two points $x, y \in \mathbb{R}^n$. We prove that

$$\left(1 - \frac{\varepsilon}{3}\right) m \|x - y\| \le \|P_{\mathcal{F}}(x) - P_{\mathcal{F}}(y)\| \le \left(1 + \frac{\varepsilon}{3}\right) m \|x - y\|$$

holds with probability $\ge 1 - n^{-2}$. Since there are $\binom{n}{2}$ pairs of points in X, it follows that with constant probability (say $> 1/3$) this holds for all pairs of points of X. In such a case, the mapping p is a D-embedding of X into $\mathbb{R}^k$ with $D \le \frac{1+\varepsilon/3}{1-\varepsilon/3} \le 1 + \varepsilon$, for $\varepsilon \le 1$.

Let $u = x - y$. We have $P_{\mathcal{F}}(u) = P_{\mathcal{F}}(x) - P_{\mathcal{F}}(y)$ since $P_{\mathcal{F}}(\cdot)$ is a linear operator. Thus, the condition becomes $\left(1 - \frac{\varepsilon}{3}\right) m \|u\| \le \|P_{\mathcal{F}}(u)\| \le \left(1 + \frac{\varepsilon}{3}\right) m \|u\|$. Again, since projection is a linear operator, for any $\alpha > 0$, the condition is equivalent to

$$\left(1 - \tfrac{\varepsilon}{3}\right) m \|\alpha u\| \le \|P_{\mathcal{F}}(\alpha u)\| \le \left(1 + \tfrac{\varepsilon}{3}\right) m \|\alpha u\|.$$

As such, we can assume that $\|u\| = 1$ by picking $\alpha = 1/\|u\|$. Namely, we need to show that

$$\Bigl| \|P_{\mathcal{F}}(u)\| - m \Bigr| \le \frac{\varepsilon}{3} m.$$

Let $f(u) = \|P_{\mathcal{F}}(u)\|$. By Lemma 19.13 (exchanging the random space with the random vector), for $t = \varepsilon m/3$, we have that the probability that this does not hold is bounded by

$$\mathbf{Pr}\bigl[|f(u) - m| \ge t\bigr] \le 4\exp\left(-\frac{t^2 n}{2}\right) = 4\exp\left(\frac{-\varepsilon^2 m^2 n}{18}\right) \le 4\exp\left(-\frac{\varepsilon^2 k}{72}\right) < n^{-2},$$

since $m \ge \frac{1}{2}\sqrt{k/n}$ and $k = 200\varepsilon^{-2}\ln n$. ∎

19.5. Bibliographical notes

Our presentation follows Matoušek [**Mat02**]. The Brunn-Minkowski inequality is a powerful inequality which is widely used in mathematics. A nice survey of this inequality and its applications is provided by Gardner [**Gar02**]. Gardner says, "In a sea of mathematics, the Brunn-Minkowski inequality appears like an octopus, tentacles reaching far and wide, its shape and color changing as it roams from one area to the next." However, Gardner is careful in claiming that the Brunn-Minkowski inequality is one of the most powerful inequalities in mathematics since, as a wit put it, "The most powerful inequality is $x^2 \ge 0$, since all inequalities are in some sense equivalent to it."

A striking application of the Brunn-Minkowski inequality is the proof that in any partial ordering of n elements, there is a single comparison that, knowing its result, reduces the number of linear extensions that are consistent with the partial ordering, by a constant fraction. This immediately implies (the uninteresting result) that one can sort n elements in $O(n \log n)$ comparisons. More interestingly, it implies that if there are m linear extensions of the current partial ordering, we can *always* sort it using $O(\log m)$ comparisons. A nice exposition of this surprising result is provided by Matoušek [**Mat02**, Section 12.3].

There are several alternative proofs of the Johnson-Lindenstrauss lemma (i.e., JL lemma); see [**IM98**] and [**DG03**]. Interestingly, it is enough to pick each entry in the dimension-reducing matrix randomly out of $-1, 0, 1$. This requires a more involved proof [**Ach01**]. This is useful when one cares about storing this dimension reduction transformation efficiently. In particular, recently, there was a flurry of work on making the JL lemma both faster to compute (per point) and sparser. For example, Kane and Nelson [**KN10**] show that one can compute a matrix for dimension reduction with $O(\varepsilon^{-1} \log n)$ non-zero entries per column (the target dimension is still $O(\varepsilon^{-2} \log n)$). See [**KN10**] and references therein for more details.

Magen [**Mag07**] observed that the JL lemma preserves angles, and in fact can be used to preserve any "k-dimensional angle", by projecting down to dimension $O(k\varepsilon^{-2} \log n)$. In particular, Exercise 19.4 is taken from there.

Surprisingly, the random embedding preserves much more structure than distances between points. It preserves the structure and distances of surfaces as long as they are low dimensional and "well behaved"; see [**AHY07**] for some results in this direction.

Dimension reduction is crucial in computational learning, AI, databases, etc. One common technique that is being used in practice is to do PCA (i.e., principal component analysis) and take the first few main axes. Other techniques include independent component analysis and MDS (multi-dimensional scaling). MDS tries to embed points from high dimensions into low dimension ($\mathsf{d} = 2$ or 3), while preserving some properties. Theoretically, dimension reduction into really low dimensions is hopeless, as the distortion in the worst case is $\Omega(n^{1/(k-1)})$, if k is the target dimension [**Mat90**].

19.6. Exercises

Exercise 19.1 (Boxes can be separated). (Easy) Let A and B be two axis parallel boxes that are interior disjoint. Prove that there is always an axis parallel hyperplane that separates the interior of the two boxes.

Exercise 19.2 (Brunn-Minkowski inequality, slight extension). Prove the following claim.

> **Corollary 19.19.** *For A and B compact sets in $\mathbb{R}^n$, we have for any $\lambda \in [0,1]$ that* $\mathrm{Vol}(\lambda A + (1-\lambda)B) \geq \mathrm{Vol}(A)^{\lambda}\,\mathrm{Vol}(B)^{1-\lambda}$.

Exercise 19.3 (Projections are contractions). (Easy) Let $\mathcal{F}$ be a k-dimensional affine subspace, and let $P_{\mathcal{F}} : \mathbb{R}^{\mathsf{d}} \to \mathcal{F}$ be the projection that maps every point $x \in \mathbb{R}^{\mathsf{d}}$ to its nearest neighbor on $\mathcal{F}$. Prove that $P_{\mathcal{F}}$ is a contraction (i.e., 1-Lipschitz). Namely, for any $\mathsf{p}, \mathsf{q} \in \mathbb{R}^{\mathsf{d}}$, we have that $\|P_{\mathcal{F}}(\mathsf{p}) - P_{\mathcal{F}}(\mathsf{q})\| \leq \|\mathsf{p} - \mathsf{q}\|$.

Exercise 19.4 (JL lemma works for angles). Show that the Johnson-Lindenstrauss lemma also $(1 \pm \varepsilon)$-preserves angles among triples of points of P (you might need to increase the target dimension however by a constant factor). [**Hint:** For every angle, construct an equilateral triangle whose edges are being preserved by the projection (add the vertices of those triangles (conceptually) to the point set being embedded). Argue that this implies that the angle is being preserved.]

CHAPTER 20

Approximate Nearest Neighbor (ANN) Search in High Dimensions

20.1. ANN on the hypercube

20.1.1. ANN for the hypercube and the Hamming distance.

Definition 20.1. The set of points $\mathcal{H}^{\mathsf{d}} = \{0, 1\}^{\mathsf{d}}$ is the d-***dimensional hypercube***. A point $\mathsf{p} = (\mathsf{p}_1, \ldots, \mathsf{p}_\mathsf{d}) \in \mathcal{H}^{\mathsf{d}}$ can be interpreted, naturally, as a binary string $\mathsf{p}_1\mathsf{p}_2\ldots\mathsf{p}_\mathsf{d}$. The ***Hamming distance*** $\mathbf{d}_H(\mathsf{p}, \mathsf{q})$ between $\mathsf{p}, \mathsf{q} \in \mathcal{H}^{\mathsf{d}}$ is the number of coordinates where p and q disagree.

It is easy to verify that the Hamming distance, being the L_1-norm, complies with the triangle inequality and is thus a metric.

As we saw previously, to solve the $(1 + \varepsilon)$-ANN problem efficiently, it is sufficient to solve the ***approximate near neighbor*** problem. Namely, given a set P of n points in $\mathcal{H}^{\mathsf{d}}$, a radius $r > 0$, and parameter $\varepsilon > 0$, we want to decide for a query point q whether $\mathbf{d}_H(\mathsf{q}, \mathsf{P}) \le r$ or $\mathbf{d}_H(\mathsf{q}, \mathsf{P}) \ge (1 + \varepsilon)r$, where $\mathbf{d}_H(\mathsf{q}, \mathsf{P}) = \min_{\mathsf{p}\in\mathsf{P}} \mathbf{d}_H(\mathsf{q}, \mathsf{p})$.

Definition 20.2. For a set P of points, a data-structure $\mathcal{D} = \mathcal{D}_{\approx\text{Near}}(\mathsf{P}, r, (1+\varepsilon)r)$ solves the ***approximate near neighbor*** problem if, given a query point q, the data-structure works as follows.

NEAR: If $\mathbf{d}_H(\mathsf{q}, \mathsf{P}) \le r$, then $\mathcal{D}$ outputs a point $\mathsf{p} \in \mathsf{P}$ such that $\mathbf{d}_H(\mathsf{p}, \mathsf{q}) \le (1 + \varepsilon)r$.
FAR: If $\mathbf{d}_H(\mathsf{q}, \mathsf{P}) \ge (1 + \varepsilon)r$, then $\mathcal{D}$ outputs "$\mathbf{d}_H(\mathsf{q}, \mathsf{P}) \ge r$".
DON'T CARE: If $r \le \mathbf{d}(\mathsf{q}, \mathsf{P}) \le (1 + \varepsilon)r$, then $\mathcal{D}$ can return either of the above answers.

Given such a data-structure, one can construct a data-structure that answers the approximate nearest neighbor query using $O\left(\log\left(\varepsilon^{-1}\log \mathsf{d}\right)\right)$ queries using an approximate near neighbor data-structure. Indeed, the desired distance $\mathbf{d}_H(\mathsf{q}, \mathsf{P})$ is an integer number in the range $0, 1, \ldots, \mathsf{d}$. We can build a $\mathcal{D}_{\approx\text{Near}}$ data-structure for distances $(1 + \varepsilon)^i$, for $i = 1, \ldots, M$, where $M = O\left(\varepsilon^{-1}\log \mathsf{d}\right)$. Performing a binary search over these distances would resolve the approximate nearest neighbor query and requires $O(\log M)$ queries.

As such, in the following, we concentrate on constructing the approximate near neighbor data-structure (i.e., $\mathcal{D}_{\approx\text{Near}}$).

20.1.2. Construction of the near neighbor data-structure.

20.1.2.1. *On sense and sensitivity.* Let $\mathsf{P} = \{\mathsf{p}_1, \ldots, \mathsf{p}_n\}$ be a subset of vertices of the hypercube in d dimensions. In the following we assume that $\mathsf{d} = n^{O(1)}$. Let $r, \varepsilon > 0$ be two prespecified parameters. We are interested in building an approximate near neighbor data-structure (i.e., $\mathcal{D}_{\approx\text{Near}}$) for balls of radius r in the Hamming distance.

Definition 20.3. A family $\mathcal{F}$ of functions (defined over $\mathcal{H}^{\mathsf{d}}$) is $\left(r, R, \widehat{\alpha}, \widehat{\beta}\right)$***-sensitive*** if for any $\mathsf{q}, \mathsf{s} \in \mathcal{H}^{\mathsf{d}}$, we have the following

(A) If $q \in \mathbf{b}(s, r)$, then $\mathbf{Pr}\big[f(q) = f(s)\big] \geq \widehat{\alpha}$.

(B) If $q \notin \mathbf{b}(s, R)$, then $\mathbf{Pr}\big[f(q) = f(s)\big] \leq \widehat{\beta}$.

In (A) and (B), f is a randomly picked function from $\mathcal{F}$, $r < R$, and $\widehat{\alpha} > \widehat{\beta}$.

Intuitively, if we can construct an (r, R, α, β)-sensitive family, then we can distinguish between two points which are close together and two points which are far away from each other. Of course, the probabilities α and β might be very close to each other, and we need a way to do amplification.

A simple sensitive family. A priori it is not even clear such a sensitive family exists, but it turns out that the family randomly exposing one coordinate is sensitive.

Lemma 20.4. *Let $f_i(p)$ denote the function that returns the ith coordinate of p, for $i = 1, \ldots, d$. Consider the family of functions $\mathcal{F} = \{f_1, \ldots, f_d\}$. Then, for any $r > 0$ and ε, the family $\mathcal{F}$ is $\big(r, (1+\varepsilon)r, \alpha, \beta\big)$-sensitive, where $\alpha = 1 - r/d$ and $\beta = 1 - r(1+\varepsilon)/d$.*

PROOF. If $q, s \in \{0, 1\}^d$ are within distance smaller than r from each other (under the Hamming distance), then they differ in at most r coordinates. The probability that a random $h \in \mathcal{F}$ would project into a coordinate that q and s agree on is $\geq 1 - r/d$.

Similarly, if $\mathbf{d}_H(q, s) \geq (1+\varepsilon)r$, then the probability that a random $h \in \mathcal{F}$ would map into a coordinate that q and s agree on is $\leq 1 - (1+\varepsilon)r/d$. ■

A family with a large sensitivity gap. Let k be a parameter to be specified shortly, and consider the family of functions $\mathcal{G}$ that concatenates k of the given functions. Formally, let

$$\mathcal{G} = \text{combine}(\mathcal{F}, k) = \left\{ g \;\middle|\; g(p) = \big(f^1(p), \ldots, f^k(p)\big), \text{ for } f^1, \ldots, f^k \in \mathcal{F} \right\}$$

be the set of all such functions.

Lemma 20.5. *Given an (r, R, α, β)-sensitive family $\mathcal{F}$, the family $\mathcal{G} = \text{combine}(\mathcal{F}, k)$ is $\big(r, R, \alpha^k, \beta^k\big)$-sensitive.*

PROOF. For two fixed points $q, s \in \mathcal{H}^d$ such that $\mathbf{d}_H(q, s) \leq r$, we have that for a random $h \in \mathcal{F}$, we have $\mathbf{Pr}[h(q) = h(s)] \geq \alpha$. As such, for a random $g \in \mathcal{G}$, we have that

$$\mathbf{Pr}\big[g(q) = g(s)\big] = \mathbf{Pr}\big[f^1(q) = f^1(s) \text{ and } f^2(q) = f^2(s) \text{ and } \ldots \text{ and } f^k(q) = f^k(s)\big]$$
$$= \prod_{i=1}^{k} \mathbf{Pr}\big[f^i(q) = f^i(s)\big] \geq \alpha^k.$$

Similarly, if $\mathbf{d}_H(q, s) > R$, then $\mathbf{Pr}\big[g(q) = g(s)\big] = \prod_{i=1}^{k} \mathbf{Pr}\big[f^i(q) = f^i(s)\big] \leq \beta^k$. ■

The above lemma implies that we can build a family that has a gap between the lower and upper sensitivities; namely, $\alpha^k/\beta^k = (\alpha/\beta)^k$ is arbitrarily large. The problem is that if α^k is too small, then we will have to use too many functions to detect whether or not there is a point close to the query point.

Nevertheless, consider the task of building a data-structure that finds all the points of $P = \{p_1, \ldots, p_n\}$ that are equal, under a given function $g \in \mathcal{G} = \text{combine}(\mathcal{F}, k)$, to a query point. To this end, we compute the strings $g(p_1), \ldots, g(p_n)$ and store them (together with their associated point) in a hash table (or a prefix tree). Now, given a query point q, we compute $g(q)$ and fetch from this data-structure all the strings equal to it that are stored in it. Clearly, this is a simple and efficient data-structure. All the points colliding with q would be the natural candidates to be the nearest neighbor to q.

By not storing the points explicitly, but using a pointer to the original input set, we get the following easy result.

Lemma 20.6. *Given a function $g \in \mathcal{G} = \text{combine}(\mathcal{F}, k)$ (see Lemma 20.5) and a set $\mathsf{P} \subseteq \mathcal{H}^{\mathsf{d}}$ of n points, one can construct a data-structure, in $O(nk)$ time and using $O(nk)$ additional space, such that given a query point q, one can report all the points in $X = \left\{\mathsf{p} \in \mathsf{P} \,\middle|\, g(\mathsf{p}) = g(\mathsf{q})\right\}$ in $O(k + |X|)$ time.*

Amplifying sensitivity. Our task is now to amplify the sensitive family we currently have. To this end, for two τ-dimensional points x and y, let $x \doteqdot y$ be the Boolean function that returns *true* if there exists an index i such that $x_i = y_i$ and *false* otherwise. Now, the regular "=" operator requires vectors to be equal in all coordinates (i.e., it is equal to $\bigcap_i (x_i = y_i)$) while $x \doteqdot y$ is $\bigcup_i (x_i = y_i)$. The previous construction of Lemma 20.5 using this alternative equal operator provides us with the required amplification.

Lemma 20.7. *Given an $\left(r, R, \alpha^k, \beta^k\right)$-sensitive family $\mathcal{G}$, the family $\mathcal{H}_{\doteqdot} = \text{combine}(\mathcal{G}, \tau)$ if one uses the $\doteqdot$ operator to check for equality is $\left(r, R, 1 - \left(1 - \alpha^k\right)^\tau, 1 - \left(1 - \beta^k\right)^\tau\right)$-sensitive.*

Proof. For two fixed points $\mathsf{q}, \mathsf{s} \in \mathcal{H}^{\mathsf{d}}$ such that $\mathbf{d}_H(\mathsf{q}, \mathsf{s}) \leq r$, we have, for a random $g \in \mathcal{G}$, that $\mathbf{Pr}[g(\mathsf{q}) = g(\mathsf{s})] \geq \alpha^k$. As such, for a random $h \in \mathcal{H}_{\doteqdot}$, we have that

$$\begin{aligned}\mathbf{Pr}\left[h(\mathsf{q}) \doteqdot h(\mathsf{s})\right] &= \mathbf{Pr}\left[g^1(\mathsf{q}) = g^1(\mathsf{s}) \text{ or } g^2(\mathsf{q}) = g^2(\mathsf{s}) \text{ or } \ldots \text{ or } g^\tau(\mathsf{q}) = g^\tau(\mathsf{s})\right] \\ &= 1 - \prod_{i=1}^{\tau} \mathbf{Pr}\left[g^i(\mathsf{q}) \neq g^i(\mathsf{s})\right] \geq 1 - \left(1 - \alpha^k\right)^\tau.\end{aligned}$$

Similarly, if $\mathbf{d}_H(\mathsf{q}, \mathsf{s}) > R$, then

$$\mathbf{Pr}\left[h(\mathsf{q}) \doteqdot h(\mathsf{s})\right] = 1 - \prod_{i=1}^{\tau} \mathbf{Pr}\left[g^i(\mathsf{q}) \neq g^i(\mathsf{s})\right] \leq 1 - \left(1 - \beta^k\right)^\tau.$$

■

To see the effect of Lemma 20.7, it is useful to play with a concrete example. Consider an $(\alpha^k, \beta^k, r, R)$-sensitive family where $\beta^k = \alpha^k/2$ and yet α^k is very small. Setting $\tau = 1/\alpha^k$, the resulting family is (roughly) $(1 - 1/e, 1 - 1/\sqrt{e}, r, R)$-sensitive. Namely, the gap shrank, but the threshold sensitivity is considerably higher. In particular, it is now a constant, and the gap is also a constant.

Using Lemma 20.7 as a data-structure to store P is more involved than before. Indeed, for a random function $h = \left(g^1, \ldots, g^\tau\right) \in \mathcal{H}_{\doteqdot} = \text{combine}(\mathcal{G}, \tau)$ building the associated data-structure requires us to build τ data-structures for each one of the functions $g^1, \ldots, g^\tau$, using Lemma 20.6. Now, given a query point, we retrieve all the points of P that collide with each one of these functions, by querying each of these data-structures.

Lemma 20.8. *Given a function $h \in \mathcal{H}_{\doteqdot} = \text{combine}(\mathcal{G}, \tau)$ (see Lemma 20.7) and a set $\mathsf{P} \subseteq \mathcal{H}^{\mathsf{d}}$ of n points, one can construct a data-structure, in $O(nk\tau)$ time and using $O(nk\tau)$ additional space, such that given a query point q, one can report all the points in $X = \left\{\mathsf{p} \in \mathsf{P} \,\middle|\, h(\mathsf{p}) \doteqdot h(\mathsf{q})\right\}$ in $O(k\tau + |X|)$ time.*

20.1.2.2. *The near neighbor data-structure and handling a query.* We construct the data-structure $\mathcal{D}$ of Lemma 20.8 with parameters k and τ to be determined shortly, for a random function $h \in \mathcal{H}_{\doteqdot}$. Given a query point q, we retrieve all the points that collide with h and compute their distance to the query point. Next, scan these points one by one and compute their distance to q. As soon as encountering a point $s \in P$ such that $\mathbf{d}_H(q, s) \le R$, the data-structures returns ***true*** together with s.

Let's assume that we know that the expected number of points of $P \setminus \mathbf{b}(q, R)$ (i.e., $R = (1 + \varepsilon)r$) that will collide with q in $\mathcal{D}$ is in expectation L (we will figure out the value of L below). To ensure the worst case query time, the query would abort after checking $4L + 1$ points and would return *false*. Naturally, the data-structure would also return false if all points encountered have distance larger than R from q.

Clearly, the query time of this data-structure is $O(k\tau + \mathsf{d}L)$.

We are left with the task of fine-tuning the parameters τ and k to get the fastest possible query time, while the data-structure has reasonable probability to succeed. Figuring the right values is technically tedious, and we do it next.

20.1.2.3. *Setting the parameters.* If there exists $p \in P$ such that $\mathbf{d}_H(q, p) \le r$, then the probability of this point to collide with q under the function h is $\phi \ge 1 - \left(1 - \alpha^k\right)^\tau$. Let us demand that this data-structure succeeds with probability $\ge 3/4$. To this end, we set

$$\tau = 4\left\lceil 1/\alpha^k \right\rceil \implies \phi \ge 1 - \left(1 - \alpha^k\right)^\tau \ge 1 - \exp\left(-\alpha^k \tau\right) \ge 1 - \exp(-4) \ge 3/4, \tag{20.1}$$

since $1 - x \le \exp(-x)$, for $x \ge 0$.

Lemma 20.9. *The expected number of points of* $P \setminus \mathbf{b}(q, R)$ *colliding with the query point is* $L = O\left(n(\beta/\alpha)^k\right)$, *where* $R = (1 + \varepsilon)r$.

Proof. Consider the points in $P \setminus \mathbf{b}(q, R)$. We would like to bound the number of points of this set that collide with the query point. Observe that in this case, the probability of a point $p \in P \setminus \mathbf{b}(q, R)$ to collide with the query point is

$$\begin{aligned}
&\le \psi = 1 - \left(1 - \beta^k\right)^\tau = \left(1 - \left(1 - \beta^k\right)\right)\left(1 + \left(1 - \beta^k\right) + \left(1 - \beta^k\right)^2 + \ldots + \left(1 - \beta^k\right)^{\tau - 1}\right) \\
&\le \beta^k \tau \le 8\left(\frac{\beta}{\alpha}\right)^k,
\end{aligned}$$

as $\tau = 4\left\lceil 1/\alpha^k \right\rceil$ and $\alpha, \beta \in (0, 1)$. Namely, the expected number of points of $P \setminus \mathbf{b}(q, R)$ colliding with the query point is $\le \psi n$. ■

By Lemma 20.8, extracting the $O(L)$ points takes $O(k\tau + L)$ time. Computing the distance of the query time for each one of these points takes $O(k\tau + L\mathsf{d})$ time. As such, by Lemma 20.9, the query time is

$$O(k\tau + L\mathsf{d}) = O\left(k\tau + n\mathsf{d}(\beta/\alpha)^k\right).$$

To minimize this query time, we "approximately" solve the equation requiring the two terms, in the above bound, to be equal (we ignore d since, intuitively, it should be small compared to n). We get that

$$k\tau = n(\beta/\alpha)^k \rightsquigarrow \frac{k}{\alpha^k} \approx n\frac{\beta^k}{\alpha^k} \implies k \approx n\beta^k \rightsquigarrow 1/\beta^k \approx n \implies k \approx \ln_{1/\beta} n.$$

Thus, setting $k = \ln_{1/\beta} n$, we have that $\beta^k = 1/n$ and, by (20.1), that

$$\tau = 4\left\lceil 1/\alpha^k \right\rceil = \exp\left(\frac{\ln n}{\ln 1/\beta} \ln 1/\alpha\right) = O(n^\rho), \quad \text{for } \rho = \frac{\ln 1/\alpha}{\ln 1/\beta}. \tag{20.2}$$

As such, to minimize the query time, we need to minimize ρ.

Lemma 20.10. *(A) For* $x \in [0, 1)$ *and* $t \ge 1$ *such that* $1 - tx > 0$ *we have* $\dfrac{\ln(1 - x)}{\ln(1 - tx)} \le \dfrac{1}{t}$.

(B) For $\alpha = 1 - r/\mathsf{d}$ *and* $\beta = 1 - r(1+\varepsilon)/\mathsf{d}$, *we have that* $\rho = \dfrac{\ln 1/\alpha}{\ln 1/\beta} \le \dfrac{1}{1+\varepsilon}$.

Proof. (A) Since $\ln(1 - tx) < 0$, it follows that the claim is equivalent to $t \ln(1-x) \ge \ln(1-tx)$. This in turn is equivalent to

$$g(x) \equiv (1 - tx) - (1 - x)^t \le 0.$$

This is trivially true for $x = 0$. Furthermore, taking the derivative, we see $g'(x) = -t + t(1-x)^{t-1}$, which is non-positive for $x \in [0, 1)$ and $t > 0$. Therefore, g is non-increasing in the interval of interest, and so $g(x) \le 0$ for all values in this interval.

(B) Indeed $\rho = \dfrac{\ln 1/\alpha}{\ln 1/\beta} = \dfrac{\ln \alpha}{\ln \beta} = \dfrac{\ln \frac{\mathsf{d}-r}{\mathsf{d}}}{\ln \frac{\mathsf{d}-(1+\varepsilon)r}{\mathsf{d}}} = \dfrac{\ln\left(1 - \frac{r}{\mathsf{d}}\right)}{\ln\left(1 - (1+\varepsilon)\frac{r}{\mathsf{d}}\right)} \le \dfrac{1}{1+\varepsilon}$, by part (A). ■

In the following, it would be convenient to consider d to be considerably larger than r. This can be ensured by (conceptually) padding the points with fake coordinates that are all zero. It is easy to verify that this "hack" would not affect the algorithm's performance in any way and it is just a trick to make our analysis simpler. In particular, we assume that $\mathsf{d} > 2(1+\varepsilon)r$.

Lemma 20.11. *For* $\alpha = 1 - r/\mathsf{d}$, $\beta = 1 - r(1+\varepsilon)/\mathsf{d}$, n *and* d *as above, we have that (i)* $\tau = O\left(n^{1/(1+\varepsilon)}\right)$, *(ii)* $k = O(\ln n)$, *and (iii)* $L = O\left(n^{1/(1+\varepsilon)}\right)$.

Proof. By (20.1), $\tau = 4\left\lceil 1/\alpha^k \right\rceil = O(n^\rho) = O\left(n^{1/(1+\varepsilon)}\right)$, by Lemma 20.10(B).

Now, $\beta = 1 - r(1+\varepsilon)/\mathsf{d} \le 1/2$, since we assumed that $\mathsf{d} > 2(1+\varepsilon)r$. As such, we have $k = \ln_{1/\beta} n = \dfrac{\ln n}{\ln 1/\beta} = O(\ln n)$.

By Lemma 20.9, $L = O\left(n(\beta/\alpha)^k\right)$. Now $\beta^k = 1/n$ and as such $L = O\left(1/\alpha^k\right) = O(\tau) = O\left(n^{1/(1+\varepsilon)}\right)$. ■

20.1.2.4. *The result.*

Theorem 20.12. *Given a set* P *of* n *points on the hypercube* $\mathcal{H}^{\mathsf{d}}$ *and parameters* $\varepsilon > 0$ *and* $r > 0$, *one can build a data-structure* $\mathcal{D} = \mathcal{D}_{\approx Near}(\mathsf{P}, r, (1+\varepsilon)r)$ *that solves the approximate near neighbor problem (see Definition 20.2). The data-structure answers a query successfully with high probability. In addition we have the following:*
(A) The query time is $O\left(\mathsf{d}n^{1/(1+\varepsilon)} \log n\right)$.
(B) The preprocessing time to build this data-structure is $O\left(n^{1+1/(1+\varepsilon)} \log^2 n\right)$.
(C) The space required to store this data-structure is $O\left(n\mathsf{d} + n^{1+1/(1+\varepsilon)} \log^2 n\right)$.

Proof. Our building block is the data-structure described above. By Markov's inequality, the probability that the algorithm has to abort because of too many collisions with points of $\mathsf{P} \setminus \mathbf{b}(\mathsf{q}, (1+\varepsilon)r)$ is bounded by $1/4$ (since the algorithm tries $4L+1$ points). Also, if there is a point inside $\mathbf{b}(\mathsf{q}, r)$, the algorithm would find it with probability $\ge 3/4$, by (20.1). As such, with probability at least $1/2$ this data-structure returns the correct answer in this case. By Lemma 20.8, the query time is $O(k\tau + L\mathsf{d})$.

This data-structure succeeds only with constant probability. To achieve high probability, we construct $O(\log n)$ such data-structures and perform the near neighbor query in each one of them. As such, the query time is

$$O\left((k\tau + L\mathsf{d})\log n\right) = O\left(n^{1/(1+\varepsilon)} \log^2 n + \mathsf{d}n^{1/(1+\varepsilon)} \log n\right) = O\left(\mathsf{d}n^{1/(1+\varepsilon)} \log n\right),$$

by Lemma 20.11 and since $\mathsf{d} = \Omega(\lg n)$ if P contains n distinct points of $\mathcal{H}^{\mathsf{d}}$.

As for the preprocessing time, by Lemma 20.8 and Lemma 20.11, it is $O(nk\tau \log n) = O\left(n^{1+1/(1+\varepsilon)} \log^2 n\right)$.

Finally, this data-structure requires $O(\mathsf{d}n)$ space to store the input points. Specifically, by Lemma 20.8, we need an additional $O(nk\tau \log n) = O\left(n^{1+1/(1+\varepsilon)} \log^2 n\right)$ space. ∎

In the hypercube case, when $\mathsf{d} = n^{O(1)}$, we can build $M = O(\log_{1+\varepsilon} \mathsf{d}) = O(\varepsilon^{-1} \log \mathsf{d})$ such data-structures such that $(1+\varepsilon)$-ANN can be answered using binary search on those data-structures which correspond to radii $r_1, \ldots, r_M$, where $r_i = (1+\varepsilon)^i$, for $i = 1, \ldots, M$.

Theorem 20.13. *Given a set* P *of* n *points on the hypercube* $\mathcal{H}^{\mathsf{d}}$ *(where* $\mathsf{d} = n^{O(1)}$*) and a parameter* $\varepsilon > 0$*, one can build a data-structure to answer approximate nearest neighbor queries (under the Hamming distance) using* $O\left(\mathsf{d}n + n^{1/(1+\varepsilon)}\varepsilon^{-1} \log^2 n \log \mathsf{d}\right)$ *space, such that given a query point* q*, one can return a* $(1+\varepsilon)$*-ANN in* P *(under the Hamming distance) in* $O(\mathsf{d}n^{1/(1+\varepsilon)} \log n \log(\varepsilon^{-1} \log \mathsf{d}))$ *time. The result returned is correct with high probability.*

Remark 20.14. The result of Theorem 20.13 needs to be oblivious to the queries used. Indeed, for any instantiation of the data-structure of Theorem 20.13 there exist query points for which it would fail.

In particular, formally, if we perform a sequence of ANN queries using such a data-structure, where the queries depend on earlier returned answers, then the guarantee of a high probability of success is no longer implied by the above analysis (it might hold because of some other reasons, naturally).

20.2. LSH and ANN in Euclidean space

20.2.1. Preliminaries.

Lemma 20.15. *Let* $X = (X_1, \ldots, X_{\mathsf{d}})$ *be a vector of* d *independent variables which have normal distribution* $\mathbf{N}$*, and let* $v = (v_1, \ldots, v_{\mathsf{d}}) \in \mathbb{R}^{\mathsf{d}}$*. We have that* $\langle v, X \rangle = \sum_i v_i X_i$ *is distributed as* $\|v\| Z$*, where* $Z \sim \mathbf{N}$*.*

Proof. By Lemma 27.11$_{\text{p338}}$ the point X has multi-dimensional normal distribution $\mathbf{N}^{\mathsf{d}}$. As such, if $\|v\| = 1$, then this holds by the symmetry of the normal distribution. Indeed, let $e_1 = (1, 0, \ldots, 0)$. By the symmetry of the d-dimensional normal distribution, we have that $\langle v, X \rangle \sim \langle e_1, X \rangle = X_1 \sim \mathbf{N}$.

Otherwise, $\langle v, X \rangle / \|v\| \sim \mathbf{N}$, and as such $\langle v, X \rangle \sim N\left(0, \|v\|^2\right)$, which is indeed the distribution of $\|v\| Z$. ∎

Definition 20.16. A distribution $\mathcal{D}$ over $\mathbb{R}$ is called ***p*-stable** if there exists $p \geq 0$ such that for any n real numbers $v_1, \ldots, v_n$ and n independent variables $X_1, \ldots, X_n$ with distribution $\mathcal{D}$, the random variable $\sum_i v_i X_i$ has the same distribution as the variable $(\sum_i |v_i|^p)^{1/p} X$, where X is a random variable with distribution $\mathcal{D}$.

By Lemma 20.15, the normal distribution is a ***2-stable distribution***.

20.2.2. Locality sensitive hashing (LSH). Let p and q be two points in $\mathbb{R}^{\mathsf{d}}$. We want to perform an experiment to decide if $\|\mathsf{p} - \mathsf{q}\| \leq 1$ or $\|\mathsf{p} - \mathsf{q}\| \geq \eta$, where $\eta = 1 + \varepsilon$. We will randomly choose a vector $\vec{v}$ from the d-dimensional normal distribution $\mathbf{N}^{\mathsf{d}}$ (which is 2-stable). Next, let r be a parameter, and let t be a random number chosen uniformly from the interval $[0, r]$. For $\mathsf{p} \in \mathbb{R}^{\mathsf{d}}$, consider the random hash function

$$h(\mathsf{p}) = \left\lfloor \frac{\langle \mathsf{p}, \vec{v} \rangle + t}{r} \right\rfloor. \tag{20.3}$$

Assume that the distance between p and q is η and the distance between the projection of the two points to the direction $\vec{v}$ is β. Then, the probability that p and q get the same hash value is $\max(1-\beta/r, 0)$, since this is the probability that the random sliding will not separate them. Indeed, consider the line through $\vec{v}$ to be the x-axis, and assume q is projected to r and s is projected to $r-\beta$ (assuming $r \geq \beta$). Clearly, q and s get mapped to the same value by $h(\cdot)$ if and only if $t \in [0, r-\beta]$, as claimed.

As such, we have that the probability of collusion is

$$\alpha(\eta, r) = \mathbf{Pr}\Big[h(\mathsf{p}) = h(\mathsf{q})\Big] = \int_{\beta=0}^{r} \mathbf{Pr}\Big[\big|\langle \mathsf{p}, \vec{v}\rangle - \langle \mathsf{q}, \vec{v}\rangle\big| = \beta\Big]\left(1 - \frac{\beta}{r}\right) d\beta.$$

However, since $\vec{v}$ is chosen from a 2-stable distribution, we have that $Z = \langle \mathsf{p}, \vec{v}\rangle - \langle \mathsf{q}, \vec{v}\rangle = \langle \mathsf{p} - \mathsf{q}, \vec{v}\rangle \sim \mathbf{N}\Big(0, \|\mathsf{p} - \mathsf{q}\|^2\Big)$. Since we are considering the absolute value of the variable, we need to multiply this by two. Thus, we have

$$\alpha(\eta, r) = \int_{\beta=0}^{r} \frac{2}{\sqrt{2\pi}\eta} \exp\left(-\frac{\beta^2}{2\eta^2}\right)\left(1 - \frac{\beta}{r}\right) d\beta,$$

by plugging in the density of the normal distribution for Z. Intuitively, we care about the difference $\alpha(1+\varepsilon, r) - \alpha(1, r)$, and we would like to maximize it as much as possible (by choosing the right value of r). Unfortunately, this integral is unfriendly, and we have to resort to numerical computation.

Now, we are going to use this hashing scheme for constructing locality sensitive hashing, as in the hypercube case, and as such we care about the ratio

$$\rho(1+\varepsilon) = \min_{r} \frac{\log(1/\alpha(1, r))}{\log(1/\alpha(1+\varepsilon, r))};$$

see $(20.2)_{p272}$.

The following is verified using numerical computations (using a computer).

Lemma 20.17 (**[DIIM04]**). *One can choose r, such that $\rho(1+\varepsilon) \leq \frac{1}{1+\varepsilon}$.*

Lemma 20.17 implies that the hash functions defined by (20.3) are $(1, 1+\varepsilon, \alpha', \beta')$-sensitive and, furthermore, $\rho = \frac{\log(1/\alpha')}{\log(1/\beta')} \leq \frac{1}{1+\varepsilon}$, for some values of α' and β'. As such, we can use this hashing family to construct an approximate near neighbor data-structure $\mathcal{D}_{\approx\mathrm{Near}}(\mathsf{P}, r, (1+\varepsilon)r)$ for the set P of points in $\mathbb{R}^{\mathsf{d}}$. Following the same argumentation of Theorem 20.12, we have the following.

Theorem 20.18. *Given a set P of n points in $\mathbb{R}^{\mathsf{d}}$ and parameters $\varepsilon > 0$ and $r > 0$, one can build a $\mathcal{D}_{\approx Near} = \mathcal{D}_{\approx Near}(\mathsf{P}, r, (1+\varepsilon)r)$, such that given a query point q, one can decide:*

(A) If $\mathbf{b}(q, r) \cap \mathsf{P} \neq \emptyset$, then $\mathcal{D}_{\approx Near}$ returns a point $u \in \mathsf{P}$, such that $\mathbf{d}_H(u, q) \leq (1+\varepsilon)r$.

(B) If $\mathbf{b}(q, (1+\varepsilon)r) \cap \mathsf{P} = \emptyset$, then $\mathcal{D}_{\approx Near}$ returns the result that no point is within distance $\leq r$ from q.

In any other case, any of the answers is correct. The query time is $O(\mathsf{d}n^{1/(1+\varepsilon)} \log n)$ and the space used is $O\Big(\mathsf{d}n + n^{1+1/(1+\varepsilon)} \log n\Big)$. The result returned is correct with high probability.

20.2.3. ANN in high-dimensional euclidean space. Unlike the binary hypercube case, where we could just do direct binary search on the distances, here we need to use the reduction from ANN to near neighbor queries. We will need the following result which is implied by Theorem 18.24_{p252} and Theorem 18.18_{p250}.

Corollary 20.19. *Given a set P of n points in $\mathbb{R}^{\mathsf{d}}$, one can construct a data-structure $\mathcal{D}$ that answers $(1+\varepsilon)$-ANN queries, by performing $O(\log(n/\varepsilon))$ $(1+\varepsilon)$-approximate near* *neighbor queries. The total number of points stored at these approximate near neighbor data-structure of $\mathcal{D}$ is $O(n\varepsilon^{-1}\log(n/\varepsilon))$.*

Constructing the data-structure of Corollary 20.19 requires building a low quality HST; to this end we can use the fast HST construction of Theorem 11.18_{p157} that builds $2\sqrt{\mathsf{d}}n^5$-approximate HST of n points in $\mathbb{R}^{\mathsf{d}}$ in $O(\mathsf{d}n\log n)$ expected time.

20.2.3.1. *The overall result.* Plugging Theorem 20.18 into Corollary 20.19, we have the following:

Theorem 20.20. *Given a set* P *of n points in $\mathbb{R}^{\mathsf{d}}$ and parameters $\varepsilon > 0$ and $r > 0$, one can build an ANN data-structure using*

$$O\left(\mathsf{d}n + n^{1+1/(1+\varepsilon)}\varepsilon^{-2}\log^3(n/\varepsilon)\right)$$

space, such that given a query point q*, one can returns a $(1+\varepsilon)$-ANN in* P *in*

$$O\left(\mathsf{d}n^{1/(1+\varepsilon)}(\log n)\log\frac{n}{\varepsilon}\right)$$

time. The result returned is correct with high probability.

The construction time is $O\left(\mathsf{d}n^{1+1/(1+\varepsilon)}\varepsilon^{-2}\log^3(n/\varepsilon)\right)$.

Proof. We compute, in $O(n\mathsf{d}\log n)$ expected time, a low quality HST using Theorem 11.18_{p157}. Using this HST, we can construct the data-structure $\mathcal{D}$ of Corollary 20.19, where we do not compute the $\mathcal{D}_{\approx\mathrm{Near}}$ data-structures. We next traverse the tree $\mathcal{D}$ and construct the $\mathcal{D}_{\approx\mathrm{Near}}$ data-structures using Theorem 20.18.

We only need to prove the bound on the space. Observe that we need to store each point only once, since other places can refer to the point by a pointer. Thus, this is the $O(\mathsf{d}n)$ space requirement. The other term comes from plugging the bound of Corollary 20.19 into the bound of Theorem 20.18. ■

20.3. Bibliographical notes

Section 20.1 follows the exposition of Indyk and Motwani [**IM98**]. Kushilevitz, Ostrovsky, and Rabani [**KOR00**] offered an alternative data-structure with somewhat inferior performance. It is quite surprising that one can perform approximate nearest neighbor queries in high dimensions in time and space polynomial in the dimension (which is sublinear in the number of points). One can reduce the approximate near neighbor in Euclidean space to the same question on the hypercube "directly" (we show the details below). However, doing the LSH directly on the Euclidean space is more efficient.

The value of the results shown in this chapter depends to a large extent on the reader's perspective. Indeed, for a small value of $\varepsilon > 0$, the query time $O(\mathsf{d}n^{1/(1+\varepsilon)})$ is very close to linear dependency on n and is almost equivalent to just scanning the points. Thus, from the low-dimensional perspective, where ε is assumed to be small, this result is slightly sublinear. On the other hand, if one is willing to pick ε to be large (say 10), then the result is clearly better than the naive algorithm, suggesting running time for an ANN query which takes (roughly) $O\left(n^{1/11}\right)$ time.

The idea of doing locality sensitive hashing directly on the Euclidean space, as done in Section 20.2, is not shocking after seeing the Johnson-Lindenstrauss lemma. Our description follows the paper of Datar et al. [**DIIM04**]. In particular, the current analysis which relies on computerized estimates is far from being satisfactory. It would be nice to have a simpler and more elegant scheme for this case. This is an open problem for further research.

Currently, the best LSH construction in $\mathbb{R}^d$ is due to Andoni and Indyk [**AI06**]. It uses space $O\left(dn + n^{1+1/(1+\varepsilon)^2+o(1)}\right)$ and query time $O\left(dn^{1/(1+\varepsilon)^2+o(1)}\right)$. This (almost) matches the lower bound of Motwani et al. [**MNP06**]. For a nice survey on LSH see [**AI08**].

From approximate near neighbor in $\mathbb{R}^d$ to approximate near neighbor on the hypercube. The reduction is quite involved, and we only sketch the details. Let P be a set of n points in $\mathbb{R}^d$. We first reduce the dimension to $k = O(\varepsilon^{-2} \log n)$ using the Johnson-Lindenstrauss lemma. Next, we embed this space into $\ell_1^{k'}$ (this is the space $\mathbb{R}^k$, where distances are the L_1 metric instead of the regular L_2 metric), where $k' = O(k/\varepsilon^2)$. This can be done with distortion $(1 + \varepsilon)$.

Let Q' be the resulting set of points in $\mathbb{R}^{k'}$. We want to solve approximate near neighbor queries on this set of points, for radius r. As a first step, we partition the space into cells by taking a grid with sidelength (say) $k'r$ and randomly translating it, clipping the points inside each grid cell. It is now sufficient to solve the approximate near neighbor problem inside this grid cell (which has bounded diameter as a function of r), since with small probability the result would be correct. We amplify the probability by repeating this a polylogarithmic number of times.

Thus, we can assume that P is contained inside a cube of sidelength $\leq k'nr$, it is in $\mathbb{R}^{k'}$, and the distance metric is the L_1 metric. We next snap the points of P to a grid of sidelength (say) $\varepsilon r/k'$. Thus, every point of P now has an integer coordinate, which is bounded by a polynomial in $\log n$ and $1/\varepsilon$. Next, we write the coordinates of the points of P using unary notation. (Thus, a point $(2, 5)$ would be written as $(00011, 11111)$ assuming the number of bits for each coordinates is 5.) It is now easy to verify that the Hamming distance on the resulting strings is equivalent to the L_1 distance between the points.

Thus, we can solve the near neighbor problem for points in $\mathbb{R}^d$ by solving it on the hypercube under the Hamming distance.

See Indyk and Motwani [**IM98**] for more details.

We have only scratched the surface of proximity problems in high dimensions. The interested reader is referred to the survey by Indyk [**Ind04**] for more information.

CHAPTER 21

Approximating a Convex Body by an Ellipsoid

In this chapter, we show that any convex body can be approximated "reasonably well" by an ellipsoid.

21.1. Ellipsoids

A reminder of linear algebra and the notation we use is provided in Section 28.1_{p347}.

Definition 21.1. Let $\mathbf{b} = \left\{ x \in \mathbb{R}^n \;\middle|\; \|x\| \leq 1 \right\}$ be the unit ball in $\mathbb{R}^n$. Let $\overline{c} \in \mathbb{R}^n$ be a vector and let $T : \mathbb{R}^n \to \mathbb{R}^n$ be an invertible linear transformation. The set

$$\mathcal{E} = T(\mathbf{b}) + \overline{c}$$

is called an ***ellipsoid*** and $\overline{c}$ is its ***center***.

Alternatively, we can write

$$\mathcal{E} = \left\{ x \in \mathbb{R}^n \;\middle|\; \left\| T^{-1}(x - \overline{c}) \right\| \leq 1 \right\}.$$

However,

$$\begin{aligned} \left\| T^{-1}(x - \overline{c}) \right\|^2 &= \left\langle T^{-1}(x - \overline{c}), T^{-1}(x - \overline{c}) \right\rangle = (T^{-1}(x - \overline{c}))^T T^{-1}(x - \overline{c}) \\ &= (x - \overline{c})^T \left(T^{-1} \right)^T T^{-1}(x - \overline{c}). \end{aligned}$$

In particular, let $Q = \left(T^{-1} \right)^T T^{-1}$, and observe that Q is symmetric and positive definite (see Claim 28.2_{p347}). Thus,

$$\mathcal{E} = \left\{ x \in \mathbb{R}^n \;\middle|\; (x - \overline{c})^T Q (x - \overline{c}) \leq 1 \right\}.$$

If we change the basis of $\mathbb{R}^n$ to be the set of unit eigenvectors of Q, then Q becomes a diagonal matrix, and we have that

$$\mathcal{E} = \left\{ (y_1, \ldots, y_n) \in \mathbb{R}^n \;\middle|\; \lambda_1(y_1 - \overline{c}_1)^2 + \cdots + \lambda_n(y_n - \overline{c}_n)^2 \leq 1 \right\},$$

where $\overline{c} = (\overline{c}_1, \ldots, \overline{c}_n)$ is the center and $\lambda_1, \ldots, \lambda_n$ are the eigenvalues of Q (all positive since Q is positive definite). In particular, this implies that the points $(\overline{c}_1, \ldots, \overline{c}_{i-1}, \overline{c}_i \pm 1/\sqrt{\lambda_i}, \overline{c}_{i+1}, \ldots, \overline{c}_n)$, for $i = 1, \ldots, n$, lies on the boundary of $\mathcal{E}$. Specifically, the eigenvectors are the main axes of the ellipsoid, and their corresponding eigenvalues specify how long the ellipse along each one of these axes is. That is, the ith axis of the ellipsoid $\mathcal{E}$ has length $1/\sqrt{\lambda_i}$. Visually, the ellipsoid is being created by taking the unit ball and stretching it by $1/\sqrt{\lambda_i}$ along its ith axis.

In particular,

$$\mathrm{Vol}(\mathcal{E}) = \frac{\mathrm{Vol}(\mathbf{b})}{\sqrt{\lambda_1} \cdots \sqrt{\lambda_n}} = \frac{\mathrm{Vol}(\mathbf{b})}{\sqrt{\det(Q)}}.$$

For a convex body K (i.e., a convex and bounded set), let $\mathcal{E}$ be a largest volume ellipsoid contained inside K. One can show that $\mathcal{E}$ is unique. Namely, there is a single ***maximum volume ellipsoid*** inside K.

The following is obvious and is stated for the sake of clarity.

Observation 21.2. *Given two bodies $X, Y \subseteq \mathbb{R}^n$ such that $X \subseteq Y$ and a non-singular affine transformation $\mathbf{M}$, then* $\mathrm{Vol}(X)/\mathrm{Vol}(Y) = \mathrm{Vol}(\mathbf{M}(X))/\mathrm{Vol}(\mathbf{M}(Y))$ *and* $\mathbf{M}(X) \subseteq \mathbf{M}(Y)$.

Theorem 21.3. *Let $\mathsf{K} \subset \mathbb{R}^n$ be a convex body, and let $\mathcal{E} \subseteq \mathsf{K}$ be its maximum volume ellipsoid. Suppose that $\mathcal{E}$ is centered at the origin. Then $\mathcal{E} \subseteq \mathsf{K} \subseteq n\,\mathcal{E} = \left\{ nx \;\middle|\; x \in \mathcal{E} \right\}$.*

PROOF. By applying an affine transformation (and by Observation 21.2), we can assume that $\mathcal{E}$ is the unit ball $\mathbf{b}$. Assume, for the sake of contradiction, that there is a point $\mathsf{p} \in \mathsf{K}$, such that $\|\mathsf{p}\| > n$. Consider the set C which is the convex hull of $\{\mathsf{p}\} \cup \mathbf{b}$. Since K is convex, we have that $C \subseteq \mathsf{K}$.

We will reach a contradiction by finding an ellipsoid $\mathcal{G}$ which has volume larger than $\mathbf{b}$ and is enclosed inside C, and thus $\mathcal{G} \subseteq C \subseteq \mathsf{K}$.

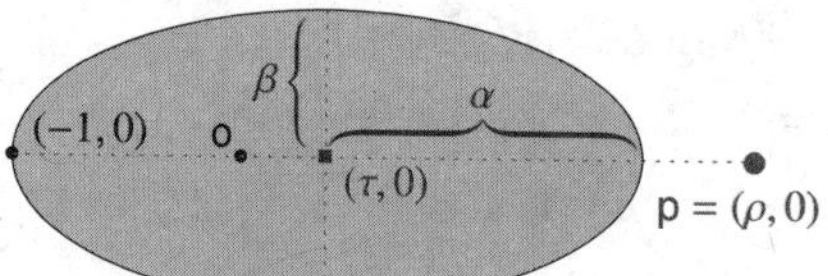

By rotating space, we can assume that the apex p of C is the point $(\rho, 0, \ldots, 0)$, for $\rho > n$. For $\alpha, \beta, \tau \geq 0$, we consider ellipsoids of the form

$$\mathcal{G} = \left\{ (y_1, \ldots, y_n) \;\middle|\; \frac{(y_1 - \tau)^2}{\alpha^2} + \frac{1}{\beta^2} \sum_{i=2}^{n} y_i^2 \leq 1 \right\}.$$

(Using the notation above, we have $\lambda_1 = 1/\alpha^2$, $\lambda_2 = \ldots = \lambda_n = 1/\beta^2$, $\bar{c}_1 = \tau$, and $\bar{c}_2 = \cdots = \bar{c}_n = 0$.) We need to pick the values of τ, α, and β such that $\mathcal{G} \subseteq C$. Observe that by symmetry, it is enough to enforce that $\mathcal{G} \subseteq C$ in the first two dimensions.

Thus, we can consider C and $\mathcal{G}$ to be in two dimensions. Now, the center of $\mathcal{G}$ is going to be on the x-axis, at the point $(\tau, 0)$. The set $\mathcal{G}$ is an ellipse with axes parallel to the x- and y-axes. In particular, we require that $(-1, 0)$ be on the boundary of $\mathcal{G}$. This implies that $(-1 - \tau)^2 = \alpha^2$. Now, the center point $(\tau, 0)$ can be anywhere on the interval between the origin and the midpoint between $(-1, 0)$ and $\mathsf{p} = (\rho, 0)$. As such, we conclude that $0 \leq \tau \leq (\rho + (-1))/2$. This implies that

$$\alpha = 1 + \tau. \tag{21.1}$$

In particular, the equation of the curve forming (the boundary of) $\mathcal{G}$ is

$$F(x, y) = \frac{(x - \tau)^2}{(1 + \tau)^2} + \frac{y^2}{\beta^2} - 1 = 0. \tag{21.2}$$

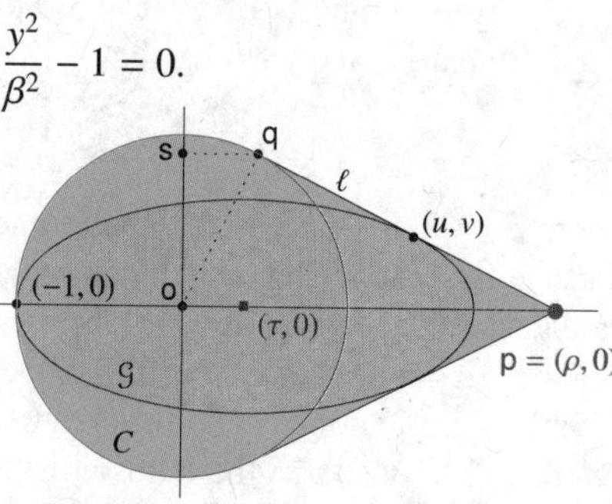

We next compute the value of β^2. Consider the tangent ℓ to the unit circle that passes through $\mathsf{p} = (\rho, 0)$. Let q be the point where ℓ touches the unit circle. (See the figure on the right.) We have $\triangle \mathsf{poq} \sim \triangle \mathsf{oqs}$, since oq forms a right angle with pq, where $\mathsf{s} = (0, s_y)$. Since $\|\mathsf{q} - \mathsf{o}\| = 1$, we have

$$x_{\mathsf{q}} = \|\mathsf{s} - \mathsf{q}\| = \frac{\|\mathsf{s} - \mathsf{q}\|}{\|\mathsf{q} - \mathsf{o}\|} = \frac{\|\mathsf{q} - \mathsf{o}\|}{\|\mathsf{o} - \mathsf{p}\|} = \frac{1}{\rho}.$$

Furthermore, since q is on the unit circle, we have

$$\mathsf{q} = \left(1/\rho, \sqrt{1 - 1/\rho^2}\right).$$

As such, the equation of the line ℓ is

$$\left\langle \mathsf{q}, (x, y) \right\rangle = 1 \implies \frac{x}{\rho} + \sqrt{1 - 1/\rho^2}\, y = 1 \implies \ell \equiv y = -\frac{1}{\sqrt{\rho^2 - 1}} x + \frac{\rho}{\sqrt{\rho^2 - 1}}.$$

Next, consider the tangent line ℓ' to $\mathcal{G}$ at some point (u, v). We will derive a formula for ℓ' as a function of (u, v) and then require that $\ell = \ell'$. (Thus, (u, v) will be the intersection point of $\partial\mathcal{G}$ and ℓ, which exists as $\mathcal{G}$ is as large as possible.) The slope of ℓ' is the slope of the tangent to $\mathcal{G}$ at (u, v). We now compute the derivative of F (see (21.2) above) at (u, v), which is

$$\frac{dy}{dx} = -\frac{F_x(u, v)}{F_y(u, v)} = -\left(\frac{2(u - \tau)}{(1 + \tau)^2}\right) \Big/ \left(\frac{2v}{\beta^2}\right) = -\frac{\beta^2(u - \tau)}{v(1 + \tau)^2},$$

by computing the derivatives of the implicit function F[①]. Since $\alpha = 1 + \tau$ (see (21.1)), the line ℓ' is

$$\frac{(u - \tau)}{\alpha^2}(x - \tau) + \frac{v}{\beta^2} y = 1,$$

as it has the required slope and it passes through $(u, v) \in \partial\mathcal{G}$ and by (21.2). Namely $\ell' \equiv y = -\frac{\beta^2(u - \tau)}{v\alpha^2}(x - \tau) + \frac{\beta^2}{v}$. Setting $\ell' = \ell$, we have that the line ℓ' passes through $(\rho, 0)$. As such,

$$\frac{(\rho - \tau)(u - \tau)}{\alpha^2} = 1 \implies \frac{(u - \tau)}{\alpha} = \frac{\alpha}{\rho - \tau} \implies \frac{(u - \tau)^2}{\alpha^2} = \frac{\alpha^2}{(\rho - \tau)^2}. \tag{21.3}$$

Since ℓ and ℓ' are the same line, we can equate their slopes; namely,

$$\frac{\beta^2(u - \tau)}{v\alpha^2} = \frac{1}{\sqrt{\rho^2 - 1}}. \tag{21.4}$$

However, by (21.3), we have

$$\frac{\beta^2(u - \tau)}{v\alpha^2} = \frac{\beta^2}{v\alpha} \cdot \frac{(u - \tau)}{\alpha} = \frac{\beta^2}{v\alpha} \cdot \frac{\alpha}{\rho - \tau} = \frac{\beta^2}{v(\rho - \tau)}.$$

Thus, plugging the above into (21.4), we have

$$\frac{\beta^2}{v} = \frac{\rho - \tau}{\sqrt{\rho^2 - 1}}.$$

Squaring and inverting both sides, we have $\frac{v^2}{\beta^4} = \frac{\rho^2 - 1}{(\rho - \tau)^2}$ and thus $\frac{v^2}{\beta^2} = \beta^2 \frac{\rho^2 - 1}{(\rho - \tau)^2}$. The point $(u, v) \in \partial\mathcal{G}$, and as such $\frac{(u - \tau)^2}{\alpha^2} + \frac{v^2}{\beta^2} = 1$. Using (21.3) and the above, we get

$$\frac{\alpha^2}{(\rho - \tau)^2} + \beta^2 \frac{\rho^2 - 1}{(\rho - \tau)^2} = 1,$$

[①]For the rusty reader (and the rustier author) – here is a quick "proof" of the formula for the derivative of an implicit function in two dimensions: Locally near (u, v) the function $F(x, y)$ behaves like the linear function $G(x, y) = F(u, v) + (x - u)F_x(u, v) + (y - v)F_y(u, v) = (x - u)F_x(u, v) + (y - v)F_y(u, v)$, as $F(u, v) = 0$. The curve is defined by $F(x, y) = 0$, which locally is equivalent to $G(x, y) = 0$. But $G(x, y) = 0$ is a line (which is the tangent line to $F(x, y)$ at (u, v)). Rearranging, we get that $y = -\left(F_x(u, v)/F_y(u, v)\right)x + \langle \textit{whatever} \rangle$, establishing the claim.

and thus, as $\alpha = \tau + 1$, we have

$$\beta^2 = \frac{(\rho-\tau)^2 - \alpha^2}{\rho^2 - 1} = \frac{(\rho-\tau)^2 - (\tau+1)^2}{\rho^2-1} = \frac{(\rho - 2\tau - 1)(\rho+1)}{\rho^2 - 1} = \frac{\rho - 2\tau - 1}{\rho - 1}$$
$$= 1 - \frac{2\tau}{\rho - 1}.$$

Namely, for $n \geq 2$ and $0 \leq \tau < (\rho - 1)/2$, we have that the ellipsoid $\mathcal{G}$ is defined by the parameters $\alpha = 1 + \tau$ and β and it is contained inside the "cone" C. We have that $\mathrm{Vol}(\mathcal{G}) = \beta^{n-1}\alpha\,\mathrm{Vol}(\mathbf{b})$, and thus

$$\mu = \ln \frac{\mathrm{Vol}(\mathcal{G})}{\mathrm{Vol}(\mathbf{b})} = (n-1)\ln\beta + \ln\alpha = \frac{n-1}{2}\ln\beta^2 + \ln\alpha.$$

For $\tau > 0$ sufficiently small, we have $\ln\alpha = \ln(1+\tau) = \tau + O(\tau^2)$, because of the Taylor expansion $\ln(1+x) = x - x^2/2 + x^3/3 - \cdots$, for $-1 < x \leq 1$. Similarly, $\ln\beta^2 = \ln\left(1 - \frac{2\tau}{\rho-1}\right) = -\frac{2\tau}{\rho-1} + O(\tau^2)$. Thus,

$$\mu = \frac{n-1}{2}\left(-\frac{2\tau}{\rho-1} + O(\tau^2)\right) + \tau + O(\tau^2) = \left(1 - \frac{n-1}{\rho-1}\right)\tau + O(\tau^2) > 0,$$

for τ sufficiently small and if $\rho > n$. Thus, $\mathrm{Vol}(\mathcal{G})/\mathrm{Vol}(\mathbf{b}) = \exp(\mu) > 1$, implying that $\mathrm{Vol}(\mathcal{G}) > \mathrm{Vol}(\mathbf{b})$, a contradiction. ■

A convex body K centered at the origin is ***symmetric*** if $p \in \mathsf{K}$ implies that $-p \in \mathsf{K}$. Interestingly, the constant in Theorem 21.3 can be improved to $\sqrt{n}$ in this case. We omit the proof, since it is similar to the proof of Theorem 21.3.

Theorem 21.4. *Let $\mathsf{K} \subset \mathbb{R}^n$ be a symmetric convex body, and let $\mathcal{E} \subseteq \mathsf{K}$ be its maximum volume ellipsoid. Suppose that $\mathcal{E}$ is centered at the origin. Then $\mathsf{K} \subseteq \sqrt{n}\,\mathcal{E}$.*

21.2. Bibliographical notes

We closely follow the exposition of Barvinok [**Bar02**]. The maximum volume ellipsoid is sometimes referred to as John's ellipsoid. In particular, Theorem 21.3 is known as John's theorem and was originally proved by Fritz John [**Joh48**]. One can approximate the John ellipsoid using interior point techniques [**GLS93**]. However, this is not very efficient in low dimensions. In the next chapter, we will show how to do this by other (simpler) techniques, which are faster (and simpler) in low dimensions.

There are numerous interesting results in convexity theory about what convex shapes look like. One of the surprising results is Dvoretsky's theorem, which states that any symmetric convex body K around the origin (in "high enough" dimension) can be cut by a k-dimensional subspace, such that the intersection of K with this k-dimensional space, contains an ellipsoid $\mathcal{E}$ and is contained inside an ellipsoid $(1+\varepsilon)\mathcal{E}$ (k here is an arbitrary parameter). Since one can define a metric in a Banach space by providing a symmetric convex body which defines the unit ball, this implies that any high enough dimensional Banach space contains (up to translation that maps the ellipsoid to a ball) a subspace which is almost Euclidean.

A survey of those results is provided by Ball [**Bal97**].

CHAPTER 22

Approximating the Minimum Volume Bounding Box of a Point Set

22.1. Some geometry

Let $\mathcal{H} = [0,1]^d$ denote the unit hypercube in $\mathbb{R}^d$. The ***width***, $\omega(K)$, of a set $K \subseteq \mathbb{R}^d$ is the minimum distance between two parallel planes that enclose K. Alternatively, the width is the minimum length of the projection of K into a line in $\mathbb{R}^d$, where the length of the projection is the minimum length interval covering the projected set on the line. The ***diameter*** of K is the maximum distance between any two points of K. Alternatively, this is the maximum length of a projection of K into a line.

Lemma 22.1. *The width of* $\mathcal{H} = [0,1]^d$ *is* 1 *and its diameter is* $\sqrt{d}$.

PROOF. The upper bound on the width follows by considering the hyperplanes $x_1 = 0$ and $x_1 = 1$ that enclose the hypercube. As for the lower bound, observe that $\mathcal{H}$ encloses a ball of radius $1/2$. Now, the projection of a ball of radius r on any line results in an interval of length $2r$. As such, the projection of $\mathcal{H}$ in any direction is of length at least 1.

The diameter is just the distance between the two points $(0,\ldots,0)$ and $(1,\ldots,1)$. ■

The following is an explicit formula for the volume of a pyramid in $\mathbb{R}^d$. We provide the easy proof for the sake of completeness.

Claim 22.2. *Let* K *be a* **pyramid** *in* $\mathbb{R}^d$*; that is, there exist a set* X *that lies on a hyperplane* h *and a point* $p \in \mathbb{R}^d$ *such that* $K = \mathcal{CH}(X \cup \{p\})$*. Then* $\mathrm{Vol}(K) = \mathrm{Vol}_{d-1}(X)\rho/d$*, where* ρ *is the distance of* p *to* h *and* $\mathrm{Vol}_{d-1}(X)$ *is the* $(d-1)$*-dimensional volume of* X.

PROOF. Assume, for the sake of simplicity of exposition, that $h \equiv (x_d = 0)$ and $p = (0,\ldots,0,\rho)$. Clearly, the intersection of $x_d = \alpha$, for $\alpha \in [0,\rho]$, with K is a translated copy of $((\rho-\alpha)/\rho)X$. Now, $f(\alpha) = \mathrm{Vol}_{d-1}\left(\frac{\rho-\alpha}{\rho}X\right) = \left(\frac{\rho-\alpha}{\rho}\right)^{d-1}\mathrm{Vol}_{d-1}(X)$. As such, we have

$$\begin{aligned}\mathrm{Vol}(K) &= \int_{x_d=0}^{\rho} f(x_d)\,dx_d = \int_{x_d=0}^{\rho}\left(\frac{\rho-x_d}{\rho}\right)^{d-1}\mathrm{Vol}_{d-1}(X)\,dx_d = \frac{\mathrm{Vol}_{d-1}(X)}{\rho^{d-1}}\int_{x_d=0}^{\rho} x_d^{d-1}\,dx_d\\ &= \frac{\mathrm{Vol}_{d-1}(X)}{\rho^{d-1}}\cdot\frac{\rho^d}{d} = \frac{\mathrm{Vol}_{d-1}(X)\rho}{d}.\end{aligned}$$

■

Lemma 22.3. *Let* K *be a convex body, and let* h *be a hyperplane cutting* K*. Let* $\mu = \mathrm{Vol}(h \cap K)$*, and let* $\vec{v}$ *be a unit vector orthogonal to* h*. Let* ρ *be the length of the projection of* K *onto the line spanned by* $\vec{v}$*. Then* $\mathrm{Vol}(K) \geq \mu \cdot \rho/d$.

PROOF. Let $K^+ = K \cap h^+$ (resp. $K^- = K \cap h^-$) and let p^+ (resp. p^-) be the point of maximum distance of K^+ (resp. K^-) from h, where h^+ (resp. h^-) denotes the positive (resp. negative) halfspace induced by h.

Let $C = h \cap \mathsf{K}$. By the convexity of K, the pyramid $R = \mathcal{CH}(C \cup \{\mathsf{p}^+\})$ is contained inside K. Let α^+ (resp. α^-) be the distance of p^+ (resp. p^-) from h. By Claim 22.2, $\mathrm{Vol}(R) = \alpha^+\mu/\mathsf{d}$, which implies that $\mathrm{Vol}(\mathsf{K}^+) \geq \alpha\mu/\mathsf{d}$. A similar argument can be applied for $\mathsf{K}^- = \mathsf{K} \cap h^-$. Thus, we have that $\mathrm{Vol}(\mathsf{K}) = \mathrm{Vol}(\mathsf{K}^-) + \mathrm{Vol}(\mathsf{K}^+) \geq (\alpha^- + \alpha^+)\mu/\mathsf{d} = \rho\mu/\mathsf{d}$. ■

Lemma 22.4. *Let h be any hyperplane. We have* $\mathrm{Vol}(h \cap [0,1]^{\mathsf{d}}) \leq \mathsf{d}$.

Proof. Since the width of $\mathcal{H} = [0,1]^{\mathsf{d}}$ is 1, the projection of $\mathcal{H}$ onto any line in the direction perpendicular to h is of length ≥ 1. As such, applying Lemma 22.3 to $\mathcal{H}$, we have $\rho \geq 1$, and if $\mathrm{Vol}(h \cap \mathcal{H}) > \mathsf{d}$, then $\mathrm{Vol}(\mathcal{H}) > 1$, a contradiction. ■

Let $r(\mathsf{K})$ denote the radius of the largest ball enclosed inside K.

Lemma 22.5. *Let $\mathsf{K} \subseteq [0,1]^{\mathsf{d}}$ be a convex body. Let $\mu = \mathrm{Vol}(\mathsf{K})$. Then for the width of K we have $\omega(\mathsf{K}) \geq \mu/\mathsf{d}$. Furthermore, K contains a ball of radius $\mu/(2\mathsf{d}^2)$; that is, $r(\mathsf{K}) \geq \mu/(2\mathsf{d}^2)$.*

Proof. By Lemma 22.4, the volume of the intersection of K with any hyperplane is bounded by d. Thus, $\mu = \mathrm{Vol}(\mathsf{K}) \leq \omega(\mathsf{K})\,\mathsf{d}$. Namely, $\omega(\mathsf{K}) \geq \mu/\mathsf{d}$.

Next, let $\mathcal{E}$ be the largest volume ellipsoid that is contained inside K. By John's theorem (Theorem 21.3$_{\mathrm{p280}}$), we have that $\mathsf{K} \subseteq \mathsf{d}\mathcal{E}$. Let α be the length of the shortest axis of $\mathcal{E}$. Clearly, $\omega(\mathsf{K}) \leq 2\mathsf{d}\alpha$, since $\omega(\mathsf{d}\mathcal{E}) = 2\mathsf{d}\alpha$. Thus $2\mathsf{d}\alpha \geq \mu/\mathsf{d}$. This implies that $\alpha \geq \mu/(2\mathsf{d}^2)$.

Thus, $\mathcal{E}$ is an ellipsoid whose shortest axis is of length $\alpha \geq \mu/(2\mathsf{d}^2)$. In particular, $\mathcal{E}$ contains a ball of radius α, which is in turn contained inside K. ■

Considerably better bounds are known on the value of $r(\mathsf{K})$. In particular, it is known that $\omega(\mathsf{K}) \,/\, \left(2\sqrt{\mathsf{d}}\right) \leq r(\mathsf{K})$ for odd dimension and $\omega(\mathsf{K}) \,/\, \left(2(\mathsf{d}+1)/\sqrt{\mathsf{d}+2}\right) \leq r(\mathsf{K})$ for even dimension. Thus, $r(\mathsf{K}) \geq \omega(\mathsf{K}) \,/\, \left(2\sqrt{\mathsf{d}+1}\right)$ [**GK92**]. Plugging this into the proof of Lemma 22.5 will give us a slightly better lower bound.

22.1.1. Approximating the diameter.

Lemma 22.6. *Given a point set P in $\mathbb{R}^{\mathsf{d}}$, one can compute in $O(n\mathsf{d})$ time a pair of points $\mathsf{p}, \mathsf{q} \in S$, such that* $\|\mathsf{p}-\mathsf{q}\| \leq \mathrm{diam}(\mathsf{P}) \leq \min\left(2, \sqrt{\mathsf{d}}\right)\|\mathsf{p}-\mathsf{q}\|$.

Proof. Let B be the minimum volume axis parallel box containing P, and consider the longest dimension of B. There must be two points, p and q in P, that lie on the two facets of B orthogonal to this dimension; namely, the distance $\|\mathsf{p}-\mathsf{q}\|$ is larger than (or equal to) the longest dimension of B. The points p and q are easily found in $O(n\mathsf{d})$ time. Now, $\left\|\mathsf{p}-\mathsf{q}\right\| \leq \mathrm{diam}(\mathsf{P}) \leq \mathrm{diam}(B) \leq \sqrt{\mathsf{d}}\left\|\mathsf{p}-\mathsf{q}\right\|$.

Alternatively, pick a point $\mathsf{s} \in \mathsf{P}$, and compute its furthest point $\mathsf{s}' \in \mathsf{P}$. Next, let $\mathsf{t}_1, \mathsf{t}_2$ be the two points realizing the diameter. We have $\mathrm{diam}(\mathsf{P}) = \|\mathsf{t}_1-\mathsf{t}_2\| \leq \|\mathsf{t}_1-\mathsf{s}\|+\|\mathsf{s}-\mathsf{t}_2\| \leq 2\|\mathsf{s}-\mathsf{s}'\|$. Thus, $\|\mathsf{s}-\mathsf{s}'\|$ is a 2-approximation to the diameter of P. ■

22.2. Approximating the minimum volume bounding box

22.2.1. Constant factor approximation.

Lemma 22.7. *Given a set P of n points in $\mathbb{R}^{\mathsf{d}}$, one can compute in $O(\mathsf{d}^2 n)$ time a bounding box $B(\mathsf{P})$ with $\mathrm{Vol}(B_{\mathrm{opt}}(\mathsf{P})) \leq \mathrm{Vol}(B(\mathsf{P})) \leq 2^{\mathsf{d}}\mathsf{d}!\,\mathrm{Vol}(B_{\mathrm{opt}}(\mathsf{P}))$, where B_{opt} is the minimum volume bounding box of P.*

Furthermore, there exists a vector $\vec{\mathsf{v}} \in \mathbb{R}^{\mathsf{d}}$, such that $c \cdot B + \vec{\mathsf{v}} \subseteq \mathcal{CH}(\mathsf{P})$, where $c = 1/\left(2^{\mathsf{d}}\mathsf{d}!\mathsf{d}^{5/2}\right)$.

Proof. Using the algorithm of Lemma 22.6, we compute in $O(\mathrm{d}n)$ time two points $s, t \in \mathsf{P}$ which form a 2-approximation of the diameter of P.

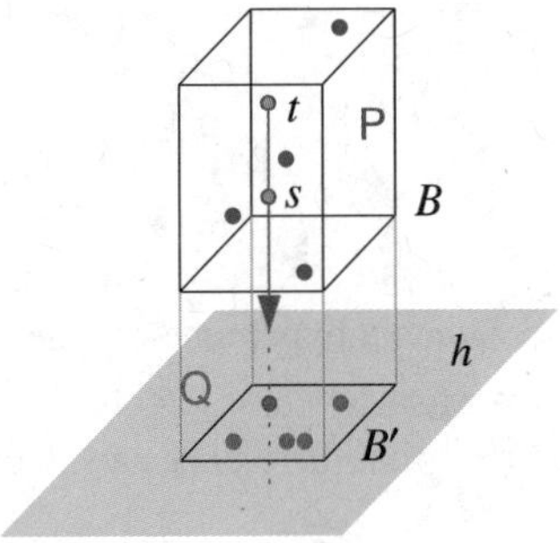

For simplicity of exposition, we assume that (i) st is on the x_d-axis, (ii) there is at least one point of P that lies on the hyperplane $h \equiv x_\mathrm{d} = 0$, and (iii) $x_\mathrm{d} \geq 0$ for all points of P. Let Q be the orthogonal projection of P into h, and let $\mathcal{I}$ be the shortest interval on the x_d-axis which contains the projection of P onto this axis. Computing $\mathcal{I}$ and projecting P to h to compute Q can be done in $O(\mathrm{d}n)$ time.

By recursion, compute a bounding box B' of Q in h, and let $B = B' \times \mathcal{I}$ be the bounding box of P resulting from "lifting" B' into the dth dimension; see the figure on the right. Note that in the bottom of the recursion, the point set is one dimensional and the minimum interval containing the points can be computed in linear time.

Clearly, $\mathsf{P} \subseteq B$, and thus we only need to bound the quality of approximation. We next show that $\mathrm{Vol}_\mathrm{d}(C) \geq \mathrm{Vol}_\mathrm{d}(B)/c_\mathrm{d}$, where $C = \mathcal{CH}(\mathsf{P})$ and $c_\mathrm{d} = 2^\mathrm{d} \cdot \mathrm{d}!$ (this implies the first part of the lemma, as $\mathrm{Vol}_\mathrm{d}\left(B_{\mathrm{opt}}\right) \geq \mathrm{Vol}_\mathrm{d}(C)$). We prove this by induction on the dimension. For $\mathrm{d} = 1$ the claim trivially holds.

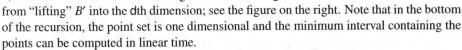

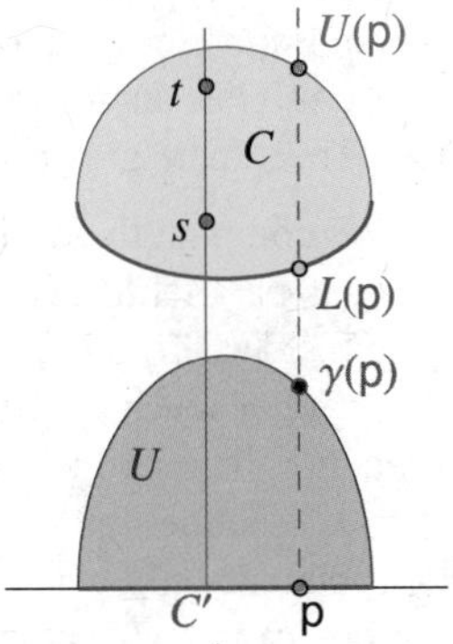

Now, by induction, we have, in the $(\mathrm{d}-1)$-dimensional case, that $\mathrm{Vol}_{\mathrm{d}-1}(C') \geq \mathrm{Vol}_{\mathrm{d}-1}(B')/c_{\mathrm{d}-1}$, where $C' = \mathcal{CH}(\mathsf{Q})$. For a point $\mathsf{p} \in C'$, let ℓ_p be the line parallel to the x_d-axis passing through p. Let $L(\mathsf{p})$ be the minimum value of x_d for the points of ℓ_p lying inside C, and similarly, let $U(\mathsf{p})$ be the maximum value of x_d for the points of ℓ_p lying inside C. That is, $\ell_\mathsf{p} \cap C = [L(\mathsf{p}), U(\mathsf{p})]$. Clearly, since C is convex, the function $L(\cdot)$ is convex and $U(\cdot)$ is concave. As such, $\gamma(\mathsf{p}) = U(\mathsf{p}) - L(\mathsf{p})$ is a concave function, being the difference between a concave and a convex function. See the figure on the right. In particular, $\gamma(\cdot)$ induces the convex body

$$U = \bigcup_{x \in C'} \Big[(x, 0), (x, \gamma(x))\Big].$$

Clearly, $\mathrm{Vol}_\mathrm{d}(U) = \mathrm{Vol}_\mathrm{d}(C)$. Furthermore, $\gamma\big((0, \ldots, 0)\big) \geq \|s - t\|$ and U contains a pyramid. Indeed, the pyramid base on the hyperplane $x_\mathrm{d} = 0$ is the set C'. Furthermore, the segment $[(0, \ldots, 0), (0, \ldots, 0, \|s - t\|)]$ is contained inside U and forms the pole of the pyramid. Thus,

$$\mathrm{Vol}_\mathrm{d}(C) = \mathrm{Vol}_\mathrm{d}(U) \geq \|s - t\| \, \mathrm{Vol}_{\mathrm{d}-1}(C')/\mathrm{d},$$

by Claim 22.2. Observe that $\|\mathcal{I}\|$ is the length of the projection of C into the x_d-axis. We have that $\|\mathcal{I}\| \leq 2\|s - t\|$, since $\|s - t\|$ is a 2-approximation to the diameter of P. Thus,

$$\begin{aligned}
\mathrm{Vol}_\mathrm{d}(B) \geq \mathrm{Vol}_\mathrm{d}\left(B_{\mathrm{opt}}\right) \geq \mathrm{Vol}_\mathrm{d}(C) &\geq \frac{\mathrm{Vol}_{\mathrm{d}-1}(C')\,\|s - t\|}{\mathrm{d}} \\
&\geq \frac{\left(\mathrm{Vol}_{\mathrm{d}-1}(B')/c_{\mathrm{d}-1}\right) \cdot (2\|s - t\|)}{2\mathrm{d}} \\
&\geq \frac{\mathrm{Vol}_{\mathrm{d}-1}(B')\,\|\mathcal{I}\|}{2c_{\mathrm{d}-1}\mathrm{d}} = \frac{\mathrm{Vol}_\mathrm{d}(B)}{2^\mathrm{d}\mathrm{d}!} = \frac{\mathrm{Vol}_\mathrm{d}(B)}{c_\mathrm{d}},
\end{aligned}$$

as $c_\mathrm{d} = 2^\mathrm{d}\mathrm{d}!$.

We are left with the task of proving the last part of the lemma. Let T be an affine transformation that maps B to the unit hypercube $\mathcal{H} = [0,1]^{\mathsf{d}}$. Observe that for any set $X \subseteq \mathbb{R}^{\mathsf{d}}$ we have that $\mathrm{Vol}_{\mathsf{d}}(T(X)) = \alpha \mathrm{Vol}_{\mathsf{d}}(X)$, where $\alpha = \mathrm{Vol}_{\mathsf{d}}\big(T\big([0,1]^{\mathsf{d}}\big)\big)$. As such, we have that

$$\mathrm{Vol}_{\mathsf{d}}(T(C)) = \alpha \mathrm{Vol}_{\mathsf{d}}(C) \geq \frac{\alpha \mathrm{Vol}_{\mathsf{d}}(B)}{c_{\mathsf{d}}} = \frac{\mathrm{Vol}_{\mathsf{d}}(T(B))}{c_{\mathsf{d}}} = \frac{\mathrm{Vol}_{\mathsf{d}}(\mathcal{H})}{c_{\mathsf{d}}} = \frac{1}{c_{\mathsf{d}}}.$$

By Lemma 22.5, there is a ball $\mathbf{b}$ of radius $r \geq 1/(2\mathsf{d}^2 c_{\mathsf{d}})$ contained inside $T(C)$. The ball $\mathbf{b}$ contains a hypercube of sidelength $2r/\sqrt{\mathsf{d}} \geq 2/\big(2\mathsf{d}^2\sqrt{\mathsf{d}}c_{\mathsf{d}}\big)$. Thus, for $c = 1/\big(2^{\mathsf{d}}\mathsf{d}!\mathsf{d}^{5/2}\big)$, there exists a vector $v' \in \mathbb{R}^{\mathsf{d}}$, such that $c\mathcal{H} + v' \subseteq T(C)$. Thus, applying T^{-1} to both sides, we have that there exists a vector $v = T^{-1}(v') \in \mathbb{R}^{\mathsf{d}}$, such that $c \cdot B + v = c \cdot T^{-1}(\mathcal{H}) + v \subseteq C$. ■

22.3. Exact algorithm for the minimum volume bounding box

In this section, we sketch exact algorithms for computing the minimum volume bounding box in two and three dimensions.

22.3.1. A sketch of an exact algorithm in two dimensions – rotating calipers. Let P be a set of points in the plane. We are interested in the problem of computing the minimum area rectangle covering P.

An important property of the minimum area rectangle is that one of the edges of the convex hull lies on one of the bounding rectangle edges. We will refer to this edge as being *flush*. Thus, there are only n possibilities we have to check.

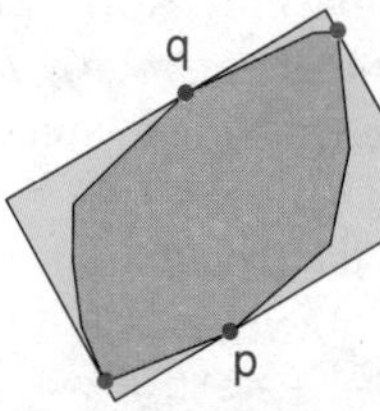

We compute the convex hull $C = \mathcal{CH}(\mathsf{P})$. Next, we rotate two lines parallel to each other, which touch the boundary of C on both sides. To this end, consider the two points where these parallel lines touch C. Clearly, rotating these lines is equivalent to rotating these two tangency points p and q around the boundary of C; see the figure on the right. Observe that the tangency points change only when one of the two lines becomes collinear with one of the edges of C. In particular, by walking around C, one can track these two tangency points. This can be easily done in linear time.

We can also rotate two parallel lines which are perpendicular to the first set. Those four lines together induce a rectangle. It is easy to observe that during this rotation, we will encounter the minimum area rectangle. The function that those lines define changes every time the lines change the vertices they rotate around. This happens $O(n)$ time. When the vertices that the lines rotate around are fixed, the area function is a constant size function, and as such its minimum/maximum can be computed in constant time.

Thus, the minimum volume rectangle of a set of n points in the plane can be computed in $O(n \log n)$ time (as this is the time to compute the convex hull of n points in the plane).

22.3.2. An exact algorithm in 3 dimensions. Let P be a set of points in $\mathbb{R}^3$. Our purpose is to compute the minimum volume bounding box B_{opt} of P. It is easy to verify that B_{opt} must contain a point of P on every one of its faces.

So, consider an edge e for the bounding box B_{opt}. The edge e is the normal of a plane h that contains a face of B_{opt}. In particular, if we project the points of P in the direction of e into the perpendicular plane h, we get a two-dimensional point set Q. Similarly, the projection of B_{opt} is a rectangle R. If the projection of $\mathcal{CH}(\mathsf{P})$ onto the line spanned by e is of length α, then the volume of B_{opt} is area$(\mathsf{R})\alpha$. In particular, any rectangle R' in

h that covers Q can be used to construct such a bounding box of volume area$(\mathsf{R}')\alpha$. We conclude that R must be a minimum area bounding rectangle for Q, since otherwise we could construct a smaller volume bounding box than B_{opt} for P, a contradiction.

As such, the projection of B_{opt} into h is a minimum area rectangle, and hence it has one flush edge. This flush edge corresponds to an edge e' of the convex hull of P. The edge e' lies on a face of B_{opt}, and we refer to it as being ***flush***.

In fact, there must be two adjacent faces of B_{opt} that have flush edges of $\mathcal{CH}(\mathsf{P})$ on them. Otherwise, consider a face f of B_{opt} that has an edge flush on it. All the four adjacent faces of B_{opt} then cannot have flush edges on them. But that is not possible, since we can project the points in the direction of the normal of f and argue that in the projection there must be a flush edge. This flush edge corresponds to an edge of $\mathcal{CH}(\mathsf{P})$ that lies on one of the faces of B_{opt} that is adjacent to f.

Lemma 22.8. *If B_{opt} is the minimum volume bounding box of P, then it has two adjacent faces which are flush (i.e., these two faces of B_{opt} each contains an edge of $\mathcal{CH}(\mathsf{P})$).*

This provides us with a natural algorithm to compute the minimum volume bounding box. Indeed, let us check all possible pairs of edges $e, e' \in \mathcal{CH}(\mathsf{P})$. We assume here that $e \neq e'$, as the degenerate case $e = e'$ is handled in a similar fashion. For each such pair, compute the minimum volume bounding box that has e and e' as flush edges.

Consider the normal $\vec{n}$ of the face of a bounding box that contains e. The normal $\vec{n}$ lies on a great circle on the sphere of directions, which are all the directions that are orthogonal to e. Computing an orthonormal basis of space having the direction of e as one of the vectors results in two orthogonal vectors v_2, v_3 perpendicular to each other and the direction of e. In particular, consider the unit circle in the plane spanned by v_2 and v_3. We parameterize this circle, in the natural way, by a real number in the range $[0, 2\pi]$. Namely, we parameterize, by a number $\alpha \in [0, 2\pi]$, the points of the great circle of directions orthogonal to e. In particular, let $\vec{n}(\alpha)$ denote this direction for the specified value of α.

Next, consider the normal $\vec{n'}(\alpha)$ of the face that is flush to e'. Clearly, we need $\vec{n'}(\alpha)$ to be orthogonal to both e' and $\vec{n}(\alpha)$. As such, we can compute this normal in constant time given e' and $\vec{n}(\alpha)$ (i.e., $\vec{n'}(\alpha)$ is a function of $\vec{n}(\alpha)$, which is in turn a function of the variable α). Similarly, we can compute the third direction of the bounding box using vector product in constant time. Thus, if e and e' are fixed, there is a one-dimensional family of bounding boxes of P that have e and e' flush on them and comply with all the requirements to be a minimum volume bounding box. Indeed, given a value α, one can easily compute the two normals $\vec{n}(\alpha)$ and $\vec{n'}(\alpha)$. This uniquely defines the bounding box of P having two faces normal to these directions. Let $B(\alpha)$ denote the resulting bounding box of P.

(Note that this bounding box might not be feasible for certain values of α or not feasible for any value of α. The algorithm skips over such values or pairs of edges. The details of exactly how to do this are easy but tedious, and we skip over them in this sketch.)

It is now easy to verify that we can compute the representation of this family of bounding boxes, by tracking what vertices of the convex hull the bounding boxes touch (i.e., this is similar to the rotating calipers algorithm, but one has to be more careful about the details).

We omit the tedious but straightforward details – one has to perform synchronized walks on $\mathcal{CH}(\mathsf{P})$ tracking the extreme vertices in the directions of the normals of $B(\alpha)$ as α changes. This can be done in linear time, and as such, one can compute the minimum volume bounding box in this family in linear time. Doing this for all pairs of edges results in an algorithm with running time $O(n^3)$, where $n = |\mathsf{P}|$.

Theorem 22.9. *Let* P *be a set of* n *points in* $\mathbb{R}^3$. *One can compute the minimum volume bounding box of* P *in* $O(n^3)$ *time.*

22.4. Approximating the minimum volume bounding box in three dimensions

Let P be a set of n points in $\mathbb{R}^3$, and let B_{opt} denote the minimum volume bounding box of P.

The algorithm. Let $B = B(\mathsf{P})$ be the bounding box of P computed by Lemma 22.7, and let B_ε be a translated copy of $(\varepsilon/c)B$ centered at the origin, where c is an appropriate constant to be determined shortly. Let $\mathcal{G} = \mathsf{G}(B_\varepsilon/2)$ denote the grid covering the whole space, where every grid cell is a translated copy of $B_\varepsilon/2$. We approximate P on $\mathcal{G}$; that is, for each point $\mathsf{p} \in \mathsf{P}$ let $\mathcal{G}(\mathsf{p})$ be the set of eight vertices of the cell of $\mathcal{G}$ that contains p, and let $\mathsf{P}_\mathcal{G} = \bigcup_{\mathsf{p}\in\mathsf{P}} \mathcal{G}(\mathsf{p})$.

Let Q be the set of vertices of $\mathcal{CH}(\mathsf{P}_\mathcal{G})$. We next apply the exact algorithm of Theorem 22.9 to Q. Let $\widehat{B}$ denote the resulting bounding box, which we return as the approximate bounding box.

Correctness. Observe that $\widehat{B}$ contains $\mathsf{P}_\mathcal{G}$ and such that it contains P, and as such, it is a valid bounding box of P.

Running time analysis. It takes linear time to compute B and $\mathsf{P}_\mathcal{G}$. One can compute $\mathcal{CH}(\mathsf{P}_\mathcal{G})$ in $O(n + (1/\varepsilon^2)\log(1/\varepsilon))$ time. Indeed, the set $\mathsf{P}_\mathcal{G}$ has at most $O(\min(n, 1/\varepsilon^3))$ points since all the vertices are from a grid having $B_\varepsilon/2$ as a basic cell, and all these cells lie inside $(1 + 2\varepsilon/c)B$. As such, there are at most $O\big((2/\varepsilon)^3\big)$ relevant grid cells, and this also bounds the number of points in $\mathsf{P}_\mathcal{G}$. Now, since we are interested in the convex hull of this set of points, we can just take the extreme points along each one of the $O(1/\varepsilon^2)$ parallel grid lines containing these points. Using hashing, this filtering process can be done in $O(n + 1/\varepsilon^2)$ time. As such, in the end of this process we are left with a set P' of $O(1/\varepsilon^2)$ points that might be vertices of $\mathcal{CH}(\mathsf{P}_\mathcal{G})$. Computing the convex hull of this set takes $O(\varepsilon^{-2}\log\varepsilon^{-1})$ time, since the convex hull of n points in $\mathbb{R}^3$ can be computed in $O(n\log n)$ time.

Since $|\mathsf{Q}| = O(1/\varepsilon^2)$, the final step (of using the algorithm of Theorem 22.9) takes $O\big(|\mathsf{Q}|^3\big) = O\big((1/\varepsilon^2)^3\big) = O\big(1/\varepsilon^6\big)$ time.

Thus, overall, the running time of the algorithm is $O(n + 1/\varepsilon^6)$.

Quality of approximation. We remind the reader that for two sets $\mathcal{A}$ and $\mathcal{B}$ in $\mathbb{R}^3$, the ***Minkowski sum*** of A and B is the set $\mathcal{A} + \mathcal{B} = \left\{ a + b \;\middle|\; a \in \mathcal{A}, b \in \mathcal{B} \right\}$. Clearly, $\mathcal{CH}(\mathsf{P}) \subseteq \mathcal{CH}(\mathsf{P}_\mathcal{G}) \subseteq \mathcal{CH}(\mathsf{P}) + B_\varepsilon$.

It remains to show that $\widehat{B}$ is a $(1+\varepsilon)$-approximation of $B_{\mathrm{opt}}(\mathsf{P})$. Let $B^\varepsilon_{\mathrm{opt}}$ be a translation of $\frac{\varepsilon}{4}B_{\mathrm{opt}}(\mathsf{P})$ that contains B_ε. (The existence of a $B^\varepsilon_{\mathrm{opt}}$ that contains B_ε is guaranteed by Lemma 22.7 if we take $c = 200$.) Thus, $\mathsf{Q} \subseteq \mathcal{CH}(\mathsf{P}) + B_\varepsilon \subseteq \mathcal{CH}(\mathsf{P}) + B^\varepsilon_{\mathrm{opt}} \subseteq B_{\mathrm{opt}}(\mathsf{P}) + B^\varepsilon_{\mathrm{opt}}$. Since $B_{\mathrm{opt}}(\mathsf{P}) + B^\varepsilon_{\mathrm{opt}}$ is a box and since it contains Q, it follows that it is a bounding box of P. Now $\widehat{B}$ is the minimum volume bounding box of Q and as such

$$\mathrm{Vol}\big(\widehat{B}\big) \le \mathrm{Vol}\big(B_{\mathrm{opt}}(\mathsf{P}) + B^\varepsilon_{\mathrm{opt}}\big) = \left(1 + \frac{\varepsilon}{4}\right)^3 \mathrm{Vol}\, B_{\mathrm{opt}}(\mathsf{P}) < (1+\varepsilon)\,\mathrm{Vol}\big(B_{\mathrm{opt}}(\mathsf{P})\big),$$

as desired. (The last inequality is the only place where we use the assumption $\varepsilon \le 1$.)

Theorem 22.10. *Let* P *be a set of* n *points in* $\mathbb{R}^3$, *and let* $0 < \varepsilon \le 1$ *be a parameter. One can compute in* $O(n+1/\varepsilon^6)$ *time a bounding box* $B(\mathsf{P})$ *with* $\mathrm{Vol}(B(\mathsf{P})) \le (1+\varepsilon)\,\mathrm{Vol}(B_{\mathrm{opt}}(\mathsf{P}))$.

Note that the box $B(\mathsf{P})$ computed by the above algorithm is most likely not minimal along its directions. The minimum bounding box of P that is a homothet (i.e., translated and scaled copy) of $B(\mathsf{P})$ can be computed in linear time from $B(\mathsf{P})$.

22.5. Bibliographical notes

Our exposition follows roughly the work of Barequet and Har-Peled [**BH01**]. However, the basic idea (of finding the diameter, projecting along it, and recursively finding a good bounding box on the projected input) is much older and can be traced back to the work of Macbeath [**Mac50**].

For approximating the diameter, one can find in linear time a $(1/\sqrt{3})$-approximation of the diameter in *any* dimension; see [**EK89**].

The rotating calipers algorithm (Section 22.3.1) is due to Toussaint [**Tou83**]. The elegant extension of this algorithm to the computation of the exact minimum volume bounding box is due to O'Rourke [**O'R85**].

Lemma 22.7 is (essentially) from [**BH01**].

The current constants in Lemma 22.7 are unreasonable, but there is no reason to believe they are tight.

Conjecture 22.11. *The constants in Lemma 22.7 can be improved to be polynomial in the dimension.*

Coresets. One alternative approach to the algorithm of Theorem 22.10 is to construct G using $B_\varepsilon/2$ as before and instead picking from each non-empty cell of G one point of P as a representative point. This results in a set $\mathcal{S}$ of $O(1/\varepsilon^2)$ points. Compute the minimum volume bounding box of $\mathcal{S}$ using the exact algorithm. Let B denote the resulting bounding box. It is easy to verify that $(1+\varepsilon)B$ contains P and that it is a $(1+O(\varepsilon))$-approximation to the optimal bounding box of P. The running time of the new algorithm is identical. The interesting property is that we are running the exact algorithm on a subset of the input.

This is a powerful technique for approximation algorithms. You first extract a small subset from the input and run an exact algorithm on this input, making sure that the result provides the required approximation. The subset $\mathcal{S}$ is referred to as the ***coreset*** of B as it preserves a geometric property of P (in our case, the minimum volume bounding box). We will see more about this notion in the following chapters.

22.6. Exercises

Exercise 22.1 (Bounding by a simplex). Given a set P of n points in $\mathbb{R}^3$ and a parameter ε, describe how to $(1+\varepsilon)$-approximate the minimum volume simplex (i.e., the simplex in three dimensions is the convex hull of four points) that contains P. What is the running time of your algorithm?

Exercise 22.2 (Covering by many boxes). Given a set P of n points in $\mathbb{R}^3$ and a parameter k, consider a cover of P by k boxes, where the volume of this cover is the sum of the volumes of the k boxes.

(A) Show how one can compute the (exact) minimum volume cover of P by k boxes in $n^{O(k)}$ time.

(B) Assuming k is a small fixed constant (say, 8), show how to compute in linear time $O(k^2)$ boxes that cover all the points of P. Furthermore, the total volume of these boxes is at most $O(k)$ larger than the volume of the minimum cover of P by k boxes. (Hint: Use sampling.)

CHAPTER 23

Coresets

In this chapter, we will introduce the concept of coresets. Intuitively, a coreset is a small subset of the data that captures the key features that we care about. Surprisingly, in many cases such small coresets can be computed quickly and as such facilitate fast approximation algorithms for various geometric optimization problems.

23.1. Coreset for directional width

Given a point set P in $\mathbb{R}^d$, the result of ***projecting*** P ***into the direction*** $u \in \mathbb{R}^d$ is the set resulting from projecting P into a line ℓ through the origin spanned by u.

Definition 23.1. Let P be a set of points in $\mathbb{R}^d$. For a vector $v \in \mathbb{R}^d$, such that $v \neq 0$, let

$$\overline{\omega}(v, \mathsf{P}) = \max_{p \in \mathsf{P}} \langle v, p \rangle - \min_{p \in \mathsf{P}} \langle v, p \rangle$$

denote the ***directional width*** of P in the direction of v.

The directional width has several interesting properties.

(A) It's translation invariant. Indeed, for any vector s in $\mathbb{R}^d$, we have that $\overline{\omega}(v, \mathsf{s} + \mathsf{P}) = \overline{\omega}(v, \mathsf{P})$.

(B) The directional width scales linearly; that is, for any constant $c \geq 0$, we have that $\overline{\omega}(v, c\mathsf{P}) = c\,\overline{\omega}(v, \mathsf{P})$, where $c\mathsf{P}$ denotes the scaling of P by c.

(C) For any set $\mathsf{P} \subseteq \mathbb{R}^d$, we have that

$$\overline{\omega}(v, \mathsf{P}) = \overline{\omega}(v, \mathcal{CH}(\mathsf{P})),$$

where $\mathcal{CH}(\mathsf{P})$ denotes the convex hull of P.

(D) For any $\mathsf{Q} \subseteq \mathsf{P}$ and any vector v, we have that $\overline{\omega}(v, \mathsf{Q}) \leq \overline{\omega}(v, \mathsf{P})$.

Definition 23.2. A subset $\mathcal{S} \subseteq \mathsf{P}$ is an ***ε-coreset for directional width*** if

$$\forall u \in \mathbb{S}^{(d-1)} \quad \overline{\omega}(u, \mathcal{S}) \geq (1 - \varepsilon)\overline{\omega}(u, \mathsf{P}), \tag{23.1}$$

where $\mathbb{S}^{(d-1)}$ is the sphere of directions in $\mathbb{R}^d$. Namely, the coreset $\mathcal{S}$ provides a concise approximation to the directional width of P.

Note that the property of (23.1) holds for all directions u if and only if it holds for all vectors in $\mathbb{R}^d$.

Coresets have the ***sketch*** and ***merge*** properties, which make them algorithmically useful. The proof of the following lemma is sufficiently easy that we delegate its proof to Exercise 23.1.

Lemma 23.3. *(A)* **Sketch***: Consider the sets $X \subseteq Y \subseteq \mathsf{P} \subseteq \mathbb{R}^d$, where Y is an ε-coreset (for directional width) of P and X is an ε'-coreset of Y. Then, X is an $(\varepsilon + \varepsilon')$-coreset of P.*

(B) **Merge***: Consider $X \subseteq \mathsf{P} \subseteq \mathbb{R}^d$ and $X' \subseteq \mathsf{P}' \subseteq \mathbb{R}^d$, where X (resp. X') is an ε-coreset of P (resp. P'). Then $X \cup X'$ is an ε-coreset for $\mathsf{P} \cup \mathsf{P}'$.*

Claim 23.4. *Let* P *be a set of points in* $\mathbb{R}^{\mathsf{d}}$, *let* $0 < \varepsilon \le 1$ *be a parameter, and let* $\mathcal{S}$ *be a* δ*-coreset of* P *for directional width, for* $\delta = \varepsilon/(8\mathsf{d})$. *Let* $\mathcal{B}_{\mathrm{opt}}(\mathcal{S})$ *denote the minimum volume bounding box of* $\mathcal{S}$. *Let* B' *be the rescaling of* $\mathcal{B}_{\mathrm{opt}}(\mathcal{S})$ *around its center by a factor of* $(1 + 3\delta)$. *Then,* $\mathsf{P} \subset B'$ *and* $\mathrm{Vol}(B') \le (1+\varepsilon)\,\mathrm{Vol}\big(\mathcal{B}_{\mathrm{opt}}(\mathsf{P})\big)$, *where* $\mathcal{B}_{\mathrm{opt}}(\mathsf{P})$ *denotes the minimum volume bounding box of* P.

Proof. For the volume claim, observe that

$$\mathrm{Vol}(B') = (1+3\delta)^{\mathsf{d}}\,\mathrm{Vol}\big(\mathcal{B}_{\mathrm{opt}}(\mathcal{S})\big) \le \exp(3\delta\mathsf{d})\,\mathrm{Vol}\big(\mathcal{B}_{\mathrm{opt}}(\mathcal{S})\big) = \exp\!\left(3\mathsf{d}\frac{\varepsilon}{8\mathsf{d}}\right)\mathrm{Vol}\big(\mathcal{B}_{\mathrm{opt}}(\mathcal{S})\big)$$
$$\le (1+\varepsilon)\,\mathrm{Vol}\big(\mathcal{B}_{\mathrm{opt}}(\mathcal{S})\big) \le (1+\varepsilon)\,\mathrm{Vol}\big(\mathcal{B}_{\mathrm{opt}}(\mathsf{P})\big).$$

We remain with the task of proving that $\mathsf{P} \subseteq B'$. To this end, let ℓ be a line through the origin parallel to one of the edges of $\mathcal{B}_{\mathrm{opt}}(\mathcal{S})$, and let u be its direction; see the figure on the right. Let $I_{\mathcal{S}}$ and I' be the projection into ℓ of $\mathcal{B}_{\mathrm{opt}}(\mathcal{S})$ and B', respectively. Let I_{P} be the interval formed by the projection of $\mathcal{CH}(\mathsf{P})$ into ℓ.

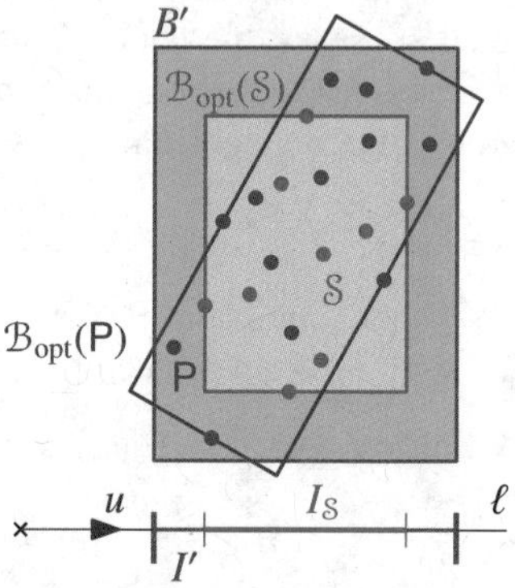

Since $I_{\mathcal{S}}$ is the interval realizing the (directional) width of $\mathcal{S}$ in the direction of ℓ (otherwise, the bounding box $\mathcal{B}_{\mathrm{opt}}(\mathcal{S})$ would not be optimal), we have that $I_{\mathcal{S}} \subseteq I_{\mathsf{P}}$. Since $\mathcal{S}$ is a δ-coreset for the directional width, we have that

$$\|I_{\mathcal{S}}\| = \overline{\omega}(u,\mathcal{S}) \ge (1-\delta)\overline{\omega}(u,\mathsf{P}) = (1-\delta)\,\|I_{\mathsf{P}}\|\,.$$

Now, since $I_{\mathcal{S}} \subseteq I_{\mathsf{P}}$, we have that the maximum distance between an endpoint of I_{P} and $I_{\mathcal{S}}$ is at most $\delta\,\|I_{\mathsf{P}}\|$. As such, the distance between the center of $I_{\mathcal{S}}$ and an endpoint of I_{P} is bounded by

$$\frac{\|I_{\mathcal{S}}\|}{2} + \delta\,\|I_{\mathsf{P}}\| \le \left(\frac{1}{2} + \frac{\delta}{1-\delta}\right)\|I_{\mathcal{S}}\| = \left(1 + \frac{2\delta}{1-\delta}\right)\frac{\|I_{\mathcal{S}}\|}{2} \le (1+3\delta)\,\frac{\|I_{\mathcal{S}}\|}{2},$$

since $\delta \le 1/3$. This implies that $(1+3\delta)I_{\mathcal{S}} = I'$ contains the segment I_{P}. Since this holds for all the directions of the edges of B', we conclude that $\mathsf{P} \subseteq B'$, as claimed. ■

It is easy to verify that a coreset for directional width also preserves (approximately) the diameter and width of the point set. Namely, it captures "well" the geometry of P. Claim 23.4 hints at a connection between coresets for directional width and the minimum volume bounding box. In particular, if we have a good bounding box, we can compute a small coreset for directional width.

Lemma 23.5. *Let* P *be a set of n points in* $\mathbb{R}^{\mathsf{d}}$, *and let* B *be a bounding box of* P, *such that* $\mathsf{s} + c_{\mathsf{d}}B \subseteq \mathcal{CH}(\mathsf{P}) \subseteq B$, *where* s *is a vector in* $\mathbb{R}^{\mathsf{d}}$, c_{d} *is a constant that depends only on* d, *and* $c_{\mathsf{d}}B$ *denotes the rescaling of* B *by a factor of* c_{d} *around its center point.*

Then, one can compute an ε*-coreset* $\mathcal{S}$ *for the directional width of* P. *The size of the coreset is* $O(1/\varepsilon^{\mathsf{d}-1})$, *and the construction time is* $O(n)$.

Proof. We partition B into a grid G, by breaking each edge of B into $M = \lceil 4/(\varepsilon c_{\mathsf{d}})\rceil$ equal length intervals (namely, we tile B with M^{d} copies of B/M). A cell in this grid is uniquely defined by a d-tuple $(i_1, \ldots, i_{\mathsf{d}})$. In particular, for a point $\mathsf{p} \in \mathsf{P}$, let $\mathrm{id}(\mathsf{p})$ denote the ID of the cell that contains it. Clearly, $\mathrm{id}(\mathsf{p})$ can be computed in constant time.

Given a $(\mathsf{d}-1)$-tuple $I = (i_1, \ldots, i_{\mathsf{d}-1})$, its ***pillar*** is the set of grid cells that have $(i_1, \ldots, i_{\mathsf{d}-1})$ as the first $\mathsf{d}-1$ coordinates of their ID. For each pillar record the highest and lowest points encountered. Here highest/lowest refers to their values in the dth direction. Let $\mathcal{S}$ be the resulting set of points.

The computation of $\mathcal{S}$ can be done in linear time by hashing the points into the ID of the pillar that contains them. Indeed, we keep a list of all the non-empty pillars encountered, and for each pillar we maintain the highest and lowest points encountered. A new point inserted into this data-structure can be handled in $O(1)$ time (considering d to be a constant).

We claim that $\mathcal{S}$ is the required coreset. To this end, let Q be the set of points formed by the union of grid cells of G that contains a point of $\mathcal{S}$. Since the coreset contains the highest and lowest points in each pillar, it follows that $\mathsf{P} \subseteq \mathcal{CH}(\mathsf{Q})$ (note, however, that is not necessarily true that $\mathsf{P} \subseteq \mathcal{CH}(\mathcal{S})$). This implies that $\overline{\omega}(v, \mathsf{Q}) = \overline{\omega}(v, \mathcal{CH}(\mathsf{Q})) \geq \overline{\omega}(v, \mathsf{P})$, for any v. Every grid cell of G is a translated copy of B/M, and we have that $\overline{\omega}(v, \mathcal{S}) + 2\overline{\omega}(v, B/M) \geq \overline{\omega}(v, \mathsf{Q}) \geq \overline{\omega}(v, \mathsf{P})$. Now,

$$\overline{\omega}(v, B/M) = \frac{\overline{\omega}(v, B)}{M} = \frac{\overline{\omega}(v, c_{\mathsf{d}}B)}{c_{\mathsf{d}}M} \leq \frac{\overline{\omega}(v, \mathsf{P})}{c_{\mathsf{d}}M} \leq \frac{\overline{\omega}(v, \mathsf{P})}{4/\varepsilon} \leq \frac{\varepsilon}{4}\overline{\omega}(v, \mathsf{P}),$$

since $\mathsf{s} + c_{\mathsf{d}}B \subseteq \mathcal{CH}(\mathsf{P})$. As such, $\overline{\omega}(v, \mathcal{S}) \geq \overline{\omega}(v, \mathsf{P}) - 2\overline{\omega}(v, B/M) \geq (1 - \varepsilon/2)\overline{\omega}(v, \mathsf{P})$. ■

Theorem 23.6. *Let P be a set of n points in $\mathbb{R}^{\mathsf{d}}$, and let $0 < \varepsilon < 1$ be a parameter. One can compute an ε-coreset $\mathcal{S}$ for the directional width of P. The size of the coreset is $O(1/\varepsilon^{\mathsf{d}-1})$, and the construction time is $O(n)$.*

Proof. Compute, in linear time, a constant approximation to the minimum volume bounding box of P using Lemma 22.7$_{\text{p284}}$. Then apply Lemma 23.5 to P. ■

23.1.1. Coresets vs random sampling.

Lemma 23.7. *For a set P of n points in $\mathbb{R}^{\mathsf{d}}$, consider an $\varepsilon/4$-coreset $\mathcal{S} \subseteq \mathsf{P}$ for directional width. Then, for any ball $\mathbf{g}$ containing $\mathcal{S}$ we have that $(1+\varepsilon)\mathbf{g}$ contains P.*

Proof. Assume this is false, and consider a point $\mathsf{p} \in \mathsf{P}$ which is outside the expanded ball $(1+\varepsilon)\mathbf{g}$. Let u be the direction of the segment $\overline{\mathsf{c}}\mathsf{p}$, where $\overline{\mathsf{c}}$ is the center of $\mathbf{g}$. Let ℓ be the line spanned by direction u.

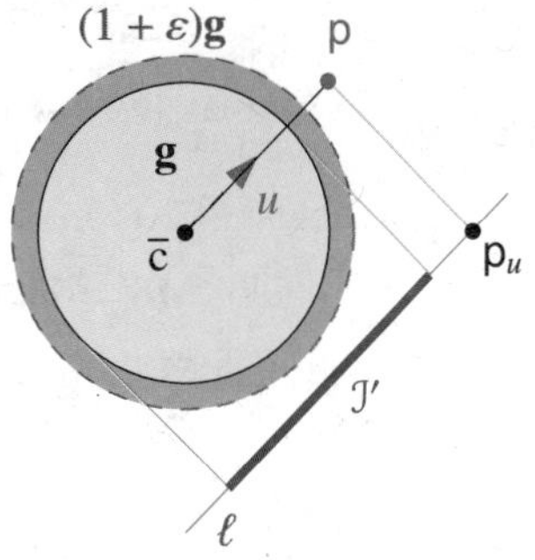

We claim that $\mathcal{S}$ fails as an $\varepsilon/4$-coreset for the directional width of P for the direction u. Indeed, let r be the radius of $\mathbf{g}$. Then the projection of $\mathcal{S}$ into the direction u results in an interval $\mathcal{I}$ of length at most $2r$, and the projection of p into the direction of u, denoted by p_u, falls outside this interval. Observe that the projection of $\mathbf{g}$ into the direction u is an interval $\mathcal{I}'$ of length $2r$ that contains $\mathcal{I}$. As such, $(1+\varepsilon)\mathcal{I} \subseteq (1+\varepsilon)\mathcal{I}'$. However, $(1+\varepsilon)\mathcal{I}'$ is the projection of $(1+\varepsilon)\mathbf{g}$ into the direction u (to see that, imagine translating ℓ such that it passes through the center $\overline{\mathsf{c}}$ of $\mathbf{g}$). See the figure on the right. Now, since the point p is outside $(1+\varepsilon)\mathbf{g}$, this implies that p_u is outside $(1+\varepsilon)\mathcal{I}'$. Namely, p is outside $(1+\varepsilon)\mathcal{I}$, and so $\mathcal{S}$ is not an $\varepsilon/4$-coreset for directional width of P. Indeed, the above implies that $\overline{\omega}(u, \mathsf{P}) - \overline{\omega}(u, \mathcal{S}) \geq (\varepsilon/2)\left\|\mathcal{I}'\right\| \geq (\varepsilon/2)\overline{\omega}(u, \mathcal{S})$. Namely,

$$\overline{\omega}(u, \mathcal{S}) \leq \frac{1}{1+\varepsilon/2}\overline{\omega}(u, \mathsf{P}) < (1 - \varepsilon/4)\,\overline{\omega}(u, \mathsf{P}),$$

since $0 < \varepsilon < 1$. However, this contradicts the fact that $\mathcal{S}$ is an $\varepsilon/4$-coreset of P for directional width (see Definition 23.2). ■

23.1.1.1. *Discussion.* Informally, coresets are small subsets of the input that capture the structure of the input that we care about. They seems similar to random samples in what they can provide. To see the difference, consider the concrete problem of finding the smallest enclosing ball of a point set $\mathsf{P} \subseteq \mathbb{R}^{\mathsf{d}}$. By the ε-net theorem a random sample R from P of size (roughly) $O(\mathsf{d}/\varepsilon \log 1/\varepsilon)$ is (with constant probability and for this discussion assume this holds) an ε-net for ranges which are the complement of balls.

Now, if any enclosing ball $\mathbf{g}$ for R has more than $\varepsilon |\mathsf{P}|$ points of P outside it, then the complement $\overline{\mathbf{g}} = \mathbb{R}^{\mathsf{d}} \setminus \mathbf{g}$ contains more than $\varepsilon |\mathsf{P}|$ points inside it. But then, since R is an ε-net for such ranges, it follows that $\overline{\mathbf{g}}$ must contain a point of P. Namely, $\mathbf{g}$ does not enclose all the points of R, a contradiction.

As such, any ball that encloses R covers all but an ε-fraction of the points of P.

On the other hand, Lemma 23.5 and Lemma 23.7 imply that there is an ε-coreset for this problem of size $O\left(1/\varepsilon^{\mathsf{d}-1}\right)$ (in fact, one can find even smaller coresets for this problem). Furthermore, for any ball covering this coreset, if we expand it by $(1+\varepsilon)$, then it covers all the input points.

Namely, coresets capture the whole structure of the input, where random samples can only capture the structure for "most" of the points. Coresets trade off geometric error (i.e., an ε fraction of the radius of the enclosing ball) while random samples trade off statistical error (i.e., hold for all except for an ε fraction of the input) to achieve this.

23.1.2. Smaller coreset for directional width.

A set $\mathsf{P} \subseteq \mathbb{R}^{\mathsf{d}}$ is ***α-fat***, for $\alpha \leq 1$, if there exists a point $\mathsf{p} \in \mathbb{R}^{\mathsf{d}}$ and a hypercube $\overline{\mathcal{H}}$ centered at the origin such that

$$\mathsf{p} + \alpha\overline{\mathcal{H}} \subset \mathcal{CH}(\mathsf{P}) \subset \mathsf{p} + \overline{\mathcal{H}}.$$

23.1.2.1. *Transforming a set into a fat set.*

Claim 23.8. *Let P be a set of n points in $\mathbb{R}^{\mathsf{d}}$, and let $\mathbf{M}$ be a full rank symmetric $\mathsf{d} \times \mathsf{d}$ matrix (i.e., $\mathbf{M}^T = \mathbf{M}$). Then, $\mathcal{S} \subseteq \mathsf{P}$ is an ε-coreset for directional width for P if and only if $\mathbf{M}(\mathcal{S})$ is an ε-coreset for $\mathbf{M}(\mathsf{P})$.*

PROOF. For any $\mathsf{p} \in \mathbb{R}^{\mathsf{d}}$, we have that $\langle v, \mathbf{M}\mathsf{p}\rangle = v^T\mathbf{M}\mathsf{p} = \left(\mathbf{M}^T v\right)^T \mathsf{p} = \left\langle \mathbf{M}^T v, \mathsf{p}\right\rangle = \langle \mathbf{M}v, \mathsf{p}\rangle$. As such, for any set $X \subseteq \mathbb{R}^{\mathsf{d}}$ and any vector v, we have that

$$\overline{\omega}(v, \mathbf{M}X) = \max_{\mathsf{p}\in X} \langle v, \mathbf{M}\mathsf{p}\rangle - \min_{\mathsf{p}\in X} \langle v, \mathbf{M}\mathsf{p}\rangle = \max_{\mathsf{p}\in X} \langle \mathbf{M}v, \mathsf{p}\rangle - \min_{\mathsf{p}\in X} \langle \mathbf{M}v, \mathsf{p}\rangle = \overline{\omega}(\mathbf{M}v, X).$$

In particular, if $\mathcal{S}$ is an ε-coreset for P, then, for any $v \in \mathbb{R}^{\mathsf{d}}$, we have

$$\overline{\omega}(v, \mathbf{M}\mathcal{S}) = \overline{\omega}(\mathbf{M}v, \mathcal{S}) \geq (1-\varepsilon)\overline{\omega}(\mathbf{M}v, \mathsf{P}) = (1-\varepsilon)\overline{\omega}(v, \mathbf{M}\mathsf{P}),$$

which implies that $\mathbf{M}\mathcal{S}$ is an ε-coreset for directional width for $\mathbf{M}\mathsf{P}$.

The other direction follows by similar argumentation. ■

Lemma 23.9. *Let P be a set of n points in $\mathbb{R}^{\mathsf{d}}$ such that the volume of $\mathcal{CH}(\mathsf{P})$ is non-zero, and let $\mathcal{H} = [-1,1]^{\mathsf{d}}$. One can compute in $O(n)$ time an affine transformation T and a vector $\vec{w}'$, such that $T(\mathsf{P})$ is an α-fat point set satisfying $\vec{w}' + \alpha\mathcal{H} \subset \mathcal{CH}(T(\mathsf{P})) \subset \mathcal{H}$, where α is a positive constant depending on d. Furthermore, for all subsets $\mathcal{S} \subseteq \mathsf{P}$, we have that $\mathcal{S}$ is an ε-coreset of P for directional width if and only if $T(\mathcal{S})$ is an ε-coreset of $T(\mathsf{P})$ for directional width.*

PROOF. Using the algorithm of Lemma 22.7$_{\text{p284}}$ compute, in $O(n)$ time, a bounding box B of P, such that there exists a vector $\vec{w}$ such that $\vec{w} + B/c \subseteq \mathcal{CH}(\mathsf{P}) \subseteq B$, where c is some constant that depends only on d.

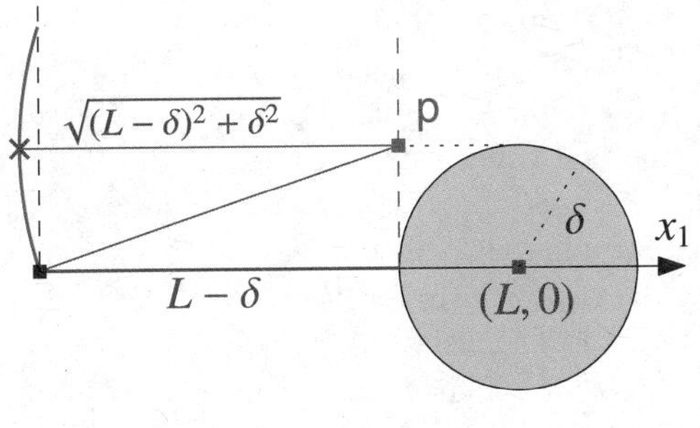

FIGURE 23.1. Illustration of the proof of Lemma 23.11.

Let T_1 be the translation of space that maps the center of B to the origin. Clearly, $\mathcal{S} \subseteq \mathsf{P}$ is an ε-coreset for directional width of P if and only if $\mathcal{S}_1 = T_1(\mathcal{S})$ is an ε-coreset for directional width of $\mathsf{P}_1 = T_1(\mathsf{P})$. Next, let T_2 be a rotation such that $B_1 = T_1(B)$ has its sides parallel to the axes. Again, $\mathcal{S}_2 = T_2(\mathcal{S}_1)$ is an ε-coreset for $\mathsf{P}_2 = T_2(\mathsf{P}_1)$ if and only if $\mathcal{S}_1$ is a coreset for P_1.

Finally, let T_3 be a scaling such that $B_2 = T_2(B_1)$ is mapped to the hypercube $\mathcal{H}$. The transformation T_3 can be written as a diagonal matrix, with the ith entry on the diagonal being the scaling in the ith dimension. By Claim 23.8, $\mathcal{S}_3 = T_3(\mathcal{S}_2)$ is an ε-coreset for $\mathsf{P}_3 = T_3(\mathsf{P}_2)$ if and only if $\mathcal{S}_2$ is an ε-coreset for P_2.

Set $T = T_3 \circ T_2 \circ T_1$, and observe that, by the above argumentation, $\mathcal{S}$ is an ε-coreset for P if and only if $T(\mathcal{S})$ is an ε-coreset for $T(\mathsf{P})$. However, note that $T(B) = \mathcal{H}$ and $T(\vec{w} + B/c) \subseteq T(\mathcal{CH}(\mathsf{P})) = \mathcal{CH}(T(\mathsf{P}))$. Namely, there exists a vector $\vec{w'}$ such that $\vec{w'} + \mathcal{H}/c \subseteq \mathcal{CH}(T(\mathsf{P})) \subseteq \mathcal{H}$. Therefore, by definition, the point set $T(\mathsf{P})$ is $\alpha = (1/c)$-fat. ■

23.1.2.2. *Computing a smaller coreset.*

Observation 23.10. *If $\mathcal{T}$ is a δ-coreset for directional width of $\mathcal{S}$ and $\mathcal{S}$ is an ε-coreset for directional width of* P*, then $\mathcal{T}$ is a $(\delta+\varepsilon)$-coreset for directional width of* P.

PROOF. For any vector v, we have $\overline{\omega}(v, \mathcal{T}) \geq (1-\delta)\overline{\omega}(v, \mathcal{S}) \geq (1-\delta)(1-\varepsilon)\overline{\omega}(v, \mathsf{P}) \geq (1-\delta-\varepsilon)\overline{\omega}(v, \mathsf{P})$. ■

This implies that even a very slow algorithm that computes a (smaller) coreset is useful. Indeed, given a point set P, we can first extract from it an $\varepsilon/2$-coreset of size $O(1/\varepsilon^{\mathsf{d}-1})$, using Lemma 23.5. Let Q denote the resulting set. We then compute a considerably smaller $\varepsilon/2$-coreset for Q. By the above observation, this results in an ε-coreset for directional width of P. Namely, since we use the slow procedure on a small set (i.e., the rough coreset), the resulting algorithm is still reasonably efficient. We next describe how to compute a smaller coreset. To this end, we need the following technical lemma.

Lemma 23.11. *Let σ, x, and y be points. Let u be the direction of the vector $\overrightarrow{\sigma x}$, and let $\delta = \|x-y\|$ and $L = \|\sigma - x\|$. If $2\delta \leq L$, then* $\displaystyle\min_{\mathsf{p} \in \mathrm{ball}(y, \|y-\sigma\|)} \langle u, \mathsf{p}-\sigma\rangle \geq -\delta^2/L$.

PROOF. By rotating and translating space, we can assume that σ is at the origin o, $x = (L, 0, \ldots, 0)$, and $u = (1, 0, \ldots, 0)$. As such, for $\mathsf{p} = y$, we need to prove that

$$\mu(\mathsf{p}) = \min_{\mathsf{s} \in \mathrm{ball}(\mathsf{p}, \|\mathsf{p}-\sigma\|)} \langle \mathsf{s}-\sigma, u\rangle = \min_{\mathsf{s} \in \mathrm{ball}(\mathsf{p}, \|\mathsf{p}-\mathsf{o}\|)} \langle \mathsf{s}-\mathsf{o}, u\rangle = \min_{\mathsf{s}=(\mathsf{s}_1, \mathsf{s}_2, \ldots, \mathsf{s}_\mathsf{d}) \in \mathrm{ball}(\mathsf{p}, \|\mathsf{p}\|)} \mathsf{s}_1 \geq -\frac{\delta^2}{L}.$$

Clearly, $\mu(\mathsf{p})$ is the value of the first coordinate of the left-most point in the ball centered at p and of radius $\|\mathsf{p}\|$.

As demonstrated in the figure on the right, if we move p away from the x_1-axis, then this quantity decreases (i.e., $p_1 \to p_2$ in the figure). Similarly, it decreases if we decrease the first coordinate of p (i.e., $p_2 \to p_3$ in the figure).

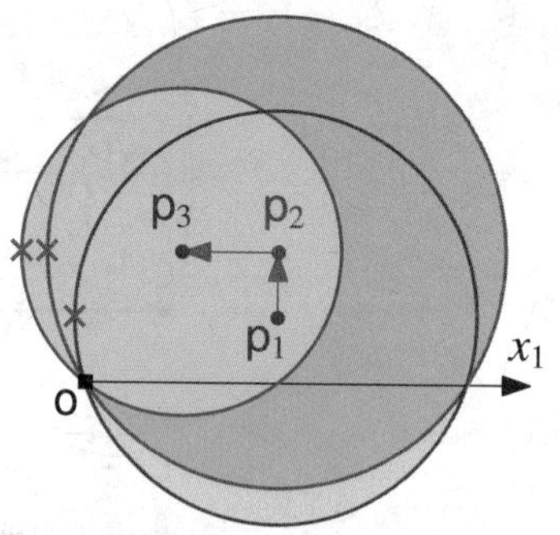

Thus, by symmetry, the minimum value of $\mu(y)$ is bounded from below by the value of $\mu(p)$ for $p = (L-\delta, \delta, 0, \ldots, 0)$. See Figure 23.1. The distance between p and the origin is $\sqrt{(L-\delta)^2+\delta^2}$, and

$$\mu(p) = (L-\delta) - \sqrt{(L-\delta)^2+\delta^2}$$
$$= \frac{(L-\delta)^2-(L-\delta)^2-\delta^2}{(L-\delta)+\sqrt{(L-\delta)^2+\delta^2}} \geq -\frac{\delta^2}{2(L-\delta)} \geq -\frac{\delta^2}{L},$$

since $L \geq 2\delta$. ■

Lemma 23.12. *Let Q be a set of m points. Then one can compute an $\varepsilon/2$-coreset for directional width of Q of size $O\left(1/\varepsilon^{(d-1)/2}\right)$, in time $O\left(m/\varepsilon^{(d-1)/2}\right)$.*

PROOF. Note that by Lemma 23.9, we can assume that Q is α-fat for some constant α and $v + \alpha[-1,1]^d \subseteq \mathcal{CH}(Q) \subseteq [-1,1]^d$ for some $v \in \mathbb{R}^d$. In particular, for any direction $u \in \mathbb{S}^{(d-1)}$, we have $\overline{\omega}(u, Q) \geq 2\alpha$.

Let $\mathbb{S}'$ be the sphere of radius $\sqrt{d}+1$ centered at the origin. Set $\delta = \sqrt{\varepsilon\alpha/4} \leq 1/4$. One can construct a set $\mathcal{I}$ of $O\left((\sqrt{d}/\delta)^{d-1}\right) = O\left(1/\varepsilon^{(d-1)/2}\right)$ points on the sphere $\mathbb{S}'$ so that for any point x on $\mathbb{S}'$, there exists a point $y \in \mathcal{I}$ such that $\|x-y\| \leq \delta$. For each point $y \in \mathcal{I}$, we compute the nearest neighbor to y in Q, denoted by $nn(y, Q)$, by scanning the points of Q and returning the nearest point to y. As such, per query, computing $nn(y, Q)$ takes $O(m)$ time. We return the set $\mathcal{S} = \left\{ nn(y, Q) \;\middle|\; y \in \mathcal{I} \right\}$ as the desired coreset; see the figure on the right. Overall, the time to compute $\mathcal{S}$ is $O(m/\varepsilon^{(d-1)/2})$.

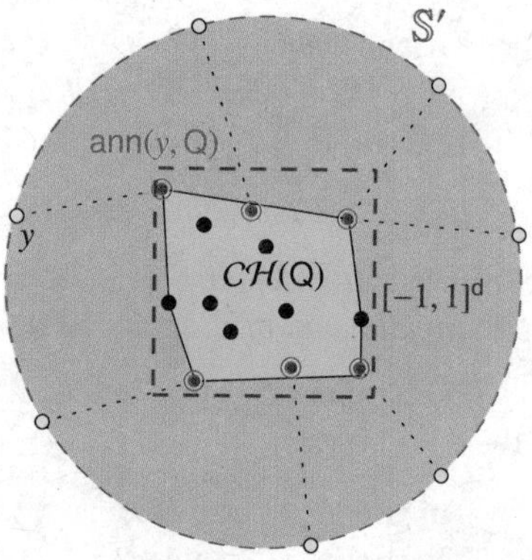

We next prove that $\mathcal{S}$ is an $(\varepsilon/2)$-coreset of Q. (The proof also works with minor modifications if we compute the ANN instead of the exact nearest neighbor when constructing $\mathcal{S}$.)

Fix a direction $u \in \mathbb{S}^{(d-1)}$ (i.e., $\|u\| = 1$). Let $\sigma \in$ Q be the point that maximizes $\langle u, p\rangle$ over all $p \in Q$. Suppose the ray emanating from σ in the direction of u hits $\mathbb{S}'$ at a point x. There exists a point $y \in \mathcal{I}$ such that $\|x-y\| \leq \delta$.

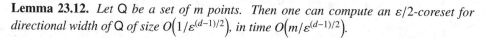

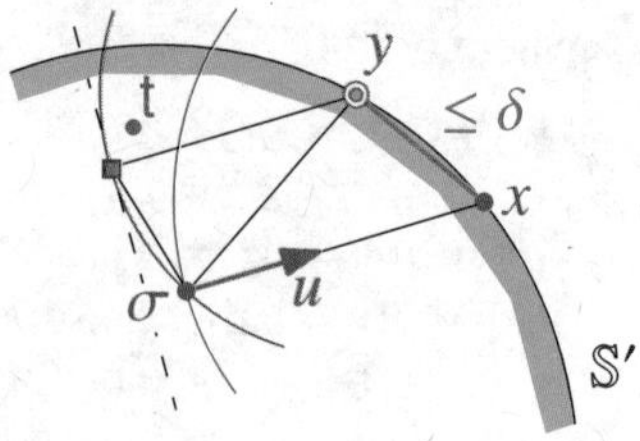

Observe that the minimum distance between any point of $\mathbb{S}'$ (which is of radius $\sqrt{d}+1$ and centered at the origin) and any point of Q (contained inside the $[-1,1]^d$ hypercube) is at least $\sqrt{d} + 1 - \text{diam}\left([-1,1]^d\right)/2 = \sqrt{d}+1-\sqrt{d}/2 \geq 1$.

Set $L = \|\sigma - x\|$. By the above argumentation, we have $\|x-y\| \leq \delta$, $L \geq 1$, and $2\delta < L$. Let $t = nn(y, Q)$, and observe that $\|y-t\| \leq \|y-\sigma\|$. As such, by Lemma 23.11, we have that

$$\langle u, \mathsf{t} - \sigma \rangle \geq \min_{\mathsf{p} \in \mathrm{ball}(y, \|y - \sigma\|)} \langle u, \mathsf{p} - \sigma \rangle \geq -\frac{\delta^2}{L}.$$

Since $\mathsf{t} \in \mathcal{S}$ and $\sigma \in \mathsf{Q}$, we have that

$$\max_{\mathsf{p} \in \mathcal{S}} \langle u, \mathsf{p} \rangle - \max_{\mathsf{p} \in \mathsf{Q}} \langle u, \mathsf{p} \rangle \geq \langle u, \mathsf{t} - \sigma \rangle \geq -\frac{\delta^2}{L} \geq -\delta^2 = -\frac{\varepsilon\alpha}{4}.$$

A similar argument implies that $\min_{\mathsf{p} \in \mathcal{S}} \langle u, \mathsf{p} \rangle - \min_{\mathsf{p} \in \mathsf{Q}} \langle u, \mathsf{p} \rangle \leq \varepsilon\alpha/4$. Thus, we conclude that $\overline{\omega}(u, \mathcal{S}) \geq \overline{\omega}(u, \mathsf{Q}) - 2(\varepsilon\alpha/4) = \overline{\omega}(u, \mathsf{Q}) - \varepsilon\alpha/2$. On the other hand, since $\overline{\omega}(u, \mathsf{Q}) \geq 2\alpha$, it follows that $\overline{\omega}(u, \mathcal{S}) \geq (1 - \varepsilon/2)\overline{\omega}(u, \mathsf{Q})$. ■

Theorem 23.13. *Let* P *be a set of* n *points in* $\mathbb{R}^{\mathsf{d}}$. *One can compute an* ε*-coreset for directional width of* P *in* $O(n + 1/\varepsilon^{3(d-1)/2})$ *time. The coreset size is* $O(\varepsilon^{(d-1)/2})$.

PROOF. We first use the algorithm of Theorem 23.6 to compute an $\varepsilon/2$-coreset Q of P of size $m = O\left(1/\varepsilon^{\mathsf{d}-1}\right)$. Next, we computing an $\varepsilon/2$-coreset $\mathcal{S}$ of Q using the algorithm of Lemma 23.12. The size of $\mathcal{S}$ is $O\left(1/\varepsilon^{(\mathsf{d}-1)/2}\right)$, and by Lemma 23.3(B) $\mathcal{S}$ is an ε-coreset for P.

As for the running time, the first step takes $O(n)$ time. The second step takes

$$O\left(m/\varepsilon^{(\mathsf{d}-1)/2}\right) = O\left(1/\varepsilon^{3(\mathsf{d}-1)/2}\right)$$

time, implying the claim. ■

23.2. Approximating the extent of lines and other creatures

23.2.1. Preliminaries. For a hyperplane $h \equiv x_{\mathsf{d}} = a_0 + a_1 x_1 + \cdots + a_{\mathsf{d}-1} x_{\mathsf{d}-1}$ in $\mathbb{R}^{\mathsf{d}}$ (which is not parallel to the x_{d}-axis) and a point $\mathsf{x} \in \mathbb{R}^{\mathsf{d}-1}$, let $h(\mathsf{x}) = a_0 + a_1 x_1 + \cdots + a_{\mathsf{d}-1} x_{\mathsf{d}-1}$. Geometrically, consider the line ℓ parallel to the x_{d}-axis and passing through x, and observe that $h(\mathsf{x})$ is the d-coordinate of the intersection point $\ell \cap h$.

Definition 23.14. Given a set of hyperplanes $\mathcal{H}$ in $\mathbb{R}^{\mathsf{d}}$, the minimization diagram of $\mathcal{H}$, known as the ***lower envelope*** of $\mathcal{H}$, is the function $\mathcal{L}_{\mathcal{H}} : \mathbb{R}^{\mathsf{d}-1} \rightarrow \mathbb{R}$, where we have $\mathcal{L}_{\mathcal{H}}(\mathsf{x}) = \min_{h \in \mathcal{H}} h(\mathsf{x})$, for $\mathsf{x} \in \mathbb{R}^{\mathsf{d}-1}$.

Similarly, the ***upper envelope*** of $\mathcal{H}$ is the function $\mathcal{U}_{\mathcal{H}}(\mathsf{x}) = \max_{h \in \mathcal{H}} h(\mathsf{x})$, for $\mathsf{x} \in \mathbb{R}^{\mathsf{d}-1}$.

The ***extent*** of $\mathcal{H}$ for $\mathsf{x} \in \mathbb{R}^{\mathsf{d}-1}$ is the vertical distance between the upper and lower envelopes at x; namely, $\mathcal{E}_{\mathcal{H}}(\mathsf{x}) = \mathcal{U}_{\mathcal{H}}(\mathsf{x}) - \mathcal{L}_{\mathcal{H}}(\mathsf{x})$.

23.2.2. Motivation – maintaining the bounding box of moving points. Let $\mathsf{P} = \{\mathsf{p}_1, \ldots, \mathsf{p}_n\}$ be a set of n points moving in $\mathbb{R}^{\mathsf{d}}$. For a given time t, let $\mathsf{p}_i(t) = (x_i^1(t), \ldots, x_i^{\mathsf{d}}(t))$ denote the position of p_i at time t. We will use $\mathsf{P}(t)$ to denote the set P at time t. We say that the motion of P has ***degree*** k if every $x_i^j(t)$ is a polynomial of degree at most k. A motion of degree one is ***linear***. Namely, $\mathsf{p}_i(t) = \mathsf{a}_i + \mathsf{b}_i t$, where $\mathsf{a}_i, \mathsf{b}_i \in \mathbb{R}^{\mathsf{d}}$ are fixed, for $i = 1, \ldots, n$. In the following, we assume that $\mathsf{P}(t)$ has linear motion; that is, every point is moving in a fixed direction with a fixed velocity.

Our purpose is to develop efficient approaches for maintaining various descriptors of the extent of P, including the smallest enclosing axis parallel bounding box of P. This measure indicates how spread out the point set P is. As the points move continuously, the quantity we are interested in changes continuously as well, though its combinatorial realization changes only at certain discrete times. For example, the smallest axis parallel bounding box containing P can be represented by a sequence of $2\mathsf{d}$ points of P, each lying

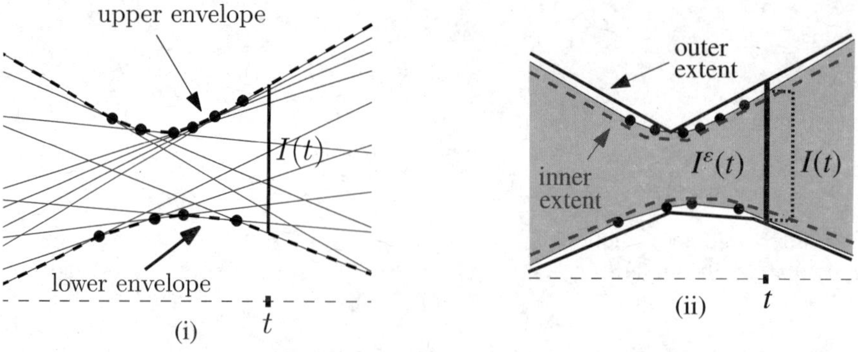

FIGURE 23.2. (i) The extent of the moving points is the length of the vertical segment connecting the lower envelope to the upper envelope. The black dots mark where the movement description of $I(t)$ changes. (ii) The approximate extent.

on one of the facets of the box. As the points move, the box also changes continuously. At certain discrete times, the points of P lying on the boundary of the box change, and we have to update the sequence of points defining the box. Similarly, our approach will be to focus on these discrete changes (or *events*) and track through time the combinatorial description of the quantity we are interested in.

Since we are computing an *axis parallel* bounding box of the moving points, we can solve the problem in each dimension separately. Thus, consider the points as moving linearly in one dimension. Thus, consistent with Definition 23.14, the ***extent*** of $\mathsf{P}(t)$, denoted by $B(t)$, is the smallest interval containing $\mathsf{P}(t)$.

It will be convenient to work in the parametric xt-plane in which a moving point $p(t) \in \mathbb{R}$ at time t is mapped to the point $(t, p(t))$. For $i = 1, \ldots, n$, we map the point $\mathsf{p}_i \in P$ to the line $\ell_i = \bigcup_t (t, \mathsf{p}_i(t))$. Let $\mathsf{L} = \{\ell_1, \ldots, \ell_n\}$ be the resulting set of lines, and let $\mathcal{A}(\mathsf{L})$ be their arrangement. Clearly, the extent $B(t)$ of $\mathsf{P}(t)$ is the vertical interval $I(t)$ connecting the upper and lower envelopes of L at time t. See Figure 23.2(i). The combinatorial structure of $I(t)$ changes at the vertices of the two envelopes of L; that is, the current pair of lines defining $I(t)$ changes only at these vertices. As such, all the different combinatorial structures of $I(t)$ can be computed in $O(n \log n)$ time by computing the upper and lower envelopes of L; see Lemma 28.7_{p348}.

For an interval $\mathcal{I} = [a, b] \subseteq \mathbb{R}$, let $\|\mathcal{I}\| = b - a$ denote the ***length*** of $\mathcal{I}$. We want to maintain a vertical interval $I_\varepsilon^+(t)$ so that $I(t) \subseteq I_\varepsilon^+(t)$ and $\left\| I_\varepsilon^+(t) \right\| \leq (1 + \varepsilon) \|I(t)\|$ for all t. In particular, the endpoints of $I_\varepsilon^+(t)$ should follow piecewise-linear trajectories, and the number of combinatorial changes in $I^\varepsilon(t)$ is small. Alternatively, we want to maintain a vertical interval $I_\varepsilon^-(t) \subseteq I(t)$ such that $\left\| I_\varepsilon^-(t) \right\| \geq (1 - \varepsilon) \|I(t)\|$. Clearly, having one approximation would imply the other by appropriate rescaling.

Geometrically, this has the following interpretation: We want to simplify the upper and lower envelopes of $\mathcal{A}(\mathsf{L})$ by convex and concave polygonal chains, respectively, so that the simplified upper (resp. lower) envelope lies below (resp. above) the original upper (resp. lower) envelope and so that for any t, the vertical segment connecting the simplified envelopes contains $(1 - \varepsilon)I(t)$. Naturally, given such an approximation, we can slightly scale it and get an outer approximation to the extent; see Figure 23.2(ii).

23.2.3. Duality and coresets for extent. In the following, we will use duality. Duality is described in detail in Chapter 25. The duality we use is the following:

$$\mathsf{p} = (p_1, \ldots, p_\mathsf{d}) \implies \mathsf{p}^\star \equiv x_\mathsf{d} = p_1 x_1 + \cdots + p_{\mathsf{d}-1} x_{\mathsf{d}-1} - p_\mathsf{d}$$

$$h \equiv x_\mathsf{d} = a_1 x_1 + \cdots + a_{\mathsf{d}-1} x_{\mathsf{d}-1} + a_\mathsf{d} \implies h^\star = (a_1, \ldots, a_{\mathsf{d}-1}, -a_\mathsf{d}).$$

The properties of this duality that we need are summarized in Lemma 25.3_{p319}.

Claim 23.15. *Let $\mathcal{H}$ be the set of hyperplanes in $\mathbb{R}^\mathsf{d}$. For any $\mathsf{p} \in \mathbb{R}^{\mathsf{d}-1}$ consider the points $\mathcal{L}_\mathcal{H}(\mathsf{p})$ and $\mathcal{U}_\mathcal{H}(\mathsf{p})$. The following properties hold:*
(A) The dual hyperplanes $\mathcal{L}_\mathcal{H}(\mathsf{p})^\star$ and $\mathcal{U}_\mathcal{H}(\mathsf{p})^\star$ are parallel and their normal is the vector $(\mathsf{p}, -1) \in \mathbb{R}^\mathsf{d}$.
(B) The points of $\mathcal{H}^\star$ lie below (resp., above) the hyperplane $\mathcal{L}_\mathcal{H}(\mathsf{p})^\star$ (resp., $\mathcal{U}_\mathcal{H}(\mathsf{p})^\star$).
(C) The hyperplanes $\mathcal{L}_\mathcal{H}(\mathsf{p})^\star$ and $\mathcal{U}_\mathcal{H}(\mathsf{p})^\star$ support $C = \mathcal{CH}(\mathcal{H}^\star)$. In fact, C is contained in the slab induced by these two hyperplanes.
(D) The extent $\mathcal{E}_\mathcal{H}(\mathsf{p})$ is the vertical distance between $\mathcal{L}_\mathcal{H}(\mathsf{p})^\star$ and $\mathcal{U}_\mathcal{H}(\mathsf{p})^\star$.

Proof. The proof is easy and is included for the sake of completeness.

(A) For a point $\mathsf{p} = (\mathsf{p}_1, \ldots, \mathsf{p}_{\mathsf{d}-1}) \in \mathbb{R}^{\mathsf{d}-1}$, let $(\mathsf{p}, \alpha_U) = \mathcal{U}(\mathsf{p}) = \mathcal{U}_\mathcal{H}(\mathsf{p})$ and $(\mathsf{p}, \alpha_L) = \mathcal{L}(\mathsf{p}) = \mathcal{L}_\mathcal{H}(\mathsf{p})$. Namely, $\mathcal{L}(\mathsf{p})\mathcal{U}(\mathsf{p})$ is a vertical segment of length $|\alpha_U - \alpha_L|$. Their two dual hyperplanes are

$$\text{(23.2)} \quad \begin{aligned} \mathcal{L}(\mathsf{p}) = (\mathsf{p}_1, \ldots, \mathsf{p}_{\mathsf{d}-1}, \alpha_L) &\implies \mathcal{L}(\mathsf{p})^\star \equiv x_\mathsf{d} = \mathsf{p}_1 x_1 + \cdots + \mathsf{p}_{\mathsf{d}-1} x_{\mathsf{d}-1} - \alpha_L, \\ \mathcal{U}(\mathsf{p}) = (\mathsf{p}_1, \ldots, \mathsf{p}_{\mathsf{d}-1}, \alpha_U) &\implies \mathcal{U}(\mathsf{p})^\star \equiv x_\mathsf{d} = \mathsf{p}_1 x_1 + \cdots + \mathsf{p}_{\mathsf{d}-1} x_{\mathsf{d}-1} - \alpha_U. \end{aligned}$$

These two hyperplanes are parallel and their normal is $(\mathsf{p}, -1)$.

(B) By definition, the point $\mathcal{U}(\mathsf{p})$ lies above (resp. below) all the hyperplanes of $\mathcal{H}$. Now, there exists $h \in \mathcal{H}$ such that $\mathcal{U}(\mathsf{p}) \in h$. As such, by Lemma 25.3_{p319} (ii), in the dual, the hyperplane $\mathcal{U}(\mathsf{p})^\star$ passes through the dual point $h^\star$. Furthermore, all the dual points $\mathcal{H}^\star$ lie above (or on) $\mathcal{U}(\mathsf{p})^\star$.

A similar argument for $\mathcal{L}(\mathsf{p})$ implies that all the points of $\mathcal{H}^\star$ lie below (or on) the hyperplane $\mathcal{L}(\mathsf{p})^\star$.

(C) By (B) we have that $C = \mathcal{CH}(\mathcal{H}^\star)$ lies between the two hyperplanes $\mathcal{L}(\mathsf{p})^\star$ and $\mathcal{U}(\mathsf{p})^\star$. Furthermore, both of these parallel hyperplanes support C.

(D) We have $\mathcal{E}_\mathcal{H}(\mathsf{p}) = \mathcal{U}_\mathcal{H}(\mathsf{p}) - \mathcal{L}_\mathcal{H}(\mathsf{p}) = |\alpha_U - \alpha_L|$. Now, the two hyperplanes $\mathcal{L}(\mathsf{p})^\star$ and $\mathcal{U}(\mathsf{p})^\star$ are parallel (see (23.2)), and their vertical distance is the difference between their two constant terms, which is $|\alpha_U - \alpha_L|$, as claimed. ■

Definition 23.16. For a set of (not vertical) hyperplanes $\mathcal{H}$, a subset $\mathcal{S} \subseteq \mathcal{H}$ is an ***ε-coreset*** of $\mathcal{H}$ for the extent if for any $\mathsf{x} \in \mathbb{R}^{\mathsf{d}-1}$ we have $\mathcal{E}_\mathcal{S}(\mathsf{x}) \geq (1 - \varepsilon)\mathcal{E}_\mathcal{H}(\mathsf{x})$.

Similarly, for a point set $\mathsf{P} \subseteq \mathbb{R}^\mathsf{d}$, a set $\mathcal{S} \subseteq \mathsf{P}$ is an ***ε-coreset for vertical extent*** of P if, for any direction $v \in \mathbb{S}^{(\mathsf{d}-1)}$, we have that $\mu_v(\mathcal{S}) \geq (1 - \varepsilon)\mu_v(\mathsf{P})$, where $\mu_v(\mathsf{P})$ is the vertical distance between the two supporting hyperplanes of P which are perpendicular to v.

Claim 23.15 implies that computing a coreset for the extent of a set of hyperplanes is by duality equivalent to computing a coreset for the vertical extent of a point set.

Lemma 23.17. *The set $\mathcal{S}$ is an ε-coreset of the point set $\mathsf{P} \subseteq \mathbb{R}^\mathsf{d}$ for the vertical extent if and only if $\mathcal{S}$ is an ε-coreset for directional width.*

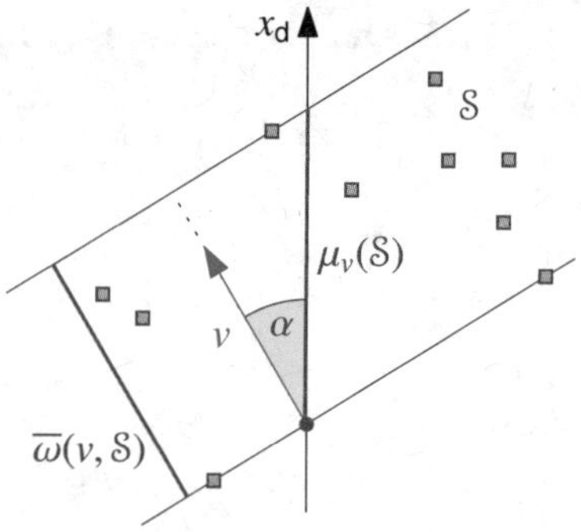

Proof. Consider any direction $v \in \mathbb{S}^{(d-1)}$, and let α be its (smaller) angle with the x_d-axis. Clearly, as demonstrated in the figure on the right, we have $\overline{\omega}(v, \mathcal{S}) = \mu_v(\mathcal{S}) \cos \alpha$. Similarly, we have $\overline{\omega}(v, \mathsf{P}) = \mu_v(\mathsf{P}) \cos \alpha$. Thus, if $\overline{\omega}(v, \mathcal{S}) \geq (1 - \varepsilon)\overline{\omega}(v, \mathsf{P})$, then $\mu_v(\mathcal{S}) \geq (1 - \varepsilon)\mu_v(\mathsf{P})$, and vice versa. ∎

Theorem 23.18. *Let $\mathcal{H}$ be a set of n hyperplanes in $\mathbb{R}^d$. One can compute an ε-coreset for extent of $\mathcal{H}$, of size $O(1/\varepsilon^{d-1})$, in $O(n)$ time. Alternatively, one can compute an ε-coreset of size $O\left(1/\varepsilon^{(d-1)/2}\right)$, in time $O(n + 1/\varepsilon^{3(d-1)/2})$.*

Proof. By Lemma 23.17, the coreset computation is equivalent to computing coreset for directional width. However, this can be done in the stated bounds, by Theorem 23.6 and Theorem 23.13. ∎

Going back to our motivation, we have the following result:

Lemma 23.19. *Let $\mathsf{P}(t)$ be a set of n points with linear motion in $\mathbb{R}^d$. We can compute an axis parallel moving bounding box $b(t)$ for $\mathsf{P}(t)$ that changes $O(d/\sqrt{\varepsilon})$ times (in other times, the bounding box moves with linear motion). The time to compute this bounding box is $O\left(d(n + 1/\varepsilon^{3/2})\right)$.*

Furthermore, for all t, we have that $\mathrm{Box}(\mathsf{P}(t)) \subseteq b(t) \subseteq (1 + \varepsilon)\mathrm{Box}(\mathsf{P}(t))$, *where* $\mathrm{Box}(\mathsf{P}(t))$ *is the minimum axis parallel bounding box of* P.

Proof. We compute the solution for each dimension separately. In each dimension, we compute an $\varepsilon/4$-coreset (for extent) of the resulting set of lines in two dimensions, using Theorem 23.18 with $d = 2$. The resulting coreset has size $O\left(1/\varepsilon^{1/2}\right)$ and it takes $O(n + 1/\varepsilon^{3/2})$ time to compute it. Next, we compute the upper and lower envelopes of the coreset $\mathcal{S}$, which takes $O\left((1/\varepsilon^{1/2}) \log(1/\varepsilon)\right)$ time. At any point in time, for this specific dimension, we have that $\mathcal{E}_{\mathsf{P}}(t) \geq \mathcal{E}_{\mathcal{S}}(t) \geq (1 - \varepsilon/4)\mathcal{E}_{\mathsf{P}}(t)$.

Next, we expand the upper and lower envelopes of the coreset appropriately so that they include the original upper and lower envelopes. Specifically, letting $\mathcal{I}_{\mathcal{S}}(t)$ denote the vertical interval realizing $\mathcal{E}_{\mathcal{S}}(t)$, we have that $\mathcal{I}_{\mathcal{S}}(t) \subseteq \mathcal{I}_{\mathsf{P}}(t) \subseteq (1 + \varepsilon)\mathcal{I}_{\mathcal{S}}(t)$. Here, $c\mathcal{I}$ denotes the scaling of the interval $\mathcal{I}$ by a factor of c around its center. ∎

23.2.4. General definition of coresets. At this point, our discussion exposes a very powerful technique for approximate geometric algorithms: (i) extract a small subset that represents the data well (i.e., a coreset) and (ii) run some other algorithm on the coreset. To this end, we need a more unified definition of coresets.

Definition 23.20 (Coresets). Given a set P of points (or geometric objects) in $\mathbb{R}^d$ and an objective function $f : 2^{\mathbb{R}^d} \to \mathbb{R}$ (for example, $f(\mathsf{P})$ is the width of P), an ***ε-coreset*** is a subset $\mathcal{S}$ of the points of P such that

$$f(\mathcal{S}) \geq (1 - \varepsilon) f(\mathsf{P}).$$

Here $\mathcal{S}$ is an ***ε-coreset of* P *for* $f(\cdot)$**.

If the function $f(\cdot)$ is parameterized (for example, $f(\mathsf{P}, v) = \overline{\omega}(v, \mathsf{P})$), namely $f(\mathcal{S}, v)$, then $\mathcal{S} \subseteq \mathsf{P}$ is a coreset if

$$\forall v \quad f(\mathcal{S}, v) \geq (1 - \varepsilon) f(\mathsf{P}, v).$$

Coresets are of interest when they can be computed quickly and have small size. Hopefully this size is independent of n, the size of the input set P. Interestingly, our current techniques are almost sufficient to show the existence of coresets for a large family of problems.

23.3. Extent of polynomials

Let $\mathcal{F} = \{f_1, \ldots, f_n\}$ be a family of d-variate polynomials and let $u_1, \ldots, u_{\mathsf{d}}$ be the variables over which the functions of $\mathcal{F}$ are defined. Each f_i corresponds to a surface in $\mathbb{R}^{\mathsf{d}+1}$: For example, any d-variate linear function can be considered as a hyperplane in $\mathbb{R}^{\mathsf{d}+1}$ (and vice versa). The upper/lower envelopes and extent for any set of functions $\mathcal{F}$ can therefore be defined similarly to the hyperplane case.

We remind the reader that a ***monomial*** is the "atomic" term that appears in polynomials. Specifically, a monomial is a product of variables (note that a variable can be included several times). For example, xy, x, and $x^{10}y^{20}z^{30}$ are all valid monomials. As another example, the polynomial $10xy + 20x^2y^{10}$ includes the monomials xy and x^2y^{10} (note that we ignore the leading constant when defining the monomial).

Example 23.21. Consider a family of polynomials $\mathcal{F} = \{f_1, \ldots, f_n\}$, where $f_i(x, y) = a_ix^2 + a_iy^2 + b_ix + c_iy + d_i$ and $a_i, b_i, c_i, d_i \in \mathbb{R}$, for $i = 1, \ldots, n$. This family of polynomials defined over $\mathbb{R}^2$ can be linearized to a family of linear functions defined over $\mathbb{R}^4$, by setting $h_i(x, y, z, w) = a_iz + a_iw + b_ix + c_iy + d_i$. Now, let $\mathcal{H} = \{h_1, \ldots, h_n\}$. Clearly, $\mathcal{H}$ is a set of hyperplanes in $\mathbb{R}^5$, and $f_i(x, y) = h_i(x, y, x^2, y^2)$. Thus, for any point $(x, y) \in \mathbb{R}^2$, instead of evaluating $\mathcal{F}$ on (x, y), we can evaluate $\mathcal{H}$ on $\eta(x, y) = (x, y, x^2, y^2)$, where $\eta(x, y)$ is the ***linearization image*** of (x, y). The advantage of this linearization is that $\mathcal{H}$, being a family of linear functions, is now easier to handle than $\mathcal{F}$.

Consider the case that $\mathcal{F}$ is a set of polynomials defined in d dimensions. Each monomial over $u_1, \ldots, u_{\mathsf{d}}$ appearing in $\mathcal{F}$ can be mapped to a distinct variable x_i. Let $x_1, \ldots, x_s$ be the resulting variables. Hence, by substituting each monomial with its corresponding variable, we can map each $f_i(u_1, \ldots, u_{\mathsf{d}}) \in \mathcal{F}$ into a (linear!) function $h_i(x_1, \ldots, x_s)$. As such $\mathcal{F}$ can be linearized into a set $\mathcal{H} = \{h_1, \ldots, h_n\}$ of linear functions over $\mathbb{R}^s$. In particular, $\mathcal{H}$ is a set of n hyperplanes in $\mathbb{R}^{s+1}$. (Note that the surface induced by f_i in $\mathbb{R}^{\mathsf{d}+1}$ corresponds only to a subset of the surface of h_i in $\mathbb{R}^{s+1}$.)

The ***linearization function*** (or ***lifting map***) $\eta : \mathbb{R}^{\mathsf{d}} \to \mathbb{R}^s$ maps a point $u = (u_1, \ldots, u_{\mathsf{d}})$ into the value of the u monomials at this point. As such, for any $\mathsf{p} \in \mathbb{R}^{\mathsf{d}}$, we have $f_i(\mathsf{p}) = h_i(\eta(\mathsf{p}))$.

The ***linearization dimension*** of this process is the target dimension of the lifting map η. Usually, one tries to come up with a mapping with linearization dimension as small as possible.

This technique thus replaces a set of polynomials by a set of linear functions and is called ***linearization***.

Example 23.22. One can linearize the functions in the above example into lower dimension. Observing that all the functions in $\mathcal{F}$ share $x^2 + y^2$ as a common term, we can group them together into a single variable. Indeed, for $f_i(x, y) = a_i(x^2 + y^2) + b_ix + c_iy + d_i$, let $g_i(x, y, z) = a_iz + b_ix + c_iy + d_i$. The resulting set of functions is $\mathcal{H}' = \{g_1, \ldots, g_n\}$. Thus, for any point $(x, y) \in \mathbb{R}^2$, instead of evaluating $\mathcal{F}$ on (x, y), we can evaluate $\mathcal{H}'$ on $\eta'(x, y) = (x, y, x^2 + y^2)$. The advantage of this linearization is that the resulting linearized functions have only three variables instead of four.

In this case, observe that $X = \eta'(\mathbb{R}^2)$ is a *subset* of $\mathbb{R}^3$ (this is the "standard" paraboloid), and we are interested in the value of $\mathcal{H}$ only on points belonging to X. In particular, the set X is not necessarily convex. The set X resulting from the linearization is a semi-algebraic set of constant complexity, and as such, basic manipulation operations of X can be performed in constant time.

Observation 23.23. *Let $\mathcal{F}$ be a set of functions, and let $\mathcal{H}$ be the set of functions resulting from linearizing $\mathcal{F}$. If $\mathcal{H}' \subseteq \mathcal{H}$ is an ε-coreset of $\mathcal{H}$ for the extent, then the corresponding subset in $\mathcal{F}$ is an ε-coreset of $\mathcal{F}$ for the extent.*

The following theorem is a restatement of Theorem 23.18 in this setting.

Theorem 23.24. *Given a family of d-variate polynomials $\mathcal{F} = \{f_1, \ldots, f_n\}$ and a parameter ε, one can compute, in $O(n)$ time, a subset $\mathcal{F}' \subseteq \mathcal{F}$ of $O(1/\varepsilon^s)$ polynomials, such that $\mathcal{F}'$ is an ε-coreset of $\mathcal{F}$ for the extent. Here s is the number of different monomials present in the polynomials of $\mathcal{F}$.*

Alternatively, one can compute an ε-coreset of $\mathcal{F}$ for the extent, of size $O(1/\varepsilon^{s/2})$, in time $O(n + 1/\varepsilon^{3s/2})$.

23.4. Roots of polynomials

We now consider the problem of approximating the extent of a family of square roots of polynomials. Note that this is considerably harder than handling polynomials because square roots of polynomials cannot be directly linearized. It turns out, however, that it is enough to $O(\varepsilon^2)$-approximate the extent of the functions inside the roots and then take the root of the resulting approximation.

Theorem 23.25. *Let $\mathcal{F} = \left\{(f_1)^{1/2}, \ldots, (f_n)^{1/2}\right\}$ be a family of k-variate functions (over $\mathsf{p} = (x_1, \ldots, x_k) \in \mathbb{R}^k$), where each f_i is a polynomial that is non-negative for every $\mathsf{p} \in \mathbb{R}^k$. Let k' be the linearization dimension of $\{f_1, \ldots, f_n\}$. Given any $\varepsilon > 0$, we can compute, in $O(n)$ time, an ε-coreset $\mathcal{G} \subseteq \mathcal{F}$ of size $O(1/\varepsilon^{2k'})$, for the extent.*

Alternatively, one can compute a set $\mathcal{G}' \subseteq \mathcal{F}$, in $O(n+1/\varepsilon^{3k'})$ time, that ε-approximates the extent of $\mathcal{F}$, where $|\mathcal{G}'| = O\left(1/\varepsilon^{k'}\right)$.

PROOF. Let $\mathcal{F}^2$ denote the family $\{f_1, \ldots, f_n\}$. Using the algorithm of Theorem 23.24, we compute a δ'-coreset $\mathcal{G}^2 \subseteq \mathcal{F}^2$ of $\mathcal{F}^2$, where $\delta' = \varepsilon^2/16$. Let $\mathcal{G} \subseteq \mathcal{F}$ denote the corresponding family $\left\{(f_i)^{1/2} \,\middle|\, f_i \in \mathcal{G}^2\right\}$.

Consider any point $\mathsf{x} \in \mathbb{R}^k$. We have that $\mathcal{E}_{\mathcal{G}^2}(\mathsf{x}) \geq (1 - \delta')\mathcal{E}_{\mathcal{F}^2}(\mathsf{x})$, and let $a = \mathcal{L}_{\mathcal{F}^2}(\mathsf{x})$, $A = \mathcal{L}_{\mathcal{G}^2}(\mathsf{x})$, $B = \mathcal{U}_{\mathcal{G}^2}(\mathsf{x})$, and $b = \mathcal{U}_{\mathcal{F}^2}(\mathsf{x})$. Clearly, we have $0 \leq a \leq A \leq B \leq b$ and

$$\mathcal{E}_{\mathcal{G}^2}(\mathsf{x}) = B - A \geq (1 - \delta')(b - a) = (1 - \delta')\mathcal{E}_{\mathcal{F}^2}(\mathsf{x}).$$

Since $(1 + 2\delta')(1 - \delta') \geq 1$, and by the above, we have that

$$(1 + 2\delta')(B - A) \geq (1 + 2\delta')(1 - \delta')(b - a) \geq b - a.$$

By Lemma 23.27 below (for $\delta = 2\delta'$), we have that $\sqrt{A} - \sqrt{a} \leq (\varepsilon/2)U$ and $\sqrt{b} - \sqrt{B} \leq (\varepsilon/2)U$, where $U = \sqrt{B} - \sqrt{A}$. Namely,

$$\sqrt{A} - \sqrt{a} + \sqrt{b} - \sqrt{B} \leq \varepsilon\left(\sqrt{B} - \sqrt{A}\right) \implies \sqrt{b} - \sqrt{a} \leq (1 + \varepsilon)\left(\sqrt{B} - \sqrt{A}\right)$$

$$\implies (1 - \varepsilon)\mathcal{E}_{\mathcal{F}}(\mathsf{x}) = (1 - \varepsilon)\left(\sqrt{b} - \sqrt{a}\right) \leq \left(1 - \varepsilon^2\right)\left(\sqrt{B} - \sqrt{A}\right) \leq \sqrt{B} - \sqrt{A} = \mathcal{E}_{\mathcal{G}}(\mathsf{x}).$$

Thus, $\mathcal{G}$ is an ε-coreset to the extent of $\mathcal{F}$.

The bounds on the size of $\mathcal{G}$ and the running time follow immediately from Theorem 23.24. ■

Corollary 23.26. *Let* $\mathcal{F} = \left\{ (f_1)^{1/2}, \ldots, (f_n)^{1/2} \right\}$ *be a set of functions in* $\mathbb{R}^d$. *Let* $\mathcal{S} = \{g_1, \ldots, g_s\}$ *be an* $\varepsilon^2/16$*-coreset of the extent for* $\mathcal{F}^2 = \{f_1, \ldots, f_n\}$. *Then, the set* $\mathcal{S}^{1/2} = \left\{ (g_1)^{1/2}, \ldots, (g_s)^{1/2} \right\}$ *is an* ε*-coreset of the extent for* $\mathcal{F}$.

Lemma 23.27. *Let* $0 \le a \le A \le B \le b$, *and let* $0 < \varepsilon \le 1$ *be given parameters, so that* $b - a \le (1+\delta)(B - A)$, *where* $\delta = \varepsilon^2/8$. *Then,* $\sqrt{A} - \sqrt{a} \le (\varepsilon/2)U$ *and* $\sqrt{b} - \sqrt{B} \le (\varepsilon/2)U$, *where* $U = \sqrt{B} - \sqrt{A}$.

PROOF. Clearly, for any $x, y \ge 0$ we have that $\sqrt{x+y} \le \sqrt{x} + \sqrt{y}$. Furthermore, $b - a \le (1+\delta)(B - A)$ implies that $b - B + A - a \le \delta(B - A)$. As such, $A - a \le b - B + A - a \le \delta(B - A) \le \delta B \le \delta b$. Thus, we have that

$$\sqrt{A} + \sqrt{B} \le \sqrt{a} + \sqrt{A - a} + \sqrt{b} \le \sqrt{a} + \sqrt{\delta b} + \sqrt{b} \le \left(1 + \sqrt{\delta}\right)\left(\sqrt{a} + \sqrt{b}\right).$$

Namely, $\frac{\sqrt{A}+\sqrt{B}}{1+\sqrt{\delta}} \le \sqrt{a} + \sqrt{b}$. On the other hand,

$$\begin{aligned} \sqrt{b} - \sqrt{a} &= \frac{b - a}{\sqrt{b} + \sqrt{a}} \le \frac{(1+\delta)(B - A)}{\sqrt{b} + \sqrt{a}} \le (1+\delta)\left(1 + \sqrt{\delta}\right)\frac{B - A}{\sqrt{B} + \sqrt{A}} \\ &= \left(1 + \frac{\varepsilon^2}{8}\right)\left(1 + \frac{\varepsilon}{4}\right)\left(\sqrt{B} - \sqrt{A}\right) \le (1 + \varepsilon/2)\left(\sqrt{B} - \sqrt{A}\right). \end{aligned}$$

■

23.4.1. Applications.

23.4.1.1. *Minimum width annulus.* Let $\mathsf{P} = \{\mathsf{p}_1, \ldots, \mathsf{p}_n\}$ be a set of n points in the plane. Let $f_i(\mathsf{q})$ denote the distance of the ith point from the point q. It is easy to verify that $f_i(\mathsf{q}) = \sqrt{\left(x_{\mathsf{q}} - x_{\mathsf{p}_i}\right)^2 + \left(y_{\mathsf{q}} - y_{\mathsf{p}_i}\right)^2}$. Let $\mathcal{F} = \{f_1, \ldots, f_n\}$. It is easy to verify that for a center point $\mathsf{x} \in \mathbb{R}^2$, the width of the minimum width annulus containing P which is centered at x has width $\mathcal{E}_{\mathcal{F}}(\mathsf{x})$. Thus, we would like to compute an ε-coreset for $\mathcal{F}$.

We can apply Theorem 23.25 directly to $\mathcal{F}$, but we can do slightly better by being more careful about the linearization. Specifically, consider the set of functions $\mathcal{F}^2$. Clearly, $f_i^2(x, y) = \left(x - x_{\mathsf{p}_i}\right)^2 + \left(y - y_{\mathsf{p}_i}\right)^2 = x^2 - 2x_{\mathsf{p}_i}x + x_{\mathsf{p}_i}^2 + y^2 - 2y_{\mathsf{p}_i}y + y_{\mathsf{p}_i}^2$. We slightly improve the linearization dimension of this set of polynomials by observing that we can remove terms that appear in all polynomials. Indeed, all the functions of $\mathcal{F}^2$ have the (additive) common factor of $x^2 + y^2$. Since we only care about the vertical extent, we can consider

$$\mathcal{H} = \left\{ -2x_{\mathsf{p}_i}x + x_{\mathsf{p}_i}^2 - 2y_{\mathsf{p}_i}y + y_{\mathsf{p}_i}^2 \;\middle|\; i = 1, \ldots, n \right\}$$

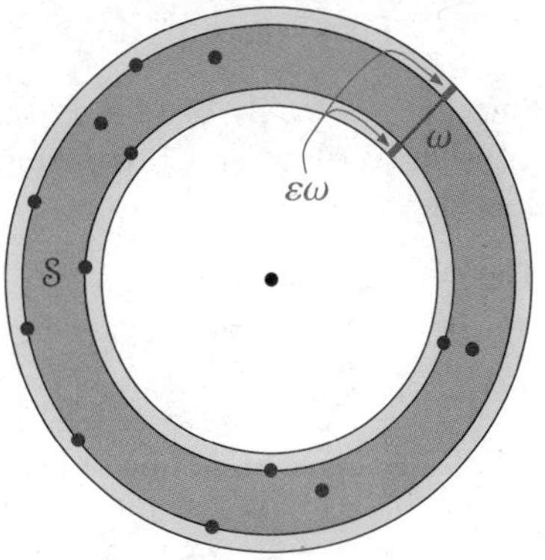

FIGURE 23.3.

which has the same extent as $\mathcal{F}^2$; formally, for any $\mathsf{x} \in \mathbb{R}^2$, we have $\mathcal{E}_{\mathcal{F}^2}(\mathsf{x}) = \mathcal{E}_{\mathcal{H}}(\mathsf{x})$. Now, $\mathcal{H}$ is a family of hyperplanes in $\mathbb{R}^3$, and so it has an $\varepsilon^2/16$-coreset $\mathcal{S}_{\mathcal{H}}$ (for the extent) of size $O\left(1/\varepsilon^2\right)$, which can be computed, in $O\left(n + 1/\varepsilon^6\right)$ time, using Theorem 23.18.

This corresponds to an $\varepsilon^2/16$-coreset $\mathcal{S}_{\mathcal{F}^2}$ of $\mathcal{F}^2$. By Corollary 23.26 (see also Theorem 23.25), this corresponds to an ε-coreset $\mathcal{S}_{\mathcal{F}}$ of $\mathcal{F}$. Finally, this corresponds to a coreset

$\mathcal{S} \subseteq \mathsf{P}$ of size $O\left(1/\varepsilon^2\right)$, such that if we expand the minimum width annulus of $\mathcal{S}$, it will contain all the points of P. Here, the expansion is done by decreasing (resp. increasing) the inner (resp. outer) radius of the ring by $\varepsilon\omega$, where ω is the width of the ring. See Figure 23.3. Thus, we can find the minimum width annulus of $\mathcal{S}$. This can be done in $O\left(|\mathcal{S}|^2\right) = O\left(1/\varepsilon^4\right)$ time using an exact algorithm. Putting everything together, we get the following.

Theorem 23.28. *Let* P *be a set of* n *points in the plane, and let* $0 < \varepsilon < 1$ *be a parameter. One can compute a* $(1+\varepsilon)$*-approximate minimum width annulus of* P *in* $O\left(n + 1/\varepsilon^6\right)$ *time.*

23.4.1.2. *Computing an* ε*-coreset for directional width of moving points.* We now show how this technique can be extended to handle various measures of moving points (width, diameter, etc.). We will demonstrate this for directional width but the basic approach extends to these other cases. Let $\mathsf{P} = \{\mathsf{p}_1, \ldots, \mathsf{p}_n\}$ be a set of n points in $\mathbb{R}^{\mathsf{d}}$, each moving independently. Let $\mathsf{p}_i(t) = (\mathsf{p}_{i,1}(t), \ldots, \mathsf{p}_{i,d}(t))$ denote the position of point p_i at time t. Set $\mathsf{P}(t) = \left\{\mathsf{p}_i(t) \,\middle|\, 1 \le i \le n\right\}$. We remind the reader that if each $\mathsf{p}_{i,j}$ is a polynomial of degree at most r, we say that the motion of P is ***algebraic*** and of ***degree*** r. We call the motion of P ***linear*** if $r = 1$.

Given a parameter $\varepsilon > 0$, we call a subset $\mathcal{S} \subseteq \mathsf{P}$ an ε-coreset of P for directional width if for any direction $u \in \mathbb{S}^{(\mathsf{d}-1)}$, we have

$$(1-\varepsilon)\,\overline{\omega}\left(u, \mathsf{P}(t)\right) \le \overline{\omega}\left(u, \mathcal{S}(t)\right) \qquad \text{for all } t \in \mathbb{R}.$$

First let us assume that the motion of P is linear; i.e., $\mathsf{p}_i(t) = a_i + b_i t$, for $1 \le i \le n$, where $a_i, b_i \in \mathbb{R}^{\mathsf{d}}$. For a direction $u = (u_1, \ldots, u_{\mathsf{d}}) \in \mathbb{S}^{(\mathsf{d}-1)}$, we define a $(\mathsf{d}+1)$-variate polynomial

$$f_i(u,t) = \langle \mathsf{p}_i(t), u\rangle = \langle a_i + b_i t, u\rangle = \sum_{j=1}^{\mathsf{d}} a_{i,j} u_j + \sum_{j=1}^{\mathsf{d}} b_{i,j} \cdot (t u_j).$$

Set $\mathcal{F} = \{f_1, \ldots, f_n\}$. Then

$$\overline{\omega}(u, \mathsf{P}(t)) = \max_i \langle \mathsf{p}_i(t), u\rangle - \min_i \langle \mathsf{p}_i(t), u\rangle = \max_i f_i(u,t) - \min_i f_i(u,t) = \mathcal{E}_{\mathcal{F}}(u,t).$$

Since $\mathcal{F}$ is a family of $(\mathsf{d}+1)$-variate polynomials, which admits a linearization of dimension $2\mathsf{d}$ (there are $2\mathsf{d}$ monomials: $u_1, \ldots, u_{\mathsf{d}}, u_1 t, \ldots, u_{\mathsf{d}} t$), using Theorem 23.24, we conclude the following.

Theorem 23.29. *Given a set* P *of* n *points in* $\mathbb{R}^{\mathsf{d}}$*, each moving linearly, and a parameter* $\varepsilon > 0$*, we can compute an* ε*-coreset of* P *for directional width of size* $O\left(1/\varepsilon^{2\mathsf{d}}\right)$*, in* $O(n)$ *time, or an* ε*-coreset of size* $O\left(1/\varepsilon^{\mathsf{d}}\right)$*, in* $O\left(n + 1/\varepsilon^{3\mathsf{d}}\right)$ *time.*

If the degree of motion of P is $r > 1$, we can write the d-variate polynomial $f_i(u,t)$ as

$$f_i(u,t) = \langle \mathsf{p}_i(t), u\rangle = \left\langle \sum_{j=0}^{r} a_{i,j} t^j, u \right\rangle = \sum_{j=0}^{r} \left\langle a_{i,j} t^j, u \right\rangle$$

where $a_{i,j} \in \mathbb{R}^{\mathsf{d}}$. A straightforward extension of the above argument shows that f_is admit a linearization of dimension $(r+1)\mathsf{d}$. Using Theorem 23.24, we obtain the following.

Theorem 23.30. *Given a set* P *of* n *moving points in* $\mathbb{R}^{\mathsf{d}}$ *whose motion has degree* $r > 1$ *and a parameter* $\varepsilon > 0$*, we can compute an* ε*-coreset for directional width of* P *of size* $O\left(1/\varepsilon^{(r+1)\mathsf{d}}\right)$ *in* $O(n)$ *time or of size* $O\left(1/\varepsilon^{(r+1)\mathsf{d}/2}\right)$ *in* $O\left(n + 1/\varepsilon^{3(r+1)\mathsf{d}/2}\right)$ *time.*

23.4.1.3. *Minimum width cylindrical shell.* Let $\mathsf{P} = \{\mathsf{p}_1, \ldots, \mathsf{p}_n\}$ be a set of n points in $\mathbb{R}^{\mathsf{d}}$, and let $0 < \varepsilon < 1$ be a parameter. Let $w^* = w^*(\mathsf{P})$ denote the width of the thinnest cylindrical shell, the region lying between two co-axial cylinders, containing P. Let $d_\ell(\mathsf{p})$ denote the distance between a point $\mathsf{p} \in \mathbb{R}^{\mathsf{d}}$ and a line $\ell \subset \mathbb{R}^{\mathsf{d}}$. (The distance between p and ℓ is realized by a segment that is orthogonal to the line.) If we fix a line ℓ, then the width of the thinnest cylindrical shell with axis ℓ and containing P is $w(\ell, \mathsf{P}) = \max_{\mathsf{p} \in \mathsf{P}} d_\ell(\mathsf{p}) - \min_{\mathsf{p} \in \mathsf{P}} d_\ell(\mathsf{p})$. A line $\ell \in \mathbb{R}^{\mathsf{d}}$ not parallel to the hyperplane $x_{\mathsf{d}} = 0$ can be represented by a $(2\mathsf{d}-2)$-tuple $(x_1, \ldots, x_{2\mathsf{d}-2}) \in \mathbb{R}^{2\mathsf{d}-2}$. Specifically, the corresponding line in $\mathbb{R}^{\mathsf{d}}$ is the set

$$\ell = \left\{ \mathsf{b} + t\mathsf{c} \;\middle|\; t \in \mathbb{R} \right\},$$

where $\mathsf{b} = (x_1, \ldots, x_{\mathsf{d}-1}, 0)$ is the intersection point of ℓ with the hyperplane $x_{\mathsf{d}} = 0$ and $\mathsf{c} = (x_{\mathsf{d}}, \ldots, x_{2\mathsf{d}-2}, 1)$ is the orientation of ℓ (i.e., c is the intersection point of the hyperplane $x_{\mathsf{d}} = 1$ with the line parallel to ℓ and passing through the origin). (The lines parallel to the hyperplane $x_{\mathsf{d}} = 0$ can be handled separately by a simpler algorithm, so let us assume this case does not happen.) The distance between ℓ and a point p is the same as the distance of the line $\ell' = \ell - \mathsf{p} = \left\{ (\mathsf{b} - \mathsf{p}) + t\mathsf{c} \;\middle|\; t \in \mathbb{R} \right\}$ from the origin. The point y on ℓ' closest to the origin satisfies $y = (\mathsf{b} - \mathsf{p}) + t\mathsf{c}$ for some t, and at the same time $\langle y, \mathsf{c} \rangle = \langle (\mathsf{b} - \mathsf{p}) + t\mathsf{c}, \mathsf{c} \rangle = 0$, as the segment $\mathsf{o}y$ is orthogonal to ℓ', where o denotes the origin. This implies that $t = -\langle (\mathsf{b} - \mathsf{p}), \mathsf{c} \rangle / \|\mathsf{c}\|^2$. Thus,

$$d_\ell(\mathsf{p}) = \|y\| = \|(\mathsf{b} - \mathsf{p}) + t\mathsf{c}\| = \left\| \mathsf{b} - \mathsf{p} - \frac{\langle (\mathsf{b} - \mathsf{p}), \mathsf{c} \rangle}{\|\mathsf{c}\|^2} \mathsf{c} \right\|.$$

For the set of points P, define $\mathcal{F} = \left\{ f_i \;\middle|\; \mathsf{p}_i \in \mathsf{P} \right\}$, where $f_i(\ell) = f_i(\mathsf{b}, \mathsf{c}) = d_\ell(\mathsf{p}_i)$. Then $w^* = \min_{x \in \mathbb{R}^{2d-2}} \mathcal{E}_{\mathcal{F}}(x)$. Let

$$f_i'(\mathsf{b}, \mathsf{c}) = \|\mathsf{c}\|^2 \cdot f_i(\mathsf{b}, \mathsf{c}) = \left\| \|\mathsf{c}\|^2 (\mathsf{b} - \mathsf{p}_i) - \langle \mathsf{b} - \mathsf{p}_i, \mathsf{c} \rangle \mathsf{c} \right\|,$$

and set $\mathcal{F}' = \left\{ f_1', \ldots, f_n' \right\}$.

Define $g_i(\mathsf{b}, \mathsf{c}) = \left(f_i'(\mathsf{b}, \mathsf{c}) \right)^2$, and let $\mathcal{G} = \{g_1, \ldots, g_n\}$. Then g_i is a $(2d-2)$-variate polynomial and has $O(\mathsf{d}^2)$ monomials. Therefore $\mathcal{G}$ admits a linearization of dimension $O(\mathsf{d}^2)$. Using the algorithm of Theorem 23.25, we compute, in $O(n)$ time, an ε-coreset of $\mathcal{F}'$ of size $O\left(1/\varepsilon^{O(\mathsf{d}^2)}\right)$. It is now easy to verify that this corresponds to an ε-coreset (for the extent) for $\mathcal{F}$. Finally, this corresponds to a subset $\mathcal{S} \subseteq \mathsf{P}$, such that $\mathcal{S}$ is a coreset of P for $w(\ell, \mathsf{P})$. Formally, a subset $\mathcal{S} \subseteq \mathsf{P}$ is an ***ε-coreset for cylindrical shell width*** if

$$w(\ell, \mathcal{S}) \geq (1 - \varepsilon) w(\ell, \mathsf{P}), \quad \text{for all lines } \ell.$$

Thus, we can compute in $O(n)$ time a set $\mathcal{S} \subseteq \mathsf{P}$ of $1/\varepsilon^{O(\mathsf{d}^2)}$ points so that for any line ℓ, $w(\ell, \mathsf{P}) \geq w(\ell, \mathcal{S}) \geq (1 - \varepsilon) w(\ell, \mathsf{P})$. Furthermore, by $(1 + \varepsilon)$-expanding this cylindrical shell (both inward and outward) of ℓ computed for the coreset $\mathcal{S}$, we get a cylindrical shell that covers P of width at most $(1 + \varepsilon) w(\ell, \mathsf{P})$. Hence, we conclude with the following.

Theorem 23.31. *Given a set* P *of n points in $\mathbb{R}^{\mathsf{d}}$ and a parameter $\varepsilon > 0$, we can compute in $O(n + 1/\varepsilon^{O(\mathsf{d}^2)})$ time a subset $\mathcal{S} \subseteq \mathsf{P}$ of size $O(1/\varepsilon^{O(\mathsf{d}^2)})$ so that for any line ℓ in $\mathbb{R}^{\mathsf{d}}$, we have $w(\ell, \mathcal{S}) \geq (1 - \varepsilon) w(\ell, \mathsf{P})$.*

Note that Theorem 23.31 does not compute the optimal cylinder; it just computes a small coreset for this problem (i.e., we still need to determine the optimal axis line ℓ). The brute force algorithm (for this problem) used on the computed coreset would take $O\left(n + 1/\varepsilon^{O(\mathsf{d}^4)}\right)$ time to compute the optimal cylinder for the coreset. By expanding this cylindrical shell, we get the desired approximate solution. (The running time of this algorithm can be further improved by being a bit more careful about the details.)

23.5. Bibliographical notes

Section 23.1 follows (roughly) the work of Agarwal et al. [**AHV04**]. The result of Section 23.1.2.2 was observed independently by Chan [**Cha06**] and Yu et al. [**YAPV04**]. It is a simplification of an algorithm of Agarwal et al. [**AHV04**], which in turn is an adaptation of a method of Dudley [**Dud74**].

The running time of Theorem 23.13 can be improved to $O(n + 1/\varepsilon^{d-1})$ by using a special nearest neighbor algorithm on a grid. Trying to use some of the other ANN data-structures will not work since it would not improve the running time over the naive algorithm. The interested reader can see the paper by Chan [**Cha06**].

Linearization is widely used in fields such as machine learning [**CS00**] and computational geometry [**AM94**]. It was probably known since the dawn of modern mathematics (i.e., it is similar to a change of variables when doing integration).

There is a general technique for finding the best possible linearization (i.e., a mapping η with the target dimension as small as possible); see [**AM94**] for details.

Section 23.4.1.2 is from Agarwal et al. [**AHV04**], and the results can be (very slightly) improved by treating the direction as a $(d-1)$-dimensional entity; see [**AHV04**] for details.

Theorem 23.31 is also from [**AHV04**]. The improved running time to compute the approximate cylinder mentioned in the text follows by a more involved algorithm, which together with the construction of the coreset also computes a compact representation of the extent of the coreset. The technical details are not trivial. In particular, the resulting running time for computing the approximate cylindrical shell is $O\left(n + 1/\varepsilon^{O(d^2)}\right)$. See [**AHV04**] for more details.

Coresets are quite useful; see the survey [**AHV05**] for more information.

23.6. Exercises

Exercise 23.1 (Sketch and merge for coresets)**.** Prove Lemma 23.3.

Exercise 23.2 (Lower bound on coreset size)**.** Prove that in the worst case, an ε-coreset for directional width has to be of size $\Omega(\varepsilon^{-(d-1)/2})$.

CHAPTER 24

Approximation Using Shell Sets

In this chapter, we define a concept (shell set) that is slightly weaker than coreset and show how a reweighting algorithm can be use a shell set to derive a fast approximation algorithm. In many cases, coresets are much smaller, in practice, than what the worst case analysis implies. As coresets are also shell sets, approximation algorithms using shell sets work quite well as they can use these considerably smaller sets to compute their approximation.

24.1. Covering problems, expansion, and shell sets

Consider a set P of n points in $\mathbb{R}^d$ that we are interested in covering by the "best" shape in a family of shapes $\mathcal{F}$. For example, $\mathcal{F}$ might be the set of all balls in $\mathbb{R}^d$, and we are looking for the minimum enclosing ball of P. An ε-coreset $\mathcal{S} \subseteq \mathsf{P}$ would guarantee that *any* ball that covers $\mathcal{S}$ will cover the whole point set if we expand it by $1 + \varepsilon$.

However, sometimes, computing the coreset is computationally expensive, the coreset does not exist at all, or its size is prohibitively large. It is still natural to look for a small subset $\mathcal{V}$ of the points, such that finding the optimal solution for $\mathcal{V}$ generates (after appropriate expansion) an approximate solution to the original problem.

Alternatively, consider a function we are trying to optimize. As a concrete example, consider the minimum radius of an enclosing ball of a point set. This function $f_{\mathsf{P}}(\overline{c})$ is defined on the given point set P and is parameterized by the center $\overline{c}$ of the ball, and we are looking for the minimum of this function (i.e., the minimum radius enclosing ball). Since this problem has a coreset, there exists a small subset $\mathcal{S} \subseteq \mathsf{P}$ such that, for all $\overline{c}$, we have that $f_{\mathsf{P}}(\overline{c}) \approx f_{\mathcal{S}}(\overline{c})$. But this is stronger than what we really need. Indeed, it is sufficient for us if there is a small subset $X \subseteq \mathsf{P}$, such that the point $\overline{c}_X$ realizing the minimum of $f_X(\cdot)$ is close to the global minimum of $f_{\mathsf{P}}(\cdot)$ (i.e., $f_{\mathsf{P}}(\overline{c}_X)$ is a $(1 + \varepsilon)$-approximation to the global minimum of $f_{\mathsf{P}}(\cdot)$). Having at hand such a subset X is already sufficient for one to be able to approximate the minimum of $f_{\mathsf{P}}(\cdot)$ efficiently. The notion of *shell set* tries to capture this.

Example 24.1. Consider the problem of computing the minimum enclosing ball for a set P of n points in $\mathbb{R}^d$. It is known that there exists a subset $\mathcal{S} \subseteq \mathsf{P}$ of size $O(1/\varepsilon)$, such that if $\mathbf{g}$ is the minimum radius ball covering $\mathcal{S}$ and if we expand it to $(1 + \varepsilon)\mathbf{g}$, then it covers the whole point set. Note that the bound here does not depend on the dimension d at all! We delegate proving a somewhat weaker bound to Exercise 24.1.

Note that any coreset for this problem has to be of size $\Omega\left(1/\varepsilon^{(d-1)/2}\right)$ as can be easily verified. As such, there is a small set of points that captures the optimal solution, which is considerably smaller than the corresponding coreset size.

Underlying our discussion here is the "continuity" of our target function (i.e., minimum radius of a ball). The ***expansion*** operation we use to get from a cover of the subset to a cover of the whole point set increases the cost of the cover only slightly.

24.1.1. Settings. So, let's assume we are given a "slow" procedure **algSlow** that can solve the covering/clustering problem we are interested in. The basic idea would be to run the algorithm on a very small subset and use this solution for the whole point set. To this end, we assume the following:

(P1) There is a cost function $f : \mathcal{F} \to \mathbb{R}$ that scores each shape. Let

$$f_{\mathrm{opt}}(\mathsf{P}) = \min_{\mathbf{r}\in\mathcal{F}, \mathsf{P}\subseteq\mathbf{r}} f(\mathbf{r})$$

be the minimum cost of a shape in $\mathcal{F}$ that covers P.

Note that we have monotonicity; that is, for any $\mathsf{Q} \subseteq \mathsf{P}$ we have that $f_{\mathrm{opt}}(\mathsf{Q}) \geq f_{\mathrm{opt}}(\mathsf{P})$.

(P2) For any subset $\mathsf{Q} \subseteq \mathsf{P}$, **algSlow**$(\mathsf{Q})$ returns a shape $\mathbf{r}$ that covers Q and has cost $f_{\mathrm{opt}}(\mathsf{Q})$. The running time of **algSlow**(Q) is $T_{\mathbf{algSlow}}(|\mathsf{Q}|)$. (For our purposes it will be sufficient if $f(\mathbf{r}) \leq (1+\varepsilon) f_{\mathrm{opt}}(\mathsf{Q})$.)

(P3) There is a natural ***expansion*** operation defined over $\mathcal{F}$. Namely, given a set $\mathbf{r} \in \mathcal{F}$, one can compute a set $(1+\varepsilon)\mathbf{r}$ which is the expansion of $\mathbf{r}$ by a factor of $1+\varepsilon$. In particular, we require that $f\big((1+\varepsilon)\mathbf{r}\big) \leq (1+\varepsilon) f(\mathbf{r})$.

(P4) The range space $(\mathsf{P}, \mathcal{F})$ has finite VC dimension, say, δ.

Definition 24.2. A subset $\mathcal{V} \subseteq \mathsf{P}$ is an ε-***shell set*** for P if for any subset Q such that $\mathcal{V} \subseteq \mathsf{Q} \subseteq \mathsf{P}$ and $\mathbf{r} =$ **algSlow**(Q), we have that $(1+\varepsilon)\mathbf{r}$ covers P.

It is beneficial to see the corresponding definition of a coreset in this setting.

Definition 24.3. A subset $\mathcal{S} \subseteq \mathsf{P}$ is an ε-***coreset*** for P if for any shape $\mathbf{r} \in \mathcal{F}$ that covers $\mathcal{S}$, we have that $(1+\varepsilon)\mathbf{r}$ covers P.

Namely, a coreset is universal – any range that covers the coreset will cover P after expansion. On the other hand, the shell set guarantees this only for ranges that are computed in a specific way by using **algSlow**.

Lemma 24.4. *Let $\mathcal{V} \subseteq \mathsf{P}$ be an ε-shell set, and consider any subset Q, such that $\mathcal{V} \subseteq \mathsf{Q} \subseteq \mathsf{P}$. Then, for $\mathbf{r} =$ **algSlow**(Q), we have that $(1+\varepsilon)\mathbf{r}$ is a $(1+\varepsilon)$-approximation to the optimal range covering P.*

Proof. Indeed, by the definition of shell set we have that $(1+\varepsilon)\mathbf{r}$ covers P. Now, by the above assumptions, we have $f_{\mathrm{opt}}(\mathsf{P}) \leq f\big((1+\varepsilon)\mathbf{r}\big) \leq (1+\varepsilon) f(\mathbf{r}) = (1+\varepsilon) f_{\mathrm{opt}}(\mathsf{Q}) \leq (1+\varepsilon) f_{\mathrm{opt}}(\mathsf{P})$. ■

The implies that ε-shell sets are considerably more restricted and weaker than coresets. Of course, an ε-coreset is automatically an ε-shell set. Also, if a problem has a shell set, then to approximate it efficiently, all we need to do is to find some set, hopefully small, that contains the shell set.

Remark 24.5. As demonstrated by Example 24.1, it might be that there is an ε-shell set for a problem, but a coreset for the same set of points might need to be so large that it has to be the whole point set. Indeed, an ε-coreset for the minimum enclosing ball is, in the worst case, of size $\Omega\big(1/\varepsilon^{(\mathsf{d}-1)/2}\big)$, and this bound gets prohibitively large if the dimension is large (e.g., $\mathsf{d} > \log n$).

24.1.2. The approximation algorithm. Here, we show how one can approximate $f_{\mathrm{opt}}(\mathsf{P})$ efficiently using shell sets. Note that we are not computing a shell set of P. Rather, we use the existence of a shell set to argue that this algorithm works.

We have two approximation parameters $\varepsilon > 0$ and $\tau > 0$, where the value of $\varepsilon > 0$ is specified in the input and the value of τ would be determined by analysis of the algorithm.

The algorithm is essentially the same as the one used for the geometric set cover (see Theorem 6.14_{p96}). Indeed, assume that there exists an ε-shell set $\mathcal{V}$ of size k for the given instance. The given instance is specified by a set P of n points in $\mathbb{R}^d$.

Initially, the points are all assigned weight one. At the ith iteration, the algorithm picks a τ-net R_i of $\mathsf{S} = (\mathsf{P}, \mathcal{F})$ of size $O(\delta/\tau \log(1/\tau))$ using random sampling (i.e., using Theorem 5.28_{p71}), for τ to be specified shortly. Note that the sampling is done according to the current weights of the points. Next, we compute the optimal range $\mathbf{r}_i$ that covers R_i by calling **algSlow**(R_i). Now, using the expansion operation, we compute the expanded range $(1 + \varepsilon)\mathbf{r}_i$.

Next, we compute the set of points U_i that fall outside the expanded range $(1 + \varepsilon)\mathbf{r}_i$; namely, $\mathsf{U}_i = \mathsf{P} \setminus (1 + \varepsilon)\mathbf{r}_i$. If U_i is empty, then we are done as we found a cover to P. Otherwise, if the total weight of U_i is smaller than $\tau\omega_{i-1}(\mathsf{P})$, then we double the weight of all the points of U_i and continue to the next iteration, where $\omega_{i-1}(\mathsf{P})$ is the total weight of the points P in the end of the $(i - 1)$st iteration.

24.1.2.1. *Analysis.* If the algorithm terminates in the ith iteration, then $(1+\varepsilon)\mathbf{r}_i$ covers P and its price is

$$f((1 + \varepsilon)\mathbf{r}_i) \leq (1 + \varepsilon)f(\mathbf{r}_i) = (1 + \varepsilon)f_{\mathrm{opt}}(\mathsf{U}_i) \leq (1 + \varepsilon)f_{\mathrm{opt}}(\mathsf{P}).$$

Namely, the solution returned is the required approximation.

Since S has VC dimension δ and, as such, so does the complement range space, it follows that with constant probability (say $> 1/2$) the sample R_i is a τ-net. Now, if R_i is a τ-net, then the total weight of P outside the range $\mathbf{r}_i$ is at most $\tau\omega_{i-1}(\mathsf{P})$. Indeed, if the weight of U_i exceeds this bound, this implies that R fails to stub the (τ-heavy) range $\mathbb{R}^d \setminus (1 + \varepsilon)\mathbf{r}$.

Thus, with constant probability, the ith iteration succeeds and doubles the weight of the points of U_i. Now, observe that the total weight of the point set P in the end of the ith *successful* iteration is $(1 + \tau)^i n$. On the other hand, if U_i is not empty, then it must contain at least one point of the ε-shell set $\mathcal{V}$. As such, the lower bound on the weight of the points of the shell set in the end of the ith iteration is $k2^{i/k}$, where $k = |\mathcal{V}|$. Thus, we have that

$$k2^{i/k} \leq (1 + \tau)^i n.$$

In particular, setting $\tau = 1/4k$, the right side of this equation grows more slowly than the left side, and it no longer holds for $i = \Omega(k \log n)$ as can be easily verified. As such, this algorithm performs $m = O(k \log n)$ successful iterations and $O(k \log n)$ iterations overall, as each iteration has at least probability $1/2$ to succeed. (Specifically, by Chernoff's inequality, with high probability, the algorithm performs $O(\log n)$ iterations till it performs m successful iterations and stops.) This implies the following lemma.

Lemma 24.6. *The algorithm described above computes a subset* $\mathsf{Q} \subseteq \mathsf{P}$ *of size* $O(k\delta \log k)$*, where* k *is a prespecified upper bound on the size of the optimal* ε*-shell set. We have that* $(1 + \varepsilon)\mathbf{r}$ *covers* P*, where* $\mathbf{r}$ *is the smallest range in* $\mathcal{F}$ *covering* Q*. With high probability, the algorithm performs* $O(k \ln n)$ *iterations.*

Theorem 24.7. *Under the settings of Section 24.1.1, one can compute a subset* $\mathsf{Q} \subseteq \mathsf{P}$ *of size* $O(k\delta \log(k))$*, where* k *is the size of the smallest* ε*-shell set of the given instance. The range* $(1 + \varepsilon)\mathbf{r}$ *covers* P*, where* $\mathbf{r}$ *is the smallest range covering* Q*. The running time of the resulting algorithm is*

$$O\Big(\big(n + T_{\mathbf{algSlow}}(k\delta \log k)\big)\, k \ln n\Big),$$

with high probability. In particular, the cost of $(1+\varepsilon)\mathbf{r}$ *is a* $(1+\varepsilon)$*-approximation to* $f_{\mathrm{opt}}(\mathsf{P})$.

PROOF. The algorithm is described above. The bound on the running time follows from the bound on the number of iterations from Lemma 24.6. The only minor technicality is that k is not known. However, we can perform an exponential search for the right value of k, starting with $k = 2$, and doubling it till successful. Clearly, the running time is dominated by the running time of the last iteration, which implies the result. ■

Note that Theorem 24.7 does compute a shell set (i.e., $\mathsf{U}_{\text{final iteration}}$). However we usually do not care about shell sets on their own, as they are just a tool to approximate the minimum cost solution.

Remark 24.8. As Remark 24.5 pointed out, a coreset, even if it exists, might be prohibitively large. Naturally, such a coreset however guarantees the existence of a shell set. It is quite believable in many cases that there is a shell set that is significantly smaller than the corresponding coreset. As such, one can use Theorem 24.7 even without having guaranteed bounds on its running time. Namely, this algorithm is a natural candidate for an approach that would work better in practice then an algorithm that directly constructs a coreset.

Remark 24.9. The algorithm in Theorem 24.7 can be modified to compute a small coreset. The idea here is at each iteration to consider the subset R_i and to consider all the ranges that cover it. If one can find a range $\mathbf{r} \in \mathcal{F}$ such that

(1) $\mathbf{r}$ covers P_i,
(2) $(1+\varepsilon)\mathbf{r}$ does not cover all the points of P,

then, one can use the same reweighting approach. Clearly, the same analysis as above implies that the resulting set Q is an ε-coreset.

However, finding a violating range at each iteration is a computationally much more challenging task than just checking the optimal one for R_i. As such, this algorithm for computing a small coreset, while theoretically interesting, is probably not practical.

24.2. Covering by cylinders

Here, we provide an example of an approximation algorithm that deploys the algorithm presented in the previous section. The problem is covering a set of n points in $\mathbb{R}^{\mathsf{d}}$ by a set of k cylinders that minimizes the maximum radius of the cylinders. In the following, we prove the existence of a small coreset for this problem. There is currently no efficient construction algorithm for this coreset. By using the above shell set approximation algorithm, one can compute the required approximation.

Before diving into this example, we need to better understand the notion of expansion when a single range is made out of several clusters.

24.2.1. Clustering and coresets.

We would like to cover a set P of n points $\mathbb{R}^{\mathsf{d}}$ by k balls, such that the radius of the maximum radius ball is minimized. We remind the reader that this is known as the ***k-center clustering*** problem (or just *k-center*). The ***price function***, in this case, $\mathrm{rd}_k(\mathsf{P})$, is the radius of the maximum radius ball in the optimal solution.

Definition 24.10. Let P be a point set in $\mathbb{R}^{\mathsf{d}}$, and let $1/2 > \varepsilon > 0$ be a parameter.

For a cluster c, let $c(\delta)$ denote the cluster resulting from expanding c by δ. Thus, if c is a ball of radius r, then $c(\delta)$ is a ball of radius $r+\delta$. For a set C of clusters, let

$$C(\delta) = \left\{ c(\delta) \,\middle|\, c \in C \right\}$$

be the ***additive expansion operator***; that is, $C(\delta)$ is a set of clusters resulting from expanding each cluster of C by δ.

Similarly,

$$(1+\varepsilon)C = \left\{ (1+\varepsilon)c \;\middle|\; c \in C \right\}$$

is the ***multiplicative expansion operator***, where $(1+\varepsilon)c$ is the cluster resulting from expanding c by a factor of $(1+\varepsilon)$. Namely, if C is a set of balls, then $(1+\varepsilon)C$ is a set of balls, where a ball $c \in C$ corresponds to a ball of radius $(1+\varepsilon)\,\mathrm{radius}(c)$ in $(1+\varepsilon)C$.

Definition 24.11. A set $\mathcal{S} \subseteq \mathsf{P}$ is an ***additive ε-coreset*** of P, in relation to a price function radius, if for any clustering C of $\mathcal{S}$, we have that P is covered by $C(\varepsilon\,\mathrm{radius}(C))$, where $\mathrm{radius}(C) = \max_{c \in C} \mathrm{radius}(c)$. Namely, we expand every cluster in the clustering by an ε-fraction of the size of the *largest* cluster in the clustering. Thus, if C is a set of k balls, then $C(\varepsilon\,\mathrm{radius}(C))$ is just the set of balls resulting from expanding each ball by εr, where r is the radius of the largest ball.

A set $\mathcal{S} \subseteq \mathsf{P}$ is a ***multiplicative ε-coreset*** of P if for any clustering C of $\mathcal{S}$, we have that P is covered by $(1+\varepsilon)C$.

It is not a priori clear why a multiplicative coreset exists, since such a coreset needs to handle cases where, for example, one of the clusters might be tiny compared to the largest cluster and this tiny cluster expands only slightly by this multiplicative expansion. We start by proving the existence of an additive coreset.

Lemma 24.12. *Let P be a set of n points in $\mathbb{R}^{\mathsf{d}}$, and let $\varepsilon > 0$ be a parameter. There exists an additive ε-coreset for the k-center problem, and this coreset has $O\left(k/\varepsilon^{\mathsf{d}}\right)$ points.*

Proof. Let C denote the optimal clustering of P. Cover each ball of C by a grid of sidelength $\varepsilon r_{\mathrm{opt}}/2\mathsf{d}$, where r_{opt} is the radius of the optimal k-center clustering of P. From each such grid cell, pick one points of P. Clearly, the resulting point set $\mathcal{S}$ is of size $O(k/\varepsilon^{\mathsf{d}})$ and it is an additive ε-coreset of P, as one can easily verify. ■

Lemma 24.13. *Let P be a set of n points in $\mathbb{R}^{\mathsf{d}}$, and let $\varepsilon > 0$ be a parameter. There exists a multiplicative ε-coreset for the k-center problem, and this coreset has $O\left(k!/\varepsilon^{dk}\right)$ points.*

Proof. For $k = 1$, the additive coreset of P is also a multiplicative coreset, and it is of size $O(1/\varepsilon^{\mathsf{d}})$.

As in the proof of Lemma 24.12, we cover the point set by a grid of radius $\varepsilon r_{\mathrm{opt}}/(5\mathsf{d})$ and let X be the set of cells (i.e., cubes) of this grid which contain points of P. Clearly, $|X| = O\left(k/\varepsilon^{\mathsf{d}}\right)$.

Let $\mathcal{S}$ be an additive $(\varepsilon/4)$-coreset of P. For each cell of X, we inductively compute an ε-multiplicative coreset of $\mathsf{P} \cap \square$, for $k-1$ balls. Let $Q_{\square,k-1}$ be this set, and let $\mathcal{Q} = \mathcal{S} \cup \bigcup_{\square \in X} Q_{\square,k-1}$.

We claim that $\mathcal{Q}$ is the required coreset.

The bound on the size of $\mathcal{Q}$ is easy as $|\mathcal{Q}| = T(k,\varepsilon) = O\left(k/\varepsilon^{d}\right) T(k-1,\varepsilon) + O\left(k/\varepsilon^{\mathsf{d}}\right) = O\left(k!/\varepsilon^{dk}\right)$.

As for the covering property, let C be any k-center clustering of $\mathcal{S}$, and let $\square$ be any cell of X. If $\square$ intersects all the k balls of C, then one of them must be of radius at least $(1-\varepsilon/2)\mathrm{rd}(\mathsf{P},k)$ since the additive coreset $\mathcal{S}$ is a subset of $\mathcal{Q}$. Let c be this ball. Clearly, when we expand c by a factor of $(1+\varepsilon)$, it would completely cover $\square$, and as such it would also cover all the points of $\mathsf{P} \cap \square$.

Thus, we can assume that $\square$ intersects at most $k-1$ balls of C. But then, by construction (and induction), $\mathcal{Q}$ contains a multiplicative coreset for $k-1$ balls for the point set $\mathsf{P} \cap \square$. As such, the multiplicative expansion of the balls of C will cover all the points of $\mathsf{P} \cap \square$. ■

24.2.2. Covering by union of cylinders. Let's assume that we want to cover P by k cylinders of such that the maximum radius is minimized (i.e., fit the points to k lines). Formally, consider $\mathcal{G}$ to be the set of all cylinders in $\mathbb{R}^{\mathsf{d}}$, and let

$$\mathcal{F}^k = \left\{ c_1 \cup c_2 \cup \ldots \cup c_k \;\middle|\; c_1, \ldots, c_k \in \mathcal{F} \right\}$$

be the set where each of its members is a union of k cylinders. For $C \in \mathcal{F}^k$, let

$$f(C) = \max_{c \in C} \operatorname{radius}(c).$$

For a point set $\mathsf{P} \subseteq \mathbb{R}^{\mathsf{d}}$, let $f_{\mathrm{opt}}(\mathsf{P}) = \min_{C \in \mathcal{F}, \mathsf{P} \subseteq C} f(C)$.

24.2.2.1. *Covering by cylinders – a slow algorithm.* Given a set $\mathsf{P} \subseteq \mathbb{R}^{\mathsf{d}}$ of n points, consider its minimum radius enclosing cylinder c. The cylinder c has (at most) $2\mathsf{d} - 1$ points of P on its boundary for which if we compute their minimum enclosing cylinder, it is c. Note that c might contain even more points on its boundary; we are only claiming that there is a defining subset of size $2\mathsf{d} - 1$. This is one of those "easy to see" but very tedious to verify facts. Let us quickly outline an intuitive explanation (but not a proof!) of this. Consider the set of lines $\mathcal{L}^{\mathsf{d}}$ of lines in $\mathbb{R}^{\mathsf{d}}$. Every member of $\ell \in \mathcal{L}^{\mathsf{d}}$ can be parameterized by the closest point p on ℓ to the origin. Consider the hyperplane h that passes through p and is orthogonal to $\mathsf{o}p$, where o is the origin. The line ℓ now can be parameterized by its orientation in h. This requires specifying a point on the $(\mathsf{d} - 2)$-dimensional unit hypersphere $\mathbb{S}^{(\mathsf{d}-2)}$. Thus, we can specify ℓ using $2\mathsf{d}-2$ real numbers. Next, define for each point $\mathsf{p}_i \in \mathsf{P}$ its distance $g_i(\ell)$ from $\ell \in \mathcal{L}^{\mathsf{d}}$. This is a messy but a nice algebraic function defined over $2\mathsf{d}-2$ variables. In particular, $g_i(\ell)$ induces a surface in $2\mathsf{d}-1$ dimensions (i.e., $\bigcup_{\ell}(\ell, g_i(\ell))$). Consider the arrangement $\mathcal{A}$ of those surfaces. Clearly, the minimum volume cylinder lies on a feature of this arrangement. That is, the point of minimum height on the upper envelope of these surfaces corresponds to the desired cylinder.

Thus, to specify the minimum radius cylinder, one needs to specify the feature (i.e., vertex, edge, etc.) of the arrangement in the parametric space that contains this point.

However, every feature in an arrangement of well-behaved surfaces in $2\mathsf{d} - 1$ dimensions, can be specified by $2\mathsf{d} - 1$ surfaces. This is intuitively clear but requires a proof – an intersection of k surfaces is going to be $(t-k)$-dimensional, where t is the dimension of the surfaces. If we add a surface to the intersection and it does not reduce the dimension of the intersection, we can reject it and take the next surface passing through the feature we care about. Clearly, after picking $2\mathsf{d} - 1$ surfaces, the intersection would be zero-dimensional (i.e., a single point).

Since every such surface is induced by some point of P, we have that if we want to specify a minimum radius cylinder induced by a subset of P, all we need to specify are the $2\mathsf{d} - 1$ points that correspond to the feature of the arrangement that contains the minimum cylinder in the parametric space. To specify k such cylinders, we need to specify $M = (2\mathsf{d} - 1)k$ points. This immediately implies that we can find the optimal cover of P by k cylinders in $O\left(n^{(2\mathsf{d}-1)k+1}\right)$ time, by enumerating all such subsets of M points and computing for each subset its optimal cover (note that the O notation hides a constant that depends on k and d).

Thus, we have a slow algorithm that can compute the optimal cover of P by k cylinders.

24.2.3. Existence of a small coreset. Since the coreset in this case is either multiplicative or additive, it is first important to define the expansion operation carefully. In particular, if C is a set of k cylinders, the $(1 + \varepsilon)$-expanded set of cylinders would be $C(\varepsilon\, \mathrm{radius}(C))$, where $\mathrm{radius}(C)$ is the radius of the largest cylinder in C.

Let P be the given set of n points in $\mathbb{R}^{\mathsf{d}}$. Let C_{opt} be the optimal cover of P by k cylinders. For each cylinder of C_{opt} place $O(1/\varepsilon^{\mathsf{d}-1})$ parallel lines inside it, so that for any point inside the union of the cylinders, there is a line in this family within distance $\leq (\varepsilon/10)r_{\mathrm{opt}}$ from it. Let L denote this set of lines.

Let Q be the point set resulting from snapping each point of P to its closest point on L. We claim that Q is an $(\varepsilon/10)$-coreset for P, as can be easily verified. Indeed, if a set C of k cylinders cover Q, then the largest cylinder must be of radius $r \geq (1 - \varepsilon/10)\mathrm{rd}(\mathsf{P}, k)$, where $\mathrm{rd}(\mathsf{P}, k)$ is the radius of the optimal cover of P by k cylinders. Otherwise, $\mathrm{rd}(\mathsf{P}, k) \leq r + (\varepsilon/10)\mathrm{rd}(\mathsf{P}, k) < \mathrm{rd}(\mathsf{P}, k)$.

The set Q lies on $O(1/\varepsilon^{\mathsf{d}-1})$ lines. Let $\ell \in \mathsf{L}$ be such a line, and consider the point set Q_ℓ. Assume for a second that there was a multiplicative ε-coreset $\mathcal{T}_\ell$ on this line. If $\mathcal{T}_\ell$ is covered by k cylinders, each cylinder intersects ℓ along an interval. Expanding each such cylinder by a factor of $1 + \varepsilon$ is equivalent to expanding each such intersecting interval by a factor of (at least) $1+\varepsilon$ (note that the interval might expand by much more than just $1+\varepsilon$, but this works in our favor, so think about the expansion as being exactly $1 + \varepsilon$). However, by Lemma 24.13, we know that such a multiplicative $(\varepsilon/10)$-coreset exists, of size $O(k!/\varepsilon^k)$. Thus, let $\mathcal{T}_\ell$ be the multiplicative $(\varepsilon/10)$-coreset for Q_ℓ for k intervals on the line. Let $\mathcal{T} = \bigcup_{\ell \in \mathsf{L}} \mathcal{T}_\ell$. We claim that $\mathcal{T}$ is an (additive) $(\varepsilon/10)$-coreset for Q. This is trivial, since being a multiplicative coreset for each line implies that the union is a multiplicative coreset and a δ-multiplicative coreset is also a δ-additive coreset. Thus, $\mathcal{T}$ is a $((1 + \varepsilon/10)^2 - 1)$-coreset for P. The only problem is that the points of $\mathcal{T}$ are not points in P. However, they correspond to points in P which are within distance at most $(\varepsilon/10)\mathrm{rd}(\mathsf{P}, k)$ from them. Let $\mathcal{S}$ be the corresponding set of points of P. It is now easy to verify that $\mathcal{S}$ is indeed an ε-coreset for P, since $((1 + \varepsilon/10)^2 - 1) + \varepsilon/10 \leq \varepsilon$. We summarize:

Lemma 24.14. *Let* P *be a set of* n *points in* $\mathbb{R}^{\mathsf{d}}$. *There exists an (additive)* ε*-coreset for* P *of size* $O\left(k \cdot k!/\varepsilon^{\mathsf{d}-1+k}\right)$ *for covering* P *by* k *cylinders of minimum (maximum) radius.*

Plugging this into Theorem 24.7 results in the following.

Theorem 24.15. *Let* P *be a set of* n *points in* $\mathbb{R}^{\mathsf{d}}$, *and let* k *and* $\varepsilon > 0$ *be parameters. One can compute a cover of* P *by* k *cylinders of radius* $(1 + \varepsilon)r_{\mathrm{opt}}$, *where* r_{opt} *is the minimum radius required to cover* P *by* k *cylinders of the same radius. The running time of this algorithm is* $O(Kn \log n)$, *where* K *is a constant that depends only on* k, $\varepsilon > 0$, *and* d.

24.3. Bibliographical notes

The observation that the reweighting technique can be used to speed up approximation algorithms by using a shell set argument is due to Agarwal et al. [**APV02**]. The discussion of shell sets is implicit in the work of Bădoiu et al. [**BHI02**]. Lemma 24.13 is based on work in [**APV02**]. Our discussion of multiplicative and additive coresets follows [**Har04**].

24.4. Exercises

Exercise 24.1 (Bounding ball made easy)**.** Let P be a set of n points in $\mathbb{R}^{\mathsf{d}}$. Assume you are given a slow algorithm that can in $O\left(\mathsf{d}m^{O(1)}\right)$ time compute the minimum enclosing ball of any subset of m points of P. Let $\varepsilon > 0$ be a parameter, and consider the incremental algorithm that starts with $\mathsf{Q}_0 = \{\mathsf{p}\}$ (here p is an arbitrary point of P). In the ith iteration, the algorithm computes the minimum enclosing ball $\mathbf{b}_{i-1}$ of Q_{i-1}. Next, the algorithm checks if $(1 + \varepsilon)\mathbf{b}_{i-1}$ covers all the points of P. If so, it returns the expanded ball as the desired approximation to the minimum enclosing ball. Otherwise, it finds a point p_i not covered by the expanded ball and generates the new set $\mathsf{Q}_i = \mathsf{P}_{i-1} \cup \{\mathsf{p}_i\}$.

(A) Prove that the number of iterations performed by this algorithm is bounded by $O(1/\varepsilon^2)$. (Hint: Give a lower bound on how much the radius of the bounding ball grows in each iteration.)

(B) What is the bound on the running time of your algorithm?

CHAPTER 25

Duality

Duality is a transformation that maps lines and points into points and lines, respectively, while preserving some properties in the process. Despite its relative simplicity, it is a powerful tool that can dualize what seem like "hard" problems into easy dual problems. There are several alternative definitions of duality, but they are essentially similar, and we present one that works well for our purposes.

25.1. Duality of lines and points

Consider a line $\ell \equiv y = ax + b$ in two dimensions. It is parameterized by two constants a and b, which we can interpret, paired together, as a point in the parametric space of the lines. Naturally, this also gives us a way of interpreting a point as defining the coefficients of a line. Thus, conceptually, points are lines and lines are points.

Formally, the ***dual point*** to the line $\ell \equiv y = ax + b$ is the point $\ell^\star = (a, -b)$. Similarly, for a point $\mathsf{p} = (c, d)$ its ***dual line*** is $\mathsf{p}^\star \equiv y = cx - d$. Namely,

$$\mathsf{p} = (a, b) \implies \mathsf{p}^\star \equiv y = ax - b,$$
$$\ell \equiv y = cx + d \implies \ell^\star = (c, -d).$$

We will consider a line $\ell \equiv y = cx + d$ to be a linear function in one dimension and let $\ell(x) = cx + d$.

A point $\mathsf{p} = (a, b)$ lies ***above*** a line $\ell \equiv y = cx+d$ if p lies vertically above ℓ. Formally, we have that $b > \ell(a) = ca + d$. We will denote this fact by $\mathsf{p} \succ \ell$. Similarly, the point p lies ***below*** ℓ if $b < \ell(a) = ca + d$, denoted by $\mathsf{p} \prec \ell$.

A line ℓ ***supports*** a convex set $S \subseteq \mathbb{R}^2$ if it intersects S but the interior of S lies completely on one side of ℓ.

Basic properties. For a point $\mathsf{p} = (a, b)$ and a line $\ell \equiv y = cx + d$, we have the following:

(P1) $\mathsf{p}^{\star\star} = (\mathsf{p}^\star)^\star = \mathsf{p}$.

Proof. Indeed, $\mathsf{p}^\star \equiv y = ax - b$ and $(\mathsf{p}^\star)^\star = (a, -(-b)) = \mathsf{p}$. ■

(P2) The point p lies above (resp. below, on) the line ℓ if and only if the point $\ell^\star$ lies above (resp. below, on) the line $\mathsf{p}^\star$. (Namely, a point and a line change their vertical ordering in the dual.)

Proof. Indeed, $\mathsf{p} \succ \ell(a)$ if and only if $b > ca + d$. Similarly, $(c, -d) = \ell^\star \succ \mathsf{p}^\star \equiv y = ax - b$ if and only if

$$-d > ac - b \iff b > ca + d,$$

and this is the above condition. ■

(P3) The vertical distance between p and ℓ is the same as that between $\mathsf{p}^\star$ and $\ell^\star$.

PROOF. Indeed, the vertical distance between p and ℓ is $|b - \ell(a)| = |b - (ca + d)|$. The vertical distance between $\ell^\star = (c, -d)$ and $\mathsf{p}^\star \equiv y = ax - b$ is $|(-d) - \mathsf{p}^\star(c)| = |-d - (ac - b)| = |b - (ca + d)|$. ■

(P4) The vertical distance $\delta(\ell, \hbar)$ between two parallel lines ℓ and $\hbar$ is the same as the length of the vertical segment $\ell^\star \hbar^\star$.

PROOF. The vertical distance between $\ell \equiv y = ax + b$ and $\hbar \equiv y = ax + e$ is $|b - e|$. Similarly, since $\ell^\star = (a, -b)$ and $\hbar^\star = (a, -e)$, we have that the segment $\ell^\star\hbar^\star$ is indeed vertical and the vertical distance between its endpoints is $|(-b) - (-e)| = |b - e|$. ■

The missing lines. Consider the vertical line $\ell \equiv x = 0$. Clearly, ℓ does not have a dual point (specifically, its hypothetical dual point has an x-coordinate with infinite value). In particular, our duality cannot handle vertical lines. To visualize the problem, consider a sequence of non-vertical lines ℓ_i that converges to a vertical line ℓ. The sequence of dual points $\ell_i^\star$ is a sequence of points that diverges to infinity.

25.1.1. Examples.

25.1.1.1. *Segments and wedges.*

Consider a segment $s = \mathsf{pq}$ that lies on a line ℓ. Observe, that the dual of a point $\mathsf{t} \in \ell$ is a line $\mathsf{t}^\star$ that passes through the point $\ell^\star$ (by (P2) above). Specifically, the two lines $\mathsf{p}^\star$ and $\mathsf{q}^\star$ define two double wedges. Let $\mathcal{W}$ be the double wedge that does not contain the vertical line that passes through $\ell^\star$; see the figure on the right.

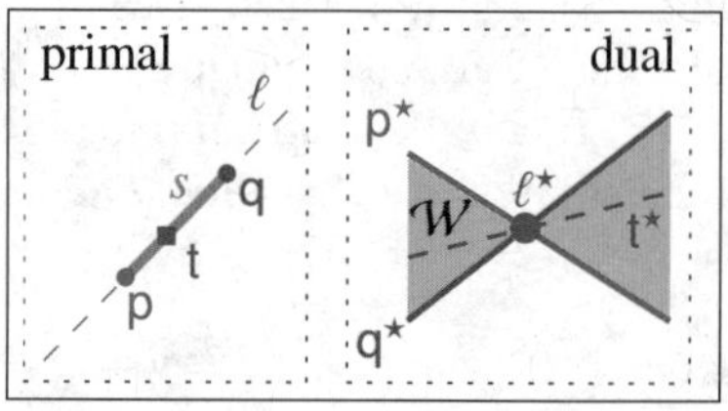

Consider now the point t as it moves along s. When it is equal to p (resp. q), then its dual line $\mathsf{t}^\star$ is the line $\mathsf{p}^\star$ (resp. $\mathsf{q}^\star$). As t moves along s from p to q, its x-coordinate changes continuously, and hence the slope of its dual changes continuously from that of $\mathsf{p}^\star$ to that of $\mathsf{q}^\star$. Furthermore, all these dual lines must all pass through the point $\ell^\star$. As such, as t moves from p to q, the dual line $\mathsf{t}^\star$ sweeps over the double wedge $\mathcal{W}$. Note that the x-coordinate of t during this process is in the interval $\left[\min(x(\mathsf{p}), x(\mathsf{q})), \max(x(\mathsf{p}), x(\mathsf{q}))\right]$; namely, it is bounded. As such, the double wedge being swept over is the one that does not include the vertical line through $\ell^\star$.

What about the other double wedge? It represents the two rays forming $\ell \setminus s$. The vertical line through $\ell^\star$ represents the singularity point at infinity where the two rays are "connected" together. Thus, as t travels along one of the rays of $\ell \setminus s$ (say starting at q), the dual line $\mathsf{t}^\star$ becomes steeper and steeper, till it becomes vertical. Now, the point t "jumps" from the "infinite endpoint" of this ray to the "infinite endpoint" of the other ray. Now, as t travels down the other ray, the dual line $\mathsf{t}^\star$ continues to rotate from its current vertical position, sweeping over the rest of the double wedge, till t reaches p.[①] (The reader who feels uncomfortable with notions like "infinite endpoint" can rest assured that the author feels the same way. As such, this should be taken as an intuitive description of what's going on and not as a formally correct one. This argument can be formalized by using the projective plane.)

[①] At this point t rests for awhile from this long trip of going to infinity and coming back.

25.1.1.2. *Convex hull and upper/lower envelopes.* Consider a set L of lines in the plane. The minimization diagram of L, known as the ***lower envelope*** of L, is the function $\mathcal{L}_{\mathsf{L}} : \mathbb{R} \to \mathbb{R}$, where we have $\mathcal{L}(x) = \min_{\ell \in \mathsf{L}} \ell(x)$, for $x \in \mathbb{R}$. Similarly, the ***upper envelope*** of L is the function $\mathcal{U}(x) = \max_{\ell \in \mathsf{L}} \ell(x)$, for $x \in \mathbb{R}$. The ***extent*** of L at $x \in \mathbb{R}$ is the vertical distance between the upper and lower envelopes at x; namely, $\mathcal{E}_{\mathsf{L}}(x) = \mathcal{U}(x) - \mathcal{L}(x)$.

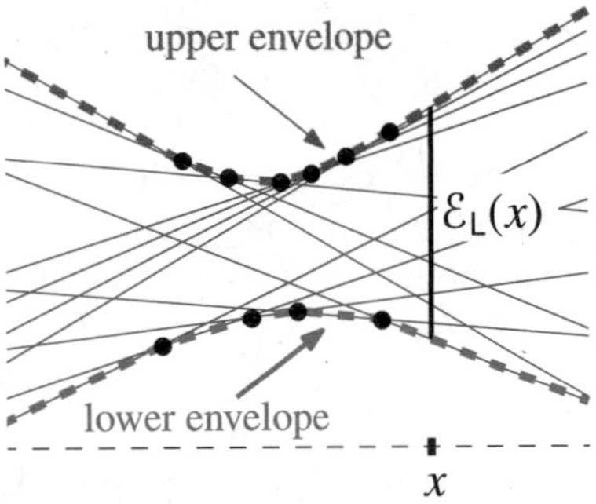

Computing the lower and/or upper envelopes can be useful. A line might represent a linear constraint, where the feasible solution must lie above this line. Thus, the feasible region is the region of points that lie above all the given lines. Namely, the region of the feasible solution is defined by the upper envelope of the lines.

The upper (and lower) envelope is a polygonal chain made out of two infinite rays and a sequence of segments, where each segment/ray lies on one of the given lines. As such, the upper envelop can be described as the sequence of lines appearing on it and the vertices where they change.

Developing an efficient algorithm for computing the upper envelope of a set of lines is a tedious but doable task. However, it becomes trivial if one uses duality.

Lemma 25.1. *Let L be a set of lines in the plane. Let $\alpha \in \mathbb{R}$ be any number, and let $\beta^- = \mathcal{L}_{\mathsf{L}}(\alpha)$ and $\beta^+ = \mathcal{U}_{\mathsf{L}}(\alpha)$. Let $\mathsf{p} = (\alpha, \beta^-)$ and $\mathsf{q} = (\alpha, \beta^+)$. Then:*

(i) The dual lines $\mathsf{p}^\star$ and $\mathsf{q}^\star$ are parallel, and they are both perpendicular to the direction $(\alpha, -1)$.

(ii) The lines $\mathsf{p}^\star$ and $\mathsf{q}^\star$ support $\mathcal{CH}(\mathsf{L}^\star)$.

(iii) The extent $\mathcal{E}_{\mathsf{L}}(\alpha)$ is the vertical distance between the lines $\mathsf{p}^\star$ and $\mathsf{q}^\star$.

Proof. (i) We have $\mathsf{p}^\star \equiv y = \alpha x - \beta^-$ and $\mathsf{q}^\star \equiv y = \alpha x - \beta^+$. These two lines are parallel since they have the same slope. In particular, they are parallel to the direction $(1, \alpha)$. But this direction is perpendicular to the direction $(\alpha, -1)$.

(ii) By property (P2), we have that all the points of $\mathsf{L}^\star$ are below (or on) the line $\mathsf{p}^\star$. Furthermore, since p is on the lower envelope of L, it follows that $\mathsf{p}^\star$ must pass through one of the points $\mathsf{L}^\star$. Namely, $\mathsf{p}^\star$ supports $\mathcal{CH}(\mathsf{L}^\star)$ and it lies above it. A similar argument applies to $\mathsf{q}^\star$.

(iii) This is a restatement of property (P4). ∎

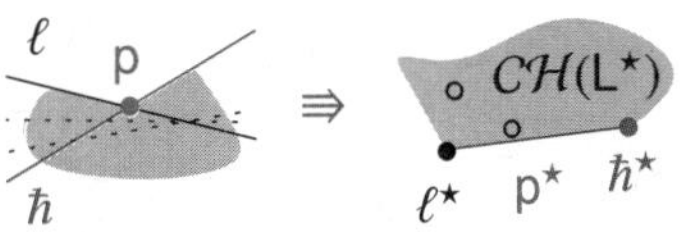

Thus, consider a vertex p of the upper envelope of the set of lines L. The point p is the intersection point of two lines ℓ and $\hbar$ of L (for the sake of simplicity of exposition, assume no other line of L passes through p). Consider the dual set of points $\mathsf{L}^\star$ and the dual line $\mathsf{p}^\star$. Since p lies above (or on) all the lines of L, by property (P2), it must be that the line $\mathsf{p}^\star$ lies below (or on) all the points of $\mathsf{L}^\star$. On the other hand (again by property (P2)), the line $\mathsf{p}^\star$ passes through the two points $\ell^\star$ and $\hbar^\star$. Namely, $\mathsf{p}^\star$ is a line that supports the convex hull of $\mathsf{L}^\star$ and it passes through its vertices $\ell^\star$ and $\hbar^\star$. (The reader should verify that $\ell^\star$ and $\hbar^\star$ are indeed vertices of the convex hull.)

The convex hull of $\mathsf{L}^\star$ is a convex polygon $\mathcal{P}$ which can be broken into two convex chains by breaking it at the two extreme points in the x direction (we are assuming here that L does not contain parallel lines, and as such the extreme points are unique). Note that such an endpoint is shared between the two chains and corresponds to a line that defines two asymptotes (one of the upper envelope and one, on the other side, for the lower envelope).

We will refer to this upper polygonal chain of the convex hull as the ***upper convex chain*** and to the lower one as the ***lower convex chain***. In particular, two consecutive segments of the upper envelope correspond to two consecutive vertices on the lower chain of the convex hull of $\mathsf{L}^\star$.

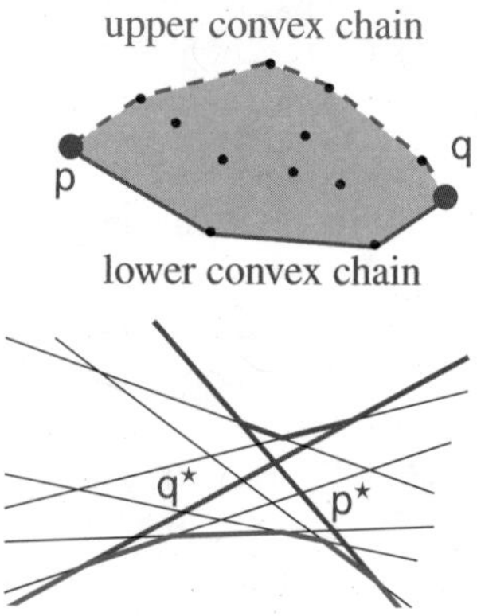

The lower chain of $\mathcal{CH}(\mathsf{L}^\star)$ corresponds to the upper envelope of L, and the upper chain corresponds to the lower envelope of L. Of special interest are the two x extreme points p and q of the convex hull. They are the dual of the two lines with the smallest/largest slopes in L. These two lines appear on both the upper and lower envelopes of the lines and they contain the four infinite rays of these envelopes.

Lemma 25.2. *Given a set L of n lines in the plane, one can compute its lower and upper envelopes in $O(n \log n)$ time.*

Proof. One can compute the convex hull of n points in the plane in $O(n \log n)$ time. Thus, computing the convex hull of $\mathsf{L}^\star$ and dualizing the upper and lower chains of $\mathcal{CH}(\mathsf{L}^\star)$ results in the required envelopes. ■

25.2. Higher dimensions

The above discussion can be easily extended to higher dimensions. We provide the basic properties without further proof, since they are easy extensions of the two-dimensional case. A hyperplane $h \equiv x_d = b_1 x_1 + \cdots + b_{d-1} x_{d-1} + b_d$ in $\mathbb{R}^d$ can be interpreted as a function from $\mathbb{R}^{d-1}$ to $\mathbb{R}$. Given a point $\mathsf{p} = (p_1, \ldots, p_d)$, let $h(\mathsf{p}) = b_1 p_1 + \cdots + b_{d-1} p_{d-1} + b_d$. In particular, a point p lies ***above*** the hyperplane h if $p_d > h(\mathsf{p})$. Similarly, p lies ***below*** the hyperplane h if $p_d < h(\mathsf{p})$. Finally, a point is on the hyperplane if $h(\mathsf{p}) = p_d$.

The ***dual*** of a point $\mathsf{p} = (p_1, \ldots, p_d) \in \mathbb{R}^d$ is a hyperplane $\mathsf{p}^\star \equiv x_d = p_1 x_1 + \cdots p_{d-1} x_{d-1} - p_d$, and the ***dual*** of a hyperplane $h \equiv x_d = a_1 x_1 + a_2 x_2 + \cdots + a_{d-1} x_{d-1} + a_d$ is the point $h^\star = (a_1, \ldots, a_{d-1}, -a_d)$. Summarizing:

$$\mathsf{p} = (p_1, \ldots, p_d) \quad \Longrightarrow \quad \mathsf{p}^\star \equiv x_d = p_1 x_1 + \cdots + p_{d-1} x_{d-1} - p_d$$
$$h \equiv x_d = a_1 x_1 + \cdots + a_{d-1} x_{d-1} + a_d \quad \Longrightarrow \quad h^\star = (a_1, \ldots, a_{d-1}, -a_d).$$

In the following we will slightly abuse notation, and for a point $\mathsf{p} \in \mathbb{R}^d$ we will refer to $(p_1, \ldots, p_{d-1}, \mathcal{L}_{\mathcal{H}}(\mathsf{p}))$ as the point $\mathcal{L}_{\mathcal{H}}(\mathsf{p})$. Similarly, $\mathcal{U}_{\mathcal{H}}(\mathsf{p})$ would denote the corresponding point on the upper envelope of $\mathcal{H}$.

The proof of the following lemma is an easy extension of the proof of Lemma 25.1 and is left as an exercise.

Lemma 25.3. *For a point $\mathsf{p} = (b_1, \ldots, b_d)$, we have the following:*

(i) $\mathsf{p}^{\star\star} = \mathsf{p}$.

(ii) The point p lies above (resp. below, on) the hyperplane h if and only if the point $h^\star$ lies above (resp. below, on) the hyperplane $\mathsf{p}^\star$.

(iii) The vertical distance between p and h is the same as that between $\mathsf{p}^\star$ and $h^\star$.

(iv) *The vertical distance $\delta(h, g)$ between two parallel hyperplanes h and g is the same as the length of the vertical segment $h^\star g^\star$.*

(v) *Computing the lower and upper envelopes of $\mathcal{H}$ is equivalent to computing the convex hull of the dual set of points $\mathcal{H}^\star$.*

25.3. Bibliographical notes

The duality discussed here should not be confused with linear programming duality [**Van97**]. Although the two topics seem to be connected somehow, the author is unaware of a natural and easy connection.

A natural question is whether one can find a duality that preserves the orthogonal distances between lines and points. The surprising answer is no, as Exercise 25.2 testifies. It is not too hard to show using topological arguments that any duality must distort such distances arbitrarily badly.

Open Problem 25.4. Given a set P of n points in the plane and a set L of n lines in the plane, consider the minimum possible distortion of a duality (i.e., the one that minimizes the distortion of orthogonal distances) for P and L. What is the minimum distortion (of the duality) possible, as a function of n?

Formally, we define the distortion of the duality as

$$\max_{\mathsf{p}\in\mathsf{P},\ell\in\mathsf{L}} \left(\frac{\mathbf{d}(\mathsf{p}, \ell)}{\mathbf{d}(\mathsf{p}^\star, \ell^\star)}, \frac{\mathbf{d}(\mathsf{p}^\star, \ell^\star)}{\mathbf{d}(\mathsf{p}, \ell)} \right).$$

A striking (negative) example of the power of duality is the work of Overmars and van Leeuwen [**OvL81**] on the dynamic maintenance of convex hulls in the plane and the maintenance of the lower/upper envelopes of lines in the plane. Clearly, by duality, the two problems are identical. However, the authors (smart people indeed) did not observe it, and the paper is twice as long as it should be since the two problems are solved separately. (In defense of the authors this is an early work in computational geometry.)

Duality is heavily used throughout computational geometry, and it is hard to imagine managing without it. Results and techniques that use duality include bounds on k-sets/k-levels [**Dey98**], partition trees [**Mat92a**], and coresets for extent measure [**AHV04**] (this is a random short list of relevant results and it is by no means exhaustive).

25.3.1. Projective geometry and duality.

Note that our duality did not work for vertical lines. This "missing lines phenomena" is inherent to all dualities, since the space of lines in the plane has the topology of an open Möbius strip which is not homeomorphic to the plane; that is, there is no continuous mapping between all lines in the plane and all points in the plane. Naturally, there are a lot of other possible dualities, and the one presented here is the one most useful for our purposes.

One way to overcome the limitations that not all lines or points are presented by the duality is to add an extra coordinate. Now a point is represented by a triplet (w, x, y), which represents the planar point $(x/w, y/w)$. Thus a point no longer has a unique representation, and for example the triplets $(1, 1, 1)$ and $(2, 2, 2)$ represent the same point $(1, 1)$. Specifically, the set of all points in three dimensions representing a point in the plane is a line in three dimensions passing through the origin.

Similarly, a line in the plane is now represented by a triplet $\langle A, B, C\rangle$ which corresponds to the (planar) line $A + Bx + Cy = 0$. Alternatively, consider this triplet as representing the plane $Aw + Bx + Cy = 0$ that passes through the origin.

Duality is now defined in a very natural way. Indeed, if the point (a, b, c) lies on the line (A, B, C), then we have that $Aa + Bb + Cc = 0$, which implies that the point represented by (A, B, C) lies on the line represented by (a, b, c).

We can now think about the plane as being the unit sphere in three dimensions. Geometrically, we can interpret a plane (i.e., two-dimensional line) as a great circle (i.e., the intersection of the respective 3D plane and the unit sphere) and a point by its representative point on the sphere. That is, a point (x, y) in the plane induces the line $\ell \equiv \left\{ c \cdot (1, x, y) \,\middle|\, c \in \mathbb{R} \right\}$, and its (say top) intersection point with the unit sphere represents the original point. Specifically, the two antipodal intersection points are considered to be the same. (That is, we "collapse" the unit sphere by treating antipodal points as the same point.) Now, we get a classical projective geometry, where any two distinct lines intersect in a single intersection point and any two distinct points define a single line.

This homogeneous representation has the beauty that all lines are now represented and duality is universal. Expressions for the intersection point of two lines no longer involve division, which makes life much easier if one wants to implement geometric algorithms using exact arithmetic. For example, this representation is currently used by the CGAL project [**FGK⁺00**] (a software library implementing basic geometric algorithms). Another advantage is that any theorem in the primal has an immediate dual theorem in the dual. This is mathematically elegant.

This duality is still not perfect, since now there is no natural definition for what a segment connecting two points is. Indeed, there are two portions of a great circle connecting a pair of points. There is an extension of this notion to add orientation. Thus $\langle 1, 1, 1 \rangle$ and $\langle -1, -1, -1 \rangle$ represent different lines. Intuitively, one of them represents one halfplane bounded by this line, and the other represents the other halfplane. Now, if one goes through the details carefully, everything falls into place and you can speak about segments (or precisely oriented segments), and so on.

This topic is presented quite nicely in the book by Stolfi [**Sto91**].

25.3.2. Duality, Voronoi diagrams, and Delaunay triangulations. Given a set P of points in $\mathbb{R}^{\mathsf{d}}$, its ***Voronoi diagram*** is a partition of space into cells, where each cell is the region closest to one of the points of P. The ***Delaunay triangulation*** of a point set is a graph which is planar (the planarity holds only for $\mathsf{d} = 2$, but a similar definition holds in any dimension) where two points are connected by a straight segment if there is a ball that touches both points and its interior is empty. It is easy to verify that these two structures are dual to each other in the sense of graph duality. Maybe more interestingly, this duality also has an easy geometric interpretation.

Indeed, given P, its Voronoi diagram boils down to the computation of the lower envelope of cones. Indeed, for a point $\mathsf{p} \in \mathsf{P}$ the function $f(x) = \|x - \mathsf{p}\|$ has an image, which is a cone in three dimensions. Clearly, the minimization diagram of all these functions is the Voronoi diagram. This set of cones can be linearized and then the computation of the Voronoi diagram boils down to computing the lower envelope of hyperplanes, one hyperplane for each point of P. Similarly, the computation of the Delaunay triangulation of P can be reduced, after lifting the points to the hyperboloid, to the computation of the convex hull of the points. In fact, the projection down of the lower part of the convex hull is the required triangulation. Thus, the two structures are dual to each other. The interested reader should check out [**dBCvKO08**].

25.4. Exercises

Exercise 25.1 (Duality of the extent)**.** Prove Lemma 25.3

Exercise 25.2 (No duality preserves orthogonal distances)**.** Show a counterexample proving that no duality can preserve (exactly) orthogonal distances between points and lines.

CHAPTER 26

Finite Metric Spaces and Partitions

We had encountered finite and infinite metric spaces earlier in the book, see Section 4.1.1 and Section 11.2.1. In this chapter we study them on their own merit, and a provide a short (partial) introduction to this fascinating topic.

26.1. Finite metric spaces

Definition 26.1. A ***metric space*** is a pair $(\mathcal{X}, \mathbf{d})$ where $\mathcal{X}$ is a set and $\mathbf{d} : \mathcal{X} \times \mathcal{X} \to [0, \infty)$ is a ***metric*** satisfying the following axioms: (i) $\mathbf{d}(x, y) = 0$ if and only if $x = y$, (ii) $\mathbf{d}(x, y) = \mathbf{d}(y, x)$, and (iii) $\mathbf{d}(x, y) + \mathbf{d}(y, z) \geq \mathbf{d}(x, z)$ (triangle inequality).

For example, $\mathbb{R}^2$ with the regular Euclidean distance is a metric space.

It is usually of interest to consider the finite case, where $\mathcal{X}$ is a set of n points. Then, the function $\mathbf{d}$ can be specified by $\binom{n}{2}$ real numbers, that is, the distance between every pair of points of $\mathcal{X}$. Alternatively, one can consider $(\mathcal{X}, \mathbf{d})$ as a weighted complete graph, where edges are assigned positive weights that comply with the triangle inequality.

Finite metric spaces rise naturally from (sparser) graphs. Indeed, let $\mathsf{G} = (\mathcal{X}, E)$ be an undirected weighted graph defined over $\mathcal{X}$, and let $\mathbf{d}_{\mathsf{G}}(x, y)$ be the length of the shortest path between x and y in G. It is easy to verify that $(\mathcal{X}, \mathbf{d}_{\mathsf{G}})$ is a finite metric space. As such if the graph G is sparse, it provides a compact representation to the finite space $(\mathcal{X}, \mathbf{d}_{\mathsf{G}})$.

Definition 26.2. Let $(\mathcal{X}, \mathbf{d})$ be an n-point metric space. We denote the ***open ball*** of radius r about $x \in \mathcal{X}$ by $\mathbf{b}(x, r) = \left\{ y \in \mathcal{X} \;\middle|\; \mathbf{d}(x, y) < r \right\}$.

Underlying our discussion of metric spaces are algorithmic applications. The hardness of various computational problems depends heavily on the structure of the finite metric space. Thus, given a finite metric space and a computational task, it is natural to try to map the given metric space into a new metric where the task at hand becomes easy.

Example 26.3. Consider the problem of computing the diameter. While it is not trivial in two dimensions, it is easy in one dimension. Thus, if we could map points in two dimensions into points in one dimension, such that the diameter is preserved, then computing the diameter becomes easy. Indeed, this approach yields an efficient approximation algorithm; see Exercise 26.5.

Of course, this mapping from one metric space to another is going to introduce error. We would be interested in minimizing this error.

Definition 26.4. Let $(\mathcal{X}, \mathbf{d}_{\mathcal{X}})$ and $(\mathcal{Y}, \mathbf{d}_{\mathcal{Y}})$ be metric spaces. A mapping $f : \mathcal{X} \to \mathcal{Y}$ is an ***embedding***. It is ***C-Lipschitz*** if $\mathbf{d}_{\mathcal{Y}}(f(x), f(y)) \leq C \cdot \mathbf{d}_{\mathcal{X}}(x, y)$, for all $x, y \in \mathcal{X}$. The mapping f is called ***K-bi-Lipschitz*** if there exists a $C > 0$ such that

$$CK^{-1} \cdot \mathbf{d}_{\mathcal{X}}(x, y) \leq \mathbf{d}_{\mathcal{Y}}\big(f(x), f(y)\big) \leq C \cdot \mathbf{d}_{\mathcal{X}}(x, y),$$

for all $x, y \in \mathcal{X}$.

The least K for which f is K-bi-Lipschitz is the ***distortion*** of f and is denoted $\mathrm{dist}(f)$. The least distortion with which $\mathcal{X}$ may be embedded in Y is denoted by $c_Y(\mathcal{X})$.

There are several powerful results that show the existence of embeddings with low distortion, and in this chapter we present some of these results. These include the following:

(A) **Probabilistic trees.** Every finite metric can be randomly embedded into a tree such that the "expected" distortion for a specific pair of points is $O(\log n)$. See Theorem 26.8$_{\text{p327}}$.

(B) **Embedding into Euclidean space.** Any n-point metric space can be embedded into (finite-dimensional) Euclidean space with $O(\log n)$ distortion. This is known as Bourgain's theorem, and we prove a weaker result with distortion $O\left(\log^{3/2} n\right)$; see Theorem 26.12$_{\text{p331}}$.

(C) **Dimension reduction.** Any n-point set in Euclidean space with the regular Euclidean distance can be embedded into $\mathbb{R}^k$ with distortion $1 + \varepsilon$, where $k = O(\varepsilon^{-2} \log n)$. This is the JL lemma that we already encountered; see Theorem 19.18$_{\text{p266}}$.

26.1.1. Examples.

What is distortion? When considering a mapping $f : \mathcal{X} \to \mathbb{R}^{\mathsf{d}}$ of a metric space $(\mathcal{X}, \mathbf{d})$ to $\mathbb{R}^{\mathsf{d}}$, it would useful to observe that since $\mathbb{R}^{\mathsf{d}}$ can be scaled, we can consider f to be an ***expansive mapping*** (i.e., no distances shrink). Furthermore, we can assume that there is at least one pair of points $x, y \in \mathcal{X}$, such that $\mathbf{d}(x, y) = \|x - y\|$. As such, we have $\mathrm{dist}(f) = \max_{x,y} \frac{\|x-y\|}{\mathbf{d}(x,y)}$.

Why distortion is necessary? Consider a graph $\mathsf{G} = (V, E)$ with one vertex s connected to three other vertices a, b, c, where the weights on the edges are all 1 (i.e., G is the star graph with three leaves). We claim that G cannot be embedded into Euclidean space with distortion $\leq 2/\sqrt{3}$. Indeed, consider the associated metric space $(V, \mathbf{d}_{\mathsf{G}})$ and an (expansive) embedding $f : V \to \mathbb{R}^{\mathsf{d}}$.

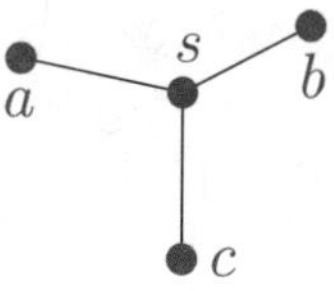

Let $\triangle$ denote the triangle formed by $a'b'c'$, where $a' = f(a), b' = f(b)$, and $c' = f(c)$. Next, consider the quantity $\max(\|a' - s'\|, \|b' - s'\|, \|c' - s'\|)$ which lower bounds the distortion of f.

This quantity is minimized when $r = \|a' - s'\| = \|b' - s'\| = \|c' - s'\|$. Namely, s' is the center of the smallest enclosing circle of $\triangle$. However, r is minimized when all the edges of $\triangle$ are of equal length and are of length $\mathbf{d}_{\mathsf{G}}(a, b) = 2$. Observe that the height of the equilateral triangle with sidelength 2 is $h = \sqrt{3}$, and the radius of its inscribing circle is $r = (2/3)h = 2/\sqrt{3}$; see the figure on the right. As such, it follows that $\mathrm{dist}(f) \geq r = 2/\sqrt{3}$.

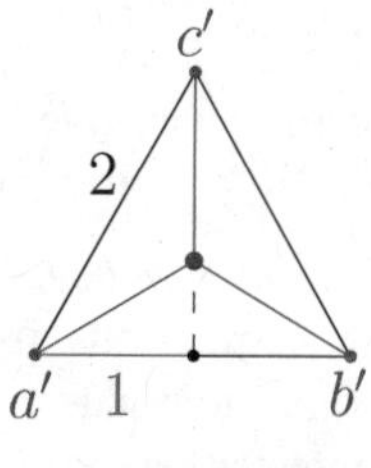

Note that the above argument is independent of the target dimension d. A packing argument shows that embedding the star graph with n leaves into $\mathbb{R}^{\mathsf{d}}$ requires distortion $\Omega\left(n^{1/\mathsf{d}}\right)$; see Exercise 26.7. It is known that $\Omega(\log n)$ distortion is necessary in the worst case when embedding a graph into Euclidean space (this is shown using expanders). A proof of distortion $\Omega(\log n / \log \log n)$ is sketched in the bibliographical notes.

26.1.2. Hierarchical tree metrics. Tree metrics, or more specifically HSTs, are quite useful in practice and nicely demonstrate why algorithmically finite metric spaces are useful. We already encountered them when considering shifted quadtrees; see Definition 11.10$_{\text{p155}}$. For convenience, we reproduce the definition here.

Definition 26.5. A ***hierarchically well-separated tree*** (***HST***) is a metric space defined on the leaves of a rooted tree $\mathcal{T}$. To each vertex $u \in \mathcal{T}$ there is associated a label $\Delta_u \geq 0$. This label is zero for all the leaves of $\mathcal{T}$, and it is a positive number for all the interior nodes. The labels are such that if a vertex u is a child of a vertex v, then $\Delta_u \leq \Delta_v$. The distance between two leaves $x, y \in \mathcal{T}$ is defined as $\Delta_{\mathrm{lca}(x,y)}$, where $\mathrm{lca}(x, y)$ is the least common ancestor of x and y in $\mathcal{T}$. An HST $\mathcal{T}$ is a ***k-HST*** if for all vertices $v \in \mathcal{T}$, we have that $\Delta_v \leq \Delta_{\overline{p}(v)}/k$, where $\overline{p}(v)$ is the parent of v in $\mathcal{T}$.

Note that an HST is a limited metric. For example, consider the cycle $\mathsf{G} = C_n$ of n vertices, with weight 1 on the edges, and consider an expansive embedding f of G into an HST H. It is easy to verify that there must be two consecutive nodes of the cycle which are mapped to two different subtrees of the root r of H. Since H is expansive, it follows that $\Delta_r \geq n/2$. As such, $\mathrm{dist}(f) \geq n/2$. Namely, HSTs fail to faithfully represent even simple metrics.

26.1.3. Clustering. One natural problem to solve on a graph (i.e., finite metric space) $(\mathcal{X}, \mathbf{d})$ is to partition it into clusters. One such natural clustering is the ***k-median clustering***, where we would like to choose a set $C \subseteq \mathcal{X}$ of k centers, such that

$$\nu_C(\mathcal{X}, \mathbf{d}) = \sum_{\mathsf{q} \in \mathcal{X}} \mathbf{d}(\mathsf{q}, C)$$

is minimized, where $\mathbf{d}(\mathsf{q}, C) = \min_{c \in C} \mathbf{d}(\mathsf{q}, c)$ is the distance of q to its closest center in C.

It is known that finding the optimal k-median clustering in a (general weighted) graph is NP-Complete. As such, the best we can hope for is an approximation algorithm. However, if the structure of the finite metric space $(\mathcal{X}, \mathbf{d})$ is simple, then the problem can be solved efficiently. For example, if the points of $\mathcal{X}$ are on the real line (and the distance between a and b is just $|a - b|$), then the k-median can be solved using dynamic programming.

Another interesting case is when the metric space $(\mathcal{X}, \mathbf{d})$ is an HST. It is not too hard to prove the following lemma. See Exercise 26.2.

Lemma 26.6. *Let $(\mathcal{X}, \mathbf{d})$ be an HST defined over n points, and let $k > 0$ be an integer. One can compute the optimal k-median clustering of $\mathcal{X}$ in $O(k^2 n)$ time.*

Thus, if we can embed a general graph G into an HST H, with low distortion, then we could approximate the k-median clustering on G by clustering the resulting HST and "importing" the resulting partition to the original space. The quality of approximation would be bounded by the distortion of the embedding of G into H.

26.2. Random partitions

Let $(\mathcal{X}, \mathbf{d})$ be a finite metric space. A ***partition*** of $\mathcal{X}$ is a set Π of disjoint subsets of $\mathcal{X}$ such that $\bigcup_{B \in \Pi} B = \mathcal{X}$. Given a partition $\Pi = \{C_1, \ldots, C_m\}$ of $\mathcal{X}$, we refer to the sets C_i as ***clusters***. We write $\mathcal{P}_{\mathcal{X}}$ for the set of all partitions of $\mathcal{X}$. For $x \in \mathcal{X}$ and a partition $\Pi \in \mathcal{P}_{\mathcal{X}}$ we denote by $\Pi(x)$ the unique cluster of Π containing x. Finally, the set of all probability distributions on $\mathcal{P}_{\mathcal{X}}$ is denoted by $\mathcal{D}_{\mathcal{X}}$.

What we want and what we can get. The target is to partition the metric space into clusters, such that each cluster would have diameter at most Δ, for some prespecified parameter Δ.

We would like to have a partition that does not disrupt distances too "badly". Intuitively, that means that a pair of points such that their distance is larger than Δ will be separated by the clustering, but points that are closer to each other would be in the same cluster. This is of course impossible, as any clustering must separate points that are close to each other. To see that, consider a set of points densely packed on the interval $[0, 10]$, and let $\Delta < 5$. Clearly, there would always be two close points that would be in two separate clusters.

As such, our strategy would be to use partitions that are constructed randomly, and the best we can hope for is that the probability of them being separated is a function of their distance t, which would be small if t is small. As an example, for the case of points on the real line, take the natural partition into intervals of length Δ (that is, all the points in the interval $[i\Delta, (i+1)\Delta)$ would belong to the same cluster) and randomly shift it by a random number x picked uniformly in $[0, \Delta)$. Namely, all the points belonging to $[x+i\Delta,\ x+(i+1)\Delta)$ would belong to the same cluster. Now, it is easy to verify that for any two points $\mathsf{p}, \mathsf{q} \in \mathbb{R}$, of distance $t = |\mathsf{p} - \mathsf{q}|$ from each other, the probability that they are in two different intervals is bounded by t/Δ (see Exercise 26.6). Intuitively, this is the best one can hope for.

As such, the clustering scheme we seek should separate two points within distance t from each other with probability $(t/\Delta) * \text{noise}$, where noise is hopefully small.

26.2.1. Constructing the partition. Let $\Delta = 2^u$ be a prescribed parameter, which is the required diameter of the resulting clusters. Choose, uniformly at random, a permutation π of $\mathcal{X}$ and a random value $\alpha \in [1/4, 1/2]$. Let $R = \alpha\Delta$, and observe that it is uniformly distributed in the interval $[\Delta/4, \Delta/2]$.

The partition is now defined as follows: A point $x \in \mathcal{X}$ is assigned to the cluster C_y of y, where y is the first point in the permutation within distance $\leq R$ from x. Formally,

$$C_y = \left\{ x \in \mathcal{X} \;\middle|\; x \in \mathbf{b}(y, R) \text{ and } \pi(y) \leq \pi(z) \text{ for all } z \in \mathcal{X} \text{ with } x \in \mathbf{b}(z, R) \right\}.$$

Let $\Pi = \{C_y\}_{y \in \mathcal{X}}$ denote the resulting partition.

Here is a somewhat more intuitive explanation: Once we fix the radius of the clusters R, we start scooping out balls of radius R centered at the points of the random permutation π. At the ith stage, we scoop out only the remaining mass at the ball centered at x_i of radius R, where x_i is the ith point in the random permutation.

26.2.2. Properties.

Lemma 26.7. *Let $(\mathcal{X}, \mathbf{d})$ be a finite metric space, let $\Delta = 2^u$ be a prescribed parameter, and let Π be the random partition of $\mathcal{X}$ generated by the above scheme. Then the following hold:*

(i) For any $C \in \Pi$, we have $\operatorname{diam}(C) \leq \Delta$.

(ii) Let x be any point of $\mathcal{X}$, and let t be a parameter $\leq \Delta/8$. For $B = \mathbf{b}(x, t)$, we have that

$$\mathbf{Pr}\Big[B \not\subseteq \Pi(x)\Big] \leq \frac{8t}{\Delta} \ln \frac{M}{m},$$

where $m = |\mathbf{b}(x, \Delta/8)|$ and $M = |\mathbf{b}(x, \Delta)|$.

PROOF. Since $C_y \subseteq \mathbf{b}(y, R)$, we have that $\operatorname{diam}\big(C_y\big) \leq \operatorname{diam}\big(\mathbf{b}(y, R)\big) = 2R \leq \Delta$, and thus the first claim holds.

Let U be the set of points $w \in \mathbf{b}(x, \Delta)$ such that $\mathbf{b}(w, \Delta/2) \cap B \neq \emptyset$. Arrange the points of U in increasing distance from x, and let $w_1, \ldots, w_{M'}$ denote the resulting order, where $M' = |U| \leq M$. For $k = 1, \ldots, M'$, let $I_k = [\mathbf{d}(x, w_k) - t, \mathbf{d}(x, w_k) + t]$ and write $\mathcal{E}_k$ for the event that w_k is the *first* point in π such that $B \cap C_{w_k} \neq \emptyset$ and yet $B \not\subseteq C_{w_k}$. Observe that if $B \not\subseteq \Pi(x)$, then one of the events $\mathcal{E}_1, \ldots, \mathcal{E}_{M'}$ must occur.

Note that if $w_k \in \mathbf{b}(x, \Delta/8)$, then $\mathbf{Pr}[\mathcal{E}_k] = 0$ since $t \leq \Delta/8$ and $B = \mathbf{b}(x, t) \subseteq \mathbf{b}(x, \Delta/8) \subseteq \mathbf{b}(w_k, \Delta/4) \subseteq \mathbf{b}(w_k, R)$. Indeed, when we "scoop" out the cluster C_{w_k}, either B would be fully contained inside C_{w_k} or, alternatively, if B is not fully contained inside C_{w_k}, then some parts of B were already "scooped out" by some other point of U, and as such $\mathcal{E}_k$ does not happen.

In particular, $w_1, \ldots, w_m$ are inside $\mathbf{b}(x, \Delta/8)$ and as such $\mathbf{Pr}[\mathcal{E}_1] = \cdots = \mathbf{Pr}[\mathcal{E}_a] = 0$. Also, note that if $\mathbf{d}(x, w_k) < R - t$, then $\mathbf{b}(w_k, R)$ contains B and as such $\mathcal{E}_k$ cannot happen. Similarly, if $\mathbf{d}(x, w_k) > R + t$, then $\mathbf{b}(w_k, R) \cap B = \emptyset$ and $\mathcal{E}_k$ cannot happen. As such, if $\mathcal{E}_k$ happens, then $R - t \leq \mathbf{d}(x, w_k) \leq R + t$. Namely, if $\mathcal{E}_k$ happens, then $R \in I_k$. We conclude that

$$\mathbf{Pr}[\mathcal{E}_k] = \mathbf{Pr}[\mathcal{E}_k \cap (R \in I_k)] = \mathbf{Pr}[R \in I_k] \cdot \mathbf{Pr}[\mathcal{E}_k \,|\, R \in I_k].$$

Now, R is uniformly distributed in the interval $[\Delta/4, \Delta/2]$, and I_k is an interval of length $2t$. Thus, $\mathbf{Pr}[R \in I_k] \leq 2t/(\Delta/4) = 8t/\Delta$.

Next, to bound $\mathbf{Pr}[\mathcal{E}_k \,|\, R \in I_k]$, we observe that $w_1, \ldots, w_{k-1}$ are closer to x than w_k and their distance to $\mathbf{b}(x, t)$ is smaller than R. Thus, if any of them appear before w_k in π, then $\mathcal{E}_k$ does not happen. Thus, $\mathbf{Pr}[\mathcal{E}_k \,|\, R \in I_k]$ is bounded by the probability that w_k is the first to appear in π out of $w_1, \ldots, w_k$. But this probability is $1/k$, and thus $\mathbf{Pr}[\mathcal{E}_k \,|\, R \in I_k] \leq 1/k$.

We are now ready for the kill. Indeed,

$$\begin{aligned}\mathbf{Pr}\Big[B \not\subseteq \Pi(x)\Big] &= \sum_{k=1}^{M'} \mathbf{Pr}\Big[\mathcal{E}_k\Big] = \sum_{k=m+1}^{M'} \mathbf{Pr}\Big[\mathcal{E}_k\Big] = \sum_{k=m+1}^{M'} \mathbf{Pr}\Big[R \in I_k\Big] \cdot \mathbf{Pr}\Big[\mathcal{E}_k \,\Big|\, R \in I_k\Big] \\ &\leq \sum_{k=m+1}^{M'} \frac{8t}{\Delta} \cdot \frac{1}{k} \leq \frac{8t}{\Delta} \ln \frac{M'}{m} \leq \frac{8t}{\Delta} \ln \frac{M}{m},\end{aligned}$$

since $\sum_{k=m+1}^{M'} \frac{1}{k} \leq \int_m^{M'} \frac{dx}{x} = \ln \frac{M'}{m}$ and $M' \leq M$. ■

26.3. Probabilistic embedding into trees

In this section, given an n-point finite metric $(\mathcal{X}, \mathbf{d})$, we would like to embed it into an HST. As mentioned above, one can verify that for any embedding into HST, the distortion in the worst case is $\Omega(n)$. Thus, we define a randomized algorithm that embeds $(\mathcal{X}, \mathbf{d})$ into a tree. Let T be the resulting tree, and consider two points $x, y \in \mathcal{X}$. Consider the *random variable* $\mathbf{d}_T(x, y)$. We constructed the tree T such that distances never shrink; i.e., $\mathbf{d}(x, y) \leq \mathbf{d}_T(x, y)$. The ***probabilistic distortion*** of this embedding is $\max_{x,y} \mathbf{E}\Big[\frac{\mathbf{d}_T(x,y)}{\mathbf{d}(x,y)}\Big]$. Somewhat surprisingly, one can find such an embedding with logarithmic probabilistic distortion.

Theorem 26.8. *Given an n-point metric $(\mathcal{X}, \mathbf{d})$, one can randomly embed it into a 2-HST with probabilistic distortion $\leq 24 \ln n$.*

Proof. The construction is recursive. Let P denote the current point set being handled, where initially $\mathsf{P} = \mathcal{X}$.

Compute a random partition of P with cluster diameter $\mathrm{diam}(\mathsf{P})/2$, using the construction of Section 26.2.1. We recursively construct a 2-HST for each cluster and hang the resulting clusters on the root node v, which is labeled with $\Delta_v = \mathrm{diam}(\mathsf{P})$. Clearly, the resulting tree $\mathcal{T}$ is a 2-HST.

For a node $v \in \mathcal{T}$, let $\mathcal{X}(v)$ be the set of points of $\mathcal{X}$ contained in the subtree of v.

For the analysis, assume $\mathrm{diam}(\mathcal{X}) = 1$, and consider two points $x, y \in \mathcal{X}$. We consider a node $v \in \mathcal{T}$ to be in level i if $\mathrm{level}(v) = \lceil \lg \Delta_v \rceil = i$ (note that i is zero for the root and negative for any other node in $\mathcal{T}$). Any two points x and y correspond to two leaves in $\mathcal{T}$, and let $\widehat{u}$ be the least common ancestor of x and y in $\mathcal{T}$. We have $\mathbf{d}_{\mathcal{T}}(x,y) = \Delta_{\widehat{u}} \leq 2^{\mathrm{level}(\widehat{u})}$. Furthermore, note that along a path the levels are strictly monotonically decreasing as we go down the tree.

Being conservative, let w be the first ancestor of x (from the root), such that $\mathbf{b} = \mathbf{b}(x, \mathbf{d}(x,y))$ is not contained (completely) in any of the sets $\mathcal{X}(u_1), \ldots, \mathcal{X}(u_m)$, where $u_1, \ldots, u_m$ are the children of w (putting it differently, $\mathbf{b}$ has a non-empty intersection with two or more clusters defined by the children of w). Clearly, $\mathrm{level}(w) \geq \mathrm{level}(\widehat{u})$, and as such $\mathbf{d}_{\mathcal{T}}(x,y) \leq 2^{\mathrm{level}(w)}$.

Consider the path σ from the root of $\mathcal{T}$ to x, and let $\mathcal{E}_i$ be the event that $\mathbf{b}$ is not fully contained in $\mathcal{X}(v_i)$, where v_i is the node of σ of level i (if such a node exists). Furthermore, let Y_i be the indicator variable which is 1 if $\mathcal{E}_i$ is the first to happened out of the sequence of events $\mathcal{E}_0, \mathcal{E}_{-1}, \ldots$. Clearly, $\mathbf{d}_{\mathcal{T}}(x,y) \leq \sum_i 2^i Y_i$.

Let $t = \mathbf{d}(x,y)$ and $j = \lfloor \lg \mathbf{d}(x,y) \rfloor$, and let $n_i = \left|\mathbf{b}\left(x, 2^i\right)\right|$ for $i = 0, \ldots, -\infty$. We have

$$\mathbf{E}\Big[\mathbf{d}_{\mathcal{T}}(x,y)\Big] \leq \mathbf{E}\left[\sum_i 2^i Y_i\right] = \sum_{i=j}^{0} 2^i\,\mathbf{E}\Big[Y_i\Big] \leq \sum_{i=j}^{0} 2^i\,\mathbf{Pr}\Big[\mathcal{E}_i \cap \overline{\mathcal{E}_{i-1}} \cap \overline{\mathcal{E}_{i-1}} \cdots \overline{\mathcal{E}_0}\Big]$$
$$\leq \sum_{i=j}^{0} 2^i \cdot \frac{8t}{2^i} \ln \frac{n_i}{n_{i-3}},$$

by Lemma 26.7. Thus,

$$\mathbf{E}\Big[\mathbf{d}_{\mathcal{T}}(x,y)\Big] \leq 8t \ln\left(\prod_{i=j}^{0} \frac{n_i}{n_{i-3}}\right) \leq 8t \ln(n_0 \cdot n_1 \cdot n_2) \leq 24 t \ln n.$$

It follows that the expected distortion for the distance between x and y is $\leq 24 \ln n$. ■

26.3.1. Application: Approximation algorithm for k-median clustering. Consider an n-point metric space $(\mathcal{X}, \mathbf{d})$, and let k be an integer number. We would like to compute the optimal k-median clustering. Namely, find a subset $C_{\mathrm{opt}} \subseteq \mathcal{X}$, such that the price of the clustering $\nu_{C_{\mathrm{opt}}}(\mathcal{X}, \mathbf{d})$ is minimized; see Section 26.1.3. To this end, we randomly embed $(\mathcal{X}, \mathbf{d})$ into an HST H using Theorem 26.8. Next, using Lemma 26.6, we compute the optimal k-median clustering of H. Let C be the set of centers computed. We return C together with the partition of $\mathcal{X}$ it induces as the required clustering.

Theorem 26.9. *Let $(\mathcal{X}, \mathbf{d})$ be an n-point metric space. One can compute, in polynomial time, a k-median clustering of $\mathcal{X}$ which has expected price $O(\alpha \log n)$, where α is the price of the optimal k-median clustering of $(\mathcal{X}, \mathbf{d})$.*

Proof. The algorithm is described above. Its running time is polynomial as can be easily verified. To prove the bound on the quality of the clustering, for any point $\mathsf{p} \in \mathcal{X}$, let $\overline{c}(\mathsf{p})$ denote the closest point in C_{opt} to p according to $\mathbf{d}$, where C_{opt} is the set of k-medians in the optimal clustering. Let C be the set of k-medians returned by the algorithm, and let H be the HST used by the algorithm. We have

$$\beta = \nu_C(\mathcal{X}, \mathbf{d}) \leq \nu_C(\mathcal{X}, \mathbf{d}_{\mathsf{H}}) \leq \nu_{C_{\mathrm{opt}}}(\mathcal{X}, \mathbf{d}_{\mathsf{H}}) = \sum_{\mathsf{p} \in \mathcal{X}} \mathbf{d}_{\mathsf{H}}(\mathsf{p}, C_{\mathrm{opt}}) \leq \sum_{\mathsf{p} \in \mathcal{X}} \mathbf{d}_{\mathsf{H}}(\mathsf{p}, \overline{c}(\mathsf{p})).$$

Thus, in expectation, we have

$$\mathbf{E}\big[\beta\big] = \mathbf{E}\left[\sum_{\mathsf{p}\in\mathcal{X}} \mathbf{d}_{\mathrm{H}}(\mathsf{p},\bar{c}(\mathsf{p}))\right] = \sum_{\mathsf{p}\in\mathcal{X}} \mathbf{E}\Big[\mathbf{d}_{\mathrm{H}}(\mathsf{p},\bar{c}(\mathsf{p}))\Big] = \sum_{\mathsf{p}\in\mathcal{X}} O\Big(\mathbf{d}(\mathsf{p},\bar{c}(\mathsf{p}))\log n\Big)$$
$$= O\left(\log n \sum_{\mathsf{p}\in\mathcal{X}} \mathbf{d}(\mathsf{p},\bar{c}(\mathsf{p}))\right) = O\Big(\nu_{C_{\mathrm{opt}}}(\mathcal{X},\mathbf{d})\ \log n\Big),$$

by linearity of expectation and Theorem 26.8. ∎

26.4. Embedding any metric space into Euclidean space

Lemma 26.10. *Let $(\mathcal{X},\mathbf{d})$ be a metric, and let $Y \subset \mathcal{X}$. Consider the mapping $f : \mathcal{X} \to \mathbb{R}$, where $f(x) = \mathbf{d}(x, Y) = \min_{y\in Y} \mathbf{d}(x, y)$. Then for any $x, y \in \mathcal{X}$, we have $|f(x) - f(y)| \le \mathbf{d}(x, y)$. Namely, f is non-expansive.*

Proof. Indeed, let x' and y' be the closest points of Y to x and y, respectively. Observe that $f(x) = \mathbf{d}(x, x') \le \mathbf{d}(x, y') \le \mathbf{d}(x, y) + \mathbf{d}(y, y') = \mathbf{d}(x, y) + f(y)$ by the triangle inequality. Thus, $f(x) - f(y) \le \mathbf{d}(x, y)$. By symmetry, we have $f(y) - f(x) \le \mathbf{d}(x, y)$. Thus, $|f(x) - f(y)| \le \mathbf{d}(x, y)$. ∎

26.4.1. The bounded spread case. Let $(\mathcal{X},\mathbf{d})$ be an n-point metric. The ***spread*** of $\mathcal{X}$, denoted by $\Phi(\mathcal{X}) = \mathrm{diam}(\mathcal{X})/\min_{x,y\in\mathcal{X},x\neq y} \mathbf{d}(x, y)$, is the ratio between the diameter of $\mathcal{X}$ and the distance between the closest pair of points of $\mathcal{X}$.

Theorem 26.11. *Given an n-point metric $\mathcal{Y} = (\mathcal{X},\mathbf{d})$, with spread Φ, one can embed it into Euclidean space $\mathbb{R}^k$ with distortion $O\big(\sqrt{\ln\Phi}\ln n\big)$, where $k = O(\ln\Phi \ln n)$.*

Proof. Assume that $\mathrm{diam}(\mathcal{Y}) = \Phi$ (i.e., the smallest distance in $\mathcal{Y}$ is 1), and let $r_i = 2^{i-2}$, for $i = 1,\ldots,\alpha$, where $\alpha = \lceil \lg\Phi\rceil + 2$. Let $\Pi_{i,j}$ be a random partition of $\mathcal{X}$ with diameter r_i, using Theorem 26.8, for $i = 0,\ldots,\alpha$ and $j = 1,\ldots,\beta$, where $\beta = \lceil c\ln n\rceil$ and c is a large enough constant.

For each cluster of $\Pi_{i,j}$ randomly toss a coin, and let $V_{i,j}$ be the set of all the points of $\mathcal{X}$ that belong to clusters in $\Pi_{i,j}$ that obtained 'T' in their coin toss. For a point $\mathsf{p} \in \mathcal{X}$, let $f_{i,j}(\mathsf{p}) = \mathbf{d}\big(\mathsf{p}, \mathcal{X}\setminus V_{i,j}\big) = \min_{\mathsf{q}\in\mathcal{X}\setminus V_{i,j}} \mathbf{d}(\mathsf{p},\mathsf{q})$, for $i = 0,\ldots,\alpha$ and $j = 1,\ldots,\beta$. Let $F : \mathcal{X} \to \mathbb{R}^{(\alpha+1)\beta}$ be the embedding, such that $F(x) = \big(f_{0,1}(x), f_{0,2}(x),\ldots, f_{0,\beta}(x), f_{1,1}(x), f_{0,2}(x),\ldots, f_{1,\beta}(x), \ldots, f_{\alpha,1}(x), f_{\alpha,2}(x),\ldots, f_{\alpha,\beta}(x)\big)$.

Next, consider two points $x, y \in \mathcal{X}$, with distance $\phi = \mathbf{d}(x, y)$. Let u be an integer such that $r_u \le \phi/2 \le r_{u+1}$. Clearly, in all the partitions $\Pi_{u,1},\ldots,\Pi_{u,\beta}$ the points x and y belong to different clusters. Furthermore, the events $x \in V_{u,j}$ and $y \notin V_{u,j}$ are independent and each has probability $1/2$, for $1 \le j \le \beta$.

Let $\mathcal{E}_j$ denote the event that $\mathbf{b}(x,\rho) \subseteq V_{u,j}$ and $y \notin V_{u,j}$, for $j = 1,\ldots,\beta$, where $\rho = \phi/(c_1 \ln n)$, where c_1 is a constant to be specified shortly. By Lemma 26.7, for c_1 sufficiently large, we have

$$\mathbf{Pr}\Big[\mathbf{b}(x,\rho) \nsubseteq \Pi_{u,j}(x)\Big] \le \frac{8\rho}{r_u}\ln n \le \frac{\phi}{8r_u} \le \frac{1}{2}.$$

Thus, we have

$$\mathbf{Pr}\Big[\mathcal{E}_j\Big] = \mathbf{Pr}\Big[\Big(\mathbf{b}(x,\rho) \subseteq \Pi_{u,j}(x)\Big) \cap \Big(x \in V_{u,j}\Big) \cap \Big(y \notin V_{u,j}\Big)\Big]$$
$$= \mathbf{Pr}\Big[\mathbf{b}(x,\rho) \subseteq \Pi_{u,j}(x)\Big]\cdot \mathbf{Pr}\Big[x \in V_{u,j}\Big]\cdot \mathbf{Pr}\Big[y \notin V_{u,j}\Big] \ge 1/8,$$

since those three events are independent. Notice that if $\mathcal{E}_j$ happens, then $f_{u,j}(x) \geq \rho$ and $f_{u,j}(y) = 0$ and as such $\left(f_{u,j}(x) - f_{u,j}(y)\right)^2 \geq \rho^2$.

Let X_j be an indicator variable which is 1 if $\mathcal{E}_j$ happens, for $j = 1, \ldots, \beta$. Let $Z = \sum_j X_j$ and $\mu = \mathbf{E}[Z] = \mathbf{E}\left[\sum_{j=1}^{\beta} X_j\right] \geq \beta/8$. The probability that only $\beta/16$ of $\mathcal{E}_1, \ldots, \mathcal{E}_\beta$ happens is

$$\mathbf{Pr}\left[Z < (1 - 1/2)\,\mathbf{E}[Z]\right] \leq \exp\left(-\frac{\mu}{2 \cdot 2^2}\right) \leq \exp\left(-\frac{\beta}{64}\right) \leq \exp\left(-\frac{c}{64}\ln n\right) \leq \frac{1}{n^{10}},$$

by the Chernoff inequality and for c sufficiently large.

Thus, with high probability, $Z \geq \mathbf{E}[Z]/2 \geq \beta/16$, and as such

$$\left\|F(x) - F(y)\right\| \geq \sqrt{\sum_{j=1}^{\beta}\left(f_{u,j}(x) - f_{u,j}(y)\right)^2} \geq \sqrt{\rho^2 Z} \geq \frac{\sqrt{\beta}\rho}{4} = \phi \cdot \frac{\sqrt{\beta}}{4c_1 \ln n},$$

since $\rho = \phi/(c_1 \ln n)$. On the other hand, by Lemma 26.10, $\left|f_{i,j}(x) - f_{i,j}(y)\right| \leq \mathbf{d}(x, y) = \phi$. Thus, as the number of coordinates is $(\alpha + 1)\beta$, we have

$$\left\|F(x) - F(y)\right\| \leq \sqrt{(\alpha + 1)\beta\phi^2} \leq 2\sqrt{\alpha\beta}\phi.$$

Thus, setting $G(x) = F(x)\frac{4c_1 \ln n}{\sqrt{\beta}}$, we get a mapping that maps two points of distance ϕ from each other to two points with distance in the range

$$\left[\phi, \phi \cdot 2\sqrt{\alpha\beta} \cdot \frac{4c_1 \ln n}{\sqrt{\beta}}\right] = \phi \cdot \left[1,\ 8c_1\sqrt{\alpha}\ln n\right].$$

Namely, $G(\cdot)$ is an embedding with distortion $O\left(\sqrt{\alpha}\ln n\right) = O\left(\sqrt{\ln \Phi}\ln n\right)$.

The probability that G fails on one of the pairs is smaller than $\binom{n}{2}/n^{10} < 1/n^8$. In particular, we can check the distortion of G for all $\binom{n}{2}$ pairs, and if any of them fail (i.e., the distortion is too big), we restart the process. ■

26.4.2. The unbounded spread case. Our next task is to extend Theorem 26.11 to the case of unbounded spread. Indeed, let $(\mathcal{X}, \mathbf{d})$ be an n-point metric, such that $\text{diam}(\mathcal{X}) \leq 1/2$. Again, we look on the different resolutions $r_1, r_2, \ldots$, where $r_i = 1/2^{i-1}$. For each one of those resolutions r_i, we can embed this resolution into β coordinates, as done for the bounded case. Then we concatenate the coordinates together.

There are two problems with this approach: (i) the number of resulting coordinates is infinite and (ii) a pair x, y might be distorted a "lot" because it contributes to all resolutions, not only to its "relevant" resolutions. Both problems can be overcome with careful tinkering.

Snapped metric. For a resolution r_i, we are going to modify the metric, so that it ignores short distances (i.e., distances $\leq r_i/n^2$). Formally, for each resolution r_i, let G_i be the complete weighted graph over $\mathcal{X}$, where the edge xy is assigned weight 0 if $\mathbf{d}(x, y) \leq r_i/n^2$ and distance $\min(\mathbf{d}(x, y), 4r_i)$ otherwise. Let $\mathbf{d}_i$ denote the shortest path distance function induced by G_i. Let $\widehat{\Pi}_i$ be the partition of $\mathcal{X}$ such that two points $x, y \in \mathcal{X}$ are in the same cluster if $\mathbf{d}_i(x, y) = 0$. Let $\mathcal{X}_i$ be a set containing a representative from each cluster of $\widehat{\Pi}_i$. For a point $\mathsf{p} \in \mathcal{X}$, we denote by $\text{rep}_i(\mathsf{p})$ the ***representative*** of the cluster $\widehat{\Pi}_i(\mathsf{p})$ in $\mathcal{X}_i$.

For the new metric space $\mathcal{Z}_i = (\mathcal{X}_i, \mathbf{d}_i)$ we have that

(A) the spread of $\mathcal{Z}_i$ is $O(n^2)$,

(B) $\forall x, y \in \mathcal{X}$, if $\mathbf{d}_i(\text{rep}_i(x), \text{rep}_i(y)) = 0$, then $\mathbf{d}(x, y) \leq r_i/n$,

(C) $\forall x, y \in \mathcal{X} \implies \mathbf{d}(x, y) - 2r_i/n \le \mathbf{d}_i(\text{rep}_i(x), \text{rep}_i(y)) \le \mathbf{d}(x, y)$, and
(D) $\forall x, y \in \mathcal{X} \implies \mathbf{d}_i(\text{rep}_i(x), \text{rep}_i(y)) \le 4r_i$.

Intuitively, the metric $\mathbf{d}_i$ cares only about distances that are roughly r_i and ignores much shorter or longer distances.

The embedding. We now embed $\mathcal{Z}_i = (\mathcal{X}_i, \mathbf{d}_i)$ into $\beta = O(\log n)$ coordinates (with $\Delta = r_i$). Next, for any point of $\mathsf{p} \in \mathcal{X}$, its embedding is the embedding of $\text{rep}_i(\mathsf{p})$. Let E_i be the resulting embedding for resolution r_i. Namely, $E_i(x) = (f_{i,1}(x), f_{i,2}(x), \ldots, f_{i,\beta}(x))$, where $f_{i,j}(x) = \mathbf{d}_i(x, \mathcal{X} \setminus V_{i,j})$. The resulting embedding is $F(x) = \bigoplus E_i(x) = (E_1(x), E_2(x), \ldots)$.

Consider two points $x, y \in \mathcal{X}$, and observe that

$$\left|f_{i,j}(x) - f_{i,j}(y)\right| = \left|\mathbf{d}_i(\text{rep}_i(x), V_{i,j}) - \mathbf{d}_i(\text{rep}_i(y), V_{i,j})\right| \le \mathbf{d}_i\big(\text{rep}_i(x), \text{rep}_i(y)\big) \le \mathbf{d}(x, y),$$

by Lemma 26.10. This implies that, for any $x, y \in \mathcal{X}$,

$$\left\|E_i(x) - E_i(y)\right\|^2 \le (\mathbf{d}_i(\text{rep}_i(x), \text{rep}_i(y)))^2 \beta \le \min\big(16\beta r_i^2, \beta\phi^2\big).$$

Bounded distortion. Consider a pair $x, y \in \mathcal{X}$ such that $\phi = \mathbf{d}(x, y) \in [2r_i, 4r_i]$. To see that $F(\cdot)$ is the required embedding (up to scaling), observe that $\mathbf{d}_i(\text{rep}_i(x), \text{rep}_i(y)) > r_i$ and arguing as in Theorem 26.11, with high probability, we have $\|F(x) - F(y)\| = \Omega\left(\phi \frac{\sqrt{\beta}}{\ln n}\right)$. To get an upper bound on this distance, observe that for i such that $r_i > \phi n^2$, we have $E_i(x) = E_i(y)$ and as such $\|E_i(x) - E_i(y)\|^2 = 0$. Thus, we have

$$\begin{aligned}
\|F(x) - F(y)\|^2 &= \sum_i \|E_i(x) - E_i(y)\|^2 = \sum_{i, r_i < \phi n^2} \|E_i(x) - E_i(y)\|^2 \\
&= \sum_{i, \phi/n^2 < r_i < \phi n^2} \|E_i(x) - E_i(y)\|^2 + \sum_{i, r_i < \phi/n^2} \|E_i(x) - E_i(y)\|^2 \\
&= \beta\phi^2 \lg\big(n^4\big) + \sum_{i, r_i < \phi/n^2} 16 r_i^2 \beta \le 4\beta\phi^2 \lg n + \frac{32\phi^2\beta}{n^4} = O\big(\beta\phi^2 \lg n\big).
\end{aligned}$$

Thus, $\|F(x) - F(y)\| = O\big(\phi\sqrt{\beta \lg n}\big)$. We conclude that, with high probability, $F(\cdot)$ is an embedding of $\mathcal{X}$ with distortion $O\Big(\phi\sqrt{\beta \lg n} \,/\, \Big(\phi \frac{\sqrt{\beta}}{\ln n}\Big)\Big) = O\big(\log^{3/2} n\big)$.

Reducing the number of coordinates. We still have to handle the infinite number of coordinates problem. However, the above proof shows that we care about a resolution r_i (i.e., it contributes to the estimates in the above proof) only if there is a pair x and y such that $r_i/n^2 \le \mathbf{d}(x, y) \le r_i n^2$. Thus, for every pair of distances there are $O(\log n)$ relevant resolutions. Thus, there are at most $\eta = O(n^2 \beta \log n) = O(n^2 \log^2 n)$ relevant coordinates, and we can ignore all the other coordinates. As such the number of coordinates in the resulting embedding is finite.

Note that all this process succeeds with high probability. If it fails, we try again. We conclude:

Theorem 26.12 (Low quality Bourgain theorem). *Given an n-point metric M, one can embed it into Euclidean space of dimension $n - 1$, such that the distortion of the embedding is at most $O(\log^{3/2} n)$.*

Using the JL lemma, the dimension can be further reduced to $O(\log n)$. Being more careful in the proof, it is possible to reduce the dimension to $O(\log n)$ directly. Also, the bound on the distortion can be improved to $O(\log n)$.

26.5. Bibliographical notes

The partitions we use are due to Calinescu et al. [**CKR01**]. The idea of embedding into spanning trees is due to Alon et al. [**AKPW95**], which showed that one can get a probabilistic distortion of $2^{O\left(\sqrt{\log n \log\log n}\right)}$. Bartal realized that by allowing trees with additional vertices, one can get a considerably better result. In particular, he showed [**Bar96**] that probabilistic embedding into trees can be done with polylogarithmic average distortion. He later improved the distortion to $O(\log n \log\log n)$ in [**Bar98**]. Improving this result was an open question, culminating in the work of Fakcharoenphol et al. [**FRT03**] which achieves the optimal $O(\log n)$ distortion.

Interestingly, if one does not care about the optimal distortion, one can get similar result (for embedding into probabilistic trees), by first embedding the metric into Euclidean space, then reducing the dimension by the JL lemma, and finally, constructing an HST by constructing a randomly shifted quadtree over the points. It is easy to verify that this yields $O(\log^4 n)$ distortion. See the survey by Indyk [**Ind01**] for more details. The random shifting of quadtrees was discussed in Chapter 11.

Our proof of Lemma 26.7 (which is originally from [**FRT03**]) is from [**KLMN05**]. The proof of Theorem 26.12 is by Gupta [**Gup00**].

A good exposition of metric spaces is available in Matoušek [**Mat02**].

Lower bound on distortion when embedding into Euclidean space. We sketch an argument showing that there are metrics that cannot be embedded into Euclidean space with distortion smaller than $O(\log n/ \log\log n)$. Our description follows Matoušek's presentation [**Mat02**], who also shows the stronger lower bound of $\Omega(\log n)$ using expanders.

Let $\mathsf{G}(n, \ell)$ denote the unweighted graph with a maximum number of edges defined over a set of n vertices V that has no cycle of length shorter than ℓ (the length of the shortest cycle in a graph is its ***girth***). It is not too hard to verify (see Exercise 26.8) that for $m(n, \ell) = |E(\mathsf{G}(n, \ell))|$ it follows that $m(n, \ell) = n^{1+\Theta(1/\ell)}$. In particular, for $\ell = \log n/(c \log\log n)$ for c a sufficiently large constant, $m(n, \ell) > n \log^2 n$.

Given a subgraph H of G, consider the metric $\widehat{\mathbf{d}_{\mathsf{H}}}(x, y) = \min(\mathbf{d}_{\mathsf{H}}(x, y), 2\ell)$, where $\mathbf{d}_{\mathsf{H}}(\cdot, \cdot)$ is the shortest path metric of H. Consider two distinct subgraphs H, Q of G and an edge $uv \in E(\mathsf{H}) \setminus E(\mathsf{Q})$. Observe that $\widehat{\mathbf{d}_{\mathsf{H}}}(u, v) = 1$ while $\widehat{\mathbf{d}_{\mathsf{Q}}}(u, v) \geq \ell - 1$. Indeed, otherwise, there would be a cycle in G that is shorter than ℓ. In particular, if we embed the graphs H and Q with distortion smaller than $D = \ell/4$, then the two embeddings must be different. Namely, every subgraph of G gives rise to a metric that must have a distinct embedding (if the distortion is at most D). Let $\mathcal{F}$ be the set of all subgraphs of G.

Now, we are embedding into Euclidean space, and by the JL lemma, we can assume the target dimension is $\mathsf{d} = O(\log n)$. Since we are interested in embeddings that are expansive and since the diameter of a metric $\widehat{\mathbf{d}_{\mathsf{Q}}}$ is 2ℓ (for any $\mathsf{Q} \in \mathcal{F}$), we can assume that the embeddings we consider map V into the ball of radius $R = 2\ell D$ in $\mathbb{R}^{\mathsf{d}}$ centered in the origin. Consider a $(1/2)$-packing P of this ball (see Definition 4.4_{p51}). Clearly, we can consider only embeddings that map V into points of P (this can increase the distortion by a factor of at most 2). An easy calculation shows that $|P| \leq (4R\mathsf{d})^{\mathsf{d}} = (8\ell D\mathsf{d})^{\mathsf{d}}$.

Observe that the number of different mappings from V to P is bounded by $M = |P|^n$. Since every two graphs of $\mathcal{F}$ must have a distinct embedding, by the above discussion, it follows that $M \geq |\mathcal{F}| = 2^{m(n,\ell)}$. In particular, as $\mathsf{d}, D, \ell = O(\log n)$, it follows that $2^{O(n \log n \log\log n)} \geq (8\ell D\mathsf{d})^{\mathsf{d}n} \geq |P|^n$. As such,

$$2^{O(n \log n \log\log n)} \geq |P|^n = M \geq 2^{m(n,\ell)} \qquad \Longrightarrow \qquad m(n, \ell) = O(n \log n \log\log n),$$

which is false, as $m(n, \ell) > n \log^2 n$. We conclude that one cannot embed all of the metrics arising out of $\mathcal{F}$ with distortion smaller than $O(\log n / \log \log n)$ into Euclidean space.

26.6. Exercises

Exercise 26.1 (L_∞ **is the mother of all metrics).** Let $(\mathcal{X}, \mathbf{d})$ be a finite metric space, where $\mathcal{X} = \{\mathsf{p}_1, \ldots, \mathsf{p}_n\}$. For a point $x \in \mathcal{X}$, consider the mapping

$$F(x) = \Big(\mathbf{d}(\mathsf{p}_1, x), \mathbf{d}(\mathsf{p}_2, x), \ldots, \mathbf{d}(\mathsf{p}_n, x)\Big).$$

Prove that for any $x, y \in \mathcal{X}$, we have that $\|F(x) - F(y)\|_\infty = \mathbf{d}(x, y)$. Namely, any n-point metric space can be embedded ***isometrically*** (i.e., without distortion) into $\mathbb{R}^n$ with the L_∞-norm.

Exercise 26.2 (Clustering for HSTs)**.** Let $(\mathcal{X}, \mathbf{d})$ be an HST defined over n points, and let $k > 0$ be an integer. Provide an algorithm that computes the optimal k-median clustering of $\mathcal{X}$ in $O(k^2 n)$ time.

[**Hint:** Transform the HST into a tree where every node has only two children. Next, run a dynamic programming algorithm on this tree.]

Exercise 26.3 (Partition induced metric)**.** (Easy) Give a counterexample to the following claim: Let $(\mathcal{X}, \mathbf{d})$ be a metric space, and let Π be a partition of $\mathcal{X}$. Then, the pair $(\Pi, \mathbf{d}')$ is a metric, where $\mathbf{d}'(C, C') = \mathbf{d}(C, C') = \min_{x \in C, y \in C'} \mathbf{d}(x, y)$ and $C, C' \in \Pi$.

Exercise 26.4 (How many distances are there, anyway?)**.** Let $(\mathcal{X}, \mathbf{d})$ be an n-point metric space, and consider the set

$$U = \left\{ i \;\middle|\; 2^i \leq \mathbf{d}(x, y) \leq 2^{i+1}, \text{ for } x, y \in \mathcal{X} \right\}.$$

Prove that $|U| = O(n)$. Namely, there are only n different resolutions that "matter" for a finite metric space.

Exercise 26.5 (Computing the diameter via embeddings)**.** (A) Let ℓ be a line in the plane, and consider the embedding $f : \mathbb{R}^2 \to \ell$, which is the projection of the plane into ℓ. Prove that f is 1-Lipschitz but it is not K-bi-Lipschitz for any constant K.

(B) Prove that one can find a family of projections $\mathcal{F}$ of size $O\big(1/\sqrt{\varepsilon}\big)$, such that for any two points $x, y \in \mathbb{R}^2$, for one of the projections $f \in \mathcal{F}$ we have $\mathbf{d}(f(x), f(y)) \geq (1-\varepsilon)\mathbf{d}(x, y)$.

(C) Given a set P of n points in the plane, give an algorithm that, in $O\big(n/\sqrt{\varepsilon}\big)$ time, outputs two points $x, y \in \mathsf{P}$, such that $\mathbf{d}(x, y) \geq (1 - \varepsilon)\mathrm{diam}(\mathsf{P})$, where $\mathrm{diam}(\mathsf{P}) = \max_{z, w \in \mathsf{P}} \mathbf{d}(z, w)$ is the diameter of P.

(D) Given P, show how to extract, in $O(n)$ time, a set $Q \subseteq \mathsf{P}$ of size $O(\varepsilon^{-2})$, such that $\mathrm{diam}(Q) \geq (1 - \varepsilon/2)\mathrm{diam}(\mathsf{P})$. (Hint: Construct a grid of appropriate resolution.)

In particular, give a $(1 - \varepsilon)$-approximation algorithm to the diameter of P that works in $O(n + \varepsilon^{-2.5})$ time. (There are slightly faster approximation algorithms known for approximating the diameter.)

Exercise 26.6 (Partitions in Euclidean space)**.** (A) For a real number $\Delta > 0$ and a random number $x \in [0, \Delta]$ consider the random partition of the real line into intervals of length Δ, such that all the points falling into the interval $[i\Delta, i\Delta + x)$ are in the same cluster. Prove that for two points $\mathsf{p}, \mathsf{q} \in \mathbb{R}$, the probability of p and q to be in two different clusters is at most $|\mathsf{p} - \mathsf{q}| / \Delta$.

(B) Consider the d-dimensional grid of sidelength Δ, and let p be a random vector in the hypercube $[0, \Delta]^d$. Shift the grid by p, and consider the partition of $\mathbb{R}^d$ induced by this grid. Formally, the space is partitioned into clusters, where all the points inside the cube $p + [0, \Delta)^d$ are one cluster. Consider any two points $q, s \in \mathbb{R}^d$. Prove that the probability that q and s are in different clusters is bounded by $d \|q - s\| / \Delta$.

(C) Strengthen (B) by showing that the probability is bounded by $\sqrt{d} \|q - s\| / \Delta$. [**Hint:** Consider the distance $t = \|q - s\|$ to be fixed, and figure out what the worst case for this partition is.]

Part (C) implies that we can partition space into clusters with diameter $\Delta' = \sqrt{d}\Delta$ such that the probability of points at distance t from each other to be separated is bounded by dt/Δ'.

Exercise 26.7 (Star distortion). Prove that the distortion of embedding the star graph with n leaves into $\mathbb{R}^d$ is $\Theta(n^{1/d})$. In addition, show that it can be embedded into $\mathbb{R}^k$ with distortion $\sqrt{2} + \delta$, for any constant $\delta > 0$, for $k = O(\log n)$.

Exercise 26.8 (Girth). Let $m(n, \ell)$ be the number of edges in a graph with n vertices and girth at least ℓ.

(A) Show that $m(n, \ell) = O(n^{1+2/\ell})$. (Hint: Throw away low degree vertices, and consider a breadth-first search (BFS) tree and its depth.)

(B) Show that there exists a graph with girth $\geq \ell$ and number of edges $m(n, \ell) = \Omega(n^{1+1/4\ell})$. (Hint: Generate a random graph by picking every edge with a certain probability. Compute the expected number of bad short cycles, and throw away an edge from each one of them. Argue that the remaining graph still has many edges.)

(By the way, finding the right bounds for $m(n, \ell)$ is still an open problem.)

CHAPTER 27

Some Probability and Tail Inequalities

"Wir müssen wissen. Wir werden wissen." (We must know. We shall know.)
– David Hilbert.

In this chapter, we prove some standard bounds on concentration of mass for random variables. See Table 27.1 for a summary of the results shown. These inequalities are crucial in analyzing many randomized and geometric algorithms.

27.1. Some probability background

Theorem 27.1 (Markov's inequality). *For a non-negative random variable X and $t > 0$, we have*

$$\mathbf{Pr}\left[X \geq t\right] \leq \frac{\mathbf{E}[X]}{t}.$$

PROOF. Assume that this is false and there exists $t_0 > 0$ such that $\mathbf{Pr}[X \geq t_0] > \frac{\mathbf{E}[X]}{t_0}$. However,

$$\mathbf{E}\left[X\right] = \sum_x x \cdot \mathbf{Pr}[X = x] = \sum_{x<t_0} x \cdot \mathbf{Pr}[X = x] + \sum_{x \geq t_0} x \cdot \mathbf{Pr}[X = x]$$
$$\geq 0 + t_0 \cdot \mathbf{Pr}[X \geq t_0] > 0 + t_0 \cdot \frac{\mathbf{E}[X]}{t_0} = \mathbf{E}\left[X\right],$$

a contraction. ■

Definition 27.2. The ***variance*** of a random variable X with expectation $\mu = \mathbf{E}[X]$ is the quantity $\mathbf{V}\left[X\right] = \mathbf{E}\left[(X-\mu)^2\right] = \mathbf{E}\left[X^2\right] - \mu^2$.

The ***standard deviation*** of X is $\sigma_X = \sqrt{\mathbf{V}[X]}$.

Theorem 27.3 (Chebychev's inequality). *Let X be a random variable with $\mu_x = \mathbf{E}[X]$ and let σ_x be the standard deviation of X. That is, $\sigma_X^2 = \mathbf{E}\left[(X-\mu_x)^2\right]$. Then, we have* $\mathbf{Pr}\left[\,|X - \mu_X| \geq t\sigma_X\right] \leq \frac{1}{t^2}$.

PROOF. Note that

$$\mathbf{Pr}\left[\,|X - \mu_X| \geq t\sigma_X\right] = \mathbf{Pr}\left[(X - \mu_X)^2 \geq t^2\sigma_X^2\right].$$

Set $Y = (X - \mu_X)^2$. Clearly, $\mathbf{E}[Y] = \sigma_X^2$. Now, apply the Markov inequality to Y. ■

Definition 27.4. The random variables X and Y are ***independent*** if, for any x, y, we have

$$\mathbf{Pr}\left[(X = x) \cap (Y = y)\right] = \mathbf{Pr}[X = x] \cdot \mathbf{Pr}[Y = y].$$

The following is easy to verify:

TABLE 27.1. Summary of Chernoff-type inequalities covered. Here we have n variables $X_1, \ldots, X_n$, $Y = \sum_i X_i$, and $\mu = \mathbf{E}[Y]$. Note that $\mathbf{Pr}[Y > (1+\delta)\mu'] \leq \mathbf{Pr}[Y > (1+\delta)\mu]$ if $\mu' \geq \mu$. As such, one can use upper bounds like μ' instead of μ in the above inequalities (that go in this direction). Similarly, one can use lower bounds on the value of μ in the bounds of the form $\mathbf{Pr}[Y < (1-\delta)\mu]$.

Values	Probabilities	Inequality	Reference
$-1, +1$	$\mathbf{Pr}[X_i = -1] = \mathbf{Pr}[X_i = 1] = \frac{1}{2}$	$\mathbf{Pr}[Y \geq \Delta] \leq e^{-\Delta^2/2n}$	Theorem 27.13
		$\mathbf{Pr}[Y \leq -\Delta] \leq e^{-\Delta^2/2n}$	Theorem 27.13
		$\mathbf{Pr}[\lvert Y\rvert \geq \Delta] \leq 2e^{-\Delta^2/2n}$	Corollary 27.14
$0, 1$	$\mathbf{Pr}[X_i = 0] = \mathbf{Pr}[X_i = 1] = \frac{1}{2}$	$\mathbf{Pr}\Big[\big\lvert Y - \frac{n}{2}\big\rvert \geq \Delta\Big] \leq 2e^{-2\Delta^2/n}$	Corollary 27.15
0,1	$\mathbf{Pr}[X_i = 0] = 1 - p_i$ $\mathbf{Pr}[X_i = 1] = p_i$	$\mathbf{Pr}[Y > (1+\delta)\mu] < \left(\frac{e^\delta}{(1+\delta)^{1+\delta}}\right)^\mu$	Theorem 27.17
	For $\delta \leq 2e - 1$	$\mathbf{Pr}[Y > (1+\delta)\mu] < \exp\big(-\mu\delta^2/4\big)$	Theorem 27.17
	For $\delta \geq 2e - 1$	$\mathbf{Pr}[Y > (1+\delta)\mu] < 2^{-\mu(1+\delta)}$	
	For $\delta \geq 0$	$\mathbf{Pr}[Y < (1-\delta)\mu] < \exp\big(-\mu\delta^2/2\big)$	Theorem 27.18
$X_i \in \big[a_i, b_i\big]$	The X_is have arbitrary independent distributions.	$\mathbf{Pr}\Big[\big\lvert Y - \mu\big\rvert \geq \eta\Big] \leq 2\exp\left(-\frac{2\eta^2}{\sum_{i=1}^n (b_i - a_i)^2}\right)$	Theorem 27.22

Claim 27.5. *If X and Y are independent, then* $\mathbf{E}[XY] = \mathbf{E}[X]\,\mathbf{E}[Y]$.

If X and Y are independent, then $Z = e^X$, $W = e^Y$ are also independent variables.

Lemma 27.6 (Linearity of expectation). *For any two random variables X and Y, we have* $\mathbf{E}[X + Y] = \mathbf{E}[X] + \mathbf{E}[Y]$.

PROOF. For the simplicity of exposition, assume that X and Y receive only integer values. We have that

$$\begin{aligned}
\mathbf{E}\big[X + Y\big] &= \sum_x \sum_y (x + y)\,\mathbf{Pr}\big[(X = x) \cap (Y = y)\big] \\
&= \sum_x \sum_y x\,\mathbf{Pr}\big[(X = x) \cap (Y = y)\big] + \sum_x \sum_y y\,\mathbf{Pr}\big[(X = x) \cap (Y = y)\big] \\
&= \sum_x x \sum_y \mathbf{Pr}\big[(X = x) \cap (Y = y)\big] + \sum_y y \sum_x \mathbf{Pr}\big[(X = x) \cap (Y = y)\big] \\
&= \sum_x x\,\mathbf{Pr}\big[X = x\big] + \sum_y y\,\mathbf{Pr}\big[Y = y\big] \\
&= \mathbf{E}\big[X\big] + \mathbf{E}\big[Y\big].
\end{aligned}$$

■

Note that linearity of expectation works even if the variables are dependent. Its obliviousness to whether or not the variables are independent makes it a useful property of expectation.

Lemma 27.7. *Let X and Y be two independent random variables. We have that $\mathbf{V}[X+Y] = \mathbf{V}[X] + \mathbf{V}[Y]$.*

PROOF. Let $\mu_x = \mathbf{E}[X]$ and $\mu_Y = \mathbf{E}[Y]$. We have that

$$\begin{aligned}
\mathbf{V}\left[X+Y\right] &= \mathbf{E}\left[(X+Y)^2\right] - \left(\mathbf{E}[X+Y]\right)^2 = \mathbf{E}\left[(X+Y)^2\right] - \left(\mu_X + \mu_Y\right)^2 \\
&= \mathbf{E}\left[X^2 + 2XY + Y^2\right] - \mu_X^2 - 2\mu_X\mu_Y - \mu_Y^2 \\
&= \left(\mathbf{E}\left[X^2\right] - \mu_X^2\right) + 2\,\mathbf{E}[X]\,\mathbf{E}[Y] - 2\mu_X\mu_Y + \left(\mathbf{E}\left[Y^2\right] - \mu_Y^2\right) \\
&= \mathbf{V}\left[X\right] + 2\mu_X\mu_Y - 2\mu_X\mu_Y + \mathbf{V}\left[Y\right] = \mathbf{V}\left[X\right] + \mathbf{V}\left[Y\right],
\end{aligned}$$

by Claim 27.5. ∎

Lemma 27.8. *For any two random variables X and Y, we have that $\mathbf{E}\left[\mathbf{E}\left[X \,\middle|\, Y\right]\right] = \mathbf{E}\left[X\right]$.*

Lemma 27.9 (Jensen's inequality)**.** *Let $f(x)$ be a convex function. Then, for any random variable X, we have $\mathbf{E}[f(X)] \geq f(\mathbf{E}[X])$.*

27.1.1. Normal distribution. The ***normal distribution*** has

$$f(x) = 1/\sqrt{2\pi}\exp\left(-x^2/2\right)$$

as its density function. We next verify that it is a valid density function.

Lemma 27.10. *We have $I = \int_{-\infty}^{\infty} f(x)dx = 1$.*

PROOF. Observe that

$$\begin{aligned}
I^2 &= \left(\int_{x=-\infty}^{\infty} f(x)\,dx\right)^2 = \left(\int_{x=-\infty}^{\infty} f(x)\,dx\right)\left(\int_{y=-\infty}^{\infty} f(y)\,dy\right) \\
&= \int_{x=-\infty}^{\infty}\int_{y=-\infty}^{\infty} f(x)f(y)\,dx\,dy \\
&= \frac{1}{2\pi}\int_{x=-\infty}^{\infty}\int_{y=-\infty}^{\infty} \exp\left(-\frac{x^2+y^2}{2}\right)\,dx\,dy.
\end{aligned}$$

Change the variables to $x = r\cos\alpha$, $y = r\sin\alpha$, and observe that the determinant of the Jacobian is

$$J = \det\begin{vmatrix} \frac{\partial x}{\partial r} & \frac{\partial x}{\partial \alpha} \\ \frac{\partial y}{\partial r} & \frac{\partial y}{\partial \alpha} \end{vmatrix} = \det\begin{vmatrix} \cos\alpha & -r\sin\alpha \\ \sin\alpha & r\cos\alpha \end{vmatrix} = r\left(\cos^2\alpha + \sin^2\alpha\right) = r.$$

As such,

$$\begin{aligned}
I^2 &= \frac{1}{2\pi}\int_{r=0}^{\infty}\int_{\alpha=0}^{2\pi} \exp\left(-\frac{r^2}{2}\right)|J|\;d\alpha\,dr = \frac{1}{2\pi}\int_{r=0}^{\infty}\int_{\alpha=0}^{2\pi} \exp\left(-\frac{r^2}{2}\right) r\,d\alpha\,dr \\
&= \int_{r=0}^{\infty} \exp\left(-\frac{r^2}{2}\right) r\,dr = \left[-\exp\left(-\frac{r^2}{2}\right)\right]_{r=0}^{r=\infty} = -\exp(-\infty) - (-\exp(0)) = 1.
\end{aligned}$$

∎

The ***multi-dimensional normal distribution***, denoted by $\mathbf{N}^d$, is the distribution in $\mathbb{R}^d$ that assigns a point $\mathsf{p} = (\mathsf{p}_1, \ldots, \mathsf{p}_d)$ the density

$$g(\mathsf{p}) = \frac{1}{(2\pi)^{d/2}}\exp\left(-\frac{1}{2}\sum_{i=1}^{d}\mathsf{p}_i^2\right).$$

It is easy to verify, using the above, that $\int_{\mathbb{R}^d} g(\mathsf{p})d\mathsf{p} = 1$. Furthermore, we have the following useful but easy properties.[①]

Lemma 27.11. *(A) The multi-dimensional normal distribution is symmetric; that is, for any two points* $\mathsf{p}, \mathsf{q} \in \mathbb{R}^d$ *such that* $\|\mathsf{p}\| = \|\mathsf{q}\|$ *we have that* $g(\mathsf{p}) = g(\mathsf{q})$*, where* $g(\cdot)$ *is the density function of the multi-dimensional normal distribution* $\mathbf{N}^d$*.*

(B) The projection of the normal distribution on any direction is a one-dimensional normal distribution.

(C) Picking d *variables* $X_1, \ldots, X_d$ *using the one-dimensional normal distribution* $\mathbf{N}$ *results in a point* $(X_1, \ldots, X_d)$ *that has multi-dimensional normal distribution* $\mathbf{N}^d$*.*

The generalized multi-dimensional distribution is a ***Gaussian***. Fortunately, we only need the simpler notion.

27.2. Tail inequalities

27.2.1. Weak concentration of mass via Chebychev's inequality. Intuitively, it is clear that if one flips a coin a sufficient number of times, we expect to get roughly the same number of heads and tails.

So, let $X_1, \ldots, X_n$ be n random independent variables that are 1 or 0 with equal probability. We have that

$$\mathbf{V}\left[X_i\right] = \mathbf{E}\left[X_i^2\right] - (\mathbf{E}[X_i])^2 = \left(\frac{1}{2}1 + \frac{1}{2}0\right) - \frac{1}{4} = \frac{1}{4}.$$

For $Y = \sum_{i=1}^n X_i$, we have that $\mathbf{E}[Y] = n/2$ and

$$\mathbf{V}\left[Y\right] = \mathbf{V}\left[\sum_{i=1}^n X_i\right] = \sum_{i=1}^n \mathbf{V}\left[X_i\right] = \frac{n}{4},$$

by Lemma 27.7. As such $\sigma_Y = \sqrt{\mathbf{V}[Y]} = \sqrt{n}/2$.

The following is now implied by plugging the above into Chebychev's inequality (Theorem 27.3) .

Lemma 27.12. *Let* $Y = \sum_{i=1}^n X_i$ *be the number of heads in n independent coin flips. We have that* $\mathbf{Pr}\left[|X - n/2| \geq t\sqrt{n}/2\right] \leq 1/t^2$*.*

This implies that the probability that we get $(\sqrt{n}/2)\sqrt{\log n}$ more heads than tails when doing n coin flips is smaller than $1/\log n$. Namely, as n grows, the distribution of Y concentrates around the expectation of Y.

Surprisingly, one can prove a considerably stronger concentration of Y around its expectation $n/2$. The proof follows from boosting the Markov inequality by a clever mapping. We next prove these stronger bounds.

27.2.2. The Chernoff bound – special case.

Theorem 27.13. *Let* $X_1, \ldots, X_n$ *be n independent random variables, such that* $\mathbf{Pr}[X_i = 1] = \mathbf{Pr}[X_i = -1] = 1/2$*, for* $i = 1, \ldots, n$*. Let* $Y = \sum_{i=1}^n X_i$*. Then, for any* $\Delta > 0$*, we have*

$$\mathbf{Pr}\left[Y \geq \Delta\right] \leq e^{-\Delta^2/2n}.$$

[①] The normal distribution has such useful properties that it seems that the only thing normal about it is its name.

PROOF. Clearly, for an arbitrary t, to be specified shortly, we have

$$\mathbf{Pr}\Big[Y \geq \Delta\Big] = \mathbf{Pr}\Big[\exp(tY) \geq \exp(t\Delta)\Big] \leq \frac{\mathbf{E}[\exp(tY)]}{\exp(t\Delta)}.$$

The first part follows as $\exp(\cdot)$ preserves ordering, and the second part follows by Markov's inequality. Observe that

$$\begin{aligned}\mathbf{E}\Big[\exp(tX_i)\Big] &= \frac{1}{2}e^t + \frac{1}{2}e^{-t} \\ &= \frac{1}{2}\left(1 + \frac{t}{1!} + \frac{t^2}{2!} + \frac{t^3}{3!} + \cdots\right) \\ &\quad + \frac{1}{2}\left(1 - \frac{t}{1!} + \frac{t^2}{2!} - \frac{t^3}{3!} + \cdots\right) \\ &= 1 + 0 + \frac{t^2}{2!} + 0 + \cdots + \frac{t^{2k}}{(2k)!} + \cdots,\end{aligned}$$

by the Taylor expansion of $\exp(\cdot)$. Note that $(2k)! \geq (k!)2^k$, and thus

$$\mathbf{E}\Big[\exp(tX_i)\Big] = \sum_{i=0}^{\infty} \frac{t^{2i}}{(2i)!} \leq \sum_{i=0}^{\infty} \frac{t^{2i}}{2^i(i!)} = \sum_{i=0}^{\infty} \frac{1}{i!}\left(\frac{t^2}{2}\right)^i = \exp\Big(t^2/2\Big),$$

again, by the Taylor expansion of $\exp(\cdot)$. Next, by the independence of the X_is, we have

$$\begin{aligned}\mathbf{E}\Big[\exp(tY)\Big] &= \mathbf{E}\left[\exp\left(\sum_i tX_i\right)\right] = \mathbf{E}\left[\prod_i \exp(tX_i)\right] = \prod_{i=1}^{n} \mathbf{E}\Big[\exp(tX_i)\Big] \\ &\leq \prod_{i=1}^{n} e^{t^2/2} = e^{nt^2/2}.\end{aligned}$$

We have

$$\mathbf{Pr}\Big[Y \geq \Delta\Big] \leq \frac{\exp\big(nt^2/2\big)}{\exp(t\Delta)} = \exp\Big(nt^2/2 - t\Delta\Big).$$

Next, to minimize the above quantity for t, we set $t = \Delta/n$. We conclude that

$$\mathbf{Pr}\Big[Y \geq \Delta\Big] \leq \exp\left(\frac{n}{2}\left(\frac{\Delta}{n}\right)^2 - \frac{\Delta}{n}\Delta\right) = \exp\left(-\frac{\Delta^2}{2n}\right).$$

■

By the symmetry of Y, we get the following:

Corollary 27.14. *Let $X_1, \ldots, X_n$ be n independent random variables, such that* $\mathbf{Pr}[X_i = 1] = 1/2$ *and* $\mathbf{Pr}[X_i = -1] = 1/2$, *for $i = 1, \ldots, n$. Let $Y = \sum_{i=1}^{n} X_i$. Then, for any $\Delta > 0$, we have* $\mathbf{Pr}\Big[|Y| \geq \Delta\Big] \leq 2\exp\Big(-\Delta^2/2n\Big)$.

Corollary 27.15. *Let $X_1, \ldots, X_n$ be n independent* coin flips, *such that* $\mathbf{Pr}[X_i = 0] = \mathbf{Pr}[X_i = 1] = 1/2$, *for $i = 1, \ldots, n$. Let $Y = \sum_{i=1}^{n} X_i$. Then, for any $\Delta > 0$, we have* $\mathbf{Pr}\Big[|Y - n/2| \geq \Delta\Big] \leq 2\exp\Big(-2\Delta^2/n\Big)$.

PROOF. Observe that $Z_i = 2X_i - 1$ is a random variable in $\{-1, +1\}$, for $i = 1, \ldots, n$. Observe that $Z_1, \ldots, Z_n$ are independent and $\mathbf{Pr}[Z_i = +1] = \mathbf{Pr}[Z_i = -1] = 1/2$. Now,

$X_i = (Z_i + 1)/2$ and

$$\mathbf{Pr}\left[\left|Y - \frac{n}{2}\right| \geq \Delta\right] = \mathbf{Pr}\left[\left|\sum_{i=1}^{n} X_i - \frac{n}{2}\right| \geq \Delta\right] = \mathbf{Pr}\left[\left|\sum_{i=1}^{n} \frac{Z_i + 1}{2} - \frac{n}{2}\right| \geq \Delta\right]$$
$$= \mathbf{Pr}\left[\left|\sum_{i=1}^{n} (Z_i + 1) - n\right| \geq 2\Delta\right] = \mathbf{Pr}\left[\left|\sum_{i=1}^{n} Z_i\right| \geq 2\Delta\right] \leq 2\exp\left(-\frac{2\Delta^2}{n}\right),$$

by Corollary 27.14. ■

Remark 27.16. Before going any further, it might be instrumental to understand what these inequalities imply. Consider the case where X_i is either 0 or 1 with probability $1/2$, and let $Y = \sum_{i=1}^{n} X_i$. In this case $\mu = \mathbf{E}[Y] = n/2$. Set $\Delta = t\sigma_Y = t\sqrt{n}/2$. We have, by Corollary 27.15, that

$$\mathbf{Pr}\left[\left|Y - \frac{n}{2}\right| \geq t\sigma_Y\right] \leq 2\exp\left(-\frac{2\Delta^2}{n}\right) = 2\exp\left(-\frac{2(t\sqrt{n}/2)^2}{n}\right) = 2\exp\left(-\frac{t^2}{2}\right).$$

Thus, Chernoff's inequality implies exponential decay on the probability that Y strays by more than t standard deviations from its expectations. In Lemma 27.12 we bounded the same probability using Chebychev's inequality and obtained a considerably weaker bound of $1/t^2$. In particular, for $t = \sqrt{10 \ln n}$, the above implies that the probability of Y straying by more than $10\sigma_Y \ln n$ from its expectation is smaller than $1/n^5$, while Lemma 27.12 implies that this probability is smaller than $1/(10 \ln n)$.

27.2.3. The Chernoff bound – general case. Here we present the Chernoff bound in a more general setting. Let $X_1, \ldots, X_n$ be n independent Bernoulli trials, where

$$\mathbf{Pr}\left[X_i = 1\right] = p_i \quad \text{and} \quad \mathbf{Pr}\left[X_i = 0\right] = q_i = 1 - p_i,$$

for $i = 1, \ldots, n$. Each X_i is known as a ***Poisson trial***. Let $X = \sum_{i=1}^{n} X_i$ and $\mu = \mathbf{E}[X] = \sum_{i=1}^{n} p_i$.

Theorem 27.17. *For any $\delta > 0$, we have $\mathbf{Pr}\left[X \geq (1+\delta)\mu\right] \leq \left(\frac{e^\delta}{(1+\delta)^{1+\delta}}\right)^\mu$, or, in a more simplified form, for any $\delta \leq 2e - 1$,*

$$\mathbf{Pr}\left[X \geq (1+\delta)\mu\right] \leq \exp\left(-\mu\delta^2/4\right) \tag{27.1}$$

and

$$\mathbf{Pr}\left[X \geq (1+\delta)\mu\right] \leq 2^{-\mu(1+\delta)}, \tag{27.2}$$

for $\delta \geq 2e - 1$.

Proof. We have $\mathbf{Pr}\left[X \geq (1+\delta)\mu\right] = \mathbf{Pr}\left[e^{tX} \geq e^{t(1+\delta)\mu}\right]$, for $t \geq 0$. By Markov's inequality, we have

$$\mathbf{Pr}\left[X \geq (1+\delta)\mu\right] \leq \frac{\mathbf{E}\left[e^{tX}\right]}{e^{t(1+\delta)\mu}}.$$

On the other hand,

$$\mathbf{E}\left[e^{tX}\right] = \mathbf{E}\left[e^{t(X_1+X_2+\cdots+X_n)}\right] = \mathbf{E}\left[e^{tX_1}\right]\cdots\mathbf{E}\left[e^{tX_n}\right].$$

Namely,

$$\mathbf{Pr}\left[X \geq (1+\delta)\mu\right] \leq \frac{\prod_{i=1}^{n} \mathbf{E}\left[e^{tX_i}\right]}{e^{t(1+\delta)\mu}} = \frac{\prod_{i=1}^{n}\left((1-p_i)e^0 + p_i e^t\right)}{e^{t(1+\delta)\mu}} = \frac{\prod_{i=1}^{n}(1 + p_i(e^t - 1))}{e^{t(1+\delta)\mu}}.$$

Let $y = p_i(e^t - 1)$. We know that $1 + y < e^y$ (since $y > 0$). Thus, settings $t = \ln(1+\delta)$, we have

$$\mathbf{Pr}\Big[X \geq (1+\delta)\mu\Big] \leq \frac{\prod_{i=1}^n \exp(p_i(e^t - 1))}{e^{t(1+\delta)\mu}} = \frac{\exp\Big(\sum_{i=1}^n p_i(e^t - 1)\Big)}{e^{t(1+\delta)\mu}}$$

$$= \frac{\exp\Big((e^t - 1)\sum_{i=1}^n p_i\Big)}{e^{t(1+\delta)\mu}} = \frac{\exp((e^t - 1)\mu)}{e^{t(1+\delta)\mu}} = \left(\frac{\exp(e^t - 1)}{e^{t(1+\delta)}}\right)^{\mu} = \left(\frac{\exp(\delta)}{(1+\delta)^{(1+\delta)}}\right)^{\mu}.$$

For the proof of the simplified form, see Section 27.2.4. ■

27.2.3.1. *The Chernoff bound – general case – the other direction.* Here we prove the other direction of the Chernoff inequality in the general case. The proof is very similar to the original proof, with a few minor technicalities, and is included for the sake of completeness.

Theorem 27.18. *Let $X_1, \ldots, X_n$ be independent Bernoulli trials, where $\mathbf{Pr}[X_i = 1] = p_i$ and $\mathbf{Pr}[X_i = 0] = q_i = 1 - p_i$, for $i = 1, \ldots, n$. Furthermore, let $X = \sum_{i=1}^b X_i$ and $\mu = \mathbf{E}[X] = \sum_i p_i$. Then we have $\mathbf{Pr}\Big[X < (1-\delta)\mu\Big] < \exp\Big(-\mu\delta^2/2\Big)$.*

Proof. We have $\mathbf{Pr}[X \leq (1-\delta)\mu] = \mathbf{Pr}\Big[e^{tX} \leq e^{t(1-\delta)\mu}\Big] = \mathbf{Pr}\Big[e^{-tX} \geq e^{-t(1-\delta)\mu}\Big]$, for $t \geq 0$. By Markov's inequality, we have

$$\mathbf{Pr}\Big[X \leq (1-\delta)\mu\Big] \leq \frac{\mathbf{E}\Big[e^{-tX}\Big]}{e^{-t(1-\delta)\mu}}.$$

As before, we have that $\mathbf{E}\Big[e^{-tX}\Big] = \mathbf{E}\Big[e^{-tX_1}\Big] \cdots \mathbf{E}\Big[e^{-tX_n}\Big]$. Namely,

$$\mathbf{Pr}\Big[X \leq (1-\delta)\mu\Big] \leq \frac{\prod_{i=1}^n \mathbf{E}\Big[e^{-tX_i}\Big]}{e^{-t(1-\delta)\mu}} = \frac{\prod_{i=1}^n\Big((1-p_i)e^{-0} + p_i e^{-t}\Big)}{e^{-t(1-\delta)\mu}} = \frac{\prod_{i=1}^n (1 + p_i(e^{-t} - 1))}{e^{-t(1-\delta)\mu}}.$$

Let $y_i = p_i(e^{-t} - 1)$ and observe that $1 + y_i \leq \exp(y_i)$ for $y_i \in [-1, 1]$. Thus, using $t = \ln \dfrac{1}{1-\delta}$, we have

$$\mathbf{Pr}\Big[X \leq (1-\delta)\mu\Big] \leq \frac{\prod_{i=1}^n \exp(p_i(e^{-t} - 1))}{e^{-t(1-\delta)\mu}} = \frac{\exp\Big(\sum_{i=1}^n p_i(e^{-t} - 1)\Big)}{e^{-t(1-\delta)\mu}}$$

$$= \frac{\exp\Big((e^{-t} - 1)\sum_{i=1}^n p_i\Big)}{e^{-t(1-\delta)\mu}} = \frac{\exp((e^{-t} - 1)\mu)}{e^{-t(1-\delta)\mu}} = \left(\frac{\exp(e^{-t} - 1)}{e^{-t(1-\delta)}}\right)^{\mu}$$

$$= \left(\frac{\exp(-\delta)}{(1-\delta)^{(1-\delta)}}\right)^{\mu}.$$

Now, Lemma 27.19 below implies that $(1-\delta)^{1-\delta} \geq \exp\Big(-\delta + \delta^2/2\Big)$. As such, we conclude that

$$\mathbf{Pr}\Big[X \leq (1-\delta)\mu\Big] \leq \left(\frac{\exp(-\delta)}{\exp(-\delta + \delta^2/2)}\right)^{\mu} = \exp\left(-\frac{\mu\delta^2}{2}\right).$$

■

Lemma 27.19. *For $\delta \in [0, 1)$, we have that $(1-\delta)\ln(1-\delta) \geq -\delta + \delta^2/2$.*

PROOF. Plugging in the Taylor expansion of $\ln(1-\delta)$ (see Lemma 28.8_{p348}), we have

$$(1-\delta)\ln(1-\delta) = (1-\delta)\sum_{i=1}^{\infty}(-1)^{i+1}\frac{(-\delta)^i}{i} = (\delta-1)\sum_{i=1}^{\infty}\frac{\delta^i}{i} = \sum_{i=1}^{\infty}\frac{\delta^{i+1}}{i} - \sum_{i=1}^{\infty}\frac{\delta^i}{i}$$

$$= \sum_{i=2}^{\infty}\frac{\delta^i}{i-1} - \sum_{i=1}^{\infty}\frac{\delta^i}{i} = -\delta + \sum_{i=2}^{\infty}\left(\frac{1}{i-1}-\frac{1}{i}\right)\delta^i$$

$$= -\delta + \sum_{i=2}^{\infty}\frac{1}{i(i-1)}\delta^i \geq -\delta + \frac{\delta^2}{2}.$$

■

27.2.4. A more convenient form.

PROOF OF THE SIMPLIFIED FORM OF THEOREM 27.17. Equation (27.2) is just Exercise 27.1. As for (27.1), we will prove this only for $\delta \leq 1/2$. For details about the case $1/2 \leq \delta \leq 2e-1$; see [**MR95**]. By Theorem 27.17, we have

$$\mathbf{Pr}\Big[X > (1+\delta)\mu\Big] < \left(\frac{e^{\delta}}{(1+\delta)^{1+\delta}}\right)^{\mu} = \exp\Big(\mu\delta - \mu(1+\delta)\ln(1+\delta)\Big).$$

The Taylor expansion of $\ln(1+\delta)$ is $\delta - \frac{\delta^2}{2} + \frac{\delta^3}{3} - \frac{\delta^4}{4} + \cdot \geq \delta - \frac{\delta^2}{2}$, for $\delta \leq 1$, see Lemma 28.8_{p348}. Thus,

$$\mathbf{Pr}\Big[X > (1+\delta)\mu\Big] < \exp\Big(\mu\Big(\delta - (1+\delta)\Big(\delta-\delta^2/2\Big)\Big)\Big) = \exp\Big(\mu\Big(\delta-\delta+\delta^2/2-\delta^2+\delta^3/2\Big)\Big)$$

$$= \exp\Big(\mu\Big(-\delta^2/2+\delta^3/2\Big)\Big) \leq \exp\Big(-\mu\delta^2/4\Big),$$

for $\delta \leq 1/2$. ■

27.3. Hoeffding's inequality

In this section, we prove a generalization of Chernoff's inequality. The proof is considerably more tedious.

Lemma 27.20. *Let X be a random variable. If* $\mathbf{E}[X] = 0$ *and $a \leq X \leq b$, then for any $s > 0$, we have* $\mathbf{E}\Big[e^{sX}\Big] \leq \exp\Big(s^2(b-a)^2/8\Big)$.

PROOF. Let $a \leq x \leq b$ and observe that x can be written as a convex combination of a and b. In particular, we have

$$x = \lambda a + (1-\lambda)b \quad \text{for } \lambda = \frac{b-x}{b-a} \in [0,1].$$

Since $s > 0$, the function $\exp(sx)$ is convex, and as such

$$e^{sx} \leq \frac{b-x}{b-a}e^{sa} + \frac{x-a}{b-a}e^{sb},$$

since we have that $f(\lambda x + (1-\lambda)y) \leq \lambda f(x) + (1-\lambda)f(y)$ if $f(\cdot)$ is a convex function. Thus, for a random variable X, by linearity of expectation, we have

$$\mathbf{E}\Big[e^{sX}\Big] \leq \mathbf{E}\left[\frac{b-X}{b-a}e^{sa} + \frac{X-a}{b-a}e^{sb}\right] = \frac{b-\mathbf{E}[X]}{b-a}e^{sa} + \frac{\mathbf{E}[X]-a}{b-a}e^{sb}$$

$$= \frac{b}{b-a}e^{sa} - \frac{a}{b-a}e^{sb},$$

since $\mathbf{E}[X] = 0$.

Next, set $p = -\frac{a}{b-a}$ and observe that $1 - p = 1 + \frac{a}{b-a} = \frac{b}{b-a}$ and

$$-ps(b-a) = -\left(-\frac{a}{b-a}\right)s(b-a) = sa.$$

As such, we have

$$\begin{aligned}\mathbf{E}\left[e^{sX}\right] &\le (1-p)e^{sa} + pe^{sb} = (1 - p + pe^{s(b-a)})e^{sa}\\ &= (1 - p + pe^{s(b-a)})e^{-ps(b-a)}\\ &= \exp\left(-ps(b-a) + \ln\left(1 - p + pe^{s(b-a)}\right)\right) = \exp(-pu + \ln(1 - p + pe^u)),\end{aligned}$$

for $u = s(b-a)$. Setting

$$\phi(u) = -pu + \ln(1 - p + pe^u),$$

we thus have $\mathbf{E}\left[e^{sX}\right] \le \exp(\phi(u))$. To prove the claim, we will show that $\phi(u) \le u^2/8 = s^2(b-a)^2/8$.

To see that, expand $\phi(u)$ about zero using Taylor's expansion. We have

$$\phi(u) = \phi(0) + u\phi'(0) + \frac{1}{2}u^2\phi''(\theta) \tag{27.3}$$

where $\theta \in [0, u]$, and notice that $\phi(0) = 0$. Furthermore, we have

$$\phi'(u) = -p + \frac{pe^u}{1 - p + pe^u},$$

and as such $\phi'(0) = -p + \frac{p}{1-p+p} = 0$. Now,

$$\phi''(u) = \frac{(1 - p + pe^u)\,pe^u - (pe^u)^2}{(1 - p + pe^u)^2} = \frac{(1-p)\,pe^u}{(1 - p + pe^u)^2}.$$

For any $x, y \ge 0$, we have $(x + y)^2 \ge 4xy$ as this is equivalent to $(x - y)^2 \ge 0$. Setting $x = 1 - p$ and $y = pe^u$, we have that

$$\phi''(u) = \frac{(1-p)\,pe^u}{(1 - p + pe^u)^2} \le \frac{(1-p)\,pe^u}{4(1-p)\,pe^u} = \frac{1}{4}.$$

Plugging this into (27.3), we get that

$$\phi(u) \le \frac{1}{8}u^2 = \frac{1}{8}(s(b-a))^2 \quad \text{and} \quad \mathbf{E}\left[e^{sX}\right] \le \exp(\phi(u)) \le \exp\left(\frac{1}{8}(s(b-a))^2\right),$$

as claimed. ■

Lemma 27.21. *Let X be a random variable. If $\mathbf{E}[X] = 0$ and $a \le X \le b$, then for any $s > 0$, we have*

$$\mathbf{Pr}\left[X > t\right] \le \frac{\exp\left(\frac{s^2(b-a)^2}{8}\right)}{e^{st}}.$$

Proof. Using the same technique we used in proving Chernoff's inequality, we have that

$$\mathbf{Pr}\left[X > t\right] = \mathbf{Pr}\left[e^{sX} > e^{st}\right] \le \frac{\mathbf{E}\left[e^{sX}\right]}{e^{st}} \le \frac{\exp\left(\frac{s^2(b-a)^2}{8}\right)}{e^{st}}.$$

■

Theorem 27.22 (Hoeffding's inequality). *Let $X_1, \ldots, X_n$ be independent random variables, where $X_i \in [a_i, b_i]$, for $i = 1, \ldots, n$. Then, for the random variable $S = X_1 + \cdots + X_n$ and any $\eta > 0$, we have*

$$\mathbf{Pr}\Big[\big|S - \mathbf{E}[S]\big| \geq \eta\Big] \leq 2\exp\left(-\frac{2\eta^2}{\sum_{i=1}^n (b_i - a_i)^2}\right).$$

Proof. Let $Z_i = X_i - \mathbf{E}[X_i]$, for $i = 1, \ldots, n$. Set $Z = \sum_{i=1}^n Z_i$, and observe that

$$\mathbf{Pr}\Big[Z \geq \eta\Big] = \mathbf{Pr}\Big[e^{sZ} \geq e^{s\eta}\Big] \leq \frac{\mathbf{E}[\exp(sZ)]}{\exp(s\eta)},$$

by Markov's inequality. Arguing as in the proof of Chernoff's inequality, we have

$$\mathbf{E}\Big[\exp(sZ)\Big] = \mathbf{E}\left[\prod_{i=1}^n \exp(sZ_i)\right] = \prod_{i=1}^n \mathbf{E}\Big[\exp(sZ_i)\Big] \leq \prod_{i=1}^n \exp\left(\frac{s^2(b_i - a_i)^2}{8}\right),$$

since the Z_is are independent and by Lemma 27.20. This implies that

$$\mathbf{Pr}\Big[Z \geq \eta\Big] \leq \exp(-s\eta)\prod_{i=1}^n e^{s^2(b_i - a_i)^2/8} = \exp\left(\frac{s^2}{8}\sum_{i=1}^n (b_i - a_i)^2 - s\eta\right).$$

The upper bound is minimized for $s = 4\eta/\left(\sum_i (b_i - a_i)^2\right)$, implying

$$\mathbf{Pr}\Big[Z \geq \eta\Big] \leq \exp\left(-\frac{2\eta^2}{\sum (b_i - a_i)^2}\right).$$

The claim now follows by the symmetry of the upper bound (i.e., apply the same proof to $-Z$). ■

27.4. Bibliographical notes

The exposition here follows more or less the exposition in [**MR95**]. The special symmetric case (Theorem 27.13) is taken from [**Cha01**]. Although the proof is only very slightly simpler than the generalized form, it does yield a slightly better constant, and it is useful when discussing discrepancy.

An orderly treatment of probability is outside the scope of our discussion. The standard text on the topic is the book by Feller [**Fel91**]. A more accessible text might be any introductory undergraduate text on probability; for example, [**MN98**] has a nice chapter on the topic.

The proof of Hoeffding's inequality follows (roughly) the proof provided in Carlos Rodriguez's class notes (which are available online). There are generalization of Hoeffding's inequality to considerably more general setups; see Talagrand's inequality [**AS00**].

Exercise 27.2 (without the hint) is from [**Mat99**].

Finally, a recent good reference on concentration of measure and tail inequalities is the book by Dubhashi and Panconesi [**DP09**].

27.5. Exercises

Exercise 27.1 (Simpler tail inequality). Prove that for $\delta > 2e - 1$, we have

$$F^+(\mu, \delta) = \left[\frac{e^\delta}{(1+\delta)^{(1+\delta)}}\right]^\mu < \left[\frac{e}{1+\delta}\right]^{(1+\delta)\mu} \leq 2^{-(1+\delta)\mu}.$$

Exercise 27.2 (Chernoff inequality is tight)**.** Let $S = \sum_{i=1}^{n} S_i$ be a sum of n independent random variables each attaining values $+1$ and -1 with equal probability. Let $P(n, \Delta) = \mathbf{Pr}[S > \Delta]$. Prove that for $\Delta \leq n/C$,

$$P(n, \Delta) \geq \frac{1}{C} \exp\left(-\frac{\Delta^2}{Cn}\right),$$

where C is a suitable constant. That is, the Chernoff bound $P(n, \Delta) \leq \exp(-\Delta^2/2n)$ is close to the truth.

[**Hint:** Use estimates for the middle binomial coefficient [**MN98**, pages 83–84]. Alternatively, a simpler approach (but less elementary) is to use Stirling's formula.]

Exercise 27.3 (Tail inequality for geometric variables)**.** Let $X_1, \ldots, X_m$ be m independent random variables with geometric distribution with probability p (i.e., $\mathbf{Pr}[X_i = j] = (1 - p)^{j-1} p$). Let $Y = \sum_i X_i$, and let $\mu = \mathbf{E}[Y] = m/p$. Prove that

$$\mathbf{Pr}\Big[Y \geq (1 + \delta)\mu\Big] \leq \exp\left(-\frac{m\delta^2}{8}\right).$$

CHAPTER 28

Miscellaneous Prerequisite

The purpose of this chapter is to remind the reader (and the author) about some basic definitions and results in mathematics used in the text. The reader should refer to standard texts for further details.

28.1. Geometry and linear algebra

A set X in $\mathbb{R}^{\mathsf{d}}$ is ***closed***, if any sequence of converging points of X converges to a point that is inside X. A set $X \subseteq \mathbb{R}^{\mathsf{d}}$ is ***compact*** if it is closed and bounded; namely; there exists a constant c, such that for all $\mathsf{p} \in X$, $\|\mathsf{p}\| \leq c$.

Definition 28.1 (Convex hull)**.** The ***convex hull*** of a set $\mathsf{R} \subseteq \mathbb{R}^{\mathsf{d}}$ is the set of all convex combinations of points of R; that is,

$$\mathcal{CH}(\mathsf{R}) = \left\{ \sum_{i=0}^{m} \alpha_i \mathsf{s}_i \;\middle|\; \forall i \;\; \mathsf{s}_i \in \mathsf{R}, \alpha_i \geq 0, \text{ and } \sum_{j=1}^{m} \alpha_i = 1 \right\}.$$

In the following, we cover some material from linear algebra. Proofs of these facts can be found in any text on linear algebra, for example **[Leo98]**.

For a matrix $\mathbf{M}$, let $\mathbf{M}^T$ denote the transposed matrix. We remind the reader that for two matrices $\mathbf{M}$ and $\mathbf{B}$, we have $(\mathbf{MB})^T = \mathbf{B}^T\mathbf{M}^T$. Furthermore, for any three matrices $\mathbf{M}, \mathbf{B}$, and $\mathbf{C}$, we have $(\mathbf{MB})\mathbf{C} = \mathbf{M}(\mathbf{BC})$.

A matrix $\mathbf{M} \in \mathbb{R}^{n \times n}$ is ***symmetric*** if $\mathbf{M}^T = \mathbf{M}$. All the eigenvalues of a symmetric matrix are real numbers. A symmetric matrix $\mathbf{M}$ is ***positive definite*** if $x^T\mathbf{M}x > 0$, for all $x \in \mathbb{R}^n$. Among other things this implies that $\mathbf{M}$ is non-singular. If $\mathbf{M}$ is symmetric, then it is positive definite if and only if all its eigenvalues are positive numbers.

In particular, if $\mathbf{M}$ is symmetric positive definite, then $\det(\mathbf{M}) > 0$. Since all the eigenvalues of a positive definite matrix are positive real numbers, the following holds, as can be easily verified.

Claim 28.2. *A symmetric matrix* $\mathbf{M}$ *is positive definite if and only if there exists a matrix* $\mathbf{B}$ *such that* $\mathbf{M} = \mathbf{B}^T\mathbf{B}$ *and* $\mathbf{B}$ *is not singular.*

For two vectors $u, v \in \mathbb{R}^n$, let $\langle u, v \rangle = u^T v$ denote their dot product.

Lemma 28.3. *Given a simplex* $\triangle$ *in* $\mathbb{R}^{\mathsf{d}}$ *with vertices* $v_1, \ldots, v_{\mathsf{d}}, v_{\mathsf{d}+1}$ *(or equivalently* $\triangle = \mathcal{CH}(v_1, \ldots, v_{\mathsf{d}+1})$*), the* **volume** *of this simplex is the absolute value of* $(1/\mathsf{d}!)|C|$*, where* C *is the value of the determinant* $C = \begin{vmatrix} 1 & 1 & \ldots & 1 \\ v_1 & v_2 & \ldots & v_{\mathsf{d}+1} \end{vmatrix}$. *In particular, for a triangle with vertices at* (x, y), (x', y'), *and* (x'', y'') *its area is the absolute value of* $\dfrac{1}{2}\begin{vmatrix} 1 & x & y \\ 1 & x' & y' \\ 1 & x'' & y'' \end{vmatrix}$.

28.1.1. Linear and affine subspaces.

Definition 28.4. The ***linear subspace*** spanned by a set of vectors $\mathcal{V} \subseteq \mathbb{R}^{\mathsf{d}}$ is the set $\mathrm{linear}(\mathcal{V}) = \left\{ \sum_i \alpha_i \vec{\mathsf{v}}_i \;\middle|\; \alpha_i \in \mathbb{R}, \vec{\mathsf{v}}_i \in \mathcal{V} \right\}$.

An ***affine combination*** of vectors $\vec{v}_1, \ldots, \vec{v}_n$ is a linear combination $\sum_{i=1}^{n} \alpha_i \cdot \vec{v}_i = \alpha_1 \vec{v}_1 + \alpha_2 \vec{v}_2 + \cdots + \alpha_n \vec{v}_n$ in which the sum of the coefficients is 1; thus, $\sum_{i=1}^{n} \alpha_i = 1$. The maximum dimension of the affine subspace in such a case is $(n-1)$-dimensions.

Definition 28.5. The ***affine subspace*** spanned by a set $\mathcal{V} \subseteq \mathbb{R}^{\mathsf{d}}$ is

$$\mathrm{affine}(\mathcal{V}) = \left\{ \sum_i \alpha_i \vec{\mathsf{v}}_i \;\middle|\; \alpha_i \in \mathbb{R}, \vec{\mathsf{v}}_i \in \mathcal{V}, \text{ and } \sum_i \alpha_i = 1 \right\}.$$

For any vector $\vec{\mathsf{v}} \in \mathcal{V}$, we have that $\mathrm{affine}(\mathcal{V}) = \vec{\mathsf{v}} + \mathrm{linear}\left(\mathcal{V} - \vec{\mathsf{v}}\right)$, where $\mathcal{V} - \vec{\mathsf{v}} = \left\{ \vec{\mathsf{v}}' - \vec{\mathsf{v}} \;\middle|\; \vec{\mathsf{v}}' \in \mathcal{V} \right\}$.

28.1.2. Computational geometry. The following are standard results in computational geometry; see [**dBCvKO08**] for more details.

Lemma 28.6. *The convex hull of n points in the plane can be computed in $O(n \log n)$ time.*

Lemma 28.7. *The lower and upper envelopes of n lines in the plane can be computed in $O(n \log n)$ time.*

Proof. Use duality and the algorithm of Lemma 28.6. ■

28.2. Calculus

Lemma 28.8. *For $x \in (-1, 1)$, we have* $\ln(1+x) = x - \frac{x^2}{2} + \frac{x^3}{3} - \frac{x^4}{4} + \cdots = \sum_{i=1}^{\infty} (-1)^{i+1} \frac{x^i}{i}$.

Lemma 28.9. *The following hold:*
(A) For all $x \in \mathbb{R}$, $1 + x \le \exp(x)$.
(B) For $x \ge 0$, $1 - x \le \exp(-x)$.
(C) For $0 \le x \le 1$, $\exp(x) \le 1 + 2x$.
(D) For $x \in [0, 1/2]$, $\exp(-2x) \le 1 - x$.

Proof. (A) Let $f(x) = 1 + x$ and $g(x) = \exp(x)$. Observe that $f(0) = g(0) = 1$. Now, for $x \ge 0$, we have that $f'(x) = 1$ and $g'(x) = \exp(x) \ge 1$. As such $f(x) \le g(x)$ for $x \ge 0$. Similarly, for $x < 0$, we have $g'(x) = \exp(x) < 1$, which implies that $f(x) \le g(x)$.

(B) This is immediate from (A).

(C) Observe that $\exp(1) \le 1 + 2 \cdot 1$ and $\exp(0) = 1 + 2 \cdot 0$. By the convexity of $1 + 2x$, it follows that $\exp(x) \le 1 + 2x$ for all $x \in [0, 1]$.

(D) Observe that (i) $\exp(-2(1/2)) = 1/e \le 1/2 = 1 - (1/2)$, (ii) $\exp(-2 \cdot 0) = 1 \le 1 - 0$, (iii) $\exp(-2x)' = -2\exp(-2x)$, and (iv) $\exp(-2x)'' = 4\exp(-2x) \ge 0$ for all x. As such, $\exp(-2x)$ is a convex function and the claim follows. ■

Lemma 28.10. *For $1 > \varepsilon > 0$ and $y \ge 1$, we have that* $\frac{\ln y}{\varepsilon} \le \log_{1+\varepsilon} y \le 2 \frac{\ln y}{\varepsilon}$.

Proof. By Lemma 28.9, $1 + x \le \exp(x) \le 1 + 2x$ for $x \in [0, 1]$. This implies that $\ln(1+x) \le x \le \ln(1+2x)$. As such, $\log_{1+\varepsilon} y = \frac{\ln y}{\ln(1+\varepsilon)} = \frac{\ln y}{\ln(1 + 2(\varepsilon/2))} \le \frac{\ln y}{\varepsilon/2}$. The other inequality follows in a similar fashion. ■

Bibliography

[AACS98] P. K. Agarwal, B. Aronov, T. M. Chan, and M. Sharir, *On levels in arrangements of lines, segments, planes, and triangles*, Discrete Comput. Geom. **19** (1998), 315–331.

[AB99] M. Anthony and P. L. Bartlett, *Neural network learning: Theoretical foundations*, Cambridge, 1999.

[ABCC07] D. L. Applegate, R. E. Bixby, V. Chvatal, and W. J. Cook, *The traveling salesman problem: A computational study*, Princeton University Press, 2007.

[AC09] P. Afshani and T. M. Chan, *On approximate range counting and depth*, Discrete Comput. Geom. **42** (2009), no. 1, 3–21.

[ACFS09] M. A. Abam, P. Carmi, M. Farshi, and M. H. M. Smid, *On the power of the semi-separated pair decomposition*, Proc. 11th Workshop Algorithms Data Struct., 2009, pp. 1–12.

[Ach01] D. Achlioptas, *Database-friendly random projections*, Proc. 20th ACM Sympos. Principles Database Syst., 2001, pp. 274–281.

[ACNS82] M. Ajtai, V. Chvátal, M. Newborn, and E. Szemerédi, *Crossing-free subgraphs*, Ann. Discrete Math. **12** (1982), 9–12.

[AD97] P. K. Agarwal and P. K. Desikan, *An efficient algorithm for terrain simplification*, Proc. 8th ACM-SIAM Sympos. Discrete Algorithms, 1997, pp. 139–147.

[AdBF+09] M. A. Abam, M. de Berg, M. Farshi, J. Gudmundsson, and M. H. M. Smid, *Geometric spanners for weighted point sets*, Proc. 17th Annu. European Sympos. Algorithms, 2009, pp. 190–202.

[AdBFG09] M. A. Abam, M. de Berg, M. Farshi, and J. Gudmundsson, *Region-fault tolerant geometric spanners*, Discrete Comput. Geom. **41** (2009), no. 4, 556–582.

[AE98] P. K. Agarwal and J. Erickson, *Geometric range searching and its relatives*, Advances in Discrete and Computational Geometry (B. Chazelle, J. E. Goodman, and R. Pollack, eds.), Amer. Math. Soc., 1998.

[AEIS99] A. Amir, A. Efrat, P. Indyk, and H. Samet, *Efficient algorithms and regular data structures for dilation, location and proximity problems*, Proc. 40th Annu. IEEE Sympos. Found. Comput. Sci., 1999, pp. 160–170.

[AES99] P. K. Agarwal, A. Efrat, and M. Sharir, *Vertical decomposition of shallow levels in 3-dimensional arrangements and its applications*, SIAM J. Comput. **29** (1999), 912–953.

[AESW91] P. K. Agarwal, H. Edelsbrunner, O. Schwarzkopf, and E. Welzl, *Euclidean minimum spanning trees and bichromatic closest pairs*, Discrete Comput. Geom. **6** (1991), no. 5, 407–422.

[AG86] N. Alon and E. Győri, *The number of small semispaces of a finite set of points in the plane*, J. Combin. Theory Ser. A **41** (1986), 154–157.

[AGK+01] V. Arya, N. Garg, R. Khandekar, K. Munagala, and V. Pandit, *Local search heuristic for k-median and facility location problems*, Proc. 33rd Annu. ACM Sympos. Theory Comput., 2001, pp. 21–29.

[AH08] B. Aronov and S. Har-Peled, *On approximating the depth and related problems*, SIAM J. Comput. **38** (2008), no. 3, 899–921.

[AH10] M. A. Abam and S. Har-Peled, *New constructions of SSPDs and their applications*, Proc. 26th Annu. ACM Sympos. Comput. Geom., 2010, `http://www.cs.uiuc.edu/~sariel/papers/09/sspd/`, pp. 192–200.

[AHK06] S. Arora, E. Hazan, and S. Kale, *Multiplicative weights method: a meta-algorithm and its applications*, manuscript. Available from `http://www.cs.princeton.edu/~arora/pubs/MWsurvey.pdf`, 2006.

[AHS07] B. Aronov, S. Har-Peled, and M. Sharir, *On approximate halfspace range counting and relative epsilon-approximations*, Proc. 23rd Annu. ACM Sympos. Comput. Geom., 2007, pp. 327–336.

[AHV04] P. K. Agarwal, S. Har-Peled, and K. R. Varadarajan, *Approximating extent measures of points*, J. Assoc. Comput. Mach. **51** (2004), no. 4, 606–635.

[AHV05] P. K. Agarwal, S. Har-Peled, and K. Varadarajan, *Geometric approximation via coresets*, Combinatorial and Computational Geometry (J. E. Goodman, J. Pach, and E. Welzl, eds.), Math. Sci. Research Inst. Pub., Cambridge, 2005.

[AHY07] P. Agarwal, S. Har-Peled, and H. Yu, *Embeddings of surfaces, curves, and moving points in Euclidean space*, Proc. 23rd Annu. ACM Sympos. Comput. Geom., 2007, pp. 381–389.

[AI06] A. Andoni and P. Indyk, *Near-optimal hashing algorithms for approximate nearest neighbor in high dimensions*, Proc. 47th Annu. IEEE Sympos. Found. Comput. Sci., 2006, pp. 459–468.

[AI08] ———, *Near-optimal hashing algorithms for approximate nearest neighbor in high dimensions*, Commun. ACM **51** (2008), no. 1, 117–122.

[AKPW95] N. Alon, R. M. Karp, D. Peleg, and D. West, *A graph-theoretic game and its application to the k-server problem*, SIAM J. Comput. **24** (1995), no. 1, 78–100.

[AM94] P. K. Agarwal and J. Matoušek, *On range searching with semialgebraic sets*, Discrete Comput. Geom. **11** (1994), 393–418.

[AM98] S. Arya and D. Mount, *ANN: library for approximate nearest neighbor searching*, `http://www.cs.umd.edu/~mount/ANN/`, 1998.

[AM02] S. Arya and T. Malamatos, *Linear-size approximate Voronoi diagrams*, Proc. 13th ACM-SIAM Sympos. Discrete Algorithms, 2002, pp. 147–155.

[AM05] S. Arya and D. M. Mount, *Computational geometry: Proximity and location*, Handbook of Data Structures and Applications (D. Mehta and S. Sahni, eds.), CRC Press LLC, 2005, pp. 63.1–63.22.

[Ame94] N. Amenta, *Helly-type theorems and generalized linear programming*, Discrete Comput. Geom. **12** (1994), 241–261.

[AMM02] S. Arya, T. Malamatos, and D. M. Mount, *Space-efficient approximate Voronoi diagrams*, Proc. 34th Annu. ACM Sympos. Theory Comput., 2002, pp. 721–730.

[AMM09] ———, *Space-time tradeoffs for approximate nearest neighbor searching*, J. Assoc. Comput. Mach. **57** (2009), no. 1, 1–54.

[AMN+98] S. Arya, D. M. Mount, N. S. Netanyahu, R. Silverman, and A. Y. Wu, *An optimal algorithm for approximate nearest neighbor searching in fixed dimensions*, J. Assoc. Comput. Mach. **45** (1998), no. 6, 891–923.

[AMS98] P. K. Agarwal, J. Matoušek, and O. Schwarzkopf, *Computing many faces in arrangements of lines and segments*, SIAM J. Comput. **27** (1998), no. 2, 491–505.

[AP02] P. K. Agarwal and C. M. Procopiuc, *Exact and approximation algorithms for clustering*, Algorithmica **33** (2002), no. 2, 201–226.

[APV02] P. K. Agarwal, C. M. Procopiuc, and K. R. Varadarajan, *Approximation algorithms for k-line center*, Proc. 10th Annu. European Sympos. Algorithms, 2002, pp. 54–63.

[Aro98] S. Arora, *Polynomial time approximation schemes for Euclidean TSP and other geometric problems*, J. Assoc. Comput. Mach. **45** (1998), no. 5, 753–782.

[Aro03] ———, *Approximation schemes for NP-hard geometric optimization problems: a survey*, Math. Prog. **97** (2003), 43–69.

[ARR98] S. Arora, P. Raghavan, and S. Rao, *Approximation schemes for Euclidean k-median and related problems*, Proc. 30th Annu. ACM Sympos. Theory Comput., 1998, pp. 106–113.

[AS00] N. Alon and J. H. Spencer, *The probabilistic method*, 2nd ed., Wiley InterScience, 2000.

[Aur91] F. Aurenhammer, *Voronoi diagrams: A survey of a fundamental geometric data structure*, ACM Comput. Surv. **23** (1991), 345–405.

[Bal97] K. Ball, *An elementary introduction to modern convex geometry*, vol. MSRI Publ. 31, Cambridge Univ. Press, 1997, `http://msri.org/publications/books/Book31/files/ball.pdf`.

[Bar96] Y. Bartal, *Probabilistic approximations of metric space and its algorithmic application*, Proc. 37th Annu. IEEE Sympos. Found. Comput. Sci., October 1996, pp. 183–193.

[Bar98] ———, *On approximating arbitrary metrics by tree metrics*, Proc. 30th Annu. ACM Sympos. Theory Comput., 1998, pp. 161–168.

[Bar02] A. Barvinok, *A course in convexity*, Graduate Studies in Mathematics, vol. 54, Amer. Math. Soc., 2002.

[BCM99] H. Brönnimann, B. Chazelle, and J. Matoušek, *Product range spaces, sensitive sampling, and derandomization*, SIAM J. Comput. **28** (1999), 1552–1575.

[BEG94] M. Bern, D. Eppstein, and J. Gilbert, *Provably good mesh generation*, J. Comput. Syst. Sci. **48** (1994), 384–409.

[BH01] G. Barequet and S. Har-Peled, *Efficiently approximating the minimum-volume bounding box of a point set in three dimensions*, J. Algorithms **38** (2001), 91–109.

[BHI02] M. Bădoiu, S. Har-Peled, and P. Indyk, *Approximate clustering via coresets*, Proc. 34th Annu. ACM Sympos. Theory Comput., 2002, pp. 250–257.

[BM58] G. E.P. Box and M. E. Muller, *A note on the generation of random normal deviates*, Ann. Math. Stat. **28** (1958), 610–611.

[BMP05] P. Brass, W. Moser, and J. Pach, *Research problems in discrete geometry*, Springer, 2005.

[Brö95] H. Brönnimann, *Derandomization of geometric algorithms*, Ph.D. thesis, Dept. Comput. Sci., Princeton University, Princeton, NJ, May 1995.

[BS07] B. Bollobás and A. Scott, *On separating systems*, Eur. J. Comb. **28** (2007), no. 4, 1068–1071.

[BVZ01] Y. Boykov, O. Veksler, and R. Zabih, *Fast approximate energy minimization via graph cuts*, IEEE Trans. Pattern Anal. Mach. Intell. **23** (2001), no. 11, 1222–1239.

[BY98] J.-D. Boissonnat and M. Yvinec, *Algorithmic geometry*, Cambridge University Press, 1998.

[Cal95] P. B. Callahan, *Dealing with higher dimensions: the well-separated pair decomposition and its applications*, Ph.D. thesis, Dept. Comput. Sci., Johns Hopkins University, Baltimore, Maryland, 1995.

[Car76] L. Carroll, *The hunting of the snark*, 1876.

[CC09] P. Chalermsook and J. Chuzhoy, *Maximum independent set of rectangles*, Proc. 20th ACM-SIAM Sympos. Discrete Algorithms, 2009, pp. 892–901.

[CCH09] C. Chekuri, K. L. Clarkson., and S. Har-Peled, *On the set multi-cover problem in geometric settings*, Proc. 25th Annu. ACM Sympos. Comput. Geom., 2009, pp. 341–350.

[CF90] B. Chazelle and J. Friedman, *A deterministic view of random sampling and its use in geometry*, Combinatorica **10** (1990), no. 3, 229–249.

[CH08] K. Chen and S. Har-Peled, *The Euclidean orienteering problem revisited*, SIAM J. Comput. **38** (2008), no. 1, 385–397.

[CH09] T. M. Chan and S. Har-Peled, *Approximation algorithms for maximum independent set of pseudo-disks*, Proc. 25th Annu. ACM Sympos. Comput. Geom., 2009, `cs.uiuc.edu/~sariel/papers/08/w_indep`, pp. 333–340.

[Cha96] T. M. Chan, *Fixed-dimensional linear programming queries made easy*, Proc. 12th Annu. ACM Sympos. Comput. Geom., 1996, pp. 284–290.

[Cha98] ______, *Approximate nearest neighbor queries revisited*, Discrete Comput. Geom. **20** (1998), 359–373.

[Cha01] B. Chazelle, *The discrepancy method: Randomness and complexity*, Cambridge University Press, New York, 2001.

[Cha02] T. M. Chan, *Closest-point problems simplified on the ram*, Proc. 13th ACM-SIAM Sympos. Discrete Algorithms, Society for Industrial and Applied Mathematics, 2002, pp. 472–473.

[Cha05] ______, *Low-dimensional linear programming with violations*, SIAM J. Comput. (2005), 879–893.

[Cha06] ______, *Faster core-set constructions and data-stream algorithms in fixed dimensions.*, Comput. Geom. Theory Appl. **35** (2006), no. 1-2, 20–35.

[Che86] L. P. Chew, *Building Voronoi diagrams for convex polygons in linear expected time*, Technical Report PCS-TR90-147, Dept. Math. Comput. Sci., Dartmouth College, Hanover, NH, 1986.

[Che06] K. Chen, *On k-median clustering in high dimensions*, Proc. 17th ACM-SIAM Sympos. Discrete Algorithms, 2006, pp. 1177–1185.

[Che08] ______, *A constant factor approximation algorithm for k-median clustering with outliers*, Proc. 19th ACM-SIAM Sympos. Discrete Algorithms, 2008, pp. 826–835.

[CK95] P. B. Callahan and S. R. Kosaraju, *A decomposition of multidimensional point sets with applications to k-nearest-neighbors and n-body potential fields*, J. Assoc. Comput. Mach. **42** (1995), 67–90.

[CKMN01] M. Charikar, S. Khuller, D. M. Mount, and G. Narasimhan, *Algorithms for facility location problems with outliers*, Proc. 12th ACM-SIAM Sympos. Discrete Algorithms, 2001, pp. 642–651.

[CKMS06] E. Cohen, H. Kaplan, Y. Mansour, and M. Sharir, *Approximations with relative errors in range spaces of finite VC dimension*, manuscript, 2006.

[CKR01] G. Calinescu, H. Karloff, and Y. Rabani, *Approximation algorithms for the 0-extension problem*, Proc. 12th ACM-SIAM Sympos. Discrete Algorithms, 2001, pp. 8–16.

[Cla83] K. L. Clarkson, *Fast algorithms for the all nearest neighbors problem*, Proc. 24th Annu. IEEE Sympos. Found. Comput. Sci., 1983, pp. 226–232.

[Cla87] ______, *New applications of random sampling in computational geometry*, Discrete Comput. Geom. **2** (1987), 195–222.

[Cla88] ———, *Applications of random sampling in computational geometry, II*, Proc. 4th Annu. ACM Sympos. Comput. Geom., 1988, pp. 1–11.

[Cla93] ———, *Algorithms for polytope covering and approximation*, Proc. 3th Workshop Algorithms Data Struct., Lect. Notes in Comp. Sci., vol. 709, Springer-Verlag, 1993, pp. 246–252.

[Cla94] ———, *An algorithm for approximate closest-point queries*, Proc. 10th Annu. ACM Sympos. Comput. Geom., 1994, pp. 160–164.

[Cla95] ———, *Las Vegas algorithms for linear and integer programming*, J. Assoc. Comput. Mach. **42** (1995), 488–499.

[CLRS01] T. H. Cormen, C. E. Leiserson, R. L. Rivest, and C. Stein, *Introduction to algorithms*, MIT Press / McGraw-Hill, 2001.

[CM96] B. Chazelle and J. Matoušek, *On linear-time deterministic algorithms for optimization problems in fixed dimension*, J. Algorithms **21** (1996), 579–597.

[CMS93] K. L. Clarkson, K. Mehlhorn, and R. Seidel, *Four results on randomized incremental constructions*, Comput. Geom. Theory Appl. **3** (1993), no. 4, 185–212.

[CRT05] B. Chazelle, R. Rubinfeld, and L. Trevisan, *Approximating the minimum spanning tree weight in sublinear time*, SIAM J. Comput. **34** (2005), no. 6, 1370–1379.

[CS89] K. L. Clarkson and P. W. Shor, *Applications of random sampling in computational geometry, II*, Discrete Comput. Geom. **4** (1989), 387–421.

[CS00] N. Cristianini and J. Shaw-Taylor, *Support vector machines*, Cambridge Press, 2000.

[CW89] B. Chazelle and E. Welzl, *Quasi-optimal range searching in spaces of finite VC-dimension*, Discrete Comput. Geom. **4** (1989), 467–489.

[dBCvKO08] M. de Berg, O. Cheong, M. van Kreveld, and M. H. Overmars, *Computational geometry: Algorithms and applications*, 3rd ed., Springer-Verlag, 2008.

[dBDS95] M. de Berg, K. Dobrindt, and O. Schwarzkopf, *On lazy randomized incremental construction*, Discrete Comput. Geom. **14** (1995), 261–286.

[dBS95] M. de Berg and O. Schwarzkopf, *Cuttings and applications*, Internat. J. Comput. Geom. Appl. **5** (1995), 343–355.

[Dey98] T. K. Dey, *Improved bounds for planar k-sets and related problems*, Discrete Comput. Geom. **19** (1998), no. 3, 373–382.

[DG03] S. Dasgupta and A. Gupta, *An elementary proof of a theorem of Johnson and Lindenstrauss*, Rand. Struct. Alg. **22** (2003), no. 3, 60–65.

[dHTT07] M. de Berg, H. J. Haverkort, S. Thite, and L. Toma, *I/o-efficient map overlay and point location in low-density subdivisions*, Proc. 18th Annu. Internat. Sympos. Algorithms Comput., 2007, pp. 500–511.

[DIIM04] M. Datar, N. Immorlica, P. Indyk, and V. S. Mirrokni, *Locality-sensitive hashing scheme based on p-stable distributions*, Proc. 20th Annu. ACM Sympos. Comput. Geom., 2004, pp. 253–262.

[DK85] D. P. Dobkin and D. G. Kirkpatrick, *A linear algorithm for determining the separation of convex polyhedra*, J. Algorithms **6** (1985), 381–392.

[DP09] D. Dubhashi and A. Panconesi, *Concentration of measure for the analysis of randomized algorithms*, Cambridge University Press, 2009.

[Dud74] R. M. Dudley, *Metric entropy of some classes of sets with differentiable boundaries*, J. Approx. Theory **10** (1974), no. 3, 227–236.

[Dun99] C. A. Duncan, *Balanced aspect ratio trees*, Ph.D. thesis, Department of Computer Science, Johns Hopkins University, Baltimore, Maryland, 1999.

[EGS05] D. Eppstein, M. T. Goodrich, and J. Z. Sun, *The skip quadtree: a simple dynamic data structure for multidimensional data*, Proc. 21st Annu. ACM Sympos. Comput. Geom., ACM, June 2005, pp. 296–305.

[EK89] O. Egecioglu and B. Kalantari, *Approximating the diameter of a set of points in the Euclidean space*, Inform. Process. Lett. **32** (1989), 205–211.

[Ele97] G. Elekes, *On the number of sums and products*, ACTA Arithmetica (1997), 365–367.

[Ele02] G. Elekes, *Sums versus products in number theory, algebra and Erdős geometry — a survey*, Paul Erdős and his Mathematics II, Bolyai Math. Soc., 2002, pp. 241–290.

[ERvK96] H. Everett, J.-M. Robert, and M. van Kreveld, *An optimal algorithm for the (≤k)-levels, with applications to separation and transversal problems*, Internat. J. Comput. Geom. Appl. **6** (1996), 247–261.

[Fel91] W. Feller, *An introduction to probability theory and its applications*, John Wiley & Sons, NY, 1991.

[FG88] T. Feder and D. H. Greene, *Optimal algorithms for approximate clustering*, Proc. 20th Annu. ACM Sympos. Theory Comput., 1988, pp. 434–444.

[FGK+00] A. Fabri, G.-J. Giezeman, L. Kettner, S. Schirra, and S. Schönherr, *On the design of CGAL a computational geometry algorithms library*, Softw. – Pract. Exp. **30** (2000), no. 11, 1167–1202.

[FH05] J. Fischer and S. Har-Peled, *Dynamic well-separated pair decomposition made easy.*, Proc. 17th Canad. Conf. Comput. Geom., 2005, pp. 235–238.

[FIS05] G. Frahling, P. Indyk, and C. Sohler, *Sampling in dynamic data streams and applications*, Proc. 21st Annu. ACM Sympos. Comput. Geom., 2005, pp. 142–149.

[FRT03] J. Fakcharoenphol, S. Rao, and K. Talwar, *A tight bound on approximating arbitrary metrics by tree metrics*, Proc. 35th Annu. ACM Sympos. Theory Comput., 2003, pp. 448–455.

[FS97] Y. Freund and R. E. Schapire, *A decision-theoretic generalization of on-line learning and an application to boosting*, J. Comput. Syst. Sci. **55** (1997), no. 1, 119–139.

[Gar82] I. Gargantini, *An effective way to represent quadtrees*, Commun. ACM **25** (1982), no. 12, 905–910.

[Gar02] R. J. Gardner, *The Brunn-Minkowski inequality*, Bull. Amer. Math. Soc. **39** (2002), 355–405.

[GK92] P. Gritzmann and V. Klee, *Inner and outer j-radii of convex bodies in finite-dimensional normed spaces*, Discrete Comput. Geom. **7** (1992), 255–280.

[GLS93] M. Grötschel, L. Lovász, and A. Schrijver, *Geometric algorithms and combinatorial optimization*, 2nd ed., Algorithms and Combinatorics, vol. 2, Springer-Verlag, Berlin Heidelberg, 1993.

[Gol95] M. Goldwasser, *A survey of linear programming in randomized subexponential time*, SIGACT News **26** (1995), no. 2, 96–104.

[Gon85] T. Gonzalez, *Clustering to minimize the maximum intercluster distance*, Theoret. Comput. Sci. **38** (1985), 293–306.

[GP84] J. E. Goodman and R. Pollack, *On the number of k-subsets of a set of n points in the plane*, J. Combin. Theory Ser. A **36** (1984), 101–104.

[GRSS95] M. Golin, R. Raman, C. Schwarz, and M. Smid, *Simple randomized algorithms for closest pair problems*, Nordic J. Comput. **2** (1995), 3–27.

[Grü03] B. Grünbaum, *Convex polytopes*, 2nd ed., Springer, May 2003, Prepared by V. Kaibel, V. Klee, and G. Ziegler.

[GT00] A. Gupta and E. Tardos, *A constant factor approximation algorithm for a class of classification problems*, Proc. 32nd Annu. ACM Sympos. Theory Comput., 2000, pp. 652–658.

[GT08] A. Gupta and K. Tangwongsan, *Simpler analyses of local search algorithms for facility location*, CoRR **abs/0809.2554** (2008).

[Gup00] A. Gupta, *Embeddings of finite metrics*, Ph.D. thesis, University of California, Berkeley, 2000.

[Har00a] S. Har-Peled, *Constructing planar cuttings in theory and practice*, SIAM J. Comput. **29** (2000), no. 6, 2016–2039.

[Har00b] ______, *Taking a walk in a planar arrangement*, SIAM J. Comput. **30** (2000), no. 4, 1341–1367.

[Har01a] ______, *A practical approach for computing the diameter of a point-set*, Proc. 17th Annu. ACM Sympos. Comput. Geom., 2001, pp. 177–186.

[Har01b] ______, *A replacement for Voronoi diagrams of near linear size*, Proc. 42nd Annu. IEEE Sympos. Found. Comput. Sci., 2001, pp. 94–103.

[Har04] ______, *No coreset, no cry*, Proc. 24th Conf. Found. Soft. Tech. Theoret. Comput. Sci., 2004.

[Hau92] D. Haussler, *Decision theoretic generalizations of the PAC model for neural net and other learning applications*, Inf. Comput. **100** (1992), no. 1, 78–150.

[Hau95] ______, *Sphere packing numbers for subsets of the Boolean n-cube with bounded Vapnik-Chervonenkis dimension*, J. Comb. Theory Ser. A **69** (1995), no. 2, 217–232.

[HI00] S. Har-Peled and P. Indyk, *When crossings count – approximating the minimum spanning tree*, Proc. 16th Annu. ACM Sympos. Comput. Geom., 2000, pp. 166–175.

[HM85] D. S. Hochbaum and W. Maas, *Approximation schemes for covering and packing problems in image processing and VLSI*, J. Assoc. Comput. Mach. **32** (1985), 130–136.

[HM04] S. Har-Peled and S. Mazumdar, *Coresets for k-means and k-median clustering and their applications*, Proc. 36th Annu. ACM Sympos. Theory Comput., 2004, pp. 291–300.

[HM05] ______, *Fast algorithms for computing the smallest k-enclosing disc*, Algorithmica **41** (2005), no. 3, 147–157.

[HM06] S. Har-Peled and M. Mendel, *Fast construction of nets in low dimensional metrics, and their applications*, SIAM J. Comput. **35** (2006), no. 5, 1148–1184.

[HS11] S. Har-Peled and M. Sharir, *Relative (p, ε)-approximations in geometry*, Discrete Comput. Geom. **45** (2011), no. 3, 462–496.

[HÜ05] S. Har-Peled and A. Üngör, *A time-optimal Delaunay refinement algorithm in two dimensions*, Proc. 21st Annu. ACM Sympos. Comput. Geom., 2005, pp. 228–236.

[HW87] D. Haussler and E. Welzl, *ε-nets and simplex range queries*, Discrete Comput. Geom. **2** (1987), 127–151.

[IM98] P. Indyk and R. Motwani, *Approximate nearest neighbors: Towards removing the curse of dimensionality*, Proc. 30th Annu. ACM Sympos. Theory Comput., 1998, pp. 604–613.

[IN07] P. Indyk and A. Naor, *Nearest neighbor preserving embeddings*, ACM Trans. Algo. **3** (2007).

[Ind99] P. Indyk, *Sublinear time algorithms for metric space problems*, Proc. 31st Annu. ACM Sympos. Theory Comput., 1999, pp. 154–159.

[Ind01] ______, *Algorithmic applications of low-distortion geometric embeddings*, Proc. 42nd Annu. IEEE Sympos. Found. Comput. Sci., 2001, Tutorial, pp. 10–31.

[Ind04] ______, *Nearest neighbors in high-dimensional spaces*, Handbook of Discrete and Computational Geometry (J. E. Goodman and J. O'Rourke, eds.), CRC Press LLC, 2nd ed., 2004, pp. 877–892.

[Joh48] F. John, *Extremum problems with inequalities as subsidary conditions*, Courant Anniversary Volume (1948), 187–204.

[Kal92] G. Kalai, *A subexponential randomized simplex algorithm*, Proc. 24th Annu. ACM Sympos. Theory Comput., 1992, pp. 475–482.

[KF93] I. Kamel and C. Faloutsos, *On packing r-trees*, Proc. 2nd Intl. Conf. Info. Knowl. Mang., 1993, pp. 490–499.

[Kle02] J. Kleinberg, *An impossibility theorem for clustering*, Neural Info. Proc. Sys., 2002.

[KLMN05] R. Krauthgamer, J. R. Lee, M. Mendel, and A. Naor, *Measured descent: A new embedding method for finite metric spaces*, Geom. funct. anal. (GAFA) **15** (2005), no. 4, 839–858.

[KLN99] D. Krznaric, C. Levcopoulos, and B. J. Nilsson, *Minimum spanning trees in d dimensions*, Nordic J. Comput. **6** (1999), no. 4, 446–461.

[KMN+04] T. Kanungo, D. M. Mount, N. S. Netanyahu, C. D. Piatko, R. Silverman, and A. Y. Wu, *A local search approximation algorithm for k-means clustering*, Comput. Geom. Theory Appl. **28** (2004), 89–112.

[KN10] D. M. Kane and J. Nelson, *A sparser Johnson-Lindenstrauss transform*, ArXiv e-prints (2010), `http://arxiv.org/abs/1012.1577`.

[KOR00] E. Kushilevitz, R. Ostrovsky, and Y. Rabani, *Efficient search for approximate nearest neighbor in high dimensional spaces*, SIAM J. Comput. **2** (2000), no. 30, 457–474.

[KPW92] J. Komlós, J. Pach, and G. Woeginger, *Almost tight bounds for ε-nets*, Discrete Comput. Geom. **7** (1992), 163–173.

[KR99] S. G. Kolliopoulos and S. Rao, *A nearly linear-time approximation scheme for the Euclidean κ-median problem*, Proc. 7th Annu. European Sympos. Algorithms, 1999, pp. 378–389.

[KRS11] H. Kaplan, E. Ramos, and M. Sharir, *Range minima queries with respect to a random permutation, and approximate range counting*, Discrete Comput. Geom. **45** (2011), no. 1, 3–33.

[KS06] H. Kaplan and M. Sharir, *Randomized incremental constructions of three-dimensional convex hulls and planar Voronoi diagrams, and approximate range counting*, Proc. 17th ACM-SIAM Sympos. Discrete Algorithms, 2006, pp. 484–493.

[KT06] J. Kleinberg and E. Tardos, *Algorithm design*, Addison-Wesley, 2006.

[Lei84] F. T. Leighton, *New lower bound techniques for VLSI*, Math. Syst. Theory **17** (1984), 47–70.

[Leo98] S. J. Leon, *Linear algebra with applications*, 5th ed., Prentice Hall, 1998.

[Lit88] N. Littlestone, *Learning quickly when irrelevant attributes abound: A new linear-threshold algorithm*, Mach. Learn. **2** (1988), no. 4, 285–318.

[LLS01] Y. Li, P. M. Long, and A. Srinivasan, *Improved bounds on the sample complexity of learning.*, J. Comput. Syst. Sci. **62** (2001), no. 3, 516–527.

[Lon01] P. M. Long, *Using the pseudo-dimension to analyze approximation algorithms for integer programming*, Proc. 7th Workshop Algorithms Data Struct., Lecture Notes Comput. Sci., vol. 2125, 2001, pp. 26–37.

[Mac50] A.M. Macbeath, *A compactness theorem for affine equivalence-classes of convex regions*, Canad. J. Math **3** (1950), 54–61.

[Mag07] A. Magen, *Dimensionality reductions in ℓ_2 that preserve volumes and distance to affine spaces*, Discrete Comput. Geom. **38** (2007), no. 1, 139–153.

[Mat90] J. Matoušek, *Bi-Lipschitz embeddings into low-dimensional Euclidean spaces*, Comment. Math. Univ. Carolinae **31** (1990), 589–600.

[Mat92a] ______, *Efficient partition trees*, Discrete Comput. Geom. **8** (1992), 315–334.

[Mat92b] ______, *Reporting points in halfspaces*, Comput. Geom. Theory Appl. **2** (1992), no. 3, 169–186.

[Mat95a] ______, *On enclosing k points by a circle*, Inform. Process. Lett. **53** (1995), 217–221.

[Mat95b] ______, *On geometric optimization with few violated constraints*, Discrete Comput. Geom. **14** (1995), 365–384.

[Mat98] ______, *On constants for cuttings in the plane*, Discrete Comput. Geom. **20** (1998), 427–448.

[Mat99] ______, *Geometric discrepancy*, Springer, 1999.

[Mat02] ______, *Lectures on discrete geometry*, Springer, 2002.

[Meg83] N. Megiddo, *Linear-time algorithms for linear programming in R^3 and related problems*, SIAM J. Comput. **12** (1983), no. 4, 759–776.

[Meg84] ______, *Linear programming in linear time when the dimension is fixed*, J. Assoc. Comput. Mach. **31** (1984), 114–127.

[MG06] J. Matoušek and B. Gärtner, *Understanding and using linear programming*, Springer-Verlag, 2006.

[Mil04] G. L. Miller, *A time efficient Delaunay refinement algorithm*, Proc. 15th ACM-SIAM Sympos. Discrete Algorithms, 2004, pp. 400–409.

[Mit99] J. S. B. Mitchell, *Guillotine subdivisions approximate polygonal subdivisions: A simple polynomial-time approximation scheme for geometric TSP, k-MST, and related problems*, SIAM J. Comput. **28** (1999), 1298–1309.

[MN98] J. Matoušek and J. Nešetřil, *Invitation to discrete mathematics*, Oxford Univ Press, 1998.

[MNP06] R. Motwani, A. Naor, and R. Panigrahi, *Lower bounds on locality sensitive hashing*, Proc. 22nd Annu. ACM Sympos. Comput. Geom., 2006, pp. 154–157.

[MP03] R. R. Mettu and C. G. Plaxton, *The online median problem*, SIAM J. Comput. **32** (2003), no. 3, 816–832.

[MR95] R. Motwani and P. Raghavan, *Randomized algorithms*, Cambridge University Press, 1995.

[MSW96] J. Matoušek, M. Sharir, and E. Welzl, *A subexponential bound for linear programming*, Algorithmica **16** (1996), 498–516.

[Mul94a] K. Mulmuley, *Computational geometry: An introduction through randomized algorithms*, Prentice Hall, 1994.

[Mul94b] ______, *An efficient algorithm for hidden surface removal, II*, J. Comp. Sys. Sci. **49** (1994), 427–453.

[MWW93] J. Matoušek, E. Welzl, and L. Wernisch, *Discrepancy and ε-approximations for bounded VC-dimension*, Combinatorica **13** (1993), 455–466.

[NZ01] G. Narasimhan and M. Zachariasen, *Geometric minimum spanning trees via well-separated pair decompositions*, J. Exp. Algorithmics **6** (2001), 6.

[O'R85] J. O'Rourke, *Finding minimal enclosing boxes*, Internat. J. Comput. Inform. Sci. **14** (1985), 183–199.

[OvL81] M. H. Overmars and J. van Leeuwen, *Maintenance of configurations in the plane*, J. Comput. Syst. Sci. **23** (1981), 166–204.

[PA95] J. Pach and P. K. Agarwal, *Combinatorial geometry*, John Wiley & Sons, 1995.

[Pol86] D. Pollard, *Rates of uniform almost-sure convergence for empirical processes indexed by unbounded classes of functions*, Manuscript, 1986.

[PS04] J. Pach and M. Sharir, *Geometric incidences*, Towards a Theory of Geometric Graphs (J. Pach, ed.), Contemporary Mathematics, vol. 342, Amer. Math. Soc., 2004, pp. 185–223.

[Rab76] M. O. Rabin, *Probabilistic algorithms*, Algorithms and Complexity: New Directions and Recent Results (J. F. Traub, ed.), Academic Press, 1976, pp. 21–39.

[RS98] S. Rao and W. D. Smith, *Improved approximation schemes for geometric graphs via "spanners" and "banyans"*, Proc. 30th Annu. ACM Sympos. Theory Comput., 1998, pp. 540–550.

[Rup93] J. Ruppert, *A new and simple algorithm for quality 2-dimensional mesh generation*, Proc. 4th ACM-SIAM Sympos. Discrete Algorithms, 1993, pp. 83–92.

[Rup95] ______, *A Delaunay refinement algorithm for quality 2-dimensional mesh generation*, J. Algorithms **18** (1995), no. 3, 548–585.

[SA95] M. Sharir and P. K. Agarwal, *Davenport-Schinzel sequences and their geometric applications*, Cambridge University Press, New York, 1995.

[SA96] R. Seidel and C. R. Aragon, *Randomized search trees*, Algorithmica **16** (1996), 464–497.

[Sag94] H. Sagan, *Space-filling curves*, Springer-Verlag, 1994.

[Sam89] H. Samet, *Spatial data structures: Quadtrees, octrees, and other hierarchical methods*, Addison-Wesley, 1989.

[Sam05] ______, *Foundations of multidimensional and metric data structures*, The Morgan Kaufmann Series in Computer Graphics and Geometric Modeling, Morgan Kaufmann Publishers Inc., 2005.

[Sei91] R. Seidel, *Small-dimensional linear programming and convex hulls made easy*, Discrete Comput. Geom. **6** (1991), 423–434.

[Sei93] ———, *Backwards analysis of randomized geometric algorithms*, New Trends in Discrete and Computational Geometry (J. Pach, ed.), Algorithms and Combinatorics, vol. 10, Springer-Verlag, 1993, pp. 37–68.

[Sha03] M. Sharir, *The Clarkson-Shor technique revisited and extended*, Comb., Prob. & Comput. **12** (2003), no. 2, 191–201.

[Smi00] M. Smid, *Closest-point problems in computational geometry*, Handbook of Computational Geometry (Jörg-Rüdiger Sack and Jorge Urrutia, eds.), Elsevier, 2000, pp. 877–935.

[SSS02] Y. Sabharwal, N. Sharma, and S. Sen, *Improved reductions of nearest neighbors search to plebs with applications to linear-sized approximate Voronoi decopositions*, Proc. 22nd Conf. Found. Soft. Tech. Theoret. Comput. Sci., 2002, pp. 311–323.

[Sto91] J. Stolfi, *Oriented projective geometry: A framework for geometric computations*, Academic Press, 1991.

[SW92] M. Sharir and E. Welzl, *A combinatorial bound for linear programming and related problems*, Proc. 9th Sympos. Theoret. Aspects Comput. Sci., Lect. Notes in Comp. Sci., vol. 577, Springer-Verlag, 1992, pp. 569–579.

[SW06] ———, *On the number of crossing-free matchings, cycles, and partitions.*, SIAM J. Comput. **36** (2006), no. 3, 695–720.

[Szé97] L. A. Székely, *Crossing numbers and hard Erdős problems in discrete geometry*, Combinatorics, Probability and Computing **6** (1997), 353–358.

[Tót01] G. Tóth, *Point sets with many k-sets*, Discrete Comput. Geom. **26** (2001), no. 2, 187–194.

[Tou83] G. T. Toussaint, *Solving geometric problems with the rotating calipers*, Proc. IEEE MELECON '83, 1983, pp. A10.02/1–4.

[Üng09] A. Üngör, *Off-centers: A new type of Steiner points for computing size-optimal quality-guaranteed Delaunay triangulations*, Comput. Geom. Theory Appl. **42** (2009), no. 2, 109–118.

[Vai86] P. M. Vaidya, *An optimal algorithm for the all-nearest-neighbors problem*, Proc. 27th Annu. IEEE Sympos. Found. Comput. Sci., 1986, pp. 117–122.

[Van97] R. J. Vanderbei, *Linear programming: Foundations and extensions*, Kluwer, 1997.

[Var98] K. R. Varadarajan, *A divide-and-conquer algorithm for min-cost perfect matching in the plane*, Proc. 39th Annu. IEEE Sympos. Found. Comput. Sci., 1998, pp. 320–331.

[Vaz01] V. V. Vazirani, *Approximation algorithms*, Springer-Verlag, 2001.

[VC71] V. N. Vapnik and A. Y. Chervonenkis, *On the uniform convergence of relative frequencies of events to their probabilities*, Theory Probab. Appl. **16** (1971), 264–280.

[Wel86] E. Welzl, *More on k-sets of finite sets in the plane*, Discrete Comput. Geom. **1** (1986), 95–100.

[Wel92] ———, *On spanning trees with low crossing numbers*, Data Structures and Efficient Algorithms, Final Report on the DFG Special Joint Initiative, Lect. Notes in Comp. Sci., vol. 594, Springer-Verlag, 1992, pp. 233–249.

[WS11] D. P. Williamson and D. B. Shmoys, *The design of approximation algorithms*, Cambridge University Press, 2011.

[WVTP97] M. Waldvogel, G. Varghese, J. Turener, and B. Plattner, *Scalable high speed IP routing lookups*, Proc. ACM SIGCOMM 97, Octeber 1997.

[YAPV04] H. Yu, P. K. Agarwal, R. Poreddy, and K. R. Varadarajan, *Practical methods for shape fitting and kinetic data structures using core sets*, Proc. 20th Annu. ACM Sympos. Comput. Geom., 2004, pp. 263–272.

[Zie94] G. M. Ziegler, *Lectures on polytopes*, Graduate Texts in Mathematics, vol. 152, Springer-Verlag, Heidelberg, 1994.

Index

Titles in This Series

174 **Lawrence S. Levy and J. Chris Robson,** Hereditary Noetherian prime rings and idealizers, 2011

173 **Sariel Har-Peled,** Geometric approximation algorithms, 2011

172 **Michael Aschbacher, Richard Lyons, Stephen D. Smith, and Ronald Solomon,** The classification of finite simple groups: Groups of characteristic 2 type, 2011

171 **Leonid Pastur and Mariya Shcherbina,** Eigenvalue distribution of large random matrices, 2011

170 **Kevin Costello,** Renormalization and effective field theory, 2011

169 **Robert R. Bruner and J. P. C. Greenlees,** Connective real K-theory of finite groups, 2010

168 **Michiel Hazewinkel, Nadiya Gubareni, and V. V. Kirichenko,** Algebras, rings and modules: Lie algebras and Hopf algebras, 2010

167 **Michael Gekhtman, Michael Shapiro, and Alek Vainshtein,** Cluster algebra and Poisson geometry, 2010

166 **Kyung Bai Lee and Frank Raymond,** Seifert fiberings, 2010

165 **Fuensanta Andreu-Vaillo, José M. Mazón, Julio D. Rossi, and J. Julián Toledo-Melero,** Nonlocal diffusion problems, 2010

164 **Vladimir I. Bogachev,** Differentiable measures and the Malliavin calculus, 2010

163 **Bennett Chow, Sun-Chin Chu, David Glickenstein, Christine Guenther, James Isenberg, Tom Ivey, Dan Knopf, Peng Lu, Feng Luo, and Lei Ni,** The Ricci flow: Techniques and applications, Part III: Geometric-analytic aspects, 2010

162 **Vladimir Maz'ya and Jürgen Rossmann,** Elliptic equations in polyhedral domains, 2010

161 **Kanishka Perera, Ravi P. Agarwal, and Donal O'Regan,** Morse theoretic aspects of p-Laplacian type operators, 2010

160 **Alexander S. Kechris,** Global aspects of ergodic group actions, 2010

159 **Matthew Baker and Robert Rumely,** Potential theory and dynamics on the Berkovich projective line, 2010

158 **D. R. Yafaev,** Mathematical scattering theory: Analytic theory, 2010

157 **Xia Chen,** Random walk intersections: Large deviations and related topics, 2010

156 **Jaime Angulo Pava,** Nonlinear dispersive equations: Existence and stability of solitary and periodic travelling wave solutions, 2009

155 **Yiannis N. Moschovakis,** Descriptive set theory, 2009

154 **Andreas Čap and Jan Slovák,** Parabolic geometries I: Background and general theory, 2009

153 **Habib Ammari, Hyeonbae Kang, and Hyundae Lee,** Layer potential techniques in spectral analysis, 2009

152 **János Pach and Micha Sharir,** Combinatorial geometry and its algorithmic applications: The Alcála lectures, 2009

151 **Ernst Binz and Sonja Pods,** The geometry of Heisenberg groups: With applications in signal theory, optics, quantization, and field quantization, 2008

150 **Bangming Deng, Jie Du, Brian Parshall, and Jianpan Wang,** Finite dimensional algebras and quantum groups, 2008

149 **Gerald B. Folland,** Quantum field theory: A tourist guide for mathematicians, 2008

148 **Patrick Dehornoy with Ivan Dynnikov, Dale Rolfsen, and Bert Wiest,** Ordering braids, 2008

147 **David J. Benson and Stephen D. Smith,** Classifying spaces of sporadic groups, 2008

146 **Murray Marshall,** Positive polynomials and sums of squares, 2008

145 **Tuna Altinel, Alexandre V. Borovik, and Gregory Cherlin,** Simple groups of finite Morley rank, 2008

144 **Bennett Chow, Sun-Chin Chu, David Glickenstein, Christine Guenther, James Isenberg, Tom Ivey, Dan Knopf, Peng Lu, Feng Luo, and Lei Ni,** The Ricci flow: Techniques and applications, Part II: Analytic aspects, 2008

143 **Alexander Molev,** Yangians and classical Lie algebras, 2007

142 **Joseph A. Wolf,** Harmonic analysis on commutative spaces, 2007

141 **Vladimir Maz′ya and Gunther Schmidt,** Approximate approximations, 2007

140 **Elisabetta Barletta, Sorin Dragomir, and Krishan L. Duggal,** Foliations in Cauchy-Riemann geometry, 2007

139 **Michael Tsfasman, Serge Vlăduţ, and Dmitry Nogin,** Algebraic geometric codes: Basic notions, 2007

138 **Kehe Zhu,** Operator theory in function spaces, 2007

137 **Mikhail G. Katz,** Systolic geometry and topology, 2007

136 **Jean-Michel Coron,** Control and nonlinearity, 2007

135 **Bennett Chow, Sun-Chin Chu, David Glickenstein, Christine Guenther, James Isenberg, Tom Ivey, Dan Knopf, Peng Lu, Feng Luo, and Lei Ni,** The Ricci flow: Techniques and applications, Part I: Geometric aspects, 2007

134 **Dana P. Williams,** Crossed products of C^*-algebras, 2007

133 **Andrew Knightly and Charles Li,** Traces of Hecke operators, 2006

132 **J. P. May and J. Sigurdsson,** Parametrized homotopy theory, 2006

131 **Jin Feng and Thomas G. Kurtz,** Large deviations for stochastic processes, 2006

130 **Qing Han and Jia-Xing Hong,** Isometric embedding of Riemannian manifolds in Euclidean spaces, 2006

129 **William M. Singer,** Steenrod squares in spectral sequences, 2006

128 **Athanassios S. Fokas, Alexander R. Its, Andrei A. Kapaev, and Victor Yu. Novokshenov,** Painlevé transcendents, 2006

127 **Nikolai Chernov and Roberto Markarian,** Chaotic billiards, 2006

126 **Sen-Zhong Huang,** Gradient inequalities, 2006

125 **Joseph A. Cima, Alec L. Matheson, and William T. Ross,** The Cauchy Transform, 2006

124 **Ido Efrat, Editor,** Valuations, orderings, and Milnor K-Theory, 2006

123 **Barbara Fantechi, Lothar Göttsche, Luc Illusie, Steven L. Kleiman, Nitin Nitsure, and Angelo Vistoli,** Fundamental algebraic geometry: Grothendieck's FGA explained, 2005

122 **Antonio Giambruno and Mikhail Zaicev, Editors,** Polynomial identities and asymptotic methods, 2005

121 **Anton Zettl,** Sturm-Liouville theory, 2005

120 **Barry Simon,** Trace ideals and their applications, 2005

119 **Tian Ma and Shouhong Wang,** Geometric theory of incompressible flows with applications to fluid dynamics, 2005

118 **Alexandru Buium,** Arithmetic differential equations, 2005

117 **Volodymyr Nekrashevych,** Self-similar groups, 2005

116 **Alexander Koldobsky,** Fourier analysis in convex geometry, 2005

115 **Carlos Julio Moreno,** Advanced analytic number theory: L-functions, 2005

114 **Gregory F. Lawler,** Conformally invariant processes in the plane, 2005

113 **William G. Dwyer, Philip S. Hirschhorn, Daniel M. Kan, and Jeffrey H. Smith,** Homotopy limit functors on model categories and homotopical categories, 2004

112 **Michael Aschbacher and Stephen D. Smith,** The classification of quasithin groups II. Main theorems: The classification of simple QTKE-groups, 2004